PLANT PHYSIOLOGY

PLANT PHYSIOLOGY

LINCOLN TAIZ
UNIVERSITY OF CALIFORNIA, SANTA CRUZ

EDUARDO ZEIGER
UNIVERSITY OF CALIFORNIA, LOS ANGELES

The Benjamin/Cummings Publishing Company, Inc.

Redwood City, California • Menlo Park, California • Reading, Massachusetts
New York • Don Mills, Ontario • Wokingham, U.K. • Amsterdam • Bonn
Sydney • Singapore • Tokyo • Madrid • San Juan

Sponsoring Editor: Edith Beard Brady

Associate Editor: Lisa Donohoe

Developmental Editor: James Funston

Photo Editor: Cecilia Mills

Editorial Assistant: Sissy Hodge

Production Supervisor: Bruce Lundquist

Interior and Cover Design: Paula Schlosser

Copyeditor: Mary Prescott

Index: E. Virginia Hobbs

Principal Illustrator: Elizabeth Morales-Denney

Illustrators: Paula McKenzie, Ben Turner Graphics

Typesetter: Polyglot Pte Ltd

Photograph and Illustration Credits appear
on pages xi–xiii

Library of Congress Cataloging-in-Publication Data
Taiz, Lincoln.
    Plant physiology/Lincoln Taiz, Eduardo Zeiger.
       p.   cm.
    Includes bibliographical references and index.
    ISBN 0-8053-0245-X
    1. Plant physiology.   I. Zeiger, Eduardo.   II. Title.
QK711.2.T35   1991   90-25748
581.1—dc20   CIP

ISBN 0–8053–0245–X
  5678910 -MU- 95 94 93

The Benjamin/Cummings Publishing Company, Inc.
390 Bridge Parkway
Redwood City, California 94065

PREFACE

DURING THE PAST DECADE the biological sciences have experienced a period of unprecedented progress, and nowhere is the excitement of this new era more apparent than in the field of plant physiology. Innovations such as the patch clamp are unlocking the mysteries of membrane transport. Recombinant DNA techniques are providing new tools for understanding how light and hormones regulate gene expression and development. X-ray crystallographic analyses of key proteins and pigment-protein complexes such as Rubisco and photosynthetic reaction center are giving us our first glimpse of the molecular mechanisms of carbon fixation and the light reactions of photosynthesis. Needless to say, this fast, near dazzling pace of discovery has permeated the making of this book. Thus, while providing students with a firm foundation in the traditional topics of plant physiology, we have also endeavored to place the major concepts in the context of contemporary biology.

At the outset, it is useful to define the boundaries of the field of plant physiology, while recognizing that there are extensive areas of overlap between related disciplines. What is plant physiology? We can define it broadly as the study of plant function, encompassing the dynamic processes of growth, metabolism and reproduction in living plants. How does plant physiology differ from its sister disciplines, such as biochemistry, biophysics, and molecular biology? Consider the example of photosynthesis, a classical topic in plant physiology. Biochemists purify photosynthetic enzymes and study their characteristics in the test tube; biophysicists isolate photosynthetic membranes and determine their spectroscopic properties in cuvettes; molecular biologists clone the genes that encode photosynthetic proteins and study their regulation during development. In contrast, plant physiologists study photosynthesis in action at different levels of organization, including the chloroplast, the cell, the leaf and the whole plant. Stated differently, biochemists, biophysicists and molecular biologists study cellular components more or less in isolation, whereas plant physiologists investigate the way in which the components interact with each other to carry out biological processes and functions.

Of course, plant physiologists frequently resort to the tools of biochemistry, biophysics, and molecular biology, and specialists in the other disciplines often conduct physiological experiments. In fact, these related disciplines represent a continuum, the borders of which are rather arbitrarily defined. Thus, familiarity with the principles of biophysics, biochemistry, and molecular biology are indispensible to the physiologist. The current emphasis on molecular biology, in particular, stems in large part from the potential applications of genetic engineering to agricultural crop improvement and plant biotechnology. Does this mean that plant physiology is no longer a central discipline in the plant sciences? Hardly. The need to understand how the components of living organisms interact with each other will not be eliminated by a detailed characterization of the genome. On the contrary, the enriched understanding of gene regulation in plants will generate a wealth of new questions concerning plant function. We hope and trust that the undergraduates of today will receive from this textbook both the knowledge and the motivation to become the scientists who will lead plant physiology into the 21st century.

Joining us in this endeavor are twenty-one outstanding scientists, each an internationally recognized authority in her or his area of expertise. The multi-author approach has been successfully applied to textbooks in other disciplines, but never before to an undergraduate text specifically designed for a course in plant physiology. The authors were selected not only for their capacity to convey the excitement and direction of contemporary research, but also for their ability to identify and clearly explain the principles of plant physiology.

The organization of the book is modular in the sense that the chapters covering a given topic are designed to be self-contained and self-explanatory, while being fully cross-referenced to related chapters. This format allows instructors to substantially alter the order of the topics without sacrificing either comprehension or flow. Fundamental concepts upon which all of the other chapters depend are covered in the first two chapters. While these chapters are intended as a review of material normally covered in introductory courses on cell and molecular biology, students may also find them useful as supplementary reading on certain topics, such as bioenergetics.

Much of the credit for integrating the diverse chapters into a comprehensive and unified whole belongs to our developmental editor, James Funston; without his wisdom, good humor and infinite patience this book would not have come into being. Saundra Lee Taiz deserves special thanks for her invaluable help in library research and in production of the manuscript. We are also indebted to our colleagues at the University of California at Santa Cruz, Harry Beevers, Barry Bowman, Jean Langenheim, Bob Ludwig, and Jane Silverthorne, for their many helpful suggestions and ideas. In fact, the important contributions of the many colleagues who reviewed all aspects of this book and who helped us with their technical knowledge and thoughtful advice make this book a truly joint project of many members of the scientific community of plant physiologists. We extend to all of them our sincere thanks. L. T. also wishes to express his appreciation to Nathan Nelson at the Roche Institute, in whose lab he spent an enjoyable sabbatical during the latter stages of production. Finally, we thank Edith Beard Brady, Jim Behnke, Andy Crowley, Robin Heyden, and Bruce Lundquist from Benjamin/Cummings for their vision and encouragement, and for the privilege of working with a publishing team of such high professional standards.

Lincoln Taiz
Eduardo Zeiger

ABOUT THE AUTHORS

Lincoln Taiz is a Professor of Biology at the University of California at Santa Cruz. He received his Ph.D. in Botany from the University of California at Berkeley. Dr. Taiz's research is on plant development and plant membrane transport, particularly on the structure and function of vacuolar ATPases.

Eduardo Zeiger is a Professor of Biology at University of California at Los Angeles. He received his Ph.D. from the University of California at Davis in Plant Genetics. Dr. Zeiger's areas of specialization are stomatal function and sensory transduction in plant cells.

CONTRIBUTORS

George W. Bates is an Associate Professor of Biological Science and Molecular Biophysics at Florida State University. His Ph.D. in Plant Physiology was earned at the University of Washington in 1977. Dr. Bates' current research interests involve membrane manipulations leading to cell fusion (somatic hybridization) and plasmid DNA uptake (transformation).

Robert E. Blankenship is a Professor of Chemistry at Arizona State University. He received his Ph.D. in Chemistry from the University of California at Berkeley in 1975. Dr. Blankenship's present professional work concerns excitation and electron transfer in photosynthetic systems, evolution of biological energy conserving systems, and mechanisms of electron transfer reactions in biological and non-biological systems.

Donald P. Briskin is an Associate Professor of Plant Physiology at the University of Illinois at Urbana-Champaign. He received his Ph.D. from the University of California at Riverside in Plant Physiology in 1981. The biochemistry of membrane transport in plant cells is the subject of Dr. Briskin's research.

Daniel J. Cosgrove is an Associate Professor of Biology at the Pennsylvania State University at University Park. His Ph.D. in Biological Sciences was earned at Stanford University. Dr. Cosgrove's research interest is primarily focused on plant growth, specifically water transport at the cellular level, the biochemistry and biophysics of cell wall expansion, and the mechanisms by which light and hormones regulate plant growth.

Peter J. Davies is a Professor of Plant Physiology at Cornell University. He earned his Ph.D. in Plant Physiology from the University of Reading in England. His present professional work is on the role of hormones in whole plant senescence, and the use of plant genotypes in elucidating the role of hormones, particularly gibberellin, in plant growth and development.

Malcolm C. Drew is an Associate Professor of Plant Physiology at Texas A&M University. His Ph.D. in Plant Nutrition was earned at the University of Oxford, England, in 1966. Dr. Drew's research interests in plant physiology include the structure, physiology, and function of plant roots, and plant responses to environmental stress, particularly excess salinity, desiccation, oxygen shortage, and nutrient deficiency investigated at the cellular, tissue, and whole plant levels.

Anthony L. Fink is Chair of the Chemistry Department and Professor of Chemistry at the University of California at Santa Cruz. He received his Ph.D. at Queen's University in Ontario, Canada. The structure and function of proteins are the focus of Dr. Fink's research. Current work concerns experimental and computational aspects of the pathway of protein folding and the relationship between protein structure and stability, the physical-chemical basis for enzyme catalysis and inhibition, and mechanism-based drug design.

Donald E. Fosket is a Professor in the Department of Developmental and Cell Biology at the University of California at Irvine. He received his Ph.D. in Biology from the University of Idaho, and subsequently did post-doctoral work in the fields of cell biology at Brookhaven National Labs and Harvard. Dr. Fosket's current areas of research involve cytoskeletal proteins and the control of microtubule formation in plant cells.

Shimon Gepstein is an Associate Professor in the Department of Biology at Technion — Israel Institute of Technology. He received his Ph.D. from Tel Aviv University in 1979. Dr. Gepstein's research interests are in abscisic acid and cell function, the molecular biology of plant senescence, and guard cell photosynthesis.

Jonathan Gershenzon is an Assistant Scientist in the Institute of Biological Chemistry at Washington State University. He received his Ph.D. from the University of Texas at Austin in 1984. Dr. Gershenzon's research interests concern the regulation of terpenoid metabolism in plants, and the role of terpenoids in plant — herbivore interactions.

Adrienne R. Hardham is a Fellow in the Plant Cell Biology Group at the Australian National University in Canberra City. She received her Ph.D. in Developmental Biology from the Research School of Biological Sciences at the Australian National University in 1978. Dr. Hardham's research interests concern plant development and morphogenesis, and the role of microtubules in cell division and differentiation.

Frank Harold is presently Research Professor of Biochemistry at Colorado State University. Dr. Harold received his Ph.D. in 1955 at the University of California at Berkeley. His research contributions center on bioenergetics, including energy coupling to membrane transport and transcellular electric currents.

George H. Lorimer is a Principal Investigator and Research Leader in the Central R&D Department of E.I. Du Pont de Nemours & Co. He graduated from Michigan State University in 1972 with a Ph.D. in Biochemistry. Dr. Lorimer's research concentrates on carboxylation reactions, enzymology, and mechanisms of protein folding, and he has a long-standing interest in the enzyme Rubisco.

John W. Radin is a Supervisory Plant Physiologist and Research Leader at the U.S. Department of Agriculture in Phoenix. He also holds positions as Adjunct Professor at both Arizona State University and the University of Arizona. His Ph.D. in Plant Physiology was earned at the University of California at Davis. Dr. Radin's present work is on environmental and developmental plant physiology, both in the field and in controlled environments; root physiology and plant water relations; plant stress, especially nutrients and water; photosynthesis (gas exchange); and biomass partitioning.

Stanley J. Roux is Chair of the Department of Botany and Professor of Botany at the University of Texas at Austin. He received his Ph.D. from Yale in 1971. Dr. Roux is interested in defining the transduction steps involved in phytochrome action that convert the light signal into changes in growth and development of plants. His current work centers on the role of calcium and calcium-binding proteins in this transduction chain.

Thomas David Sharkey is an Associate Professor in the Department of Botany at the University of Wisconsin at Madison. His Ph.D. in Botany and Plant Pathology was earned at Michigan State University in 1980. Dr. Sharkey studies the interactions of the various components of photosynthesis, and the effects of the physiology of the whole plant on the process of photosynthesis.

James N. Siedow is a Professor of Botany at Duke University. He received his Ph.D. from Indiana University in 1972. Dr. Siedow's research interests focus on physiology and biochemical studies of plant oxidative processes, and electron transfer in plant mitochondria.

Susan Sovonick-Dunford is an Associate Professor of Biological Sciences at the University of Cincinnati. She received her Ph.D. from the University of Dayton in 1973 with a specialization in plant and cell physiology. Dr. Sovonick-Dunford's research is on transport processes in higher plants, including long-distance sugar transport in herbaceous and woody species and water transport, especially as it relates to carbohydrate translocation.

Richard G. Stout is an Associate Professor of Biology at Montana State University. He earned his Ph.D. in Plant Physiology in 1980 from the University of Washington. Dr. Stout's present research interests primarily focus on plant development and cell-to-cell communication.

Daphne Vince-Prue is presently a Visiting Professor in the Department of Botany at the University of Reading, England. She was previously Head of the Division of Physiology at Glasshouse Crops Research Institute, a position from which she retired in 1986. Dr. Vince-Prue's research interests concern the control of flowering and circadian rhythm in plants.

Stephen M. Wolniak is an Associate Professor in the Department of Botany at the University of Maryland at College Park. He earned his Ph.D. in Botany at the University of California at Berkeley. Dr. Wolniak's research interests include the regulation of mitotic events, the regulation of cell cycle progressions, cell motility, and signal transduction.

REVIEWERS

Donald J. Armstrong
Oregon State University

David W. Becker
Pomona College

Harry Beevers
University of California, Santa Cruz

Wade L. Berry
University of California, Los Angeles

Krystyna Bialek
Beltsville Agriculture Research Center

Arnold Bloom
University of California, Davis

Martyn M. Caldwell
Utah State University

Gaylon Campbell
Washington State University

Raymond Chollet
University of Nebraska, Lincoln

Morris G. Cline
Ohio State University

Jerry D. Cohen
Beltsville Agriculture Research Station

Peter L. Conrad
SUNY Plattsburgh

Rodney Croteau
Washington State University

Jack Dainty
University of Toronto

Alan G. Darville
University of Georgia

Gerald F. Deitzer
University of Maryland

Gerald E. Edwards
Washington State University

Jack Einset
NSF — University of California, Irvine

Emanuel Epstein
University of California, Davis

Robert C. Evans
Rutgers University

Lewis J. Feldman
University of California, Berkeley

James French
Rutgers University

Donald R. Geiger
University of Dayton

Arthur Gibson
University of California, Los Angeles

Paul B. Green
Stanford University

James A. Guikema
Kansas State University

Anthony E. Hall
University of California, Riverside

John J. Harada
University of California, Davis

Tuan-hua David Ho
Washington University

Raymond W. Holton
University of Tennessee

Russell L. Jones
University of California, Berkeley

Noel T. Keen
University of California, Riverside

Clifford LaMotte
Iowa State University

J. Levitt
University of Missouri

David E. Lincoln
University of North Carolina

Craig Martin
University of Kansas

Irving B. McNulty
University of Utah

Anastasios Melis
University of California, Berkeley

James D. Metzger
USDA Agriculture Research Service

Russell K. Monson
University of Colorado

Donald R. Ort
University of Illinois

Carl S. Pike
Franklin & Marshall College

Keith M. Pomeroy
Agricultural Research Center, Canada

Ronald J. Poole
McGill University

Herbert B. Posner
SUNY Binghamton

John W. Radin
USDA Agriculture Research Service

Dave Rayle
Oregon State University

Albert Ruesink
Indiana University

Roy M. Sachs
University of California, Davis

Rick Sandstrom
Portland State University

Ruth Satter
University of Connecticut

Elba Serrano
University of California, Los Angeles

Robert E. Sharp
University of Missouri

Wendy K. Silk
University of California, Davis

Theophanes A. Solomos
University of Maryland

Roger M. Spanswick
Cornell University

Cecil R. Stewart
Iowa State University

Gregory J. Taylor
University of Alberta

Robert Thornton
University of California, Davis

Michael P. Timko
University of Virginia

Elaine M. Tobin
University of California, Los Angeles

Leslie R. Towill
Arizona State University

Linda L. Walling
University of California, Riverside

Daniel C. Walton
New York State University

William E. Winner
Oregon State University

Roger Wyse
Rutgers University

Thomas E. Wynn
North Carolina State University

Shang Fa Yang
University of California, Davis

Jan A. D. Zeevaart
Michigan State University

CONTENTS

UNIT I

TRANSPORT AND TRANSLOCATION OF WATER AND SOLUTES *59*

CHAPTER 3   Water and Plant Cells *61*

CHAPTER 7   Phloem Translocation   *145*

UNIT **II**

BIOCHEMISTRY AND METABOLISM   177

CHAPTER **8**   Photosynthesis: The Light Reactions   *179*

CHAPTER **9**   **Photosynthesis: Carbon Metabolism** *219*

CHAPTER 10   Photosynthesis: Physiological and Ecological Considerations   249

CHAPTER 11   **Respiration and Lipid Metabolism**   *265*

UNIT **III**

GROWTH AND DEVELOPMENT   *371*

CHAPTER **15**   **The Cellular Basis of Growth and Morphogenesis**   *373*

CHAPTER 16   Auxins: Growth and Tropisms   *398*

CHAPTER 19   Ethylene and Abscisic Acid   *473*

PLANT
PHYSIOLOGY

OVERVIEW OF ESSENTIAL CONCEPTS

CHAPTER 1

Plant and Cell Architecture

THE TERM "CELL" IS derived from the latin *cella*, meaning storeroom or chamber. It was first used in biology in 1665 by the English botanist Robert Hooke to describe the individual units he observed in cork tissue under a compound microscope. The cork cells Hooke observed were really the empty lumens of dead cells, but the term is an apt one because cells are the basic building blocks that define the plant structure.

This book will emphasize the physiological and biochemical processes of plants, but it is important to recognize that these functions are determined by structures, whether the function is gas exchange in the leaf, water conduction in the xylem, photosynthesis in the chloroplast or ion transport across the plasma membrane. At every level, structure and function represent different frames of reference of a biological unity.

This chapter provides an overview of the general anatomy and ultrastructure of plants and their cells, elements of which will be treated in greater detail in the following chapters.

Plant Life: Unifying Principles

The spectacular diversity of plant size and form is familiar to everyone. Plants range in size from less than a centimeter to greater than 100 meters. Plant morphology or shape is also surprisingly diverse. At first glance, the tiny *Lemna* plant (duckweed) would seem to have little in common with a giant saguaro cactus or a redwood tree. Yet, regardless of their specific adaptations, all plants carry out fundamentally similar processes and are based on the same architectural plan. We can summarize the major design elements of plants as follows.

1. As the earth's primary producers, green plants are the ultimate solar collectors. They harvest the energy of sunlight by converting light energy to chemical energy, which they store in the bonds formed when they synthesize carbohydrates from carbon dioxide and water.

2. Plants are nonmotile. As a substitute for motility, they have evolved an indeterminate growth habit which allows them to compete effectively for essential resources such as light, water, and mineral nutrients.

3. Terrestrial plants are structurally reinforced in order to grow toward sunlight against the force of gravity.

4. Terrestrial plants lose water continuously by evaporation and have evolved mechanisms for avoiding desiccation.

5. Plants have mechanisms for moving water and minerals from the soil to the sites of photosynthesis and growth and mechanisms for moving the products of photosynthesis to nonphotosynthetic organs and tissues.

The Plant Kingdom

According to a widely used classification scheme, all living organisms can be divided into five kingdoms: Monera, Protista, Fungi, Plantae (plants), and Animalia. The **Monera** consist of the prokaryotes, whose major characteristic is lack of a true nucleus. The prokaryotes are divided into two subgroups: the eubacteria and the archaebacteria. The **eubacteria** include photosynthetic forms, such as the cyanobacteria, as well as nonphotosynthetic (heterotrophic) forms. The **archaebacteria** are often adapted to extreme environments, such as sulfur hot springs or high-salt ponds.

Protists constitute the most diverse kingdom. For the most part they are unicellular eukaryotes, but the classification also includes multicellular eukaryotes with relatively simple structures. The photosynthetic members of the protists are termed the **algae**. Six divisions of the algae are recognized: they differ from each other in their photosynthetic pigments, cell walls, and food reserves. Algae typically grow in marine or aquatic environments where water and minerals are abundant. In earlier classification schemes, algae were included in the plant kingdom. Indeed, much biochemical evidence indicates that green land plants probably evolved from the green algae. The current placement of algae among the protists, however, emphasizes their evolutionary relationship with other kingdoms rather than their biochemical affinities with plants.

Fungi, were formerly classified with the plants. However, it is clear that fungi are sufficiently different from both plants and protists to warrant treating them as a separate kingdom. Fungi are all heterotrophic; that is, they depend on other organisms for their food and satisfy their nutritional needs by absorbing inorganic ions and organic molecules from the external environment. Most species are filamentous and possess cell walls made of chitin, the same substance that is found in insect exoskeletons.

The plant kingdom includes the bryophytes, which mainly reproduce by spores, vascular plants, and the seed plants. The **bryophytes** are very simple land plants and the least abundant in terms of number of species and overall population. Bryophytes do not appear to be in the direct line of evolution leading to the vascular plants but rather to constitute a separate minor branch of their own. They include the mosses, liverworts, and hornworts. These small plants, rarely greater than 2 cm in height, have life cycles that depend on water during the sexual phase. This facilitates **fertilization**, the fusion of gametes to produce a diploid zygote, a feature also seen in their algal precursors. Bryophytes are like algae in other respects as well: they have neither true roots nor true leaves, they lack a vascular system, and they do not produce any hard tissues for structural support. Unlike algae, bryophytes are terrestrial rather than aquatic, and the absence of these design elements that are important for growth on land greatly restricts their potential size.

The ferns represent the largest group of **spore-bearing vascular plants**. In contrast to the bryophytes, **ferns** have true roots, leaves, and vascular tissues, and they produce hard tissues for support. These architectural features enable ferns to grow up to the size of small trees. Although ferns are better adapted to the desiccating conditions of terrestrial life than bryophytes, they are still dependent on water as a medium for the movement of sperms to the egg. This dependence on water during a critical stage of their life cycle restricts the ecological range of ferns to relatively moist habitats.

The most successful terrestrial plants are the **seed plants**. The embryo, protected and nourished inside the seed, is able to survive in a dormant state during unfavorable growing conditions such as drought. Seed dispersal also facilitates the dissemination of the embryos away from the parent plant. Another important feature of seed plants is their mode of fertilization. Fertilization in seed plants is brought about by wind- or insect-mediated transfer of **pollen**, the male gamete-producing structure, to the female sexual structure, the **ovule**, which houses the egg. Pollination is independent of external water, a distinct advantage in terrestrial environments. Many seed plants produce copious amounts of woody tissues, which enable them to grow to extraordinary heights. These features of seed plants have contributed to their success and account for their wide range.

There are two categories of seed plants: gymnosperms (from the Greek words meaning "naked seed") and angiosperms (based on the Greek for "vessel seed," or seeds contained in a vessel). **Gymnosperms** are the less advanced type, of which about 700 species are known. The largest group of gymnosperms are the conifers ("cone-bearers"), which include such commercially important forest trees as pine, fir, spruce, and redwood. Two types of cones are present: male cones, which produce pollen, and female cones, which bear the ovules. The ovules are located on the surfaces of specialized

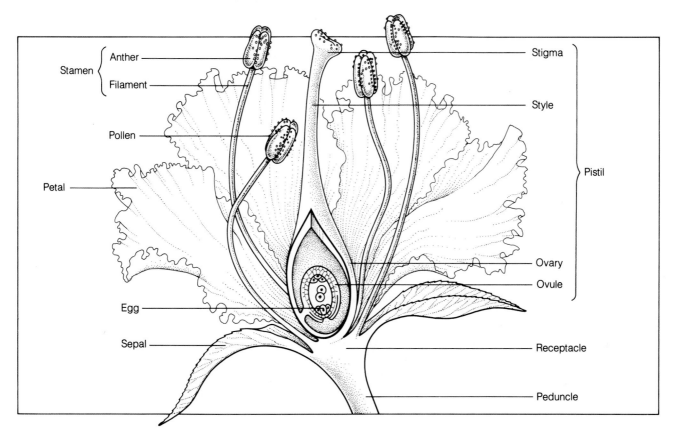

FIGURE 1.1. Schematic representation of an idealized flower of the angiosperms.

structures called **cone scales.** Following wind-mediated pollination, the sperm reaches the egg via a **pollen tube,** and the fertilized egg develops into an embryo. Upon maturation, the cone scales, which are appressed during early development, separate from each other, allowing the naked seeds to fall to the ground.

Angiosperms first became abundant during the Cretaceous period, about 100 million years ago. Today, they dominate the landscape, easily outcompeting their cousins, the gymnosperms. About 250,000 species are known, but many more remain to be characterized.

The major innovation of the angiosperms is the flower; hence they are referred to as *flowering plants.* There are other anatomical differences between angiosperms and gymnosperms, but none so crucial and far-reaching as the mode of reproduction. The flower consists of a number of leaflike structures attached to a specialized region of the stem called the **receptacle** (Fig. 1.1). **Sepals** and **petals** are the most leaf-like. The **stamen** is the male sexual structure, and the **pistil** is the female sexual structure. The **ovary,** the hollow, basal portion of the pistil, completely encloses one or more ovules. Petals have the primary function of attracting insects to serve as pollinators, which accounts for their often showy and brightly colored appearance. (The ar-

chitectural design of petals is based mainly on insect aesthetics!) After landing on the **stigma,** or tip of the pistil, the pollen germinates to form a long pollen tube, which penetrates the tissues of the **style** and ultimately enters the cavity of the **ovary.**

Within the ovary, the pollen tube enters the ovule and deposits two haploid sperm cells in the **embryo sac.** One sperm cell fuses with the egg to produce the zygote, while the other typically fuses with the two **polar nuclei** to produce a specialized storage tissue termed the **endosperm,** which provides nutrients to the growing embryo. Endosperm tissue also provides the bulk of the world's food supply in the form of cereal grains.

As in the case of the conifers, the outer tissues of the ovule typically harden into a resistant seed coat. Angiosperm seeds have a second layer of protective tissues, the fruit. The **fruit** consists of the ovary wall and, in some cases, receptacle tissue.

Angiosperms are divided into two major groups, **dicotyledons** (dicots) and **monocotyledons** (monocots). This distinction is based primarily on the number of **cotyledons,** or seed leaves, present in the embryo. In addition, the two groups differ with respect to other anatomical features such as the arrangement of their vascular tissues and floral structure.

As the dominant plant group on earth, and because of their great economic and agricultural importance, angiosperms have been studied much more intensively than other type of plants and are covered extensively in this book. Inevitably, scientists have focused their attention on a relatively small number of species that are convenient as experimental systems. However, it is important not to lose sight of the tremendous diversity of form and function to be found not only among the angiosperms but among the entire plant kingdom as well.

The Plant: An Overview of Structure

Despite their apparent diversity, seed plants all exhibit the same basic body plan (Fig. 1.2). The vegetative body is composed of three organs: **leaf, stem,** and **root.** The primary function of a leaf is photosynthesis, that of the stem is support, and that of the root is anchorage and absorption of water and minerals. Leaves are attached to the stem at the **node,** and the region of the stem between

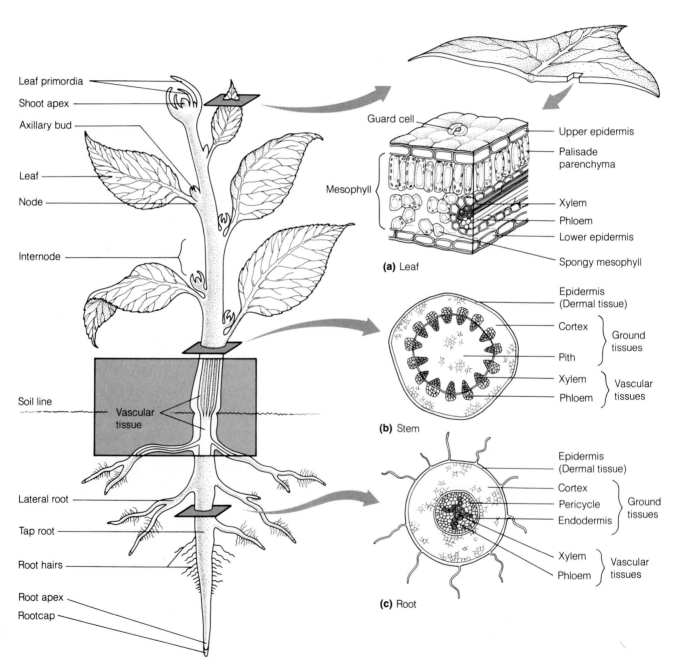

FIGURE 1.2. Schematic representation of the body of a seed plant. Cross sections of (a) the leaf, (b) the stem, and (c) the root, are also shown.

two nodes is termed the **internode**. The stem together with its leaves is commonly referred to as the **shoot**.

A fundamental difference between plants and animals is that each plant cell is surrounded by a rigid **cell wall**. In animals, embryonic cells can migrate from one location to another, resulting in the development of tissues and organs containing cells that originated in different parts of the organism. In plants, each walled cell and its neighbor are cemented together by a **middle lamella**, like bricks held together with mortar. This architecture prevents any cell migration, and, as a consequence, plant development depends solely on the pattern of cell division and cell enlargement.

Plant cells have two types of walls: primary and secondary (Fig. 1.3). **Primary cell walls** are typically thin (<1 μm) and are characteristic of young, growing cells. **Secondary cell walls** are thicker and stronger than primary walls and are deposited when most of cell enlargement has ended.

New Cells Are Produced by Dividing Tissues Called Meristems

Plants have localized regions of cell division called **meristems**. Nearly all nuclear division (mitosis) and cell division (cytokinesis) occur in these meristematic regions. In a young plant, the most active meristems are typically found at the tips of the stem and the root and are called **apical meristems** (Fig. 1.4). At the nodes, **axillary buds** contain the apical meristems for new shoots. Lateral roots arise from the **pericycle**, and *internal* meristematic

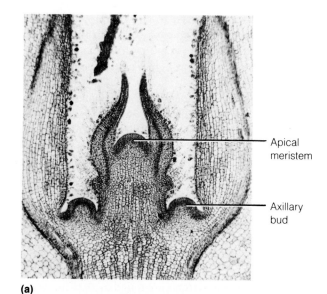

(a)

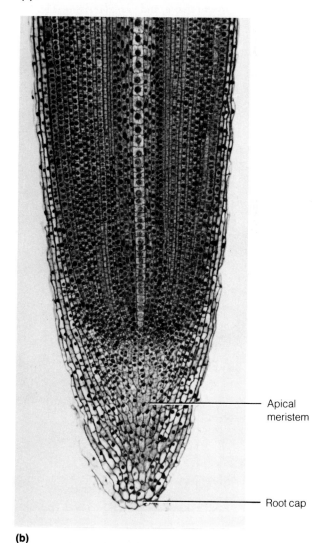

(b)

FIGURE 1.4. Longitudinal sections of (a) a shoot tip (46×) and (b) a root (100×) from flax (*Linum usitatissimum*), showing the apical meristems. (From Esau, 1960.)

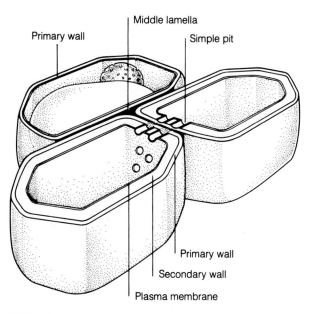

FIGURE 1.3. Schematic representation of primary and secondary cell walls and their relationship to the rest of the cell.

tissue. Immediately distal to and overlapping the meristematic regions are zones of cell elongation in which cells dramatically increase in length and width. Differentiation into specialized cell types usually occurs after cell elongation.

The phase of plant development that gives rise to new organs and to the basic plant form is called **primary growth**. Primary growth results from the activity of apical meristems (Fig. 1.4), in which cell division is followed by progressive cell enlargement, typically elongation. After elongation in a given region is essentially complete, **secondary growth** may occur. Secondary growth involves two lateral meristems, the **vascular cambium** and the **cork cambium**.

Plants Are Composed of Three Major Tissues

Three major tissue systems are found in all plant organs: *dermal tissue*, *ground tissue*, and *vascular tissue* (Fig. 1.2).

Dermal Tissue. The **epidermis** is the **dermal tissue** of young plants undergoing primary growth (Fig. 1.5a). It is generally composed of specialized, flattened polygonal cells that occur on all plant surfaces. Shoot surfaces are usually coated with a waxy **cuticle** to prevent water loss and are often covered with hairs or trichomes, which are epidermal cell extensions. Pairs of specialized epidermal cells, the **guard cells**, are found surrounding microscopic pores in all leaves. The guard cells and pores are called **stomata** (singular stoma), and they permit gas exchange (water loss and CO_2 uptake) between the atmosphere and the interior of the leaf. The root epidermis is adapted for absorption of water and minerals, and its outer wall surface typically does not have a waxy cuticle. Extensions from the root epidermal cells, the **root hairs**, increase the surface area over which absorption can take place.

Ground Tissue. Making up the bulk of the plant are cells termed the **ground tissue**. There are three types of ground

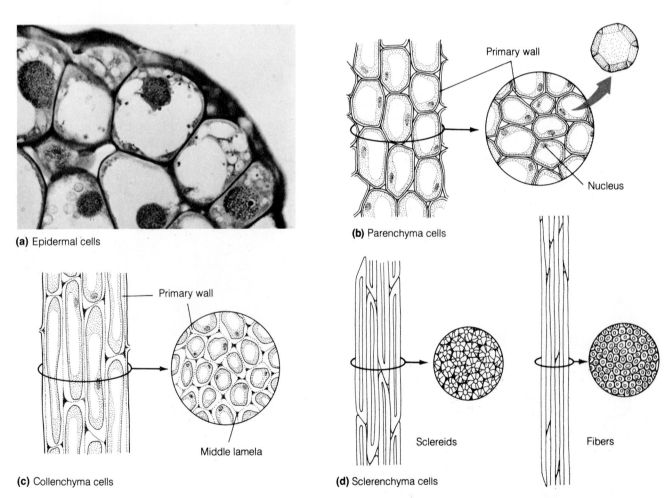

(a) Epidermal cells

(b) Parenchyma cells

Primary wall

Nucleus

(c) Collenchyma cells

Primary wall

Middle lamela

(d) Sclerenchyma cells

Sclereids

Fibers

FIGURE 1.5. (a) The outer epidermis of a wheat coleoptile, seen in transverse section (750 ×). (From O'Brien and McCully, 1969.) Diagrammatic representations of (b) parenchyma, (c) collenchyma, and (d) sclerenchyma cells. (From Esau, 1977.)

tissue: parenchyma, collenchyma, and sclerenchyma (Fig. 1.5b–d). **Parenchyma**, the most abundant ground tissue, consists of thin-walled, metabolically active cells that carry out a variety of functions in the plant, including photosynthesis and storage. **Collenchyma** tissue is composed of narrow, elongated cells with thick primary walls. Collenchyma cells provide structural support to the growing plant body, particularly shoots, and their thickened walls are nonlignified, so they can stretch as the organ elongates. Collenchyma cells are typically arranged in bundles or layers near the periphery of stems or leaf petioles. **Sclerenchyma** consists of two types of cells, sclereids and fibers. Both have thick secondary walls and are frequently dead at maturity. **Sclereids** occur in a variety of shapes, ranging from roughly spherical to branched, and are widely distributed throughout the plant. In contrast, **fibers** are narrow, elongated cells that are commonly associated with vascular tissues. The main function of sclerenchyma is to provide mechanical support, particularly to parts of the plant that are no longer elongating.

In the stem, the pith and cortex make up the ground tissue (Fig. 1.2). The **pith** is located within the cylinder of vascular tissue, where it often exhibits a spongy texture due to the presence of large intercellular air spaces. If the growth of the pith fails to keep up with that of the surrounding tissues, the pith may degenerate, producing a hollow stem. In general, roots lack piths, although there are exceptions to this rule. In contrast, the **cortex**, which is located between the epidermis and the vascular cylinder, is present in both stems and roots.

At the boundary between the ground tissue and the vascular tissue in roots, and occasionally in stems, is a specialized layer of cortex known as the **endodermis** (Fig. 1.2). This single layer of cells originates from cortical tissue at the innermost layer of the root cortex and forms a cylinder that surrounds the central vascular tissue or **stele**. Early in root development, a narrow waxy band is formed on the walls that are perpendicular to the surface of the root (the anticlinal and radial walls) circumscribing each endodermal cell. These waxy deposits, called **casparian strips**, form a barrier in the endodermal walls to the intercellular movement of water, ions, and other solutes to the vascular cells.

Leaves have two interior layers of ground tissue that are collectively known as the **leaf mesophyll** (Fig. 1.2). The **palisade parenchyma** consists of closely spaced, columnar cells located beneath the upper epidermis. There is usually one layer of palisade parenchyma in the leaf. Palisade parenchyma cells are rich in chloroplasts and are a primary site of photosynthesis in the leaf. Below the palisade parenchyma are irregularly shaped, widely spaced **spongy mesophyll** cells. The mesophyll cells are also photosynthetic, and the large spaces between these cells allow diffusion of carbon dioxide. The spongy mesophyll also contributes to leaf flexibility in the wind, and this flexibility facilitates the movement of gases within the leaf.

Vascular Tissues: Xylem and Phloem. The **vascular tissue** is composed of two major conducting systems. The **xylem** transports water and mineral ions from the root to the rest of the plant. The **phloem** distributes the products of photosynthesis and a variety of other solutes throughout the plant (Fig. 1.2).

The **tracheids** and **vessel elements** are the conducting cells of the xylem (Fig. 1.6). Both of these cell types have elaborate secondary wall thickenings and lose their cytoplasm at maturity; that is, they are dead when functional. Vessel members are also arranged end to end to form a larger unit called a **vessel**. Other cell types present in the xylem include parenchyma cells, important for storage of energy-rich molecules and phenolic compounds, and sclerenchyma fibers.

The **sieve tube elements** and **sieve cells** are responsible for sugar translocation in the phloem (Fig. 1.6). The former are found in angiosperms, and the latter perform the same function in gymnosperms. Like vessel members, sieve tube elements are often stacked in vertical rows, forming larger units called **sieve tubes**. Both types of conducting cells are living when functional but have relatively few cytoplasmic organelles. Sieve tube elements are associated with densely cytoplasmic parenchyma cells called **companion cells**. The analogous cells adjacent to the sieve cells of gymnosperms are the **albuminous cells**. In addition, the phloem frequently contains storage parenchyma and fibers that provide mechanical support.

The Plant Cell

Plants are multicellular organisms made of millions of cells with specialized functions. All plant cells, however, have a common organization; they contain a nucleus, a cytoplasm, and subcellular organelles, and they are enclosed in a membrane that defines their boundaries (Fig. 1.7). The following sections provide an overview of the components of the cell.

Biological Membranes Are Phospholipid Bilayers Containing Proteins

All cells are enclosed in a membrane that serves as their outer boundary, separating the cytoplasm from the external environment (Fig. 1.7). This **plasma membrane** allows the cell to take up and retain certain substances while excluding others. Various transport proteins embedded in the plasma membrane are responsible for this

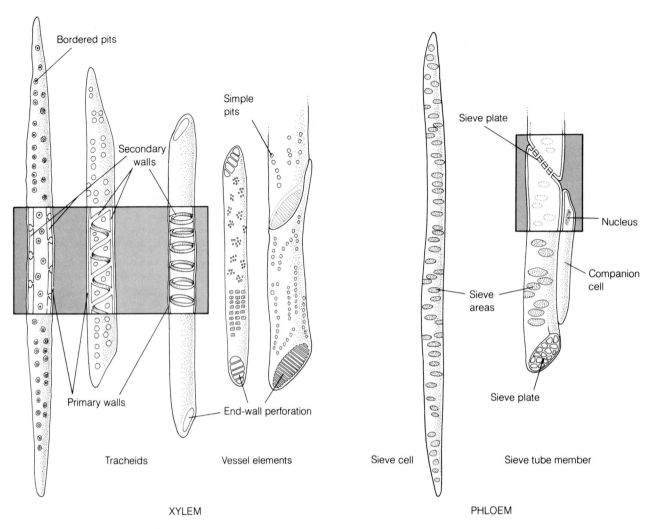

FIGURE 1.6. Diagrammatic representation of xylem and phloem conducting cells.

selective traffic of solutes across the membrane. The accumulation of ions or molecules in the cytosol through the action of transport proteins consumes metabolic energy.

All biological membranes have the same basic molecular organization. They consist of a double layer (bilayer) of phospholipids interspersed with proteins (Fig. 1.8). The composition of the phospholipids and the properties of the proteins vary from membrane to membrane, conferring on each membrane its unique functional characteristics.

Phospholipids are a class of lipids in which two fatty acids are covalently linked to glycerol, which is covalently linked to a phosphate group. Also attached to this phosphate group is a variable component such as serine, choline, glycerol, or inositol (Fig. 1.8c). In contrast to the fatty acids, these variable components are highly polar; consequently, phospholipid molecules display both hydrophilic and hydrophobic properties. The nonpolar hydrocarbon chains of the fatty acids form a region that is exclusively hydrophobic and that excludes water. The polar portions of the molecules interact with the polar water molecules surrounding the hydrophobic core area (Fig. 1.8).

The proteins associated with the lipid bilayer are of two types, integral and peripheral (they have also been called intrinsic or extrinsic, respectively). **Integral proteins** are embedded in the lipid bilayer. Most integral proteins span the entire width of the phospholipid bilayer, so that one part of the protein interacts with the outside of the cell, another part interacts with the hydrophobic core of the membrane, and a third part interacts with the interior of the cell, the cytosol.

Peripheral proteins are attached to the membrane surface by noncovalent bonds, such as ionic bonds or hydrogen bonds. Peripheral proteins have several roles in membrane function. For example, they are involved in interactions between the plasma membrane and components of the cytoskeleton, such as microtubules and actin microfilaments, which are discussed later in this chapter.

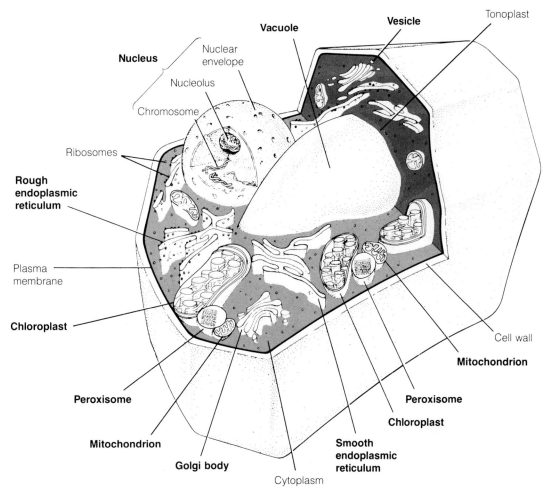

FIGURE 1.7. Diagrammatic representation of a plant cell. Note the intracellular compartments defined by the tonoplast, the nuclear envelope, and the membranes of the other organelles, labeled in boldface type.

The Nucleus Contains the Genetic Material of the Cell

The nucleus is the organelle that contains most of the cell's DNA, collectively referred to as the **genome.** The membrane system that surrounds the nucleus, called the **nuclear envelope,** is actually two discrete membranes (Fig. 1.9a). The space between the two membranes of the nuclear envelope is called the **perinuclear space,** and the two membranes of the nuclear envelope are fused at sites called **nuclear pores** (Fig. 1.9b), which consist of polygonal arrays of globular subunits rich in protein and RNA. There can be very few to many thousands of pores on an individual nuclear envelope. Macromolecules from the nucleus, including ribosomal subunits, are able to pass through the nuclear pores into the cytosol and vice versa.

The nucleus is the site of storage and replication of the **chromosomes,** composed of DNA and its associated proteins. Collectively, this DNA-protein complex is known as **chromatin,** which appears dispersed during

interphase. The linear length of all of the DNA within any plant genome is usually millions of times greater than the diameter of the nucleus in which it is found. To solve the problem of packaging this chromosomal DNA within the nucleus, segments of the linear double helix of DNA are coiled twice around a solid cylinder of eight **histone** protein molecules, forming a **nucleosome.** Nucleosomes are arranged like beads on a string along the length of each chromosome. During mitosis, the chromatin undergoes condensation by further coiling and folding processes that depend on protein-nucleic acid interactions.

The Genome Size of Higher Plants Is Quite Variable

The nuclear DNA content varies tremendously in higher plants. Whereas all mammals have approximately 3×10^9 base pairs (bp) of double-stranded DNA per haploid genome (a haploid genome is the DNA content of a

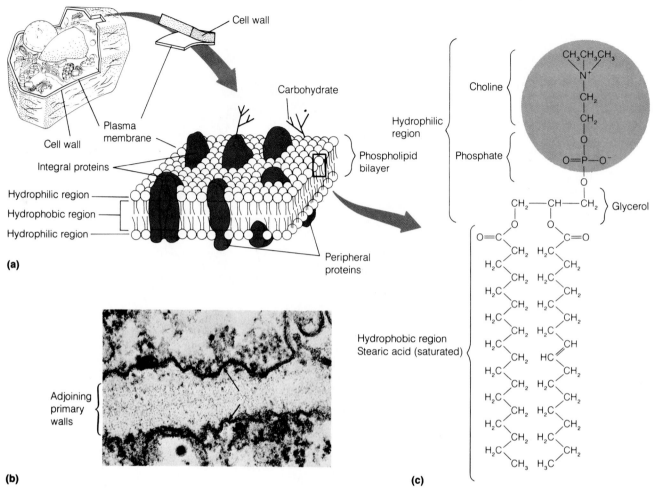

FIGURE 1.8. (a) The plasma membrane of plant cells consists of proteins embedded in a phospholipid bilayer. (b) The transmission electron micrograph shows plasma membranes (arrows) in a parenchyma cell from wheat. The overall thickness of the plasma membrane, viewed as two dense lines and an intervening space, is 8 nanometers. (From Ledbetter and Porter, 1970.) (c) Chemical structure of a typical phospholipid, phosphatidylcholine.

sperm or an egg), the higher plants have haploid genome DNA contents ranging from 7×10^7 bp, for the angiosperm *Arabidopsis thaliana*, to 2×10^{11} bp for certain gymnosperms! Even closely related beans of the genus *Vicia* exhibit genomic DNA contents that vary over a 20-fold range.

Studies of plant molecular biology have shown that most of the DNA in plants that have large genomes is **repetitive** DNA. In contrast to **unique** DNA, which codes for genes that are translated into proteins, the repetitive DNA seldom codes for actual genes but appears to play a role in plant genome organization, gene expression, and development.

Ribosomes Are Formed in the Nucleolus and Are the Sites of Protein Synthesis

Nuclei contain a densely granular region called the **nucleolus** (Fig. 1.9a) that is the site of ribosome synthesis.

The nucleolus encloses portions of one or more chromosomes where ribosomal RNA (rRNA) genes are clustered, called the **nucleolus organizer.** Typical cells have one or more nucleoli per nucleus. Each 80S ribosome is made of a large and a small subunit, and each subunit is a complex aggregate of rRNA and specific proteins. The two subunits exit the nucleus separately, through the nuclear pore, and then unite in the cytoplasm to form a complete ribosome (Fig. 1.10a).

Ribosomes are the sites of protein synthesis. This complex process starts with **transcription**—the synthesis of an RNA polymer bearing a base sequence complementary to a specific gene. The RNA transcript is processed to become messenger RNA (mRNA), which moves from the nucleus to the cytoplasm and there attaches first to the small ribosomal subunit and then to the large subunit to initiate translation. **Translation** is the process whereby a specific protein is synthesized from amino acids, using the sequence information encoded by the mRNA.

Mitochondria   Chloroplasts

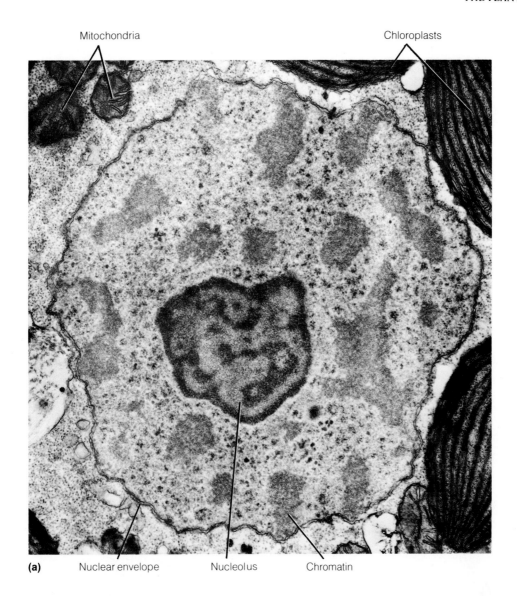

(a)   Nuclear envelope   Nucleolus   Chromatin

(b)

FIGURE 1.9. (a) Transmission electron micrograph of a plant cell showing the nucleous and the nuclear envelope (18,000 ×). (b) A freeze-etched preparation of nuclear pores from a cell of an onion root tip (28,000 ×). (Courtesy D. Branton.)

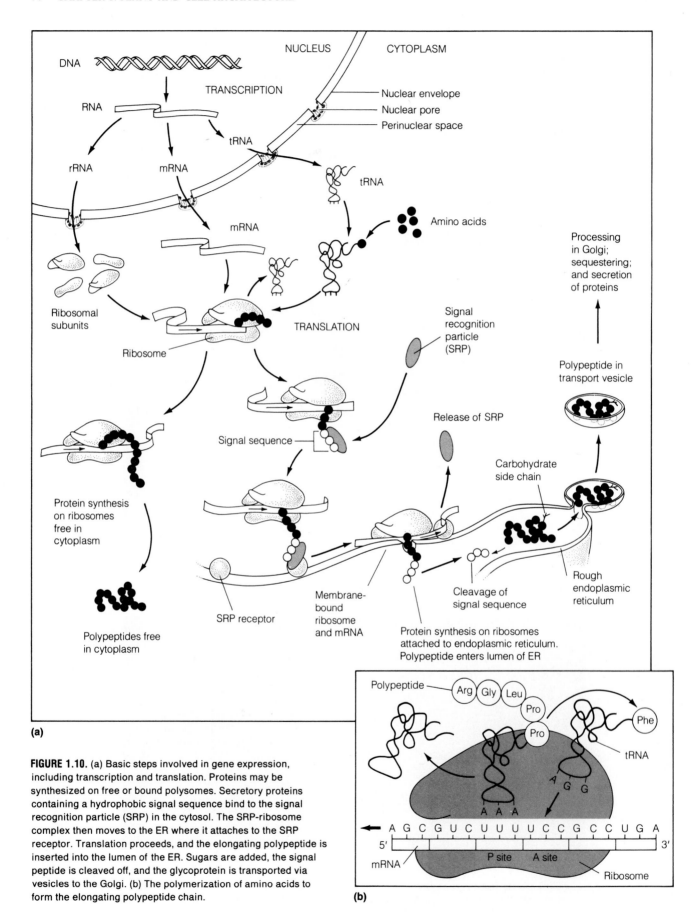

(a)

(b)

FIGURE 1.10. (a) Basic steps involved in gene expression, including transcription and translation. Proteins may be synthesized on free or bound polysomes. Secretory proteins containing a hydrophobic signal sequence bind to the signal recognition particle (SRP) in the cytosol. The SRP-ribosome complex then moves to the ER where it attaches to the SRP receptor. Translation proceeds, and the elongating polypeptide is inserted into the lumen of the ER. Sugars are added, the signal peptide is cleaved off, and the glycoprotein is transported via vesicles to the Golgi. (b) The polymerization of amino acids to form the elongating polypeptide chain.

The ribosome travels along the entire length of the mRNA and serves as the site for the sequential bonding of amino acids as specified by the base sequence of the mRNA (Fig. 1.10b).

The Endoplasmic Reticulum Is the Major Network of Internal Membranes in the Cell

Cells have an elaborate network of internal membranes: the endoplasmic reticulum (ER) and the Golgi body. The membranes present in both the ER and the Golgi body are typical lipid bilayers with interspersed integral and peripheral proteins. These membranes form flattened or tubular sacs known as **cisternae**.

Structural studies have shown that the ER is continuous with the outer membrane of the nuclear envelope. There are two types of ER, smooth ER and rough ER (Fig. 1.11), and the two types are interconnected. Rough ER differs from smooth ER in that it is covered with ribosomes; in addition, rough ER tends to be lamellar, while smooth ER tends to be tubular, although a gradation for each type can be observed in almost any cell. The structural differences between the two forms of ER are accompanied by functional differences. Smooth ER functions as a major site of lipid synthesis and membrane assembly. Rough ER is the site for synthesis of membrane proteins and proteins to be secreted outside the cell or into the vacuoles. These proteins enter the secretory pathway through the lumen of the ER after crossing the ER membrane.

The mechanism of transport across the membrane is a complex one involving the ribosomes, the mRNA coding for the secretory protein, and a special receptor in the ER membrane. All secretory proteins and most integral membrane proteins have been shown to have a hydrophobic sequence of 18 to 30 amino acid residues at the amino terminal end of the chain. During translation, this hydrophobic leader, called the **signal peptide** sequence, is recognized by a **signal recognition particle** (**SRP**), made up of protein and RNA, which facilitates binding of the free ribosome to **SRP-receptor** proteins (or "docking proteins") on the ER (Fig. 1.10). The signal peptide then mediates the transfer of the elongating polypeptide across the ER membrane. (In the case of integral membrane proteins, a portion of the completed polypeptide remains embedded in the membrane.) Once inside the lumen of the ER, the signal sequence is cleaved off by a **signal peptidase**, and sugars, especially N-acetylglucosamine, mannose and glucose, are added to the free amino group of a specific asparagine side chain. The modified

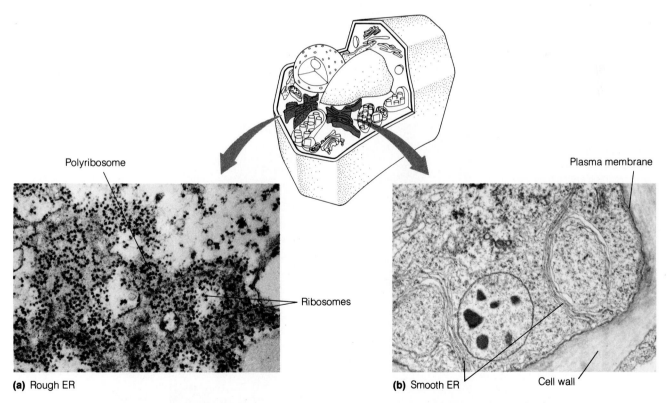

FIGURE 1.11. The endoplasmic reticulum. (a) Rough ER can be seen in a micrograph from a radish root cell. The ribosomes in the rough ER are clearly visible. The polyribosomes are aggregates of ribosomes bound by messenger RNA (36,900 ×). (b) Smooth ER can be seen in a transmission electron micrograph from a corn cell (14,700 ×). (From R. Warmbrodt.)

glycoprotein is then transported to the Golgi via vesicles for further processing.

Secretory Proteins and Polysaccharides Are Processed and Sorted in the Golgi

Each Golgi body in a plant cell is a stack of flattened membrane cisternae that is separate from the ER network (Fig. 1.12). The old term for the Golgi body, **dictyosome**, may still be found in the plant literature. The Golgi body is a polarized structure: the cisternae closest to the plasma membrane are called the *trans* face and the cisternae closest to the center of the cell are called the *cis* face. The Golgi body functions in the processing of proteins that are transferred to it, enclosed in membrane-bounded vesicles, from the lumen of the ER. These vesicles form on the ER, move through the cytosol, and fuse with the *cis* face of the Golgi (Fig. 1.12). Within the lumen of the Golgi, the glycoproteins are enzymatically modified. Certain sugars, such as mannose, are removed, and others are added to the oligosaccharide chain. Glycosylation of the OH−groups of serine, threonine and tyrosine residues (O−linked oligosaccharides) also takes place in the Golgi. After being processed within the Golgi, the glycoproteins leave the organelle in other vesicles, typically from the *trans* side of the stack. All of this processing appears to confer a specific tag or marker on each protein specifying its ultimate destination inside or outside the cell.

In plant cells, the Golgi body is important in cell wall formation. Noncellulosic cell wall polysaccharides (hemicellulose and pectins) and a cell wall protein called ex-

tensin are all synthesized and processed within the Golgi. **Secretory vesicles** derived from the Golgi carry the polymers to the plasma membrane, where the vesicles fuse with the plasma membrane and empty their contents into the cell wall region (Fig. 1.12). Two types of vesicles are present: **smooth secretory vesicles** and **coated vesicles**. Unlike the smooth vesicles, coated vesicles are surrounded by the protein **clathrin**. Clathrin subunits are arranged in the form of three-pronged "triskelions" which associate with each other to form cages around the vesicles (Fig. 1.13). In animal cells, coated vesicles function in endocytosis and in the transfer of materials within the endomembrane system. In plants, coated vesicles have been implicated in the transport of storage protein to specialized protein-storing vacuoles.

The Vacuole Occupies Most of the Volume of the Mature Plant Cell

Vacuoles are conspicuous organelles of plant cells (Fig. 1.7) and are surrounded by a membrane called the **tonoplast**. In immature cells or in meristematic tissue, vacuoles are less prominent, though they are always present, sometimes as small provacuoles. Provacuoles appear to be produced by a system of Golgi tubes called the **trans Golgi network**. As the cell begins to mature, the provacuoles fuse together to produce the large central vacuoles characteristic of most mature plant cells. In such cells, the cytoplasm occurs as a thin layer surrounding the vacuole. Some cells have cytoplasmic strands that traverse the vacuole, but each transvacuolar strand is

FIGURE 1.12. Electron micrograph of a Golgi body in a green alga, *Bulbochaete*. Vesicles from the endoplasmic reticulum at the top of the micrograph are on their way to the Golgi body. Mature vesicles are visible at the *trans* side of the stack, in the lower part of the micrograph (59,500×). (From T. W. Fraser.)

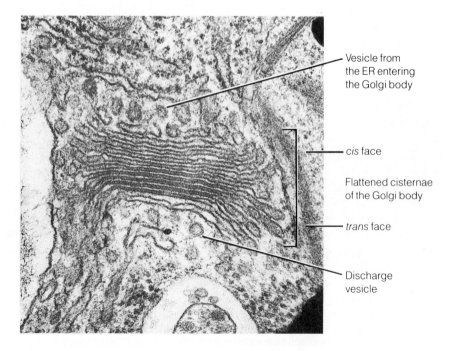

Vesicle from the ER entering the Golgi body

cis face

Flattened cisternae of the Golgi body

trans face

Discharge vesicle

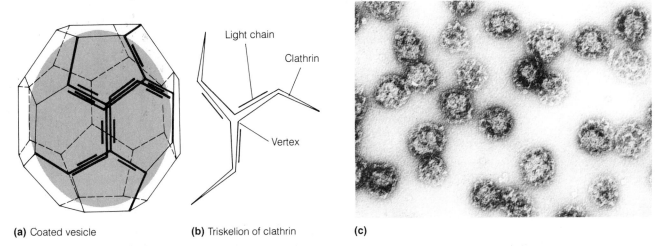

(a) Coated vesicle **(b)** Triskelion of clathrin **(c)**

FIGURE 1.13. Coated vesicles. (a) The organization of triskelions into the clathrin coat. (b) Individual triskelion, composed of three clathrin heavy chains and three light chains. (c) Preparation of coated vesicles isolated from bean leaves (102,000 ×). (From Depta et al., 1987.)

surrounded by the tonoplast. The vacuole contains inorganic ions, organic acids, sugar, enzymes, and a variety of secondary metabolites, which often play roles in plant defense. Solute accumulation provides the osmotic driving force for water uptake by the vacuole, which is necessary for plant cell enlargement. Vacuoles are rich in hydrolytic enzymes, including proteases, ribonucleases, and glycosidases, which, upon release into the cytosol, participate in the degradation of the cell during senescence. These hydrolytic enzymes may also participate in the turnover of cellular constituents throughout the life of the cell. Specialized protein-storing vacuoles, called **protein bodies**, are abundant in seeds. During germination the storage proteins are hydrolyzed and exported to the cytosol for use in protein synthesis.

Mitochondria and Chloroplasts Are Sites of Energy Conversion

A typical plant cell has two types of energy-producing organelles: mitochondria and chloroplasts. Both types of organelles are separated from the cytosol by a double membrane (an outer and an inner membrane). **Mitochondria** are the cellular sites of respiration, a process in which the energy released from sugar metabolism is used for the synthesis of ATP from ADP and inorganic phosphate (P_i). **Chloroplasts** belong to a group of double membrane-bounded organelles called **plastids**. Chloroplasts contain chorophyll and its associated proteins and are the sites of photosynthesis. Plastids that contain high concentrations of carotenoid pigments rather than chlorophyll are called **chromoplasts,** and they are the source of the yellow, orange, or red colors of many fruits and flowers, as well as autumn leaves.

Mitochondria can vary in shape from spherical to tubular, but they all have a smooth outer membrane and a highly convoluted inner membrane (Fig. 1.14). The infoldings of the inner membrane are called **cristae.** The compartment enclosed by the inner membrane is the mitochondrial **matrix,** and it contains the enzymes of the pathway of intermediary metabolism called the Krebs cycle. In contrast to the mitochondrial outer membrane and all other membranes in the cell, the inner membrane of mitochondria is almost 70% protein and contains some phospholipids that are unique to the organelle (e.g., cardiolipin). The proteins present in and on the inner membrane have special enzymatic and transport capacities. The inner membrane is highly impermeable to the passage of H^+ and serves as a barrier to the movement of protons. This important feature underlies the formation of electrochemical gradients. Dissipation of such gradients by the controlled movement of H^+ through the transmembrane protein, H^+-ATPase, is coupled to the phosphorylation of ADP to produce ATP. ATP can then be released to other cellular sites where energy is needed to drive specific reactions.

Chloroplasts possess a third system of membranes, called **thylakoids** (Fig. 1.15). A stack of thylakoids forms a **granum** (plural grana). Proteins and pigments (chlorophylls and carotenoids) that function in the photochemical events of photosynthesis are part of the thylakoid. The fluid compartment surrounding the thylakoids is called the **stroma** and is analogous to the matrix of the mitochondrion. Adjacent grana are connected by nonstacked thylakoids called **stroma lamellae.** The different components of the photosynthetic apparatus are localized in different areas of the grana and the stroma lamellae.

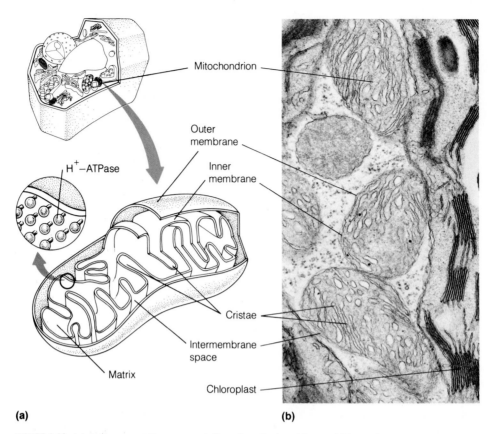

(a)   **(b)**

FIGURE 1.14. (a) A diagrammatic representation of a mitochondrion and (b) an electron micrograph of mitochondria from a leaf cell of Bermuda grass (31,600 ×). (Photo by M. E. Doohan, Courtesy of E. H. Newcomb.)

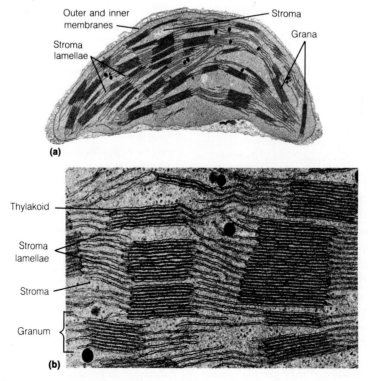

FIGURE 1.15. (a) An electron micrograph of a chloroplast from a leaf of timothy grass, *Phleum pratense* (12,800 ×). (b) The same preparation at higher magnification (44,400 ×). (Micrographs by W. P. Wergin; photos provided by E. H. Newcomb.)

Most Mitochondrial and Chloroplast Proteins Are Imported from the Cytosol During Organelle Assembly

The stroma of chloroplasts and the matrix of mitochondria contain a small circular piece of double-helical DNA, similar to the circular chromosome of bacteria and very different from the chromatin found in the plant cell nucleus. DNA replication in both mitochondria and chloroplasts is clearly independent of DNA replication in the nucleus. On the other hand, the numbers of these organelles within a given cell type remain approximately constant, which suggests that some aspects of organelle replication are under cellular regulation.

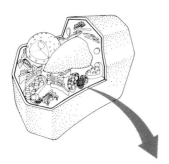

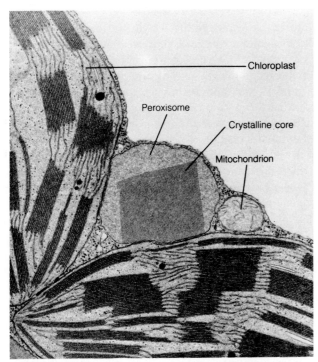

Chloroplast

Peroxisome

Crystalline core

Mitochondrion

FIGURE 1.16. An electron micrograph of a peroxisome from a mesophyll cell, showing a crystalline core (17,600 ×). The peroxisome is seen in close association with a chloroplast and a mitochondrion, probably reflecting the cooperative role of these organelles in photorespiration. The crystalline structure is catalase. (Photo provided by E. H. Newcomb.)

The mitochondrial genome ranges from 16,500 bp in humans to 78,000 bp in yeast. Most of the proteins encoded by the mitochondrial genome are 70S ribosomal proteins or components of the electron transport system. The majority of mitochondrial proteins, including Krebs cycle enzymes, are imported from the cytosol.

The chloroplast genome is larger than that of the mitochondrion, about 145,000 bp, and is present in multiple copies. Chloroplast DNA encodes rRNA; transfer RNA (tRNA); the large subunit of the enzyme that fixes CO_2, ribulose-1,5-bisphosphate carboxylase/oxygenase (Rubisco); and several of the proteins that participate in photosynthesis. Nevertheless, the majority of chloroplast proteins are synthesized in the cytosol and transported to the chloroplast.

Meristem cells contain **proplastids**, which have few or no internal membranes, no chlorophyll, and an incomplete complement of the enzymes necessary to carry out photosynthesis. Chloroplast development from proplastids is triggered by light. Upon illumination, enzymes are formed inside the proplastid or imported from the cytosol, light-absorbing pigments are produced, and membranes proliferate rapidly, giving rise to stroma lamellae and grana stacks.

Seeds usually germinate in the soil away from light, and chloroplasts develop only when the young shoot is exposed to light. If seeds are germinated in the dark, the proplastids differentiate into **etioplasts**, which contain semicrystalline tubular arrays of membrane known as **prolamellar bodies**. Instead of chlorophyll, the etioplast contains a pale yellow-green precursor pigment, protochlorophyll.

Within minutes after exposure to light, the etioplast undergoes a process of differentiation, converting the prolamellar body into thylakoids and stroma lamellae and the protochlorophyll into chlorophyll. The maintenance of chloroplast structure is dependent on the presence of light, because mature chloroplasts can revert to etioplasts during extended periods of darkness.

Microbodies Play Specialized Metabolic Roles in Leaves and Seeds

Plant cells also have **microbodies**, which are compartments bounded by a single membrane and specialized in a particular metabolic pathway. **Peroxisomes** are spherical organelles specialized in oxidation reactions (Fig. 1.16). One oxidative enzyme, catalase, constitutes up to 40% of the total protein of the peroxisome (Fig. 1.16). Along with mitochondria, peroxisomes are a major site of O_2 consumption in the cell. The oxygen is used in oxidation reactions that convert harmful metabolic by-products into nontoxic substances such as water. Peroxisomes also break down fats and participate in **photorespiration**, an important metabolic pathway coupled to photosynthesis.

Another type of microbody, the **glyoxysome**, is present in oil-storing seeds. Glyoxysomes contain the glyoxylate cycle enzymes, which help to convert stored fatty acids into sugars that can be translocated throughout the young plant to provide energy for growth.

The Cytoskeleton

The cytosol is organized into a three-dimensional network of fibrous proteins called the **cytoskeleton**. This network provides a spatial organization for the organelles and contributes to their movement in the cytosol. It also plays fundamental roles in mitosis, meiosis, cytokinesis, wall deposition, the maintenance of cell shape, and cell differentiation.

Plant cells have two types of cytoskeletal elements: microtubules and microfilaments. Each element is filamentous, having a fixed diameter and a variable length, up to many micrometers. **Microtubules** are hollow cylinders with an outer diameter of 25 nm. **Microfilaments** are solid, with a diameter of 7 nm. Both microtubules and microfilaments are macromolecular assemblies of globular proteins. Microtubules are composed of the protein **tubulin** and microfilaments are composed of the protein **actin**. A single microtubule consists of hundreds of thousands of tubulin subunits, usually arranged in 13 columns called protofilaments (Fig. 1.17). A microfilament consists of two actin chains that intertwine in a helical fashion (Fig. 1.18).

Microtubules and Microfilaments Can Assemble and Disassemble

The tubulin subunit of microtubules is composed of two similar polypeptide chains (α- and β-tubulin), each having an apparent molecular mass of 55,000 daltons (Fig. 1.17). Each actin molecule is composed of a single polypeptide with a molecular mass of approximately 42,000 daltons.

Under physiological conditions, tubulin polymerizes into a microtubule and actin polymerizes into a microfilament. Under appropriate cellular conditions, each polymer can also disassemble into its subunits. The attachments between subunits in the polymer are noncovalent but are of sufficient strength to render the structure stable. The overall *rate* of assembly and disassembly of these structures is affected by the relative concentrations of free or assembled subunits. Other factors also affect the assembly and stability of cytoskeletal components, such as the cytosolic calcium concentration (high levels of calcium promote microtubule disassembly).

Microtubules Function in Mitosis and Cytokinesis

Mitosis is the process by which previously replicated chromosomes are aligned, separated, and distributed in an orderly fashion to daughter cells (Fig. 1.19). Microtubules are an integral part of mitosis. Prior to the onset of mitosis, microtubules in the cytoplasm depolymerize, breaking down to their constituent subunits. The subunits then repolymerize early in prophase to form the microtubules characteristic of the **spindle apparatus**. At each end of the spindle, a spindle pole region contains a **microtubule organizing center** (MTOC), which is the site of microtubule formation. Some of the microtubules become attached to the chromosomes at their **kinetochores** (kinetochore microtubules), while others remain unattached (nonkinetochore microtubules). The kinetochores are located in the **centromeric** regions of the chromosomes. Some of the unattached microtubules overlap with microtubules from the opposite polar region in the spindle midzone.

Cytokinesis is the process whereby a cell is partitioned into two progeny cells. Cytokinesis usually begins late in mitosis. The precursor of the new wall that forms between incipient daughter cells is called the **cell plate**, and it is rich in pectins. Cell plate formation in higher plants is a two-step process. In the first step, Golgi vesicles and ER cisternae aggregate in the spindle midzone area (Fig. 1.20). The process of vesicle aggregation in the spindle midzone is organized by the **phragmoplast**, a

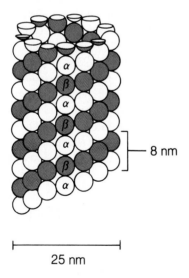

```
                8 nm
         25 nm
```

FIGURE 1.17. Drawing of a microtubule in longitudinal view. Each microtubule is arranged in 13 protofilaments. The organization of the α and β subunits is shown. (From Avers, 1986.)

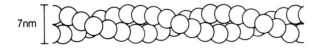

7nm

FIGURE 1.18. Diagrammatic representation of a microfilament, showing two strands of actin subunits.

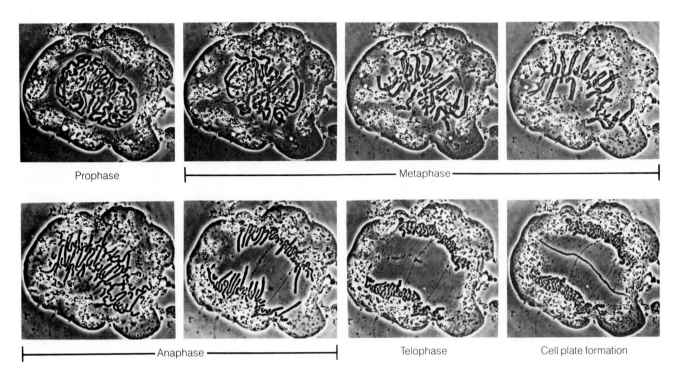

Prophase

Metaphase

Anaphase

Telophase

Cell plate formation

FIGURE 1.19. Mitosis in a living endosperm cell of the African blood lily, *Haemanthus katherinae*, recorded with time-lapse light microscopy. (Photo by W. T. Jackson.)

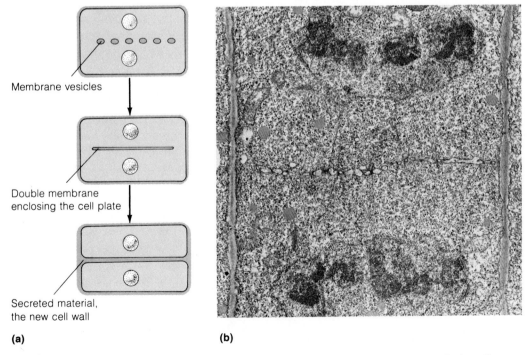

Membrane vesicles

Double membrane enclosing the cell plate

Secreted material, the new cell wall

(a)

(b)

FIGURE 1.20. Cell plate formation, shown diagrammatically in (a). (b) Electron micrograph of a cell plate forming in a root tip of soybean (11,000 ×). (Photo provided by E. H. Newcomb.)

complex of microtubules and ER that forms during late anaphase or early telophase from dissociated spindle subunits. In the second step, the membranes of the vesicles fuse with each other and with the lateral plasma membrane to become the new plasma membrane separating the progeny cells (Fig. 1.20). The contents of the vesicles and cisternae are the precursors from which the new middle lamella and primary wall are assembled outside the cell. At the end of mitosis, the phragmoplast disappears, the cell enters interphase, and microtubules reappear in the cytosol near the plasma membrane, where they play a role in the deposition of cellulose microfibrils during cell wall growth.

The Preprophase Band of Microtubules Determines the Plane of Cytokinesis

The plane in which the cell plate forms establishes the relationship of the progeny cells, and microtubules play a role in this process. Following the breakdown of cytoplasmic microtubules but before the spindle forms and mitosis begins, a prominent array of 20 to 100 microtubules is formed in the cytosol. This array is called the **preprophase band**, and it appears as a narrow belt in the region where the cell plate will form following the completion of mitosis. Although the preprophase band is no longer present when the cell plate actually forms, the cell plate attaches to the parental cell wall at the site previously occupied by the preprophase band.

Microfilaments Are Involved in Cytoplasmic Streaming and in Tip Growth

Cytoplasmic streaming is the coordinated movement of particles and organelles through the cytosol in a helical path down one side of a cell and up the other side. Cytoplasmic streaming occurs in most plant cells and has been extensively studied in the large cells of the green algae *Chara* and *Nitella*, where speeds up to 75 μm per second have been measured.

The mechanism of cytoplasmic streaming involves bundles of microfilaments that are arranged parallel to the longitudinal direction of particle movement. The generation of the shear forces necessary for movement may involve an interaction of the microfilament protein actin with the protein myosin in a fashion comparable to what occurs in muscle contraction.

Microfilaments also participate in the growth of the pollen tube. Upon germination, a pollen grain forms a tubal extension that grows down the style to accomplish fertilization, and new cell wall is rapidly formed at the tip of this tube. A network of microfilaments appears to guide vesicles containing wall precursors from their site of formation through the cytosol to the site of new wall formation at the tip. Fusion of these vesicles with the plasma membrane deposits wall precursors outside the cell, where they are assembled into wall material.

The Plant Cell Wall

The cell wall provides rigidity and protection to the plant cell without preventing diffusion of water and ions from the environment to the plasma membrane, which is the true permeability barrier of the cell.

The typical primary cell wall of a dicotyledonous plant consists of 25–30% cellulose, 15–25% hemicellulose, 35% pectin, and 5–10% protein on a dry weight basis. The precise molecular composition and structure of the cell wall depend on the cell, tissue, and plant species.

The Primary Cell Wall Is Composed of Cellulose Microfibrils and Matrix Material

Primary walls owe their unique strength and yielding properties to the composition and organization of their constituents. The primary wall consists of two phases, a crystalline microfibrillar phase embedded in a noncrystalline amorphous matrix. This arrangement accounts for the remarkable strength of the primary cell wall and its ability to grow.

Cellulose is composed of linear chains of glucose residues covalently linked to form a flattened ribbon structure that can be 0.25 to 5.0 μm long (Fig. 1.21). About 40 to 70 of these chains are held together by hydrogen bonds between the sugar OH groups to form a crystalline structure called a **microfibril**, which is about 3.0 nm in diameter (Fig. 1.21). Cellulose is very stable chemically and is extremely insoluble. The high tensile strength of the microfibrils reinforces the wall by introducing shear interactions with the matrix in the same way that glass fibers reinforce the resin matrix of fiberglass and related complex composite materials. These shear interactions are maximal when the tensile stress is parallel to the orientation of the microfibrils.

The matrix materials of primary walls can be divided into two major fractions: pectins and hemicelluloses. The **pectic polysaccharides** or **pectins** (Fig. 1.22c) can be extracted by treatment with chelating agents. The **hemicelluloses** (Fig. 1.22b) are a heterogeneous mixture of neutral and acidic polysaccharides that can be extracted with hot, dilute alkali. Many of the hemicellulosic polysaccharides appear to coat the surfaces of the cellulose microfibrils and are oriented parallel to them.

If all the polysaccharides of the wall, including cellulose, are digested away, the remaining fraction is a glycoprotein rich in the rare amino acid hydroxyproline. This glycoprotein component, called **extensin**, represents about 5–10% of the dry weight of most primary walls. Extensin appears to form a structural network that strengthens the wall. In cells of carrot roots, extensin

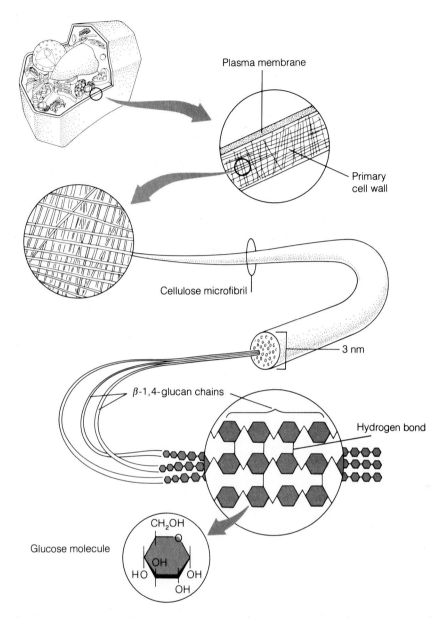

FIGURE 1.21. Cellulose microfibrils in the cell wall are made up of crystalline arrays of β-1, 4-linked glucan chains held together by intermolecular hydrogen bonds.

forms a narrow rodlike structure, 80 nm long, composed of 35% protein and 65% carbohydrate.

Secondary Walls Differ in Their Composition and Mechanical Properties from Primary Walls

The secondary cell wall contains a higher proportion of cellulose than the primary cell wall. Secondary cell walls usually have fewer pectins and therefore bind less water and are more dense than primary walls. A critical component of secondary cell walls is lignin, which constitutes 15–35% of the dry weight of woody tissues. Lignin is located in the wall matrix and plays an important role in the attainment of wall rigidity. Lignin is insoluble

in most solvents, has a high molecular weight, and has aromatic alcohols among its chemical components. Extensin has also been identified as a significant component of the secondary walls of sclerenchyma cells, such as sclereids and fibers. Plant fibers, whose tensile strength is typically 15–20 kg mm^{-2}, roughly equivalent to that of steel wire of equal diameter, may owe their great strength to both lignin and extensin.

The Wall Matrix Is Secreted Via Golgi Vesicles While Cellulose Is Synthesized on the Plasma Membrane

The constituents of the matrix are exported to the cell wall via Golgi vesicles. Polysaccharides are synthesized

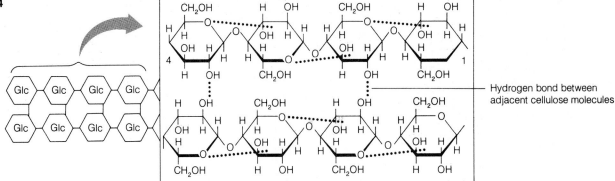

(a) Cellulose

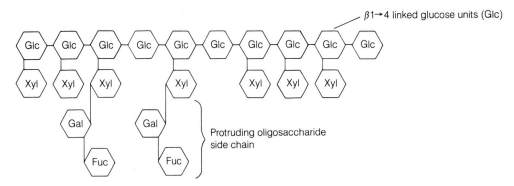

(b) Hemicellulose (Xyloglucan)

(c) Pectin

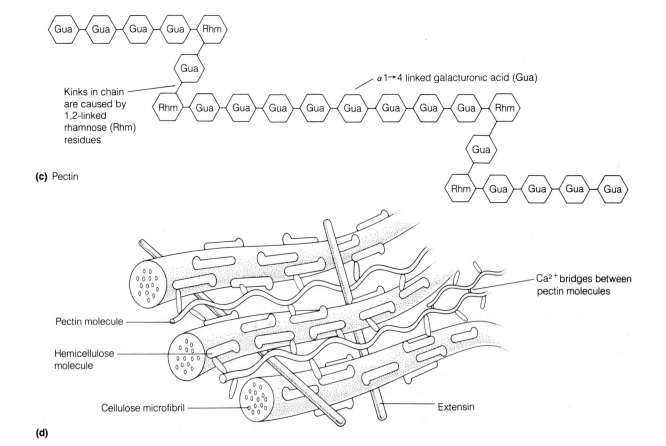

(d)

FIGURE 1.22. Chemical structures of (a) cellulose, (b) hemicellulose, and (c) pectin. (d) A model of the interconnection between the different components of the cell wall.

in Golgi bodies, while extensin is synthesized on the ER and processed in Golgi bodies. Processing includes the addition of carbohydrate and the conversion of proline residues to hydroxyproline. In contrast, the synthesis of cellulose and the formation of microfibrils are thought to occur simultaneously at **rosettes**, complexes of six molecules of the enzyme cellulose synthetase that span the plasma membrane (Fig. 1.23). As integral protein complexes embedded in the phospholipid bilayer, the rosettes can move laterally in the plane of the membrane in response to the displacement caused by extension of the anchored cellulose microfibril. In the absence of any coordination among rosettes, the orientation of the microfibrils is random, as found in primary cell walls. However, during cell elongation, a predominantly transverse alignment of the microfibrils is usually observed. Transverse microtubules within the cytosol are thought to interact with groups of rosettes, coordinating their movement during synthesis and thereby generating the microfibril alignment patterns.

Plant Cells Are Interconnected by Membrane-Lined Channels Called Plasmodesmata

During cytokinesis, Golgi-derived vesicles containing cell wall precursors fuse to form the cell plate. This cell plate, however, is not a continuous uninterrupted sheet. It is penetrated by numerous pores, about 60 nm in diameter, where remnants of the spindle apparatus, consisting of ER and microtubules, have disrupted vesicle fusion. Further deposition of wall polymers increases wall thickness, generating a membrane-lined channel called a **plasmodesma** (plural plasmodesmata) (Fig. 1.24). The plasma membrane is continuous from one cell to the next through the plasmodesmata. Many plasmodesmata also contain a central structure called a **desmotubule** (Fig. 1.24).

Plasmodesmata are small enough to prevent organelle movements between cells but large enough to allow the free exchange of small molecules dissolved in

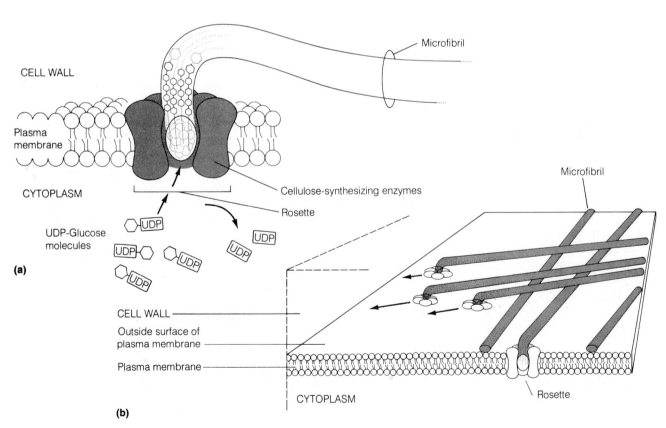

FIGURE 1.23. (a) Diagrammatic representation of a cellulose-synthesizing rosette embedded in the plasma membrane. Cellulose precursors are absorbed in the cytoplasmic side of the membrane (bottom) and the cellulose microfibril is extruded on the cell wall side. (b) View of the cellulose-synthesizing rosette as seen from the outside of the plasma membrane. Each new microfibril is attached to the preexisting cell wall, not shown in the diagram. (From Giddings et al., 1980)

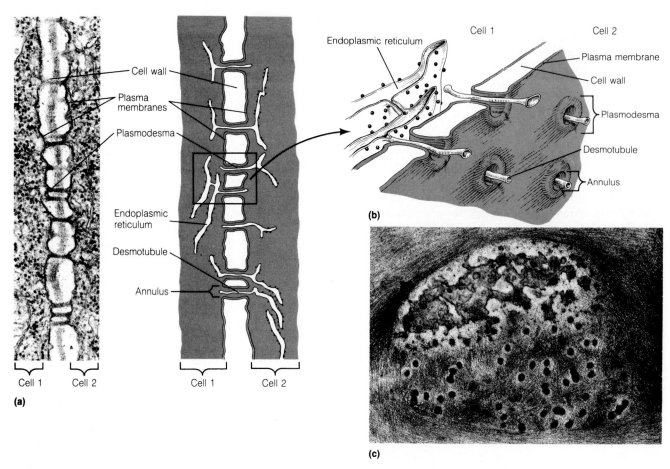

FIGURE 1.24. Plasmodesmata between cells. (a) Electron micrograph of a wall separating two adjacent cells of a timothy grass root, showing the plasmodesmata, schematically depicted on the right. (Micrograph by W. P. Wergin; photo provided by E. H. Newcomb.) (b) Schematic view of a cell wall with plasmodesmata. (c) Electron micrograph showing plasmodesmata in cross section (25,400 ×). (Courtesy of E. H. Newcomb.)

the cytosol. Studies with fluorescent molecules of different molecular weights have shown that only molecules with molecular weights of less than 800 to 1000 are free to move through plasmodesmata, although the effective pore size may, to some extent, be regulated. In this respect, plasmodesmata are much like the gap junctions of animal cells which, however, are much smaller in diameter (7 nm).

Summary

Despite their great diversity in form and size, all plants carry out similar physiological processes. As the earth's primary producers, plants convert solar energy to chemical energy. Being nonmotile, plants must grow toward the light, and they must have efficient vascular systems for the movement of water, mineral nutrients, and photosynthetic products throughout the plant body. Green land plants must also have mechanisms for avoiding desiccation.

The major vegetative organ systems of seed plants are the shoot and the root. The shoot consists of two types of organs: stems and leaves. Unlike the development of animals, plant growth is indeterminate due to the presence of permanent meristem tissue at the shoot and root apices, which give rise to new tissues and organs during the entire vegetative phase of the life cycle. Lateral meristems, vascular cambium and the cork cambium, produce growth in girth, or secondary growth.

Three major tissue systems are recognized: dermal, ground, and vascular. Each of these tissues contains a variety of cell types specialized for different functions, such as support (sclerenchyma, collenchyma), photosynthesis (mesophyll parenchyma), storage (parenchyma), gas exchange (guard cells), water conduction (xylem tracheids and vessel elements), and sugar translocation (phloem sieve tube elements).

Plants are eukaryotes and have the typical eukaryotic cell organization consisting of nucleus and cytoplasm. The plant genome, which is variable in size due to its large amounts of highly repetitive DNA, is contained in the nucleus. The cytoplasm is bounded by a plasma membrane and contains numerous membrane-bounded organelles, including plastids, mitochondria, microbodies, and a large central vacuole. The cytoskeletal components, microtubules and microfilaments, participate in a variety of processes involving intracellular movements, such as mitosis, cytoplasmic streaming, secretory vesicle transport, and cellulose microfibril deposition.

The plant cell wall plays a key role in regulating plant cell expansion. Primary walls are typical of growing cells, while secondary walls rich in lignin are found in nongrowing cells. The structure of the wall consists of a crystalline microfibrillar phase embedded in an amorphous matrix. The cytosol of adjacent cells is continuous through the cell walls because of the presence of membrane-lined channels called plasmodesmata, which play a role in cell-to-cell communication.

GENERAL READING

Alberts, B., Bray, D., Lewis, J., Raff, M., Roberts K., and Watson, J. D. (1983) *Molecular Biology of the Cell.* Garland Publishing, New York.

Avers, C. J. (1986) *Molecular Cell Biology.* Addison-Wesley, Reading, Mass.

Burgess, J. (1985) *An Introduction to Plant Cell Development.* Cambridge University Press, Cambridge, England.

Cutter, E. G. (1971) *Plant Anatomy: Experiment and Interpretation*, Vols. 1 and 2. Addison-Wesley, Reading, Mass.

Esau, K. (1960) *Anatomy of Seed Plants.* Wiley, New York.

Esau, K. (1977) *Anatomy of Seed Plants*, 2d ed. Wiley, New York.

Gunning, B. E. S., and Steer, M. W. (1975) *Ultrastructure and the Biology of Plant Cells.* Arnold Press, London.

Karp, G. (1979) *Cell Biology*, 2d ed. McGraw-Hill: New York.

Ledbetter, M. C., and Porter, K. R. (1970) *Introduction to the Fine Structure of Plant Cells.* Springer, Berlin.

Meylan, B. A., and Butterfield, B. G. (1972) *Three-Dimensional Structure of Wood: A Scanning Electron Microscope Study.* Syracuse University Press, Syracuse, N.Y.

O'Brien, T. P., and McCully, M. E. (1969) *Plant Structure and Development.* Collier-Macmillan, London.

Prescott, D. M. (1988). *Cells: Principles of Molecular Structure and Function.* Jones and Bartlett, Boston.

CHAPTER REFERENCES

Depta, H., Freundt, H., Hartmann, D., and Robinson, D. G. (1987) Preparation of a homogeneous coated vesicle fraction from bean leaves. *Protoplasm* 136:154–160.

Giddings, T. H., Jr., Brower, D. L., and Staehelin, L. A. (1980) Visualization of particle complexes in the plasma membrane of *Micvasterias denticulata* associated with the formation of cellulose fibrils in primary and secondary cell walls. *J. Cell Biol.* 84:327–339.

Hepler, P. K., and Jackson, W. T. (1969) Isopropyl *N*-phenylcarbamate affects spindle microtubule orientation in dividing endosperm cells of *Haemanthus katherinae* Baker. *J. Cell. Sci.* 5:727–743.

Thair, B. W., and Wardrop, A. B. (1971). The structure and arrangement of nuclear pores in plant cells. *Planta* 100: 1–17.

CHAPTER 2

Energy, Enzymes, and Gene Expression

The force that through the green fuse drives the flower
Drives my green age; that blasts the roots of trees
Is my destroyer.
And I am dumb to tell the crooked rose
My youth is bent by the same wintry fever.

The force that drives the water through the rocks
Drives my red blood; that dries the mouthing streams
Turns mine to wax.
And I am dumb to mouth unto my veins
How at the mountain spring the same mouth sucks.

I N THESE OPENING STANZAS from Dylan Thomas's famous poem, the poet proclaims the essential unity of the forces that propel animate and inanimate objects alike, from their beginnings to their ultimate decay. Scientists call this force "energy." Energy transformations play a key role in all the physical and chemical processes that occur in living systems. But energy alone is insufficient to drive the growth and development of organisms. Protein catalysts called enzymes are required to ensure that the rates of biochemical reactions will be rapid enough to support life. In this chapter we will examine basic concepts about energy, the way in which cells transform energy to perform useful work (bioenergetics), the structure and function of enzymes, and the regulation of enzyme activity and synthesis. We will also review basic mechanisms of enzyme induction in prokaryotes and eukaryotes.

Energy Flow Through Living Systems

The flow of matter through individual organisms and biological communities is part of everyday experience; the flow of energy is not, even though it is central to the very existence of living things. What makes concepts such as energy, work, and order so elusive is their insubstantial nature: we find it far easier to visualize the dance of atoms and molecules than the forces and fluxes that determine the direction and extent of natural processes. The branch of physical science that deals with such matters is thermodynamics, an abstract and demanding discipline that most biologists are content to skim over lightly. Yet bioenergetics is so shot through with concepts and quantitative relationships derived from thermodynamics that it is scarcely possible to discuss the subject without frequent reference to free energy, potential, entropy, and the second law. The purpose of this chapter is to collect and explain, as simply as possible, the fundamental thermodynamic concepts and relationships that recur throughout this book. Readers who prefer a more extensive treatment of the subject should consult either the introductory texts by Klotz (1967) and by Nichols (1982) or the advanced texts by Morowitz (1978) and by Edsall and Gutfreund (1983).

Thermodynamics evolved during the nineteenth century out of efforts to understand how a steam engine works and why heat is produced when one bores cannon. The very name thermodynamics, and much of its language, recalls these historical roots, but it would be more appropriate to speak of energetics, for the principles involved are universal. Living plants, like all other natural phenomena, are constrained by the laws of thermodynamics. By the same token, thermodynamics supplies an indispensable framework for the quantitative description of biological vitality.

Energy and Work

Let us begin with the meaning of "energy" and "work." Energy is defined in elementary physics, as in daily life, as the capacity to do work. The meaning of work is harder to come by and rather more narrow. Work, in the mechanical sense, is the displacement of any body against an opposing force. The work done is the product of the force and the distance displaced, or

$$W = f \, \Delta l \qquad (2.1)$$

(We may note in passing that the dimensions of work are complex, $m \, l^2 \, t^{-2}$, where m denotes mass, l distance, and t time, and that work is a scalar quantity, that is, the product of two vectorial terms.) Mechanical work appears in chemistry because whenever the final volume of a reaction mixture exceeds the initial volume, work must be done against the pressure of the atmosphere; conversely, the atmosphere performs work when a system contracts. This work is given by $P \, \Delta V$ (where P stands for pressure

and V for volume), a term that appears frequently in thermodynamic formulas. *In biology, work is employed in a broader sense to describe displacement against any of the forces that living things encounter or generate: mechanical, electrical, osmotic, or even chemical potential.*

A familiar mechanical illustration may help to clarify the relationship of energy to work. The spring in Figure 2.1 can be extended by applying to it a force over some particular distance, that is, by doing work on the spring. This work can be recovered by an appropriate arrangement of pulleys and used to lift a weight onto the table. The extended spring can thus be said to possess energy that is numerically equal to the work it can do on the weight (neglecting friction). The weight on the table, in turn, can be said to possess energy by virtue of its position in the gravitational field of the earth, which can be utilized to do some other work, such as turning a crank. The weight thus illustrates the concept of **potential energy**, a capacity to do work that arises from the position of an object in a field of force, and the sequence as a whole illustrates the conversion of one kind of energy into another, or **energy transduction**.

The First Law: The Total Energy Is Always Conserved

It is common experience that mechanical devices involve both the performance of work and the production or absorption of heat. We are at liberty to vary the amount of work done by the spring, up to some maximum, by using various weights, and the amount of heat produced will also vary. But much experimental work has shown that, under ideal circumstances, the sum of the work done and of the heat evolved is constant and depends only on the initial and final extensions of the spring. We can thus envisage a property, the internal energy of the spring, with the characteristic that

$$\Delta U = \Delta Q + \Delta W \qquad (2.2)$$

Here Q is the amount of heat absorbed by the system and W is the work done on the system. In the present illustration the work is mechanical, but it could just as

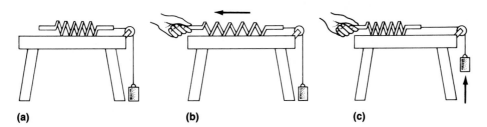

FIGURE 2.1. Energy and work in a mechanical system. In (a), a weight resting on the floor is attached to a spring via a string. Pulling on the spring (b) places the spring under tension. In (c), the potential energy stored in the extended spring performs the work of raising the weight when the spring contracts.

(a)   (b)   (c)

well be electrical, chemical, or any other kind of work.[1] Thus ΔU is the net amount of energy put into the system, either as heat or as work; conversely, both the performance of work and the evolution of heat entail a decrease in the internal energy. We cannot specify an absolute value for the energy content; only changes in internal energy can be measured. Note that Equation 2.2 assumes that heat and work are equivalent; its purpose is to stress that, under ideal circumstances, ΔU depends only on the initial and final states of the system, whereas the partitioning between heat and work is variable.

Equation 2.2 is a statement of the first law of thermodynamics, which is the principle of energy conservation. If a particular system exchanges no energy with its surroundings, its energy content remains constant; if exchange occurs, the change in internal energy will be given by the difference between the energy gained from the surroundings and that lost to the surroundings. The change in internal energy depends only on the initial and final states of the system, not on the pathway or mechanism of energy exchange. Energy and work are interconvertible; even heat is a measure of the kinetic energy of the molecular constituents of the system. To put it as simply as possible, Equation 2.2 states that no machine, including the chemical machines that we recognize as living, can do work without an energy source.

An example of the application of the first law to a biological phenomenon is the energy budget of a leaf. Leaves absorb energy from their surroundings in two ways, as direct incident irradiation from the sun and as infrared irradiation from the surroundings. Some of the energy absorbed by the leaf is radiated back to the surroundings as infrared irradiation and heat, while a fraction of the absorbed energy is stored, as either photosynthetic products or leaf temperature changes. Thus we can write the equation

$$\begin{matrix} \text{Total energy} \\ \text{absorbed by leaf} \end{matrix} = \begin{matrix} \text{energy emitted} \\ \text{from leaf} \end{matrix} + \begin{matrix} \text{energy stored} \\ \text{by leaf} \end{matrix}$$

Note that while the energy absorbed by the leaf has been transformed, the total energy remains the same, in accordance with the first law.

The Change in the Internal Energy of a System Represents the Maximum Work It Can Do

The equivalence of energy and work must be qualified by invoking "ideal conditions," that is, by requiring that

the process be carried out reversibly. The meaning of "reversible" in thermodynamics[2] is a special one: the term describes conditions under which the opposing forces are so nearly balanced that an infinitesimal change in one or the other would reverse the direction of the process. Under these circumstances the process yields the maximum possible amount of work. It will be obvious that reversibility in this sense does not often hold in nature, as in the example of the leaf above. Ideal conditions differ so little from a state of equilibrium that any process or reaction would require infinite time and would therefore not take place at all. Nonetheless, the concept of thermodynamic reversibility is a useful one: if we measure the change in internal energy that a process entails, we have an upper limit to the work that it can do; for any real process the maximum work will be less.

In the study of plant biology we encounter several sources of energy, notably light and chemical transformations, as well as a variety of work functions, including mechanical, osmotic, electrical, and chemical work. The meaning of the first law in biology stems from the certainty, painstakingly achieved by nineteenth-century physicists, that the various kinds of energy and work are measurable, equivalent, and, within limits, interconvertible. Energy is to biology what money is to economics: the means by which living things purchase useful goods and services.

Each Type of Energy Is Characterized by a Size Factor and a Potential Factor

It will be obvious that the amount of work that can be done by a system, be it mechanical or chemical, is a function of the size of the system. Work can always be defined as the product of two factors—force and distance, for example. One is a potential or intensity factor, which is independent of the size of the system; the other is a capacity factor and is directly proportional to the size (Table 2.1).

In biochemistry, energy and work have traditionally been expressed in calories, the calorie being the amount of heat required to raise the temperature of one gram of water from 15.0 to 16.0°C. In principle, one can carry out the same process by doing the work mechanically with a paddle; such experiments led to the establishment of the mechanical equivalent of heat as 4.186 joules per calorie ($J\ cal^{-1}$). In current standard usage based on the meter, kilogram, and second, the fundamental unit of energy is the joule (1 J = 0.24 cal) or the kilojoule (1 kJ = 1000 J). We shall also have occasion to use the equivalent electrical units, based on the volt: a volt is the potential difference

---

1. Equation 2.2 is more commonly encountered in the form $\Delta U = \Delta Q - \Delta W$, which results from the convention that Q is the amount of heat absorbed by the system from the surroundings and W is the amount of work done by the system on the surroundings. This convention affects the sign of W but does not alter the meaning of the equation.

---

2. In biochemistry, reversibility has a different meaning. Usually the term refers to a reaction whose pathway can be reversed, often with an input of energy.

TABLE 2.1. Potential and capacity factors in energetics

Type of energy	Potential factor	Capacity factor
Mechanical (*PV*)	Pressure	Volume
Electrical	Electric potential	Charge
Chemical	Chemical potential	Mass
Osmotic	Concentration	Mass
Thermal	Temperature	Entropy

between two points when one joule of work is involved in the transfer of a coulomb of charge from one point to the other. [A coulomb is the amount of charge carried by a current of one ampere (A) flowing for one second. Transfer of one mole (mol) of charge across a potential of one volt (V) involves 96,500 joules of energy or work.] The difference between energy and work is often a matter of the sign. Work must be done to bring a positive charge closer to another positive charge, but the charges thereby acquire potential energy, which in turn can do work.

The Direction of Spontaneous Processes

Left to themselves, events in the real world take a predictable course. The apple falls from the branch. A mixture of hydrogen and oxygen gases is converted into water. The fly trapped in a bottle is doomed to perish, the pyramids to crumble into sand; things fall apart. But there is nothing in the principle of energy conservation that forbids the apple to return to its branch with absorption of heat from the surroundings or that prevents water from dissociating into its constituent elements in a like manner. The search for the reason why neither of these things ever happens led to profound philosophical insights and also generated useful quantitative statements about the energetics of chemical reactions and the amount of work that can be done thereby. Since living things are, in many respects, chemical machines, we must examine these matters in some detail.

The Second Law: The Total Entropy Always Increases

From daily experience with weights falling and warm bodies growing cold, one might expect spontaneous processes to proceed in the direction that lowers the internal energy, that is, the direction in which ΔU is negative. But there are too many exceptions for this to be a general rule. The melting of ice is one: an ice cube placed in water at 1°C will melt, yet measurements show that liquid water (at any temperature above 0°C) is in a state of higher energy than ice; evidently, some spontaneous processes are accompanied by an increase in internal energy. Our melting ice cube does not violate the first law, for heat is absorbed as it melts. This suggests that there is a relationship between the capacity for spontaneous heat absorption and the criterion determining the direction of spontaneous processes, and that is the case. The thermodynamic function we seek is called entropy, mathematically the capacity factor corresponding to temperature, Q/T. We may state the answer to our question, as well as the second law of thermodynamics, thus: The direction of all spontaneous processes is such as to increase the entropy of a system plus its surroundings.

Few concepts are so basic to a comprehension of the world we live in, yet so opaque, as entropy—presumably because entropy is not intuitively related to our sense perceptions, as mass and temperature are. The explanation given here follows the particularly lucid exposition by Atkinson (1977), who states the second law in a form bearing, at first sight, little resemblance to that given above:

> We shall take [the second law] as the concept that any system not at absolute zero has an irreducible minimum amount of energy that is an inevitable property of that system at that temperature. That is, a system requires a certain amount of energy just to be at any specified temperature.

The molecular constitution of matter supplies a ready explanation: some energy is stored in the thermal motions of the molecules and in the vibrations and oscillations of their constituent atoms. We can speak of it as "isothermally unavailable energy," since the system cannot give up any of it without a drop in temperature (assuming that there is no physical or chemical change). The isothermally unavailable energy of any system increases with temperature, since the energy of molecular and atomic motions increases with temperature. Quantitatively, the isothermally unavailable energy for a particular system is given by ST, where T is the absolute temperature and S is the entropy.

But what is this thing, entropy? Reflection on the nature of the isothermally unavailable energy suggests that, for any particular temperature, the amount of such energy will be greater the more atoms and molecules are free to move and to vibrate—that is, the more chaotic is the system. By contrast, the orderly array of atoms in a crystal, with a place for each and each in its place, corresponds to a state of low entropy. Indeed, at absolute zero, when all motion ceases, the entropy of a pure substance is likewise zero; this is sometimes called the third law of thermodynamics.

A large molecule, a protein for example, within which many kinds of motion can take place, will have considerable amounts of energy stored in this fashion—more than would, say, an amino acid molecule. But the entropy of the protein molecule will be less than that of the constituent amino acids into which it can dissociate, because of the constraints placed on the motions of those amino acids as long as they are part of the larger structure. Any process leading to the release of these constraints increases freedom of movement, and hence entropy. This is the universal tendency of spontaneous processes as expressed in the second law; it is why the costly enzymes stored in the refrigerator tend to decay and why ice melts into water. The increase in entropy as ice melts into water is paid for, as it were, by the absorption of heat from the surroundings. As long as the net change in entropy of the system plus its surroundings is positive, the process can take place spontaneously. That does not necessarily mean that the process will in fact take place: the rate is usually determined by kinetic factors quite separate from the entropy change. All the second law mandates is that the fate of the pyramids is to crumble into sand, while the sand will never reassemble itself into a pyramid; the law does not tell how quickly this must come about.

A Process Is Spontaneous if ΔS for the System and Its Surroundings Is Positive

There is nothing mystical about entropy; it is a thermodynamic quantity like any other, measurable by experiment and expressed in entropy units. One method of quantifying it is through the heat capacity of a system, the amount of energy required to raise the temperature by one degree Celsius. In some cases the entropy can even be calculated from first principles, though only for simple molecules. For our purposes, what matters is the sign of the entropy change, ΔS: A process can take place spontaneously when ΔS for the system and its surroundings is positive; a process for which ΔS is negative cannot take place spontaneously, but the opposite process can; and for a system at equilibrium the entropy of the system plus its surroundings is maximal and ΔS is zero.

Equilibrium is another of those familiar words that are easier to use than to define. Its everyday meaning implies that the forces acting on a system are equally balanced, so that there is no net tendency to change; this is the sense in which equilibrium will be used here. A mixture of chemicals may be in the midst of rapid interconversion, but if the rates of the forward reaction and the back reaction are equal, there will be no net change in composition, and equilibrium obtains.

The second law has been stated in many versions over the years. One version forbids perpetual motion machines: because energy is, by the second law, perpetually degraded into heat and rendered isothermally unavailable ($\Delta S > 0$), continued motion requires an input of energy from the outside. The most celebrated yet perplexing version of the second law was provided by R. J. Clausius: "The energy of the universe is constant; the entropy of the universe tends towards a maximum." How can entropy increase forever, created out of nothing? The root of the difficulty is verbal, as Klotz (1967) neatly explains. Had Clausius defined entropy with the opposite sign (corresponding to order rather than to chaos), its universal tendency would be to diminish; it would then be obvious that spontaneous changes proceed in the direction that decreases the capacity for further spontaneous change. Solutes diffuse from a region of higher concentration to one of lower concentration; heat flows from a warm body to a cold one. Sometimes these changes can be reversed by an outside agency so as to reduce the entropy of the system under consideration, but then that external agency must change in such a way as to reduce its own capacity for further change. In sum, "entropy is an index of exhaustion; the more a system has lost its capacity for spontaneous change, the more this capacity has been exhausted, the greater is the entropy" (Klotz, 1967). Conversely, the farther a system is from equilibrium, the greater is its capacity for change and the less its entropy. Obviously, living things fall into the latter category: a cell is the very epitome of a state remote from equilibrium.

Free Energy and Chemical Potential

Many energy transactions that take place in living things are chemical in nature; we therefore need a quantitative expression for the amount of work a chemical reaction can do. For this purpose, relationships that involve the entropy change in the system plus its surroundings are unsuitable. We need a function that does not depend on the surroundings but that, like ΔS, attains a minimum under conditions of equilibrium and so can serve both as a criterion of the feasibility of a reaction and as a measure of the energy available from it for the performance of work. The function universally employed for this purpose is the free energy, abbreviated G in honor of the nineteenth-century physical chemist J. Willard Gibbs, who first introduced it.

ΔG Is Negative for a Spontaneous Process at Constant Temperature and Pressure

In the preceding section we spoke of the isothermally unavailable energy, ST. Free energy is defined as the energy that *is* available under isothermal conditions, and by the relationship

$$\Delta H = \Delta G + T \Delta S \qquad (2.3)$$

The term H, enthalpy or heat content, is not quite equivalent to U, the internal energy. To be exact, ΔH is a measure of the total energy change, including work that may result from changes in volume during the reaction, whereas ΔU excludes this term. (We will return to the concept of enthalpy in a later section.) However, in the biological context we are usually concerned with reactions in solution, for which volume changes are negligible. For most purposes, then,

$$\Delta U \simeq \Delta G + T \Delta S \qquad (2.4)$$

and

$$\Delta G \simeq \Delta U - T \Delta S \qquad (2.5)$$

What makes this a useful relationship is the demonstration that *for all spontaneous processes at constant temperature and pressure, ΔG is negative.* The change in free energy is thus a criterion of feasibility. Any chemical reaction that proceeds with a negative ΔG can take place spontaneously; a process for which ΔG is positive cannot take place, but the reaction can go in the opposite direction; and a reaction for which ΔG is zero is at equilibrium and no net change will occur. For a given temperature and pressure, ΔG depends only on the composition of the reaction mixture; hence the alternative term "chemical potential" is particularly apt. Again, nothing is said about rate, only about direction. Whether a reaction having a given ΔG will proceed, and at what rate, is determined by kinetic rather than thermodynamic factors.

There is a close and simple relationship between the free energy change of a chemical reaction and the work that the reaction can do. Provided that the reaction is carried out reversibly,

$$\Delta G = -W_{max} \qquad (2.6)$$

That is, for a reaction taking place at constant temperature and pressure, −ΔG is a measure of the maximum work the process can perform. More precisely, −ΔG is the maximum work possible, exclusive of pressure-volume work, and thus is a quantity of great importance in bioenergetics. Any process going toward equilibrium can, in principle, do work. We can therefore describe processes for which ΔG is negative as "energy-releasing," or exergonic. Conversely, for any process moving away from equilibrium ΔG is positive, and we speak of an "energy-consuming," or endergonic, reaction. Actually, of course, an endergonic reaction cannot occur: all real processes go toward equilibrium, with a negative ΔG. The concept of endergonic reactions is nevertheless a useful abstraction, for many biological reactions appear to move away from equilibrium. A prime example is the synthesis of ATP during oxidative phosphorylation, whose apparent ΔG is as high as 67 kJ mol^{-1} (16 kcal mol^{-1}). Clearly, the cell must do work to render the reaction exergonic overall. The occurrence of an endergonic process in nature thus implies that it is coupled to a second, exergonic process. Much of cellular and molecular bioenergetics is concerned with the mechanisms by which energy coupling is effected.

The Standard Free Energy Change, ΔG°, Is the Free Energy Change When the Concentration of Reactants and Products Is 1 M

Free energy changes can be measured experimentally by calorimetric methods. They have been tabulated in two forms: as the free energy of formation of a compound from its elements and as ΔG for a particular reaction. It is of the utmost importance to remember that, by convention, the numerical values refer to a particular set of conditions. *The standard free energy change, ΔG°, refers to conditions such that all reactants and products are present at a concentration of 1 M;* in biochemistry it is more convenient to employ ΔG°′, which is defined in the same way except that the pH is taken to be 7. The conditions obtained in the real world are likely to be very different from these, particularly with respect to the concentrations of the participants. To take a familiar example, ΔG°′ for the hydrolysis of ATP is about −33 kJ mol^{-1} (−8 kcal mol^{-1}). In the cytoplasm, the actual nucleotide concentrations are approximately 3 mM ATP, 1 mM ADP, and 10 mM P$_i$. As we will see, free energy changes depend strongly on concentrations, and ΔG for ATP hydrolysis under physiological conditions is much more negative than ΔG°′, about −50 to −65 kJ mol^{-1} (−12 to −15 kcal mol^{-1}). *Thus, whereas values of ΔG°′ for many reactions are easily accessible, they must not be used uncritically as guides to what happens in cells.*

The Value of ΔG Is a Function of the Displacement of the Reaction from Equilibrium

From the preceding discussion of the concept of free energy, it is apparent that there must be a relationship between ΔG and the equilibrium constant of a reaction: at equilibrium ΔG is zero, and the farther a reaction is from equilibrium, the larger is ΔG and the more work the reaction can do. The quantitative statement of this relationship is

$$\Delta G° = -RT \ln K = -2.3RT \log K \qquad (2.7)$$

where R is the gas constant, T the absolute temperature, and K the equilibrium constant of the reaction. This equation is one of the most useful links between thermodynamics and biochemistry and has a host of applications. For example, the equation is easily modified to allow computation of the free energy change for concentrations other than the standard ones. For the reactions shown in Equation 2.8,

$$A + B \rightleftharpoons C + D \qquad (2.8)$$

the actual free energy change ΔG is given by

$$\Delta G = \Delta G^{\circ} + RT \ln \frac{[C][D]}{[A][B]} \qquad (2.9)$$

where the terms in brackets refer to the actual concentrations at the time of the reaction. Strictly speaking, one should use activities, but these are usually not known for cellular conditions and so concentrations must do.

Equation 2.9 can be rewritten to make its import a little plainer. Let q stand for the mass-action ratio $[C][D]/[A][B]$. Substitution of Equation 2.7 into Equation 2.9, followed by rearrangement, then yields Equation 2.10:

$$\Delta G = -2.3RT \log \frac{K}{q} \qquad (2.10)$$

In other words, the value of ΔG is a function of the displacement of the reaction from equilibrium. In order to displace a system from equilibrium, work must be done on it and ΔG is positive. Conversely, a system displaced from equilibrium can do work on another system, provided the kinetic parameters allow the reaction to proceed and a mechanism exists that couples the two systems. Quantitatively, a reaction mixture at 25°C whose composition is one order of magnitude away from equilibrium ($\log K/q = 1$) corresponds to a free energy change of 5.7 kJ mol^{-1} (1.36 kcal mol^{-1}). The value of ΔG is negative if the actual mass-action ratio is less than the equilibrium ratio and positive if the mass-action ratio is greater.

The point that ΔG is a function of the displacement of a reaction (indeed, of any thermodynamic system) from equilibrium is central to an understanding of biological energetics. Figure 2.2 illustrates this relationship diagrammatically for the chemical interconversion of substances A and B, and it will shortly reappear in other guises.

The Enthalpy Change Measures the Energy Transferred as Heat

Chemical and physical processes are almost invariably accompanied by generation or absorption of heat, which reflects the change in the internal energy of the system. The amount of heat transferred and the sign of the reaction are related to the change in free energy as set out in Equation 2.3. The energy absorbed or evolved as heat under conditions of constant pressure is designated as the change in heat content or enthalpy, ΔH. Processes that generate heat, such as combustion, are said to be exothermic; those in which heat is absorbed, such as melting or evaporation, are referred to as endothermic. Oxidation of glucose to CO_2 and water is an exergonic reaction ($\Delta G^{\circ} = -2858$ kJ mol^{-1} [-686 kcal mol^{-1}]); when this reaction takes place during respiration, part of the free energy is conserved through coupled reactions

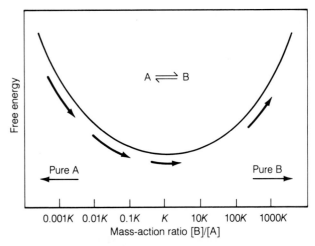

FIGURE 2.2. Free energy of a chemical reaction as a function of displacement from equilibrium. Imagine a closed system containing components A and B at concentrations [A] and [B]. The two components can be interconverted by reaction A ⇌ B, which is at equilibrium when the mass-action ratio, [B]/[A], equals unity. The curve shows qualitatively how the free energy G of the system varies when the total [A] + [B] is held constant but the mass-action ratio is displaced from equilibrium. The arrows represent schematically the free energy change ΔG for a small conversion of [A] into [B] occurring at different mass-action ratios. (Modified from Nicholls, 1982, with permission of Academic Press.)

that generate ATP. Combustion of glucose dissipates the free energy of reaction, releasing most of it as heat ($\Delta H = -2804$ kJ mol^{-1} [-673 kcal mol^{-1}]). Bioenergetics is preoccupied with energy transduction and therefore gives pride of place to free energy transactions, but at times heat transfer may also carry biological significance. For example, water has a high heat of vaporization, 44 kJ mol^{-1} (10.5 kcal mol^{-1}) at 25°C, which plays an important role in the regulation of leaf temperature. During the day, the evaporation of water from the leaf surface (transpiration) dissipates heat to the surroundings and helps to cool the leaf. Conversely, condensation of water vapor as dew heats the leaf, since water condensation is the reverse of the process of evaporation and is therefore exothermic. The abstract enthalpy function is a direct measure of the energy exchanged in the form of heat.

Redox Reactions

Oxidation and reduction refer to the transfer of one or more electrons from a donor to an acceptor, usually to another chemical species; an example is the oxidation of ferrous iron by oxygen with the formation of ferric iron and water. Reactions of this kind require special consideration, for they play a central role in both respiration and photosynthesis.

The Free Energy Change of an Oxidation-Reduction Reaction Is Expressed as the Standard Redox Potential in Electrochemical Units

Redox reactions can be quite properly described in terms of their free energy changes. However, the participation of electrons make it convenient to follow the course of the reaction with electrical instrumentation and encourages the use of an electrochemical notation. It also permits one to dissect the chemical process into separate oxidative and reductive half-reactions. For the oxidation of iron, we can write

$$2Fe^{2+} \rightleftharpoons 2Fe^{3+} + 2e^- \qquad (2.11)$$

$$\frac{1}{2}O_2 + 2H^+ + 2e^- \rightleftharpoons H_2O \qquad (2.12)$$

$$\overline{2Fe^{2+} + \frac{1}{2}O_2 + 2H^+ \rightleftharpoons 2Fe^{3+} + H_2O} \qquad (2.13)$$

The tendency of a substance to donate electrons, its "electron pressure," is measured by its standard reduction (or redox) potential, E_0, with all components present at a concentration of 1 M. In biochemistry, it is more convenient to employ E'_0, which is defined in the same way except that the pH is 7. By definition, then, E'_0 is the electromotive force given by a half-cell in which the reduced and oxidized species are both present at 1.0 M, 25°C, and pH 7, in equilibrium with an electrode that can reversibly accept electrons from the reduced species. By convention, the reaction is written as a reduction. The standard reduction potential of the hydrogen electrode[3] serves as reference: at pH 7, it equals −0.42 V. The standard redox potential as defined above is often referred to in the bioenergetics literature as the midpoint potential, E_m. A negative midpoint potential marks a good reducing agent; oxidants have positive midpoint potentials.

The redox potential for the reduction of oxygen to water is +0.82 V, and that for the reduction of Fe^{3+} to Fe^{2+} (the direction opposite to Eq. 2.11) is +0.77 V. We can therefore predict that, under standard conditions, the Fe^{2+}-Fe^{3+} couple will tend to reduce oxygen to water rather than the reverse. A mixture containing Fe^{2+}, Fe^{3+}, and oxygen will probably not be at equilibrium, and the extent of its displacement from equilibrium can be expressed in terms of either the free energy change for Equation 2.13 or the difference in redox potential, $\Delta E'_0$, between the oxidant and the reductant couples (+0.05 V in the case of iron oxidation). In general,

$$\Delta G^{\circ\prime} = -nF \Delta E'_0 \qquad (2.14)$$

where n is the number of electrons transferred and F is Faraday's constant (23.06 kcal V^{-1} mol^{-1}). In other words, the standard redox potential is a measure, in electrochemical units, of the free energy change of an oxidation-reduction process.

As with free energy changes, the actual redox potential measured under conditions other than the standard ones depends on the concentrations of the oxidized and reduced species according to Equation 2.15 (note the similarity in form to Eq. 2.9):

$$E_h = E'_0 + \frac{2.3RT}{nF} \log \frac{[\text{oxidant}]}{[\text{reductant}]} \qquad (2.15)$$

Here E_h is the measured potential in volts and the other symbols have their usual meanings. It follows that the redox potential under biological conditions may differ substantially from the standard reduction potential.

The Electrochemical Potential

In the preceding section we introduced the concept that a mixture of substances whose composition diverges from the equilibrium state represents a potential source of free energy (Fig. 2.2). Conversely, a similar amount of work must be done on an equilibrium mixture in order to displace its composition from equilibrium. In this section, we shall examine the free energy changes associated with another kind of displacement from equilibrium, namely gradients of concentration and of electrical potential.

Transport of an Uncharged Solute Against Its Concentration Gradient Decreases the Entropy of the System

Consider a vessel divided by a membrane into two compartments that contain solutions of an uncharged solute at concentrations C_1 and C_2, respectively. The work required to transfer one mole of solute from the first compartment to the second is given by

$$\Delta G = 2.3RT \log \frac{C_2}{C_1} \qquad (2.16)$$

This expression is analogous to that for a chemical reaction (Eq. 2.10) and has the same meaning. If C_2 is greater than C_1, ΔG is positive, and work must be done to transfer the solute. The free energy change for the transport of one mole of solute against a tenfold gradient of concentration is again 5.7 kJ, or 1.36 kcal. The reason why work must be done to move a substance from a region of lower concentration to one of higher concentration is that the process entails a change to a less probable state and therefore a decrease in the entropy of the system. Conversely, diffusion of the solute from the region of higher concentration to that of lower concen-

---

3. The standard hydrogen electrode consists of platinum, over which hydrogen gas is bubbled at a pressure of 1 atmosphere. The electrode is immersed in a solution containing hydrogen ions. When the activity of hydrogen ions is 1, approximately 1 M H^+, the potential of the electrode is taken to be 0.

tration takes place in the direction of greater probability; it results in an increase in the entropy of the system and can proceed spontaneously. The sign of ΔG turns negative and the process can do the amount of work specified by Equation 2.16, provided a mechanism exists that couples the exergonic diffusion process to the work function.

The Membrane Potential Is the Work That Must Be Done to Move an Ion from One Side of the Membrane to the Other

Matters become a little more complex if the solute in question bears an electric charge. Transfer of positively charged solute from compartment 1 to compartment 2 will then cause a difference in charge to develop across the membrane, with the second compartment becoming electropositive relative to the first. Since like charges repel one another, the work done by the agent that moves the solute from compartment 1 to compartment 2 is a function of the charge difference; more precisely, it depends on the electric potential difference across the membrane. This term, called membrane potential for short, will appear again in later pages. The **membrane potential**, ΔE_m,[4] is defined as the work that must be done by an agent to move a test charge from one side of the membrane to the other. When one joule of work must be done to move one coulomb of charge, the potential difference is said to be one volt. The absolute electric potential of any one phase cannot be measured, but the potential difference between two phases can be. By convention, the membrane potential is always given in reference to the movement of a positive charge. It states the intracellular potential relative to the extracellular one, which is defined as zero.

The work that must be done to move one mole of an ion against a membrane potential of ΔE_m volts is given by

$$\Delta G = zF \, \Delta E_m \tag{2.17}$$

where z is the valence of the ion and F is Faraday's constant. The value of ΔG for the transfer of cations into a positive compartment is positive and so calls for work. Conversely, the value of ΔG is negative when cations move into the negative compartment, so work can be done. The electric potential is negative across the plasma membrane of the great majority of cells; therefore cations tend to leak in but have to be "pumped" out.

The Electrochemical Potential Difference, $\Delta\bar{\mu}$, Includes Both Concentration and Electrical Potentials

In general, ions moving across a membrane are subject to gradients of both concentration and electric potential. Consider, for example, the situation depicted in Figure 2.3, which corresponds to a major event in energy transduction during photosynthesis. A cation of valence z moves from compartment 1 to compartment 2, against both a concentration gradient ($C_2 > C_1$) and a gradient of membrane electric potential (compartment 2 is electropositive relative to compartment 1). The free energy change involved in this transfer is given by Equation 2.18; ΔG is positive, and the transfer can proceed only if coupled to a source of energy, in this instance the absorption of light:

$$\Delta G = zF \, \Delta E_m + 2.3RT \log \frac{C_2}{C_1} \tag{2.18}$$

As a result of this transfer, cations in compartment 2 can be said to be at a higher electrochemical potential than the same ions in compartment 1. The electrochemical potential for a particular ion is designated $\bar{\mu}_{ion}$. Ions tend to flow from a region of high electrochemical potential to one of low potential and can in principle do work thereby. The maximum amount of this work, neglecting friction, is given by the change in free energy of the ions that flow from compartment 2 to compartment 1 (Eq. 2.6) and is numerically equal to the difference in electrochemical potential, $\Delta\bar{\mu}_{ion}$. This principle underlies much of biological energy transduction.

The electrochemical potential difference $\Delta\bar{\mu}_{ion}$ is properly expressed in kilojoules per mole or kilocalories

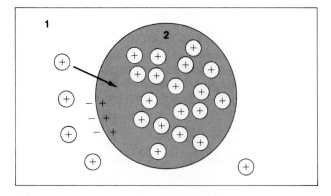

FIGURE 2.3. Transport against an electrochemical potential gradient. The agent that moves the charged solute $\oplus$ from compartment 1 to compartment 2 must do work to overcome both the electrochemical potential gradient and the concentration gradient. As a result, cations in compartment 2 have been raised to a higher electrochemical potential than those in compartment 1. Neutralizing anions have been omitted.

---

4. Many texts use the term $\Delta\psi$ for the membrane potential difference. However, to avoid confusion with the use of $\Delta\psi$ to indicate water potential (see end of chapter), the term ΔE_m will be used here.

per mole. However, it is frequently convenient to state the driving force for ion movement in electrical terms, with the dimensions of volts or millivolts. To convert $\Delta\bar{\mu}_{ion}$ into millivolts (mV), all the terms in Equation 2.18 are divided by F:

$$\frac{\Delta\bar{\mu}_{ion}}{F} = z\,\Delta E_m + \frac{2.3RT}{F}\log\frac{C_2}{C_1} \qquad (2.19)$$

An important case in point is the proton motive force, or protonic potential, which will be considered at length in Chapter 6.

Equations 2.18 and 2.19 have proved to be of central importance in bioenergetics. First, they measure the amount of energy that must be expended on the active transport of ions and metabolites, a major function of biological membranes. Second, since the free energy of chemical reactions is often transduced into other forms via the intermediate generation of electrochemical potential gradients, they play a major role in descriptions of biological energy coupling. It should be emphasized that the electrical and concentration terms may either add, as in Equation 2.18, or subtract and that the application of the equations to particular cases requires careful attention to the sign of the gradients. We should also note that free energy changes in chemical reactions (Eq. 2.10) are scalar, whereas transport reactions have direction; this turns out to be a subtle but critical aspect of the biological role of ion gradients.

Ion distribution at equilibrium is an important special case of the general electrochemical equation (Eq. 2.18). Figure 2.4 shows a membrane-bound vesicle (compartment 2) that contains a high concentration of the salt K_2SO_4, surrounded by a medium (compartment 1) containing a lower concentration of the same salt; it is stipulated that the membrane is impermeable to anions but freely passes cations. Potassium ions will therefore tend to diffuse out of the vesicle into the solution, whereas the anions are retained. Diffusion of the cations generates a membrane potential, with the vesicle interior negative, which restrains further diffusion. At equilibrium, ΔG and $\Delta\bar{\mu}_{K^+}$ equal zero (by definition). Equation 2.18 can then be arranged to give

$$\Delta E_m = \frac{-2.3RT}{zF}\log\frac{C_2}{C_1} \qquad (2.20)$$

where C_2 and C_1 are the concentrations of K^+ ions in the two compartments; z, the valence, is unity; and ΔE_m is the membrane potential in equilibrium with the potassium concentration gradient.

This is one form of the celebrated **Nernst equation.** It states that at equilibrium, a permeant ion will be so distributed across the membrane that the chemical driving force (outward in this instance) will be balanced by the electric driving force (inward). For a univalent cation at

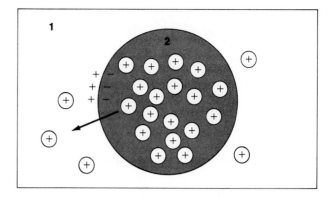

FIGURE 2.4. Generation of an electrical potential by ion diffusion. Compartment 2 has a higher salt concentration than compartment 1 (anions are not shown). If the membrane is permeable to the cations but not to the anions, the cations will tend to diffuse into compartment 1, generating a membrane potential with compartment 2 negative.

25°C, each tenfold increase in concentration factor corresponds to a membrane potential of 59 mV; for a divalent ion the value is 29.5 mV.

The preceding discussion of the energetic and electrical consequences of ion translocation illustrates a point that must be clearly understood, namely that an electrical potential across a membrane may arise by two distinct mechanisms. The first mechanism, illustrated by Figure 2.4, is the diffusion of charged particles down a preexisting concentration gradient, an exergonic process. A potential generated by such a process is described as a diffusion potential, or as a Donnan potential. (Donnan potential is defined as the diffusion potential that occurs in the limiting case where the counterion is completely impermeant or fixed, as in Fig. 2.4.) Many ions are unequally distributed across biological membranes and differ widely in their rates of diffusion across the barrier; therefore diffusion potentials always contribute to the observed membrane potential. But in most biological systems the measured electric potential differs from that which would be expected on the basis of passive ion diffusion. In these cases one must invoke electrogenic ion pumps, transport systems that carry out the exergonic process indicated in Figure 2.3 at the expense of an external energy source. Transport systems of this kind transduce the free energy of a chemical reaction into the electrochemical potential of an ion gradient and play a leading role in biological energy coupling.

The Water Potential Is the Chemical Potential of Water Expressed in Units of Pressure

Water is the most abundant constituent of living organisms. The transport processes considered above involve the movement of solute from one aqueous compartment to another; we now turn to the behavior of the solvent, water. Water relations are especially prominent in the physiology of plants, which must absorb water from the soil and maintain turgor.

In order to describe the movement of water, it is necessary to enlarge somewhat the concept of electrochemical potential, $\bar{\mu}$. Like any other substance, water flows downhill from a region of high potential to one where its potential is lower. The factors that determine the potential ($\bar{\mu}_j$) of any species j appear as terms in Equation 2.21.

$$\mu_j = \mu_j^* + 2.3RT \log C_j + z_j FE + \bar{V}_j P + m_j gh \quad (2.21)$$

The meaning of the terms is as follows. μ_j^* is a constant that gives the potential of species j in the reference state. In practice, one is chiefly interested in the difference between the chemical potentials of j at two locations, so that μ_j^* cancels out.

The second term, $2.3RT \log C_j$, specifies the contribution that the concentration (C) (more precisely, the activity) of j makes to its chemical potential. In the case of water, the concentration of the solvent is reduced by the addition of a solute; when two solutions are separated by a membrane permeable to water but not to the solute, water tends to flow from the dilute compartment to the more concentrated one.

In the third term, $z_j FE$, z_j is the valence of species j, F is Faraday's constant, and E is the electrical potential. Since water is uncharged, its movements are unaffected by the electrical potential; for that reason the plant literature speaks of the chemical (not electrochemical) potential of water.

The fourth term, $\bar{V}_j P$, takes account of the general fact that substances tend to flow from a region of high pressure to one of lower pressure. $\bar{V}_j$ is the partial molal volume of species j, the volume occupied by one mole of species j (18 ml per mole in the case of water), and P is the pressure. This term was neglected in earlier sections, which implicitly ignored differences in pressure, but it enters prominently into water relations in plants. Because of their rigid walls, microorganisms and plants can develop large hydrostatic pressures.

Finally, $m_j gh$ takes account of the gravitational field: m_j is the mass of species j, g the gravitational acceleration, and h the height. This term, negligible on the cellular scale, can be large on the organismic level. It is obvious, for instance, that gravitational work must be done to transport water to the top of a tree.

The movement of solutes across biological membranes is often mediated by specialized proteins that couple metabolic energy to solute transport. This is not the case for water: as far as we know, water moves by a combination of diffusion and pressure-driven bulk flow (see Chapter 3). The direction of new flow is determined by the sum of the terms in Equation 2.21: water moves from a region of high chemical potential to one of lower potential.

The interplay of driving forces is illustrated by the genesis of osmotic pressure, a major element in the physiology of walled cells and organisms. When solutes are dissolved in water, the mole fraction of water decreases and so does its concentration (activity) and therefore its chemical potential. If the solution makes contact (through a semipermeable membrane) with a second compartment containing pure water, water tends to flow into the solution, diluting it and increasing its volume. The osmotic pressure, π, is the hydrostatic pressure that, when applied to the solution, just prevents the influx of water. This, of course, is what happens in walled cells: water tends to flow into the cytoplasm, thanks to its content of solutes, but enlargement of the cell is prevented by the rigid cell wall. The plasma membrane exerts substantial pressure on the wall, as much as 1 megapascal (MPa)[5] and sometimes more. Strictly speaking, an isolated solution has no osmotic pressure, but it does have an osmotic potential. We shall neglect this subtlety here and define the osmotic pressure of any solution as a function of its water concentration (more precisely, water activity) by the relationship

$$2.3RT \log C_w = -\bar{V}_w \pi \quad (2.22)$$

Here, RT and π are expressed in pressure units (RT at 20°C is 24.37 liter-bars mol^{-1}). Even dilute solutions have substantial osmotic pressures: 0.1 M sucrose has an osmotic pressure of approximately 0.24 MPa.

By the use of Equation 2.22, one can replace the concentration term in Equation 2.21 with the osmotic pressure term:

$$\bar{\mu}_w = \mu_w^* - \bar{V}_w \pi + \bar{V}_w P + m_w gh \quad (2.23)$$

As a matter of experimental convenience, plant physiologists prefer to employ a related term, the water potential ψ_w. The water potential is defined as follows:

$$\psi_w = \frac{\bar{\mu}_w - \mu_w^*}{\bar{V}} = P - \pi + \rho gh \quad (2.24)$$

where π is the osmotic pressure and ρ is the density of water. Whereas the dimensions of $\bar{\mu}$ and μ^* give the energy content per mole (joules per mole or kilocalories per mole), the water potential is expressed in units of pressure (megapascals or bars). Note than an increase in the hydrostatic pressure raises the water potential, whereas an increase in the solute content lowers it; also, water at treetop level has a higher potential than water at ground level. We shall examine the uses of the water potential in detail in Chapters 3 and 4.

---

5. 1 MPa = 10 bars $\approx$ 10 atm.

Enzymes: The Agents of Life

Proteins constitute about 30% of the total dry weight of typical plant cells. If we exclude inert materials, such as the cell wall and starch, which can account for up to 90% of the dry weight of some cells, proteins and amino acids represent about 60–70% of the dry weight of the living cell. As we have seen in Chapter 1, cytoskeletal structures such as microtubules and microfilaments are composed of protein. Proteins can also occur as storage forms, particularly in seeds. However, the major function of proteins in metabolism is to serve as enzymes, biological catalysts that greatly increase the rates of biochemical reactions, making life possible. Enzymes participate in these reactions but are not themselves fundamentally changed in the process (Mathews and van Holde, 1990).

Enzymes have been called the "agents of life"—a very apt term, since essentially all life processes are controlled by them. A typical cell has several thousand different enzymes, which carry out a wide variety of reactions. The most remarkable features of enzymes are their *specificity*, which permits them to distinguish among very similar molecules, and their *catalytic efficiency*, which is far greater than that of ordinary catalysts. The stereospecificity of enzymes is quite amazing, allowing them to distinguish not only between enantiomers, for example, but even between apparently identical atoms or groups of atoms (Creighton, 1983).

This ability to discriminate between similar molecules results from the fact that the first step in enzyme catalysis is the formation of a tightly bound, noncovalent complex, the **enzyme-substrate complex**, between the enzyme and the substrate(s). Enzyme-catalyzed reactions exhibit unusual kinetic properties that are also related to the formation of these very specific complexes. Another distinguishing feature of enzymes is that they are subject to various kinds of regulatory control, ranging from subtle effects on the catalytic activity by effector molecules (inhibitors or activators) to regulation of enzyme synthesis and destruction by the control of gene expression and protein turnover.

Enzymes are unique in the large rate enhancements they bring about, orders of magnitude greater than those due to other catalysts. Typical rate enhancements of enzyme-catalyzed reactions over the corresponding uncatalyzed reactions are 10^8 to 10^{12}. Many enzymes will convert about a thousand molecules of substrate to product in 1 second. Some will convert as many as a million!

Unlike most other catalysts, enzymes function at ambient temperature and atmospheric pressure and usually in a narrow pH range near neutrality (there are exceptions; for instance, vacuolar proteases and ribonucleases are most active at pH 4–5). A few enzymes are able to function under extremely harsh conditions; examples are pepsin, the protein-degrading enzyme of the stomach, which has a pH optimum around 2.0, and the hydrogenase of the hyperthermophilic ("extreme heat-loving") archaebacterium *Pyroccus furiosus*, which oxidizes H_2 with a temperature optimum greater than 95°C (Bryant and Adams, 1989). The presence of such remarkably heat-stable enzymes enables *Pyrococcus* to grow optimally at around 100°C!

Enzymes are usually named after their substrates by adding the suffix *-ase*—for example, α-amylase, malate dehydrogenase, β-glucosidase, phosphoenolpyruvate carboxylase, horseradish peroxidase. Many thousands of enzymes have already been discovered, and new ones are being found all the time. Each enzyme has been named in a systematic fashion, based on the reaction it catalyzes, by the International Union of Biochemistry. In addition, many enzymes have common, or trivial, names. Thus the common name Rubisco refers to D-ribulose 1,5-bisphosphate carboxylase/oxygenase (EC 4.1.1.39). The Enzyme Commission (EC) number indicates the class (4 = lyase) and subclasses (4.1 = carbon-carbon cleavage; 4.1.1 = cleavage of C—COO − bond).

The versatility of enzymes reflects their properties as proteins. It is the nature of proteins that permits both the exquisite recognition by an enzyme of its substrate and the catalytic apparatus necessary to carry out diverse and rapid chemical reactions (Stryer, 1988).

Proteins Are Chains of Amino Acids Joined by Peptide Bonds

Proteins are composed of long chains of amino acids (Fig. 2.5) linked by amide bonds, known as **peptide bonds** (Fig. 2.6). The 20 different amino acid side chains endow proteins with a large variety of groups with different chemical and physical properties, including hydrophilic (polar, water-loving) and hydrophobic (nonpolar, water-avoiding) groups, charged and neutral polar groups, and acidic and basic groups. This diversity, in conjunction with the relative flexibility of the peptide bond, allows for the tremendous variation in protein properties, ranging from the rigidity and inertness of structural proteins to the reactivity of hormones, catalysts, and receptors. The three-dimensional aspect of protein structure provides for precise discrimination in the recognition of **ligands**, the molecules that interact with proteins, as shown by the ability of enzymes to recognize their substrates and of antibodies to recognize antigens, for example.

All molecules of a particular protein have the same sequence of amino acid residues, determined by the sequence of nucleotides in the gene coding for that protein. Although the protein is synthesized as a linear chain on the ribosome, upon release it folds spontaneously into a specific three-dimensional shape, the **native** state. The chain of amino acids is called a polypeptide.

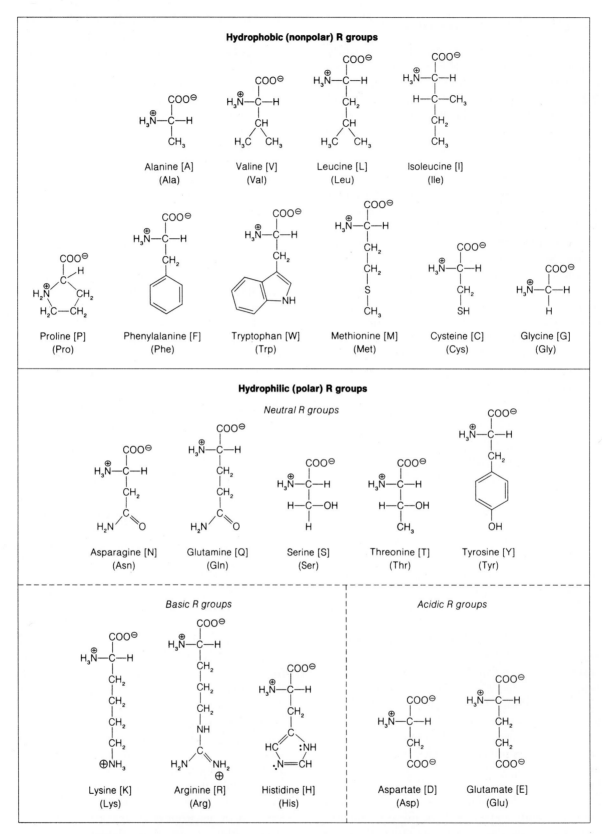

FIGURE 2.5. The structures, names, single-letter codes (in square brackets), abbreviations, and classification of the amino acids.

(a)

Peptide bond

(b)

Rigid unit

FIGURE 2.6. (a) The peptide bond links two amino acids by an amide bond. (b) Sites of free rotation, within the limits of steric hindrance, about the N—C_α and C_α—C bonds (ϕ and ψ); there is no rotation about the amide bond because of its double-bond character.

HYDROGEN BONDS

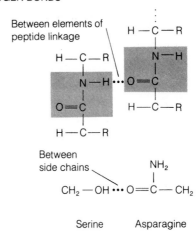

Between elements of peptide linkage

Between side chains

Serine   Asparagine

ELECTROSTATIC ATTRACTIONS

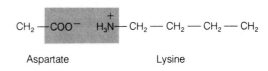

Aspartate   Lysine

VAN DER WAALS INTERACTIONS

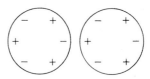

FIGURE 2.7. Examples of noncovalent interactions in proteins. Hydrogen bonds are weak electrostatic interactions involving a hydrogen atom between two electronegative atoms. In proteins the most important hydrogen bonds are those between the peptide bonds. Electrostatic interactions are essentially ionic bonds between positively and negatively charged groups. The van der Waals interactions are short-range transient dipole interactions. Hydrophobic interactions (not shown) involve restructuring of the solvent water around nonpolar groups, minimizing the exposure of nonpolar surface area to polar solvent, and are entropy-driven.

The three-dimensional arrangement of the atoms in the molecule is referred to as the **conformation**. Changes in conformation do *not* involve breaking of covalent bonds. The process of *denaturation* involves the loss of this unique three-dimensional shape and results in the loss of catalytic activity.

The forces that are responsible for the shape of a protein molecule are noncovalent in nature (Fig. 2.7). These noncovalent interactions involve hydrogen bonds; electrostatic interactions (also known as ionic bonds or salt bridges); van der Waals interactions (dispersion forces), which are transient dispoles between spatially close atoms; and hydrophobic "bonds"—the tendency of nonpolar groups to avoid contact with water and thus to associate with themselves. In addition, covalent disulfide bonds are found in many proteins. Although each of these types of noncovalent interaction is weak, there are so many noncovalent interactions that in total they contribute a large amount of free energy to stabilizing the native structure.

Protein Structure Is Hierarchical

Proteins are built up with increasingly complex organizational units. The primary structure of a protein refers to the sequence of amino acid residues. The secondary structure refers to regular, local structural units, usually held together by hydrogen bonding. The most common of these units are the α helix and β strands forming parallel and antiparallel β-pleated sheets and turns (Fig. 2.8). The tertiary structure—the final three-dimensional structure of the polypeptide—results from the packing together of the secondary structure units and the exclusion of solvent. The quaternary structure refers to the association of two or more separate three-dimensional polypeptides to form complexes. When associated in this manner, the individual polypeptides are called **subunits**.

A protein molecule consisting of a large single polypeptide chain is composed of several independently folding units known as domains. Typically, domains have molecular mass of about 10^4 daltons. The active site of an enzyme—that is, the region where the substrate binds

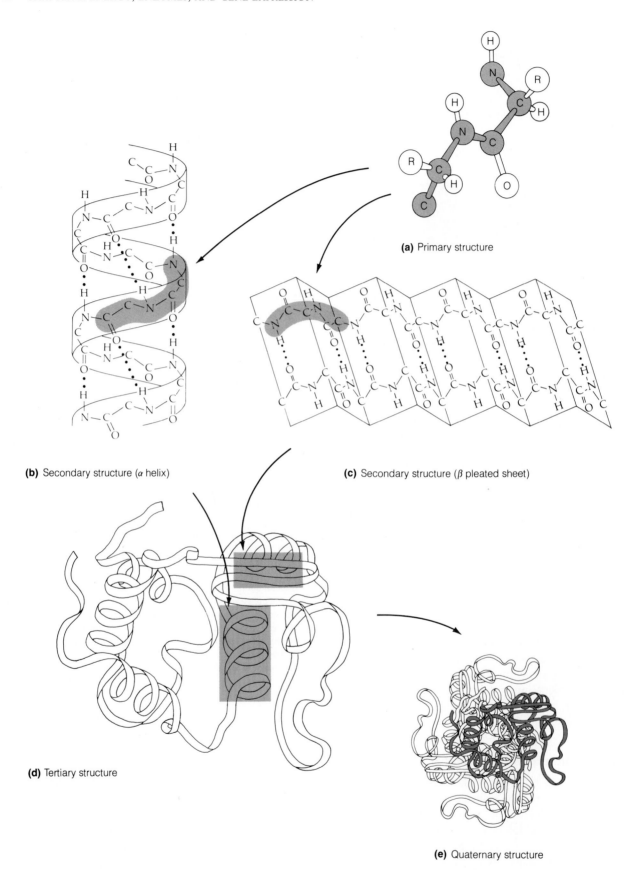

(a) Primary structure

(b) Secondary structure (α helix)

(c) Secondary structure (β pleated sheet)

(d) Tertiary structure

(e) Quaternary structure

FIGURE 2.8. Hierarchy of protein structure. (a) Primary structure: peptide bond. (b and c) Secondary structure: α helix and antiparallel β-pleated sheet. (d) Tertiary structure: α helices, β-pleated sheets, and random coils. (e) Quaternary structure: four subunits.

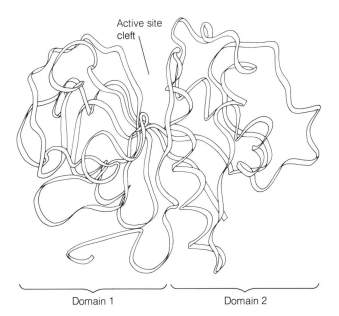

FIGURE 2.9. The backbone structure of papain, showing the two domains and the active-site cleft running between them.

and the catalytic reaction occurs—is often located at the interface between two domains. For example, in the enzyme papain (a vacuolar protease that is found in papaya and is representative of a large class of plant thiol proteases), the active site lies at the junction of two domains (Fig. 2.9). A number of helices, turns, and β sheets contribute to the unique three-dimensional shape of this enzyme.

Determinations of the conformation of proteins have revealed that there are families of proteins with common three-dimensional folds, as well as common patterns of super-secondary structure, such as β-α-β.

Enzymes Are Highly Specific Protein Catalysts

All enzymes are proteins, although recently some small ribonucleic acids and protein-RNA complexes have been found to exhibit enzyme-like behavior in the processing of RNA. Proteins have molecular masses ranging from 10^4 to 10^6 daltons, and they may be a single folded polypeptide chain (subunit, or protomer) or oligomers of several subunits (most commonly oligomers are dimers or tetramers). Normally, enzymes have only one type of catalytic activity associated with the same protein. **Isoenzymes** or **isozymes** are enzymes with similar catalytic function that have different structures and catalytic parameters and are encoded by different genes. For example, a number of different isozymes have been found for peroxidase, an enzyme in plant cell walls that is involved in the synthesis of lignin. An isozyme of peroxidase has also been localized in vacuoles. Isozymes may exhibit tissue specificity and show developmental regulation.

Enzymes frequently contain a nonprotein **prosthetic group** or **cofactor** that is necessary for biological activity.

The association of a cofactor with an enzyme depends on the three-dimensional structure of the protein. Once bound to the enzyme, the cofactor contributes to the specificity of catalysis. Typical examples of cofactors are metal ions (e.g., zinc, iron, molybdenum), heme groups or iron-sulfur clusters (especially in oxidation-reduction enzymes), and coenzymes (e.g., nicotinamide adenine dinucleotide [NAD$^+$/NADH], flavin adenine dinucleotide [FAD/FADH$_2$], flavin mononucleotide [FMN], and pyridoxal phosphate [PLP]). Coenzymes are usually vitamins or are derived from vitamins and act as carriers. For example, NAD$^+$ and FAD carry hydrogens and electrons in redox reactions, biotin carries CO_2, and tetrahydrofolate carries one-carbon fragments. Peroxidase has both heme and Ca^{2+} prosthetic groups and is glycosylated; that is, it contains carbohydrates covalently added to asparagine, serine, or threonine side chains. Such proteins are called **glycoproteins.**

A particular enzyme will catalyze only one type of chemical reaction for only one class of molecule—in some cases, for only one particular compound. Enzymes are also very stereospecific and produce no by-products. For example, β-glucosidase catalyzes the hydrolysis of β-glucosides, compounds formed by a glycosidic bond to D-glucose. The substrate must have the correct anomeric configuration: it must be β-, not α-. Furthermore, it must have the glucose structure; no other carbohydrates such as xylose or mannose can act as substrates for β-glucosidase. Finally, the substrate must have the correct stereochemistry, in this case the D absolute configuration. Rubisco (D-ribulose 1,5-bisphosphate carboxylase/oxygenase) catalyzes the addition of carbon dioxide to D-ribulose 1,5-bisphosphate to form two molecules of 3-phospho-D-glycerate, the initial step in the C3 photosynthetic carbon reduction cycle, and is the world's most abundant enzyme. It has very strict specificity for the carbohydrate substrate, but it also catalyzes an oxygenase reaction in which O_2 replaces CO_2, as will be discussed further in Chapter 9.

Enzymes Lower the Free Energy Barrier Between Substrates and Products

Catalysts speed the rate of a reaction by lowering the energy barrier between substrates (reactants) and products and are not themselves used up in the reaction but are regenerated. Thus a catalyst increases the rate of a reaction but does not affect the equilibrium ratio of reactants and products, because the rates of the reaction in both directions are increased to the same extent. It is important to realize that enzymes cannot make a nonspontaneous (energetically uphill) reaction occur. However, many energetically unfavorable reactions in cells proceed because they are coupled to an energetically more favorable reaction usually involving ATP hydrolysis (Fig. 2.10).

Enzymes act as catalysts because they lower the free energy of activation for a reaction. They do this by a combination of raising the ground state ΔG of the substrate and lowering the ΔG of the **transition state** (TS) of the reaction, thereby decreasing the barrier for the reaction to occur (Fig. 2.11). The presence of the enzyme leads to a new reaction pathway that is different from that of the uncatalyzed reaction.

$$A + B \rightarrow C \qquad\qquad \Delta G = +4.0 \text{ kcal mol}^{-1}$$

$$ATP + H_2O \rightarrow ADP + P_i + H^+ \qquad \Delta G = -7.3 \text{ kcal mol}^{-1}$$

$$A + B + ATP + H_2O \rightarrow C + ADP + P_i + H^+ \quad \Delta G = -3.3 \text{ kcal mol}^{-1}$$

$$A + ATP \rightarrow A{-}P + ADP$$

$$A{-}P + B + H_2O \rightarrow C + H^+ + P_i$$

$$A + B + ATP + H_2O \rightarrow C + ADP + P_i + H^+$$

FIGURE 2.10. Coupling of the hydrolysis of ATP to drive an energetically unfavorable reaction. The reaction A + B → C is thermodynamically unfavorable, whereas the hydrolysis of ATP to form ADP and inorganic phosphate (P_i) is thermodynamically very favorable (it has a large negative ΔG). Through appropriate intermediates, such as A-P, the two reactions are coupled so as to yield an overall reaction that is the sum of the individual reactions and has a favorable free energy change.

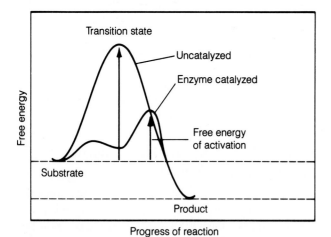

FIGURE 2.11. Free energy curves for the same reaction, either uncatalyzed or enzyme-catalyzed. As a catalyst, an enzyme lowers the free energy of activation of the transition state between substrates and products compared with the uncatalyzed case. It does this by forming various complexes and intermediates such as enzyme-substrate (ES) and enzyme-product (EP) complexes. The ground state free energy of the ES complex in the enzyme-catalyzed reaction may be higher than that of the substrate in the uncatalyzed reaction, and the transition state free energy of the enzyme-bound substrate will be significantly less than that in the corresponding uncatalyzed case.

Catalysis Occurs at the Active Site

The **active site** of an enzyme molecule is usually a cleft or pocket on or near the surface of the enzyme that involves only a small fraction of the enzyme surface. It is convenient to consider the active site as consisting of two components: the **binding site** for the substrate (which attracts and positions the substrate) and the **catalytic groups** (the reactive side chains of amino acids or cofactors, which carry out the bond-breaking and bond-forming reactions involved).

Binding of substrate at the active site initially involves noncovalent interactions between the substrate and either side chains or peptide bonds of the protein. The rest of the protein structure provides a means of positioning the substrate and catalytic groups, flexibility for conformational changes, and regulatory control. The shape and polarity of the binding site account for much of the specificity of enzymes, and there is complementarity between the shape and polarity of the substrate and those of the active site. In some cases, binding of the substrate induces a conformational change in the active site of the enzyme. This is particularly common where there are two substrates. Binding of the first substrate sets up a conformational change of the enzyme that results in formation of the binding site for the second substrate. Hexokinase is a good example of this type of reaction (Fig. 2.12).

The catalytic groups are usually the amino acid side chains and/or cofactors that can function as catalysts. Common examples of catalytic groups are acids (—COOH from the side chains of aspartic acid or glutamic acid, imidazolium from the side chain of histidine), bases (—NH_2 from lysine, imidazole from histidine, —S^- from cysteine), nucleophiles (imidazole from histidine, —S^- from cysteine, —OH from serine), and electrophiles (often metal ions, such as Zn^{2+}). The acidic catalytic groups function by donating a proton, the basic ones by accepting a proton. Nucleophilic catalytic groups form a transient covalent bond to the substrate.

The decisive factor in catalysis is the direct interaction between the enzyme and the substrate. In many cases, there is an intermediate involving a covalent bond between the enzyme and the substrate. Although the details of the catalytic mechanism differ from one type of enzyme to another, a limited number of features are involved in all enzyme catalysis. These include acid-base catalysis, electrophilic or nucleophilic catalysis, and ground state distortion through electrostatic or mechanical strains on the substrate.

A Simple Kinetic Equation Describes an Enzyme-Catalyzed Reaction

Enzyme-catalyzed systems often exhibit a special form of kinetics, called Michaelis-Menten kinetics, which are characterized by a hyperbolic relationship between re-

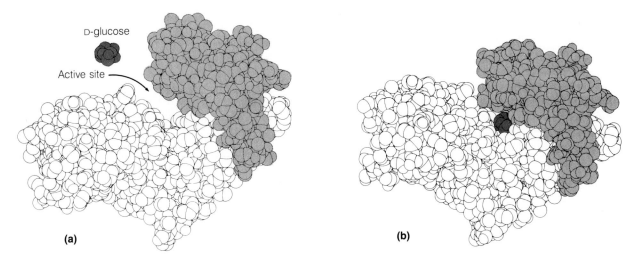

FIGURE 2.12. Hexokinase conformational change induced by its first substrate, D-glucose. (a) Before glucose binding. (b) After glucose binding. The binding of glucose to hexokinase induces a conformational change in which the two major domains come together to close the cleft containing the active site. This change sets up the binding site for the second substrate, ATP. In this manner the enzyme prevents unproductive hydrolysis of ATP by shielding the substrates from the aqueous solvent. The overall reaction is the phosphorylation of glucose and formation of ADP.

action velocity, v, and substrate concentration, [S] (Fig. 2.13). This type of plot is known as a saturation plot because when the enzyme becomes saturated with substrate (i.e., each enzyme molecule has a substrate molecule associated with it), the rate becomes independent of substrate concentration. Saturation kinetics imply that there is an equilibrium process preceding the rate-limiting step:

$$E + S \underset{}{\overset{\text{fast}}{\rightleftharpoons}} ES \xrightarrow{\text{slow}} E + P$$

where E represents the enzyme, S the substrate, P the product, and ES the enzyme-substrate complex. Thus, as

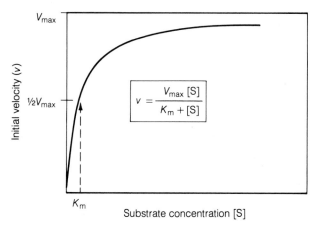

FIGURE 2.13. Plot of initial velocity, V, versus substrate concentration, [S], for an enzyme-catalyzed reaction. The curve is hyperbolic. The maximal rate, V_{max}, occurs when all the enzyme molecules are fully occupied by substrate. The value of K_m, defined as the substrate concentration at $\frac{1}{2}V_{max}$, is a reflection of the affinity of the enzyme for the substrate. The smaller the value of K_m, the tighter the binding.

the substrate concentration is increased, a point will be reached at which all the enzyme molecules are in the form of the ES complex, and the enzyme is saturated with substrate. Since the rate of the reaction depends on the concentration of ES, the rate will not increase further since there can be no higher concentration of ES.

When an enzyme is mixed with a large excess of substrate, there will be an initial very short time period (usually milliseconds) during which the concentrations of enzyme-substrate complexes and intermediates build up to certain levels; this is known as the pre-steady-state period. Once the intermediate levels have been built up, they remain relatively constant until the substrate is depleted; this period is known as the **steady state**.

Normally enzyme kinetics are measured under steady-state conditions, and such conditions usually prevail in the cell. For many enzyme-catalyzed reactions the kinetics under steady-state conditions can be described by a simple expression known as the Michaelis-Menten equation:

$$v = \frac{V_{max}[S]}{K_m + [S]} \tag{2.25}$$

where v is the observed rate or velocity (in units such as moles per liter per second), V_{max} is the maximum velocity (at infinite substrate concentration), and K_m (usually measured in units of molarity) is a constant characteristic of the particular enzyme-substrate system and is related to the association constant of the enzyme for the substrate (Fig. 2.13). K_m represents the concentration of substrate required to half-saturate the enzyme and thus is the substrate concentration at $V_{max}/2$. In many cellular systems the usual substrate concentration is in the vicinity of K_m. The smaller the value of K_m, the more strongly

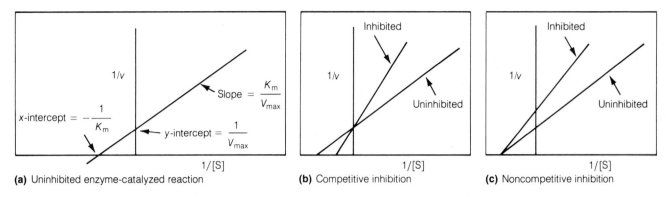

(a) Uninhibited enzyme-catalyzed reaction

(b) Competitive inhibition

(c) Noncompetitive inhibition

FIGURE 2.14. Lineweaver-Burk double-reciprocal plots. A plot of $1/v$ versus $1/[S]$ yields a straight line. (a) Uninhibited enzyme-catalyzed reaction showing the calculation of K_m from the x-intercept and of V_{max} from the y-intercept. (b) The effect of a competitive inhibitor on the parameters K_m and V_{max}. The apparent K_m is increased but the V_{max} is unchanged. (c) A noncompetitive inhibitor reduces V_{max} but has no effect on K_m.

the enzyme binds the substrate. Typical values for K_m are in the range 10^{-6} to 10^{-3} M.

The parameters V_{max} and K_m are readily obtained by fitting experimental data to the Michaelis-Menten equation, either by computerized curve fitting or by using a linearized form of the equation. An example of a linearized form of the equation is the Lineweaver-Burk double-reciprocal plot shown in Figure 2.14a. When divided by the concentration of enzyme, the value of V_{max} gives the **turnover number**, the number of molecules of substrate converted to product per unit of time per molecule of enzyme. Typical turnover number values range from 10^2 to 10^3 s^{-1}.

Enzymes Are Subject to Various Kinds of Inhibition

Any agent that decreases the velocity of an enzyme-catalyzed reaction is called an inhibitor. Inhibitors may exert their effects in many different ways. Generally, if the inhibition is irreversible the compound is called an **inactivator**. There are also agents that can increase the efficiency of an enzyme, and these are called **activators**. Inhibitors and activators are very important in the cellular regulation of enzymes. Many agriculturally important insecticides and herbicides are enzyme inhibitors. The study of enzyme inhibition can provide useful information about kinetic mechanisms, the nature of enzyme-substrate intermediates and complexes, the chemical mechanism of catalytic action, and the regulation and control of metabolic enzymes. In addition, the study of inhibitors of potential target enzymes is essential to the rational design of herbicides.

Inhibitors can be classified as reversible or irreversible. **Irreversible inhibitors** form covalent bonds with an enzyme or denature it. For example, iodoacetate (ICH_2COOH) irreversibly inhibits thiol proteases such as

papain by alkylating the active-site —SH group. One class of irreversible inhibitors is called affinity labels, or active site-directed modifying agents, because their structure directs them to the active site. An example is tosyl-lysine chloromethyl ketone (TLCK), which irreversibly inactivates papain. The tosyl-lysine part of the inhibitor resembles the substrate structure and so binds in the active site. The chloromethyl ketone part of the bound inhibitor reacts with the active-site histidine side chain. Such compounds are very useful in mechanistic studies of enzymes but have limited practical use as herbicides because of their chemical reactivity, which can be harmful to the plant.

Reversible inhibitors form weak, noncovalent bonds with the enzyme and may be competitive, noncompetitive, or mixed. For example, the widely used broad-spectrum herbicide glyphosate (Roundup®) works by competitively inhibiting a key enzyme in the biosynthesis of aromatic amino acids, 5-enolpyruvylshikimate-3-phosphate (EPSP) synthase. Resistance to glyphosate has recently been achieved by genetically engineering plants so that they are capable of overproducing EPSP synthase (Weising et al., 1988).

Competitive Inhibition. This is the simplest and most common form of reversible inhibition. Competitive inhibition usually arises from binding of the inhibitor to the active site with an affinity similar to or stronger than that of the substrate. Thus the effective concentration of the enzyme is decreased by the presence of the inhibitor, and the catalytic reaction will be slower than if the inhibitor were absent. Competitive inhibition is usually based on the fact that the structure of the inhibitor resembles that of the substrate, hence the strong affinity of the inhibitor for the active site. Competitive inhibition may also occur in **allosteric enzymes**, where the inhibitor binds to a distant site on the enzyme, causing a conformational change that alters the active site and prevents normal substrate binding. Such a binding site is called an **allosteric site**. In this case, the competition between substrate and inhibitor is indirect.

Competitive inhibition results in an apparent increase in K_m and no effect on V_{max} (Fig. 2.14b). By measuring the apparent K_m as a function of inhibitor concentration, one can calculate K_i, the inhibitor constant, which reflects the affinity of the enzyme for the inhibitor.

Noncompetitive Inhibition. In this case the inhibitor does not compete with the substrate for binding to the active site. Instead, it may bind to another site on the protein and obstruct the substrate's access to the active site, thereby changing the enzyme's catalytic properties, or it may bind to the ES complex and thus alter catalysis. Noncompetitive inhibition is frequently observed in the regulation of metabolic enzymes. The diagnostic property of this type of inhibition is that K_m is unaffected, whereas V_{max} decreases in the presence of increasing amounts of inhibitor (Fig. 2.14c).

Mixed Inhibition. This form of inhibition is characterized by effects on both V_{max} (which decreases) and K_m (which increases). Mixed inhibition is very common and results from the formation of a complex between the enzyme, the substrate, and the inhibitor that does not break down to products.

pH and Temperature Affect the Rate of Enzyme-Catalyzed Reactions

Enzyme catalysis is very sensitive to pH. This sensitivity is easily understood when one considers that the essential catalytic groups are usually ionizable ones (imidazole, carboxyl, amino) and that they are catalytically active in only one of their ionization states. For example, imidazole acting as a base will be functional only at pH values above 7. Plots of the rates of enzyme-catalyzed reactions versus pH are usually bell-shaped, corresponding to two sigmoidal curves, one for an ionizable group acting as an acid and the other for the group acting as a base (Fig. 2.15). Although the effects of pH on enzyme catalysis usually reflect the ionization of the catalytic group, they may also reflect a pH-dependent conformational change in the protein that leads to loss of activity due to disruption of the active site.

The temperature dependence of most chemical reactions also applies to enzyme-catalyzed reactions. Thus, most enzyme-catalyzed reactions show an exponential increase in rate with increasing temperature. However, because the enzymes are proteins, another major factor comes in to play, namely denaturation. After a certain temperature is reached, enzymes show a very rapid decrease in activity due to the onset of denaturation

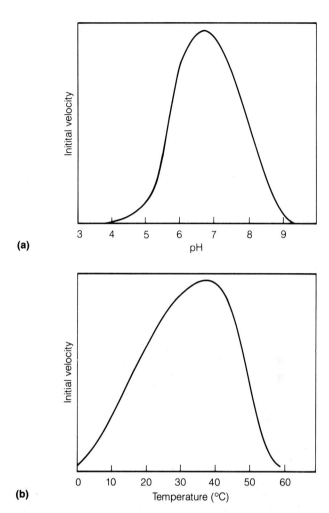

(a)

(b)

FIGURE 2.15. pH and temperature curves for typical enzyme reactions. (a) Many enzyme-catalyzed reactions show bell-shaped profiles of rate versus pH. The inflection point on each shoulder corresponds to the pK_a of an ionizing group in the active site. (b) Temperature causes an exponential increase in the reaction rate until the optimum is reached. Beyond the optimum, thermal denaturation dramatically decreases the rate.

(Fig. 2.15b). The temperature at which denaturation begins, and hence at which catalytic activity is lost, varies with the particular protein as well as the environmental conditions, such as pH. Frequently, denaturation begins at about 40–50°C and is complete over a range of about 10 degrees.

Cooperative Systems Increase the Sensitivity to Substrates and Are Usually Allosteric

Cells control the concentrations of most metabolites very closely. To keep such tight control, the enzymes controlling metabolite interconversion must be very sensitive. From the plot of velocity versus substrate concentration (Fig. 2.13), we can see that the velocity of an enzyme-catalyzed reaction increases with increasing substrate concentration up to V_{max}. However, we can calculate from the Michaelis-Menten equation (Eq. 2.25) that raising the velocity of an enzyme-catalyzed reaction from $0.1V_{max}$ to $0.9V_{max}$ requires an enormous (81-fold) increase in the substrate concentration:

$$0.1V_{max} = \frac{V_{max}[S]}{K_m + [S]}, \qquad 0.9V_{max} = \frac{V_{max}[S]'}{K_m + [S]'}$$

$$0.1K_m = 0.9[S], \qquad 0.9K_m = 0.1[S]'$$

$$\frac{0.1}{0.9} = \frac{0.9}{0.1} \times \frac{[S]}{[S]'}$$

$$\frac{[S]}{[S]'} = \left(\frac{0.1}{0.9}\right)^2 = \frac{0.01}{0.81}$$

The calculation shows that reaction velocity is rather insensitive to small changes in substrate concentration. The same factor applies in the case of inhibitors and inhibition. In **cooperative systems**, on the other hand, a small change in one parameter, such as inhibitor concentration, brings about a *large change* in velocity. A consequence of a cooperative system is that the plot of v versus [S] is no longer hyperbolic but becomes *sigmoidal* (Fig. 2.16). The advantage of cooperative systems is that a small change in the concentration of the critical effector (substrate, inhibitor, activator) will bring about a large change in the rate. In other words, the system behaves like a switch.

Cooperativity is typically observed in allosteric enzymes that contain multiple active sites located on multiple subunits. Such oligomeric enzymes usually exist in two major conformational states, one active and one inactive (or relatively inactive). Binding of ligands (substrates, activators, or inhibitors) to the enzyme perturbs the position of the equilibrium between the two conformations. For example, an inhibitor will favor the inactive form; an activator will favor the active form. The cooperative aspect comes in as follows: a positive co-

operative event is one in which binding of the first ligand makes binding of the next one easier. Similarly, negative cooperativity means that the second ligand will bind less readily than the first. Cooperativity in substrate binding (homoallostery) occurs when the binding of substrate to a catalytic site on one subunit increases the substrate affinity of an identical catalytic site located on a different subunit. Effector ligands (inhibitors or activators), in contrast, bind to sites other than the catalytic site (heteroallostery). This fits nicely with the fact that the end products of metabolic pathways, which frequently serve as feedback inhibitors, usually bear no structural resemblance to the substrates of the first step.

The Kinetics of Some Membrane Transport Processes Can Also Be Described by the Michaelis-Menten Equation

Membranes contain proteins that speed up the movement of specific ions or organic molecules across the lipid bilayer. Some membrane transport proteins are enzymes, such as ATPases, which use the energy of hydrolysis of ATP to pump ions across the membrane. When run in the reverse direction, the ATPases of mitochondria and chloroplasts can synthesize ATP. Other types of membrane proteins function as carriers, binding their substrate on one side of the membrane and releasing it on the other side. The kinetics of carrier-mediated transport can be described by the Michaelis-Menten equation in the same manner as the kinetics of enzyme-catalyzed reactions. Instead of a biochemical reaction with a substrate and product, however, the carrier binds to the solute and transfers it from one side of a membrane to the other. Letting X be the solute, we can write

$$X_{out} + carrier \xrightarrow{\text{fast}} [X\text{-}carrier] \xrightarrow{\text{slow}} X_{in} + carrier$$

Since the carrier can bind to the solute more rapidly than it can transport the solute to the other side of the membrane, solute transport exhibits saturation kinetics. That is, a concentration is reached beyond which adding more solute does not result in a more rapid rate of transport (Fig. 2.17). As indicated in Fig. 2.17, V_{max} is the maximum rate of transport of X across the membrane; K_m is equivalent to the binding constant of the solute for the carrier. Like enzyme-catalyzed reactions, carrier-mediated transport requires a high degree of structural specificity of the protein. The actual transport of the solute across the membrane apparently involves conformational changes, also similar to those in enzyme-catalyzed reactions.

Enzyme Activity Is Often Regulated

Cells can control the flux of metabolites by regulating the concentration of enzymes and their catalytic activ-

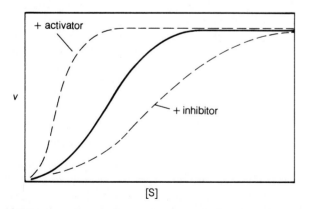

FIGURE 2.16. Allosteric systems exhibit sigmoidal plots of rate versus substrate concentration. The addition of an activator shifts the curve to the left; the addition of an inhibitor shifts it to the right.

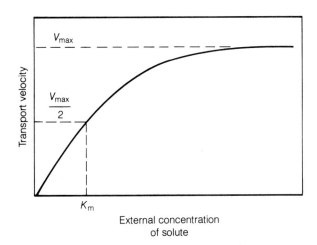

FIGURE 2.17. The kinetics of carrier-mediated transport of a solute across a membrane are analogous to those of enzyme-catalyzed reactions. Thus plots of transport velocity against concentration of solute are hyperbolic, becoming asymptotic to the maximal velocity at high solute concentration.

ity. By using allosteric activators or inhibitors, cells can modulate enzymatic activity and obtain very carefully controlled expression of catalysis.

Control of Enzyme Concentration. The amount of enzyme in a cell is determined by the relative rates of synthesis and degradation of the enzyme. The rate of synthesis is regulated at the genetic level by a variety of mechanisms, which are discussed in greater detail in the last section of this chapter.

Compartmentalization. Different enzymes or isozymes with different catalytic properties (e.g., substrate affinity) may be localized in different regions of the cell, such as mitochondria and cytosol. Similarly, enzymes associated with special tasks are often compartmentalized; for example, the enzymes involved in photosynthesis are found in chloroplasts. Vacuoles contain many hydrolytic enzymes such as proteases, ribonucleases, glycosidases, and phosphatases, as well as peroxidases. The cell walls contain glycosidases and peroxidases. The mitochondria are the main locations of the enzymes involved in oxidative phosphorylation and energy metabolism, including the enzymes of the tricarboxylic acid (TCA) cycle.

Covalent Modification. Control by covalent modification of enzymes is quite common and usually involves their phosphorylation or adenylylation, such that the phosphorylated form, for example, is active and the nonphosphorylated form is inactive. These control mechanisms are normally energy-dependent and usually involve ATP.

Proteases are normally synthesized as inactive precursors known as zymogens or proenzymes. For example, papain is made as an inactive precursor called "propapain" and becomes activated later by cleavage (hydrolysis) of a peptide bond. This type of covalent modification avoids premature proteolytic degradation of cellular constituents by the newly synthesized enzyme.

Feedback Inhibition. Consider a typical metabolic pathway with two or more end products such as that shown in Figure 2.18. Control of the system requires that if too much of the end products build up, their rate of formation is decreased. Similarly, if too much reactant A builds up, the rate of conversion of A to products should be increased. Regulation of the process is usually achieved by control of the flux at the first step of the pathway and at each branch point. The final products, G and J, which might bear no resemblance to the substrate A, inhibit the enzymes at A → B and at the branch point.

By having two enzymes at A → B, each being inhibited by one of the end metabolites but not by the other, it is possible to exert finer control than with just one enzyme. The first step in a metabolic pathway is usually called the *committed step*. It is at this step that enzymes are subject to major control.

Fructose 2,6-bisphosphate plays a central role in the regulation of carbon metabolism in plants. It functions as an activator in glycolysis (the breakdown of sugars to generate energy) and an inhibitor in gluconeogenesis (the synthesis of sugars). Fructose 2,6-bisphosphate is synthesized from fructose 6-phosphate in a reaction requiring ATP and catalyzed by the enzyme fructose 6-phosphate, 2-kinase. It is degraded in the reverse reaction catalyzed by fructose-2,6-bisphosphatase, which releases inorganic phosphate (P_i). Both of these enzymes are subject to metabolic control by fructose 2,6-bisphosphate, as well as ATP, P_i, fructose 6-phosphate, dihydroxyacetone phosphate, and 3-phosphoglycerate. The role of fructose 2,6-bisposphate in plant metabolism will be discussed further in Chapters 9 and 11.

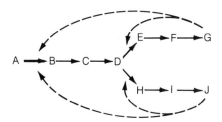

FIGURE 2.18. Feedback inhibition in a hypothetical metabolic pathway. The uppercase letters (A, B, C, etc.) represent metabolites, and each arrow represents an enzyme-catalyzed reaction. The double arrow for the first reaction indicates that two different enzymes with different inhibitor susceptibilities are involved. Broken lines indicate metabolites that inhibit particular enzymes. The first step in the metabolic pathway and the branch points are particularly important sites for feedback control.

Gene Expression and Protein Turnover

Enzyme concentrations in cells must be regulated to allow programmed growth and development of the organism. Enzymes, like other types of proteins, are encoded by specific genes, many of which are under developmental control. Differential gene expression alters the cell's complement of enzymes, thereby altering its function (Murphy and Thompson, 1988). The first step in gene expression is **transcription**, the synthesis of an mRNA copy of the DNA template encoding the enzyme (Darnell et al., 1990; Alberts et al., 1989). Developmental studies have shown that each plant organ contains large numbers of organ-specific mRNAs. For example, anthers contain at least 11,000 different mRNAs that are not found in other plant structures, while roots contain up to 7000 root-specific transcripts (Goldberg, 1986). As we shall see in Unit III, light and hormones elicit the synthesis of specific mRNAs in their target tissues. A large body of evidence suggests that transcription is controlled by DNA-binding proteins, which in turn can be regulated by small effector molecules. A number of steps leading from transcription to **translation**, the actual synthesis of the protein on the ribosome, can also affect gene expression. Finally, the rate of enzyme degradation and turnover can play a critical role in determining the steady-state level of the enzyme in the cell.

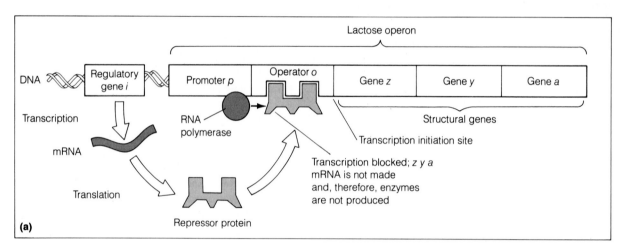

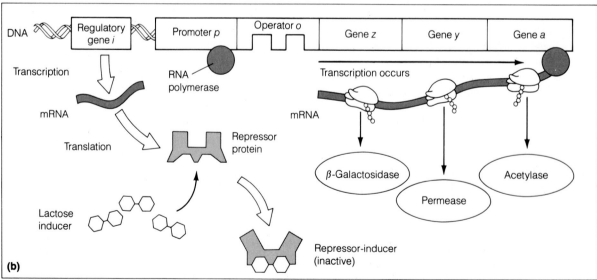

FIGURE 2.19. The *lac* operon of *E. coli*. (a) The regulatory gene *i* located upstream of the operon is transcribed to produce an mRNA encoding a repressor protein. The repressor protein binds to the operator gene *o*. The operator is a short stretch of DNA located between the promoter sequence *p* (the site of RNA polymerase attachment to the DNA) and the three structural genes, *z*, *y*, and *a*. Upon binding to the operator, the repressor prevents RNA polymerase from interacting with the transcription initiation site. (b) When lactose (inducer) is added to the medium, it binds to the repressor and inactivates it. The inactivated repressor is unable to bind to *o*, and transcription and translation can proceed.

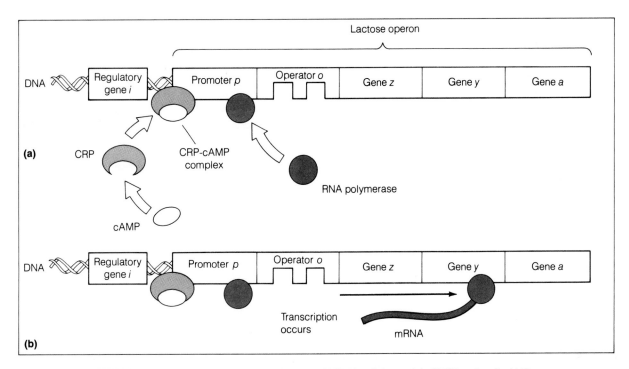

FIGURE 2.20. Stimulation of transcription by the catabolite regulator protein (CRP) and cyclic AMP (cAMP). CRP has no effect on transcription until cAMP binds to it. (a) the CRP-cAMP complex binds to a specific DNA sequence near the promoter region of the *lac* operon. (b) Binding of the CRP-cAMP complex alters the promoter region, making it more accessible to RNA polymerase. Transcription rates are enhanced.

DNA-Binding Proteins Regulate Transcription in Prokaryotes

Much of our understanding of the elements of transcription is based on bacterial systems, especially *E. coli.* In prokaryotes, genes are arranged in **operons**, sets of contiguous genes that include structural genes and regulatory sequences. A famous example is the *E. coli* lactose (*lac*) operon.

The *lac* operon is responsible for the production of three proteins involved in the utilization of the disaccharide lactose. This operon consists of three structural genes and three regulatory genes. The structural genes (*z*, *y*, and *a*) code for the sequence of amino acids in three proteins: β-galactosidase, the enzyme that catalyzes the hydrolysis of lactose to glucose and galactose; permease, which is a carrier protein for the membrane transport of lactose into the cell; and transacetylase, whose significance is unknown (Fig. 2.19).

The three regulatory sequences (*i*, *p*, and *o*) control the transcription of mRNA for the synthesis of these proteins. Gene *i* is responsible for the synthesis of a **repressor protein** that recognizes and binds to a specific nucleotide sequence, the **operator**. The operator, *o*, is located downstream from the **promoter** sequence, *p*, where RNA polymerase attaches to the operon to initiate transcription, and immediately upstream from the transcription start site, where transcription actually begins. In

the absence of lactose, the lactose repressor forms a tight complex with the operator sequence and blocks the interaction of RNA polymerase with the transcription start site, effectively preventing transcription. When present, lactose binds to the repressor, causing it to detach from the operator. When the operator sequence is unobstructed, the RNA polymerase can move along the DNA, synthesizing a continuous mRNA. The translation of this mRNA yields the three proteins, and lactose is said to induce their synthesis.

The lactose operon genes are also regulated by an activator protein, the catabolite regulator protein (CRP), which binds to a different nucleotide sequence just upstream from the operator and promoter sites (Fig. 2.20). In contrast to the behavior of the lactose repressor protein, when the CRP protein is complexed with its effector, cyclic AMP, its affinity for its DNA binding site is dramatically *increased*. In this case, the ternary complex formed between CRP, cyclic AMP, and the lactose operon DNA sequences activates transcription of the lactose operon structural genes by increasing the affinity of RNA polymerase for the neighboring promoter site. Bacteria synthesize cyclic AMP when glucose in their growth media is exhausted. The lactose operon genes are thus under opposing regulation by glucose absence (high cyclic AMP) and lactose presence, since glucose is a catabolite of lactose.

In bacteria, metabolites can also serve as *corepres-*

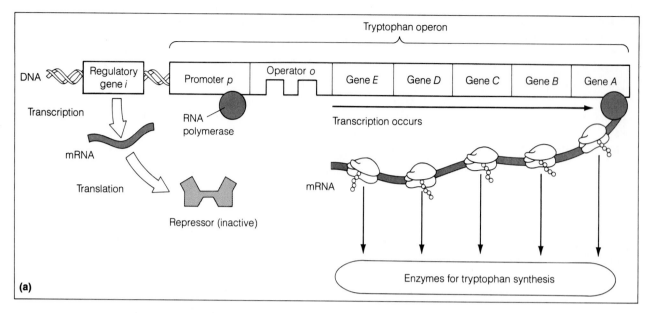

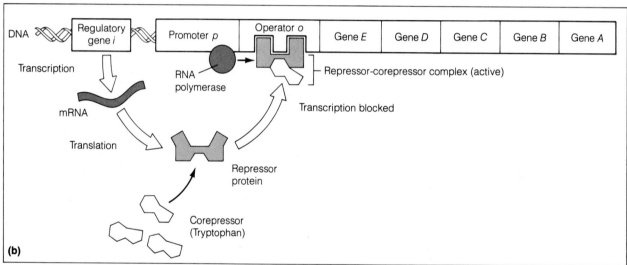

FIGURE 2.21. The *trp* operon of *E. coli*. Tryptophan (Trp) is the end product of the pathway catalyzed by tryptophan synthetase and other enzymes. Transcription of the repressor gene results in the production of a repressor protein, which, however, is inactive until it forms a complex with its corepressor, Trp. In the absence of Trp, transcription and translation proceed (a). In the presence of Trp, the activated repressor-corepressor complex blocks transcription by binding to the operator sequence (b).

sors, activating a repressor protein that blocks transcription. Repression of enzyme synthesis is often involved in regulation of biosynthetic pathways in which one or more enzymes are synthesized only if the end product of the pathway—an amino acid, for example—is not available from the environment. In such a case the amino acid acts as a corepressor; it complexes with the repressor protein, and this complex attaches to the operator DNA, preventing transcription and, of course, protein synthesis. The tryptophan (*trp*) operon in *E. coli* is an example of an operon that works by corepression (Fig. 2.21).

Gene Expression in Eukaryotes Is Regulated at Many Levels

The study of bacterial gene expression has provided models that can be tested in eukaryotes. However, it has been found that the details of the process are quite different in eukaryotes. Of foremost importance is the presence of the nuclear membrane separating the genome from the translational machinery in eukaryotes. In prokaryotes, translation is coupled to transcription. Even as the mRNA transcripts elongate, they bind to ribosomes and begin synthesizing proteins. In eukaryotes, the transcripts must first be transported to the cytosol, adding another level of control. Eukaryotes also differ from prokaryotes in the organization of their genomes. Each gene encodes a single polypeptide; there are no operons in the eukaryotic genome. Furthermore, eukaryotic genes are divided into coding regions called exons and noncoding regions called introns. Since the initial mRNA transcript, or pre-mRNA, contains both exon and intron sequences, the pre-mRNA must be processed or spliced by spliceosome complexes, composed of proteins and small nuclear RNAs, to remove the introns (McCorkle and Altman, 1987). Prior to splicing, the unprocessed pre-mRNA is modified in two important ways. First it is "capped" by the addition of 7-methylguanylate to the 5' end of the transcript via a 5' → 5' linkage. Next, a poly(A) tail is added at the 3' end. After splicing, the mature

transcript exits the nucleus as a nucleoprotein complex (Fig. 2.22).

Each of the steps involved in eukaryotic gene expression can potentially regulate the rate. In addition to transcription initiation, splicing may be regulated; different transcripts seem to form spliceosome complexes with different affinities. The stabilities (turnover rates) of translatable mRNAs are regulated and may vary from tissue to tissue, depending on the physiological conditions. The translatability of mRNA molecules is also regulated; RNAs fold into molecules with secondary and tertiary structure that can influence the accessibility of the translation initiation codon (the first AUG sequence) to the ribosome.

Because there is redundancy in mRNA codons that specify a given amino acid during translation, codon usage also affects the translatability of mRNAs. Plant

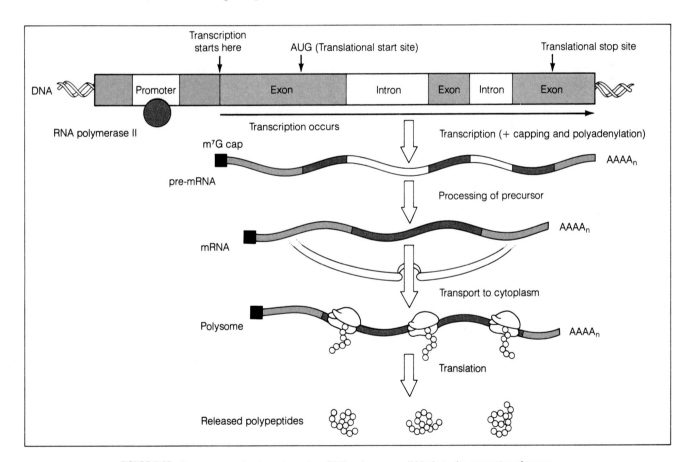

FIGURE 2.22. Gene expression in eukaryotes. RNA polymerase II binds to the promoter of genes encoding proteins. RNA polymerases I and III transcribe ribosomal genes and tRNA genes, respectively. Unlike prokaryotic genes, eukaryotic genes are not clustered in operons, and each is divided into introns and exons. Transcription from the template strand proceeds in the 3' → 5' direction at the transcription start site. Translation begins with the first AUG encoding methionine, as in prokaryotes, and ends with the stop codon. The unprocessed pre-mRNA transcript is first "capped" by the addition of GTP to the 5' end, and methyl groups are attached to carbon 7 of the terminal G. The 3' end is shortened slightly by cleavage at a specific site, and a poly(A) tail is added. The capped and polyadenylated pre-mRNA is then spliced by a spliceosome complex, and the introns are removed. The mature mRNA exits the nucleus through the pores and initiates translation on polysomes in the cytosol.

cells seem to have dramatically different populations of amino-acylated tRNAs available for complementary binding to different but redundant mRNA codons during translation. Finally, the cellular location at which translation occurs seems to affect rates of gene expression. Free polysomes may translate mRNAs at very different rates than ribosomes bound to the rough endoplasmic reticulum; even within the endoplasmic reticulum, differential translation rates may occur.

Transcription in Eukaryotes Is Also Regulated by DNA-Binding Proteins

Synthesis of many eukaryotic proteins is under transcriptional control. Both positive- and negative-acting DNA-binding regulatory proteins have been identified.

For example, in yeast, which offers a number of experimental advantages for molecular biological studies, many of the enzymes involved in galactose metabolism and transport are inducible and coregulated, despite the fact that the genes are located on different chromosomes. Three of the genes encode enzymes that convert galactose to glucose 1-phosphate; a fourth gene encodes a galactose transport protein; a fifth gene encodes the enzyme that converts melibiose to galactose, α-galactosidase; a sixth gene, GAL3, encodes an enzyme that metabolizes galactose to an as yet unidentified product.

The genes encoding the enzymes along this metabolic pathway are under both positive and negative control (Fig. 2.23). The GAL4 gene encodes a protein that binds to a specific regulatory sequence adjacent to the promoters of all six genes, about 200 bases upstream of the

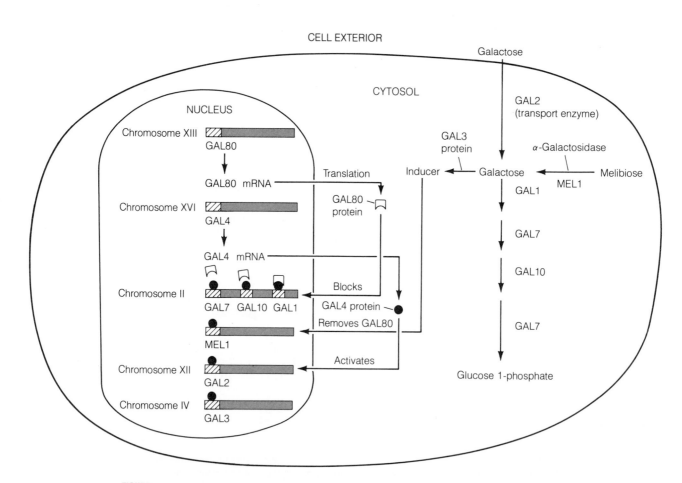

FIGURE 2.23. A model for eukaryotic gene induction: the galactose metabolism pathway of yeast. A number of enzymes involved in galactose transport and metabolism are induced by a metabolite of galactose. The genes GAL7, GAL10, GAL1, and MEL1 are located on chromosome II; GAL2 is on chromosome XII; GAL3 is on chromosome IV. GAL4 and GAL80, located on two other chromosomes, encode positive and negative regulatory proteins, respectively. The GAL4 protein binds to an activating sequence located upstream of each of the genes in the pathway, indicated by the hatched lines. The GAL80 protein forms an inhibitory complex with the GAL4 protein. In the presence of galactose, the metabolite formed by the GAL3 gene product diffuses to the nucleus and stimulates transcription by causing dissociation of the GAL80 protein from the complex. (Modified from Fig. 11-19(a) in Darnell et al., 1990.)

transcription start sites. Binding of the GAL4 protein to these genes activates their transcription. Another gene on a different chromosome, GAL80, encodes a protein that forms a complex with the GAL4 protein when it is bound to the upstream regulatory sequence. When the GAL80 protein is complexed with the GAL4 protein at the regulatory site on the gene, transcription is blocked. In the presence of galactose, the metabolite formed by the GAL3 gene product acts as an inducer by causing dissociation of the GAL80 protein from the GAL4 protein (Johnston, 1987; Mortimer et al., 1989).

While most DNA-binding regulatory proteins regulate gene expression by interacting with sites adjacent to the promoter, in eukaryotes a second class of activator regulates gene expression by interacting with DNA binding sites at a distance from promoter sites. These remote DNA-binding sites, enhancers, were first characterized in several animal virus DNA genomes and are a common feature of all eukaryotes. Unlike operator sequences, enhancer sequences function when positioned either upstream or downstream from promoters and as far as 10,000 bp from the transcription start site (Atchison, 1988).

How do enhancer DNA sequences activate gene expression at a distance? Like operator sequences, enhancer sequences complex with specific DNA-binding proteins. These activator-enhancer complexes then seem to engage RNA polymerase-promoter site complexes by bending intervening DNA sequences between the two complexes so as to bring them into proximity (alternatively, the activator-enhancer complex may engage a nucleotide sequence adjacent to a promoter rather than an RNA polymerase-promoter complex itself). This DNA bending-mediated interaction between the enhancer complex and the promoter complex somehow activates gene expression. Enhancers can activate whole groups of genes, in contrast to typical regulatory proteins, which have only local effects on specific genes.

DNA-binding sequences that control gene expression by regulating initiation of transcription, including enhancer sequences, seem to be highly prevalent among plant nuclear genes, which, as we have seen, are not clustered into operons (Kuhlemeier et al., 1987). Each individual plant gene must be developmentally and environmentally regulated, which in some cases, requires interactions between a number of specific DNA-binding sites and specific activator and repressor proteins. This complexity introduces numerous variables into the regulation of a particular plant gene. As a consequence, simple on/off conditions for plant nuclear genes may be rare.

Protein Turnover Is an Important Factor Affecting Enzyme Levels in the Cell

An enzyme molecule, once synthesized, has a finite lifetime in the cell ranging from a few minutes to several hours. Hence, steady-state levels of cellular enzymes are attained as the result of an equilibrium between enzyme synthesis and enzyme degradation, or turnover. In animal cells there are two distinct pathways of protein turnover, one occurring in specialized digestive vacuoles called lysosomes and the other occurring in the cytosol. Proteins destined to be digested in lysosomes appear to be specifically targeted to these organelles. Upon entering the lysosomes, the proteins are rapidly degraded by lysosomal proteases. Lysosomes are also capable of engulfing and digesting entire organelles by an autophagic process. The recently elucidated nonlysosomal pathway of protein turnover involves the ATP-dependent formation of a covalent bond to a small, 76-amino-acid polypeptide called ubiquitin. Ubiquitination of an enzyme molecule apparently marks it for destruction by a proteolytic complex that specifically recognizes the "tagged" molecule.

The central vacuole, rich in proteases, is the plant equivalent of lysosomes, but as yet there is no clear evidence that plant vacuoles either engulf organelles or participate in the turnover of cytosolic proteins, except during senescence. In contrast, ubiquitin has been shown to be present in plant cells, and there is evidence that it participates in the turnover of cellular proteins (Shanklin et al., 1987).

Summary

Living organisms, including green plants, are governed by the same physical laws of energy flow that apply everywhere in the universe. These laws of energy flow have been encapsulated in the laws of thermodynamics.

Energy is defined as the capacity to do work, which may be mechanical, electrical, osmotic, or chemical work. The first law of thermodynamics states the principle of energy conservation: energy can be converted from one form to another, but the total energy of the universe remains the same. The second law of thermodynamics describes the direction of spontaneous processes. A spontaneous process is one that results in a net increase in the total entropy (ΔS), or randomness, of the system plus its surroundings. Processes involving heat transfer, such as the cooling due to water evaporation from leaves, are best described in terms of the change in heat content, or enthalpy (ΔH), defined as the amount of energy absorbed or evolved as heat under constant pressure.

The free energy change, ΔG, is a convenient parameter for determining the direction of spontaneous processes in chemical or biological systems without reference to their surroundings. The value of ΔG is negative for all spontaneous processes at constant temperature and pressure. The standard free energy change of a reaction, $\Delta G°$, is the free energy change when all the reactants and products are at a concentration of 1.0 M. In biochemistry, it is more convenient to use $\Delta G°'$, which is defined in the same way as $\Delta G°$ but with the pH taken to be 7. The actual ΔG of a reaction is a function of its displacement from equilibrium. The greater the displacement from equilibrium, the more work the reaction can do. Living systems have evolved to maintain their biochemical reactions as far from equilibrium as possible.

The redox potential represents the free energy change of an oxidation-reduction reaction expressed in electrochemical units. The standard redox potential, E_0, is the tendency of a substance to donate electrons when all components are at 1.0 M. As with free energy changes, the actual redox potential of a system depends on the concentrations of the oxidized and reduced species.

The establishment of ion gradients across membranes is an important aspect of the work carried out by living systems. The membrane potential is a measure of the work required to transport an ion across a membrane. The electrochemical potential difference, $\Delta \bar{\mu}$, includes both concentration and electric potentials. Transport of water across membranes is best described by the water potential, which is the chemical potential of water expressed in pressure units.

The laws of thermodynamics predict whether and in which direction a reaction can occur, but they say nothing about how fast a reaction will be. Life depends on highly specific protein catalysts called enzymes to speed up the rates of reactions. All proteins are composed of amino acids linked together by peptide bonds. Protein structure is hierarchical and can be classified into primary, secondary, tertiary, and quaternary structure. The forces responsible for the shape of a protein molecule are noncovalent in nature and are easily disrupted by heat, chemicals, or pH, leading to loss of conformation, or denaturation.

Enzymes function by lowering the free energy barrier between the substrates and products of a reaction. Catalysis occurs at the active site of the enzyme. Enzyme-mediated reactions exhibit saturation kinetics and can be described by the Michaelis-Menten equation in which V_{max} is the maximum velocity of the enzyme-catalyzed reaction and K_m is the substrate concentration at $\frac{1}{2}V_{max}$. The K_m values are inversely related to the affinity of an enzyme for its substrate. The Lineweaver-Burk plot is a linearized version of the Michaelis-Menten equation that facilitates determination of K_m and V_{max}. Since reaction velocity is relatively insensitive to substrate concentration, many enzymes exhibit cooperativity. Typically, such enzymes are allosteric, containing two or more active sites which interact with each other and which may be located on different subunits.

Enzymes are subject to reversible and irreversible inhibition. Irreversible inhibitors typically form covalent bonds with the enzyme; reversible inhibitors form non-covalent bonds with the enzyme and may be competitive, noncompetitive, or mixed.

Enzyme activity is often regulated in cells. Regulation may be effected by compartmentation of enzymes and/or substrates; covalent modification; feedback inhibition, in which the end products of metabolic pathways inhibit the enzymes involved in earlier steps; and control of the enzyme concentration in the cell by gene expression and protein degradation.

Most of our current understanding of gene expression is based on prokaryotic systems. In prokaryotes, structural genes involved in related functions are organized into operons, such as the *lac* operon. Regulatory genes encode DNA-binding proteins, which may be repressors or activators. When repressors or activators bind to the DNA near the transcription start site, they may either repress or activate transcription. In inducible systems, the regulatory proteins are themselves activated or inactivated by binding to small effector molecules. Similar systems are present in eukaryotic genomes, which, however, are organized quite differently in that related genes are not clustered in operons and are subdivided into exons and introns. Pre-mRNA transcripts must be processed by splicing, capping, and adding poly(A) tails to produce the mature mRNA, and the mature mRNA must then exit the nucleus to initiate translation in the cytosol. Despite being scattered throughout the genome, many eukaryotic genes are both inducible and coregulated. An example is the group of enzymes involved in galactose metabolism in yeast.

Finally, enzyme concentration is regulated by protein degradation, or turnover. As yet there is no evidence that plant vacuoles function like animal lysosomes in protein turnover, except during senescence, when the contents of the vacuole are released. However, protein turnover via the covalent attachment of the short polypeptide ubiquitin and subsequent proteolysis may be an important mechanism for regulating the cytosolic protein concentration in plants.

GENERAL READING

Alberts, B., Bray, D., Lewis, J., Raff, M., Roberts, K., and Watson, J. D. (1989) *Molecular Biology of the Cell*, 2d ed. Garland Publishing, New York.

Atchison, M. L. (1988) Enhancers: Mechanisms of action and cell specificity. *Annu. Rev. Cell Biol.* 4:127–153.

Atkinson, D. E. (1977) *Cellular Energy Metabolism and Its Regulation*. Academic Press, New York.

Creighton, T. E. (1983) *Proteins: Structures and Molecular Principles*. W. H. Freeman, New York.

Darnell, J., Lodish, H., and Baltimore, D. (1990) *Molecular Cell Biology*, 2d ed. Scientific American Books. W. H. Freeman, New York.

Edsall, J. T., and Gutfreund, H. (1983) *Biothermodynamics: The Study of Biochemical Processes at Equilibrium*. Wiley, New York.

Fersht, A. (1985) *Enzyme Structure and Mechanism*, 2d ed. W. H. Freeman, New York.

Johnston, M. (1987) A model fungal gene regulatory mechanism: The GAL genes of *Saccharomyces cerevisiae*. *Microbiol. Rev.* 51:458–476.

Klotz, I. M. (1967) *Energy Changes in Biochemical Reactions*. Academic Press, New York.

Mathews, C. K., and van Holde, K. E. (1990) *Biochemistry*. Benjamin/Cummings, Redwood City, Calif.

Morowitz, H. J. (1978) *Foundations of Bioenergetics*. Academic Press, New York.

Mortimer, R. K., Schild, D., Contopoulou, C. R., and Kans, J. A. (1989) Genetic map of *Saccharomyces cerevisiae*, Edition 10. *Yeast* 5:321–403.

Murphy, T. M., and Thompson, W. F. (1988) *Molecular Plant Development*. Prentice-Hall, Englewood Cliffs, N.J.

Nicholls, D. G. (1982) *Bioenergetics: An Introduction to the Chemiosmotic Theory*. Academic Press, New York.

Nobel, P. S. (1983) *Biophysical Plant Phsiology and Ecology*. W. H. Freeman, San Francisco.

Stryer, L. (1988) *Biochemistry*, 3d ed. W. H. Freeman, New York.

Walsh, C. T. (1979) *Enzymatic Reaction Mechanisms*. W. H. Freeman, New York.

Webb, E. (1984) *Enzyme Nomenclature*. Academic Press, Orlando, Fla.

Weising, K., Schell, J., and Kahl, G. (1988) Foreign genes in plants: Transfer, structure, expression, and applications. *Annu. Rev. Genet.* 22:421–477.

CHAPTER REFERENCES

Bennett, W. S., Jr., and Steitz, T. A. (1978) Glucose-induced conformational change in yeast hexokinase. *Proc. Natl. Acad. Sci. USA* 75:4848–4852.

Bryant, F. O., and Adams, M. W. W. (1989) Characterization of hydrogenase from the hyperthermophilic archaebacterium, *Pyrococcus furiosus*. *J. Biol. Chem.* 264:5070–5079.

Cech, T. R. (1987) The chemistry of self-splicing RNA and RNA enzymes. *Science* 236:1532–1539.

Goldberg, R. B. (1986) Regulation of plant gene expression. *Philos. Trans. R. Soc. London Ser. B* 314:343–353.

Kraut, J. (1988) How do enzymes work? *Science* 242:533–540.

Kuhlemeier, C., Green, P. J., and Chua, N. -H. (1987) Regulation of gene expression in higher plants. *Annu. Rev. Plant Physiol. Plant Mol. Biol.* 38:221–257.

McCorkle, G. M., and Altman, S. (1987) RNA's at catalysts. *Concepts Biochem.* 64:221–226.

Shanklin, J., Jabben, M., and Vierstra, R. D. (1987) Red light-induced formation of ubiquitin-phytochrome conjugates: Identification of possible intermediates of phytochrome degradation. *Proc. Natl. Acad. Sci. USA* 84:359–363.

UNIT

I

TRANSPORT AND TRANSLOCATION OF WATER AND SOLUTES

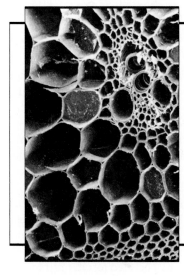

CHAPTER 3

Water and Plant Cells

B Y FAR THE MOST abundant constituent of living plant cells is water. Water typically constitutes 80–95% of the mass of growing tissues. Common vegetables such as carrots and lettuce may contain 85–95% water. Wood, which is composed of dead cells, generally has a lower water content; sapwood, which has active xylem transport, contains 35–75% water, while heartwood, which does not transport water, has a slightly lower water content. Seeds, with a water content of 5–15%, are among the driest of plant tissues, yet before germinating they must absorb a considerable amount of water.

Water serves several unique functions in the life of the plant. It is the most abundant and the best solvent known. As a solvent, it makes up the medium for movement of molecules within and between cells and greatly influences the molecular structures and properties of proteins, membranes, nucleic acids, and other cell constituents. Water is the environment in which most of the biochemical reactions in cells occur, and it participates in a number of essential reactions, such as those involving hydrolysis and dehydration.

Unlike many other substances in the plant cell, a water molecule is only a temporary resident. Plants continuously absorb and lose water. On a warm, sunny, dry day a leaf will exchange up to 100% of its water in a single hour. During the plant's lifetime, water equivalent to 100 times the mass of the plant may be lost through the leaf surfaces. Such loss of water from the surface of the plant is called transpiration, and it is an important means of cooling the plant. The uptake of water by the roots is an important means of bringing dissolved soil minerals to the root surface for absorption. Furthermore, water has an extraordinarily high capacity for holding heat, and its presence ensures that temperature fluctuations in plants occur rather slowly.

Of all the resources that plants need to grow and function, water is the most abundant and at the same time the most limiting for agricultural productivity (Fig. 3.1). The fact that water is limiting is the reason for the practice of crop irrigation. Water availability also limits the productivity of natural ecosystems (see Fig. 3.2). Thus an understanding of the uptake and loss of water by plants is of particular importance.

The Structure and Properties of Water

We will begin our study of water by considering how its structure gives rise to some of its unique properties. This will serve as a basis for examining various mechanisms by which water is transported from the soil through the plant to the atmosphere.

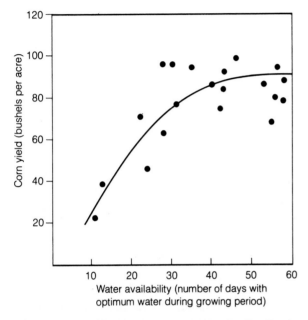

FIGURE 3.1. Corn yield as a function of water availability. The data plotted here were gathered at an Iowa farm over a 4-year period. Water availability was assessed as the number of days without water stress during a 9-week growing period. (Data from CAED Report 20.)

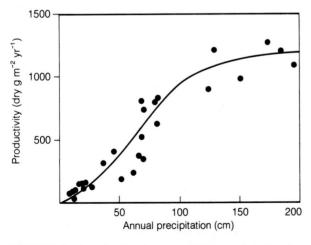

FIGURE 3.2. Productivity of various ecosystems as a function of annual precipitation. Productivity was estimated as net above ground accumulation of organic matter through growth and reproduction. (After Whittaker, 1970; cited in R. E. Ricklefs, 1970.)

The Polarity of Water Molecules Gives Rise to Extensive Intermolecular Attractions Called Hydrogen Bonds

The water molecule consists of an oxygen atom covalently bonded to two hydrogen atoms. The two O—H bonds form an angle of 105° (see Fig. 3.3). Because the oxygen atom is more **electronegative** than hydrogen, it tends to attract the electrons of the covalent bond. This results in a partial negative charge at the oxygen end of the molecule and a partial positive charge at each hydrogen. These partial charges are equal, so the water molecule carries no *net* charge. Nevertheless, this partial charge separation, together with the shape of the water molecule, makes water a *polar molecule*, and the opposite partial charges between neighboring water molecules tend to attract each other. The electrostatic attraction between water molecules is known as a **hydrogen bond**, and it is responsible for many of the unusual physical properties of water. Hydrogen bonding between water molecules leads to local aggregations of water in aqueous solutions. It is thought that quasi-crystalline networks or clusters of water continually form, break up, and re-form (Fig. 3.4).

The Polarity of Water Makes It an Excellent Solvent

The physical properties of water make it uniquely suitable as a medium for life. First, water is an excellent solvent: it dissolves greater amounts of a wider variety of substances than do other related solvents. This is due in part to the small size of water molecules and in part to their polar nature, which makes water a particularly good solvent for ionic substances. The water molecules orient themselves around ions and polar solutes in solution and effectively shield their electric charges. This shielding decreases the electrostatic interaction between the

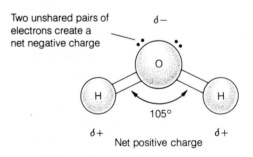

FIGURE 3.3. Diagram of the water molecule. The two intramolecular hydrogen-oxygen bonds form an angle of 105°. The opposite partial charges on the water molecule result in the formation of intermolecular hydrogen bonds with other water molecules. The average distances involved in intermolecular hydrogen bonds and intramolecular hydrogen-oxygen bonds are expressed in nanometers (1 nm = 10^{-9} m).

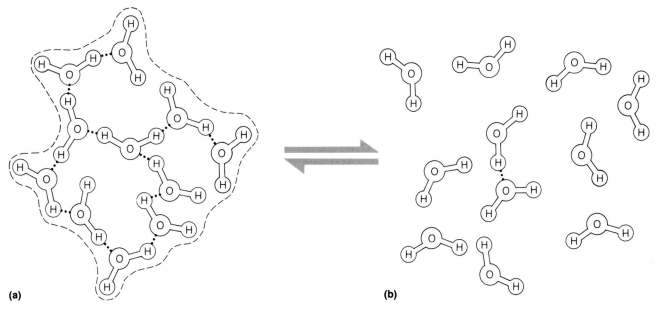

FIGURE 3.4. Hydrogen bonding between water molecules results in local aggregations of ordered, quasi-crystalline water (a). Because of the continuous thermal agitation of the water molecules, these aggregations are very short-lived; they break up rapidly to form much more random configurations (b).

charged substances and thereby increases their solubility. Furthermore, the polar ends of water molecules can orient themselves next to charged or partially charged groups in macromolecules, forming **shells of hydration**. Hydrogen bonding between macromolecules and water reduces the interaction between the macromolecules and helps to draw them into solution.

The Ability of Water to Form Hydrogen Bonds Gives Rise to Its Thermal, Cohesive, and Adhesive Properties

The extensive hydrogen bonding between water molecules results in unusual thermal properties, such as high specific heat and high latent heat of vaporization. **Specific heat** refers to the heat energy required to raise the temperature of a substance by a specific amount. When the temperature of water is raised, the molecules must vibrate faster, so a great deal of energy must be put into the system to break the hydrogen bonds between water molecules. Thus, compared with other liquids, water requires a relatively large energy input to raise its temperature. For plants this is important because it reduces potentially harmful temperature fluctuations. **Latent heat of vaporization** refers to the energy needed to separate molecules from the liquid phase and move them into the gas phase at constant temperature—a process that occurs during transpiration. For water at 25°C the heat of vaporization is 10.5 kJ mol^{-1}, which is the highest value known for any liquid.[1] Most of this energy is used to

break hydrogen bonds between water molecules. As we saw in Chapter 2, the high heat of vaporization of water enables plants to cool themselves by evaporating water from leaf surfaces, which are prone to heat up because of the radiant input from the sun. Transpiration is an important means of temperature regulation for plants.

In addition, the extensive hydrogen bonding in water gives rise to the property of **cohesion**, which refers to the mutual attraction between molecules. A related property, known as **adhesion**, refers to the attraction of water to a solid phase such as a cell wall or glass surface.

Water molecules at an air-water interface are more strongly attracted to neighboring water molecules than to the gas phase on the other side of the surface. As a consequence of this unequal attraction, the air-water interface assumes a shape that minimizes its surface area. In effect, the water molecules exert a force at the air-water interface. This force not only influences the shape of the surface but also may create a pressure in the rest of the liquid. The condition that exists at the interface is known as **surface tension** and, as we will see later, is very important in generating the forces needed to pull water to the top of trees.

Cohesion, adhesion, and surface tension give rise to a phenomenon known as **capillarity**. As shown in Figure 3.5, water tends to move up a glass capillary tube. The upward movement of the water volume is due to attraction of water at the periphery to the polar surface of a clean glass tube (adhesion) and to the surface tension of water, which tends to minimize the surface area. Together, adhesion and surface tension exert a tension on the water molecules at and just below the surface, causing

1. 1 kJ = 0.24 kcal.

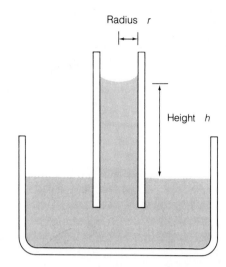

LIFTING FORCE = circumference × surface tension
= $2\pi r$ × 0.0728 Nm^{-1}

DOWNWARD FORCE = height × area × density × acceleration due to gravity
= h × $2\pi r^2$ × 998 kg-m^{-3} × 9.8 ms^{-2}

At equilibrium: LIFTING FORCE = DOWNWARD FORCE

Solving for height: $h = \dfrac{1.49 \times 10^{-5}\ \text{m}^2}{r\ (\text{in meters})}$

Examples:	Capillary radius (μm)	Height of rise (m)
	1	1.49
	10	0.149
	100	0.0149
	1000	0.00149
	Typical vessel (75μm)	0.02

FIGURE 3.5. Capillary rise of water up a tube due to the lifting force of adhesion and surface tension. The lifting force is given by the circumference of the tube × the surface tension of water. The weight of the column of water is given by the height × area × density × acceleration due to gravity. The maximum capillary rise depends inversely on the tube radius, as tabulated here: the smaller the tube, the higher the capillary rise. For a typical vessel with a radius of, say, 75 μm, the capillary rise amounts to only 0.02 m. Because this rise is much less than the height of trees, it is evident that capillarity cannot account for water movement up tall trees. The units for surface tension are Newtons per meter (Nm^{-1}). One Newton equals 0.01 Joules per cm.

them to move up the tube until the force of adhesion is balanced by the weight of the water column. The conducting elements of the xylem form small tubes that can, under certain artificial conditions (e.g., in a dry log), draw water up by capillary action. However, water in functional xylem elements forms a continuous column from the root to the leaf—there are no air-water interfaces. Thus, water movement in the xylem differs fundamentally from water movement in capillary tubes.

Besides small tubes, fibrous materials can act like wicks to draw water up by capillarity. The basis for this action is similar to that in small tubes, except that the air-water interfaces occur within the network of microscopic interstices of the material. The plant cell wall can act as an effective wick for capillary water movement. Capillarity ensures that cell wall surfaces directly exposed to the air, such as those in leaf mesophyll, remain wetted and do not dry out.

Water Has a High Tensile Strength

Cohesion also gives water a high **tensile strength**, defined as the ability to resist a pulling force. We do not usually think of liquids as having tensile strength; however, such a property must exist for water to be pulled up a capillary tube.

The tensile strength of water can also be demonstrated by placing it in a capped syringe as shown in Figure 3.6. If you *push* on the plunger, the water is compressed and a positive **hydrostatic pressure** builds up. The pressure is measured in units called **Pascals** (Pa) or, more conveniently, **megapascals** (MPa).[2] Table 3.1 compares various units of pressure. If instead of pushing you *pull* on the plunger in Figure 3.6, a tension or negative hydrostatic pressure develops in the water to resist the pull. How hard must you pull on the plunger before the water molecules are torn away from each other and the water column breaks? To break the water column requires sufficient energy to overcome the forces attract-

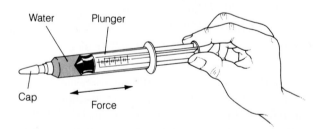

FIGURE 3.6. A sealed syringe can be used to create positive and negative pressures in a fluid like water. Pushing on the plunger compresses the fluid, and a positive pressure builds up. If a small air bubble is trapped within the syringe, it can be seen to shrink as the pressure increases. Pulling on the plunger causes the fluid to develop a tension, or negative pressure. If the syringe contains any air bubbles, they will expand as the pressure is reduced from atmospheric pressure. With a gas bubble in the syringe, the pressure never goes below a pure vacuum because the gas phase can expand indefinitely. If the syringe is filled with a degassed solution and lacks bubbles, pressures below vacuum (i.e., negative pressures) can develop because the intermolecular bonds holding the fluid together can withstand considerable tension before breaking.

---

2. 1 MPa = 9.87 atmospheres.

TABLE 3.1. Comparison of various units of pressure

1 atmosphere = 14.7 pounds per square inch
= 760 mm Hg (at sea level, 45° latitude)
= 1.013 bar
= 0.1013 MPa
= 1.013 × 10⁵ Pa

A car tire is typically inflated to about 0.2 MPa.

The water pressure in home plumbing is typically 0.2–0.3 MPa.

The water pressure under 15 feet (5 m) of water is about 0.05 MPa.

ing water molecules to one another. Experimental studies have demonstrated that water in small capillaries can resist tensions more negative than -30 MPa (the negative sign indicates a tension, as opposed to a compression). This value is about one-third of the theoretical strength computed on the basis of the strength of hydrogen bonds. It is about 10% of the tensile strength of copper or aluminum wire and is thus quite substantial.

The presence of dissolved gases reduces the tensile strength of water. This is the case because as the pressure in an aqueous solution is reduced, gases that are dissolved in water tend to come out of solution and form a *gas bubble*. The lowest absolute pressure possible in a gas phase is 0 MPa (pure vacuum) because the molecular forces of attraction needed to resist a negative pressure (or tension) do not exist in gases. In contrast, in solids and liquids the intermolecular attractions can resist tensile forces. Therefore, if even a tiny gas bubble forms in a column of water under tension, the gas bubble will expand indefinitely, with the result that the tension in the liquid phase will collapse, a phenomenon known as **cavitation**. As we will see later, cavitation can have a devastating effect on water transport through the xylem of trees.

Transport Processes

When water moves from the soil through the plant to the atmosphere, it travels through a widely variable medium, and the mechanism of transport also varies with the type of medium (cell wall, cytoplasm, lipid bilayer). We will now consider the fundamental processes, and their driving forces, that lead to water transport.

Diffusion Is the Movement of Molecules Along a Concentration Gradient by Random Thermal Agitation

Water molecules in a solution are not static; they are in continuous motion, colliding with one another and exchanging kinetic energy. **Diffusion** refers to the process by which molecules intermingle as a result of their random thermal agitation. Such agitation gives rise to the random but progressive movement of substances from regions of high concentration to regions of low concentration, that is, down a concentration gradient (Fig. 3.7).

Fick discovered that the rate of solute transport by diffusion is directly proportional to the concentration gradient ($\Delta C/\Delta x$), where the proportionality constant is the diffusion coefficient of the substance. In symbols, we write this relation as

$$J_s = -D_s \frac{\Delta C_s}{\Delta x} \tag{3.1}$$

The rate of solute transport, or the **flux** (J_s), is the amount of substance s crossing a unit area per unit time (e.g., J_s may have units of mol m^{-2} s^{-1}). The **diffusion coefficient** (D_s) is a proportionality constant that measures how easily substance s moves through a particular medium. The **concentration gradient** (ΔC_s) is the difference in concentration of substance s between two points separated by the distance Δx. The negative sign in the equation indicates that the flux moves down a concentration gradient.

The Rate of Diffusion Is Rapid over Short Distances but Extremely Slow over Long Distances

From Fick's law (Eq. 3.1), one can derive (with difficulty!) an expression for the time it takes for a substance to diffuse a particular distance. If the initial conditions are such that all the solute molecules are concentrated at the starting position (see Fig. 3.8), then the concentration front moves away from the starting position as shown for discrete time points in Figure 3.8. As the substance diffuses away from the starting point, the concentration gradient becomes less steep, and thus net movement becomes slower. The time it takes for the substance at any given distance from the starting point to reach one-half of the concentration at the starting point ($t_{C=1/2}$) is given by

$$t_{C=1/2} = \frac{(\text{distance})^2}{D} K \tag{3.2}$$

where K is a constant that depends on the geometry of the system (for convenience, we will use a K value of 1 in the following calculations) and D is the diffusion coefficient. Equation 3.2 shows that the time required for a substance to diffuse a given distance depends on the *square* of the distance.

We can see what this means by considering two numerical examples. First, let us ask how long it would take a small molecule to diffuse across a typical cell. The diffusion coefficient for a small molecule might be about

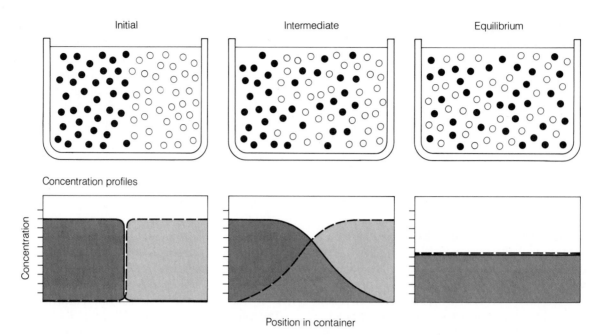

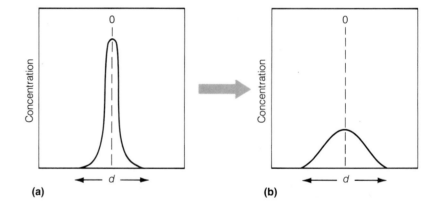

Initial   Intermediate   Equilibrium

Concentration profiles

Position in container

FIGURE 3.7. Thermal motion of molecules leads to *diffusion*—the gradual mixing of molecules and eventual dissipation of concentration differences. Initially, two materials containing different molecules are brought into contact. The materials may be gas, liquid, or solid. Diffusion is fastest in gases, intermediate in liquids, and slowest in solids. The initial separation of the molecules is depicted graphically in the upper panels, and the corresponding concentration profiles are shown in the lower panels as a function of position. With time, the mixing and randomization of the molecules lead to thin net movement. At equilibrium the two types of molecules are randomly (evenly) distributed.

FIGURE 3.8. Graphical representation of the concentration gradient of a solute diffusing according to Fick's law. The solute molecules were initially located in the plane indicated on the x-axis (0). The first graph (a) shows the distribution of solute molecules shortly after placement at the plane of origin. Note how sharply the concentration drops off as the distance, d, from the origin increases. The solute distribution at a later time point is shown in the second graph (b). The average distance of the diffusing molecules from the origin has increased, and the slope of the gradient has flattened out. (Modified from Nobel, 1983.)

(a)   **(b)**

10^{-9} m^2 s^{-1}, and the cell size might be 50 μm. Thus,

$$t = \frac{(50 \times 10^{-6} \text{ m})^2}{10^{-9} \text{ m}^2 \text{ s}^{-1}} = 2.5 \text{ s}$$

This calculation shows that the diffusion of small molecules over cellular dimensions is rapid. What about diffusion over longer distances? Calculating the time needed for the same substance to move a distance of 1 m (e.g., the length of a corn leaf), we find

$$t = \frac{(1 \text{ m})^2}{10^{-9} \text{ m}^2 \text{ s}^{-1}} = 10^9 \text{ s} \approx 24 \text{ years}$$

a value that exceeds by orders of magnitude the life span

of a plant such as corn, which lives only a few months.

From these numerical examples we see that diffusion of small molecules in aqueous solutions can be effective at cellular dimensions but is far too slow for mass transport over long distances. As will be shown in Chapter 4, diffusion is of great importance during loss of water vapor from leaves. It is also important in water uptake during growth and for solute movement in metabolism.

Long-Distance Water Transport in the Plant Occurs by Bulk Flow

A second process leading to water movement is known as **bulk flow** or **mass flow** and refers to the movement of groups of molecules in response to a pressure gradient. Among many common examples of bulk flow are convection currents, water moving through a garden hose, a river flowing, and rain falling.

If we consider bulk flow through a pipe, the rate of volume flow depends on the radius (r) of the pipe, the viscosity (η) of the liquid, and the size of the pressure gradient ($\Delta P/\Delta x$) that drives the flow. This relation is given by the **Poiseuille equation**:

$$\text{Volume flow rate} = \frac{\pi r^4}{8\eta}\frac{\Delta P}{\Delta x} \qquad (3.3)$$

Equation 3.3 demonstrates that bulk flow is very sensitive to the radius of the pipe. If the radius is doubled, the volume flow rate increases by a factor of 16 (2^4).

Pressure-driven bulk flow of water is the predominant mechanism responsible for long-distance transport of water in the plant via the xylem. It may also account for much of the water flow that occurs through the soil and through the cell walls of plant tissues. In contrast to diffusion, pressure-driven bulk flow is independent of solute concentration gradients, as long as viscosity changes are negligible.

Osmosis, the Movement of Water Through a Selectively Permeable Membrane, Involves Both Bulk Flow and Diffusion

A third process leading to water transport is **osmosis**, which refers to movement of a solvent such as water across a membrane. In all living cells, membranes are the great dividers: they separate different parts of the cell from each other and greatly impede movement of substances between compartments. The membranes of plant cells are **selectively permeable**; that is, they permit the movement of water and other small uncharged substances such as carbon dioxide across them while greatly restricting the movement of larger solutes and especially charged substances. The movement of many substances found in cells is virtually blocked by membranes; for such substances to cross the various cell membranes, special transport proteins are required. This aspect of membrane transport will be discussed in Chapter 6. Here we restrict our discussion to the movement of water across membranes, that can occur without the help of transport proteins. Note, however, that some proteins form membrane channels that allow both ions and water to move across the membrane (see Finkelstein, 1987, and Stein, 1986).

Like diffusion and bulk flow, osmosis occurs spontaneously in response to a driving force. In diffusion, transport is driven by a concentration gradient; in bulk flow, it is driven by a pressure gradient; in osmosis, both types of gradients influence transport. The direction and rate of water flow across a membrane are determined not solely by the concentration gradient of water or by the pressure gradient, but by the sum of these two driving forces. This observation leads to the development of the concept of a composite or total driving force, representing the free energy gradient of water. In practice, this driving force is expressed as a chemical potential gradient or, more commonly by plant physiologists, as a water potential gradient.

The Chemical Potential of Water or "Water Potential" Represents the Free Energy Status of Water

The **chemical potential** of water is a quantitative expression of the free energy associated with the water (see Chapter 2). In thermodynamics, free energy represents a potential for performing work. All living things, including plants, require a continuous input of free energy in order to maintain and repair their structures and their organized states, as well as to grow and reproduce. Processes such as biochemical reactions, solute accumulation, and long-distance transport are all driven by an input of free energy into the plant.

The concepts of free energy and chemical potential derive from thermodynamics, the study of the transformations of energy. As we discussed in Chapter 2, thermodynamics provides a useful framework for studying the energetics of many processes essential to life. Here we will restrict ourselves to a discussion of the thermodynamic basis for transport of water in plants.

Osmosis is an energetically spontaneous process. That is, water moves down a chemical potential gradient, from a region of high chemical potential to a region of low chemical potential. No work is involved in osmosis; rather, free energy is released. To understand how this comes about, we need to examine more closely what influences the chemical potential of water.

In general, the chemical potential of any substance may be influenced by four factors: *concentration, pressure, gravity,* and *electrical potential*. Since water is a neutral molecule, the electrical potential has no influence on its chemical potential. Thus, only concentration, pressure, and gravity affect the chemical potential of water.

One important point to note is that chemical potential is a relative quantity: it is always expressed as the difference between the potential of a substance in a given state and the potential of the same substance in a standard state. The unit of chemical potential is energy per mole of substance (joules per mole).

For historical and practical reasons, it turns out that the unit of chemical potential is inconvenient for most work in plant physiology. For this reason plant physiologists have defined another parameter called the **water potential**, defined as the chemical potential of water divided by the partial molal volume of water ($\overline{V}_w$, the volume of 1 mole of water, 18 cm³ mol⁻¹). When this parameter is used, the free energy of water is expressed in pressure units such as megapascals, which are more convenient units in practice. We will now consider more fully the important concept of water potential.

The Major Factors Contributing to Water Potential Are Represented by the Equation $\psi = P - \pi$

Water potential is symbolized by the Greek letter ψ and is directly related to the chemical potential of water. Like chemical potential, water potential is a relative quantity and depends on concentration, pressure, and gravity. This relation may be written as

$$\psi = \psi^* + f(\text{concentration}) + f(\text{pressure}) + f(\text{gravity})$$
$$(3.4)$$

where ψ^* is the water potential in the standard state and $f(\text{concentration})$, $f(\text{pressure})$, and $f(\text{gravity})$ denote the effects of these three factors on the water potential. Let us now consider each of these terms.[3]

Standard State of Water. The reference or standard state of water is taken to be that of pure water at ambient pressure and at the same temperature as the sample of water. The water potential in this standard state (ψ^*) is arbitrarily assigned the value of 0 MPa. In principle, we could assign any value to ψ^*, but by giving ψ^* a value of zero we can drop it from the equation. This arbitrary assignment makes it algebraically simpler to calculate ψ for a sample, but it should not be forgotten that ψ is always defined relative to the water potential of pure water.

Concentration. The term $f(\text{concentration})$ represents the effect of the concentration of water on ψ. The higher the

concentration of water, (more precisely, the activity of water), the higher the water potential. By convention, concentration is defined as the mole fraction. The mole fraction of a substance, s, is the number of moles of s divided by the total number of moles of all species (including s) in the system.

Because the solutions in most plant cells are fairly dilute, the water concentration (mole fraction) is very high, and it is easier in practice to define $f(\text{concentration})$ in terms of the concentration of solutes rather than of the solvent, water, as in Equation 2.22 of the previous chapter. The effect of dissolved solutes on ψ of a dilute solution is given approximately by $-RTC_s$, where R is the gas constant (0.0083143 L MPa mol⁻¹ K⁻¹), T is the absolute temperature (degrees kelvin, or K), and C_s is the osmolality of the solution. The minus sign indicates that dissolved solutes reduce the water potential of a solution, by reducing the concentration of water. Osmolality is a measure of the concentration of total dissolved solutes in a solution, regardless of the molecular species or mass of the dissolved solutes. Osmolality may be expressed as moles of total dissolved solutes per kilogram (or liter) of water.

Historically, RTC_s has been expressed in pressure units (e.g., MPa) because this allows us to sum the effects of solutes and hydrostatic pressure on the water potential. RTC_s is often called the **osmotic pressure** of a solution and is represented by the Greek letter π. Do not consider π to be a physical pressure, and do not confuse this use of π with the geometric use of π. Thus:

$$f(\text{concentration}) = -\pi = -RTC_s \qquad (3.5)$$

Table 3.2 shows the values of RT at various temperatures and the values of osmotic pressure associated with solutions of different solute concentrations. Note that π is a positive quantity; its effect on ψ, however, is negative. That is, π reduces the water potential of a solution.

It is important to realize that osmotic pressure and osmolality are measures of the total number of dissolved particles, regardless of type. Thus, if we dissolve 1 mole of sucrose in 1 kg of water, we obtain a solution with an osmolality of 1 mol kg⁻¹. If instead we dissolve 1 mole of salt (NaCl) in the same mass of water, we obtain an osmolality of 2 mol kg⁻¹ because salt dissociates into two particles.

Expression 3.5, defining the osmotic pressure as RTC_s, is valid for "ideal" solutions at dilute concentration. Real solutions frequently deviate from the ideal, especially at high concentrations—say, greater than 0.1 mol kg⁻¹. In our treatment of water potential, we will assume that we are dealing with ideal solutions. Readers are referred to more advanced treatments for further discussion of nonideal behavior (e.g., see Nobel, 1983; Friedman 1986).

---

3. Students planning further study of plant water relations should note that the components of water potential defined above are sometimes given different names and symbols. In particular, ψ is often defined as the sum of *component potentials*; thus:

$$\psi_t = \psi_s + \psi_p + \psi_g$$

The symbol ψ_t is called the total water potential and is identical to ψ. ψ_s is called the *solute or osmotic potential*; it is a *negative* quantity and equals the negative of the osmotic pressure (i.e., $\psi_s = -\pi$). To add to the confusion, the symbol π is sometimes also called osmotic potential and has been used variously to mean the osmotic pressure and the negative of the osmotic pressure. It is important to stay alert to these varying definitions of π to avoid confusion. The symbol ψ_p is called the *pressure potential*; it is identical to P, the hydrostatic pressure in excess of ambient pressure. The *gravitational potential*, ψ_g, is equal to $\rho_w g h$, where ρ_w is the density of water, g the gravitational acceleration, and h the height of water above the reference state.

TABLE 3.2. Values of *RT* and osmotic pressures of solutions at various temperatures

Temperature (°C)	RT* (L MPa mol^{-1})	Osmotic pressure (MPa) of solution with solute concentration			Osmotic pressure of seawater (MPa)
		0.01 M	0.10 M	1.00 M	
0	2.271	0.0227	0.227	2.27	2.6
20	2.436	0.0244	0.244	2.44	2.8
25	2.478	0.0248	0.248	2.48	2.8
30	2.519	0.0252	0.252	2.52	2.9

*$R = 0.0083143$ L MPa mol^{-1} K^{-1}, L = liter.

Pressure. The term f(**pressure**) expresses the effect of hydrostatic pressure on the water potential of a solution. Because plant cells have rigid walls, they can build up a large positive internal hydrostatic pressure, usually called **turgor pressure** by plant physiologists. Moreover, in the xylem and in the walls between cells, a **tension** or a **negative hydrostatic pressure** can develop. As we shall see, negative pressures outside cells are very important in moving water long distances through the plant.

In water potential studies, f(pressure) is symbolized by the letter P and is defined as the hydrostatic pressure in excess of ambient atmospheric pressure. That is,

$$f(\text{pressure}) = P$$

where P = absolute pressure − atmospheric pressure (atmospheric pressure = 0.1 MPa). P is sometimes called the gauge pressure. Note that water in the reference state is at ambient pressure, so by definition $P = 0$ MPa for water in the standard state. Thus, the value of P for pure water in an open beaker is 0 MPa even though its absolute pressure is 0.1 MPa (1 atmosphere). Water under a perfect vacuum would have a P value of -0.1 MPa, whereas its absolute pressure would be 0 MPa. It is important to keep in mind the distinction between absolute pressure and P.

Gravity. Gravity causes water to move downward, unless the force of gravity is opposed by an equal and opposite force. The potential for water movement thus depends on height. The effect of gravity on water potential, f(**gravity**), depends on the height (h) of the water above the reference-state water, the density of water (ρ_w), and the acceleration due to gravity (g). In symbols,

$$f(\text{gravity}) = \rho_w g h$$

$\rho_w g$ has a value of 0.01 MPa m^{-1}.

For a standing column of water, as in Figure 3.9, the gravity and pressure components of water potential exactly balance one another. For this reason there is no net water movement in such a column, even though there are opposing gradients in both pressure and gravity effects.

Total Water Potential. From the foregoing, it follows that we can write the **total water potential** of a solution as

$$\psi = \psi^* - \pi + P + \rho_w g h \tag{3.6}$$

In practice, the reference state ψ^* is omitted because it is defined to have a value of 0 MPa. Thus we have

$$\psi = -\pi + P + \rho_w g h \tag{3.7}$$

For transport over a short vertical height (say, less than 5 or 10 m) or between adjacent cells, the gravitational term is small and is commonly omitted, giving

$$\psi = P - \pi \tag{3.8}$$

This equation states that the water potential is affected by two principal factors: the hydrostatic pressure and the osmotic pressure. Note again that the osmotic pressure (π) is defined as a positive quantity. The negative sign is there to account for the reduction in water potential due to dissolved solutes.

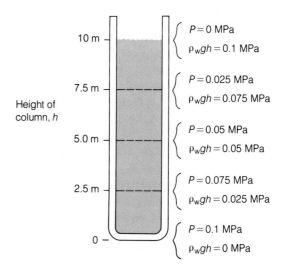

FIGURE 3.9. In a standing column of water, the pressure and gravity components of the water potential are related and complementary. At the top of a 10 m water column, $P = 0$ MPa, and $\rho_w g h = 0.1$ MPa. At the bottom the reverse pattern is found. At the midpoint, the two components are equal.

Water Enters the Cell Along a Water Potential Gradient Until the Water Potential Inside Equals the Water Potential Outside

What are typical values for the water potential and its components? These are best illustrated by examples. First, imagine an open beaker full of pure water (Fig. 3.10a). Since the water is open to the atmosphere, the hydrostatic pressure of the water is the same as atmospheric pressure ($P = 0$ MPa). There are no solutes in the water, so the osmotic pressure is 0 MPa ($\pi = 0$ MPa); therefore the water potential is 0 MPa ($\psi = P - \pi$). Now imagine dissolving sucrose in the water to a concentration of 0.1 mol per liter of water (0.1M) (Fig. 3.10b). This addition increases the osmotic pressure, π, to 0.244 MPa and decreases the water potential, ψ, to -0.244 MPa.

Second, consider a flaccid plant cell (i.e., a cell without any turgor pressure) with a total internal solute concentration of 0.3 M (Fig. 3.10c). This solute concentration gives an osmotic pressure of 0.732 MPa. Because the cell is flaccid, the internal pressure is the same as ambient pressure, so the hydrostatic pressure, P, is 0 MPa and the water potential of the cell is -0.732 MPa.

What will happen if this cell is placed in the beaker containing 0.1 M sucrose (Fig. 3.10c)? The sucrose solution has an osmotic pressure, π, of 0.244 MPa and a water potential, ψ, of -0.244 MPa. Initially, there is a large difference between the ψ value of the external solution and that of cell. The difference in ψ, symbolized by $\Delta\psi$, is ψ outside minus ψ inside and equals 0.488 MPa. This $\Delta\psi$ leads to an influx of water from a region of high ψ (less negative) to region of low ψ (more negative); the volume of the cell thus increases. Because plant cells are surrounded by rigid cell walls, even a slight increase in cell volume causes a large increase in the hydrostatic pressure within the cell. As water enters the cell, the cell wall is stretched by the contents of the cell. The wall resists such stretching by pushing back on the cell contents. This is analogous to inflating a basketball with air, except that air is compressible, whereas water is nearly incompressible. Thus, as water moves into the cell because of the $\Delta\psi$ across the plasma membrane, the hydrostatic pressure, P, of the cell increases. Consequently, the cell ψ increases and $\Delta\psi$ is reduced. Eventually, P increases enough to raise the cell ψ to the same value as ψ of the external solution. At this point, equilibrium is reached ($\Delta\psi = 0$ MPa), and net water transport ceases. The tiny amount of water taken up by the cell does not significantly affect the solute concentration of the external solution, whose volume is very much larger than that of the cell. Hence, ψ, π, and P of the external solution are not altered. Therefore, at equilibrium $\psi_{cell} = \psi_{solution} = -0.244$ MPa. Calculating P and π of the cell is straightforward. If we assume that the cell has a very rigid cell wall, then very little water will enter, so π remains nearly unchanged during the equilibration process; thus, π remains at $+0.732$ MPa. Cell hydrostatic pressure can be obtained by calculation: $P = \psi + \pi = -0.244 + 0.732 = 0.488$ MPa.

Because of the Rigid Cell Wall, Small Changes in Cell Volume Cause Large Changes in the Turgor Pressure

The constancy of the cell osmotic pressure, π, during water uptake is approximately true for cells with very rigid cell walls. In principle, the water that is absorbed during equilibration dilutes the solutes in the cell somewhat and thus reduces π. For many plant cells, this dilution is so small that we can ignore it. However, if the cell walls stretch substantially as water is absorbed, the dilution of the solutes in the cell may be substantial. In such a case, the effect of the dilution on π must be taken into account. The change in cell volume with any given change in pressure is determined by the rigidity of the wall, which is usually measured as a volumetric **modulus of elasticity**, symbolized by ε. This wall property is given by the change in pressure divided by the relative change in volume ($\Delta V / V$); that is,

$$\varepsilon = \frac{\Delta P}{\Delta V / V}$$

The elastic modulus has units of pressure, with typical values on the order of 10 MPa. From the above equation it can be seen that a 1% increase in volume increases turgor pressure by 0.1 MPa, or about 1 atmosphere. This is equivalent to a 20% increase in turgor pressure for a typical cell with $P \cong 0.5$ MPa. In contrast, a 1% increase in volume would result in only a 1% decrease in cell osmotic pressure. Thus, when the water potential of turgid cells increases due to water uptake, most of this change is effected through an increase in P rather than a decrease in π.

Water Can Also Leave the Cell in Response to a Water Potential Gradient

Water can also leave the cell by osmosis. If, in the previous example, we remove our plant cell from the 0.1 M sucrose solution and place it in a 0.3 M sucrose solution (Fig. 3.10d), $\psi_{outside}$ (-0.732 MPa) is more negative than ψ_{cell} (-0.244 MPa), and water will move from the cell to the solution. As water leaves the cell, the cell volume decreases. As the cell volume decreases, P decreases also until $\psi_{cell} = \psi_{outside}$. From the water potential equation we can calculate that at equilibrium $P = 0$ MPa.

If we squeeze the cell by pressing it between two plates (Fig. 3.10e), we effectively raise the cell P, consequently raising the cell ψ and creating a $\Delta\psi$ such that

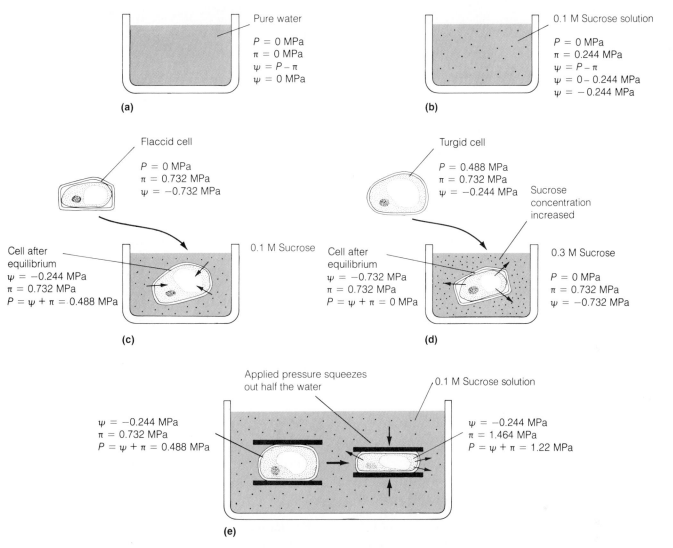

FIGURE 3.10. Five examples to illustrate the concept of water potential and its components. In (a), the water potential and its components are illustrated for pure water. (b) shows a solution containing 0.1 M sucrose. In (c), a flaccid cell is dropped in the 0.1 M sucrose solution. Because the cell's starting water potential is less than the solution water potential, the cell takes up water. After equilibration, the cell's water potential rises to equal the water potential of the solution, and the result is a cell with a positive turgor pressure. (d) illustrates how the cell is made to lose water by increasing the concentration of sucrose in the solution. The increased sucrose concentration lowers the solution water potential, draws out water from the cell, and thereby reduces the cell's turgor pressure. Another way to make the cell lose water is by slowly pressing it between two plates (e). In this case, half of the cell water is removed, so cell osmotic pressure doubles and turgor pressure increases correspondingly.

water now flows *out* of the cell. If we continue squeezing until half of the cell water is removed and then hold the cell in this condition, the cell will reach a new equilibrium. As in the previous example, at equilibrium $\Delta\psi = 0$ MPa, and the amount of water added to the external solution is so small that it can be ignored. The cell will thus return to the ψ value that it had before the squeezing procedure, but the components of ψ will be quite different. Because

half of the water was squeezed out of the cell while the solutes remained inside the cell (the membrane is selectively permeable), the cell solution at equilibrium is concentrated twofold, and thus π is twice its starting value ($0.732 \times 2 = 1.464$ MPa). Knowing the equilibrium values of ψ and π, we can readily calculate the turgor pressure from Equation 3.8 as $P = \psi + \pi = -0.244 + 1.464 = 1.22$ MPa.

How Water Potential Is Measured

CELL GROWTH, PHOTOSYNTHESIS, and crop productivity are all strongly influenced by water potential and its components. Like the body temperature of humans, water potential is a good overall indicator of plant health. Planet scientists have thus expended considerable effort in devising accurate and reliable methods for evaluating a plant's water status. Four instruments have been used extensively to measure ψ, π, and P, as described below.

PSYCHROMETER

Psychrometry (psychro comes from the Greek word for " to cool") is the estimation of ψ by measuring the change in temperature during evaporation. It is based on the principle that evaporation of water from a surface cools the surface. Cooling occurs because the water molecules that escape into the atmosphere are those that have a higher-than-average energy — enough energy to break the bonds holding them in the liquid. When they escape, they leave behind a mass of molecules with lower-than-average energy, and thus a cooler body of water.

One psychrometric technique, known as *isopiestic psychrometry*, has been extensively employed by John Boyer and co-workers (Boyer and Knipling, 1965) and is illustrated in Figure 3.A. To make a measurement, a piece of tissue is sealed inside a chamber containing a temperature sensor (in this case, a thermocouple) in contact with a small droplet of water. Initially, water evaporates from both the tissue and the water drop, raising the humidity of the air inside the sealed chamber. Evaporation continues until the air becomes saturated or nearly saturated with water vapor. At this point, if the plant tissue and the water drop have the same water potential, net movement of water from the droplet stops, and the temperature of the drop, measured with the temperature sensor, is the same as the ambient temperature. However, if the tissue has a lower water potential than the droplet, then water evaporates from the drop, diffuses through the air, and is absorbed by the tissue. This slight evaporation of water cools the drop. The larger the difference in water potential between the tissue and the drop, the higher the rate of water transfer and hence the cooler the drop. Rather than placing pure water on the temperature sensor, one may instead place a standard solution of known

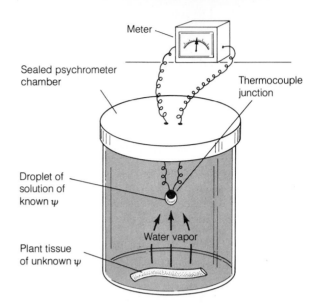

FIGURE 3.A. Diagram illustrating the use of isopiestic psychrometry to measure the water potential of a plant tissue.

solute concentration (known π, and thus known ψ). If the standard solution has a lower water potential than the sample to be measured, water will diffuse from the tissue to the drop, causing a warming of the drop. By measuring the change in temperature of the drop for several solutions of known ψ, it is possible to match exactly the water potential of the solution and that of the sample (Fig. 3.B). When the match is perfect, the change in temperature of the drop is zero.

Psychrometers have been used to measure the water potentials of excised pieces of tissue and of intact tissue. Moreover, the method is applicable to solutions, wherein the water potential equals $-\pi$. Thus psychrometry can measure both the water potential of living tissue and the osmotic pressure of a solution. Frequently, the ψ value of a tissue is measured with a phychrometer, and then the tissue is crushed and the π value of the expressed cell sap is measured with the same instrument. By combining the two measurements, it is possible to estimate the hydrostatic pressure that existed in the cells before the tissue was crushed ($P = \psi + \pi$).

This method is very useful, but it is very sensitive to temperature fluctuations. For example, a change in temperature of $0.01°C$ may correspond to a change in water potential of about 0.1 MPa (the value varies with the type of temperature sensor). Since it is often

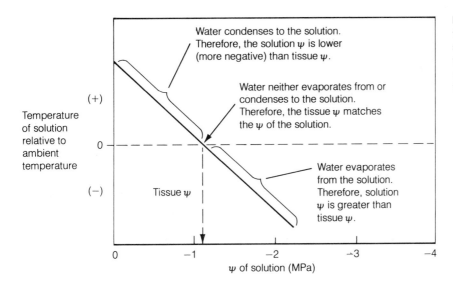

FIGURE 3.B. The temperature of the sensor in the psychrometer shown in Figure 3.A depends on the water potential of the solution relative to the water potential of the tissue sample.

desirable to have a resolution of 0.01 MPa, the instrument must be kept under very stringent conditions of constant temperature. For this reason, the method is used primarily in laboratory settings and has found only limited use for field work, where temperature cannot easily be controlled. There are many variations in psychrometric technique, and interested readers should consult Brown and Van Haveren (1972) and Slavík (1974).

CRYOSCOPIC OSMOMETER

This device measures the osmotic pressure of a solution by measuring its freezing point. One of the **colligative properties of solutions** is the decrease in the freezing point as the solute concentration increases. Other colligative properties include increase in boiling point, decrease in vapor pressure, and increase in osmotic pressure. These properties of solutions are called colligative because they all occur together, or collectively. Colligative properties depend on the number of dissolved particles and not on the nature of the solute. As an example of freezing point depression, a solution containing 1 mole of solutes per kilogram of water has a freezing point of −1.86°C, compared with 0°C for pure water.

Various instruments are used to measure the freezing point depression of solutions (for two examples, see Prager and Bowman, 1963, and Bearce and Kohl, 1970). With one of these instruments, samples of solution as small as 1 nanoliter are placed in an oil medium located on a temperature-controlled stage (Fig. 3.C). The very small sample size allows sap from single cells to be measured and permits

rapid thermal equilibration with the stage. To prevent evaporation, the samples are suspended in oil-filled wells in a silver plate (silver has high thermal conductivity). The temperature of the stage is rapidly dropped to about −30°C, which causes the sample to freeze. The temperature is then raised very slowly, and the melting process in the sample is observed through a microscope. When the last ice crystal in the sample melts, the temperature of the stage is recorded (note that the melting and freezing points are the same). It is a straightforward job to calculate the solute concentration from the freezing point depression; and from the solute concentration (C_s), π is calculated as RTc_s.

PRESSURE PROBE

If a cell were as large as a watermelon or even a grape, it would be relatively easy to measure its hydrostatic pressure. Because of the small size of plant cells, however, the development of methods for direct measurement of turgor pressure has been slow. Paul Green at the University of Pennsylvania developed the first direct method for measuring turgor pressure in plant cells (Green and Stanton, 1967). In this technique, an air-filled glass tube sealed at one end is inserted into a cell (Fig. 3.D). The high pressure in the cell compresses the trapped gas, and from the change in volume one can readily calculate the pressure of the cell from the ideal gas law (pressure × volume = constant). This method works only for cells of relatively large volume, such as the giant cell of the filamentous green alga *Nitella*. For smaller cells, the loss of cell sap into the glass tube is suffi-

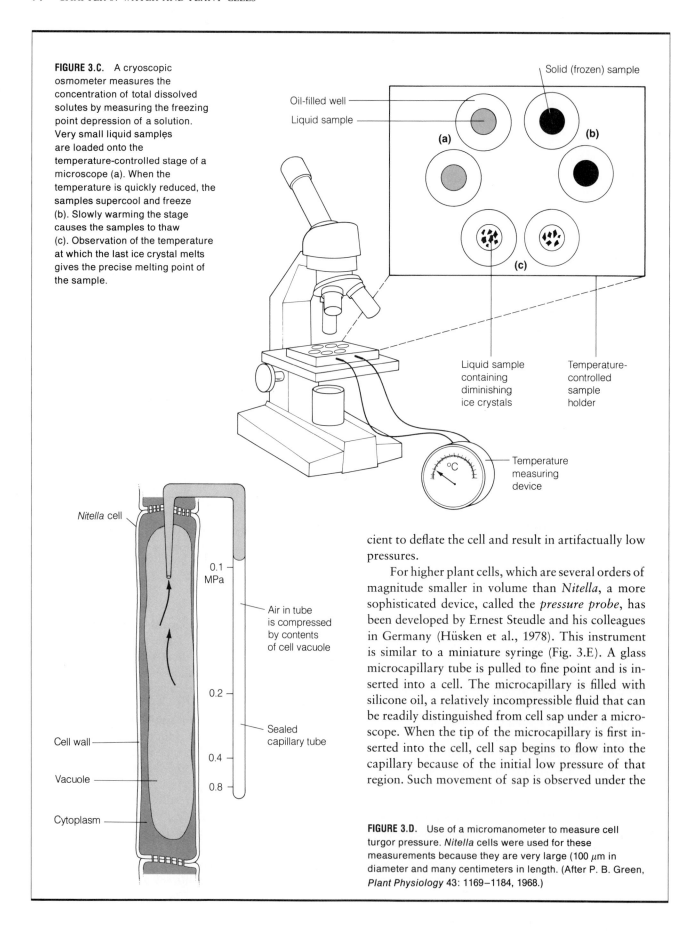

FIGURE 3.C. A cryoscopic osmometer measures the concentration of total dissolved solutes by measuring the freezing point depression of a solution. Very small liquid samples are loaded onto the temperature-controlled stage of a microscope (a). When the temperature is quickly reduced, the samples supercool and freeze (b). Slowly warming the stage causes the samples to thaw (c). Observation of the temperature at which the last ice crystal melts gives the precise melting point of the sample.

Solid (frozen) sample

Oil-filled well

Liquid sample

(a)   (b)

(c)

Liquid sample containing diminishing ice crystals

Temperature-controlled sample holder

Temperature measuring device

°C

Nitella cell

0.1 MPa

Air in tube is compressed by contents of cell vacuole

0.2

Sealed capillary tube

Cell wall

Vacuole

Cytoplasm

0.4

0.8

cient to deflate the cell and result in artifactually low pressures.

For higher plant cells, which are several orders of magnitude smaller in volume than *Nitella*, a more sophisticated device, called the *pressure probe*, has been developed by Ernest Steudle and his colleagues in Germany (Hüsken et al., 1978). This instrument is similar to a miniature syringe (Fig. 3.E). A glass microcapillary tube is pulled to fine point and is inserted into a cell. The microcapillary is filled with silicone oil, a relatively incompressible fluid that can be readily distinguished from cell sap under a microscope. When the tip of the microcapillary is first inserted into the cell, cell sap begins to flow into the capillary because of the initial low pressure of that region. Such movement of sap is observed under the

FIGURE 3.D. Use of a micromanometer to measure cell turgor pressure. *Nitella* cells were used for these measurements because they are very large (100 μm in diameter and many centimeters in length. (After P. B. Green, *Plant Physiology* 43: 1169–1184, 1968.)

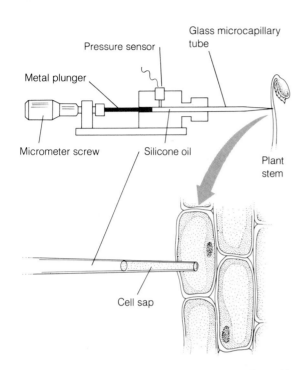

FIGURE 3.E. Diagram of the simplest pressure probe, not to scale. The primary advantage of this method over the one shown in Figure 3.D is that cell volume is minimally disturbed. This is of great importance for the tiny cells typical of higher plants, where loss of even a few picoliters (10^{-12} liter) can substantially reduce turgor pressure.

microscope and is counteracted by pushing on the plunger of the device, thus building up a pressure. In such a fashion the boundary between the oil and the cell sap can be pushed back to the tip of the microcapillary. When the boundary is returned to the tip and is held in a constant position, the initial volume of the cell is restored and the pressure inside the cell is exactly balanced by the pressure in the capillary. This pressure is measured by a pressure sensor in the device. Thus the hydrostatic pressure of individual cells may be measured directly. This method has been used to measure P in cells of both excised and intact tissues of a variety of plant species. Its primary limitation is that some cells are too small for current instruments to measure. Furthermore, some cells tend to leak after being stabbed with the capillary, and others plug up the tip of the capillary, thereby preventing valid measurements.

PRESSURE CHAMBER

A relatively quick method for estimating the water potential of large pieces of tissues, such as whole leaves and shoots, is with the **pressure chamber**. This method was pioneered by Henry Dixon at Trinity College, Dublin, at the beginning of this century, but

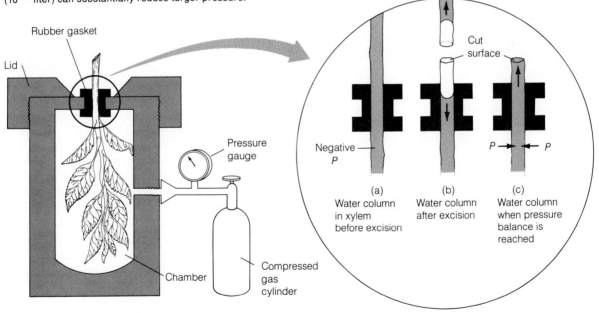

FIGURE 3.F. The pressure chamber method for measuring plant water potential. The diagram at the left shows a shoot sealed into a chamber, which may be pressurized with compressed gas. The diagrams at right show the state of the water columns within the xylem at three points in time. In (a), the xylem is uncut and under a negative pressure, or tension. In (b), the shoot is cut, causing the water to pull back into the tissue, away from the cut surface, in response to the tension in the xylem. In (c), the chamber is pressurized, bringing the xylem sap back to the cut surface.

it did not come into widespread use until P. Scholander and co-workers at Scripps Institute improved the instrument design and showed its practical use (Scholander et al., 1965). The pressure chamber measures the negative hydrostatic pressure (tension) that exists in the xylem of most plants. It is assumed that the water potential of the xylem is fairly close to the average water potential of the whole organ, which is probably valid because (1) in many cases the osmotic pressure of the xylem solution is low, so the major component of the water potential in the xylem is the (negative) hydrostatic pressure in the xylem column, and (2) the xylem is in intimate contact with most cells and the plant.

In this technique, the organ to be measured is excised from the plant and is partly sealed in a pressure chamber (Fig. 3.F). Before excision, the water column in the xylem is under some tension. When the water column is broken by excision of the organ, the water is pulled into the xylem capillary by the now unopposed tension. The cut surface consequently appears dull and dry. To make a measurement, the chamber is pressurized with compressed gas until the water in the xylem is brought back to the cut surface. The pressure needed to bring the water back to the surface is called the *balance pressure* and is readily detected by the change in the appearance of the cut surface, which becomes wet and shiny when this pressure is attained. The balance pressure is equal in magnitude (but opposite in sign) to the negative pressure that existed in the xylem column before the plant material was excised. For example, if a balance pressure of 0.5 MPa is found, then P in the xylem before excision was -0.5 MPa. If we know π of the xylem sap from other measurements, we may calculate the water potential of the xylem, which, as stated above, is assumed to be close to the water potential of the whole organ.

In many outdoor plants, P in the xylem may be -1 to -2 MPa, whereas π may be only 0.1 or 0.2 MPa. Therefore in many situations pressure chamber measurements by themselves provide an adequate estimate of the water potential of the plant.

Because the pressure chamber method is rapid and does not require delicate instrumentation or elaborate temperature control, it has been extensively used under field conditions to estimate water potentials.

The "Matric Potential" Is the Overall Reduction in Water Potential Caused by Insoluble Materials Such as Soil Colloids or Cell Walls

In discussions of soils and cell walls one sometimes finds reference to a **matric potential**. The matric potential is a parameter used to account for the reduction in water potential when water interacts with the surface of a solid phase such as a cell wall or a soil clay particle. Such interactions reduce the tendency of water molecules to react chemically or to evaporate.

This reduction in water potential stems from three different effects. First, there may be attractive forces between water and the soil surface that tend to bind water in a microscopic film around the surface of the solid. This film is analogous to the shells of hydration around individual molecules. Second, for solids that have fixed charges on their surfaces, swarms of ions with opposite charge, or counterions, congregate near the fixed charges. Examples of such fixed charges are the negatively charged carboxyl groups of wall pectins and the negatively charged silicon and aluminum oxides of clay particles. The counterions remain near the fixed charges because of the need to maintain electrical neutrality. As a result, they increase the local osmotic pressure of the solution near the particle surface. Third, concave menisci develop at air-water interfaces, causing a tension (negative hydrostatic pressure) through the surface tension properties described earlier in relation to capillarity. Since it is difficult to separate out the individual matric effects experimentally, an overall *matric potential* is sometimes measured, for example, in soils. Soil water potential will be discussed in Chapter 4.

The matric potential does not represent a new type of contribution to the water potential of the cell (Passioura, 1980; Tyree and Jarvis, 1982). In principle, these interfacial effects can be grouped within the π and P components of water potential. Moreover, the location of the cell wall outside the protoplast means that matric effects in the wall cannot contribute to the water potential of the cell. Because matric potential within plant cells can be adequately represented by P and π, we can ignore it in our discussion of plant water relations.

The Water Potential Concept Is Useful for Evaluating the Water Status of a Plant

The concept of water potential has two principal uses. First, the water potential is the quantity that governs the direction of water flow across cell membranes. Specifically, the difference in water potential ($\Delta\psi$) across a membrane is the driving force that leads to transport of water by osmosis. With some important limitations, a water potential difference is also the driving force for water movement across multicellular tissues. As we shall see in Chapter 7, however, the water potential gradient does not always predict the direction of pressure-driven bulk flow.

The second important use of the water potential is as a measure of the *water status* of a plant. Water deficits lead to inhibition of plant growth and photosynthesis, as well as to other effects. Figure 3.11 lists some of the physiological changes that occur as plants become drier. The process that is most affected by water deficit is cell growth. More severe water stress leads to inhibition of cell division, inhibition of wall and protein synthesis, accumulation of solutes, closing of stomata, and inhibition of photosynthesis. Water potential is one measure of how "wet" or "dry" a plant is and thus provides a relative index of the *water stress* the plant is experiencing (see Chapter 14).

Figure 3.11 also shows representative values for ψ at various stages of water stress. Leaves of well-watered plants have ψ values in the range of -0.2 to -0.6 MPa, whereas the leaves of plants in arid climates can have much lower values, perhaps -2 to -5 MPa under extreme conditions. Because water transport is a passive process, plants can take up water only when the plant ψ value is below that of the soil. As the soil becomes drier, the plant similarly becomes drier (attains a lower ψ). If this were not the case, the soil would begin to extract water from the plant.

Like ψ, values of π also vary considerably, depending on the growing conditions and the type of plant. Well-watered garden plants may have π values inside their cells as low as 0.5 MPa. Examples include lettuce, cucumber seedlings, and bean leaves. The lower limit for π is probably set by the minimum concentration of dissolved ions, metabolites, and proteins in the cytoplasm of living cells. At the other extreme, plants sometimes build up a much higher π. For instance, water stress typically leads to an accumulation of solutes, which permits maintenance of turgor pressure in the presence of low water potentials. Plant tissues that store high concentrations of sucrose or other sugars, such as sugar beet roots, sugarcane stems, or grape berries, necessarily have high values of π. Values as high as 2.5 MPa are not unusual. **Halophytes**, plants that grow in saline environments, typically have very high internal solute concentrations; by this means they reduce cell ψ enough to extract water from salt water. Most crop plants cannot survive in seawater because it has a lower water potential, due to the dissolved salts, than the tissues of the plant can attain and still function properly.

Values for P are normally lower than those for π. This follows directly from Equation 3.8 and from the fact that water potentials of plants are always negative. Within cells of well-watered garden plants, the hydrostatic pressure may range from 0.1 to perhaps 1 MPa, depending on the value of π inside the cell.

A positive hydrostatic pressure in plant cells is important for two principal reasons. First, the growth of plant cells requires a hydrostatic pressure to stretch the walls. As a plant loses water, P falls very rapidly because the cell wall is rather rigid and water is incompressible. Such a decrease in P as ψ decreases may explain why cell growth is so sensitive to water stress. Second, the hydrostatic pressure within cells increases the mechanical rigidity of cells and tissues. This function of cell turgor pressure is particularly important for young nonlignified tissues, which cannot support themselves mechanically without a high internal pressure. **Wilting** occurs when the

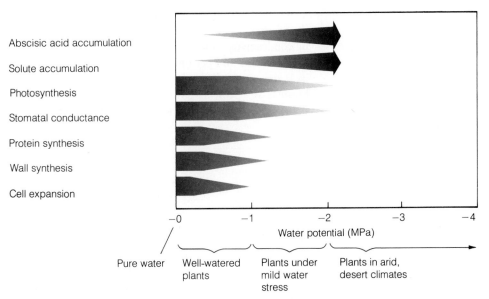

Abscisic acid accumulation

Solute accumulation

Photosynthesis

Stomatal conductance

Protein synthesis

Wall synthesis

Cell expansion

Water potential (MPa)

Pure water

Well-watered plants

Plants under mild water stress

Plants in arid, desert climates

FIGURE 3.11. Water potential of plants under various growing conditions, and sensitivity of various physiological processes to water potential. Thickness of a bar corresponds to the magnitude of the process. For example, cell expansion decreases as water potential falls (becomes more negative). Abscisic acid is a hormone that induces stomatal closure during water stress (Chapter 19). (Adapted from T. C. Hsiao, *Agricultural Meteorology* 14: 59–84, 1974.)

hydrostatic pressure inside the cells of such tissues falls toward zero.

In the xylem of rapidly transpiring plants, P is negative and may attain values of -1 MPa or lower. The magnitude of P in the xylem varies considerably, depending on the rate of transpiration and the height of the plant.

The Rate of Water Transport Depends on the Magnitude of the Driving Force and the Hydraulic Conductivity of the Transport Pathway

So far, we have seen that water moves in response to a driving force, which may be a concentration gradient in the case of diffusion, a pressure gradient in the case of bulk flow, or a water potential gradient in the case of osmosis. The direction of flow is determined by the direction of the appropriate gradient.

What is it that determines the *rate* at which the water moves? The rate of flow depends on both the magnitude of the driving force and the physical characteristics of the medium through which water is transported. One common way to relate the rate of flow to the driving force is via a **resistance to flow**; thus:

$$\text{Flow rate} = \frac{\text{driving force}}{\text{resistance}} \qquad (3.9)$$

The larger the resistance, the smaller the flow rate for any given driving force. A second common transport equation relates the flow rate to the driving force via a **conductance**; thus:

$$\text{Flow rate} = \text{driving force} \times \text{conductance} \qquad (3.10)$$

This relation is analogous to Ohm's law. It is readily seen that conductance = 1/resistance.

Resistances and conductances can have various units, depending on how the flow rate and the driving force are measured. Two examples will serve to illustrate the concepts behind these terms.

First, consider two compartments, one containing pure water and the other containing a sucrose concentration of 1.0 M, as shown in Figure 3.12. The compartments are separated by a membrane that is permeable to water but impermeable to the solute. This is an ideal condition for osmosis, so we would predict that water would tend to flow from the compartment containing pure water to that containing the solute. Since osmosis is the mechanism of transport, the driving force is the difference in water potentials across the membrane, which is readily calculated to be 2.44 MPa (at 20°C). This calculation tells us how large the driving force is but gives us no information about flow rate. How fast will water move across the membrane? The rate depends on the area of the membrane (the greater the surface area, the more

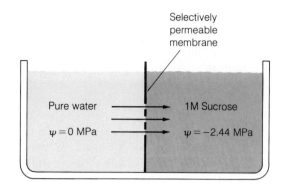

Volume flow rate $= L(\Delta\psi)$
Assume $L = 10^{-6}$ cm^3 s^{-1} MPa^{-1}
$\Delta\psi = 0 - (-2.44$ MPa$)$
$\qquad = 2.44$ MPa
Therefore: Volume flow rate $= 2.44 \times 10^{-6}$ cm^3 s^{-1}

FIGURE 3.12. The rate of water flow between two compartments separated by an ideal osmotic membrane depends on the hydraulic conductance of the membrane (L) and on the water potential difference ($\Delta\psi$). Hydraulic conductance is the product of membrane cross-sectional area (A) and the membrane hydraulic conductivity (Lp), ($L = A \times Lp$). Membrane hydraulic conductivity is a measure of the ease with which water crosses the membrane. For a giant algal cell such as *Nitella*, Lp would typically have a value of 10^{-4} cm s^{-1} Mpa^{-1}. In this example, we have arbitrarily assigned a value of 10^{-6} cm^3 s^{-1} MPa^{-1}. Note that the resulting flow is given in units of volume per time.

water will flow) and on the permeability of the membrane material to water, a property usually called the **hydraulic conductivity, Lp**, of the membrane. The hydraulic conductivity expresses how readily water can move across a membrane and has units of volume of water per unit area of membrane per unit time per unit driving force—for instance, cm^3 cm^{-2} s^{-1} MPa^{-1} or cm s^{-1} MPa^{-1}. The larger the hydraulic conductivity, the larger the flow rate. The total **hydraulic conductance, L**, of the membrane is the product of the area of the membrane and the hydraulic conductivity (units: cm^3 s^{-1} MPa^{-1}, that is, a volume per unit time and per unit driving force). In our example in Figure 3.12, the hydraulic conductance of the membrane is 10^{-6} cm^3 s^{-1} MPa^{-1}. The volume flow rate can then be computed from the following equation:

$$\text{Volume flow rate} = L(\Delta\psi) \qquad (3.11)$$

resulting in a flow rate of 2.44×10^{-6} cm^3 s^{-1}. If we doubled the area of the membrane, the flow rate would double because the total hydraulic conductance would double.

The second example illustrates the concept of hydraulic resistance, the reciprocal of hydraulic conductivity. Consider Figure 3.13, which shows two tanks containing water at different heights and connected by a long, thin pipe. In this case, water will move primarily by bulk flow in response to the pressure difference across

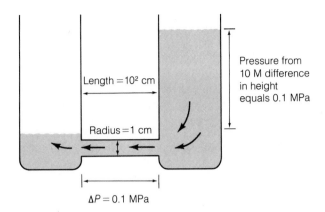

FIGURE 3.13. Illustration of bulk flow of water in response to a pressure difference. Two tanks of water are connected by a narrow pipe. The flow rate depends on the driving force and the conductance of the transporting pipe. The driving force is the pressure difference at the two ends of the pipe that arises from the difference in water heights in the two tanks. The conductance of the pipe depends on its length and radius. Viscosity of the water also influences flow rate.

the two ends of the pipe. If we measure the rate of water flow, as in the last example, in units of volume per unit time ($cm^3\ s^{-1}$), then the total resistance to water flow, due to frictional drag of the water against the walls of the pipe, will have units of $cm^{-3}\ s\ MPa$ (to confirm this, check the units by using Eq. 3.9). On what does the total resistance depend? Common experience indicates that the length and diameter of the pipe and the viscosity of the water will determine the resistance. Indeed, from the Poiseuille equation (Eq. 3.3), we can calculate the hydraulic resistance of the pipe as

$$\text{Resistance} = \frac{8}{\pi}\ \frac{\text{length}}{(\text{radius})^4}\ \frac{\text{viscosity}}{1} \quad (3.12)$$

$$= \frac{8}{\pi}\ \frac{10^2\ cm}{(1\ cm)^4}\ \frac{10^{-9}\ MPa\ s}{1}$$

$$= 2.5 \times 10^{-7}\ cm^{-3}\ s\ MPa$$

So with a driving force of 0.1 MPa (the difference in heights of the water is 10 m), the flow rate is 0.4×10^7 $cm^3\ s^{-1}$. If we double the length of the pipe, the flow rate decreases to one-half of this value. If we shrink the radius of the pipe by one-half, the flow rate would decrease to $\frac{1}{16}$ of the former value ($[\frac{1}{2}]^4 = \frac{1}{16}$).

Sometimes it is more useful to measure flow rate as a velocity ($cm^3\ cm^{-2}\ s^{-1}$, or $cm\ s^{-1}$) rather than as a volume of per unit time ($cm^3\ s^{-1}$). In this case, the hydraulic conductances and resistances will be defined slighly differently and will have slightly different units. Therefore, it is necessary to pay attention to units when calculating flow rates. Application of the concept of hydraulic conductivity to cells and tissues will be discussed in Chapter 4.

Summary

Water is important in the life of plants because it makes up the matrix and medium in which most biochemical processes essential for life occur. The structure and properties of water strongly influence the structure and properties of proteins, membranes, nucleic acids, and other cell constituents.

In most land plants, water is continually lost to the atmosphere and taken from the soil. The movement of water may occur by diffusion, bulk flow, osmosis, or some combination of these three fundamental transport mechanisms. Diffusion moves water from regions of high water concentration (low solute concentration) to regions of low water concentration (high solute concentration). It occurs because molecules are in constant thermal agitation, which tends to even out any concentration differences. Bulk flow occurs in response to a pressure difference, whenever there is a suitable pathway for bulk movement of water. Osmosis is the process by which water moves across membranes. It depends on the chemical potential of water, or water potential.

Solute concentration and pressure are the two major factors affecting water potential, although gravity is also important when large vertical distances are involved. In addition to its importance in osmosis, water potential serves as a useful measure of the water status of plants.

The rate of water transport depends on the magnitude of the driving force and the hydraulic conductivity of the transport pathway. For diffusion, the hydraulic conductance simply depends on the diffusion coefficient. For bulk flow, the path geometry and the solution viscosity are important. For osmosis, the membrane permeability is of prime importance. As we will see in the following chapter, diffusion, bulk flow, and osmosis all take part in the movement of water from the soil through the plant to the atmosphere.

GENERAL READING

Boyer, J. S. (1985) Water Transport. *Annu. Rev. Plant Physiol.* 36: 473–516.

Brown, R. W., and Van Haveren, P., eds. *Psychrometry in Water Relations Research.* Utah Agricultural Experiment Station, Utah State University, Logan.

Cosgrove, D. (1986) Biophysical control of plant cell growth. *Annu. Rev. Plant Physiol.* 37: 377–405.

Finkelstein, A. (1987) *Water Movement Through Lipid Bilayers, Pores, and Plasma Membranes: Theory and Reality.* Wiley, New York.

Friedman, M. H. (1986) *Principles and Models of Biological Transport.* Springer-Verlag, Berlin.

Kramer, P. J. (1983) *Water Relations of Plants.* Academic Press, New York.

Milburn, J. A. (1979) *Water Flow in Plants.* Longman Group, London.

Nobel, P. (1983) *Biophysical Plant Physiology and Ecology.* W. H. Freeman, San Francisco.

Preston, R. D. (1974) *The Physical Biology of Plant Cell Walls.* Chapman and Hall, London.

Ricklefs, R. E. (1970) *Ecology.* Chiron Press, Newton, Mass.

Slatyer, R. O. (1967) *Plant-Water Relationships.* Academic Press, London and New York.

Slavík, B. (1974) *Methods of Studying Plant Water Relations.* Academia, Prague.

Stein, W. D. (1986) *Transport and Diffusion Across Cell Membranes.* Academic Press, Orlando, Fla.

Sutcliffe, J. (1968) *Plants and Water.* St. Martin's, New York.

Tyree, M. T., and Jarvis, P. G. (1982) Water in tissues and cells. In: *Physiological Plant Ecology. II. Water Relations and Carbon Assimilation, Encyclopedia of Plant Physiology, New Series,* pp. 35–77. Lange, O. L., Nobel, P. S., Osmond, C. B., and Ziegler, H., eds. Springer, Berlin.

CHAPTER REFERENCES

Bearce, B. C., and Kohl, H. C., Jr. (1970) Measuring osmotic pressure of sap within live cells by means of a visual melting point apparatus. *Plant Physiol.* 46: 515–519.

Boyer, J. S., and Knipling, E. B. (1965) Isopiestic technique for measuring leaf water potentials with a thermocouple psychrometer. *Proc. Natl. Acad. Sci. USA* 54: 1044–1051.

Green, P. B., and Stanton, F. W. (1967) Turgor pressure: Direct manometric measurement in single cells of *Nitella. Science* 155: 1675–1676.

Hüsken, D., Steudle, E., and Zimmermann, U. (1978) Pressure probe technique for measuring water relations of cells in higher plants. *Plant Physiol.* 61: 158–163.

Passioura, J. B. (1980) The meaning of matric potential. *J. Exp. Bot.* 31: 1161–1169.

Prager, D. J., and Bowman, R. L. (1963) Freezing-point depression: New method for measuring ultramicro quantities of fluids. *Science* 142: 237–239.

Scholander, P. F., Hammel, H. T., Bradstreet, E. D., and Hemmingsen, E. A. (1965) Sap pressure in vascular plants, *Science* 148: 339–346.

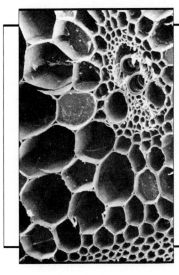

CHAPTER 4

Water Balance of the Plant

EARTH'S ATMOSPHERE PRESENTS A dilemma to land plants. On the one hand, it is the source of carbon dioxide, which is needed for carbon fixation during photosynthesis. Plants therefore need ready access to the atmosphere so that they can photosynthesize. On the other hand, excessive exposure to the atmosphere poses a danger, because it is relatively dry and can dehydrate the plant. To meet the contradictory demands of maximizing carbon dioxide uptake while limiting water loss, plants have evolved adaptations that minimize water loss from the leaf, allow some degree of control over water loss, and permit rapid transport of water from the soil to replace water lost to the atmosphere.

In this chapter we will examine how water moves from the soil through the plant to the atmosphere. This pathway can be thought of as continuous, but it is not homogeneous. In traveling along this pathway, water moves as a liquid through soil, across membranes, through cells, along hollow conduits, and through cell walls; it then escapes as a vapor into the air spaces in the leaf and diffuses to the outside atmosphere. The passage of water as a vapor from the leaf to the atmosphere is called **transpiration**. Different mechanisms of transport are called into play along various parts of this pathway.

Water in the Soil

The **water content** and rate of water movement in soils depend to a large extent on soil type. Table 4.1 shows that the physical characteristics of different soils can vary greatly. At one extreme is sand, in which the soil particles may be a millimeter or more in diameter. Sandy soils have a relatively low surface area per gram of soil and have large spaces or channels between particles. At the other extreme is clay, in which particles are smaller than 2 μm in diameter. Clay soils have much greater surface areas and smaller channels between particles. The spaces between soil particles may be filled with air or with water.

When a soil is heavily watered by rain or by irrigation, the water percolates downward by gravity through the spaces between soil particles, partially displacing, and and in some cases trapping, air in these channels. Water in the soil may exist as a film adhering to the surface of soil particles or may fill the entire channel between particles. In sandy soils, the spaces between particles are so large that water tends to drain from them and remain only on the particle surfaces and at interstices between

TABLE 4.1. Physical characteristics of different soils

Soil	Particle diameter (μm)	Surface area per gram (m^2)
Coarse sand	2000–200 ⎫	
Fine sand	200–20 ⎬	10 to less than 1
Silt	20–2	10–100
Clay	below 2	100–1000

particles. In clay soils, the channels are small enough that water does not freely drain from them but is held more tightly. This phenomenon is reflected in the moisture-holding capacity, or **field capacity**, of soils, which is large for clay soils and soils with a high humus content and much lower for sandy soils. Field capacity refers to the water content of a soil after it has been saturated with water and excess water has been allowed to drain away.

Soil Water Potential Is a Function of the Osmotic Pressure of the Soil Water and the Negative Hydrostatic Pressure Caused by Adhesion and Surface Tension

The soil water potential, like the cellular water potential, depends on two components. The first component, the osmotic pressure (π) of soil water, is generally low; a typical value might be 0.01 MPa. For saline soils that contain a substantial concentration of salts, π is much higher, perhaps 0.2 MPa or greater. The second component of soil water potential, hydrostatic pressure (P), depends largely on the water content of the soil. It is always less than or equal to zero; that is, soil water is under a tension. For wet soils, P is very close to zero. As a soil dries out, P decreases, and for arid soils P can attain values of about -3 MPa.

Where does the negative pressure in soil water come from? From our discussion of capillarity in Chapter 3, we know that water has a high surface tension that tends to minimize air-water interfaces. As a soil dries out, water is first removed from the middle of the largest spaces between particles. Because of adhesive forces, water tends to cling to the surfaces of soil particles, so that a large surface area between soil water and soil air develops (see Fig. 4.1). As the water content of the soil decreases, the water recedes into the interstices between soil particles and the air-water surface stretches and develops curved menisci. Water under these curved surfaces develops a tension or negative pressure, which is given by the approximate formula

$$P = \frac{-2T}{r} \qquad (4.1)$$

where T is the surface tension of water (7.28×10^{-8} MPa m) and r is the radius of curvature of the meniscus.

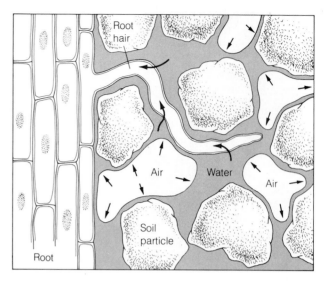

FIGURE 4.1. Root hairs make intimate contact with soil particles and amplify the surface area needed for water absorption by the plant. The soil is a mixture of particles (both mineral and organic), water, dissolved solutes, and air. As water is absorbed by the plant, the soil solution recedes into smaller pockets, channels, and crevices between the soil particles. This recession causes the surface of the soil solution to develop concave menisci (curved interfaces between air and water), which brings the solution into tension, or negative pressure, by surface tension. As more water is removed from the soil, more acute menisci are formed, resulting in greater tensions (more negative pressures).

Such tension in soil water can become quite large because the radius of curvature of air-water surfaces may become very small in dry soils. Tension values may easily reach -1 or -2 MPa. In comparison, the osmotic pressure π is rather small. Thus the tension in soil water caused by its adhesion to soil particles is the predominant component of the water potential in nonsaline soils.

Water Moves Through the Soil by Bulk Flow at a Rate Governed by the Pressure Gradient and the Soil Hydraulic Conductivity

Water movement through soils is predominantly by bulk flow driven by a pressure gradient, although diffusion also accounts for some water movement. As a plant absorbs water from the soil, it depletes the soil of water near the surface of the roots. This depletion reduces the value of P in the water near the root surface and establishes a pressure gradient with respect to neighboring regions of soil with higher P values. Because the water-filled pore spaces in the soil are interconnected, water moves through these channels down the pressure gradient to the root surface by bulk flow.

The rate of water flow through soils depends on the size of the pressure gradient through the soil and on the hydraulic conductivity of the soil. Soil hydraulic conductivity is a measure of the ease with which water moves through the soil, and it varies with the type of soil. Sandy

Irrigation

PEOPLE HAVE USED irrigation to provide water for crops almost as long as they have practiced agriculture. In the harsh, hot deserts of southern Arizona, the prehistoric Hohokam Indian communities dug extensive networks of irrigation canals to divert water from the nearby Salt River to their crops. Thus their society thrived where it would otherwise have perished. Aerial photography reveals the location of these ancient canals, which follow many of the same routes as today's concrete-lined waterways that carry water throughout the area of Phoenix, Arizona. In the Tigris and Euphrates valleys of Mesopotamia, which have been called the "cradle of civilization," crops were irrigated with similar results. It is possible that civilization declined there when the irrigated fields were "salted out"—made useless by the continuing addition of salts to the soil from poor-quality water. The Incas of ancient Peru also practiced irrigation extensively, including some innovative methods involving raised beds and provision of water from below.

Irrigation as practiced today in world agriculture is governed by the same general principles as in ancient times. Water is expensive and therefore must be used as efficiently as possible. If surface water is used, dams, and waterways must be constructed and maintained. If water is pumped from the ground, energy must be expended to raise it to the surface. Perhaps surprisingly, most of the inefficiency in the use of water is encountered before the water ever reaches the plant in the field! Evaporation from lakes, seepage from canals, transpiration from aquatic weeds that grow in the canals, and uneven application in the fields are just some of the problems faced by irrigation specialists. A system is considered efficient if 50% of the water is delivered to the root zone of the crop it is supporting.

Various methods have been developed for delivering water to fields. If the land can be graded and sloped appropriately, furrows with a very slight downhill slope can be placed between rows of plants. Water is applied at the uphill end and allowed to flow through the furrows. Some water is wasted, because the soil at the uphill end becomes saturated before enough has been delivered in the middle of the field. Water may also "puddle" at the downhill end. Level-basin irrigation is being used increasingly to overcome these problems. Large basins are leveled to within 1 to 2 cm by laser-directed machinery. If the soil is uniformly permeable throughout the basin, this allows water to be evenly applied. If the land cannot be leveled or if the amounts applied are small, sprinklers are often used. However, sprinklers not only require energy to pump and pressurize the water but also produce small droplets from which evaporation is excessive. In addition, if the water is of poor quality, sprinkling will deposit salts directly on the leaves, where they can injure the plants. In some cases, sprinkling at night can prevent much of the injury because the salt-laden droplets do not evaporate as fast.

Some soils with very high clay content expand as they are wetted and shrink and crack as they dry. When they are wet they tend to become sealed, and water enters very slowly. For cracking clay soils, a technique known as surge irrigation has been developed. Water is put into the field extremely rapidly so that it can flow down the cracks and enter the root zone before the soil swells and the cracks disappear.

In recent years a technique known as drip irrigation (also called trickle irrigation or micro-irrigation) has come into use in some areas of the world. Water is pumped directly to the base of a plant by plastic tubing and bled through an emitter at a slow rate that just meets the plant's needs. This approach is very efficient, but it is also very expensive and requires diligent maintenance of the hardware to keep the system working. For instance, the emitters tend to become plugged by both mineral deposits and slime produced by microorganisms. Periodically, the system must be flushed out with acid or with disinfectant. So far, drip irrigation has been used mostly for high-value crops whose quality (and price) depends strongly on a reliable supply of water. Under these conditions the profit from drip irrigation will pay for the extra costs. Fresh fruits such as blueberries and strawberries are extensively irrigated by this method in the United States.

With the availability of extremely efficient irrigation systems such as drip irrigation, more attention can be shifted to inefficient use of water by the crops themselves and to possible improvements in plant water use efficiency. The physiology of water in plants is an exciting area of research because many problems wait to be solved. For example, grain crops are most sensitive to drought when they are flowering, when abortion of the very young embryos can result in a barren plant. When problems such as this are solved, agronomists will know how to make grain yields reliable even without resorting to irrigation!

soils, with their large spaces between particles, have large hydraulic conductivities, whereas clay soils, with their minute spaces between particles, have appreciably smaller hydraulic conductivities. Besides soil type, the water content also affects the hydraulic conductivity of soil. Figure 4.2 shows that as the water content (and hence the water potential) of a soil decreases, the hydraulic conductivity decreases drastically. This decrease in soil hydraulic conductivity is due primarily to the replacement of water in the soil spaces by air. When air moves into a soil channel previously filled with water, water movement through that channel is restricted to its periphery. As more of the soil spaces become filled with air, there are in effect fewer channels through which water can easily flow, and the hydraulic conductivity falls.

When a soil dries out sufficiently, the water potential (ψ), defined as the chemical potential of water expressed in pressure units, may fall below what is called the **permanent wilting point**. This is the point at which the water potential of the soil is so low that plants could not regain turgor pressure even if all water loss through

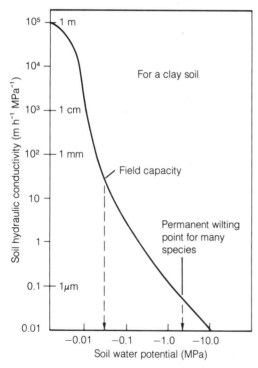

FIGURE 4.2. Soil hydraulic conductivity as a function of the water potential of the soil. Conductivity measures the ease with which water moves through the soil. The decrease in conductivity as the soil dries is due primarily to the movement of air into the soil to replace the water. As the air moves in, the pathways for water flow between soil particles become smaller and more tortuous and flow becomes more difficult. The overall shape of this curve is representative of many soils, but the actual shape for a particular soil may be influenced by its particle size distribution and organic matter content. The *field capacity* refers to the amount of water the soil is able to retain against gravitational forces.

transpiration were stopped. The plants remain wilted (i.e., $P = 0$) even at night, when transpiration ceases. Because plant cells have a positive P only when $\psi > -\pi$, this means that the soil ψ is less than or equal to $-\pi$ of the plant. Since cell π varies with plant species, the permanent wilting point is clearly not a unique property of the soil but depends on the plant species as well (Turner and Kramer, 1980).

Water Absorption by the Root

Intimate contact between the surface of the root and the soil is essential for effective water absorption by the root. Such intimate contact is maximized by the growth of root hairs into the soil. **Root hairs** are microscopic extensions of root epidermal cells which greatly increase the surface area of the root, thus providing greater capacity for absorption (Fig. 4.1). When 4-month-old rye (*Secale*) plants were examined, it was found that root hairs constituted more than 60% of the surface area of the roots. Measurements have shown that water enters the root most readily in the region just behind the zone of elongation, that is, in the region where root hairs are developing.

Root hairs are delicate cells that are easily ruptured when the soil is disturbed. This is the reason why newly transplanted seedlings and plants need to be protected from water loss for the first few days after transplantation. Thereafter, new root hairs grow into the soil and the plant can better withstand water stress.

Water Crosses the Root Radially Via the Apoplast, Transmembrane, and Symplast Pathways Until It Reaches the Endodermis

When water comes in contact with the root surface, the nature of water transport changes. In the soil, water is transported predominantly by bulk flow (Kramer, 1984). Figure 4.3 shows that from the epidermis to the endodermis of the root, there are multiple pathways through which water can flow. In the **apoplast pathway**, water moves exclusively through the cell wall without crossing any membranes. The term **apoplast** refers to the continuous system of cell walls and intercellular air spaces in plant tissues. Water may also move via the **cellular pathway**, which has two components. The first component, the **transmembrane** pathway, is the route followed by water that sequentially enters a cell on one side, exits the cell on the other side, enters the next in the series, and so on. In this pathway, water crosses at least two membranes for each cell in its path (the plasma membrane on entering and on exiting). Transport across the tonoplast may also be involved. Before entering each cell, the water finds itself temporarily in the cell wall space. In the second component of the cellular pathway, water moves through the **symplast**, traveling from one cell to the next via the

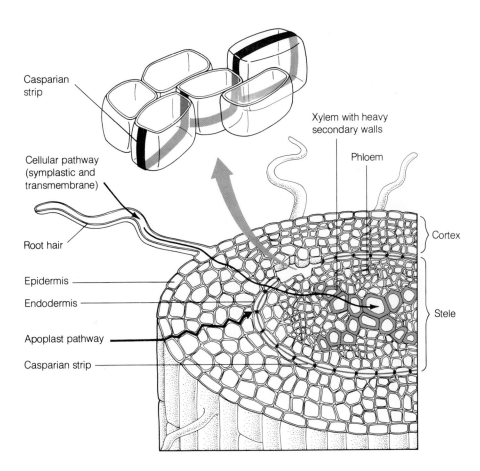

Casparian strip

Cellular pathway (symplastic and transmembrane)

Root hair

Epidermis

Endodermis

Apoplast pathway

Casparian strip

Xylem with heavy secondary walls

Phloem

Cortex

Stele

FIGURE 4.3. Pathways for water uptake by the root. Through the cortex, water may travel via the apoplast pathway or the cellular pathway, which includes the transmembrane and symplast pathways. In the symplast pathway, water flows between cells through the plasmodesmata without crossing the plasma membrane. In the transmembrane pathway, water moves across the plasma membranes, with a short visit to the cell wall space. At the endodermis, the apoplast is blocked by the Casparian strip. Water entering the root's vascular system must cross the plasma membrane of the endodermis.

plasmodesmata (see Chapter 1). The symplast consists of the entire continuous network of cell cytoplasms interconnected by plasmodesmata.

Although the relative importance of the apoplast, transmembrane, and symplast pathways is not yet clearly established, water transport across the root always occurs through some combination of these three pathways.

At the endodermis, water movement through the apoplast pathway is blocked by the **Casparian strip**. The Casparian strip is a band of radial cell walls in the endodermis impregnated with the waxlike, hydrophobic substance **suberin**. Suberin is an effective barrier to water and solute movement, so all molecules transported across the endodermis must cross a plasma membrane and enter the cytoplasm of the endodermal cells.

From the foregoing discussion, it should be evident that radial movement of water across the root is a complex process. One way to simplify the process conceptually is to treat the whole multicellular pathway—from the root hairs to the root xylem—as if it were a single membrane with a single resistance (or its reciprocal, conductance) to water flow. This simplification appears to work reasonably well, and it can be demonstrated that the whole root behaves like a single selectively permeable membrane.

Many studies have characterized the **hydraulic conductance** of roots. Interestingly, root hydraulic conductance decreases whenever the rate of respiration of the root tissues decreases. This effect can be shown experimentally by using low temperature, respiration inhibitors (such as CN^-), or anaerobic conditions to inhibit respiration; in each case, less water is transported by the roots. This decrease in root hydraulic conductance provides an explanation for the wilting of plants in waterlogged soils: submerged roots soon run out of oxygen, which is normally provided by diffusion through the air spaces in the soil (diffusion through gas is 10^4 times faster than diffusion through water). The anaerobic roots transport less water to the shoots, which consequently suffer net water loss and begin to wilt.

When Transpiration Is Prevented, Solute Accumulation in the Xylem Can Generate Positive Hydrostatic Pressure or "Root Pressure"

Plants sometimes exhibit a phenomenon referred to as **root pressure**. For example, if the stem of a young seedling is cut off just above the soil line, the stump will often

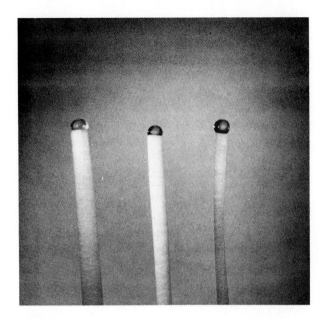

FIGURE 4.4. Exudation from cut stems of zucchini (left), soybean (center), and cucumber (right). The photograph was taken about 30 minutes after excision of the zucchini and soybean stems and about 15 minutes after excision of the cucumber stem. The exudation is a manifestation of the positive pressure (referred to as root pressure) in the xylem of such plants. (Photograph courtesy of Daniel Cosgrove, Penn State University.)

exude sap from the cut xylem for many hours (see Fig. 4.4). If a manometer is sealed over the stump, positive pressures can be measured. These pressures can be as high as 0.5 MPa, although they are typically less than 0.1 MPa.

Root pressures can be understood as follows. The root absorbs ions from the dilute soil solution and transports them into the xylem. The buildup of solutes in the xylem sap leads to an increase in the xylem osmotic pressure, π, and thus a decrease in the xylem ψ. This lowering of the xylem ψ provides a driving force for water absorption, which in turn leads to a positive hydrostatic pressure in the xylem. In effect, the whole root acts like a cell; the multicellular root tissue behaves like an osmotic membrane, building up a positive hydrostatic pressure in the xylem in response to the accumulation of solutes. Root pressure is most prominent in well-hydrated plants under humid conditions where there is little transpiration. Under more arid conditions, when transpiration rates are high, water is taken up so rapidly by the leaves and lost into the atmosphere that a positive pressure never develops in the xylem.

Plants that develop root pressure frequently exhibit exudation of liquid from the leaves, a phenomenon known as **guttation**. Positive xylem pressure causes exudation of xylem sap through **hydathodes**, structures that are located near terminal tracheids of the bundle ends around the margin of leaves. The "dewdrops" that can be seen on the tips of grass leaves in the morning are actually guttation droplets exuded from such specialized

pores. Guttation is most notable when transpiration is suppressed and the relative humidity is high, as at night.

Water Is Conducted Through Hollow Tracheids and Vessels of the Xylem

In most plants the xylem constitutes the longest part of the pathway of water transport. In a plant 1 m tall, more than 99.5% of the water transport pathway through the plant is within the xylem, and in tall trees the xylem represents an even greater fraction of the pathway. Compared with the complex pathway across the root tissue, the xylem is a simple pathway of low resistance.

Conducting cells in the xylem have a specialized anatomy that enables them to transport large quantities of water with great efficiency. There are two important types of **tracheary elements** in the xylem: tracheids and vessel members (see Fig. 4.5). Both of these elements are dead when functional: they have no membranes or organelles. In effect, they are like hollow tubes reinforced by lignified secondary walls. **Tracheids** are elongated, spindle-shaped cells that communicate with adjoining tracheids by means of the numerous pits in their lateral walls. These **pits** are microscopic regions where the secondary wall is absent and the primary wall is thin and porous. Pits of one tracheid are typically located opposite pits of an adjoining tracheid, forming **pit pairs**. Pit pairs, constitute a low-resistance path for water movement between tracheids. The porous layer between pit pairs, consisting of two primary walls and a middle lamella, is called the **pit membrane**.

Vessel members tend to be shorter and wider than tracheids and have perforated end walls that form a **perforation plate** at each end of the cell. Like tracheids, vessel members also have pits on their lateral walls. Unlike tracheids, which are arranged in overlapping vertical files, vessel members are stacked end to end to form a larger unit called a **vessel**. Because of their open cross-walls, vessels provide a very efficient low-resistance pathway for water movement. The vessel members at the extreme ends of a vessel have imperforate end walls and communicate with neighboring vessels via pit pairs.

Vessels are considered to be more evolutionarily advanced than tracheids and are found only in angiosperms and a small group of gymnosperms called the Gnetales. Tracheids are present in both angiosperms and gymnosperms.

The Pressure Gradient Required to Move Water Through the Xylem Is Much Less Than That Required to Transport Water Through Living Cells

The lack of cell membranes in the tracheary elements and the perforations in the walls of vessel members allow water to move rather freely through these water-filled

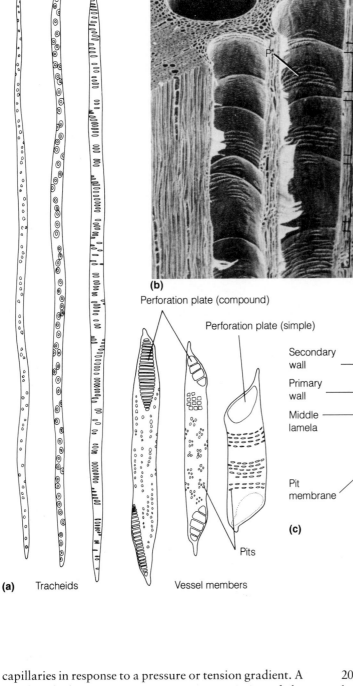

(b)

Perforation plate (compound)

Perforation plate (simple)

Secondary wall

Primary wall

Middle lamela

Pit membrane

Pit-pair

Pit canal

(c)

Pits

(a) Tracheids   Vessel members

FIGURE 4.5. (a) Structural comparison of tracheids and vessel members, two classes of tracheary elements involved in xylem transport of water. Tracheids (left) are elongate, hollow, dead cells with highly lignified walls. The walls contain numerous pits—regions where secondary wall is absent but primary wall remains. The shape and pattern of wall pitting vary with species and organ type. Tracheids are present in all vascular plants. Vessels consist of a stack of two or more vessel members (right). Vessel members are also dead cells and are connected to one another through perforation plates—regions of the wall where a pore or hole in the wall has developed. Vessels are connected to other vessels and to tracheids through pits. Vessels are found in most angiosperms and are lacking in most gymnosperms. (b) Scanning electron micrograph of red oak wood showing stacks of individual vessel members (VM) comprising a portion of a vessel. Large pits (P) are visible on the side walls (100×). (Photograph courtesy of W. A. Côté.) (c) Diagram of a simple pit pair.

capillaries in response to a pressure or tension gradient. A simple calculation can give us some appreciation of the extraordinary efficiency of a vessel. We will calculate the driving force required to move water through the xylem at a typical velocity and compare it with the driving force that would be needed to move water through a comparable cell-to-cell pathway.

Let us begin by noting that the velocity with which water travels up the trunk of a tree depends on both the type of tree and the transpirational demand placed on the xylem. For trees with wide vessels (radii of 100 to

200 μm), peak velocities of 16 to 45 m h^{-1} (4–13 mm s^{-1}) have been measured. Trees with smaller vessels (radii 25–75 μm) have lower peak velocities, from 1 to 6 m h^{-1} (0.3–1.7 mm s^{-1}). For our calculation we will use a figure of 4 mm s^{-1} for the xylem transport velocity and 40 μm as the vessel radius. This is a rather high velocity for such a narrow vessel and so will tend to exaggerate the pressure gradient required to support water flow in the xylem. Poiseuille's law can be used to estimate the pressure gradient ($\Delta P/\Delta x$) needed to move water at this velocity (4×10^{-3} m s^{-1}) through a pipe of radius 40 μm.

By dividing Equation 3.3 by the cross-sectional area of the pore (πr^2), we find that the velocity of transport is given by

$$J_v = \frac{(\text{radius})^2}{8(\text{viscosity})} \frac{\Delta P}{\Delta x} \qquad (4.2)$$

Taking the viscosity of xylem sap to be similar to that of water (10^{-3} Pa s), we find that the pressure gradient required is 2×10^4 Pa m^{-1} (or 0.02 MPa m^{-1}). This is the pressure gradient needed to overcome the viscous drag that arises as water moves through an *ideal* vessel at a rate of 4 mm s^{-1}. *Real* vessels have rough inner wall surfaces and constrictions, such as perforation plates, at the points where vessel members meet. Tracheids with their imperforate walls offer even greater resistance to water flow. Such deviations from an ideal pipe will increase the frictional drag above that calculated from the Poiseuille equation, but since we selected a rather low value for vessel radius, our estimate of 0.02 MPa m^{-1} should be in the correct range for pressure gradients found in real trees.

Let us now compare this value of 0.02 MPa m^{-1} with the driving force that would be necessary to move water at the same velocity through a layer of *living cells*. We will ignore water movement in the apoplast pathway in this example and focus on water moving from cell to cell, crossing the plasmalemma each time. The velocity (J_v) of water flow across a membrane depends on the membrane hydraulic conductivity (Lp) and on the difference in water potential ($\Delta\psi$) across the membrane:

$$J_v = Lp\,\Delta\psi \qquad (4.3)$$

A high value for the Lp of higher plant cells is about 4×10^{-7} m s^{-1} MPa^{-1}. Thus, to move water across a membrane at 4×10^{-3} m s^{-1} would require a driving force ($\Delta\psi$) of 10^4 MPa (4×10^{-3} m s^{-1} divided by 4×10^{-7} m s^{-1} MPa^{-1}). This is the driving force needed to move the water across a single membrane. To move through a cell, water must cross at least two membranes, so the total driving force across one cell would be 2×10^4 MPa. If we estimate the cell length as 100 μm (10^{-4} m, a generous estimate), then we can calculate that the water potential gradient needed for water to move at a velocity of 4 mm s^{-1} through a layer of cells would be 2×10^4 MPa/10^{-4} m = 2×10^8 MPa m^{-1}. This is an exceedingly large driving force and serves to illustrate that water flow through the xylem is indeed very much easier than water flow across the membranes of cells. Comparing the two driving forces, we see that there is a factor of 10^{10} difference between the two pathways.

A Pressure Gradient of About 3 MPa Is Needed to Lift Water to the Top of a 100-Meter Tree

With the foregoing example in mind, let us now ask how large a pressure gradient is needed to move water up the tallest tree. Redwoods are among the tallest trees in the world, and the tallest redwood is approximately 100 m high. If we think of the stem of a tree as a long pipe, we can estimate that the pressure difference, from the ground to the top of the tree, needed to overcome the frictional drag of water moving through the xylem would be about 2 MPa (0.02 MPa m^{-1} × 100 m). Most gymnosperms, such as redwoods, have tracheids rather than vessels in their xylem. Tracheids have end walls that increase the resistance to water flow through the xylem. Thus, our estimate of 2 MPa is probably low, but it will serve as a reasonable order-of-magnitude estimate. In addition to frictional resistance, gravity must be considered. The weight of a standing column of water 100 m tall creates a pressure of 1 MPa at the bottom of the water column (100 m × 0.01 MPa m^{-1}). This pressure gradient due to gravity must be added to that required to cause water movement through the xylem. Thus we find that a pressure difference of about 2 + 1 = 3 MPa, from the base to the top branches, is needed to carry water up the tallest trees.

The Conducting Cells of the Xylem Are Adapted for the Transport of Water Under Tension

Does this pressure difference arise because the tree builds up a positive pressure at its base, or because the tree develops a tension (negative pressure) at its top? We mentioned previously that some roots develop a positive hydrostatic pressure in their xylem, a so-called root pressure. But the root pressure is typically less than 0.1 MPa and disappears when the transpiration rate is high, so it is clearly inadequate to move water up a tall tree. Rather, the water at the top of trees develops a large tension (a negative hydrostatic pressure), and it is this tension that *pulls* water up the water columns in the xylem. It is straightforward to demonstrate the existence of such tensions by puncturing the xylem with an ink-filled hypodermic needle. When this is done, the ink is quickly and visibly drawn into the xylem vessels. The origin of tension in the water will be described below.

The large tensions that develop in the xylem of trees and other plants create special problems. First, the water under tension transmits an inward force to the walls of the xylem. If the cell walls were weak or pliant, they would collapse under the influence of this tension. The secondary wall thickenings of tracheids and vessel members are adaptations that offset this tendency to collapse.

A second, and more serious, problem is that water under such tensions is in a *physically unstable state*. We mentioned in Chapter 3 that the experimentally determined breaking strength of degassed water (i.e., water that has been boiled to remove gases) is greater than 30 MPa. This value is much larger than the estimated 3 MPa tension needed to pull water up the tallest trees.

However, water in the xylem contains dissolved gases. As the tension in water increases, there is an increased tendency for the dissolved gases to escape into the vapor phase, forming a gas bubble. Once a gas bubble has formed within the water column under tension, it will expand, since gases cannot resist tensile forces. This phenomenon of bubble formation is known as **cavitation**. Cavitation of the xylem breaks the continuity of the water column and stops water transport.

Such breaks in the water columns in plants are not unusual. With the proper equipment, one can "hear" the water columns break. John Milburn from the University of Glasgow, Melvin Tyree from the University of Toronto, and others have attached acoustical detectors to the stems of plants to detect cavitation. When the plants are deprived of water, clicks are detected. The clicks correspond to the formation and rapid expansion of vapor bubbles in the xylem, resulting in high-frequency acoustic shock waves through the rest of the plant.

These breaks in xylem water continuity, if not repaired, would be disastrous to the plant. By blocking the main transport pathway of water to the leaves, such bubbles would cause the dehydration and death of the leaves.

The impact of xylem cavitation on the plant is minimized by several means. Because the tracheary elements in the xylem are interconnected, one gas bubble might, in principle, expand to fill the whole network. In practice, gas bubbles do not spread far because they are stopped at the pitted walls between overlapping tracheids and vessels. Their expansion is stopped because the gas cannot squeeze through the small pores of the pit membranes, an effect again due to the surface tension of water. Since the capillaries in the xylem are interconnected, one gas bubble does not completely stop water flow through a vessel or column of tracheids. Instead, water can detour around the blocked point by traveling through neighboring, connected conduits (Fig. 4.6). Thus the presence of imperforate walls, besides increasing the resistance to water flow, also restricts cavitation.

Plants also have a means of getting rid of gas bubbles in the xylem. At night, when transpiration is low, the tension in the xylem decreases and the water vapor and gases may simply go back into the solution of the xylem and the free space. Moreover, as we have seen, some plants develop positive pressures (root pressures) in the xylem. Such pressures shrink the gas bubble and cause the gases to redissolve. Finally, many plants have secondary growth in which new xylem is formed each year. The new xylem becomes functional before the old xylem ceases to function because of occlusion by gas bubbles or by substances secreted by the plant.

The Source of Tension in the Xylem Is the Negative Pressure That Develops in the Leaf Cell Walls When Water Evaporates

We mentioned earlier that an actively transpiring leaf may exchange all of its water in the course of 1 hour. Such a leaf would very quickly wilt and die if the water lost by evaporation were not replaced. In the intact plant, water is brought to the leaves via the **xylem** of the leaf vascular bundle, which branches and ramifies into a very fine and sometimes intricate network of **veins** throughout the leaf (Fig. 4.7). So finely divided does the **venation pattern** become that in a typical leaf most cells are within 0.5 mm of a minor vein. From the xylem, water is drawn into the cells of the leaf and along the cell walls.

It is at the surface of the cell walls in the leaf that the negative pressure which causes water to move up through

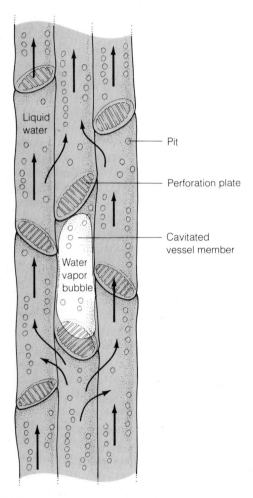

FIGURE 4.6. Detours around a vapor-locked vessel member. Tracheids and vessels constitute multiple, parallel, interconnected pathways for water movement. Cavitation in this example blocks water movement within the cavitated vessel member. However, because these water conduits are interconnected through wall pits, cavitation of a vessel or tracheid does not completely stop water movement in the cell file. Water can detour around the block by moving through adjacent tracheary elements. The spread of the vapor bubble throughout the xylem is eventually stopped by an imperforate end wall.

Labels on figure: Liquid water / Pit / Perforation plate / Cavitated vessel member / Water vapor bubble

(a)

(b)

FIGURE 4.7. Intricate venation brings the xylem water supply close to every cell in the leaf. (a) A leaf of *Coleus blumei* which has been "cleared" by treating with bleach, and then stained with a lignin-specific dye which stains the vascular system. The image shown is a negative. The petiole, midrib, secondary veins that branch from the midrib, and numerous smaller tertiary veins, quarternary veins, etc., appear white in the photograph (1.0 ×). (Photograph from W. A. Russin and R. Evert.) (b) A paradermal (parallel to the surface) section through a soybean leaf showing a vein ending (84 ×). Because of the extensive ramification of the leaf vascular system, no leaf mesophyll cell is more than a fraction of a millimeter from some part of the xylem. (Photograph ©Jack Bostrack.)

the xylem is developed. The situation is analogous to that in the soil. The cell wall acts like a very fine capillary wick soaked with water. Water adheres to the cellulose microfibrils and other hydrophilic components of the wall. The mesophyll cells within the leaf are in direct contact with the atmosphere through an extensive system of intercellular air spaces. Initially, water evaporates from a thin film lining these air spaces. As water is lost to the air, the surface of the remaining water is drawn into the interstices of the cell wall, that is, microscopic cracks and crevices between cells and between fibrils in the wall (see Fig. 4.8). The air-water surface becomes curved, forming microscopic menisci, and the surface tension induces a tension, or negative pressure, in the water. As more water is removed from the wall, the water surface develops more acute menisci, and the pressure of the water becomes more negative (see Eq. 4.1). Thus, the motive force for xylem transport is generated at the air-water interfaces within the leaf.

Water Vapor Moves from the Leaf to the Atmosphere by Diffusion Through Stomata

When water has evaporated from the cell surface into the intercellular air space, further movement of the water out of the leaf is primarily by diffusion. The waxy cuticle covering the leaf surface is a very effective barrier to water movement. It has been estimated that only about 5% of the water lost from leaves escapes through the cuticle. Almost all of the water loss from typical leaves occurs by diffusion of water vapor through the tiny pores of the stomatal apparatus, which are usually most abundant on the lower surface of the leaf. In Figure 4.9 we can see that the pathway of water vapor loss to the atmosphere can be arbitrarily divided into three regions: the air space inside the leaf, the stomatal pore, and the layer of still air next to the surface of the leaf, the so-called boundary layer. Movement of water vapor through these regions is predominantly diffusional and so is controlled by the *concentration gradient* of water vapor. In principle, water vapor loss could also be described in terms of water potential gradients, but these equations become nonlinear because of the logarithmic relationship between water vapor concentration and water potential (see Eq. 4.4).

In Chapter 3 we pointed out that diffusion in liquids is so slow that it is effective only at cellular dimensions. How long would it take for water vapor to diffuse from the cell wall surfaces inside a leaf to the outside atmosphere? To answer this question, we will again make use of Equation 3.2 relating distance to the time required for the concentration of a diffusing substance to reach a certain value, say half of the starting concentration. The distance through which water vapor must diffuse from the surface inside the leaf to the outside turbulent air is 1 mm.

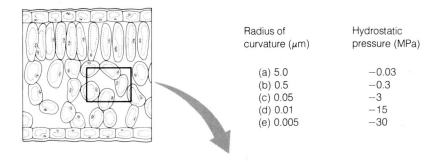

Radius of curvature (µm)	Hydrostatic pressure (MPa)
(a) 5.0	−0.03
(b) 0.5	−0.3
(c) 0.05	−3
(d) 0.01	−15
(e) 0.005	−30

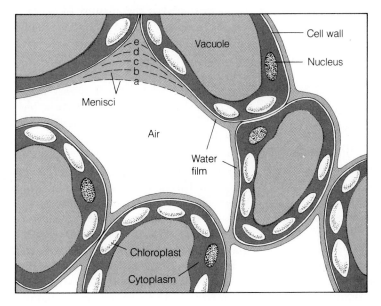

FIGURE 4.8. The origin of tensions or negative pressures in cell wall water of the leaf. As water evaporates from the surface film covering the cell walls of the mesophyll, water withdraws farther into the interstices between neighboring cells, and surface tension effects result in a negative pressure in the liquid phase. As the water potential decreases, liquid water remains only in the smaller cracks and crevices in and between cell walls, and the radius of curvature of the meniscus progressively decreases. As the radius of curvature decreases (dashed lines), the pressure decreases (becomes more negative), as calculated from the equation $P = -2T/r$, where T is the surface tension of water and r is the radius.

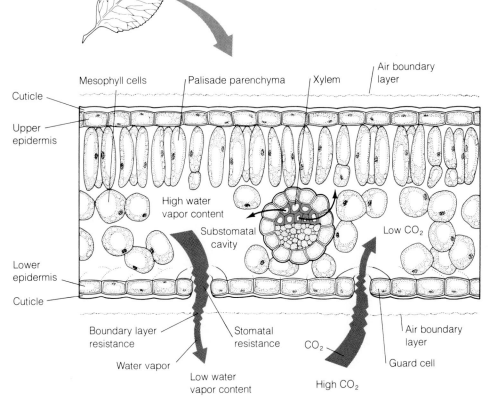

FIGURE 4.9. The water pathway through the leaf. Water is drawn from the xylem into the cell walls of the mesophyll, where it evaporates into the air spaces within the leaf. By diffusion, water vapor then moves through the leaf air space, through the stomatal pore, and across the boundary layer of still air that adheres to the outer leaf surface. CO_2 also diffuses into the leaf through stomata along a concentration gradient.

The diffusion coefficient of water vapor in air is 2.4×10^{-5} m^2 s^{-1}. Therefore the diffusion time is

$$t_{1/2} = \frac{(\text{distance})^2}{\text{diffusion coefficient}} = \frac{(10^{-3}\ \text{m})^2}{2.4 \times 10^{-5}\ \text{m}^2\ \text{s}^{-1}}$$
$$= 0.042\ \text{s}$$

This time is so short because diffusion is very much more rapid in a gas than in a liquid. Thus we see that diffusion is adequate to move water vapor through the gas phase of the leaf.

Loss of water from the leaf depends on two factors: (1) the gradient in water vapor concentration from the leaf air spaces to the external air and (2) the diffusional resistance of the pathway to the external air. We will examine each of these factors in turn.

The Driving Force for Water Loss from the Leaf to the Atmosphere Is the Absolute Concentration Gradient of Water Vapor

To evaluate the **gradient in water vapor concentration**, we need to know the water vapor concentration in the outside air and in the air spaces of the leaf. The first parameter can be readily measured, but the water concentration in the intercellular space of the leaf is more difficult to assess. We can estimate this value by assuming that the air space in the leaf is close to water potential equilibrium with the cell wall surfaces. This approximation is not strictly true because water is diffusing away from these surfaces. However, it introduces little error because the major resistance to vapor loss is at the stomatal pore. Moreover, the air space volume inside the leaf is rather small, whereas the wet surface from which water evaporates is comparatively large. Air space volume is about 5% of the total leaf volume for pine needles, 10% for corn leaves, 30% for barley, and 40% for tobacco leaves. The internal surface area from which water evaporates may be from 7 to 30 times the external leaf area. This high ratio of surface area to volume makes for rapid vapor equilibration inside the leaf.

Making the assumption of equilibrium, we can calculate the water vapor concentration in the leaf air spaces if we know (1) the water potential of the leaf (which is the same as the water potential of the wall surfaces from which water is evaporating) and (2) the leaf temperature. Let us take as an example a leaf with a water potential of -1.0 MPa. To reach vapor equilibrium, water evaporates from the cell wall surfaces until the water potential of the air inside the leaf equals the water potential of the leaf. The water potential of the air is given by

$$\psi = \frac{RT}{\overline{V}_{\text{w}}} \ln(\text{RH}) \tag{4.4}$$

where RH is the relative humidity of the air. **Relative humidity** is a measure of the water vapor concentration in air, expressed as a fraction of the **saturation water vapor concentration**. If we use the symbols c_{wv} for the actual water vapor concentration and c_{wv}(sat.) for the water vapor concentration of air saturated with water vapor at $\psi = 0$ MPa, then relative humidity is given by

$$\text{RH} = \frac{c_{\text{wv}}}{c_{\text{wv}}(\text{sat.})} \tag{4.5}$$

c_{wv} is expressed as the quantity of water per volume of air (e.g., mol m^{-3}) and RH is between 0 and 1. RH multiplied by 100 is the percent relative humidity.

Table 4.2 shows values of the relative humidity of air as a function of water potential calculated from Equation 4.4. From this table it is easy to see that the air spaces of living leaves must have a high RH, nearly 1 (100%), and thus a high water potential. Moreover, outside air, with an RH of say 0.5 (50%), has a remarkably low water potential.

To convert from RH to c_{wv}, we need to know c_{wv}(sat.). This quantity is *strongly dependent on temperature*. As the air temperature rises, the water-holding capacity of air increases sharply, as shown in Figure 4.10. In the range of 10 to 35°C, an increase in air temperature of 12°C doubles the water vapor concentration of saturated air. This is an important observation. If our leaf with a water potential of -1.0 MPa warms up abruptly from 20°C to 32°C, the relative humidity in the leaf air space drops abruptly from 99.3% to almost 50%. This drop in RH occurs because the water-holding capacity of the air, c_{wv}(sat.), doubles. In response, water evaporates in the air space until RH returns to the value of 99.3% and the air is again in water potential equilibrium with the leaf. As a consequence, c_{wv} of the leaf air space

TABLE 4.2. Relation between relative humidity and water potential of air, calculated from Equation 4.4*

Relative humidity	Water potential (MPa)
1.0	0.0
0.999	−0.14
0.995	−0.68
0.990	−1.36
0.980	−2.73
0.950	−6.92
0.900	−14.22
0.750	−38.84
0.500	−93.57
0.200	−217.27
0.100	−310.85

*Assuming a temperature of 20°C (293 °K); at which $RT/\overline{V}_{\text{w}}$ = 135 MPa.

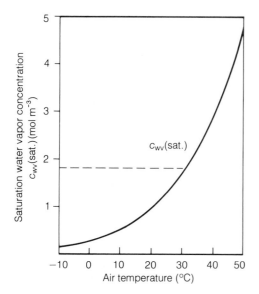

FIGURE 4.10. Concentration of water vapor in saturated air as a function of air temperature. As its temperature increases, the air holds more water.

increases from 0.95 to 1.87 mol m^{-3}, which makes for a steeper concentration gradient driving the diffusional loss of water from the leaf.

Table 4.3 illustrates what we have said about RH, c_{wv}, and water potential at various points in the transpiration pathway. We see that c_{wv} decreases along each step of the pathway from the cell wall surface to the bulk air outside the leaf. Note that it is possible for RH to increase along part of this pathway because the external air temperature may be lower than the temperature of the leaf. The important point is that the driving force for water loss from the leaf is the *absolute* concentration gradient (gradient in c_{wv}), not the relative humidity.

TABLE 4.3. Representative values for relative humidity, absolute water vapor concentration, and water potential for four points in the pathway of water loss from a leaf.*

Location	Relative humidity	Water vapor Concentration (mol m^{-3})	Potential (MPa)
Inner air spaces (25°C)	0.99	1.27	−3.18
Just inside stomatal pore (25°C)	0.95	1.21	−16.2
Just outside stomatal pore (25°C)	0.47	0.60	−239
Bulk air (20°C)	0.50	0.50	−215

*Refer to Figure 4.9.
Adapted from P. S. Nobel, *Biophysical Plant Physiology and Ecology*, W. H. Freeman, San Francisco, 1983, p. 413.

Transpiration Is Also Regulated by the Leaf Stomatal Conductance and the Boundary Layer Resistance

After the gradient in water vapor concentration, the second important factor governing water loss from the leaf is the **diffusional resistance** of the transpiration pathway. There are two important resistances in this pathway, both of them variable. The first, and more important, is the resistance associated with diffusion through the stomatal pore. The second is the resistance due to the layer of unstirred air next to the leaf surface through which water vapor must diffuse to reach the turbulent air of the atmosphere. This second resistance is usually called the leaf **boundary layer resistance**.

Because the cuticle covering the leaf is so impermeable to water, most of the leaf transpiration is restricted to loss of water vapor by diffusion through the stomatal pore. Water loss through the stomatal pore will be discussed in the next section. For now, it is important to realize that the microscopic stomatal pores provide a *low-resistance pathway* for diffusional movement of gases across the epidermis and cuticle. That is, they increase the **stomatal conductance** of the leaf. Changes in stomatal conductance are important in regulating water loss by the plant and controlling the rate of carbon dioxide uptake necessary for sustained CO_2 fixation during photosynthesis.

Once water vapor has diffused through the stomatal pore, the last step is diffusion across the **boundary layer** on the surface of the leaf. The boundary layer is a thin film of still air hugging the surface of the leaf, and its resistance to water vapor diffusion is proportional to its thickness. The boundary layer thickness is determined primarily by wind speed. When the air surrounding the leaf is very still, the layer of unstirred air on the surface of the leaf may be so thick that it is the primary deterrent to water vapor loss from the leaf. As shown in Figure 4.11, increasing the stomatal aperture under such conditions has little effect on transpiration rate (although closing the stomata completely will still reduce transpiration). When wind velocity is high, the drag of the moving air reduces the thickness of the boundary layer at the leaf surface and so effectively reduces the resistance of this layer. Under such conditions, the stomatal resistance has the largest control over water loss by the leaf (Bange, 1953). This situation is more representative of conditions in this field, where even on windless days there are small local convection currents that reduce boundary layer resistances.

Various anatomical and morphological aspects of the leaf can alter the effect of wind on the thickness of the boundary layer. Hairs on the surface of leaves can serve as microscopic windbreaks. Some plants have sunken stomata that provide a dead-volume region, outside the

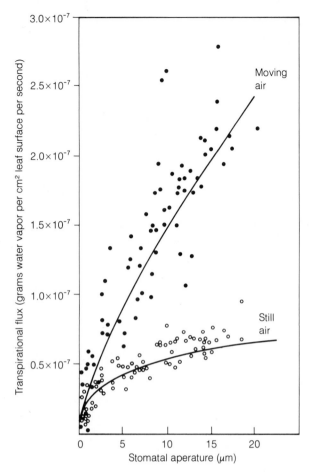

FIGURE 4.11. Dependence of transpiration flux on the stomatal aperture of zebra plant (*Zebrina pendula*) in still air and in moving air. The boundary layer is larger and more rate-limiting in still air than in moving air. As a result, the stomatal aperture has less control over transpiration in still air. (From Bange, 1953.)

stomatal pore, protected from wind. The size and shape of leaves also influence the way the wind sweeps across the leaf surface. Although these and other factors may influence the boundary layer, they are not characteristics that can be altered on an hour-to-hour or even day-to-day basis. For such control, the guard cells play the most crucial role by regulating the stomatal conductance of the leaf.

Stomatal Control Serves to Maximize Photosynthesis While Minimizing Transpiration

Land plants are faced with competing demands. The atmosphere is so far from water saturation that the plant is in danger of lethal dehydration. The cuticle covering the plant surfaces exposed to the atmosphere serves as an effective barrier to water loss and thus protects the plant from desiccation. However, a complete barrier to water vapor loss would also block the exchange of O_2 and CO_2

that is essential for respiration and photosynthesis. Plants often need to discriminate between different substances present in their environment, a task that can be accomplished by use of semipermeable membranes. However, the option of preventing water loss while allowing CO_2 to freely permeate the cuticle has not been available because there are no natural substances that could be included in the cuticle to discriminate between water and CO_2. Because of this limitation, plants cannot prevent outward diffusion of water without simultaneously excluding CO_2 from the leaf.

The functional solution for this dilemma is the *temporal* regulation of stomatal apertures. At night, when there is no photosynthesis and thus no demand for CO_2 inside the leaf, stomatal apertures are kept small, preventing unnecessary loss of water. On a sunny morning, when the supply of water is abundant and the solar radiation incident on the leaf favors high photosynthetic activity, the demand for CO_2 inside the leaf is large, and the stomatal pores are wide open, decreasing the stomatal resistance to CO_2 diffusion. Water loss by transpiration is also substantial under those conditions, but since the water supply is plentiful it is advantageous for the plant to trade water for the products of photosynthesis, which are essential for growth and reproduction. On the other hand, on another sunny morning later in the season, when the water in the soil has been depleted, the stomata will open less or even remain closed. By keeping its stomata closed, the plant avoids potentially lethal dehydration. The absolute concentration gradient of water vapor and the boundary layer resistance are physical parameters not readily amenable to biological control. Stomatal resistance, the third factor controlling leaf transpiration, is under biological control and is the plant's primary means of regulating the exchange of gases across the leaf surfaces. This biological control is exerted by a pair of specialized epidermal cells, the **guard cells**, which surround the stomatal pore.

The Radial Orientation of Cellulose Microfibrils in Guard Cell Walls Is Required for Pore Opening

Guard cells can be found in leaves of all higher plants and are also present in organs from more primitive plants such as the liverworts and the mosses (Ziegler, 1987). Guard cells show considerable morphological diversity, but in higher plants we can distinguish two main types: one is typical of grasses and a few other monocots such as the palms; the other is found in all dicots, in many monocots, and in mosses, ferns, and gymnosperms (Fig. 4.12). In grasses (Fig. 4.12a), guard cells have a characteristic dumbbell shape, with bulbous ends. The pore proper is a long slit located between the two "handles" of the dumbbells. These guard cells are always flanked by a pair

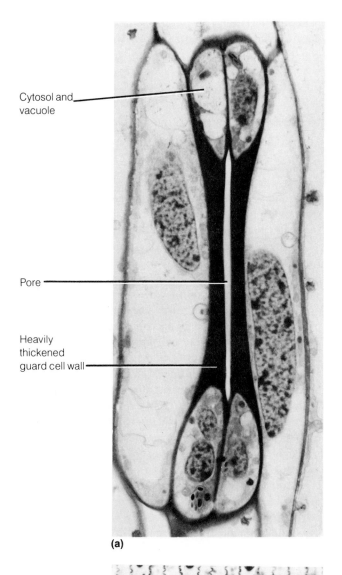

Cytosol and vacuole

Pore

Heavily thickened guard cell wall

(a)

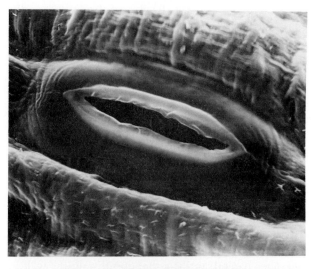

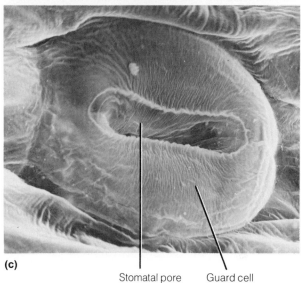

(c)

Stomatal pore   Guard cell

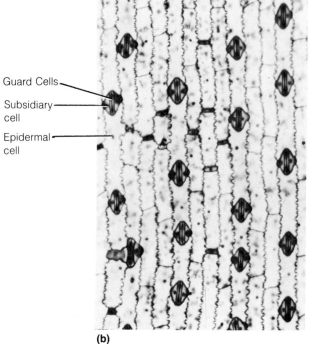

Guard Cells

Subsidiary cell

Epidermal cell

(b)

FIGURE 4.12. Electron micrographs of stomata. (a) Stomata from a grass (2750 ×). The bulbous ends of each guard cell show their cytosolic content and are joined by the heavily thickened walls. The stomatal pore separates the two midportions of the guard cells. (From Palevitz, 1981.) (b) A thin section of the epidermis from a corn leaf, showing the arrangement of the stomatal complexes interspaced between ordinary epidermal cells (290 ×). Each pair of guard cells encloses a pore and is surrounded by two subsidiary cells. (From Srivastava and Singh, 1972.) (c) Scanning electron micrographs of the onion epidermis. The top panel shows the outside surface of the leaf, with a stomatal pore inserted in the cuticle (1750 ×). The bottom panel shows a pair of guard cells facing the stomatal cavity, toward the inside of the leaf (1750 ×). (Top, from Zeiger and Hepler 1976. Bottom, from E. Zeiger and N. Burnstein.)

of differentiated epidermal cells called **subsidiary cells,** which help the guard cells to control the stomatal apertures (Fig. 4.12b). The pore, guard cells, and subsidiary cells are collectively called the **stomatal complex.** In dicot plants and non-grass monocots, kidney-shaped guard cells have an elliptical contour with the pore at its center (Fig. 4.12c). Although subsidiary cells are not uncommon in species with kidney-shaped stomata, they are often absent, and the guard cells are surrounded by ordinary epidermal cells.

A distinctive feature of the specialized organization of the guard cells is the structure of their walls. Portions of these walls are substantially thickened (Fig. 4.13) and may be up to 5 μm across, in contrast to the 1–2 μm typical of epidermal cells. In kidney-shaped guard cells, a differential thickening pattern results in very thick inner and outer (lateral) walls, a thin dorsal wall (the wall in contact with epidermal cells), and a somewhat thickened ventral (pore) wall (Fig. 4.13). The portions of the walls facing the atmosphere extend into well-developed ledges, which form the pore proper. This thickening pattern is an essential element of guard cell mechanics associated with the alignment of their **cellulose microfibrils,** which reinforce the walls and are an important determinant of cell shape (see Chapter 1). In ordinary cells with a cylindrical shape, cellulose microfibrils are oriented transverse to the long axis of the cell. As a result, the cell expands in the direction of its long axis, since the cellulose reinforcement offers the least resistance at right angles to its orientation. In guard cells the microfibril organization is different. Kidney-shaped guard cells have cellulose microfibrils fanning out radially from the pore (Fig. 4.14a). Thus the cell girth is reinforced like a steel-belted radial tire, and the guard cell curves outward during stomatal opening (Sharpe et al., 1987). In the grasses, the dumbbell-shaped guard cells function like beams with inflatable ends. As the bulbous ends of the cells increase in volume and swell, the beams are separated from each other and the slit between them widens (Fig. 4.14b).

Stomatal Opening Is Caused by an Increase in the Turgor Pressure of the Guard Cells

Guard cells function as multisensory hydraulic valves. Environmental factors such as light intensity and quality, temperature, relative humidity, and intracellular CO_2 concentrations are sensed by guard cells, and these signals are integrated into well-defined stomatal responses. If leaves kept in the dark are suddenly illuminated, the light stimulus is perceived by the guard cells as an opening signal, triggering a series of responses that result in the opening of the stomatal pore. The early aspects of this process involve ion uptake and other metabolic changes in the guard cells, which will be discussed in detail in Chapter 6. Here we will note the effect of increases in osmotic pressure, π, resulting from ion uptake and from biosynthesis of organic molecules in the guard cells. Water relations in guard cells follow the same rules as in other cells. As π increases, the water potential becomes more negative and water moves into the guard cells. As water enters the cell, turgor pressure increases. However, because of the elastic properties of their walls, guard cells can reversibly increase their volume by 40% to 100%, depending on the species. The deformation of the guard cell walls imposed by this increase in volume is a central aspect of stomatal movements.

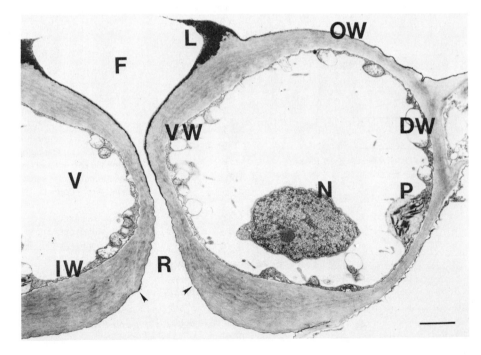

FIGURE 4.13. Electron micrograph showing a pair of guard cells from the dicot *Nicotiana tabacum,* the tobacco plant. The section was made perpendicular to the main surface of the leaf. The pore, facing the atmosphere, is on the top; the bottom faces the substomatal cavity inside the leaf. L, Ledges flanking the stomatal forechamber, F; V, vacuole; N, nucleus; P, plastid; OW, IW, VW, and DW, outer, inner, ventral, and dorsal walls. Note the uneven thickening pattern of the walls, which determines the asymmetric deformation of the guard cells when their volume increases during stomatal opening. Bar = 2μm (From Sack, 1987.)

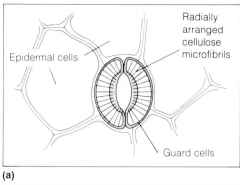

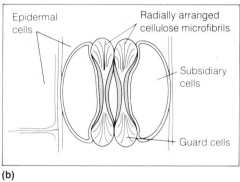

FIGURE 4.14. The radial alignment of the cellulose microfibrils in guard cells and epidermal cells of a kidney-shaped stoma (a) and a grass-like stoma (b). (From Meidner and Mansfield, 1968.)

The Transpiration Ratio Is a Measure of the Effectiveness of the Stomata in Maximizing Photosynthesis While Minimizing Water Loss

The effectiveness of plants in moderating water loss while allowing sufficient CO_2 uptake for photosynthesis can be assessed by a parameter called the **transpiration ratio**. This is defined as the amount of water transpired by the plant divided by the amount of carbon dioxide assimilated by photosynthesis:

$$\text{Transpiration ratio} = \frac{\text{moles of } H_2O \text{ transpired}}{\text{moles of } CO_2 \text{ fixed}} \quad (4.6)$$

For typical plants in which the first stable product of carbon fixation is a three-carbon compound (called C3 plants; see Chapter 9), about 500 molecules of water are lost for every molecule of CO_2 fixed by photosynthesis, giving a transpiration ratio of 500.

The large ratio of H_2O efflux to CO_2 influx depends on two factors. First, the concentration gradient driving water loss is about 50 times larger than that driving CO_2 influx. In large part, this is due to the low concentration of CO_2 in air ($\sim 0.03\%$) and the relatively high concentration of water vapor within the leaf air spaces. Second, CO_2 diffuses more slowly through air than water does

(the CO_2 molecule is larger than H_2O and has a smaller diffusion coefficient). Furthermore, CO_2 has a longer diffusion path, because it must cross the plasma membrane, the cytoplasm, and the chloroplast envelope before it is assimilated in the chloroplast. These membranes add substantially to the resistance of the CO_2 diffusion pathway.

Plants with C4 photosynthesis (i.e., in which a four-carbon compound is the first stable product of photosynthesis; see Chapter 9) generally transpire less water per CO_2 fixed; a typical transpiration ratio for C4 plants is about 250. Desert-adapted plants with CAM (Crassulacean acid metabolism) photosynthesis, in which CO_2 is initially fixed into four-carbon organic acids at night, have even lower transpiration ratios; values of about 50 are not unusual.

Sometimes the reciprocal of the transpiration ratio, called the **water use efficiency**, is cited. Plants with a transpiration ratio of 500 would have a water use efficiency of $1/500$ or 0.002.

Overview: The Soil-Plant-Atmosphere Continuum

We have seen that movement of water from the soil through the plant to the atmosphere involves different mechanisms of transport. In the soil and the xylem, water moves by bulk flow in response to a pressure gradient. In the vapor phase, movement is primarily by diffusion, at least until water reaches the outside air, where convection (a form of bulk flow) becomes dominant. When water transport occurs across membranes, the driving force is the water potential gradient across the membrane. Such osmotic flow occurs when cells absorb water and when roots transport water from the soil to the xylem.

One point deserves emphasis: *Water flow is a passive process. That is, water moves in response to physical forces.* There are no metabolic "pumps" (i.e., reactions driven by ATP hydrolysis) that push water from one place to another. As a consequence, water tends to move from regions of high water potential such as the soil and the root to regions of low water potential such as the leaves and the air, although there are exceptions to this rule. This phenomenon is illustrated in Figure 4.15, which shows representative values for water potential and its components at various points along the water transport pathway. From the soil to the leaves there is a consistent decrease in water potential. Also, the components of water potential can be quite different at different parts of the pathway. For example, inside the leaf cells, such as the mesophyll, the water potential is approximately the same as that in the neighboring xylem, yet the components of ψ are quite different. The dominant component of ψ in the

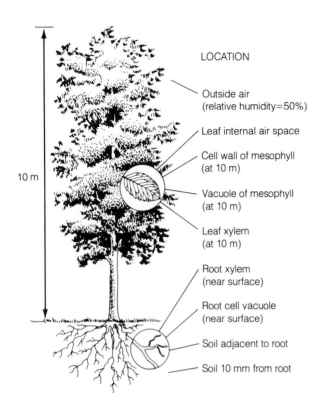

LOCATION

- Outside air (relative humidity=50%)
- Leaf internal air space
- Cell wall of mesophyll (at 10 m)
- Vacuole of mesophyll (at 10 m)
- Leaf xylem (at 10 m)
- Root xylem (near surface)
- Root cell vacuole (near surface)
- Soil adjacent to root
- Soil 10 mm from root

10 m

Water Potential and Its Components (in MPa)				
ψ	P	$-\pi$	$\rho_w gh$	$\dfrac{RT}{\overline{V}_w}\ln(RH)$
−95.1			0.1	−95.2
−0.8			0.1	−0.9
−0.8	−0.7	−0.3	0.1	
−0.8	0.2	−1.1	0.1	
−0.8	−0.8	−0.1	0.1	
−0.6	−0.5	−0.1	0.0	
−0.6	0.5	−1.1	0.0	
−0.5	−0.4	−0.1	0.0	
−0.3	−0.2	−0.1	0.0	

FIGURE 4.15. Representative overview of water potential and its components at various points in the transport pathway from the soil through the plant to the atmosphere. Water potential ψ can be measured through this continuum, but the components vary. In the liquid part of the pathway, pressure (P), osmotic pressure (π), and gravity ($\rho_w gh$) determine ψ, whereas in the air, only the effects of gravity and relative humidity—$RT\ln(RH)/\overline{V}_w$—are important. Note that although the water potential is the same in the vacuole of a mesophyll cell and in the surrounding cell wall, the components of ψ can differ greatly (e.g., in this example P inside the mesophyll cell is 0.2 MPa and outside the cells is −0.7 MPa). (After Nobel, 1983.)

xylem is the negative pressure (P), whereas in the leaf cell P is generally positive. This large difference in P occurs across the plasmalemma of the leaf cells. Within the leaf cells, water potential is reduced by a high concentration of dissolved solutes (high π). Finally, in the vapor phase we cannot dissect ψ into π and P; instead, the relative humidity is the significant factor determining ψ.

Summary

Water is the essential medium of life. Land plants are faced with potentially lethal desiccation by water loss to the atmosphere. This problem is aggravated by the large surface area of leaves, their high radiant energy gain, and the need for leaves to have an open pathway for CO_2 uptake. Thus there is a conflict between the need for water conservation and the need for CO_2 assimilation. The need to resolve this vital conflict determines much of the structure of land plants: an extensive root system to extract water from the soil, a low-resistance pathway through the xylem vessels and tracheids to bring water to the leaves, a hydrophobic cuticle covering the surfaces of the plant to reduce evaporation, microscopic stomata on the leaf surface to allow gas exchange, and guard cells to regulate the diameter (and diffusive resistance) of the stomatal aperture.

The result is an organism that transports water from the soil to the atmosphere purely in response to physical forces; no energy is expended directly by the plant to translocate water, although development and maintenance of the structures needed for efficient and controlled water transport require considerable energy input.

The mechanism of water transport includes diffusion, bulk flow, and osmosis, and each of these types of transport is coupled to a different driving force. Water in the plant can be considered as a continuous hydraulic system, connecting the water in the soil with the water vapor in the atmosphere. Overall, water moves from regions of high water potential to regions of low water potential. This chapter has been concerned with understanding how water moves through various parts of this continuum and how plants regulate water flow.

GENERAL READING

Boyer, J. S. (1985) Water transport. *Annu. Rev. Plant Physiol.* 36: 473–516.

Dainty, J. (1976) Water relations of plant cells. In: *Encyclopedia of Plant Physiology*, New Series, Vol. 2, Part A, pp. 12–35. Luttge, U., and Pitman, M., eds. Springer-Verlag, Berlin.

Hillel, D. (1971) *Soil and Water: Physical Principles and Processes*. Academic Press, New York.

Kozlowski, T. T., ed. (1968) *Water Deficits and Plant Growth*, Vols. I and II. Academic Press, New York.

Kozlowski, T. T., ed. (1972) *Water Deficits and Plant Growth*, Vol. III. Academic Press, New York.

Kramer, P. J. (1984) *Water Relations of Plants*. Academic Press, New York.

Milburn, J. A. (1979) *Water Flow in Plants*. Longman, London.

Nobel, P. S. (1983) *Biophysical Plant Physiology and Ecology*. W. H. Freeman, San Francisco.

Turner, N. C., and Kramer, P. J., eds. (1980) *Adaptation of Plants to Water and High Temperature Stress*. Wiley, New York.

Tyree, M. T., and Jarvis, P. G. (1982) Water in tissues and cells. In: *Encyclopedia of Plant Physiology*, New Series, Vol. 12B: *Physiological Plant Ecology II: Water Relations and Carbon Assimilation*, pp. 35–77. Springer-Verlag, Berlin. (This volume also contains many other review articles of interest, including one by J. P. Passioura.)

Weatherly, P. E. (1982) Water uptake and flow in roots. In: *Encyclopedia of Plant Physiology*, New Series, Vol. 12B: *Physiological Plant Ecology*, pp. 79–109. Lange, O. L., Nobel, P. S., Osmond, C. B., and Ziegler, H., eds. Springer-Verlag, Berlin.

Zeiger, E. (1983) The biology of stomatal guard cells. *Annu. Rev. Plant Physiol.* 34: 441–475.

Zeiger, E., Farquhar, G., and Cowan, I., eds. (1987) *Stomatal Function*. Stanford University Press, Stanford, Calif.

Ziegler, H. (1987) The evolution of stomata. In: *Stomatal Function*, pp. 29–57. Zeiger, E., Farquhar, G., and Cowan, I., eds. Stanford University Press, Stanford, Calif.

Zimmermann, M. H. (1983) *Xylem Structure and the Ascent of Sap*. Springer-Verlag, Berlin.

CHAPTER REFERENCES

Bange, G. G. J. (1953) On the quantitative explanation of stomatal transpiration. *Acta Bot. Neerlandia* 2: 255–296.

Cosgrove, D. J. (1987) Wall relaxation in growing stems: comparison of four species and assessment of measurement techniques. *Planta* 171: 266–278.

Hsiao, T. C., and Acevedo, E. (1974) Plant responses to water deficits, water-use efficiency, and drought resistance. *Agric. Meteorol.* 14: 59–84.

Meidner, H. and Mansfield, D. (1968) *Stomatal Physiology*. McGraw-Hill, London.

Palevitz, B. A. (1981) The structure and development of guard cells. In: *Stomatal Physiology*, T. A. Mansfield and P. G. Jarvis, eds. pp. 1–23. Cambridge University Press, Cambridge.

Sack, F. D. The development and structure of stomata. In *Stomatal Function*, pp. 59–89. Zeiger, E., Farquhar, G. and Cowan, I., eds. Stanford University Press, Stanford, Calif.

Sharpe, P. J. H., Wu, H., and Spence, R. D. (1987) Stomatal mechanics. In: *Stomatal Function*, pp. 91–114. Zeiger, E., Farquhar, G., and Cowan, I., eds. Stanford University Press, Stanford, Calif.

Srivastava, L. and Singh, A. P. (1972) Stomatal structure in corn leaves. *J. Ultrastructure Research* 39: 345–363.

Zeiger, E. and Hepler, P. K. (1976) Production of guard cell protoplasts from onion and tobacco. *Plant Physio.* 58: 492–498.

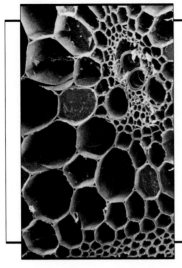

CHAPTER 5

Mineral Nutrition

MINERAL NUTRIENTS ARE CONTINUALLY cycled through organisms and their environment. Uptake and assimilation by plants represent the key step in the incorporation of minerals into the biosphere (Delwiche, 1983). It has often been stated that plants act as the "miners" of the earth's crust, supplying other organisms with inorganic nutrients essential for their growth and proliferation. Mineral nutrients enter the biosphere at the plant's root system, where the large surface area of roots and their ability to absorb mineral nutrients at low concentrations in the soil make mineral absorption a very effective process. Following absorption at the roots, the mineral nutrients are translocated to the various parts of the plant for utilization in important biological functions. Other organisms, such as mycorrhizal fungi, often work in conjunction with the root system to aid in this nutrient acquisition process.

Understanding how plants acquire and assimilate mineral nutrients essential for their growth and development is central to the study of plant physiology. This body of knowledge, called **mineral nutrition**, has had a major impact on the development of modern agriculture and has also made major contributions to our understanding of how plants function. Because of the complex nature of the mineral nutrition of higher plants, studies in this area involve not only plant physiologists but also biochemists, inorganic chemists, soil scientists, microbiologists, and ecologists.

This chapter discusses the acquisition of mineral nutrients by plants. We will examine the nutritional needs of plants, the various symptoms of nutritional deficiencies, and the use of fertilizers to ensure proper plant nutrition. Further aspects of nutrient assimilation will be discussed in Chapter 12.

The Plant Root System and Its Interaction with the Soil

The ability of plants to absorb both water and mineral nutrients from the soil is related to their capacity to develop an extensive root system. This was clearly illustrated by a detailed study of winter rye by H. J. Dittmer in 1937. When the root system of a rye plant that had grown for 16 weeks was examined, it was found to consist of 13×10^6 root axes and lateral roots. Of the total surface area of roots, root hairs contribute about 67% of the total root surface area (see Fig. 4.1). The volume of the soil in contact with the roots is also important for the uptake of mineral nutrients. In studies of wheat by J. E. Weaver in 1926, it was found that in some environments the roots would grow to depths of 100 to 200 cm and extend laterally to distances of 30 to 60 cm. Maximum depths of about 50 m have been reported for *Prosopis* trees (from the Leguminosae family). The total length of the root systems of trees planted 0.5 m apart has been estimated to reach 12 to 18 km.

Below the Soil Surface, Plant Roots Are Continuously Growing and Decaying in Response to Changes in the Soil Environment

An important aspect of root growth and development is that it is dynamic and highly dependent on the soil environment. Plant roots grow continuously. Their proliferation, however, depends on the availability of water and minerals in their microenvironment. In regions of the soil that are rich in mineral nutrients and moisture, root growth and proliferation are extensive. In soil regions poor in nutrients, proliferation is inhibited or the root system may die back.

The *form* of the root system is markedly dependent on the plant species. In monocots, root development starts with the emergence of three to six **primary** (or seminal) root axes from the germinating seed. With further growth, new adventitious roots called **nodal** or brace roots are formed. With time, the primary and nodal root axes grow and branch extensively to form a complex

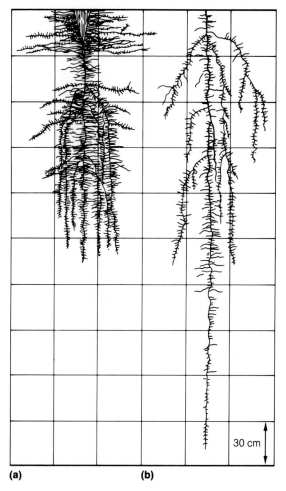

(a)   (b)

FIGURE 5.2. Taproot system of two dicots, sugar beet (left) and alfalfa (right). The sugar beet root system is typical of 5 months of growth, while the alfalfa root system is typical of 2 years of growth. In both dicots, the root system shows a major vertical root axis. In the case of sugar beet, the upper portion of the taproot system is thickened because of its function as storage tissue. Both plants were grown under adequate water conditions. The grid lines are 30 cm apart. (Adapted from Weaver, 1926.)

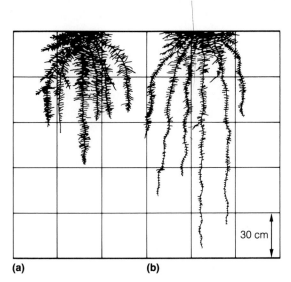

(a)   (b)

FIGURE 5.1. Fibrous root system of a wheat plant (monocot). The root system of a mature (3 months) wheat plant growing in dry soil (a) is compared to the root system of a wheat plant growing in irrigated soil (b). In the fibrous root system, the primary root axes are no longer distinguishable. It is also apparent that the morphology of the root system is affected by the amount of water present in the soil. The grid lines are 30 cm apart. (Adapted from Weaver, 1926.)

fibrous root system (Fig. 5.1). In the fibrous root system, the roots generally have the same diameter (except where environmental conditions or pathogenic interactions modify the root structure), so it is difficult to distinguish a main root axis. In dicots, the root system develops along a main single root axis, called a **taproot**, which thickens as a result of secondary cambial activity. From this main root axis, lateral roots develop to form an extensively branched root system (Fig. 5.2).

The development of the root system in both monocots and dicots depends on the activity of the root apical meristem and the production of lateral root meristems. A generalized diagram of the apical region of a plant root indicating the location of the meristem is shown in Figure 5.3.

Cell division occurs in the meristematic zone both in

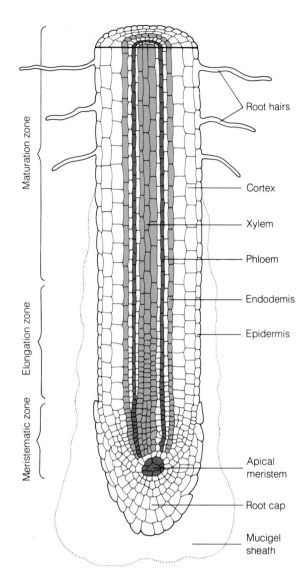

Maturation zone

Elongation zone

Meristematic zone

Root hairs

Cortex

Xylem

Phloem

Endodemis

Epidermis

Apical meristem

Root cap

Mucigel sheath

FIGURE 5.3. Diagrammatic longitudinal section of the apical region of the root. The meristematic cells are located near the tip of the root. These cells generate the root cap and the upper tissues of the root. In the elongation zone, cells differentiate to produce the xylem, phloem, and cortex. Root hairs, formed in epidermal cells, are located in the maturation zone. (Modified from Russell, 1977.)

the direction of the root apex to form the root cap and in the direction of the root base to form cells that will differentiate into the tissues of the functional root. The root cap protects the delicate meristematic cells as the root moves through the soil and also secretes a gelatinous material called **mucigel**, which commonly surrounds the root tip. The precise function of the mucigel is uncertain, but it has been suggested that it may protect the root apex from desiccation, promote the transfer of nutrients to the root, or affect root interaction with soil microorganisms (Russell, 1977). Another function thought to occur in the root cap is the perception of gravity and the correspond-

ing signal for directing the downward growth of roots. This process is termed the **gravitropic response**. As cells begin to differentiate back from the root apex, two regions of the root, the **cortex** and the **stele**, begin to be distinguished. The stele contains the vascular elements of the root: the **xylem**, which functions in water and solute conduction to the shoot, and the **phloem**, which functions in the transport of metabolites from the shoot to the root (Fig. 4.3). The cortex contains several layers of cells that surround the stele. The innermost layer of cells in the cortex adjacent to the stele forms the **endodermis**. The walls of this endodermal cell layer become thickened, and the deposition of suberin on the radial walls forms the **Casparian strip**. This hydrophobic structure is a barrier to the radial movement of water or solutes through the cell wall to the stele. As we have seen in Chapter 4, solute movement at the endodermis must occur within living cells following transport across the plasma membrane. The distinction between solute translocation within the cell wall space (**apoplast**) and translocation within the continuum of living plant cells (**symplast**) has proved to be essential for the understanding of transport in the plant (see Chapter 4).

Some Minerals Enter the Roots at the Apical Region; Others Can Be Taken Up Along the Entire Root Surface

The precise point of entry of minerals into the root system has been a topic of considerable interest. Some researchers have claimed that nutrient absorption takes place only at the apical regions of the root axes or branches (Bar-Yosef et al., 1972), whereas others believe that absorption takes place over the entire root surface (Greenwood et al., 1974; Nye and Tinker, 1977). Experimental evidence has supported both possibilities, depending on the nutrient being investigated. In barley, the absorption of potassium, phosphate, and ammonium can take place freely at all locations of the root surface (Clarkson and Hanson, 1980). In contrast, the uptake of calcium in this species appears to be restricted to the apical region. Iron uptake may take place either at the apical region, as in barley (Clarkson and Sanderson, 1978), or over the entire root surface, as in corn (Kashirad et al., 1973).

Within the soil, nutrient movement to the root surface can occur both by **bulk flow** and by **diffusion** (see Chapter 3). **Bulk flow** occurs when nutrients are carried in the convective flow of water moving through the soil toward the root. The amount of nutrient provided to the root by bulk flow is dependent on the rate of water flow through the plant and the amount of nutrient in the soil solution. Under conditions of high rates of water flow and high concentrations of nutrients in the soil solution, bulk flow can play an important role in nutrient supply.

Looking at Roots Face to Face

STUDIES OF ROOT growth below the soil surface requires some means of directly observing the root system. As early as 1873, Sachs studied root systems by using simple soil-filled boxes with one glass wall. Since that time, facilities for studying root growth in the soil have become much more complex. Large laboratories with subterranean chambers for the observation of root growth have been constructed and allow the analysis of root growth while the aerial parts of the plant are exposed to natural field conditions (Huck and Taylor, 1984). These laboratories are called **rhizotrons**, from the Greek *rhizos*, meaning root, and *tron*, meaning a device for studying (Fig. 5.A). In a rhizotron, roots grow in glass-walled chambers that line underground passageways. The water and soil conditions in the chambers are kept as similar as possible to those in the areas surrounding the chambers (guard rows). Comparisons of the growth of plants inside the chambers with the growth of plants in the guard rows allow researchers to detect if growth in the chambers has any consequence for the physiology of the plant. Grid lines on the glass walls of the root chambers indicate soil depth. Details of root morphology (root size and distribution) under natural growing conditions can be observed with specially designed microscopes, mounted adjacent to the glass walls of the root chambers. In addition, the actual growth of the roots over a period of time can be measured with time-lapse photography. Rhizotron studies also help plant physiologists to understand some important aspects of root function. Measurements of the

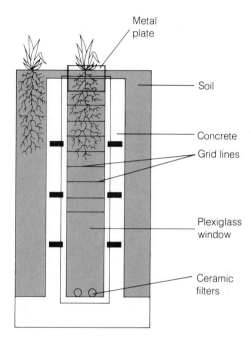

FIGURE 5.A. Diagram of a rhizotron used in the analysis of root growth.

mineral content of the root tissue and of the soil solution in the chamber provide information about mineral movement in the soil and uptake by the root. In the most advanced rhizotrons, computer-assisted data acquisition systems record automatically such parameters as soil water potential, temperature, and oxygen tension. This information can then be correlated with data on root growth and function (Huck and Taylor, 1984). Types of measurements that can be made through the use of the rhizotron are summarized in Table 5.A.

TABLE 5.A. Types of measurements that can be made through the use of a rhizotron*

General topic	Special measurements
Root growth and development	(a) Variation in growth habit as influenced by genetic factors. (b) Variation in growth habit as related to changes in soil microenvironment (temperature, oxygen water content, etc.) (c) Partitioning of carbon compounds between roots and other plant organs
Root functioning	(a) Water uptake (b) Mineral uptake and ion movement (c) Biological oxygen demand of the roots
Characterization of soil properties	(a) Soil water content and evaporation (b) Infiltration of water and minerals applied as fertilizers (c) Groundwater recharge and ion migration through the soil

*Adapted from Huck and Taylor, 1984.

Diffusion occurs when mineral nutrients move from a region of higher concentration to a region of lower concentration. Nutrient uptake by the roots lowers the concentration of nutrients at the root surface, generating concentration gradients in the soil solution surrounding the root. Diffusion of nutrients down their concentration gradient can increase nutrient availability at the root surface.

When nutrient uptake by the root is high and the nutrient concentration in the soil is low, bulk flow can supply only a small fraction of the total nutrient requirement (Mengel and Kirkby, 1979). Under these conditions, movement of most nutrients to the root surface is limited by diffusion. When diffusion is too slow to maintain high nutrient concentrations near the root, a **nutrient depletion zone** adjacent to the root surface is formed (Fig. 5.4). Nutrient depletion decreases as the distance to the root surface increases. Actual nutrient uptake, however, will depend on the relationship between the affinity of the root uptake mechanism for a given nutrient and the prevailing concentration of that nutrient at the root surface. This aspect of nutrient uptake is discussed in Chapter 6.

The formation of a depletion zone tells us something important about mineral nutrition. Because roots deplete the mineral supply in the surrounding soil, their effectiveness in mining minerals from the soil is not determined simply by the rate at which they can remove nutrients from the soil solution. Without their capacity for continuous growth, the roots would rapidly deplete the soil adjacent to their surface. Optimal nutrient acquisition therefore depends both on the capacity for nutrient uptake and on the growth characteristics of the root system.

Soil and Minerals

The soil surrounding plant roots can be considered as a heterogeneous material containing a solid phase, a liquid phase, and a gaseous phase (Chapter 4). All of these phases are involved in the supply of nutrients to the root surface. The inorganic particles of the solid phase act as a reservoir of nutrients such as potassium, calcium, magnesium, and iron. Also associated with this solid phase are organic particles containing nitrogen, phosphorus, and sulfur. The liquid phase of the soil constitutes the soil solution, which contains dissolved mineral ions and acts as the medium for ion movement to the root surface. Gases such as oxygen and carbon dioxide may be dissolved in the soil solution, but their exchange in respiring root cells generally occurs in the gaseous phase present in the air spaces between soil particles. The supply of oxygen to the roots is essential for cellular respiration, the source of metabolic energy that drives mineral uptake processes.

Negative Charges on the Surface of Soil Particles Affect the Adsorption of Mineral Cations and Anions

Soil particles, both inorganic and organic, have negative charges on their surfaces. Inorganic clay particles contain crystal lattices consisting of arrangements of the cationic forms of aluminum and silicon (Al^{3+}, Si^{4+}). These clay particles can become negatively charged by replacement of the Al^{3+} and Si^{4+} in the lattice by cations of lesser charge, a process termed isomorphous replacement. The silicate-containing clay materials have been classified into three major groups—the kaolinite group, the smectite group, and the illite group—based on differences in their structure and physical properties (Table 5.1). The kaolinite group is generally found in well-weathered soils, while the smectite and illite groups are found in younger soils. The organic particles found in humus originate from the products of the decomposition of dead tissue from plants

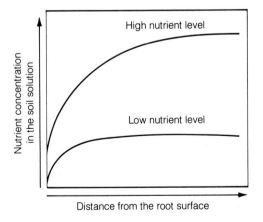

FIGURE 5.4. Formation of a nutrient depletion zone in the region of the soil adjacent to the plant root. A nutrient depletion zone is formed when the rate of nutrient uptake by the cells of the root exceeds the rate of replacement of the nutrient by diffusion in the soil solution. This depletion causes a localized decrease in the nutrient concentration in the area adjacent to the root surface. (Adapted from Mengel and Kirkby, 1979.)

TABLE 5.1. Comparative properties of three major types of silicate clays found in the soil

Property	Type of clay		
	Smectite	Illite	Kaolinite
Size (μm)	0.01–1.0	0.1–2.0	0.1–5.0
Shape	Irregular flakes	Irregular flakes	Hexagonal crystals
Cohesion	High	Medium	Low
Water swelling capacity	High	Medium	Low
Cation exchange capacity (milliequivalents 100 g^{-1})	80–100	15–40	3–15

Adapted from Brady, 1974.

and animals by the microbes of the soil. The negative surface charges of organic particles result from the dissociation of hydrogen ions from the carboxylic acid and phenolic groups present in this component of the soil.

Negative surface charges of soil particles are important in the adsorption of mineral cations to the surface of the particles. This adsorption is significant because of the high ratio of surface area to volume of both inorganic and organic soil particles, and it is an important factor in soil fertility. Mineral cations adsorbed on the surface of soil particles are not easily lost when the soil is leached by water, and they provide a nutrient reserve available to plant roots. Mineral nutrients adsorbed in this way can be replaced by other cations in a process known as **cation exchange** (Figure 5.5). The degree to which a soil can adsorb and exchange ions is termed its **cation exchange capacity** and is highly dependent on the particle type. A soil with higher cation exchange capacity will generally supply more minerals to the roots.

In contrast to mineral cations, which are adsorbed on the surface of soil particles, mineral anions are usually repelled by the negative charge of the soil particles and may remain dissolved in the soil solution. Thus, the anion exchange capacity of most agricultural soils is small compared to their cation exchange capacity. Among the most commonly required anions, nitrate and chloride are generally not adsorbed by soil particles and remain in the soil solution, where they are susceptible to leaching by water moving through the soil. On the other hand, chloride deficiency is rarely observed under natural conditions. Phosphate ions may bind to soil particles containing aluminum or iron. In this case, the positively charged

iron and aluminum ions will have hydroxyl (OH^-) groups that can be exchanged with sulfate, phosphate, and other anions. Sulfate is fairly soluble in the absence of Ca^{2+}, but in the presence of calcium its solubility is reduced by precipitation as $CaSO_4$.

Nutrient Availability, Soil Microbes, and Root Growth Are Strongly Dependent on Soil pH

Another important property of soils is the hydrogen ion concentration or soil pH. The soil pH can affect the growth of plant roots and soil microorganisms. Root growth is generally favored at slightly acidic pH values (5.5 to 6.5). Fungi generally predominate in the soil adjacent to the roots in the acid pH range, whereas at higher pH values bacteria may become more prevalent. Soil pH also determines the availability of plant nutrients (Fig. 5.6). A low soil pH favors the weathering of rocks and the release of ions such as K^+, Mg^{2+}, Ca^{2+}, and

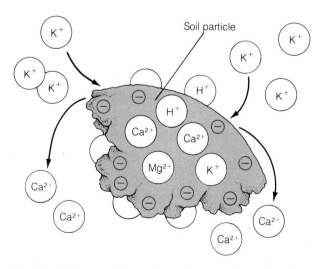

FIGURE 5.5. The principle of cation exchange on the surface of a soil particle. Cations are bound to the surface of soil particles because the surface is negatively charged. Addition of a cation such as potassium can displace another cation such as calcium from its binding on the surface of the soil particle and make it available for uptake by the root.

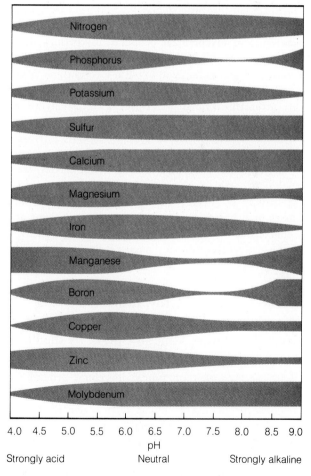

FIGURE 5.6. Influence of soil pH on the availability of nutrient elements in organic soils. The width of the shaded areas indicates the degree of nutrient availability to the plant root. (From Lucas and Davis, 1961.)

Mn^{2+}. At low pH values, the salts present in the soil as carbonates, sulfates, and phosphates are more soluble. Increasing solubility facilitates absorption by the root. Rainfall levels and decomposition of organic matter in soils are major factors in lowering the soil pH. Carbon dioxide produced as a result of the decomposition of organic material equilibrates with soil water in the reaction

$$CO_2 + H_2O \rightleftharpoons H^+ + HCO_3^-$$

forming hydrogen ions. Microbial decomposition of organic material also produces ammonia and hydrogen sulfide, which can be oxidized in the soil to form strong acids. On the other hand, in arid regions, the weathering of rock can cause the release of minerals that act as bases in the soil and result in an increase in soil pH.

Mycorrhizal Fungi and Their Association with Plant Roots

Our discussion thus far has centered on the direct acquisition of mineral nutrients by the root, but it is important to realize that this process may be modified by the association of mycorrhizal fungi with the root system. The occurrence of **mycorrhizae** (singular, mycorrhiza; from the Greek words for "fungus" and "root") is not unusual; in fact, they are widespread in natural conditions. Much of the world's vegetation appears to have roots associated with mycorrhizal fungi. Mycorrhizae are usually absent in roots from laboratory-grown plants, but such roots may not be typical of natural growing conditions (Rovira et al., 1983; Tinker and Gildon, 1983).

Mycorrhizal fungi occur in two major classes: ectotrophic mycorrhizae and vesicular-arbuscular mycorrhizae (Rovira et al., 1983; Sanders et al., 1975). Minor classes of mycorrhizal fungi include the ericaceous and orchidaceous mycorrhizae, which may have limited importance in terms of mineral nutrient uptake (Tinker and Gildon, 1983).

Mycorrhizal Fungi Grow Inside the Plant Roots and in the Surrounding Soil, Often Facilitating Mineral Uptake by the Plant

The **ectotrophic mycorrhizal** fungi typically show a thick sheath or "mantle" of fungal mycelium around the roots, with some of the mycelium penetrating between the cortical cells (Fig. 5.7). The cortical cells themselves are not penetrated by the fungal hyphae but instead are surrounded by a network of hyphae called the **Hartig net**. Often, the amount of fungal mycelium is so extensive that its total mass is comparable to that of the roots themselves. The fungal mycelium also extends into the soil, away from this compact mantle, forming single hyphal rhizomorphs and hyphal strands containing fruiting

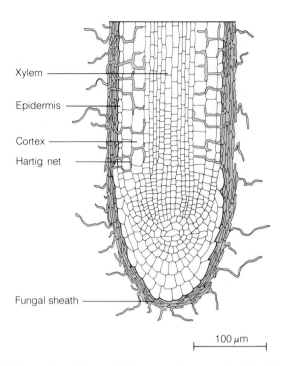

FIGURE 5.7. A root infected with ectotrophic mycorrhizal fungi. In the infected root, the fungal hyphae surround the root to produce a dense fungal sheath and penetrate the intercellular spaces of the cortex to form the Hartig net. It is apparent that the total mass of fungal hyphae is comparable to the root mass itself. (From Rovira et al., 1983.)

bodies (Tinker and Gildon, 1983). The capacity of the root system to absorb nutrients is improved by the presence of external fungal hyphae that can reach beyond the areas of nutrient-depleted soil near the roots (Clarkson, 1985). Ectotrophic mycorrhizal fungi exclusively infect tree species, including gymnosperms and woody angiosperms.

Unlike the ectotrophic mycorrhizal fungi, **vesicular-arbuscular mycorrhizal** fungi do not produce a compact mantle of fungal mycelium around the root. Instead, the hyphae grow in a less dense arrangement, both within the root itself and extending outward from the root into the surrounding soil (Fig. 5.8) After entering the root through either the epidermis or a root hair, the hyphae not only extend through the regions between cells but also penetrate individual cells of the cortex. Within the cells, the hyphae can form ovoid structures called **vesicles** and branched structures called **arbuscules**. The arbuscules appear to be sites of nutrient transfer between the fungus and the host plant. Outside the root, the external mycelium can extend several centimeters away from the root and may contain spore-bearing structures. Unlike the ectotrophic mycorrhizae, vesicular-arbuscular mycorrhizae make up only a small mass of fungal material, which is unlikely to exceed 10% of the root weight (Tinker, 1978). Vesicular-arbuscular mycorrhizae are found in association with the roots of almost all species of herba-

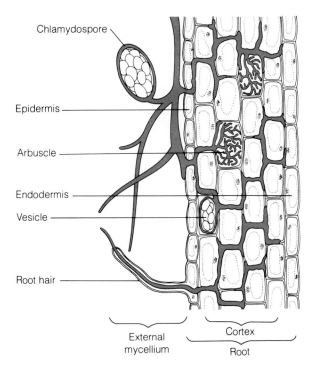

Chlamydospore

Epidermis

Arbuscle

Endodermis

Vesicle

Root hair

External mycellium

Cortex

Root

FIGURE 5.8. The association of vesicular-arbuscular mycorrhizal fungi with a section of a plant root. The external mycelium can bear reproductive chlamydospores and extend out from the root into the surrounding soil. The fungal hyphae grow into the intercellular wall spaces of the cortex and penetrate individual cortical cells. As they extend into the cell, they do not break the plasma membrane or the tonoplast of the host cell. Instead, the hypha is surrounded by these membranes as it occupies intracellular space. In this process, the fungal hyphae may form ovoid structures known as vesicles or branched structures known as arbuscules. The arbuscules participate in nutrient ion exchange between the host plant and the fungus. Arbuscules develop and proliferate following the penetration of the hyphae into the root cortical cells. In later stages, the arbuscules separate from the hyphae and degenerate. (From Mauseth, 1988.)

ceous angiosperms with the notable exception of some species of the Chenopodiaceae (spinach family) and the Cruciferae (cabbage family) (Tinker and Gildon, 1983).

The association of vesicular-arbuscular mycorrhizae with plant roots facilitates the uptake of phosphorus and trace metals such as zinc and copper. By extending beyond the depletion zone for phosphorus around the root, the external mycelium improves phosphorus absorption. Calculations by Sanders and Tinker (1971) show that a root infected with mycorrhizal fungi can transport phosphate at a rate more than four times higher than that of an uninfected root. The external mycelium of the ectotrophic mycorrhizae can also absorb phosphate and make it available to the plant. In addition, it has been suggested that ectotrophic mycorrhizae may proliferate in the organic litter of the soil and hydrolyze organic phosphorus for transfer to the root (see Tinker and Gildon, 1983, and references therein).

Leakage and Death of the Fungal Hyphae Are Probably Involved in the Transfer of Nutrients from the Mycorrhizal Fungi to the Plant

Little is known about the mechanism by which the mineral nutrients absorbed by mycorrhizal fungi are transferred to the cells of plant roots. In the ectotrophic mycorrhizae, inorganic phosphate may simply leak from the hyphae in the Hartig net and be absorbed by the root cortical cells. In the vesicular-arbuscular mycorrhizae, the situation may be more complex. Nutrients may leak from intact arbuscules within an infected cell and then diffuse to root cortical cells. Another possibility is related to the observation that some root arbuscules are continually degenerating while new ones are being formed. A degenerating arbuscule filled with nutrients could release its internal contents to the host cell, from which they would eventually be distributed to the entire plant.

A key factor in the extent of mycorrhizal infection of the plant root is the nutrient status of the host plant. Deficiency of a nutrient such as phosphorus tends to promote infection, whereas plant growth in a well-fertilized soil may tend to suppress mycorrhizal infection. The mycorrhizal infection is usually a mutualistic relationship, with the fungi receiving sugars from the host plant in exchange for increasing the plant's mineral nutrient uptake efficiency.

Essential Elements

Of the elements present in the biosphere and detected in plant tissue, only certain elements have been determined to be essential. In the absence of an essential element, a plant will demonstrate deficiency symptoms and die without completing its life cycle. Availability of these essential elements and of energy from sunlight allows plants to synthesize all the compounds they need for normal growth. Table 5.2 lists the elements currently known to be essential to higher plants. In addition to the elements listed in the table, some plants may require sodium or silicon. In plant species adapted to saline environments, sodium is taken up in high amounts and is required for growth. Silicon is essential for rice and may also be essential for *Equisetum* or "scouring rush" (Mengel and Kirkby, 1979). Certain plants in the genus *Astraglus* also accumulate selenium, although it is not known whether they have a specific requirement for this element (Marschner, 1986).

Essential elements are usually classified as macronutrients or micronutrients, according to their relative concentration in plant tissue. However, some researchers have argued that such a classification is difficult to justify physiologically. In some cases, the differences in tissue

TABLE 5.2. Adequate concentrations of nutrient elements in plant tissue

Element	Chemical symbol	Atomic weight	Concentration in dry matter		Relative number of atoms with respect to molybdenum
			μmol g^{-1}	ppm or %	
Micronutrients				ppm	
Molybdenum	Mo	95.95	0.001	0.1	1
Copper	Cu	63.54	0.10	6	100
Zinc	Zn	65.38	0.30	20	300
Manganese	Mn	54.94	1.0	50	1,000
Iron	Fe	55.85	2.0	100	2,000
Boron	B	10.82	2.0	20	2,000
Chlorine	Cl	35.46	3.0	100	3,000
Macronutrients				%	
Sulfur	S	32.07	30	0.1	30,000
Phosphorus	P	30.98	60	0.2	60,000
Magnesium	Mg	24.32	80	0.2	80,000
Calcium	Ca	40.08	125	0.5	125,000
Potassium	K	39.10	250	1.0	250,000
Nitrogen	N	14.01	1,000	1.5	1,000,000
Oxygen	O	16.00	30,000	45	30,000,000
Carbon	C	12.01	40,000	45	40,000,000
Hydrogen	H	1.01	60,000	6	60,000,000

From Epstein, 1972.

content of macronutrients and micronutrients are not as great as those shown in Table 5.2. For example, some plant tissues have almost as much iron or manganese as they do sulfur or magnesium. Many elements may be present in concentrations greater than the plant's minimal requirements, and plants may accumulate elements that are not essential nutrients.

Mengel and Kirkby (1979) have proposed that the essential elements should instead be classified according to their biochemical role and physiological function. Such a classification is shown in Table 5.3, in which plant nutrients have been divided into four basic groups. The first group includes the elements that form the organic compounds of the plant. The gaseous forms of the elements in the first group are acquired from the atmosphere, and the ionic forms are acquired from the soil solution. These nutrients (such as nitrogen, carbon, or sulfur) are assimilated in the plant in biochemical reactions involving carboxylation and oxidation-reduction. Elements of the second group can often be found in plant tissues as phosphate, borate, and silicate esters in which the elemental group is bound to the hydroxyl group of an organic molecule (i.e., sugar-phosphate), and they are important in energy transfer reactions. They are all taken up as either inorganic anions or acids from the soil solution. The third group of elements are taken up by the plant in their ionic forms. These elements are present in plant tissue as either free ions or ions bound to substances such as the pectic acids present in the plant cell wall. Of particular importance are their roles as enzyme cofactors

and in the regulation of osmotic potentials. The fourth group of essential elements are absorbed from the soil as ions or chelates and have important roles in reactions involving electron transfer.

An Essential Element Is Needed for Completion of the Life Cycle of the Plant, Causes a Specific Deficiency When Unavailable, and Has a Defined Role in Plant Metabolism

In order to demonstrate that an element is essential to a plant, certain criteria must be met. According to the criteria established originally by Arnon and Stout (1939), an element could be considered essential if (1) its deficiency prevents the plant from completing its life cycle, (2) its deficiency is specific to the element and can be corrected or prevented only by supplying that element, and (3) the element has a nutritional role apart from correcting any unfavorable microbial or chemical condition of the soil. The demonstration of the first criterion involves growing plants in nutrient solutions containing all essential elements except the element in question. Under these conditions, the plant will exhibit deficiency symptoms and will not complete its life cycle. Only the addition of that element to the nutrient solution will either alleviate or prevent the deficiency symptom (criterion 2).

Since their original consideration, it has become apparent that these criteria for essentiality may be too

TABLE 5.3. Classification of plant nutrients according to biochemical function

Nutrient element	Uptake	Biochemical functions
Group 1		
C, H, O, N, S	H_2O. Ions from soil solution: HCO_3^-, NO_3^-, NH_4^+, SO_4^{2-}. Gases from the atmosphere: CO_2.	Major constituent of all organic material. Assimilated by carboxylation and oxidation-reduction processes.
Group 2		
P, B, Si	In the form of phosphates, boric acid, or borate. Silicate from the soil solution.	Esterification with alcohol groups in plants. Phosphate esters are involved in energy transfer reactions.
Group 3		
K, Na, Mg, Ca, Mn, Cl	Ions from the soil solution.	Nonspecific functions establishing osmotic potentials. Specific contributions to the structure and function of enzyme protein. Balancing non-diffusible and diffusible anions.
Group 4		
Fe, Cu, Zn, Mo	In the form of ions or chelates from the soil solution.	Present predominantly in prosthetic groups. Permit electron transport by valence changes.

From Mengel and Kirkby, 1979.

strict. This is especially true of the second criterion, which requires that an essential element cannot be replaced by another element. For instance, in some plants bromide may partially substitute for chloride (Broyer et al., 1954), even though chloride is the essential halogen under natural conditions. According to the criteria of Arnon and Stout, this substitution would indicate that chloride is not essential. In current investigations, the first criterion is the one generally used to demonstrate essentiality, and the results are further verified by a clear demonstration of a physiological role for the element.

Another difficulty in establishing the essentiality of a micronutrient element is that it may be required at such a low concentration that trace contamination associated with the chemicals used to make macronutrient solutions may satisfy the requirement. In addition, contamination can occur in the water supply used to make up nutrient solutions. Thus, extreme care must be taken when conducting experiments to demonstrate nutrient essentiality.

Techniques for Growing Plants in Nutritional Studies

Demonstration of the essentiality of an element requires that plants be grown under controlled conditions in which only the element in question is absent. This is extremely difficult to achieve with plants grown in a complex me-

dium such as soil. In the nineteenth century, J. Sachs and W. Knop, working independently, approached this problem by growing plants in **nutrient solutions.** This approach opened the way for the development of modern techniques for the culture of plants without soil.

In the work of Sachs and Knop, plants were grown with their roots immersed in a nutrient solution containing inorganic salts, without any soil or organic matter. Their demonstration that plants could grow normally under these conditions proved unequivocally that plants can supply all their needs from only inorganic elements and sunlight.

The technique of growing plants with their roots immersed in nutrient solutions without soil is called **hydroponics** (Fig. 5.9). Successful hydroponic culture requires that the roots be supplied with adequate amounts of oxygen. Studies have shown that inadequate oxygenation of the root system or removal of oxygen by purging with nitrogen gas reduces nutrient absorption by root cells. The absorption is reduced because anoxia inhibits cell respiration (Chapter 11) and prevents an adequate supply of metabolic energy to drive absorption processes. Sufficient oxygenation of the root system can be achieved by either bubbling air through the medium or changing the nutrient solutions frequently. In experimental studies, addition of nutrients or frequent changes of nutrient solutions are also necessary because of the changes in nutrient concentrations and medium pH that occur as a

FIGURE 5.9. Hydroponic and aeroponic systems for growing plants in nutrient solutions in which composition and pH can be automatically controlled. In a hydroponic system (a), the roots are immersed in the nutrient solution, and air is bubbled through the solution. An alternative hydroponic system (b) often used in commercial production is the nutrient film growth system, in which the nutrient solution is pumped as a thin film down a shallow trough surrounding the plant roots. In this system, the composition and pH of the nutrient solution can be automatically controlled. In the aeroponic system (c), the roots are suspended over the nutrient solution, which is whipped into a mist by a driver attached to a motor shaft. (Aeroponic system adapted from Zobel et al., 1976.)

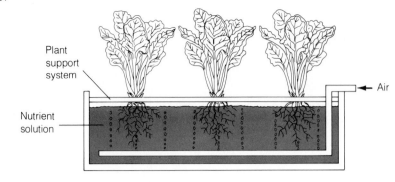

(a) Hydroponic growth system

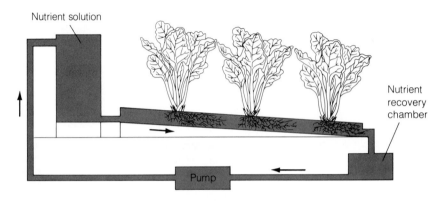

(b) Nutrient film growth system

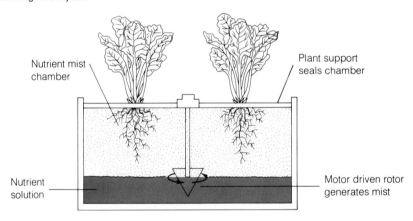

(c) Aeroponic growth system

result of nutrient uptake by the roots. Hydroponics can be used for the commercial growth of plants such as lettuce. In commercial hydroponic culture, nutrient solutions are often recirculated in a thin layer that flows through a trough surrounding the plant roots (Asher and Edwards, 1983). With this method, called the "nutrient film technique" (Fig. 5.9b), nutrient solution composition and pH can be monitored continually and adjusted automatically.

One alternative to growing plants with their roots immersed in a nutrient solution is to grow them in a support material such as sand, gravel, or vermiculite. Nutrient solutions can then be added and old solutions can be removed by leaching. Another alternative is to grow the plants **aeroponically** (Zobel et al., 1976). In this technique, plants are grown with their roots suspended in

air while being sprayed continuously with a nutrient solution (Fig. 5.9c). This approach ensures that the roots receive an ample supply of oxygen and avoids limitations on gas exchange.

Plant Physiologists Have Formulated Nutrient Solutions That Can Sustain Optimal Plant Growth

Over the years, many formulations have been used for nutrient solutions. Early formulations developed by Knop included only KNO_3, $Ca(NO_3)_2$, KH_2PO_4, $MgSO_4$, and an iron salt. At the time, this nutrient solution was believed to contain all the components required by the plant, but we know now that these experiments were carried out with impure salts that contained as contaminants other elements now known to be essential (such as boron or molybdenum). An example of a more modern formulation for a nutrient solution is shown in Table 5.4. This formulation is called a modified **Hoagland's solution**, named after Dennis R. Hoagland, a researcher prominent in the development of modern mineral nutrition research in the United States. This solution has all the elements know to be needed for optimal plant growth. An important property of this formulation is that its nitrogen is supplied as both ammonium and nitrate. Supplying nitrogen in a balanced mixture of cations and anions tends to reduce the rapid rises in the pH of the medium that are commonly observed when the nitrogen is supplied solely as nitrate anion (Asher and Edwards, 1983).

A significant problem in growing plants in nutrient solutions is that of supplying sufficient quantities of iron. When iron is supplied as an inorganic salt, such as $FeSO_4$ or $Fe(NO_3)_2$, it can precipitate out of solution as iron hydroxide. If phosphate salts are present, insoluble iron phosphate will also form. This precipitation of the iron out of solution makes it unavailable for the plant, unless iron salts are added at frequent intervals. Initial attempts to alleviate this problem involved adding iron together with citric acid or tartaric acid to the nutrient solution. These compounds, which form a soluble complex with cations such as iron and calcium, are called **chelating agents**. More modern formulations have used the chemical ethylenediaminetetraacetic acid (EDTA). The structure of EDTA is shown in Figure 5.10.

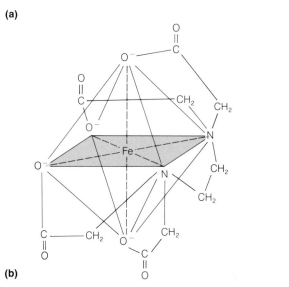

FIGURE 5.10. Chemical structure of the chelator ethylenediaminetetraacetic acid (EDTA) by itself (a) and chelated to an Fe^{3+} ion (b). Iron binds to EDTA through interaction with the two nitrogen atoms and the four ionized oxygen atoms of the carboxylate groups. The resulting ring structure clamps the metallic ion and effectively neutralizes its reactivity in solution. During the uptake of iron at the root surface, Fe^{3+} is reduced to Fe^{2+}, which is released from the EDTA-iron complex. The chelator can then bind other available Fe^{3+} ions.

TABLE 5.4. Composition of a modified Hoagland's nutrient solution for growing plants*

Compounds added to nutrient solution	Nutrient elements	Final concentration	
		μm	ppm
Macronutrients			
KNO_3, $Ca(NO_3)_2 \cdot 4H_2O$,	N	16,000	224
$NH_4H_2PO_4$, $MgSO_4 \cdot 7H_2O$	K	6,000	235
	Ca	4,000	160
	P	2,000	62
	S	1,000	32
	Mg	1,000	24
Micronutrients			
KCl, H_3BO_4, $MnSO_4 \cdot H_2O$,	Cl	50	1.77
$ZnSO_4 \cdot 7H_2O$, $CuSO_4 \cdot 5H_2O$,	B	25	0.27
H_2MoO_4, Fe-EDTA	Mn	2.0	0.11
	Zn	2.0	0.131
	Cu	0.5	0.032
	Mo	0.5	0.05
	Fe	20	1.12

Adapted from Epstein, 1972.

*The compounds are added separately from stock solutions to prevent nutrient precipitation during the preparation of the nutrient solution (see Epstein, 1972, for details).

Chelating agents contain negatively charged carboxyl groups or electron-donating nitrogen groups that can bind the iron in a coordination complex. Iron-EDTA complexes are commonly used in nutrient solutions for plants grown hydroponically, aeroponically, or in the soil. The fate of the chelation complex during iron uptake by the root cells is not clear; it has been suggested that during the absorption of iron by the roots, the iron is released from the chelator when it is reduced from Fe^{3+} to Fe^{2+} at the root surface. The chelator can then diffuse back into the nutrient (or soil) solution and react with another Fe^{3+} ion. Following uptake into root cells, iron is kept soluble by chelation with organic compounds present in plant cells. It appears that citric acid may play a major role in the chelation of iron and its long-distance transport in the xylem (Olsen et al., 1981).

Roles of Essential Elements and Nutrient Disorders

Inadequate supply of an essential element results in a nutritional disorder manifested by characteristic deficiency symptoms. In hydroponic culture a given symptom can be readily correlated with the removal of one essential element. In soil-grown plants, diagnosis can be more complex, because several elements may be deficient at once and some virus-induced plant diseases may produce symptoms similar to those of nutrient deficiencies.

Deficiency of an Essential Element Causes Typical Symptoms Resulting from the Disruption of Metabolic Processes in the Plant

Nutrient deficiency symptoms in a plant are the expression of metabolic disorders resulting from an insufficient supply of an essential element. These disorders are, in turn, related to the roles played by essential elements in normal plant metabolism and function. A partial listing of the roles of essential elements is shown in Table 5.5. This table does not include carbon, oxygen, and hydrogen, which are acquired primarily from carbon dioxide and water during photosynthesis. Despite the fact that each essential element participates in a specific set of metabolic reactions, it is possible to make some general statements about the functions of essential elements in plant metabolism. In general, the essential elements

function (1) as constituents of compounds, (2) in the activation of enzymes, and (3) in contributing to the osmotic potential of plant cells. There may be other roles related to the ability of divalent cations such as calcium or magnesium to modify the permeability of plant membranes. In addition, research is still uncovering specific roles of these elements in plant metabolism; for example, calcium appears to regulate key enzymes in the cytosol (Hepler and Wayne, 1985). Thus, many essential elements may have multiple roles in plant metabolism.

When relating deficiency symptoms to the role of an essential element (Table 5.6), it is also important to consider the extent to which an element can be recycled from older to younger leaves. Some elements such as nitrogen, phosphorus, and potassium can readily move from leaf to leaf, whereas others such as boron, iron, and calcium are relatively immobile. If an essential element is mobile, deficiency symptoms will occur first in older leaves. On the other hand, a deficiency of an immobile essential element will first become evident in younger leaves. Although the precise mechanism of the mobilization process is not well understood, plant hormones such as cytokinins appear to be involved. The specific

TABLE 5.5. Roles of essential mineral elements

Element	Role(s)
N	Constituent of amino acids, amides, proteins, nucleic acids, nucleotides and coenzymes, hexosamines, etc.
P	Component of sugar phosphates, nucleic acids, nucleotides, coenzymes, phospholipids, phytic acid, etc. Has a key role in reactions in which ATP is involved.
K	Required as a cofactor for 40 or more enzymes. Has a role in stomatal movements. Maintains electroneutrality in plant cells.
S	Component of cysteine, cystine, methionine, and thus proteins. Constituent of lipoic acid, coenzyme A, thiamine pyrophosphate, glutathione, biotin, adenosine-5′-phosphosulfate, and 3-phosphoadenosine.
Ca	Constituent of the middle lamella of cell walls. Required as a cofactor by some enzymes involved in the hydrolysis of ATP and phospholipids. Acts as a "second messenger" in metabolic regulation.
Mg	Required nonspecifically by a large number of enzymes involved in phosphate transfer. A constituent of the chlorophyll molecule.
Fe	Constituent of cytochromes and of nonheme iron proteins involved in photosynthesis, N_2 fixation, and respiration.
Mn	Required for activity of some dehydrogenases, decarboxylases, kinases, oxidases, and peroxidases, and required nonspecifically by other cation-activated enzymes. Required for photosynthetic evolution of O_2.
B	Indirect evidence for involvement in carbohydrate transport. Borate forms complexes with certain carbohydrates. Natural borate complexes in plants have not been identified.
Cu	An essential component of ascorbic acid oxidase, tyrosinase, monoamine oxidase, uricase, cytochrome oxidase. Component of plastocyanin in spinach.
Zn	Essential constituent of alcohol dehydrogenase, glutamic dehydrogenase, carbonic anhydrase, and other enzymes.
Mo	A constituent of nitrate reductase. Essential to N_2 fixation.
Cl	Required for the photosynthetic reactions involved in O_2 evolution.

Adapted from Evans and Sorger, 1966.

TABLE 5.6. Key to nutrient deficiency symptoms on tomato plants

A$_1$ Symptoms appearing first or most severely on youngest leaves . . . . B

  B$_1$ Interveinal chlorosis present on young leaves C

    C$_1$ Black spots appear adjacent to veins. Smallest veins remain green. In older leaves the necrosis may appear as interveinal brown necrotic spots 2 to 4 mm in diameter located near the main veins . . . . Manganese

    C$_2$ Black spots do not appear. Smallest veins do not remain green. Necrotic areas, if present, are not associated with any particular part of the lamina . . . . Iron

  B$_2$ Young leaves do not show *interveinal* chlorosis, but young leaflets may show chlorosis toward the central basal portion of the leaflets. . . . . D

    D$_1$ Dorsal sides of young leaflets show marked purpling. This purpling includes both veins and interveinal areas. The ventral surfaces of young leaflets are very dark green. Leaflets are small and curled downward. Oldest leaves may show slight interveinal chlorosis and necrosis . . . . Phosphorus

    D$_2$ Dorsal sides of young leaflets are not purple. Ventral surfaces of young leaflets often show central basal chlorosis. Growth may be distorted. Interveinal necrosis often appears at bases of young leaflets. Very young growing tissue shows necrosis . . . . E

      E$_1$ Plant tissues are very brittle, especially under conditions of low stress for water. Growth accompanying recovery from this deficiency is usually twisted, asymmetrical, and otherwise distorted . . . . Boron

      E$_2$ Plant tissues are soft and often flaccid even under conditions of low stress for water. Leaflets developing after onset of the deficiency are narrow and cupped downward, but there is not usually much twisting either under deficiency conditions or upon recovery . . . . Calcium

A$_2$ Symptons neither appearing first nor being most severe on youngest leaves. Symptoms about equal over entire plant, or most severe on oldest or on recently matured leaves . . . . F

  F$_1$ Interveinal chlorosis present, possibly only as a mild mottling. . . . . G

    G$_1$ Oldest leaves most chlorotic. . . . . H

      H$_1$ Chlorosis definitely interveinal, so that at least the main veins remain green. Plants are not usually spindly I

        I$_1$ Chlorosis is the first visible symptom. Leaf edges curl upward in severe deficiencies . . . . J

          J$_1$ Necrosis appears as sunken necrotic spots which appear shiny from the back of the leaf. These spots have no particular location with respect to veins. Bright yellow and orange colors of the chlorotic leaves are common . Magnesium

          J$_2$ Necrosis occurs as gradual drying of interveinal areas followed by drying of the remaining tissues. Bright coloration is not common in chlorotic leaves . . . . Molybdenum

        I$_2$ Chlorosis is not the first visible symptom. Leaf edges do not usually curl upward . . . . K

          K$_1$ Tip and marginal necrosis present on older leaves of mildly affected plants, appearing as a "scorch". Old or recently matured leaves may show interveinal chlorosis. Leaflets sometimes show small black interveinal necrotic dots. Neither excessive guttation nor "water soaking" is present. . . . . Potassium

          K$_2$ Tip and marginal necrosis is absent. Only the oldest leaves of severely damaged plants show chlorosis. Necrosis appears as irregular sunken necrotic spots which may be veinal, interveinal, or adjacent to veins. Young leaflets sometimes show small black interveinal necrotic spots. At times of low stress for water, excessive guttation and "water soaking" of the leaf tissues may be observed. . . . . Zinc

      H$_2$ Chlorosis general, so that veins do not remain green. Plants are often spindly. L

        L$_1$ Veins become bright red. Petioles and petiolules tend to be twisted and/or vertically disposed. . . . . Sulfur

        L$_2$ Veins are yellow or possibly somewhat pink. Petioles and petiolules do not show twisting or vertical disposition Nitrogen

    G$_2$ Oldest leaves not most chlorotic. Chlorosis, when present, appears as interveinal mottling on recently matured leaves. Small black necrotic dots appear on young or recently matured leaves . . . . M

      M$_1$ Tip and marginal necrosis is present on older leaves of mildly affected plants, appearing as a "scorch". Old or recently matured leaves may show interveinal chlorosis. Leaflets sometimes show small black interveinal necrotic dots . . . . Potassium

      M$_2$ Neither tip nor marginal necrosis is present on older leaves until the entire plant is severely affected. Some tip necrosis is often present on the youngest leaves. Younger leaflets often show small black interveinal necrotic dots, especially adjacent to the main veins. Necrosis is usually confined to interveinal tissues adjacent to main veins . . Manganese

  F$_2$ Interveinal chlorosis is not present. . . . . N

    N$_1$ Leaf margins and tips wilt. Leaves do not show excessive guttation . . . . O

      O$_1$ Necrotic spots, when present, are sharply delimited and sunken. Old leaflet margins roll upward stiffly. Veinal necrosis is often present. Petioles and petiolules often bend abruptly and stiffly downward . . . . Copper

      O$_2$ Necrosis appears as bronze-colored areas of cellular necrosis. These areas are neither sunken nor sharply delimited. *Old* leaflet margins are not usually rolled upward. Veinal necrosis is not present. Petioles and petiolules do not show abrupt downward bending . . . . Chlorine

    N$_2$ Leaves show no excessive wilting except in cases of petiole necrosis. Wilting, if present, is not confined to leaflet tips and margins. Excessive guttation occurs under conditions of low stress for water. This is often accompanied by the appearance of water-soaked areas on the backs of the leaves. Necrosis usually appears as irregular spots which may or may not have any particular relationship to the veins. . . . . Zinc

From Woolley and Broyer, 1957.

deficiency symptoms for the essential elements and synopses of their functional roles follow.

Nitrogen. Since nitrogen is associated with many plant cell components, such as amino acids and nucleic acids, it is not surprising that a characteristic deficiency symptom is stunted growth. In addition, nitrogen-deficient plants may have markedly slender and often woody stems. This woodiness may be due to a buildup of excess carbohydrates because they cannot be used in the synthesis of amino acids or other nitrogen compounds. Carbohydrate backbone structures not used in nitrogen metabolism can be used in anthocyanin synthesis, leading to accumulation of the pigment. This is seen as a purple coloration in leaves, petioles, and stems of some plants such as tomato and certain varieties of corn. In many plants, the first symptom of nitrogen deficiency is **chlorosis** (yellowing of the leaves), especially in the older leaves near the base of the plant. Under severe nitrogen deficiency, these leaves become completely yellow (or tan) and then fall off the plant. Younger leaves may not show these symptoms initially because nitrogen can be mobilized from older leaves. Thus, a nitrogen-deficient plant may have light green upper leaves and yellow or tan lower leaves.

Phosphorus. Phosphorus (as phosphate) is an integral component of a number of important compounds present in plant cells, including the sugar-phosphates used in respiration and photosynthesis and the phospholipids that make up plant membranes. It is also a component of nucleotides used in plant energy metabolism and in the molecules of DNA and RNA. Characteristic symptoms of phosphorus deficiency include stunted growth in young plants and a dark green coloration of the leaves, which may be malformed and contain **necrotic spots** (small spots of dead tissue). As in nitrogen deficiency, anthocyanins may be formed in excess, giving a hint of a purple coloration to the leaves. In contrast to nitrogen deficiency, however, this purple coloration is not associated with chlorosis. In fact, a dark greenish-purple coloration of the leaves may occur. Additional symptoms of phosphorus deficiency include the production of slender (but not woody) stems and the death of older leaves. Maturation of the plant may also be delayed.

Potassium. Potassium has an important role in regulation of the osmotic potential of plant cells (Chapters 3 and 6). It is also an activator of many enzymes involved in respiration and photosynthesis. The first observable symptom of potassium deficiency is mottled or marginal chlorosis, which then develops into necrosis that occurs primarily at the leaf tips and margins and between veins. In many monocots these necrotic lesions may initially form at the leaf tips and margins and then extend basipetally toward the leaf base. Since potassium can be mobilized to the younger leaves, these symptoms appear initially on the older, more mature leaves at the base of the plant. Curling and crinkling of the leaves may also occur. The stems of potassium-deficient plants may be slender and weak with abnormally short internodal regions. In potassium-deficient corn, the roots may have increased susceptibility to root-rotting fungi present in the soil, and this, together with effects on the stem, results in an increased tendency for the plant to be easily bent to the ground (lodging).

Sulfur. In plants that are deficient in sulfur, many of the symptoms are similar to those of nitrogen deficiency, including chlorosis, stunting of growth, and anthocyanin accumulation. This is not surprising, since sulfur and nitrogen are both components of proteins. However, the chlorosis caused by sulfur deficiency generally occurs initially on the young leaves, rather than on the more mature leaves as in nitrogen deficiency. This occurs because, unlike nitrogen, sulfur is not easily remobilized to the younger leaves in most species. However, in some plant species chlorosis may occur at once in all the leaves or even initially on the older leaves.

Calcium. Calcium is used in the synthesis of new cell walls, particularly in the middle lamallae separating newly divided cells. It is also used in the mitotic spindle during cell division. Calcium is required for the normal functioning of plant membranes and has been implicated as a second messenger for a number of plant responses to both environmental and hormonal signals (Hepler and Wayne, 1985). In its function as a second messenger, calcium may bind to **calmodulin**, a protein found in the cytosol of plant cells. The calmodulin-calcium complex may be involved in the regulation of many metabolic processes. Characteristic symptoms of calcium deficiency include necrosis of the tips and margins of young leaves, followed by necrosis of the terminal buds. These symptoms occur in the young meristematic regions of the plant, where cell division is occurring and new walls are forming. These symptoms are often preceded by a general chlorosis and downward hooking of the young leaves. The young leaves may also be deformed in appearance. If the root system of a calcium-deficient plant is examined, it may appear brownish, short, and highly branched. Severe stunting may occur when the meristematic regions of the plant die prematurely.

Magnesium. In plant cells, magnesium has a specific role in the activation of enzymes involved in respiration, photosynthesis, and the synthesis of DNA and RNA. Magnesium is also a part of the porphyrin component of chlorophyll. A characteristic symptom of magnesium deficiency is an interveinal chlorosis that occurs first in the older leaves due to the high mobility of this element. This pattern of chlorosis occurs because chlorophyll in the vascular bundles remains unaffected for longer periods than that in the cells between the bundles. If the deficiency is extensive, the leaves may become yellow or white. An additional symptom of magnesium deficiency may be premature leaf abscission.

Iron. Iron has an important role as a component of enzymes involved in the transfer of electrons (redox reactions), such as cytochromes. In this role it is reversibly oxidized from Fe^{2+} to Fe^{3+} during electron transfer. As in magnesium deficiency, characteristic symptoms of iron deficiency involve interveinal chlorosis. However, because iron cannot be readily mobilized from older leaves, these symptoms occur initially on the younger leaves. Under conditions of extreme or prolonged deficiency, the veins may become chlorotic so that the whole leaf takes on a white color. Chlorosis of the leaves occurs because iron is required for the synthesis of chlorophyll. However, the precise role of iron in chlorophyll synthesis is still a subject of research. The low mobility of iron is probably due to precipitation in the older leaves as insoluble oxides or phosphates or to the formation of complexes with phytoferritin, an iron-binding protein found in the leaf (Bienfait and Van der Mark, 1983). The precipitation of iron reduces subsequent mobilization of the metal to the phloem for long-distance translocation.

Copper. Like iron, copper is associated with enzymes involved in redox reactions. An example of such an enzyme is plastocyanin, which is involved in electron transfer during the light reactions of photosynthesis (Haehnel, 1984). The initial symptom of copper deficiency is the production of dark green leaves, which may contain necrotic spots. The necrotic spots appear first at the tips of the young leaves and then extend downward along the leaf margins. The leaves may also be twisted or malformed. Under extreme copper deficiency, premature loss of leaves (leaf abscission) may occur.

Boron. Although the precise function of boron in plant metabolism is unknown, there is evidence for roles in nucleic acid synthesis, hormone responses, and membrane function (Parr and Loughman, 1983). Boron-deficient plants may exhibit a wide variety of symptoms, depending on the species and the age of the plant. A characteristic symptom is black necrosis of the young leaves and terminal buds. The necrosis of the young leaves occurs primarily at the base of the leaf blade. There may also be loss of apical dominance so that the plant becomes highly branched; however, the terminal apices of the branches soon become necrotic due to inhibition of cell division. Structures such as the fruit, fleshy roots, and tubers may exhibit necrosis or abnormalities related to the breakdown of internal tissues.

Manganese. Manganese activates a number of enzymes in plant cells. In particular, decarboxylases and dehydro-genases involved in the tricarboxylic acid cycle are specifically activated by this divalent cation. The best-defined function of manganese is in the photosynthetic reaction in which oxygen is produced from water (Marschner, 1986, and references therein). The major symptom of manganese deficiency is interveinal chlorosis associated with the development of small necrotic spots. This chlorosis may occur on the younger or older leaves, depending on the plant species.

Zinc. Many enzymes require zinc for their activity, and zinc may be required for chlorophyll biosynthesis in some plants. Zinc deficiency is characterized by a reduction in internodal growth, with the result that plants display a rosette habit of growth. The leaves may also be small and distorted, with leaf margins having a puckered appearance. These symptoms may result from loss of the capacity to produce sufficient amounts of the hormone indoleacetic acid. In some species (corn, sorghum, beans), there may be interveinal chlorosis of the older leaves, followed by the development of white necrotic spots. This chlorosis may be an expression of a zinc requirement for chlorophyll biosynthesis.

Molybdenum. A well-characterized role of molybdenum is as a component of nitrate reductase, the enzyme that catalyzes the reduction of nitrate to nitrite during its assimilation by the plant cell. The first indication of a molybdenum deficiency is general interveinal chlorosis and necrosis of the older leaves. In some plants, such as cauliflower or broccoli, the leaves may not become necrotic but instead may appear twisted and subsequently die (whiptail disease). Flower formation may be prevented, or the flowers may be abscised prematurely. Since molybdenum is a component of the nitrate reductase system, a molybdenum deficiency may bring about a nitrogen deficiency if the nitrogen source is primarily nitrate (Bouma, 1983). Molybdenum deficiencies occur in some regions of the eastern United States, where it causes whiptail disease in broccoli and cauliflower crops.

Chlorine. Chlorine is required for the water-splitting reaction of photosynthesis in which oxygen is produced (Haehnel, 1984). In addition, chlorine may be required for cell division in both leaves and roots (Terry, 1977). Plants deficient in chlorine develop wilting of the leaf tips followed by general leaf chlorosis and necrosis. The leaves may also exhibit reduced growth. Eventually, the leaves may take on a bronzelike color ("bronzing"). Roots of chlorine-deficient plants may appear stunted and thickened near the root tips. Because chlorine (as chloride) is extremely soluble and generally available in soils, chlorine deficiency is unknown in plants grown in their native habitats. Most plants generally take up chlorine to concentrations much higher than those required for normal functioning.

Soil and Plant Tissue Analysis as Indicators of Plant Nutritional Status

It is often important to evaluate the nutrient status of plants, particularly for the development of fertilizer schedules in agriculture. It may be desirable to increase or decrease certain nutrients during the various stages of growth to enhance the yield of the economically important part of the crop plant (tuber, grain, etc.). Two techniques are available for estimating the nutrient supply and content of plants growing under particular soil conditions: soil analysis and plant tissue analysis.

Soil analysis involves chemical determination of the content of a nutrient in a soil sample. This technique depends on careful soil sampling methods and interpretation of soil parameters such as soil density. A major limitation of soil analysis is that it reflects the levels of nutrients *potentially* available to the plant roots but fails to evaluate the uptake conditions and the amounts of nutrients actually taken up by plants. To obtain this information, plant tissue analysis must be done.

Adequate use of **plant tissue analysis** requires an understanding of the relationship between plant growth (or yield) and the mineral content of plant tissue samples (Bouma, 1983). Such a relationship is shown in Figure 5.11. When the nutrient content in a tissue sample is low, growth is reduced. In this **deficiency zone** of the curve, an increase in the tissue concentration of the mineral is correlated with an increase in growth or yield. As the nutrient level in the tissue sample increases further, a point is reached at which additional increases in mineral content are no longer correlated with increases in growth or yield. This region of the plot is often called the **adequate zone.** The narrow transition between the deficiency and adequate zones of the plot reveals the **critical concentration** of the nutrient, which may be defined as the minimal tissue content of the nutrient that is correlated with maximal growth or yield. As the tissue nutrient content increases beyond the adequate zone, growth or yield begins to decline due to toxicity.

To elucidate the relationship between growth and the tissue content of a nutrient, plants are grown in soil or nutrient solution in which all of the nutrients are present in adequate amounts except the nutrient under consideration. This nutrient is then added in increasing concentrations to different sets of plants, and the tissue content of the nutrient and growth characteristics are measured. In general, several curves are established, one for each element. For many agricultural conditions, establishing curves for nitrogen, phosphorus, and potassium will suffice.

Plant tissue analysis has been applied to many different plant tissues and organs, including leaves of different ages and positions, stems, roots, petioles, seed, fruit, and grain. It has been generally observed that changes in nutrient content brought about by changes in the nutrient supply are closely correlated with the nutritional status of leaves (Bouma, 1983). Leaf analysis may therefore provide a better indicator of nutrient deficiency than the morphological deficiency symptoms discussed in the previous section. If the critical concentration level (Fig. 5.11) for a mineral nutrient can be established, it will represent the tissue content below which nutrient disorders will occur. Plant tissue analysis should therefore make it possible to correct for a deficiency in a crop before it results in a nutrient disorder that could cause a reduction in growth or yield. Plant analysis has been useful in the development of fertilizer programs in sugarcane production and in the production of forage crops in pastures, where plant nutrient content is important for animal nutrition.

Chemical Fertilizers, Organic Farming, and Foliar Nutrition

Farming methods used in early agricultural practices usually resulted in the recycling of mineral nutrients taken from the soil by the plants. The crops were consumed locally by humans and animals, and the nutrients were often returned to the soil in the form of manure. The major loss of nutrients from the soil was due to leaching, which could be reduced to some extent by adding lime. With the onset of modern, high-production agricultural systems, nutrient removal from the soil by crops has become a significant factor. The realization

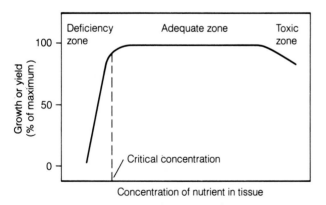

FIGURE 5.11. Relationship between yield (or growth) and the nutrient content of the plant tissue. The yield parameter may be expressed in terms of shoot dry weight or height. The deficient, adequate, and toxic zones are indicated on the graph. In order to obtain data of this type, plants are grown under conditions in which the concentration of one essential nutrient is varied while all others are in adequate supply. The effect of varying the concentration of this nutrient during plant growth is reflected in the growth or yield. The critical concentration for that nutrient is obtained by determining the concentration below which a reduction in yield or growth occurs.

that plants synthesize all their components from basic inorganic substances and sunlight has led to the modern practice of using chemical fertilizers to restore to the soil the mineral nutrients that are removed by crop plants.

Crop Yields Can Be Improved by Addition of Chemical or Organic Fertilizers

Chemical fertilizers applied to the soil generally contain salts of the macronutrients (see Table 5.2), especially those of nitrogen, phosphorus, and potassium. Fertilizers that contain only one of these three nutrients are termed "straight" fertilizers. Some examples of straight fertilizers are "superphosphate," ammonium nitrate, and muriate of potash (containing potassium). Fertilizers that contain two or more of these key mineral nutrients are called "compound" or "mixed" fertilizers. With increased agricultural production and with fertilization limited to the addition of macronutrients, consumption of micronutrients can increase to the point at which they, too, must be added to the soil as fertilizers. Of course, micronutrients may also be added to the soil to correct a preexisting deficiency. For example, some soils in the United States are deficient in boron or copper (Mengel and Kirkby, 1979) and can benefit from micronutrient supplementation. Chemical fertilizers may also be applied to the soil to modify soil pH. As shown in Figure 5.6, soil pH can affect the availability of various mineral nutrients. Lime addition to the soil can raise the pH of acidic soils, and addition of elemental sulfur can decrease the pH of alkaline soils. In the latter case, acidification occurs following the uptake of sulfur compounds by microorganisms with concomitant release of sulfate and hydrogen ions.

In contrast to chemical fertilizers that contain inorganic salts, **organic fertilizers** contain mineral nutrients in the form of complex organic molecules. Organic fertilizers originate from the wastes and residues of plant and animal life. They are often added to the soil mainly to improve its physical structure, rather than for their contribution of mineral elements. Organic fertilizers such as manures can improve soil water retention during drought and increase drainage in wet weather. Most of the nutrients in organic fertilizers are bound in organic molecules. Before they can be absorbed by the plant, the mineral nutrients must be released from the organic compounds, usually by the action of soil microorganisms. Therefore, the nutrients may become slowly available to the plant over a period of weeks or months. This slow release can reduce nutrient loss due to leaching by water moving through the soil.

Organic farming is the practice of growing crops by relying only on organic fertilizers and avoiding the use of chemical fertilizers and pesticides. Its proponents claim that crops grown in this way are of higher quality than those grown with chemical fertilizers and are more healthy for human consumption. Organically grown crops are often very vigorous, probably because of the improvement of the soil quality (soil water retention) and the supply of nutrients. There is, however, no scientific evidence that similar nutrients added as inorganic salts in chemical fertilizers or released by organic matter breakdown result in crops of different quality.

Application of Mineral Nutrients to the Leaves Can Enhance Absorption

In addition to being added as fertilizers to the soil, mineral nutrients can be applied to the leaves as sprays, a process known as **foliar application**. Leaves can absorb the applied nutrients, and this method may often have agronomic advantages over the application of nutrients to the soil. Foliar nutrition can reduce the lag time between application and uptake by the plant, which could be important during a phase of rapid growth. It can also circumvent the problem of restricted uptake of a nutrient from the soil. For example, foliar application of mineral nutrients such as iron, manganese, and copper may be more efficient than application through the soil, where they are absorbed by soil particles and hence less available to the root system.

Nutrient uptake by plant leaves is most effective when the nutrient solution remains on the leaf as a thin film (Mengel and Kirkby, 1979). To promote the production of a thin film, the nutrient solutions are often supplemented with surfactant chemicals, such as the detergent Tween 80, that reduce surface tension. Nutrient movement into the plant seems to involve diffusion through the cuticle and uptake by leaf cells. Although movement through the stomata could provide a pathway into the leaf, the architecture of the stomatal pore largely prevents liquid penetration (Ziegler, 1987). For foliar nutrient application to be successful, damage to the leaves must be prevented. If foliar sprays are applied on a hot day, when evaporation is high, salts may accumulate on the leaf surface and cause burning or scorching. This problem can be prevented by applying the salts in low concentrations and spraying on cool days or in the evening. Addition of lime to the spray reduces nutrient solubility and prevents toxicity. Foliar application has proved practical with cereals such as wheat. Seed protein content in this crop, which is important for baking and animal feeding, can be enhanced by the foliar application of nitrogen during later stages of growth.

Salt Stress and Halophytes

When excess minerals are present in the soil, the soil is said to be saline, and plant growth may be restricted if these salts (and nutrients) are at toxic levels. Excess salt in soils can be a major problem in arid and semiarid regions. In these regions, rainfall is insufficient to promote leaching of salts from the soil. In addition, irrigation water can contain 100 to 1000 g of salt per cubic meter, and with an average annual application of about 4000 cubic meters of water per acre, 400 to 4000 kilograms of salt can be added to the soil (Marschner, 1986). In saline soil, plants encounter **salt stress**. Some plants are affected adversely by the presence of excess salts; other plants can survive (**salt tolerant plants**) or even thrive (**halophytes**). The mechanisms by which plants tolerate salinity are complex (Chapter 14), involving interactions between molecular synthesis, enzyme induction, and membrane transport. In some species, excess salts are not taken up; in others, uptake occurs but the salt is excreted from the plant by salt glands associated with the leaves. To prevent toxic buildup of salts in the cytosol, many plants may sequester these ions in the vacuole (Stewart and Ahmad, 1983).

Summary

Plants are autotrophic organisms capable of synthesizing all their components by using sunlight and inorganic elements. Carbon dioxide from the atmosphere and water and minerals from the soil represent a reservoir of nutrients available to plants. To obtain mineral nutrients from the soil, plants develop an extensive root system that often interacts with microorganisms such as mycorrhizal fungi. Plant root systems can be directly studied with rhizotrons, which allow examination of roots under field conditions.

Studies of plant nutrition have shown that specific elements are essential for plant life. When a plant is deficient in one of these elements, a nutritional disorder with characteristic deficiency symptoms occurs. Nutritional disorders occur because essential elements have key roles in plant metabolism. Mineral nutrition can be studied by using solution culture techniques that allow the characterization of specific nutrient deficiencies.

Soil and plant tissue analysis can provide information on the nutrient status of plants. When crop plants are grown under modern high-production conditions, substantial amounts of mineral nutrients are removed from the soil. To prevent the development of nutrient deficiencies, nutrients can be added back either as fertilizers to the soil or as foliar sprays to the leaves. If mineral salts (essential or nonessential) are present in excess in the soil, plant growth may be adversely affected, resulting in salt stress. Certain plants are able to tolerate excess salts, and other plants, known as halophytes, may even thrive in this condition.

GENERAL READING

Asher, C. J., and Edwards, D. G. (1983) Modern solution culture techniques. In: *Inorganic Plant Nutrition, Encyclopedia of Plant Physiology*, New Series, Vol. 15A, pp. 94–119.

Lauchli, A., and Bieleski, R. L., eds. Springer-Verlag, Berlin.

Bouma, D. (1983) Diagnosis of mineral deficiencies using plant tests. In: *Inorganic Plant Nutrition, Encyclopedia of Plant Physiology*, New Series Vol. 15A, pp. 120–146. Lauchli, A., and Bieleski, R. L. eds. Springer-Verlag, Berlin.

Brady, N. C. (1974) *The Nature and Properties of Soils*, 8th ed. Macmillan Publishing, New York.

Clarkson, D. T. (1985). Factors affecting mineral nutrient acquisition by plants. *Annu. Rev. Plant Physiol.* 36: 77–115.

Clarkson, D. T., and Hanson, J. B. (1980) The mineral nutrition of higher plants. *Annu. Rev. Plant Physiol.* 31: 239–298.

Delwiche, C. C. (1983) Cycling of elements in the biosphere. In: *Inorganic Plant Nutrition, Encyclopedia of Plant Physiology*, New Series, Vol. 15A, pp. 212–238. Lauchli, A., and Bieleski, R. L., eds. Springer-Verlag, Berlin.

Epstein, E. (1972) *Mineral Nutrient of Plants: Principles and Perspectives*. Wiley, New York.

Evans, H. J., and Sorger, G. J. (1966) Role of mineral elements with emphasis on the univalent cations. *Annu. Rev. Plant Physiol.* 17: 47–76.

Feldman, L. J. (1984) Regulation of root development. Annu. Rev. Plant Physiol. 35: 223–242.

Greenway, H., and Munns, R. (1980) Mechanisms of salt tolerance in non-halophytes. *Annu. Rev. Plant Physiol.* 31: 149–190.

Haehnel, W. (1984) Photosynthetic electron transport in higher plants. *Annu. Rev. Plant Physiol.* 35: 659–693.

Hepler, P. K., and Wayne, R. O. (1985) Calcium and plant development. *Annu. Rev. Plant Physiol.* 36: 397–439.

Hewitt, E. J. (1983) A perspective of mineral nutrition: Essential and functional metals in plants. In: *Metals and Micronutrients: Uptake and Utilization by Plants*, pp. 277–323. Robb, D. A., and Pierpoint, W. S., eds. Academic Press, New York.

Lauchli, A., and Bieleski, R. L., eds. (1983) *Inorganic Plant Nutrition, Encyclopedia of Plant Physiology*, New Series, Vol. 15A. Springer-Verlag, Berlin.

Marks, G. C. (1973) *Ectomycorrhizae: Their Ecology and Physiology*. Academic Press, New York.

Marschner, H. (1986) *Mineral Nutrition of Higher Plants*. Academic Press, London.

Mauseth, J. D. (1988) *Plant Anatomy*. Benjamin/Cummings, Redwood City, Calif.

Mengel, K., and Kirkby, E. A. (1979) *Principles of Plant Nutrition*. International Potash Institute, Bern, Switzerland.

Morgan, J. M. (1984) Osmoregulation and water stress in higher plants. *Annu. Rev. Plant Physiol.* 35: 299–319.

Nye, P. H., and Tinker, P. B. (1977) *Solute Movement in the Soil-Root System*. University of California Press, Berkeley.

Russell, R. S. (1977) *Plant Root Systems. Their Function and Interaction with the Soil*. McGraw Hill, London.

Sanders, F. E., Mosse, B., and Tinker, P. B., eds. (1975) *Endomycorrhizae*. Academic Press, London.

Smith, P. F. (1962) Mineral nutrition of plant tissues. *Annu. Rev. Plant Physiol.* 13: 81–108.

Weaver, J. E. (1926) *Root Development of Field Crops*. McGraw-Hill, New York.

CHAPTER REFERENCES

Arnon, D. I., and Stout, P. R. (1939). The essentiality of certain elements in minute quantity for plants with special reference to copper. *Plant Physiol.* 14: 371–375.

Bar-Yosef, B., Kafkafi, U., and Bresler, E. (1972) Uptake of phosphorus by plants growing under field conditions. I. Theoretical model and experiment determination of its parameters. *Soil Sci.* 36: 783–800.

Bienfait, H. F., and Van der Mark, F. (1983) Phytoferritin and its role in iron metabolism. In: *Metals and Micronutrients: Uptake and Utilization by Plants*, pp. 111–123 Robb, D. A., and Pierpoint, W. S., eds. Academic Press, New York.

Broyer, T. C., Carlton, A. B., Johnson, C. M., and Stout, P. R. (1954) Chlorine—a micronutrient element for higher plants. *Plant Physiol.* 29: 526–532.

Chaney, R. L., and Bell, P. F. (1987) Complexity of iron nutrition: Lessons for plant-soil interaction research. *J. Plant Nutr.* 10: 963–994.

Clarkson, D. T., and Sanderson, J. (1978). Sites of absorption and translocation of iron in barley roots. Tracer and microautoradiographic studies. *Plant Physiol.* 61: 731–736.

Greenwood, D. J., Wood, J. T., and Cleaver, T. J. (1974) A dynamic model for the effects of soil and weather conditions on nitrogen response. *J. Agric. Sci.* 82: 455–467.

Huck, M. G., and Taylor, H. M. (1984) The rhizotron as a tool for root research. *Adv. Agron.* 35: 1–35.

Johnson, C. M., Stout, P. R., Brouer, T. G., and Carlton, A. B. (1957) Comparative chlorine requirements of different plant species. *Plant Soil* 8: 337–353.

Juniper, B. E., Groves, S., Landau-Schachar, B., and Audus, L. J. (1966) Root cap and the perception of gravity. *Nature.* 209: 93–94.

Kashirad, A., Marschner, H., and Richter, C. H. (1973) Absorption and translocation of ^{59}Fe from various parts of the corn plant. *Z. Pflanzenernaehr. Bodenkd.* 134: 136–147.

Lucas, R. E., and Davis, J. F. (1961) Relationships between pH values of organic soils and availabilities of 12 plant nutrients. *Soil Sci.* 92: 177–182.

Olsen, R. A., Clark, R. B., and Bennett, J. H. (1981) The enhancement of soil fertility by plant roots. *Amer. Sci.* 69: 378–384.

Parr, A. J., and Loughman, B. C. (1983) Boron and membrane function in plants. In: *Metals and Micronutrients: Uptake and Utilization by Plants*, pp. 87–107. Robb. D. A., and Pierpoint, W. S., eds. Academic Press, New York.

Rovira, A. D., Bowen, C. D., and Foster, R. C. (1983) The significance of rhizosphere microflora and mycorrhizas in plant nutrition. In: *Inorganic Plant Nutrition, Encyclopedia of Plant Physiology, New Series Vol. 15A*, pp. 61–93. Lauchli, A., and Bieleski, R. L., eds. Springer-Verlag, Berlin.

Sanders, F. E., and Tinker, P. B. (1971) Mechanism of absorption of phosphate from soil by endomycorrhizas. *Nature* 233: 278–279.

Stewart, G. R., and Ahmad, I. (1983) Adaptation to salinity in angiosperm halophytes. In: *Metals and Micronutrients: Uptake and Utilization by Plants*, pp. 33–50. Robb, D. A., and Pierpoint, W. S., eds. Academic Press, New York.

Terry, N. (1977) Photosynthesis, growth and the role of chloride. *Plant Physiol.* 60: 69–75.

Tinker, P. B. (1978) Effects of vesicular-arbuscular mycorrhizas on plant nutrition and plant growth. *Physiol. Veg.* 16: 743–751.

Tinker, P. B., and Gildon, A. (1983) Mycorrhizal fungi and ion uptake. In: *Metals and Micronutrients: Uptake and Utilization by Plants*, pp. 21–32. Robb, D. A., and Pierpoint, W. S., eds. Academic Press, New York.

Woolley, J. T., and Broyer, T. C. (1957) Foliar symptoms of deficiencies of inorganic elements in tomato. *Plant Physiol.* 32: 148–151.

Ziegler, H. (1987) The evolution of stomata. In: *Stomatal Function*, pp. 29–57. Zeiger, E., Farquhar, G., and Cowan, I., eds. Stanford University Press, Stanford, Calif.

Zobel, R. W., Tredici, P. D., and Torrey, J. G. (1976) Method for growing plants aeroponically. *Plant Physiol.* 57: 344–346.

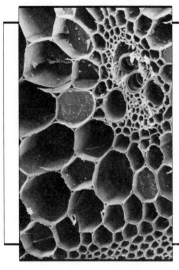

CHAPTER 6

Solute Transport

Plants share with other living organisms a capacity to maintain an internal environment with a composition different from that of their surroundings. This internal environment remains more or less constant even when the outside environment undergoes rapid or extreme changes. There is, however, a continuous movement of molecules and ions between the plant and its environment that allows the plant to accumulate nutrients and eliminate metabolic by-products. Precise regulation of the movement of molecules and ions in and out of the cell depends on the properties of the plasma membrane (Chapter 1).

Molecular and ionic movement between different compartments in biological systems is known as **transport**. Transport can occur between the outside and the inside of a plant, between different points within the plant (as in transport by the xylem or the phloem), and between a cell and the extracellular environment. Ultimately, however, transport at the cellular level is responsible for all higher-order transport events. For example, phloem transport involves movement of sucrose out of a photosynthetic cell, uptake by a sieve tube member, movement through the sieve tube, and finally uptake by a nonphotosynthetic cell.

Transport is a highly selective process; some compounds are accumulated to concentrations far greater than their concentrations in the natural environment, while other compounds are specifically excluded. Transport between the cell and its environment is controlled by the plasma membrane. This membrane, which separates the cytoplasm from the external environment, determines the types of molecules that move into and out of the cell and the direction and rate of transport. In this chapter we consider the physical, chemical, and biological bases of molecular transport across plant cell membranes. First, the physical and chemical principles that govern the movement of molecules in solution are described. Then we show how these principles apply to biological systems and discuss the molecular mechanisms of transport in living cells. Finally, we consider how, within the plant, transport between cells is organized and controlled.

Passive and Active Transport

For many years, most research activity in the field of biological membrane transport sought to determine whether the transport of a particular ion or solute was active or passive. **Passive transport** was understood to encompass movements that occurred without the cell's doing any work (i.e., without energy being expended). **Active trans-**

port, on the other hand, referred to ion and solute movements that required some direct input of cellular energy. As we have come to a better understanding of the mechanisms by which substances move into and out of cells, the differences between active and passive transport have become less clear-cut. Nonetheless, active transport and passive transport remain central concepts in this field and must be understood before many basic aspects of cell physiology can be appreciated.

An understanding of transport requires a discussion of the physical and chemical principles that govern the movement of molecules in free solution. In Chapter 3 it was shown that whenever a concentration gradient exists for a given solute, the gradient decays spontaneously over time as a result of **diffusion.** Consider the experiment depicted in Figure 6.1. Initially, the bag, which contains the dissolved solute sucrose, is placed in a medium lacking sucrose. Since the bag is somewhat permeable to sucrose—that is, sucrose can slowly pass through the bag—the sucrose inside the bag begins to diffuse out (down its concentration gradient) as a result of the random thermal motions of the individual sucrose molecules. Diffusion will continue until, at equilibrium, the sucrose concentrations inside and outside the bag are equal (Fig. 6.1b).

Fick's law (see Eq. 3.1) indicates that diffusion will always proceed spontaneously, down a concentration or chemical gradient, until equilibrium is reached. At equilibrium, no further *net* movements of solute can occur without the application of some driving force. Movements of substances against a concentration or chemical gradient are not spontaneous and require that work be done on the system by the application of free energy. In the example in Figure 6.1, the driving force for net sucrose diffusion is the potential energy gradient established at the beginning of the experiment because of the concentration difference. As we stated in Chapter 3, biological transport can also be driven by other forces: hydrostatic pressure, gravity, and electrical fields. (In biological systems, however, gravity seldom contributes substantially to the force driving transport.) Along with concentration gradients, these sources of potential energy define the **chemical potential** of any solute (Chapter 2) according to the following relation:

$$\bar{\mu}_j \quad = \quad \mu_j^* \quad + \quad RT \ln C_j \quad + \quad z_j FE \quad + \quad \bar{V}_j P$$

| Chemical potential for a given solute, j | Chemical potential of j under standard conditions | Concentration (activity) component | Electrostatic charge component | Hydrostatic pressure component |

$$(6.1)$$

Here $\bar{\mu}_j$ is the chemical potential of solute species j in joules per mole, μ_j^* is its chemical potential under standard conditions (a correction factor that will cancel out in future equations and so can be ignored), R is the universal gas constant, T is the absolute temperature, and C_j is the concentration (more accurately the activity) of j. The electrical term, $z_j FE$, applies only to dissolved ions; z is the electrostatic charge of the ion (+1 for monovalent cations, −1 for monovalent anions, and so on), F is Faraday's constant (equivalent to the electrical charge on 1 mol of protons), and E is the overall electrical potential of the solution (with respect to ground). The final term, $\bar{V}_j P$, expresses the contribution of the volume (V) and pressure (P) of j to the chemical potential of j. This final term makes a much smaller contribution to $\bar{\mu}_j$ than do the concentration and electrical terms except in the very important case of osmotic water movements. As discussed in Chapter 3, the chemical potential of water (i.e., the water potential) depends on the concentration of dissolved solutes and the hydrostatic pressure on the system.

The importance of the concept of chemical potential (Nobel, 1983) is that it sums all the forces that may act on a molecule to drive net transport. In general, diffusion (or passive transport) always moves molecules from areas of higher chemical potential downhill to areas of lower chemical potential. Movement against a chemical potential gradient is indicative of active transport (Fig. 6.2).

In the example of sucrose diffusion (Fig. 6.1), the chemical potential of sucrose in each compartment is approximated quite accurately by the concentration term alone (unless a solution is very concentrated, causing hydrostatic pressure to build up). The chemical potential of sucrose inside a cell can be described as follows (in the next three equations, the subscript "s" stands for sucrose and the superscripts "i" and "o" stand for inside and outside, respectively):

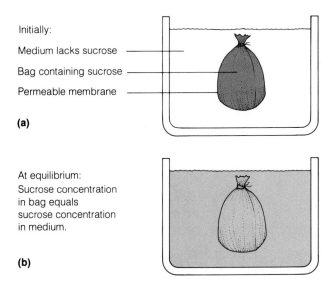

Initially:

Medium lacks sucrose ————

Bag containing sucrose ————

Permeable membrane ————

(a)

At equilibrium:
Sucrose concentration in bag equals sucrose concentration in medium.

(b)

FIGURE 6.1. Diffusion across a permeable barrier. When a bag with a sucrose solution is placed in a sucrose-free medium, sucrose diffuses out from the bag into the medium. Net movement of sucrose molecules stops when the concentration of sucrose in the medium equals that in the bag.

FIGURE 6.2. Relationship between the chemical potential, $\bar{\mu}$, and the transport of molecules across a permeability barrier. The net movement of molecular species j between compartments 1 and 2 depends on the relative magnitude of the chemical potential in each compartment, represented here by the height of the bars. Movement down a chemical gradient occurs spontaneously; movement against a gradient requires energy and is called active transport.

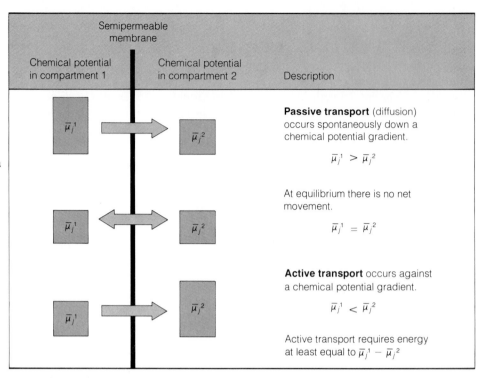

The chemical potential of sucrose inside the cell is

$$\bar{\mu}_s^i = \mu_s^* + RT \ln C_s^i$$

| Chemical potential of sucrose solution inside the cell | Chemical potential under standard conditions | Concentration component |

The chemical potential of sucrose outside the cell is

$$\bar{\mu}_s^o = \mu_s^* + RT \ln C_s^o$$

The driving force for diffusion is simply the difference in chemical potential between the solutions inside and outside the cell. The gradient is proportional to the difference in chemical potential:

$$\Delta\bar{\mu}_s = \bar{\mu}_s^o - \bar{\mu}_s^i$$
$$= (\mu_s^* + RT \ln C_s^o) - (\mu_s^* + RT \ln C_s^i)$$
$$= RT(\ln C_s^o - \ln C_s^i) \quad (6.2)$$
$$= RT \ln \frac{C_s^o}{C_s^i}$$

In other words, the driving force ($\Delta\bar{\mu}_s$) for solute diffusion is proportional to the magnitude of the concentration gradient (C_s^o/C_s^i). However, if the solute carries an electrical charge (as does the potassium ion), the electrical component of the chemical potential must also be considered. Suppose the bag in Figure 6.1 contains a KCl solution rather than sucrose. Because the ionic species (K^+ and Cl^-) diffuse independently, each has its own chemical potential. Thus the driving force for K^+ diffusion would be

$$\Delta\bar{\mu}_K = \bar{\mu}_K^o - \bar{\mu}_K^i$$
$$= RT \ln \frac{[K^+]^o}{[K^+]^i} + zF(E^o - E^i)$$

and because the electrostatic charge of K^+ is $+1$, $z = +1$ and

$$\Delta\bar{\mu}_K = RT \ln \frac{[K^+]^o}{[K^+]^i} + F(E^o - E^i) \quad (6.3)$$

A similar expression can be written for Cl^- (but remember that $z_{Cl^-} = -1$).

Equation 6.3 shows that ions, such as potassium ions, diffuse in response to both their concentration gradients ($[K^+]^o/[K^+]^i$) *and* any electrical potential difference between the two compartments ($E^o - E^i$). One very important implication of this equation is that ions can be driven *against* their concentration gradients if one applies a voltage (electrical field) between the two compartments. Because of the importance of electrical fields in biological transport, $\bar{\mu}$ is often called the **electrochemical potential** and $\Delta\bar{\mu}$ is the difference in electrochemical potential between two compartments.

Transport of Solutes Across a Membrane Barrier

If the two KCl solutions in the previous example are separated by a real biological membrane, diffusion is complicated by the fact that the ions must move through the membrane as well as across the open solutions. The extent to which a membrane permits or restricts the movement of a substance is called **membrane permeability**. As will be discussed later, permeabilities depend on the membrane's composition as well as the chemical nature of the solute. In a loose sense, permeability can be expressed in terms of a diffusion coefficient for the solute in the membrane. However, permeabilities are influenced by several additional and difficult-to-measure factors, such as membrane thickness. Despite their theoretical complexity, permeabilities can readily be measured by determining the rate at which a solute passes through a membrane under a specific set of conditions. Unless the permeability of the membrane to a particular solute is zero, the membrane will slow diffusion and thus reduce the speed with which equilibrium is reached. The membrane itself, however, cannot alter the final equilibrium conditions.

A Diffusion Potential Develops When Oppositely Charged Ions Are Transported Across a Cell Membrane at Different Rates

When salts diffuse across a membrane, an electrical membrane potential (voltage) can develop. Consider the two KCl solutions separated by a membrane in Figure 6.3. The K^+ and Cl^- will permeate the membrane independently as they diffuse down their respective electrochemical gradients. Unless the membrane is very porous, its permeabilities for the two ions will differ. Consequently, K^+ and Cl^- will initially diffuse across the membrane at different rates. The result will be a slight separation of charges, which creates an electrical potential across the membrane. In biological systems, membranes are usually more permeable to K^+ than to Cl^-. Therefore, K^+ will diffuse out of the cell (compartment 1 in Fig. 6.3) faster than Cl^-, causing the cell to develop a negative electrical charge with respect to the medium. The membrane potential that develops as a result of diffusion is called the **diffusion potential**.

An important principle that must always be kept in mind when the movement of ions across membranes is considered is the **principle of electrical neutrality**. Bulk solutions always contain equal numbers of anions and cations. The existence of a membrane potential implies that there is an uneven distribution of charges across the membrane; however, the actual number of unbalanced ions is negligible in chemical terms. For example, a membrane potential of -100 mV, like that found across the plasma membranes of many plant cells, results from the presence of only one extra anion out of every 100,000 within the cell! As depicted in Figure 6.3, all of these extra anions are found immediately adjacent to the membrane's surface; there is no charge imbalance at all throughout the bulk of the cell. In our example of KCl diffusion across a membrane, electrical neutrality is preserved because, as

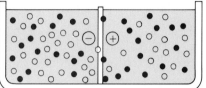

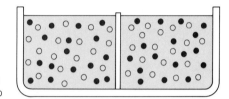

FIGURE 6.3. Development of a diffusion potential and charge separation between two compartments separated by a membrane that is preferentially permeable to cations. If the concentration of potassium chloride is higher in compartment 1 ($[KCl]_1 > [KCl]_2$), potassium and chloride will diffuse at a higher rate into compartment 2, and a diffusion potential will be established. When membranes are more permeable to potassium than to chloride, potassium ions will diffuse faster than chloride ions, and charge separation, indicated in the figure by ($+$) and ($-$), will occur.

Initial conditions: $[KCl]_1 > [KCl]_2$

Diffusion potential exists until chemical equilibrium is reached

Equilibrium conditions: $[KCl]_1 = [KCl]_2$

At chemical equilibrium diffusion potential equals zero

K^+ moves ahead of Cl^- in the membrane, the resulting diffusion potential retards the movement of K^+ and speeds that of Cl^-. Ultimately, both ions actually diffuse at the same rate, but the diffusion potential persists. As the system moves toward equilibrium and the concentration gradient collapses, the diffusion potential also collapses.

The Nernst Equation Describes the Relationship Between the Voltage Difference Across a Membrane and the Distribution of a Given Ion Under Equilibrium Conditions

When transport across a membrane reaches equilibrium, the flux (J) (Chapter 3) of any solute is the same in the two directions, outside to inside and inside to outside:

$$J_{o \to i} = J_{i \to o}$$

We have already seen that fluxes are proportional to the driving forces; thus, at equilibrium, the electrochemical potentials will be the same:

$$\bar{\mu}_j^o = \bar{\mu}_j^i$$

and for any given ion (the ion is symbolized here by the subscript j)

$$\mu_j^* + RT \ln C_j^o + z_j F E^o = \mu_j^* + RT \ln C_j^i + z_j F E^i$$

By rearranging this equation, the difference in electrical potential between the two compartments ($E^i - E^o$) is obtained:

$$E^i - E^o = \frac{RT}{z_j F} \left(\ln \frac{C_j^o}{C_j^i} \right)$$

This electrical potential difference is known as the **Nernst potential** (ΔE_n) for that ion:

$$\Delta E_{n_j} = E^i - E^o$$

and

$$\Delta E_{n_j} = \frac{RT}{z_j F} \left(\ln \frac{C_j^o}{C_j^i} \right)$$

or

$$\Delta E_{n_j} = \frac{2.3 RT}{z_j F} \log \frac{C_j^o}{C_j^i} \tag{6.4}$$

This relationship, known as the **Nernst equation**, states that at equilibrium the difference in concentration of an ion between two compartments is balanced by the voltage difference between the compartments. The Nernst equation can be further simplified for a univalent ion at 25°C:

$$\Delta E_n = 59 \log \frac{C_j^o}{C_j^i} \tag{6.5}$$

Note that a tenfold difference in concentration corresponds to a Nernst potential of 59 mV. That is, a membrane potential of 59 mV would maintain a tenfold concentration gradient of an ion that is transported by passive diffusion. Similarly, if a tenfold concentration gradient of an ion existed across the membrane, passive diffusion of that ion down its concentration gradient would result in a 59-mV difference across the membrane. All living cells exhibit a membrane potential due to the asymmetric ion distribution between the inside and outside of the cell. These membrane potentials can readily be determined by inserting a microelectrode into the cell and measuring the voltage difference between the inside of the cell and the external bathing medium (Fig. 6.4).

In the case of a membrane selectively permeable to one ion, the Nernst equation can be used to determine whether that ion is distributed actively or passively across a membrane. If the Nernst potential, ΔE_n, for an ion equals the measured membrane potential, one can conclude that the ion under consideration is distributed passively across the membrane. Therefore, the first and simplest experimental test for biological active transport is to determine the concentration gradient for an ion inside and outside a cell; then the calculated Nernst

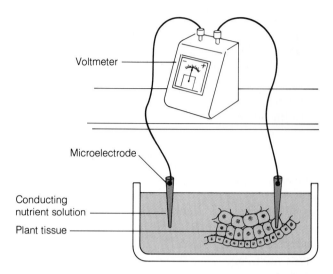

FIGURE 6.4. Diagram of a pair of microelectrodes used to measure membrane potentials across cell membranes. One of the glass microelectrodes is inserted into the cell compartment under study (usually the vacuole or the cytoplasm), while the other is kept in an electrolytic solution that serves as a reference. The microelectrodes are connected to a voltmeter, which records the potential difference between the cell compartment and the solution. Typical membrane potentials across plant membranes range from −60 to −240 mV.

potential for the ion is compared with the experimentally measured value of the membrane potential. Any significant disparity between ΔE_n and the measured membrane potential (ΔE_m) suggests that the ion is transported actively.

The Goldman Equation Describes the Relationship Between the Diffusion Potential and Prevailing Ion Gradients Across a Membrane

Many different ions permeate the membranes of living cells simultaneously; therefore, the Nernst potential for any single ionic species seldom describes the membrane's diffusion potential. Instead, all the diffusion potentials across the membrane are described by the **Goldman equation**:

$$\Delta E_m = \frac{RT}{F} \ln \frac{P_K C_K^o + P_{Na} C_{Na}^o + P_{Cl} C_{Cl}^i}{P_K C_K^i + P_{Na} C_{Na}^i + P_{Cl} C_{Cl}^o} \tag{6.6}$$

This equation relates the standing ion gradients across a membrane to the diffusion potential that develops. P_K, P_{Na}, and P_{Cl} represent the membrane permeabilities for K^+, Na^+, and Cl^-, respectively. Although the equation should include terms for all ions passing through the membrane, K^+, Na^+, and Cl^- have the largest membrane permeabilities and highest concentrations in plant cells and therefore dominate the equation.

The Goldman equation applies to conditions in which ions are moving passively and ion fluxes are at steady state. **Steady state** is the condition in which ion concentrations and the membrane potential are constant with respect to time. Steady state is *not* the same as equilibrium (Fig. 6.2); the diffusive ion fluxes of living cells never reach equilibrium because of the continuous presence of active transport.

The relationship between the Goldman equation and the Nernst equation can be seen if one imagines a membrane that is permeable to only one ion, say K^+, so that both P_{Cl} and $P_{Na} = 0$. Under these conditions, the Goldman equation reduces to the Nernst equation for K^+. Biological membranes are never permeable to only a single ionic species; however, artificial membranes can approximate this situation and are used in the manufacture of pH electrodes and other ion-specific electrodes.

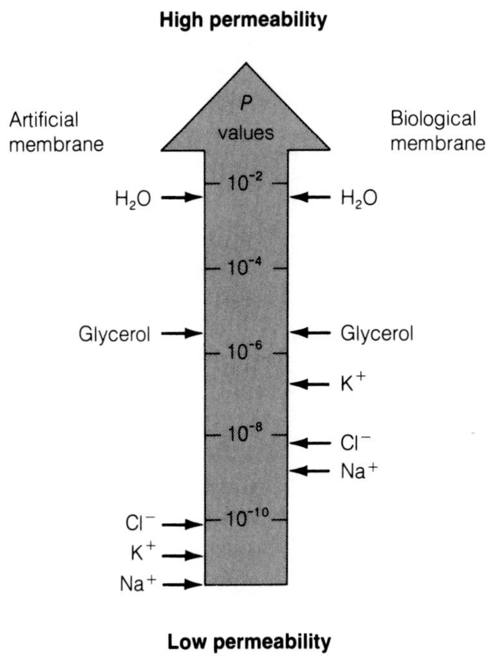

Transport Across Biological Membranes

Artificial bilayers, such as black lipid membranes, have been used extensively to study the permeability of pure phospholipid membranes. Because of the nonpolar nature of the membrane's interior, pure phospholipid bilayers are highly impermeable to ions or polar molecules. Figure 6.5 shows the permeabilities of artificial bilayers for a variety of substances. Nonpolar molecules such as O_2 pass through these membranes very rapidly, and the membrane permeabilities for small polar molecules (water, CO_2, glycerol, etc.) are slightly lower but still significant. Among polar molecules, water has an unusually high membrane permeability, probably because of its strong interactions with the polar head groups of the lipids. In general, as molecules increase in size and polarity, their membrane permeabilities decrease. Ions penetrate through pure phospholipid bilayers so slowly that these artificial membranes may be considered impermeable to them.

When the permeabilities of artificial bilayers for ions and solutes are compared with those of natural biological membranes, important similarities and differences emerge (Fig. 6.5). For nonpolar molecules and many small polar molecules, both types of membranes have similar permeabilities. On the other hand, for ions and large polar molecules such as sugars, natural membranes are much more permeable than artificial bilayers. The reason is that, unlike artificial bilayers, natural membranes contain **transport proteins** that facilitate the passage of ions and other polar molecules. The molecules for which natural and artificial membranes exhibit similar permeabilities apparently diffuse directly through the lipid phase of the membrane without the assistance of transport proteins.

FIGURE 6.5. Permeability values, *P*, in centimeters per second, for some substances diffusing across an artificial barrier, such as a black lipid membrane, and across a biological membrane. For uncharged substances, such as water, *P* values are similar in both systems, but for charged substances, such as potassium or sodium, *P* values are higher in biological membranes, reflecting the role of transport proteins. Note the logarithmic scale.

Two Main Types of Membrane Proteins Enhance the Movements of Substances Across Membranes

Transport specificity is a function of the various transport proteins found in the membrane. A particular transport protein is usually highly specific for the kinds of substances it will transport; however, it will generally also transport a family of related substances. For example, in plants the K^+ transporter on the plasma membrane may transport Rb^+ and Na^+ in addition to K^+, but K^+ is preferred. At the same time, it is completely ineffective in transporting anions such as Cl^- or uncharged solutes such as sucrose. Similarly, a protein involved in the transport of neutral amino acids may move glycine, alanine, and valine with equal ease but will not accept aspartic acid or asparagine.

There are two types of membrane transporters: **channels** and **carriers** (Figure 6.6). In general, channels are transmembrane proteins that function as selective pores in the membrane. The size of a pore and the density of surface charges on its interior lining determine its transport specificity. Molecules that can move through channels do so through the pore. In carrier-mediated transport, the substance being transported is initially bound by an active site on the carrier protein. Binding

FIGURE 6.6. Two types of specialized membrane proteins can facilitate transport across membranes. Channel proteins act as membrane pores, and their specificity is determined primarily by the biophysical properties of the channel. Carrier proteins actually bind the transported molecule on one side of the membrane and release it on the other side. Passive transport through channels and carriers mediating facilitated diffusion does not require metabolic energy. Active transport through carriers uses energy, usually from ATP hydrolysis.

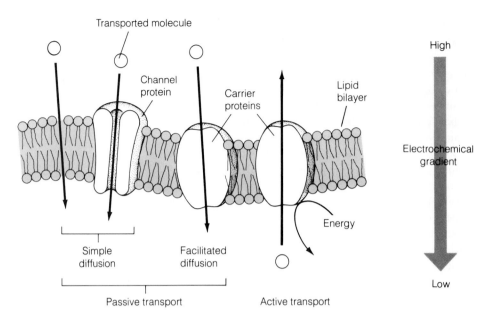

leads to a conformational change of the protein, which exposes the substance to the solution on the other side of the membrane. Transport is complete when the substance dissociates from the carrier's binding site. Carrier-mediated transport can be either passive or active. Passive transport on a carrier (down the substance's electrochemical gradient) is sometimes called **facilitated diffusion**. Carrier-mediated active transport systems that move solutes against a chemical or electrochemical gradient are often called **pumps**.

The binding and release of a molecule at a specific site on a protein that occur in carrier-mediated transport are similar to the binding and release of molecules from an enzyme in an enzyme-catalyzed reaction. In fact, the kinetic analyses that are used to study enzyme reactions can also be used experimentally to distinguish between channel-mediated diffusion and carrier-mediated transport (Chapter 2).

Measurements of Concentrations and Membrane Potentials Can Distinguish Between Active and Passive Transport

In the case of nonionized solutes, it is possible to establish whether they are taken up (or extruded) actively or passively simply by measuring their internal and external concentrations. However, in the case of charged ions, that assessment also requires consideration of the magnitude of the membrane potential. For cells at steady state, whether transport is active or passive can be determined by using the Nernst equation (Eq. 6.4). If the Nernst potential for a particular ion is equal to the membrane potential, the ion is distributed passively across the membrane. Large differences between calculated Nernst potentials and measured membrane potentials indicate

active transport. This approach works for many plant cells and tissues because their ion distributions approximate a steady state. For cells that are not at steady state, a more complex approach that is beyond the scope of this chapter is required.

Table 6.1 shows how the experimentally measured ion gradients of pea root cells compare with predicted values calculated from the Nernst equation (Higinbotham et al., 1967). In this example, the external concentration of each ion in the solution bathing the tissue and the measured membrane potential were substituted into the Nernst equation, and a predicted internal concentration was calculated for that ion. Notice that, of all the ions, only K^+ is at or near equilibrium. The anions all have higher internal concentrations than predicted, indicating their active uptake. The cations Na^+, Ca^{2+}, and Mg^{2+} have lower internal concentrations than predicted; these

TABLE 6.1. Comparison of observed and predicted ion concentrations in pea root tissue. The membrane potential was measured as −110 mV.

Ion	Concentration in external medium (mmol L^{-1})	Predicted internal concentration (mmol L^{-1})	Observed internal concentration (mmol L^{-1})
K^+	1	74	75
Na^+	1	74	8
Mg^{2+}	0.25	1,350	1.5
Ca^{2+}	1	5,400	1
NO_3^-	2	0.0272	28
Cl^-	1	0.0136	7
$H_2PO_4^-$	1	0.0136	21
SO_4^{2-}	0.25	0.000047	9.5

Data from Higinbotham et al. (1967)

ions enter the cell by diffusion down their electrochemical gradients and then are actively extruded.

The example shown in Table 6.1 is actually an oversimplification because plant cells have several internal compartments, which can differ in their ionic composition. The cytosol and the vacuole are the most important intracellular compartments determining the ionic relations of plant cells. In mature plant cells, the central vacuole often occupies 90% or more of the cell's volume and the cytosol is restricted to a thin layer around the cell's periphery. Because of its small volume, the cytosol of most angiosperm cells is difficult to assay chemically. For this reason, much of the work on the ionic relations of plants has focused on certain green algae, such as *Chara* and *Nitella*, whose cells are several inches long and can contain an appreciable volume of cytosol. The conclusions from these studies and from related work with higher plants are shown diagrammatically in Figure 6.7. Potassium is accumulated passively by both the cytosol and the vacuole except when extracellular K^+ concentrations are very low, in which case it is taken up actively. Sodium is pumped actively out of the cytosol into the extracellular spaces and vacuole. Excess protons,

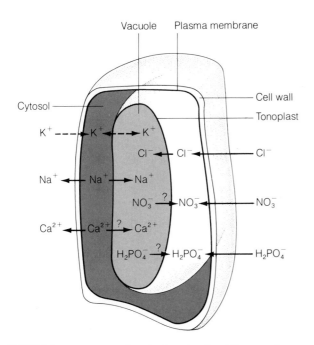

FIGURE 6.7. Ion concentrations in the cytosol and the vacuole are controlled by passive (dashed arrows) and active (solid arrows) transport processes. The arrows indicate the direction of net transport. In most plant cells the vacuole occupies up to 90% of the cell's volume and contains the bulk of the cell solutes. Control of the ionic concentrations in the cytosol is important for the enzymatic activity involved in the cell's metabolism. The cell wall surrounding the plasma membrane does not represent a permeability barrier and hence is not a factor in solute transport. Question marks indicate uncertainties about the nature of the transport processes.

generated by intermediary metabolism, are also actively extruded from the cytosol. This process helps maintain the cytosolic pH near neutrality, while the vacuole and the extracellular medium are generally more acidic by one to two pH units. All the anions are taken up actively into the cytosol. The situation of divalent cations, especially Ca^{2+}, is more complex and will be considered in more detail later.

Electrogenic Proton Transport Is a Major Determinant of the Membrane Potential

Regulation of ion fluxes at the cell membrane is essential to ensure an adequate supply of nutrients for metabolic needs and to provide a driving force for ion fluxes against diffusion gradients. Because of the differential permeability of the membrane to the prevalent anions and cations and the ensuing ion gradients arising from passive ion fluxes, all cells have an electrical potential across the plasma membrane. In most plant cells, K^+ has both the greatest internal concentration and the highest membrane permeability. This condition makes the membrane potential inside negative because K^+ tends to diffuse out of the cell faster than any anion.

When permeabilities and ion gradients are known, it is possible to calculate a diffusion potential for the membrane from the Goldman equation (Eq. 6.6). In plants and fungi, experimentally measured membrane potentials are usually much more negative than those calculated from the Goldman equation. For example, cells in stems and roots of young seedlings generally have membrane potentials of -110 to -130 mV, whereas their calculated diffusion potentials are usually only -50 to -80 mV. Thus, in addition to the diffusion potential, the membrane potential has a second component. Where does this excess voltage come from?

Whenever an ion moves into or out of a cell unbalanced by a counterion of opposite charge, a voltage is created across the membrane. Such transport is called **electrogenic** and is common in living cells. We usually think of electrogenic transport as occurring against an electrochemical gradient. This process requires energy, which is often supplied by the hydrolysis of ATP. Thus electrogenic transport is active, and the carrier that mediates it is called an **electrogenic pump**. Figure 6.8 shows how a particular type of electrogenic pump might work. At rest, the protein forms an occluded pore in the membrane with sites on its cytoplasmic side for binding both the transportable ion and ATP. When these substances are bound, the ATP is hydrolyzed and the protein becomes phosphorylated. A conformational change then occurs in the protein, opening the transport pathway to the outside and simultaneously closing it on the cytoplasmic side. The ion then leaves its binding site, and the protein reverts to its original conformation. Finally,

FIGURE 6.8. Hypothetical steps in the transport of a cation (M^+) against its chemical gradient by an electrogenic pump. The protein, embedded in the membrane, binds the cation on the inside of the cell (a) and is phosphorylated by ATP (b). This phosphorylation leads to a conformational change that exposes the cation to the outside of the cell and makes it possible for the cation to diffuse away (c). Release of the phosphate ion (P) from ATP to the inside of the cell (d) restores the initial configuration of the membrane protein and allows a new pumping cycle to begin.

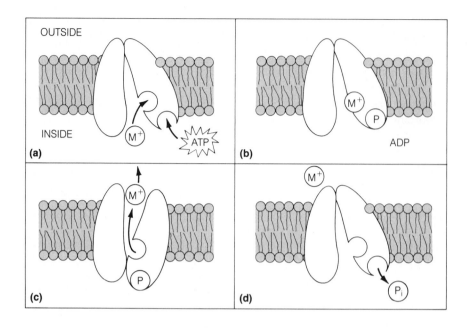

the phosphate group leaves and the cycle can start again. Phosphorylation of the transport protein coupled to ATP hydrolysis is typical of plasma membrane ATPases in general, including the proton pump in the plasma membrane of plant cells. Because of this feature, the plasma membrane ATPase is strongly inhibited by orthovanadate, a phosphate analog that competes with the phosphate from ATP for the phosphorylation site on the enzyme.

In principle, any ion could be transported by an electrogenic ATPase, the transport specificity being determined by the protein's ion binding site. However, only a few electrogenic pumps have been identified. In animal cells, an Na^+/K^+-transporting ATPase has binding sites for both Na^+ and K^+ and pumps three Na^+ out of the cell for every two K^+ pumped in. This imbalance leads to a slight negative charge inside the cell. However, because of the relatively high passive fluxes of Na^+ and K^+ across most animal plasma membranes, the Na^+/K^+ pump contributes little to the membrane potential. Because of these properties, the membrane potential of animal cells calculated from the Goldman equation approximates measured potentials reasonably well, indicating that membrane potentials in these cells are primarily a result of passive diffusion of ions down their concentration gradients.

Plants do not contain an Na^+/K^+-ATPase. Instead, H^+ extrusion appears to be the principal electrogenic ion flux across the plasma membrane (see box on chemiosmosis). The H^+-ATPase drives protons from the cytosol to the external medium, creating both a pH gradient and a large inside-negative membrane potential.

In plants the dependence of the membrane potential on ATP can be demonstrated by observing the effect of cyanide (CN^-) on the membrane potential (Fig. 6.9). Cyanide rapidly poisons the mitochondria, and the cell's ATP is depleted. As ATP synthesis is inhibited, the membrane potential falls to the level of the Goldman diffusion potential, which is due to the passive movements of K^+, Cl^-, and Na^+. Thus, the membrane potentials of

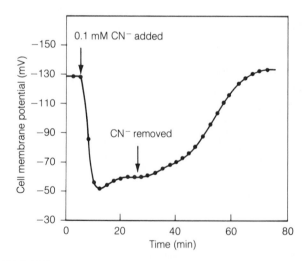

FIGURE 6.9. The membrane potential of a pea cell collapses when cyanide (CN^-) is added to the bathing solution. Cyanide blocks ATP production in the cells by poisoning the mitochondria. The collapse of the membrane potential upon addition of cyanide indicates that an ATP supply is necessary for the maintenance of the potential. Washing the cyanide out of the tissue results in slow recovery of ATP production and restoration of the membrane potential. (From Higinbotham et al., 1970.)

Chemiosmosis in Action

THE BRITISH NOBEL laureate Peter Mitchell coined the term **chemiosmosis** to explain the coupling between electrochemical gradients of protons across selectively permeable membranes and the performance of cellular work (Harold, 1986). In plants, proton gradients play a role in transport across the plasma membrane and tonoplast and drive both transport and ATP synthesis on the inner membranes of chloroplasts and mitochondria. The passage of protons along their electrical and chemical gradients can be coupled to cellular work because the free energy change is negative. This process can be represented by the equation

$$\Delta G = \Delta \bar{\mu}_{H^+} = F \Delta E_m + 2.3RT \log \frac{[H^+]_i}{[H^+]_o}$$

where $\Delta \bar{\mu}_{H^+}$ is the proton electrochemical gradient, F is the Faraday constant, ΔE_m is the membrane potential, R is the gas constant, T is the absolute temperature, and the subscripts "i" and "o" refer to the inside and outside of the cell or other membrane-bound compartment. Mitchell introduced the term **proton motive force (Δp)** for the difference in electrochemical potential between protons inside and outside a cellular compartment, $\bar{\mu}_{H^+}$. It is convenient to express Δp in units of electrical potential; that can be accomplished by dividing both sides of the foregoing equation by the Faraday constant, F, which changes the units to millivolts:

$$\Delta p = \frac{\Delta \bar{\mu}_{H^+}}{F} = \Delta E_m + \frac{2.3RT}{F} \log \frac{[H^+]_i}{[H^+]_o}$$

Since $pH = -\log[H^+]$, the term $\log([H^+]_i/[H^+]_o)$ simplifies to $(pH_i - pH_o)$. If ΔpH is defined as $pH_i - pH_o$, this yields the general expression for proton motive force:

$$\frac{\Delta \tilde{\mu}_{H^+}}{F} = \Delta p = \Delta E_m - \frac{2.3RT}{F} \Delta pH$$

At 25°C, $2.3RT/F = 59$ mV, and substituting this value into the foregoing expression results in the most commonly used equation:

$$\Delta p = \Delta E_m - 59 \, \Delta pH$$

Let us consider the example of a cell bathed in a solution of 1 mM KCl and 1 mM sucrose. Proton pumping by an H^+-ATPase results in a membrane potential of -120 mV and a pH difference between the inside and the outside of the cell of two pH units. Thus, from Equation 6.7,

$$\Delta p = -120 \text{ mV} - 59(2) \text{ mV} = -238 \text{ mV}$$

How much K^+ can the cell take up by using that Δp? Because potassium is positively charged and the membrane potential of the cell is inside negative, potassium will be taken up by the electrical component of the Δp, which equals -120 mV. Using the Nernst equation (Eq. 6.5), we can calculate that an external K^+ concentration of 1 mM and an E_m of -120 mV will equilibrate with an internal K^+ concentration of 100 mM. Thus, using the electrical component of the Δp, the cell can generate a 100-fold K^+ gradient across the membrane.

The Δp can also be used to take up sucrose against a concentration gradient, usually via a proton/sucrose symporter. Because sucrose is co-transported with a proton, both components of the electrochemical gradient (238 mV in our example) can be used for sucrose uptake (Harold, 1986). From Equation 6.2, we can calculate that an external sucrose concentration of 1 mM and a Δp of 238 mV would equilibrate with an internal sucrose concentration of 10 M! In real life, however, such concentration gradients would not exist: sucrose would back-diffuse out of the cell, and regulatory mechanisms at the membrane would repress the function of the symporter after certain critical concentrations were attained.

plant cells have two components: a diffusion potential and a component resulting from electrogenic transport (Spanswick, 1981). When cyanide inhibits electrogenic ion transport, the pH of the external medium increases while the cytosol becomes acidic because H^+ remains inside the cell. This is one piece of evidence that H^+ is the electrogenically transported ion.

Electrogenic H^+ transport occurs not only in plants but also in bacteria, algae, fungi, and some animal cells, such as those of the gastric epithelium. ATP synthesis in mitochondria and chloroplasts also depends on an H^+-ATPase. In mitochondria and chloroplasts, this transport protein is sometimes called an ATP synthase because it forms ATP rather than hydrolyzing it. The mitochondrial ATP synthase is structurally different from the H^+-ATPases located in plasma membranes.

Electrogenic Calcium Transport by an ATPase Regulates Intracellular Calcium Concentrations

Calcium is another important ion that may be transported electrogenically (Kauss, 1987). The wall and apoplastic spaces are rich in calcium, but cytosolic Ca^{2+} concentrations are maintained at very low levels (in the micromolar range) despite a strong electrochemical gradient driving Ca^{2+} diffusion into the cell. Small fluctuations in cytosolic Ca^{2+} concentration drastically alter the activities of many enzymes. Some crucial metabolic reactions are regulated through changes in cytosolic Ca^{2+} levels. One way in which plant cells achieve this regulation is by modulating the activity of Ca^{2+}-ATPases that drive Ca^{2+} out of the cytoplasm back into the extracellular spaces. In certain animal membranes Ca^{2+} transport is electrogenic. The mechanism and regulation of Ca^{2+} transport in plants are still somewhat unclear and are active areas of current research.

Not all transport ATPases are electrogenic. Some, such as the K^+/H^+-ATPase of bacteria, are electrically neutral. This ATPase drives both K^+ and H^+ against their electrochemical gradients, but because equal numbers of K^+ and H^+ ions are moving in opposite directions, their charges cancel out. It is not known whether such electroneutral transport ATPases exist in plants; however, one must always keep that possibility in mind, especially as new transport systems are studied. For example, the plant Ca^{2+}-ATPase mentioned above might turn out to catalyze an electroneutral exchange of two H^+ for each Ca^{2+}.

Cotransport Processes Use the Energy Stored in the Proton Motive Force

Why do cells expend ATP on electrogenic transport when electrically neutral transport would be metabolically less expensive? One reason is that electrogenic transporters help regulate metabolism by keeping the cytosolic concentrations of H^+, Na^+, K^+, and Ca^{2+} within permissible limits. However, the consequences of electrogenic transport are much more far-reaching. When protons are extruded from the cell electrogenically, both a membrane potential and a pH gradient are created at the expense of ATP hydrolysis. This electrochemical H^+ gradient, which is often termed the **proton motive force** or **Δp**, actually represents stored free energy in the form of the H^+ gradient.

Pure phospholipid bilayers are quite impermeable to H^+. However, plant cell membranes contain special transport proteins that allow H^+ to diffuse back into the cell if it moves with another ion or solute. In this way the proton motive force generated by electrogenic H^+ trans-

port is used to drive the transport of many other substances against their electrochemical gradients in a process known as **cotransport**. A model of how a cotransporter may operate is shown in Figure 6.10. The carrier is often a transmembrane protein with a site on the outside of the membrane that can bind a proton. Proton binding causes a second site to be exposed. This site binds the ion or solute undergoing cotransport. With both molecules bound, the transporter undergoes a conformational change that exposes the binding sites to the cytoplasmic side of the membrane. The cycle is completed by diffusion of the proton and the cotransported molecule away from their binding sites, causing the transporter to regain its original or "relaxed" conformation.

Two major types of cotransport are recognized. The example shown in Figure 6.10 is called a **symport** because the two substances are moving in the same direction through the membrane. **Antiports** (Fig. 6.11) move the two transported molecules in opposite directions. In both types of cotransport, the ion or solute undergoing cotransport is moving against its electrochemical gradient, so its transport is active. However, the energy driving this transport is provided by the proton motive force rather than by ATP hydrolysis and is captured by the cell through the inward movement of H^+ through the cotransporter, which dissipates the proton motive force. Because of its hybrid energetic character, cotransport is sometimes called **secondary active transport** and the term **primary active transport** is reserved for ATP-driven electrogenic ion pumps or active exchange pumps like the K^+/H^+ exchanger of bacteria, where *both* substrates are moving against their electrochemical gradients. These are the complexities in interpreting active versus passive transport that were mentioned earlier.

In plants and fungi, sugars and amino acids are taken up by cotransport with protons, that is, by symports. Some of the experimental evidence for this conclusion is shown in Figure 6.12. When glucose is supplied to a plant cell bathed in a simple solution of mineral salts, a reduction in the membrane potential, an increase in external pH, and uptake of glucose occur simultaneously (Novacky et., 1980). The decrease in membrane potential is due to the positive charges (H^+) that move into the cell along with glucose. This membrane depolarization is transitory, however, because the reduced membrane voltage allows the H^+ pump to work faster and thereby restore the membrane voltage and pH gradient in the presence of continuing glucose uptake.

Current research (Tazawa et al., 1987) has led to the view that most of the ionic gradients across cell membranes of higher plants are generated and maintained by the electrochemical H^+ gradient. In turn, this H^+ gradient is generated by the plasma membrane H^+-ATPase.

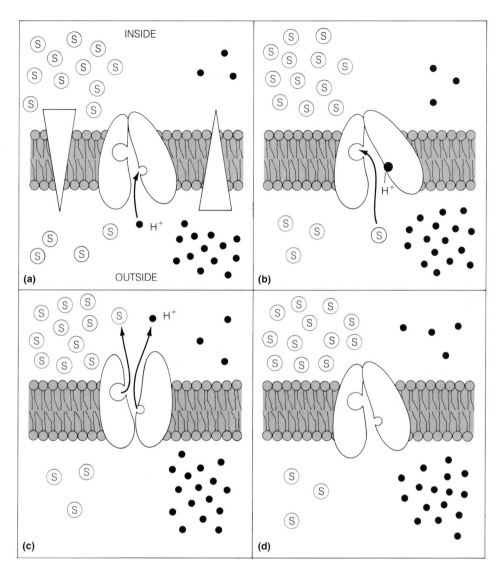

FIGURE 6.10. Hypothetical model for cotransport. The energy driving the process has been stored in a proton gradient (symbolized by the triangle on the right in (a)) and is being used to take up a substrate S against its concentration gradient (triangle on the left). In its initial conformation, the membrane protein is exposed to the outside environment and can bind a proton (a). This results in a conformational change that permits a molecule of S to be bound (b). The binding of S causes another conformational change that exposes the protein and its substrates to the inside of the cell (c). Release of a proton and a molecule of S to the cell's interior restores the original conformation of the symporter and allows a new pumping cycle to begin (d).

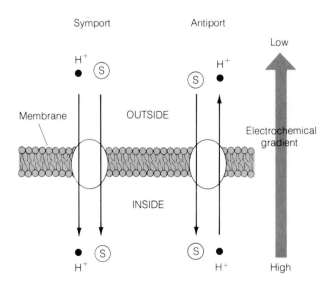

FIGURE 6.11. Two examples of cotransport coupled to a primary proton gradient. In the case of an antiport, the pumping of one proton out of the cell is energetically coupled to the uptake of one molecule of a substrate (S) from the outside medium. In the case of a symport, the energy dissipated by a proton moving back into the cell is coupled to the uptake of one molecule of a substrate. In both cases, the substrate under consideration is moving against its electrochemical gradient. Both neutral and charged substrates can be taken up in these cotransport processes.

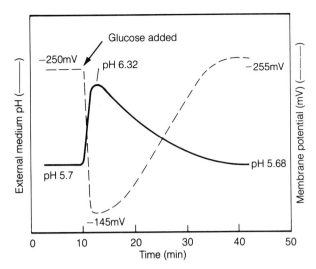

FIGURE 6.12. Evidence for a glucose/proton symport is shown by simultaneous measurements of the pH of the medium bathing the surface of the aquatic plant duckweed (*Lemna gibba*) and the membrane potential of one cell. The early portions of the curves show steady values of pH and membrane potential, conditions that change when 50 mM glucose is added to the solution. The observed increase in pH indicates that protons are disappearing from the medium at the same time that the membrane potential of the cell is decreasing. These observations are predicted by a glucose/proton symport if glucose is cotransported with a proton into the cell. With time, an increase in pump activity results in restoration of the initial pH and membrane potential values. (Adapted from Novacky et al. 1980.)

Although further research will undoubtedly reveal additional complexity, this view fits much of our present data. There is evidence that Na^+ is transported out of the cell by an Na^+/H^+ antiporter and that Cl^- and sucrose enter the cell via specific proton symporters. Regarding NO_3^- and $H_2PO_4^-$, there is some controversy about whether they enter the cell via symporters or are driven by primary pumps. Potassium often appears to be at or near electrochemical equilibrium and thus enters the cell by passive diffusion. However, even these passive K^+ movements are driven by the H^+-ATPase in the sense that K^+ uptake is in equilibrium with the membrane potential, which in turn is largely generated by the H^+ pump.

Much of the current research on membrane transport in plants centers around understanding how electrogenic transport and cotransport work at the molecular level (Sze, 1985). This research involves studying the physical properties of the ATPases and other transporters, how they interact with their substrates during transport, and how they respond to regulatory signals. On a practical level, two areas are particularly important: understanding sucrose and amino acid transport in and out of the phloem, which is thought to be crucial to understanding the determinants of seed formation and hence of crop yields, and understanding the mechanism of ion accumulation by roots, which is also crucial to understanding plant growth.

Solute Accumulation in the Vacuole Is Driven by the Tonoplast H^+-ATPase

Since plant cells enlarge primarily by the uptake of water into vacuoles, the osmotic pressure of the vacuole must be maintained sufficiently high for water to enter from the cytoplasm. The tonoplast serves as a selective barrier for the accumulation of solutes in the vacuole, just as the plasma membrane regulates uptake into the cell. Recently, a mechanism of active transport across the tonoplast has been elucidated (Sze, 1985; Boller and Wiemken, 1986). A new type of proton-pumping ATPase, which pumps protons into the vacuole, has been discovered. The tonoplast H^+-ATPase differs both structurally and functionally from the plasma membrane H^+-ATPase (Nelson and Taiz, 1989). Unlike the plasma membrane ATPase, it is insensitive to vanadate, which indicates that it does not have a phosphorylated intermediate during ATP hydrolysis, as shown in the reaction model in Figure 6.10. In contrast, the tonoplast H^+-ATPase is inhibited by nitrate, which does not inhibit the plasma membrane ATPase. Considering the large amounts of nitrate taken up by plant cells from the soil and accumulated in the vacuole, nitrate inhibition would seem a major disadvantage. How the tonoplast ATPase avoids toxicity under high-nitrate conditions remains an open question.

The tonoplast H^+-ATPase, like the plasma membrane H^+-ATPase, is an electrogenic proton pump that generates a proton motive force across the vacuolar membrane. This accounts for the fact that the vacuole is typically 20 to 30 mV more positive than the cytoplasm, although it is still negative relative to the external medium. To maintain bulk electrical neutrality, anions such as Cl^- or $malate^{2-}$ are transported from the cytoplasm into the vacuole through channels in the membrane (Bennett et al., 1984). Without the simultaneous movement of anions along with the pumped protons, the charge buildup across the tonoplast would make the pumping of additional protons energetically difficult.

In the presence of ongoing anion transport, the tonoplast H^+-ATPase can generate a large concentration (pH) gradient of protons across the tonoplast. This gradient accounts for the fact that the pH of the vacuolar sap is typically about 5.5, whereas the cytoplasmic pH is always 7.0 to 7.5. Even steeper pH gradients can develop in certain species. As anyone who has bitten into a lemon knows, lemon juice is quite acidic. In fact, the pH of lemon juice, which consists mainly of vacuolar sap, is about 2.5. The most extreme case is that of the marine brown alga *Desmerestia*, which has a vacuolar pH of < 1.0! Obviously, the tonoplast of *Desmerestia* must have remarkable properties to maintain such a steep pH gradient.

Whereas the electrical component of the proton motive force drives the uptake of anions into the vacuole, the pH gradient is harnessed to drive the uptake of cations and sugars into the vacuole via antiporters. For example, an antiporter that exchanges protons for Ca^{2+} has been characterized in red beet roots (Blumwald and Poole, 1986), and a proton/hexose antiporter has been demonstrated in roots of sugar beet (Briskin et al., 1985). Halophytes, plants adapted to saline conditions, appear to have tonoplast antiporters specific for Na^+ ions that enable the cell to sequester sodium in the vacuole and thus avoid Na^+ toxicity in the cytoplasm (Blumwald and Poole, 1985).

The various transport processes located on the tonoplast and plasma membrane are illustrated in Figure 6.13.

The tonoplast H^+-ATPase belongs to a general class of vacuolar ATPases present on the endomembrane systems of all eukaryotes. Recently, sequence analyses of the two major subunits have shown that the eukaryotic

vacuolar ATPases (V-ATPases) are distantly related to the ATP synthases of mitochondria and chloroplasts, which are thought to have evolved from eubacterial endosymbionts. Remarkably, vacuolar H^+-ATPases are even more closely related to the plasma membrane ATP synthases of archaebacteria (Gogarten et al. 1989). How the gene for an archaebacterial ATPase came to reside in the eukaryotic nucleus and gave rise to the vacuolar ATPase is an exciting area for future evolutionary studies.

Proton Pumping in Guard Cells Generates Ion Gradients That Regulate Guard Cell Turgor and Stomatal Apertures

Stomatal movements, the basis of gas exchange in leaves, provide an example of how processes that regulate solute transport in cells are used in a specialized plant function.

The German botanist H. Von Mohl, working at

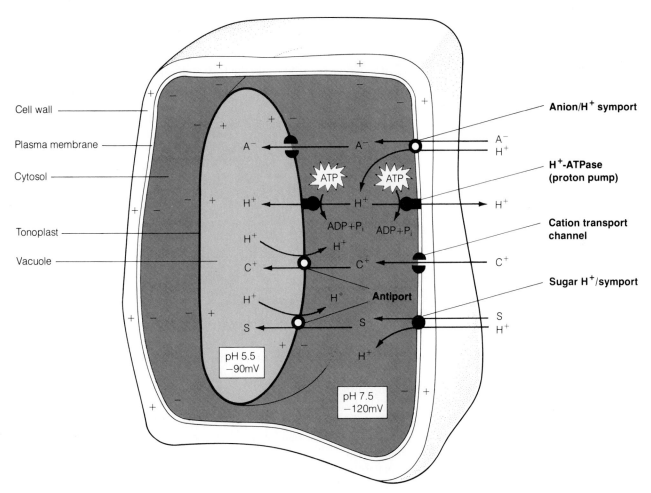

FIGURE 6.13. Transport reactions across the plasma membrane and the tonoplast. Protons are pumped out of the cell by the plasma membrane H^+-ATPase and into the vacuole by the tonoplast H^+-ATPase. Anions (A^-) can cross the plasma membrane by cotransport with a proton via a symporter. Cations (C^+) can enter the cell through channels driven by the inside-negative membrane potential. Sugars (S) appear to enter the cell via sugar/proton symporters. Once in the cell, both sugars and cations can be taken up into the vacuole by exchange with protons via specific antiporters. Anions enter the vacuole through channels, driven by the inside-positive membrane potential.

Tübingen, proposed in 1856 that turgor changes in guard cells provide the driving force for stomatal movements (Meidner, 1987). F. E. Lloyd, at the Carnegie Institution in Stanford, California, hypothesized in 1908 that these turgor changes depend on starch-sugar interconversions, a concept that became the basis of the *starch-sugar hypothesis* of stomatal movements. By that time, it had been observed that guard cell chloroplasts contain large, prominent starch grains and that the amount of starch decreases during stomatal opening and increases during stomatal closing. Starch, an insoluble, high-molecular-weight polymer of glucose, does not contribute to the cell's osmotic potential (Chapter 4). The starch-sugar hypothesis postulated that stomatal opening occurs when starch is hydrolyzed to glucose monomers, causing an increase in the osmotic potential of the guard cells. In the reverse process, starch synthesis would decrease the concentration of free sugars in the cell, resulting in lower osmotic potentials and stomatal closing.

Ion Concentrations in Guard Cells Increase During Stomatal Opening and Decrease During Stomatal Closing

The starch-sugar hypothesis was a central concept in stomatal physiology during the first decades of this century, but it was seriously questioned after the discovery of potassium fluxes in guard cells. These K^+ movements were first reported in 1943 by S. Imamura in Japan. Work in the late 1960s by M. Fujino in Nagasaki and R. A. Fischer at the University of California at Davis stimulated a large number of studies on the ionic relationships in guard cells, which showed the central role of ions in stomatal movements.

Potassium concentrations in guard cells increase severalfold when stomata open—from about 100 mM in the closed state to 400 to 800 mM in the open state, depending on the plant species and the experimental conditions. Because potassium is a cation, these vast fluctuations in its concentration require a balancing of the electrical charges in the cell. In species of the genus *Allium*, such as onion (*Allium cepa*), electroneutrality is maintained by parallel fluxes of Cl^- anions. In most species, however, the negative charges are provided by varying amounts of Cl^- and the organic anion malate^{2-} (for structural formula, see Fig. 6.19). Like potassium, chloride is taken up during stomatal opening and extruded during stomatal closing. In contrast, malate is synthesized in the guard cell cytosol, in a metabolic pathway that uses carbon skeletons resulting from the hydrolysis of starch. Hence, the decrease in starch observed during stomatal opening is functionally coupled to potassium uptake by the supply of substrate for malate biosynthesis. The fate of malate during stomatal closing is not known. Malate concentrations in guard cells

must decrease as potassium leaves the cells; that could be achieved in mitochondrial respiration, which would convert malate into CO_2, or by the back-conversion of malate into sugars and subsequent polymerization into starch. These aspects of carbon metabolism in guard cells are under investigation.

Accumulation of potassium and chloride in guard cells requires active uptake mechanisms. In theory, it could be achieved by separate transport systems pumping potassium and chloride into the cells or by the establishment of a primary gradient supporting secondary transport processes, including potassium and chloride uptake. The observation that protons are extruded into the apoplast during stomatal opening led to the hypothesis that a primary *proton* gradient in guard cells generates a proton motive force that provides the driving force for ion uptake (Zeiger, 1983).

Specific Mechanisms in Guard Cells Mediate the Stomatal Response to Light

Photosynthesis and stomatal movements are tightly coupled in the leaf, and both processes are very sensitive to light. Studies of the stomatal responses to light have characterized some of the properties of the guard cell proton pump. Figure 6.14 shows the changes in light fluence rates, leaf photosynthesis, and stomatal conductance (a measure of the water lost by the leaf per unit area per unit time) during the course of a day in a leaf of the tree *Fagus sylvatica* growing in the field. **Fluence rates** are measured amounts of light reaching the leaf surface per unit time (mol m^{-2} s^{-1}). Note how both leaf photosynthesis and stomatal conductance change in parallel with the prevailing levels of solar radiation throughout the day. We can then ask, how is the light signal transduced into stomatal apertures that match the prevailing levels of solar radiation? It is very difficult to address that question in studies of the intact leaf because many other external and internal factors besides light affect stomatal responses. However, in some species, such as the dicot broad bean (*Vicia faba*) or the monocot *Commelina communis*, portions of the leaf epidermis containing intact guard cells can readily be separated from the rest of the leaf. This preparation is called an epidermal peel. When epidermal peels floating in a solution containing potassium chloride are illuminated, stomatal opening can be observed. Many studies in which stomatal apertures have been measured microscopically have demonstrated that apertures increase as a function of light fluence rates (Fig. 6.15). These experiments indicate that guard cells do not require the presence of mesophyll to respond quantitatively to light.

Photobiological studies of stomata have shown that their light response has two components: one component matches the absorption spectrum of chlorophyll and is

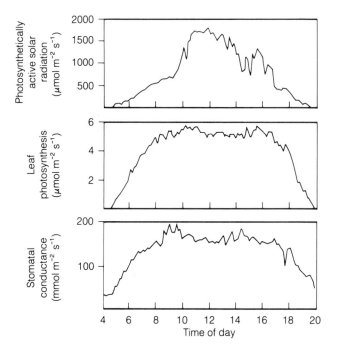

FIGURE 6.14. Photosynthetically active solar radiation, leaf photosynthesis, and stomatal conductance in a leaf of the tree *Fagus sylvatica*. Measurements of solar radiation (moles of photons per square meter of leaf per second, that is, mol m^{-2} s^{-1}) were made with a photometer; photosynthesis and stomatal conductance were measured in an intact attached leaf that was enclosed in a gas exchange chamber through which air of known CO_2 concentration and relative humidity was circulated. Photosynthesis and stomatal conductance increase in parallel with the amount of light reaching the leaf surface and decrease later in the day as the solar radiation declines. (From Schulze, 1970.)

mediated by the guard cell chloroplast, and the second component is activated specifically by blue light. The identity of the blue-light photoreceptor has yet to be established, but its activity is not restricted to stomata. **Blue-light responses** are rather common photoresponses in plants; they include bending of shoots toward the light (phototropism), tracking of the sun by leaves (solar tracking, Fig. 10.4b), and chloroplast movement within photosynthetic cells (Fig. 10.4a).

The dual-photoreceptor nature of the stomatal responses to light is illustrated by an experiment with epidermal peels of *Commelina* (Fig. 6.16). At the onset of the experiment, the peels were illuminated with red light at fluence rates that had previously been shown to saturate the opening response (that is, further increases in fluence rate did not cause any additional opening). From the absorption properties of chlorophyll, showing peaks in the red and the blue, and from studies with metabolic inhibitors, it is known that the observed opening under red-light irradiation is a result of sensory transduction by the guard cell chloroplasts. On completion of the opening in response to red light, blue-light irradiation was added. Figure 6.16 shows that addition of blue light caused substantial additional stomatal opening. The additional opening could not have been mediated by the guard cell

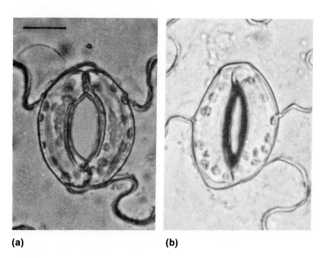

(a)   **(b)**

FIGURE 6.15. Stomata in detached epidermis of the broad bean *Vicia faba*, showing a wide-open stomatal pore (a) and a nearly closed pore (b). Stomatal apertures are measured under a microscope by recording aperture widths. Measurements of the change in aperture as a function of experimental conditions allow characterization of stomatal responses to different environmental stimuli. Bar = 20 μm. (Photos courtesy of G. Tallman and E. Zeiger.)

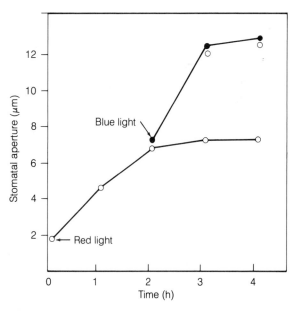

FIGURE 6.16. Changes in stomatal aperture as a function of time in stomata from detached epidermis of *Commelina communis* treated with red light at saturating fluence rates (open circles). In a parallel experiment, stomata exposed to red light also received blue light, added at the time indicated by the arrow. The increase in stomatal apertures above the level reached in the presence of saturating red light indicates that a second photosystem, responding to blue light, stimulates the aperture changes observed in the second phase of the experiment. (From Schwartz and Zeiger, 1984.)

chloroplasts because that photoreceptor system was already saturated by red light; instead, the additional response showed that a blue light-dependent system was activated. Why two different photoreceptor systems are needed in guard cells is not yet known, but their characterization has been helpful in understanding the regulation of stomatal apertures by light.

Light Activates a Proton Pump at the Guard Cell Plasma Membrane

Because of the low density of stomata in the leaf epidermis (the underside of *Vicia* leaves has a stomatal density of about 60 guard cell pairs mm^{-2}), it is difficult to use epidermal peels for studies of ion transport or guard cell biochemistry. This problem has been solved by using guard cell protoplasts, which are guard cells devoid of their walls. Protoplasts are made by treating epidermal peels with cellulolytic enzymes, which digest the cell walls (Fig. 6.17). Properly stabilized, protoplasts retain many of the properties of intact guard cells and provide a means of concentrating several million guard cells in a test tube for quantitative studies.

In recent investigations, suspensions of *Vicia* guard cell protoplasts were irradiated with red light at saturating fluence rates while the pH of the suspension medium was monitored. Under these conditions, pulses of blue light caused acidification of the suspension medium, which became more acidic as the fluence rate of blue light increased (Fig. 6.18; Shimazaki et al., 1986). These results are the expression of a light-activated proton pump at the guard cell plasma membrane, with proton extrusion detected as acidification of the suspension medium.

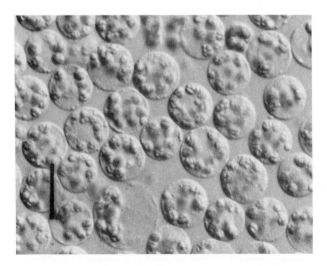

FIGURE 6.17. Guard cell protoplasts from *Nicotiana glauca*. When the cell wall is digested by appropriate enzymes in the presence of 0.35 m mannitol, the protoplasts become spherical and can be isolated as a cell suspension. Each protoplast contains 10 to 12 chloroplasts. Bar = 20 μm. (Photo courtesy of W. Copples and G. Tallman.)

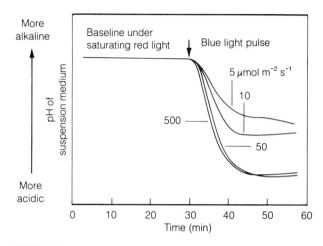

FIGURE 6.18. Guard cell protoplasts of *Vicia faba* acidify their suspension medium when exposed to pulses of blue light given under saturating fluence rates of red light. The background of saturating red light is used to eliminate interference from chlorophyll, a second photoreceptor system in the protoplasts that is sensitive to both red and blue light. Blue-light pulses 30 s long and of different fluence rates (5, 10, 50, and 500 μmol m^{-2} s^{-1}) were given as indicated by the arrow. The fluence rates of blue light were measured with a photometer calibrated to measure moles of photons. The acidification results from the stimulation by blue light of an H^+-pumping ATPase at the plasma membrane. The rate of H^+ pumping depends on the number of blue photons reaching the cells, which suggests a mechanism that can regulate stomatal apertures as a function of light levels in the leaf environment. (From Shimazaki et al., 1986.)

This interpretation has been substantiated by experiments performed with the electrophysiological technique called *patch clamping*, which permits measurement of electrical currents flowing across the plasma membrane (see box on patch clamp studies). When guard cell protoplasts were studied with this technique, blue light was found to stimulate electrical currents at the guard cell plasma membrane (Fig. 6.B; Assmann et al., 1985). Together, the acidification and patch clamp studies show that blue light activates a proton pump in guard cells and that the rate of proton pumping depends on the fluence rate. This dependence of the proton pump on fluence can account for the coupling between light fluences in the leaf environment and stomatal responses to light.

The finding of a blue light-activated proton pump in guard cells raises a question about the other component of the stomatal response to light, which depends on chlorophyll excitation. How is the chloroplast signal transduced? Another patch clamp study with guard cell protoplasts has shown that, under appropriate experimental conditions, red light also stimulates proton pumping at the guard cell plasma membrane (Serrano et al., 1988). Red light-induced proton pumping is blocked by photosynthetic inhibitors, which indicates that the response is modulated by the guard cell chloroplast. These studies show that proton pumping provides a functional link between the two photoreceptor systems regulating the stomatal response to light.

Patch Clamp Studies in Plant Cells

T HE USE OF the patch clamp method in plant physiology is generating information about the properties of plant membranes that is beyond the scope of other techniques. Conventional electrophysiological studies with intracellular electrodes (Fig. 6.4) are valuable for the measurement of membrane potentials and other electrical properties of plant cells but have limitations for the characterization of ion pumps and single ion channels. Patch clamping, on the other hand, is very well suited for that purpose (Satter and Moran, 1988; Serrano and Zeiger, 1989).

During intracellular recording, cells are impaled with a glass electrode, and the electrode tip is located inside the cell, usually in the vacuole. In patch clamp experiments, the tip of a glass electrode is brought in contact with the plasma membrane of a protoplast, and gentle suction is applied to facilitate the formation of a tight seal between the electrode and the membrane. Upon attainment of a good seal, several options become available to the investigator. Further suction can remove the portion of the membrane delimited by the opening in the electrode tip (Fig. 6.A), exposing the interior of the cell to the microelectrode solution. In this configuration, measured electrical events reflect movement of electrical charges at the plasma membrane of the whole cell (Fig. 6.B). Electrogenic pumps are best studied in this configuration. Pulling the electrode away from the cell produces a membrane patch that becomes

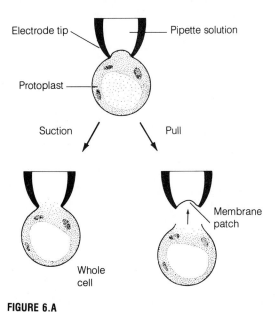

FIGURE 6.A

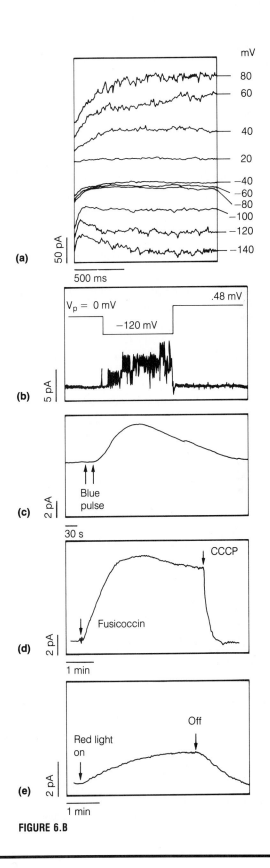

FIGURE 6.B

exposed to both the electrode and the bathing medium (Fig. 6.A). This configuration makes it possible to study the opening and closing of single ion channels (Fig. 6.6 and Fig. 6.B). Because protoplasts are spherical, their volume is easy to measure, and their lack of electrical connections with other cells facilitates calculations of charge fluxes per unit area of membrane. Other advantages of patch clamping are the ability to distinguish between electrical events at the plasma membrane and the tonoplast and the possibility of controlling the composition of both the cytosol and the bathing medium.

Most of the patch clamp studies on plant membranes have been done with guard cells. Figure 6.B shows some of the results obtained by measuring electrical current at controlled membrane potentials (voltage clamp). Whole cell currents are shown in (a). These data are obtained by applying voltage pulses to the protoplast (as indicated on the right in the figure) and measuring the electrical current in picoamperes, leaving the cell at a given voltage. These current-voltage relationships are valuable

for understanding the regulation of ion transport across the membrane. The opening of potassium channels is shown in (b). The experimenter changes the imposed voltage across the membrane, as shown at the top of the panel. In the appropriate medium conditions, the measured current, seen in the lower part of panel (b), shows the opening and closing of potassium channels (Schroeder et al., 1987). Electrical currents generated by an electrogenic pump are shown in panels (c–e). A blue-light pulse given on a background of high fluence rates of red light (c) stimulates a transient electrical current by activating a blue light-dependent proton pump (Assmann et al., 1985). In the dark, treatment of guard cell protoplasts with the fungal toxin fusicoccin, which causes stomatal opening, stimulates an electrical current (d). CCCP, a metabolic inhibitor that makes the membrane highly permeable to protons, abolishes the response. Currents stimulated by red light are shown in (e) (Serrano et al., 1988). The high resolution of the patch clamp technique is attracting increasing attention from many investigators.

A scheme of the primary and secondary transport processes in guard cells is shown in Figure 6.19. Proton pumping at the membrane results in charge separation and a proton gradient. Proton extrusion makes the membrane potential of guard cells more negative; light-dependent changes in membrane potential of up to 50 mV have been measured. In addition, proton pumping generates a pH gradient of about 0.5 to 1 pH unit. The electrical component of the gradient can drive potassium uptake through potassium channels in the membrane, which have been characterized (Schroeder et al., 1987). Chloride uptake can be driven by the pH component of the proton gradient and might take place through a Cl^-/H^+ symporter.

Kinetic Analysis of Transport Processes Can Reveal How Cells Regulate Their Intracellular Solute Concentrations

Thus far, we have described cellular transport in terms of its energetics. However, cellular transport can also be studied by using enzyme kinetics, because transport involves the binding and dissociation of molecules at active sites on transport proteins (see Chapter 2). One advantage of the kinetic approach is that it gives new insights into the regulation of transport.

Kinetic experiments involve measurements of the effects of external ion (or other solute) concentrations

on transport rates, and they lead to a different view of transport. In simple diffusion, the rate of inward transport is proportional to the external concentration of the transported molecule. In carrier-mediated transport, there is always a concentration at which uptake saturates. This saturation occurs when the concentration of the free molecule undergoing transport appreciably exceeds its carrier's affinity for it and the binding site is occupied essentially all the time. The concentration of carrier, not the concentration of solute, becomes rate limiting.

As seen in Figure 6.20, transport kinetics gives numerical values for the carrier's binding affinity (K_m) and the maximum rate of transport (V_{max}). Such kinetic analysis also distinguishes between simple diffusion (such as movement through a channel) and carrier-mediated transport. Determining whether carrier-mediated transport is active or passive requires further tests, such as measuring the effect of respiratory poisons like CN^- (Fig. 6.9) on transport rates.

Ordinarily, transport displays both active and passive properties when a wide range of solute concentrations is studied. Figure 6.21 shows sucrose uptake by soybean cotyledon protoplasts as a function of the external sucrose concentration (Lin et al., 1984). Uptake increases sharply with concentration and begins to saturate at about 10 mM. At concentrations above 10 mM, uptake becomes linear and nonsaturable. Inhibition of ATP synthesis with metabolic poisons blocks the satu-

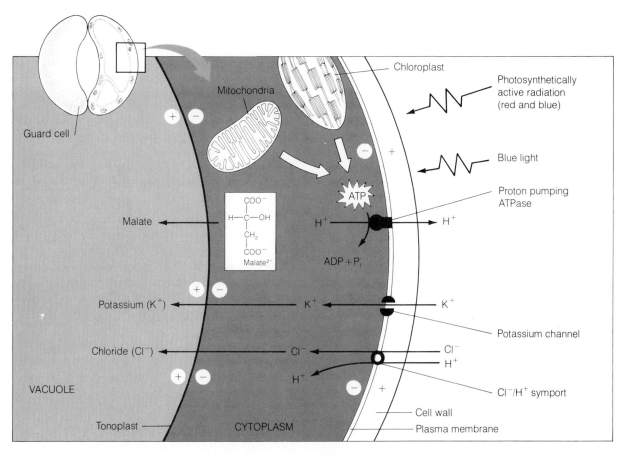

FIGURE 6.19. Stomatal guard cells have a light-activated H^+-pumping ATPase. Proton pumping is stimulated by photosynthetically active radiation via responses of the guard cell chloroplasts. The pump is also activated by a different photoreceptor system that is sensitive to blue light. The ATPase can be driven by ATP produced by the chloroplasts or by the mitochondria. The electrochemical proton gradient is used for uptake of potassium via K^+-specific channels. Chloride might be taken up via a Cl^-/H^+ symport. Malate is synthesized in the cytoplasm from carbon skeletons produced during starch hydrolysis in the chloroplast. Malate, K^+, and Cl^- are transported into the vacuole.

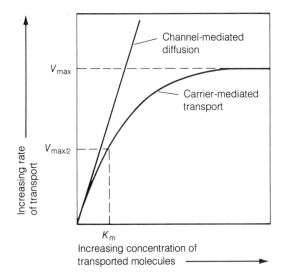

FIGURE 6.20. Transport mediated by channels and transport mediated by carriers can be distinguished by kinetic analysis. Transport via channels occurs by simple diffusion and does not show saturation in the range of concentrations typical of plant cells. In contrast, carrier-mediated transport depends on binding of the solute to a specific site on the carrier. Because the number of such sites is limited, carrier-mediated transport does exhibit saturation. The binding affinity (K_m) of the carrier for the solute under consideration can be calculated by computing the solute concentration at which the rate of transport is 50% ($V_{max/2}$) of the maximum rate (V_{max}). (Adapted from Alberts et al., 1983).

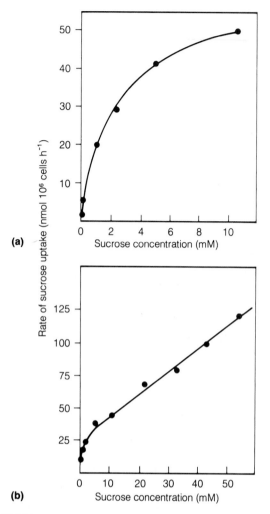

(a)

(b)

FIGURE 6.21. The transport properties of a solute can change at different solute concentrations. For example, at low concentrations (1 to 10 mm), the rate of uptake of sucrose by soybean cells shows saturation kinetics, typical of a carrier (a). At higher concentrations, however, the uptake rate increases linearly over a broad range of concentrations, indicating that uptake was occurring by simple diffusion through a channel (b). (From Lin et al., 1984.)

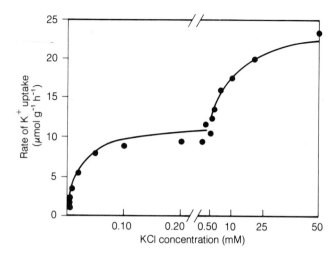

FIGURE 6.22. The transport of potassium into barley roots shows two different phases. Both phases exhibit saturation kinetics, indicative of carrier-mediated transport. This biphasic kinetics of potassium uptake has been interpreted on the basis of two classes of carriers: a high-affinity one, having a K_m of 0.02 to 0.03 mM, and a low-affinity one, with a K_m of about 11 mM, operating at the higher concentration ranges. (Adapted from Epstein, 1972.)

rable component but not the linear one. The interpretation is that sucrose uptake at low concentrations is an active carrier-mediated process (sucrose/H^+ cotransport). At higher concentrations, sucrose enters the cells by passive diffusion down its concentration gradient and is therefore not sensitive to metabolic poisons.

Sometimes kinetic analysis of transport reveals multiple saturable components for the same solute (Fig. 6.22). This result has often been interpreted on the basis of the presence in the tissue of separate classes of carriers, each of which has a different K_m for the transported solute (Epstein, 1972). These carriers may be located in the same cells or in separate cells of the tissue. For example, it has been suggested that in roots the high-affinity carrier may be located in epidermal cells and the low-affinity carrier in cortical cells. An alternative explanation is that there is

only one class of carrier for the same solute and that the K_m of the carrier changes as the internal concentration of the solute increases. Studies of K_m values have shown a feedback inhibition (Chapter 2) of carrier-mediated transport that allows the cell to maintain stable cytoplasmic concentrations over a wide range of external concentrations. For example, the K_m of the high-affinity K^+ carrier in barley roots (Fig. 6.22) changes with the plant's demand for K^+. Barley plants starved for K^+ have high-affinity K_m values in the range of 0.02 to 0.03 mM, whereas well-fertilized plants have K_m values that are four- to fivefold greater (remember, an increase in K_m reflects a decrease in the carrier's affinity for K^+; see Chapter 2). Thus, by regulating carrier K_m values, plant cells can match nutrient uptake to cellular demand.

A second way plant cells regulate their internal ion concentrations is by altering the number of carriers present in the membrane. Kinetically, such a change appears as a change in V_{max} without a change in K_m. Plants that have been starved and then resupplied with sulfate or phosphate often respond in this way. As the internal sulfate and phosphate levels recover, the V_{max} values for their uptake drop to the usual values as excess carriers are removed from the membrane.

Transcellular Transport

Mineral nutrients absorbed by the root are carried to the shoot by the transpiration stream moving through the xylem. Just as the initial uptake of nutrients is a highly specific, well-regulated process, so too is the subsequent movement of mineral ions from the root surface across the cortex and into the xylem. Ion transport across the root obeys the same biophysical laws that govern cellular transport. However, as we have seen in the case of water movement (Chapter 4), the anatomy of roots imposes some special constraints on the pathway of ion movement.

The Apoplast and the Symplast Are the Major Pathways for Solute Movement Within the Plant

Thus far, our discussion of cellular ion transport has not included the cell wall. In terms of the transport of small molecules, the cell wall is an open lattice of polysaccharides through which mineral nutrients diffuse readily. Because all plant cells are separated by cell walls, it is possible for ions to diffuse across the root entirely through the cell wall space without ever entering a living cell. This continuum of cell walls is called the free space or apoplast (see Chapter 4). The apoplastic volume of a piece of plant tissue can be determined by comparing the uptake of ^{3}H-labeled water and ^{14}C-labeled mannitol. Mannitol is a nonpermeating sugar alcohol that equilibrates with the free space but cannot enter the cells. Water, on the other hand, freely penetrates both the cells and the cell walls. Measurements of this type usually show that 5 to 20% of the plant tissue volume is occupied by cell walls.

Just as the cell walls form a continuous phase, so too do the cytoplasms of neighboring cells. Plant cells are interconnected by cytoplasmic bridges called **plasmodesmata** (Chapter 1). Plasmodesmata are cylindrical pores 20 to 60 nm in diameter (Fig. 1.24). Each plasmodesma is lined with plasma membrane and contains a central tubule, the **desmotubule**, that is a continuation of the endoplasmic reticulum. In tissues in which significant amounts of intercellular transport occur, neighboring cells contain large numbers of plasmodesmata, up to 15 per square micrometer of cell surface (Fig. 6.23). Specialized secretory cells, such as nectar and salt glands, appear to have high densities of plasmodesmata; so do the cells near root tips, where most of the nutrient absorption occurs.

By injecting dyes or by making electrical resistance measurements on cells containing large numbers of plasmodesmata, investigators have shown that ions and small solutes can move from cell to cell through these pores. As discussed in Chapter 4, water also moves through plasmodesmata. Because each plasmodesma is partially occluded by a desmotubule, large molecules such as proteins cannot pass through the plasmodesmata. Ions, on the other hand, can move from cell to cell through the entire plant by simple diffusion through the symplast (Chapter 4).

Ions Moving Through the Root Cross Both Symplastic and Apoplastic Spaces

Ion absorption by the roots (Chapter 5) is more pronounced in the area distal to the meristem, where the root hairs are located (Figs. 5.3 and 15.2). Cells in this region have completed their elongation but have not yet begun secondary growth. The root hairs are simply ex-

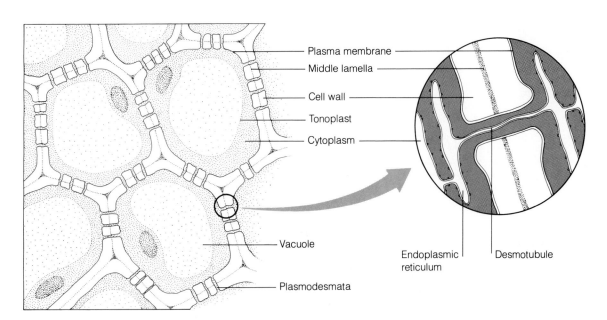

FIGURE 6.23. Diagram illustrating how plasmodesmata connect the cytoplasm of neighboring cells. Plasmodesmata are about 40 nm in diameter and allow free diffusion of water and small molecules from one cell to the next. Desmotubules provide a direct connection between endoplasmic reticula of adjacent cells.

tensions of some epidermal cells and serve to increase greatly the surface area available for ion absorption.

The apoplast forms a continuous phase from the root surface through the cortex (Fig. 4.3). At the boundary between the vascular cylinder (the **stele**) and the cortex is a layer of specialized cells, the **endodermis**. As discussed in Chapters 4 and 5, a suberized cell layer in the endodermis, known as the **Casparian strip**, acts like a gasket, sealing the apoplast of the cortex off from the apoplast of the stele. An ion that enters a root may be taken up into the cytoplasm of an epidermal cell, or it may diffuse through the apoplast into the cortex and actually first enter the symplast through uptake by a cortical cell. In all cases, ions must enter the symplast before they can penetrate the stele because of the presence of the Casparian strip. Once an ion has entered the stele through the symplastic connections across the endodermis, it continues to diffuse from cell to cell into the xylem. Finally, the ion reenters the apoplast as it diffuses into a xylem tracheid or vessel. Again, the Casparian strip prevents the ion from diffusing back out of the root through the apoplast. Because of the presence of the Casparian strip, it is possible for the plant to maintain a higher ionic concentration in the xylem than exists in the soil water surrounding the roots.

Uptake of Ions into the Xylem Could Involve One or Two Active Pump Sites

When ions have been actively taken up by the symplast and have diffused into the stele, they could enter the tracheids and vessels of the xylem by passive diffusion. In this case, the movement of ions from the root surface to the xylem would require only a single active-transport step. The site of this active uptake would be the plasma membrane surface of the root epidermal and cortical cells. Support for this model is shown in Figure 6.24. Bowling and co-workers used ion-specific microelectrodes to measure the electrochemical potentials of various ions across maize roots (Dunlop and Bowling, 1971). Data from this and other studies indicate that K^+, Cl^-, Na^+, SO_4^{2-}, and NO_3^- are all taken up actively by the epidermal and cortical cells and are maintained in the xylem against an electrochemical gradient when compared with the external medium (Lüttge and Higinbotham, 1979). However, none of these ions is at a higher electrochemical potential in the xylem than in the cortex or living portions of the stele. Therefore, the final movement of ions into the xylem could be due to passive diffusion.

Other observations have led to the view that this final step of xylem loading may also involve active transport (Lüttge and Higinbotham, 1979). With the type of apparatus shown in Figure 6.25, it is possible to make simultaneous measurements of ion uptake into the epidermal/cortical cytoplasm and ion loading into the xylem. By using treatments with inhibitors and plant hormones, it can be shown that ion uptake by the cortex and ion loading into the xylem operate independently. For example, treatment with the protein-synthesis inhibitor cycloheximide or the cytokinin benzyladenine inhibits xylem loading without affecting uptake by the cortex. If

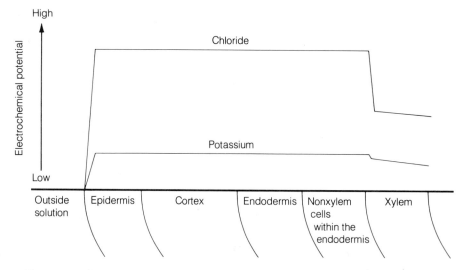

FIGURE 6.24. Diagram showing electrochemical potentials of K^+ and Cl^- across a maize root. To determine the electrochemical potentials, the root was bathed in a solution containing 1 mM KCl and 0.1 mM $CaCl_2$. A reference electrode was positioned in the bathing solution and an ion-sensitive measuring electrode was inserted in different cells of the root. The horizontal axis shows the different tissues found across the root (see Fig. 4.3). The substantial increase in electrochemical potential for both K^+ and Cl^- between the bathing medium and the epidermis indicates that ions are taken up into the root by an active transport process. In contrast, the potentials decrease at the xylem vessels, suggesting that transport of ions into the xylem is by passive diffusion down the electrochemical gradient. (From Dunlop and Bowling, 1971.)

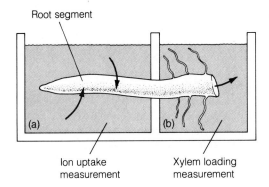

Ion uptake measurement   Xylem loading measurement

FIGURE 6.25. The relationship between ion uptake into the root and xylem loading can be measured by placing a root segment across two compartments and adding a radioactive tracer to compartment (a). The rate of disappearance of the tracer from compartment a gives a measurement of ion uptake, and the rate of appearance in compartment (b) provides a measurement of xylem loading. (From Lüttge and Higinbotham, 1979.)

only one regulatory site were involved in ion uptake by the epidermal and cortical cells, changes in the rate of xylem loading would always reflect changes in the rate of ion uptake. Clearly, this is not the case. However, whether xylem loading is active or passive and whether one or two regulatory sites are involved is not yet clear, and this remains an area of ongoing research.

Summary

Molecular movement between different compartments of biological systems is known as transport. Plants exchange solutes with their environment and among their tissues and organs. Transport at the cellular level is responsible for all higher-order activities. Transport between cells is specifically controlled by their plasma membranes.

Forces that drive biological transport, which include concentration gradients, electrostatic charges, and hydrostatic pressures, are integrated by an expression called the chemical potential. Transport of solutes down chemical gradients is known as diffusion, or passive transport. Movement of solutes against a chemical potential gradient is known as active transport and requires energy input.

The extent to which a membrane permits or restricts the movement of a substance is called membrane permeability. The composition of the membrane and the chemical properties of the solute are major determinants of membrane permeability. When cations and anions move across a membrane at different rates, the electrical potential that develops is called the diffusion potential. For each ion, the relationship between the voltage difference across the membrane and the distribution of the ion at equilibrium is described by the Nernst equation. This equation shows that the difference in concentration of an ion between two compartments is balanced by the voltage difference between the compartments. That voltage difference, or membrane potential, is seen in all living cells because of the asymmetric ion distributions between the inside and outside of the cells. All diffusion potentials across a cell membrane are summed by the Goldman equation.

Membranes contain specialized proteins that facilitate solute transport. Channels are transport proteins that form pores in the membrane; solutes diffuse through these pores down their chemical gradients. Carriers bind a solute on one side of the membrane and release it on the other side. Transport specificity is largely determined by the properties of channels and carriers. Because diffusion and chemical binding are markedly different processes, channels can be distinguished from carriers by kinetic analysis of solute transport.

Electrogenic transport occurs when an ion moves across a membrane unbalanced by a counterion of opposite charge. This process requires energy, which is usually supplied by the hydrolysis of ATP. In plants, an H^+-pumping ATPase is the primary electrogenic transporter in both the plasma membrane and the tonoplast. Proton pumping is a major component of the membrane potentials across these compartments. Plant cells also have a calcium-pumping ATPase that participates in the regulation of intracellular calcium concentrations. Stomatal guard cells have a light-stimulated H^+-pumping ATPase that regulates stomatal movements. The electrochemical gradient generated by H^+ pumping is also used to drive the transport of other substances in a process called secondary transport.

Solutes move between cells either through the extracellular spaces, or apoplast, or from cytoplasm to cytoplasm, via the symplast. Cytoplasms from neighboring cells are connected by plasmodesmata, which facilitate symplast transport. When an ion enters the root, it may be taken up into the cytoplasm of an epidermal cell or it may diffuse through the apoplast into the root cortex and enter the symplast through a cortical cell. From the symplast, the ion finds its way into the xylem and the rest of the plant.

GENERAL READING

Alberts, B., Bray, D., Lewis, J., Raff, M., Roberts, K., and Watson, J. D. (1983) *Molecular Biology of the Cell.* Garland, New York.

Boller, T., and Wiemken, A. (1986) Dynamics of vacuolar compartmentation. *Annu. Rev. Plant Physiol.* 37:137–164.

Epstein, E. (1972) *Mineral Nutrition of Plants: Principles and Perspectives.* Wiley, New York.

Glass, A. D. M. (1983) Regulation of ion transport. *Annu. Rev. Plant Physiol.* 34:311–326.

Harold, F. M. (1986) *The Vital Force: A Study of Bioenergetics,* Chap. 3. Freeman, New York.

Kauss, H. (1987) Some aspects of calcium-dependent regulation in plant metabolism. *Annu. Rev. Plant Physiol.* 38:47–72.

Lüttge, U., and Higinbotham, N. (1979) *Transport in Plants*. Springer-Verlag, Berlin.

Nobel, P. S. (1983) *Biophysical Plant Physiology and Ecology*. Freeman, San Francisco.

Satter, R. L., and Moran, N. (1988) Ionic channels in plant cell membranes. *Plant Physiol.* 72:816–820.

Sze, H. (1985) H^+-translocating ATPase: Advances using membrane vesicles. *Annu. Rev. Plant Physiol.* 36:175–208.

Tazawa, M., Shimen, T. and Mimura, T. (1987) Membrane control in the Characeae. *Annu. Rev. Plant Physiol.* 38:95–117.

Zeiger, E. (1983) The biology of stomatal guard cells. *Annu. Rev. Plant Physiol.* 34:441–475.

CHAPTER REFERENCES

Assmann, S. M., Simoncini, L., and Schroeder, J. I. (1985) Blue light activates electrogenic ion pumping in guard cell protoplasts of *Vicia faba*. *Nature* 318:285–287.

Bennett, A. B., O'Neill, S. D., and Spanswick, R. M. (1984) H^+-ATPase activity from storage tissue of *Beta vulgaris*. *Plant Physiol.* 74:538–544.

Blumwald, E., and Poole, R. J. (1985) Na^+/H^+ antiport in isolated tonoplast vesicles from storage beet of *Beta vulgaris*. *Plant Physiol.* 78:163–167.

Blumwald, E., and Poole, R. J. (1986) Kinetic of Ca^{2+}/H^+ antiport in isolated tonoplast vesicles from storage tissue of *Beta vulgaris* L. *Plant Physiol.* 80: 727–731.

Briskin, D. P., Thornley, W. R., and Wyse, R. E. (1985) Membrane transport in isolated vesicles from sugarbeet taproot. II. Evidence for a sucrose/H^+-antiport. *Plant Physiol.* 78:871–875.

Dunlop, J., and Bowling, D. J. F. (1971) The movement of ions to the xylem exudate of maize roots. *J. Exp. Bot.* 22: 453–464.

Gogarten, J. P., Kibak, H., Dittrich, P., Taiz, L., Bowman, E. J., Bowman, B. J., Manolson, M. F., Poole, R. J., Date, T., Oshima, T., Konishi, J., Denda, K., and Yoshida, M. (1989) Evolution of the vacuolar H^+-ATPase: Implications for the origins of eukaryotes. *Proc. Nat'l Acad. Sci.* 86: 6661–6665.

Higinbotham, N., Etherton, B., and Foster, R. J. (1967) Mineral ion contents and cell transmembrane electropotentials of pea and oat seedling tissue. *Plant Physiol.* 42:37–46.

Higinbotham, N., Graves, J. S., and Davis, R. F. (1970) Evidence for an electrogenic ion transport pump in cells of higher plants. *J. Membr. Biol.* 3:210–222.

Lin, W., Schmitt, M. R., Hitz, W. D., and Giaquinta, R. T. (1984) Sugar transport into protoplasts isolated from developing soybean cotyledons. *Plant Physiol.* 75:936–940.

Meidner, H. (1987) Three hundred years of research into stomata. In: *Stomatal Function*, pp. 7–27. Zeiger, E., Farquhar, G., and Cowan, I., eds. Stanford University Press, Stanford, Calif.

Nelson, N. and Taiz, L. (1989) The evolution of H^+-ATPases. *Trends in Biol. Sciences* 14: 113–116.

Novacky, A., Ullrich-Ebercus, C. I., and Lüttge, U. (1980) pH and membrane potential changes during glucose uptake in *Lemna gibba* G1 and their response to light. *Planta* 149:321–326.

Schroeder, J. I., Raschke, K., and Neher, E. (1987) Voltage dependence of K^+ channels in guard-cell protoplasts. *Proc. Natl. Acad. Sci. USA* 84: 4108–4112.

Schulze, E.-D. (1970) Der CO_2-Gaswechsel der Buche (*Fagus selvatica* L.) in Abhängigkeit von den Klimafaktoren im Freiland. *Flora* 159: 177–232.

Schwartz, A., and Zeiger, E. (1984) Metabolic energy for stomatal opening. Roles of photophosphorylation and oxidative phosphorylation. *Planta* 161:129–136.

Serrano, E. E., and Zeiger, E. (1989) Sensory transduction and electrical signaling in guard cells. *Plant Physiol.* 91:795–799.

Serrano, E. E., Zeiger, E., and Hagiwara, S. (1988) Red light stimulates an electrogenic proton pump in *Vicia* guard cell protoplasts. *Proc. Natl. Acad. Sci. USA* 85:436–440.

Shimazaki, K., Iino, M., and Zeiger, E. (1986) Blue-light dependent proton extrusion by guard-cell-protoplasts of *Vicia faba*. *Nature* 319: 324–326.

Spanswick, R. M. (1981) Electrogenic ion pumps. *Annu. Rev. Plant Physiol.* 32:267–289.

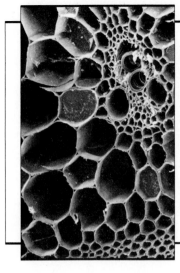

CHAPTER 7

Phloem Translocation

A S WE HAVE SEEN in Chapter 4, survival on land posed a new set of problems for plants newly emerged from an aquatic environment—the need to acquire and retain water being primary among them. In response to these environmental pressures, plants evolved roots (which anchor the plant and absorb water and nutrients) and leaves (which absorb light and exchange gases). As plants increased in size, the roots and leaves became increasingly separated from each other in space. Thus, systems evolved for long-distance transport—that is, translocation—that allowed the shoot and the root to efficiently exchange the products of absorption and assimilation. The **xylem** is the tissue that transports water and minerals from the root system to the aerial portions of the plant (see Chapters 1, 4, 6). The **phloem**, which is the subject of this chapter, is the tissue that translocates the products of photosynthesis from mature leaves to areas of growth and storage, including the roots. As we shall see, the phloem also serves to redistribute water and various compounds throughout the plant body. These compounds, some of which initially arrive in the mature leaves via the xylem, can be either transferred out of the leaves without modification or metabolized before redistribution.

In the discussion that follows, the emphasis will be on phloem translocation in angiosperms, since most of the research has been conducted on that group of species. Gymnosperms will be briefly compared to the angiosperms in terms of the anatomy of their conducting cells and possible differences in their mechanism of translocation.

Pathways of Translocation

As mentioned above, two long-distance transport pathways, the phloem and the xylem, extend throughout the plant body. The phloem is generally found on the outer side of both primary and secondary vascular tissues (Figs. 7.1 and 7.2); in plants with secondary growth the phloem constitutes the inner bark. The cells of the phloem that conduct sugars and other organic materials throughout the plant are called **sieve elements**. "Sieve element" is a comprehensive term that includes both the highly differentiated **sieve-tube members** typical of the angiosperms and the relatively unspecialized **sieve cells** of gymnosperms. In addition to sieve elements, the phloem tissue contains companion cells, parenchyma cells, and, in some cases, fibers, sclereids, and latex-containing cells (laticifers). However, only the sieve elements are directly involved in translocation. The small veins of leaves and the primary vascular bundles of

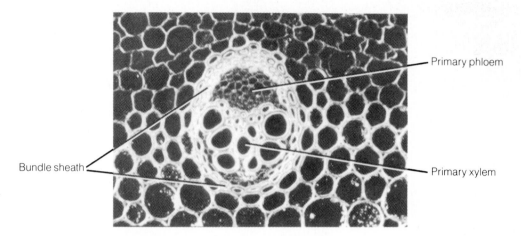

FIGURE 7.1. Transverse section of vascular bundle of buttercup (*Ranunculus*) (160 ×). The primary phloem is toward the outside of the stem. Both the primary phloem and the primary xylem are surrounded by a bundle sheath of thick-walled sclerenchyma cells, which isolates the vascular tissue from the ground tissue.

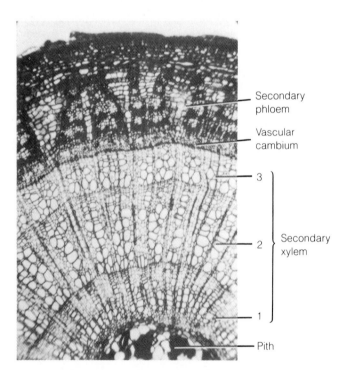

FIGURE 7.2. Transverse section of a linden (*Tilia americana*) tree stem. (50 ×). The stem is three years old. The numbers on the figure indicate growth rings in the secondary xylem. The old secondary phloem has been crushed by the expansion of the xylem. Only the most recent layer of secondary phloem is functional. (Photo © Richard Gross.)

stems are often surrounded by a **bundle sheath**, which consists of one or more layers of compactly arranged cells (Fig. 7.1). In the vascular tissue of leaves, the bundle sheath surrounds the small veins all the way to their ends, isolating the veins from the intercellular spaces of the leaf.

In some leaves, cells similar to the bundle sheath cells extend to the upper and lower epidermal layers and are thought to aid in water conduction throughout the leaf.

We will begin our discussion with the experimental evidence demonstrating that the sieve elements are the conducting cells in the phloem. This will be followed by an examination of the structure and physiology of these unusual plant cells.

Experiments with Radioactive Labels Show That Sugar Is Translocated in Phloem Sieve Elements

In classical experiments on the translocation of organic solutes performed by Malpighi in 1686, the bark of a tree was removed in a ring around the trunk (Fig. 7.3). In 1928 Mason and Maskell observed that this treatment, called **girdling**, has no immediate effect on transpiration, since water movement occurs in the xylem, interior to the bark. However, sugar transport in the trunk is blocked at the site where the bark has been removed. Sugars accumulate above the girdle, that is, on the side toward the leaves, and are depleted below the treated region. Eventually the bark below the girdle dies, while the bark above swells and remains healthy. Mason and Maskell concluded that sugar transport occurs in the bark of the tree and, further, that the sieve elements are the cellular channels of sugar transport. The latter conclusion was based on an observed high correlation between leaf and bark sucrose contents and on the high sucrose concentration calculated to be present in the sieve elements.

More sophisticated experiments on phloem translocation became possible in the 1940s when radioactive isotopes became available for scientific research. Labeled

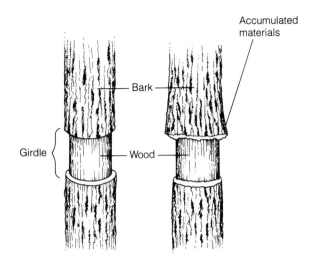

FIGURE 7.3. Tree trunk immediately after girdling (left) and after a longer period of time (right). Girdling is the removal of the bark of a tree in a ring around the trunk. Materials translocated from the leaves have accumulated in the region above the girdle and caused it to swell.

organic compounds can be introduced into the plant in a number of ways. For example, carbon dioxide can be labeled[1] with ^{14}C or ^{11}C and supplied to an intact mature leaf in a small sealed chamber. The labeled CO_2 is incorporated into organic compounds via the CO_2-fixing reactions of photosynthesis. Some of the sugar-phosphates formed in photosynthesis undergo further reactions to form the translocated sugars, such as sucrose and stachyose. In another type of experiment, the photosynthetic reactions are bypassed by supplying labeled transport sugar, usually sucrose, directly to the leaf. In this case, a solution containing the radioactive sugar must be applied in a way that allows it to reach cells in the leaf interior. The solution can be applied to the leaf surface after the epidermis has been peeled off or the cuticle has been abraded in a small area. Alternatively, the solution can be supplied through an isolated vein. Compounds labeled with other isotopes, such as ^{32}P, can be introduced in the same way.

In order to identify the pathway of sugar translocation, the tissue or cellular location of the isotope must be determined, usually by means of a technique called **autoradiography**. In tissue autoradiography, parts of

labeled plants are rapidly frozen, freeze-dried, embedded in paraffin or resin, and sliced into thin sections. The sections are then coated with a film of photographic emulsion. During a period of exposure, radiation from the label exposes the film, so that when the film is developed, silver grains appear wherever the label was present in the tissue. A comparison of the tissue section and the pattern of silver grains reveals the location of the label in the tissue. In terms of the transport pathway for sugars, the label initially appears in the sieve elements of the phloem, confirming the results of the earlier experiments (Fig. 7.4).

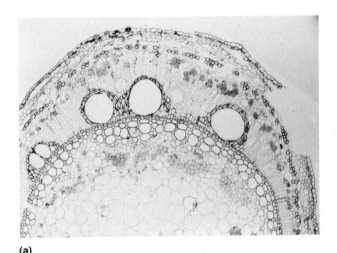

(a)

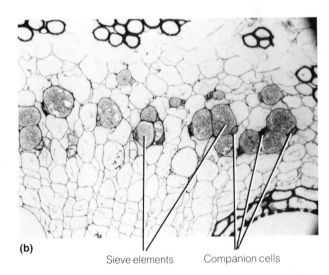

(b)

Sieve elements   Companion cells

FIGURE 7.4. Autoradiographs of stem tissue of broad bean (*Vicia faba*). $^{14}CO_2$ was supplied to the source leaf. During photosynthesis, the ^{14}C was incorporated into labeled sugars, which were then transported to other parts of the plant. The location of the label is revealed by the presence of dark grains on the film. Comparison of the film with the underlying tissue section reveals that label is confined almost entirely to the sieve elements of the phloem. (a) Cross section (50 ×). (b) Cross section (325 ×). (From Christy and Fisher, 1978.)

---

1. These two isotopes of carbon, ^{11}C and ^{14}C, have quite different half-lives (the half-life is the time required for the radioactivity to decrease by one-half). The half-life for ^{11}C is 20 minutes, while that for ^{14}C is 5,600 years. For this reason, the isotopes are used in different types of experiments. Carbon-14 is useful when the researcher wishes to analyze the compounds in which the radioactivity is present. Carbon-11 would disappear before such an analysis could be made. It is most useful when the researcher wishes to make repeated measurements on the same plant, since only a few hours are needed for the isotope in the plant to disappear.

Monitoring Traffic on the Sugar Freeway

SUGAR MOLECULES CAN be thought of as cars speeding along a freeway on their way to various destinations, called sinks. To monitor this traffic we need to count the number of cars per unit time or, in physiological terms, the mass transfer rate. Measurements of mass transfer rates were among the earliest quantitative determinations made in phloem physiology. The early studies measured either gain in weight by developing fruits or storage organs (sinks) or loss of weight by mature leaves (sources) over a known time interval; the sink experiments are generally the more accurate. For example, Mason and Lewin in 1926 determined weights of individual yam (*Dioscorea alata*) tubers over a growing season. The increase in dry weight during the growing season was divided by the time interval to obtain an average mass transfer rate. This rate, divided by the average sieve-element area per stem, yields a specific mass transfer rate of 5.7 g h^{-1} cm^{-2} of sieve elements. A number of similar studies obtained mass transfer rates of 0.2 to 4.8 g h^{-1} cm^{-2} of *phloem*. If it is assumed that sieve elements occupy approximately one-third of the phloem cross-sectional area (a reasonable estimate), these results yield specific mass transfer rates of 0.6 to 14.5 g h^{-1} cm^{-2} of *sieve elements*.

A more recent technique for determining mass transfer rates by use of radioactive tracers gives rates in excellent agreement with the earlier measurements (Geiger and Swanson, 1965). In this method the plant is usually pruned to a single mature source leaf and a single immature sink leaf. Carbon dioxide labeled with ^{14}C is supplied to the source leaf in such a way that the rate of supply is equal to the rate of fixation. The rate of arrival of label in the sink tissue is monitored over the length of the experiment. The supplied radioactive CO_2 contains a predetermined amount of radioactivity per unit weight of carbon (specific activity). After a certain period of time, the sugars in transit from the source to the sink reach the same specific activity as the supplied CO_2. When this state, called isotopic saturation, has been reached, the mass transfer rate can be calculated from the rate of arrival of label in the sink tissue, as follows:

$$\frac{Bq}{second} \div \frac{Bq}{\mu g \text{ of carbon}} = \frac{\mu g \text{ of carbon}}{second}$$

Arrival rate in sink	Specific activity of supplied label	Mass transfer

where Bq stands for becquerel, a unit of radioactivity.* The rate determined in this way, corrected for the cross-sectional area of sieve elements in the transit pathway, provides an estimate of specific mass transfer. For sugar beet, which transports sucrose, the value obtained is equivalent to a rate of 4.8 g h^{-1} cm^{-2} of sieve elements, in excellent agreement with the values determined in earlier investigations.

---

*A bequerel (Bq) is equivalent to one disintegration per second. Another unit of radioactivity that has been used quite often is the curie (Ci), equivalent to 3.7×10^{10} disintegrations per second.

Mature Sieve Elements Are Living Cells That Are Highly Specialized for Translocation

As we will see later, detailed knowledge of the ultrastructure of sieve elements is critical to any discussion of the mechanism of phloem translocation. Mature sieve elements are indeed unique among living plant cells (Figs. 7.5 and 7.6). They lack many structures normally found in living cells, including the undifferentiated cells from which mature sieve elements are formed. For example, sieve elements lose their nucleus and tonoplast (vacuolar membrane) during development. Microfilaments, microtubules, Golgi bodies, and ribosomes are also absent from the mature cells. In addition to the plasma membrane, organelles that are retained include somewhat modified mitochondria, plastids, and smooth endoplasmic reticulum. The walls are nonlignified, though they are secondarily thickened in some cases. Thus, the sieve elements are unlike the tracheary elements of the xylem, which are dead at maturity, lack a plasma membrane, and have lignified secondary walls. As we shall see, this difference is critical to the mechanism of translocation in the phloem.

The Most Characteristic Feature of the Sieve Elements Is the Presence of the Sieve Areas

Sieve elements are characterized by **sieve areas**, portions of the cell wall where pores interconnect the conducting cells. The sieve-area pores range in diameter from less than 1 μm to approximately 15 μm. The sieve areas of sieve-tube members (angiosperms) are more specialized

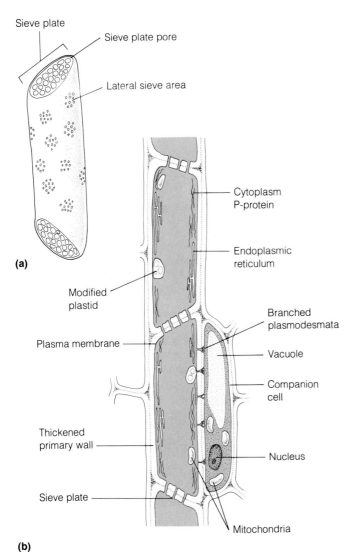

(a)

(b)

FIGURE 7.5. Schematic drawings of mature sieve elements (sieve-tube members): External view showing sieve plates and lateral sieve areas (a). Longitudinal section showing two sieve-tube members joined together to form a sieve tube (b). The pores in the sieve plates between the sieve-tube members are open channels for transport through the sieve tube. Each sieve-tube member is associated with one or more companion cells, which assume some of the essential metabolic functions that were reduced or lost during the differentiation of the sieve elements. Note that the companion cell is enriched in cytoplasmic organelles, whereas the sieve-tube member has relatively few organelles. The plasma membrane of a sieve-tube member is continuous with that of its neighboring sieve-tube member.

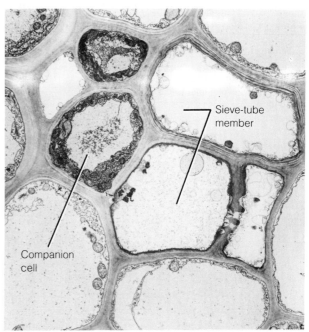

FIGURE 7.6. Electron micrograph showing a transverse section of companion cells and mature sieve-tube members (3600 ×). The cellular components are distributed along the walls of the sieve-tube members. (From R. Warmbrodt, 1985.)

TABLE 7.1. Characteristics of the two types of sieve elements.

Sieve-tube members

1. Found in angiosperms.
2. Some sieve areas are differentiated into sieve plates; individual sieve-tube members are joined together into a sieve tube.
3. Sieve-plate pores are open channels.
4. P-protein is present in dicots and a few monocots.
5. Companion cells are sources of ATP and perhaps other compounds and, in some species, are intermediary cells.

Sieve cells

1. Found in gymnosperms.
2. No sieve plates; all sieve areas are similar.
3. Pores in sieve areas appear blocked with membranes.
4. No P-protein.
5. Albuminous cells function as companion cells.

than those of sieve cells (gymnosperms). For instance, some of the sieve areas of sieve-tube members are differentiated into **sieve plates** (Fig. 7.7 and Table 7.1). Sieve plates have larger pores than the other sieve areas in the cell and are generally found on the end walls of sieve-tube members, where the individual cells are joined together to form a longitudinal series called a **sieve tube**. Furthermore, the sieve-plate pores of sieve-tube members

are essentially open channels that allow transport between cells (Fig. 7.7). In sieve cells (gymnosperms), on the other hand, all of the sieve areas are more or less the same. The pores of sieve cells are relatively unspecialized and appear to be filled with numerous membranes, which meet in large median cavities in the middle of the wall and are continuous with the smooth endoplasmic reticulum opposite the sieve areas (Fig. 7.8).

FIGURE 7.7. Electron micrograph showing a portion of two mature sieve elements (sieve-tube members), sectioned longitudinally, in the hypocotyl of winter squash (*Cucurbita maxima*). The wall between the sieve elements is a sieve plate. The insert shows sieve-plate pores in face view. In both cases, the sieve-plate pores are open, that is, unobstructed by P-protein. (From Evert, 1982.)

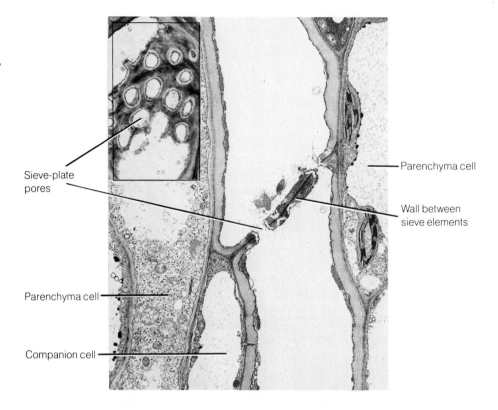

Sieve-plate pores

Parenchyma cell

Wall between sieve elements

Parenchyma cell

Companion cell

FIGURE 7.8. Electron micrograph showing a sieve area of a mature sieve cell in *Pinus resinosa*. Note that the pores are filled with membranes, which meet in large median cavities in the middle of the wall and are continuous with smooth endoplasmic reticulum on either side of the sieve-area pores. How this arrangement of the pores affects translocation is unclear. The sieve area pores are continuous from cell to cell; in this micrograph they appear to be blocked by the cell wall because of the angle of the section. (From Neuberger and Evert, 1975.)

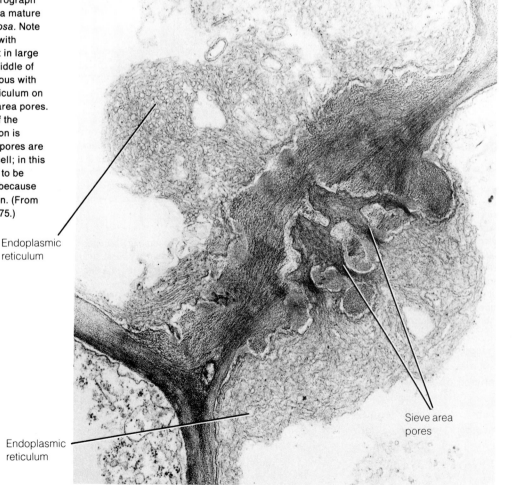

Endoplasmic reticulum

Endoplasmic reticulum

Sieve area pores

P-Protein and Callose Deposition Seal Off Damaged Sieve Elements

Sieve-tube members of the dicots are usually rich in a phloem protein called **P-protein** (Fig. 7.5). (In the past, P-protein was called "slime.") P-protein is found rarely in monocot species and never in gymnosperms. It occurs in a number of different forms (tubular, fibrillar, granular, and crystalline) depending on the species and maturity of the cell. In immature cells, P-protein is most evident as discrete bodies in the cytosol known as **P-protein bodies**. P-protein bodies may be spheroidal, spindle-shaped, or twisted and coiled. They generally disperse into tubular or fibrillar forms during cell maturation.

P-protein appears to function in sealing off damaged sieve elements by plugging up the sieve-plate pores. Sieve tubes are under high internal turgor pressure, and the sieve elements in a sieve tube are connected to each other through open sieve-plate pores. When a sieve tube is cut or punctured, the release of pressure causes the contents of the sieve elements to surge toward the cut end, from which the plant could lose much sugar-rich phloem sap if some sealing mechanism did not exist (Fig. 7.9). ("Sap" is a general term used to refer to the fluid contents of plant cells.) However, when surging occurs, P-protein and other cellular inclusions are trapped on the sieve-plate pores, helping to seal the sieve element and to prevent further loss of sap.

A longer-term solution to the damage problem in most species is the production of **callose** in the sieve pores (Fig. 7.9). Callose is a β-$(1 \rightarrow 3)$ glucan, which is identified by its reaction with aniline blue stain. Callose is synthesized by an enzyme in the plasma membrane and is deposited between the plasma membrane and the cell wall. The enzyme responsible for callose synthesis (callose synthetase) appears to be arranged vectorially in the plasma membrane, with substrate being supplied from the cytoplasmic side and product being deposited on the wall surface. It is now generally recognized that callose is synthesized in functioning sieve elements in response to damage and other stresses, such as mechanical stimulation and high temperatures. The deposition of such **wound callose** in the sieve pores efficiently seals off damaged sieve elements from surrounding intact tissue. As the sieve elements recover from damage, the callose disappears from these pores.

Callose is also found in sieve elements under circum-

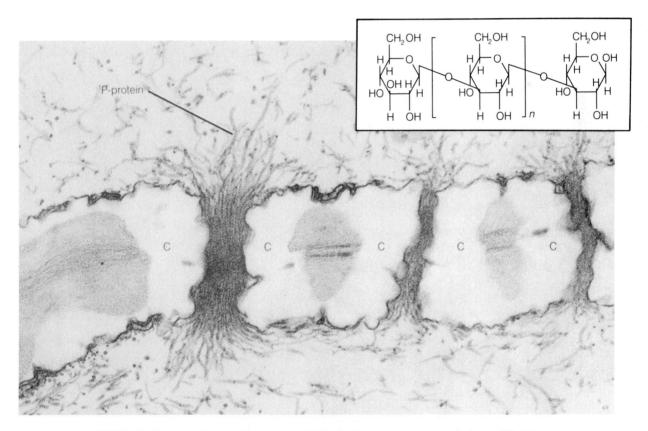

FIGURE 7.9. Electron micrograph of callose formation in the sieve-plate pores of tobacco (*Nicotiana tabacum*) (40,000 ×). Callose deposition occurred in response to wounding. In intact plants, formation of wound callose seals off damaged sieve elements and prevents loss of sugar-rich phloem sap. Contrast this to the open pores visible in Figure 7.7. The material in the pores is P-protein, which was probably more widely dispersed initially, being compressed as the callose formed in the pores. The structural formula for callose is shown in the insert above. (Photo from Susan A. Sovonick-Dunford.)

stances other than wounding, although its function always seems to be one of sealing. **Definitive callose** is deposited in dying cells or sieve elements undergoing obliteration due to the formation of secondary tissues. Callose associated with dormancy is found in many overwintering plants, that is, perennial plants that have become dormant in the winter. Such **dormancy callose** is redissolved in the spring in preparation for resumption of transport and growth.

The Highly Specialized Sieve Elements Are Functionally Supported by the Companion Cells

Each sieve-tube member is associated with one or more **companion cells** (Figs. 7.5, 7.6, and 7.7), the two cell types arising by the division of a single mother cell. Numerous intercellular connections, the **plasmodesmata**, penetrate the walls between sieve-tube members and their companion cells, suggesting a close functional relationship

and ease of transport between the two cells. The plasmodesmata are often complex and branched on the companion-cell side. In the gymnosperms the companion-cell function is assumed by **albuminous cells**, which do not arise from the same mother cells as the sieve cells.

Companion cells are thought to take over some of the critical metabolic functions, such as protein synthesis, that are reduced or lost during the differentiation of the sieve elements. The numerous mitochondria in companion cells may supply energy as ATP to the sieve elements. In some species, photosynthetic products flow from the producing cells in the mesophyll through the companion cells to the sieve elements.

The transfer of materials to the sieve elements is further facilitated in some herbaceous dicots by the development of wall ingrowths in some companion cells or phloem parenchyma cells (Fig. 7.10). Cells with such wall ingrowths are called **transfer cells**. The wall ingrowths greatly increase the surface area of the plasma membrane, thus increasing the potential for solute trans-

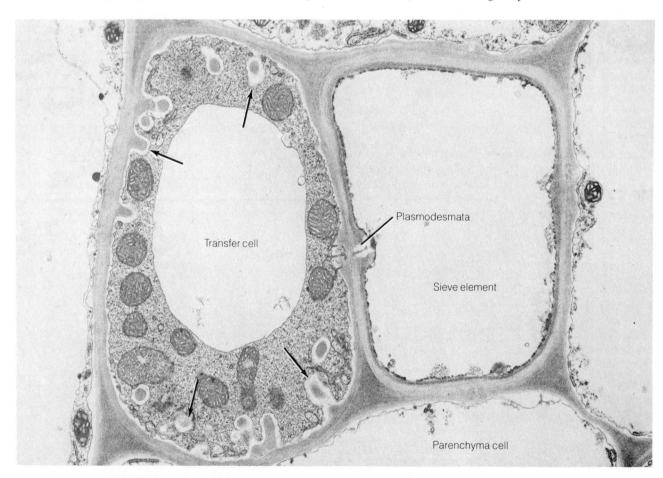

FIGURE 7.10. Electron micrograph of transfer cell and sieve element (sieve-tube member) from a source leaf of pea (*Pisum sativum*) (15,700 ×). Note the numerous wall ingrowths (arrows) on the walls adjacent to the parenchyma cells but not on the wall opposite the sieve element. Such ingrowths greatly increase the surface area of the transfer cell membrane, increasing the potential for solute transport and ultimately facilitating the transfer of materials from the mesophyll to the sieve elements. Note the presence of plasmodesmata between the transfer cell and the sieve element. (Brentwood and Cronshaw, 1978).

fer across the membrane. Xylem parenchyma cells can also be modified as transfer cells, probably serving to retrieve and reroute solutes moving in the xylem.

Patterns of Translocation

The direction of phloem transport is determined by the relative locations of the areas of supply and utilization of the products of photosynthesis. Phloem transport does not occur exclusively either in an upward or a downward direction and is not defined with respect to gravity. Translocation occurs from areas of supply (**sources**) to areas of metabolism or storage (**sinks**). Sources include any exporting organ, typically a mature leaf, that is capable of producing photosynthate in excess of its own needs. Another type of source is a storage organ during the exporting phase of its development. For example, the storage root of the biennial wild beet (*Beta maritima*) is a sink during the first growing season when it accumulates sugars received from the source leaves. During the second growing season the same root becomes a source; the sugars are remobilized and utilized to produce a new shoot, which ultimately becomes reproductive. In contrast, roots of the cultivated sugar beet (*Beta vulgaris*) can increase in dry mass during both the first and second growing seasons, so the leaves serve as sources during both flowering and fruiting stages. This occurs in cultivated varieties of beets because they have been selected for the capacity of their roots to act as sinks during all phases of development. Sinks include any nonphotosynthetic organs of the plant and organs that do not produce enough photosynthetic products to support their own growth or storage needs. Roots, tubers, developing fruits, and immature leaves, which must import carbohydrate for normal development, are all examples of sink tissues. Both girdling and labeling studies support the source-to-sink pattern of translocation in the phloem.

Source-to-Sink Pathways Follow Anatomical and Developmental Rules

While the overall pattern of phloem transport can be simply stated as source-to-sink movement, the specific pathways involved are often more complex, following anatomical and developmental rules. Sinks are not equally supplied by all source leaves on a plant; rather, certain sources preferentially supply specific sinks. In the case of herbaceous plants, the following generalizations can be made:

1. The proximity of the source to the sink is a significant factor. Thus, the upper mature leaves on a plant usually transport assimilates to the growing shoot tip and young, immature leaves, while the lower leaves predominantly supply the root system. Intermediate leaves export in both directions,

bypassing the intervening mature leaves.

2. The importance of various sinks may shift during plant development. Whereas the root and shoot apices are usually the major sinks during vegetative growth, fruits generally become the dominant sinks during reproductive development, particularly for adjacent and other nearby leaves.

3. Source leaves supply sinks with which they have direct vascular connections. A given leaf is generally connected via the vascular system to other leaves directly above or below it on the stem. Such a vertical row of leaves is called an *orthostichy*. The number of internodes between leaves on the same orthostichy varies with species. Figure 7.11a illustrates this pattern in sugar beet.

4. Interfering with the translocation pathway by wounding or pruning can alter the patterns outlined above. In the absence of direct connections between source and sink, vascular interconnections (*anastomoses*) can provide an alternative pathway. In sugar beet, for example, removing source leaves from one side of the plant can bring about cross-transfer of assimilates to young leaves (sink leaves) on the pruned side (Fig. 7.11b). Upper source leaves on a plant can be forced to translocate to the roots by removing the lower source leaves, and vice versa. However, the flexibility of the translocation pathway is dependent on the extent of the interconnections between vascular bundles and thus on the species and organs studied. In sweet orange (*Citrus sinensis* L.), half of the fruit receives no photosynthate when the entire fruit is dependent on a single source leaf (Koch and Avigne, 1984). In chickpea (*Cicer arietinum* L.), the leaves on a depodded branch cannot transport photosynthate to the pods on an adjacent defoliated branch (Singh and Pandey, 1980), but in soybean plants (*Glycine max* [L.] Merrill), cross-transfer occurs readily from a partially depodded side to a partially defoliated side (Gent, 1982).

Materials Translocated in the Phloem

Water is quantitatively the most abundant substance transported in the phloem. Dissolved in the water are the translocated solutes, which consist mainly of carbohydrates in most plants (Table 7.2). Sucrose is the sugar most commonly transported in sieve elements, in which it can reach concentrations of 0.3 to 0.9 M. Some other organic solutes also move in the phloem. Nitrogen occurs in the phloem largely in the form of amino acids and amides, especially glutamate/glutamine and aspartate/

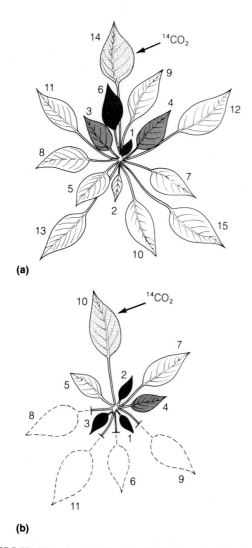

(a)

(b)

FIGURE 7.11. Diagrams showing the distribution of radioactivity from a single labeled source leaf in an intact plant and in a pruned plant. The degree of radioactive labeling is indicated by the intensity of the shading of the leaves. (a) The distribution of radioactivity in leaves of an intact sugar beet plant (*Beta vulgaris*) one week after $^{14}CO_2$ was supplied for 4 hours to a single source leaf (arrow). Leaves are numbered according to their age, with the youngest, newly emerged leaf being designated number 1. The ^{14}C label was translocated mainly to the sink leaves directly above the source leaf (that is, sink leaves on the same orthostichy as the source). (b) same as (a) except that all source leaves on the side of the plant opposite the labeled leaf were removed 24 hours prior to labeling. Sink leaves on both sides of the plant now receive ^{14}C-labeled assimilates from the source. (Based on data from Joy, 1964.)

asparagine. Reported levels of amino acids and organic acids vary widely, even for the same species, but are usually low compared with carbohydrates. Almost all of the endogenous plant hormones, including auxin, gibberellins, cytokinins, and abscisic acid, have been found in sieve elements and are thought to be transported over long distances at least partly in these cells. Nucleotide-phosphates and enzymes have also been found in phloem

TABLE 7.2. The composition of phloem sap from castor bean (*Ricinus communis*) and squash (*Cucurbita maxima* Duchesne), collected as an exudate from cuts in the phloem.

Component	Concentration (mg ml^{-1})	
	Castor bean*	Squash[†]
Sugars	80.0–106.0	0.5–12.0
Amino acids	5.2	5.0–30.0
Organic acids	2.0–3.2	3.0–5.0
Protein	1.45–2.20	76.2–112.2
Chloride	0.355–0.675	0.041–0.176
Phosphate	0.350–0.550	0.028–0.083
Potassium	2.3–4.4	2.1–4.6
Magnesium	0.109–0.122	0.016–0.033

*From Hall and Baker (1972). Like most other species, castor bean translocates mainly carbohydrate, largely in the form of sucrose. Exudate from members of the squash family (cucurbits), however, contains comparatively low levels of carbohydrate and most of that as stachyose. Cucurbits may mobilize a high proportion of their transport carbon in the form of amino acids and organic acids. Stachyose is the main carbohydrate translocated in the cucurbits.

[†]From Richardson et al. (1982).

sap. Inorganic solutes that move in the phloem include potassium, magnesium, phosphate, and chloride. In contrast, nitrate, calcium, sulfur, and iron are almost completely excluded from the phloem.

The discussion that follows begins with the methods used to identify materials translocated in the phloem. We will then examine the translocated sugars and the complexities of nitrogen transport in the plant.

Phloem Sap Can Be Collected and Analyzed

Since phloem tissue, like xylem tissue, consists of a mixture of conducting and nonconducting cells, identification of the material translocated in the sieve elements presents a formidable challenge. Methods based on labeling with radioactive isotopes and direct extraction of the tissue do not distinguish between compounds in transit and those outside the translocation pathway. However, the tendency of some species to exude phloem sap from wounds that sever sieve elements has been exploited to obtain relatively pure samples of the translocated material. The driving force for exudation is the high positive pressure in the sieve elements.

Most species do not exude detectable amounts of phloem sap because of the sealing mechanisms mentioned earlier. However, exudation rates can be increased by treating the cut surface with the chelating agent EDTA (ethylenediaminetetraacetic acid) (King and Zeevaart, 1974). Chelating agents inhibit callose formation by forming soluble complexes with divalent cations, particularly calcium, which is required by callose synthetase.

The major disadvantage of the wound-exudation method is that the fluid collected may not represent the

(a)

(b)

true composition of the translocated material. Contaminants can originate from damaged parenchyma cells or even from the sieve elements themselves. Furthermore, the abrupt lowering of the turgor pressure in the sieve elements will, as discussed in Chapter 4, cause a decrease in the water potential of these cells. As a result, water from the surroundings will enter the sieve elements along a water potential gradient, causing a dilution of the phloem sap.

The ideal way to collect phloem sap would be to tap into a single sieve element by using a tiny syringe. Fortunately, nature has provided us with just such a probe, the aphid stylet! Aphids are small insects that feed by inserting their mouthparts, consisting of four tubular stylets, into a single sieve element of a leaf or stem (Fig. 7.12). The high turgor pressure in the sieve

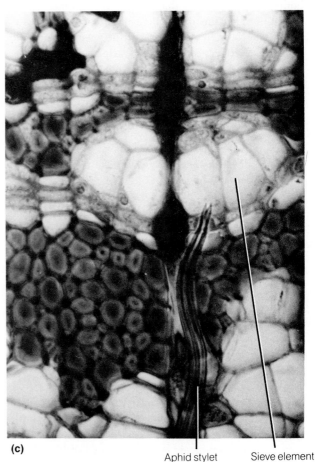

(c)

Aphid stylet   Sieve element

FIGURE 7.12. Collection of phloem sap using aphids, which tap single sieve elements. (a) An aphid (*Longistigma caryae*) feeding on a branch of linden. The insect is just releasing a droplet of honeydew, which occurs about once every 30 min. The honeydew excreted by the aphid consists of sieve-element sap from which selected solutes have been removed in the gut of the insect. The high turgor pressure in the sieve element forces the cell contents through the food canal of the aphid and into the gut. (b) Exudation from a stylet of *Longistigma* that was cut from the aphid's body after anesthetizing the insect with CO_2. Exudation continues from the cut stylet provided it has not been moved from its original position in the sieve element during the experimental procedure. (c) Transverse section through the bark of linden, showing the tips of aphid stylets that had been exuding prior to sectioning. The tips of the stylets are inside a single sieve element, while the sheath of saliva secreted by the aphid ends outside the cell (625 ×). (From Zimmerman and Brown, 1971.)

element forces the cell contents through the insect's food canal and into the gut, where amino acids are selectively removed and some sugars are metabolized. The excess sap, still rich in carbohydrates, is then excreted as "honeydew," which can be collected for chemical analysis. Sap can also be collected from aphid stylets cut from the body of the insect after it has been anesthetized with CO_2. Exudate from severed stylets provides a more accurate picture of the substances present in the sieve elements than does honeydew, whose composition has been altered by the insect.

Aphid techniques have substantial advantages in the study of phloem physiology. Of particular importance, of course, is the fact that aphids tap a single sieve element, so there is no problem of contamination from other cell types. Exudation from severed stylets can continue for hours, which suggests that the aphid prevents normal sealing mechanisms from operating or that the pressure drop that occurs on penetration of the sieve element is not sufficient to trigger sealing mechanisms.

The disadvantages of using aphids include the fact that siting the insects at a desired location and severing the stylets without disrupting them require considerable patience and skill. In addition, aphids may induce reactions in the host plant by secreting saliva into the plant tissues. However, the magnitude of these reactions over the period of time necessary to collect sap for analysis is probably negligible.

The use of aphid stylets is thus the preferred technique for many laboratory experiments on herbaceous plants, for which a large, readily manipulated aphid species, *Longistigma caryae*, is available.

Sugars Are Translocated in Nonreducing Form

Results from analyses of sap collected by the techniques outlined above indicate that the translocated carbohydrates are all nonreducing sugars. Reducing sugars, such as glucose and fructose, contain an exposed aldehyde or ketone group (Fig. 7.13) and can be assayed colorimetrically because of their ability to reduce Cu^{2+} to Cu^+ in solution. In a nonreducing sugar the ketone or aldehyde group is reduced to an alcohol or combined with a similar group on another sugar. For example, the disaccharide sucrose is a nonreducing sugar composed of one glucose molecule (an aldose sugar) bound to one fructose molecule (a ketose sugar). Most researchers believe that the nonreducing sugars are the major translocates simply because they are less reactive than their reducing components.

Sucrose is the most commonly translocated sugar; many of the other mobile carbohydrates contain sucrose bound to varying numbers of galactose molecules. Raffinose consists of sucrose and one galactose molecule; stachyose, sucrose and two galactose molecules; and verbascose, sucrose and three galactose molecules

(Fig. 7.13). Translocated sugar alcohols include mannitol and sorbitol.

Transport Patterns of Nitrogenous Compounds in the Phloem and Xylem Are Interrelated

Nitrogen is transported throughout the plant in either inorganic or organic form; the form that predominates depends on several factors, including the transport pathway. While nitrogen is transported in the phloem almost entirely in organic form, it can be transported in the xylem either as nitrate or as part of an organic molecule (see Chapter 12). In the majority of species studied, the same group of organic molecules carries nitrogen in both the xylem and the phloem.

The form in which nitrogen is transported in the xylem depends on the species studied. Species that do not form a symbiotic association with nitrogen-fixing microorganisms depend on soil nitrate as their major nitrogen source (see Chapter 12). In the xylem of these species, nitrogen is usually present as both nitrate and nitrogen-rich organic molecules, particularly the amides, asparagine and glutamine (Fig. 7.13). Species with nitrogen-fixing nodules on their roots depend on atmospheric nitrogen, rather than soil nitrate, as their major nitrogen source. Following conversion to an organic form, this nitrogen is transported in the xylem to the shoot, usually as amides or ureides such as allantoin, allantoic acid, or citrulline (Fig. 7.13). In all cases in which nitrogen is assimilated into organic compounds in the roots, both the energy and the carbon skeletons required for assimilation are derived from photosynthates transported to the roots via the phloem. Nitrogen levels in mature leaves are quite stable, which indicates that at least some of the excess nitrogen continuously arriving via the xylem must be redistributed in the phloem to fruits or younger leaves.

One of the most widely studied species in terms of nitrogen transport is the economically important soybean, a legume with nitrogen-fixing root nodules. Research has shown that 12 to 48% of the nitrogen needs of the developing seeds of this species are supplied from compounds newly synthesized in the leaves and 35 to 52% from compounds synthesized in the roots and transferred from xylem to phloem (Layzell and LaRue, 1982). The remainder probably comes from the breakdown of specialized storage proteins in the soybean leaf. White lupin (*Lupinus albus*) is another legume in which nitrogen transport has been extensively investigated. Phloem sap arriving in the shoot apex of this species is often enriched in nitrogen, compared to phloem sap in the stem or in the petioles of the source leaves (Layzell et al., 1981). This enrichment is due to transfer of nitrogenous compounds in the stem from the xylem to the phoem.

Finally, levels of nitrogenous compounds in the phloem are quite high during leaf senescence. In woody

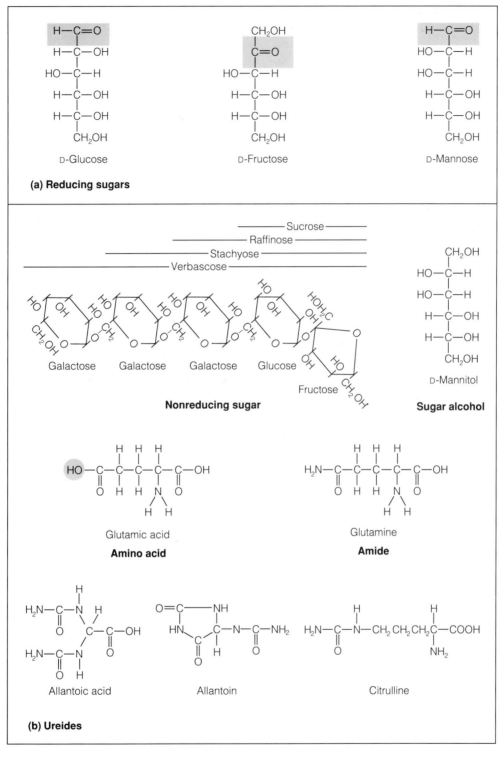

FIGURE 7.13. Structures of compounds not normally translocated in the phloem and those commonly translocated in the phloem. (a) The reducing sugars glucose, fructose, and mannose are generally not found in phloem sap. The reducing groups are aldehyde (glucose and mannose) and ketone (fructose) groups. (b) Compounds commonly translocated in the phloem. Sucrose is a disaccharide made up of one glucose and one fructose molecule. Raffinose, stachyose, and verbascose contain sucrose bound to one, two, or three galactose molecules. Mannitol is a sugar alcohol formed by the reduction of the aldehyde group of mannose. All are nonreducing. Glutamic acid, an amino acid, and glutamine, its amide, are important nitrogenous compounds in the phloem, in addition to aspartate and asparagine. Species with nitrogen-fixing nodules also utilize ureides as transport forms of nitrogen.

species, senescing leaves mobilize and export nitrogenous compounds to the woody tissues for storage, whereas in herbaceous plants the export is generally to the seeds. Other solutes, such as mineral ions, are redistributed from senescing leaves in the same manner.

Rates of Movement

The rate of movement of material in the sieve elements can be expressed in two ways: as **velocity**, the linear distance traveled per unit time, or as **mass transfer rate**, the quantity of material passing through a given cross section of phloem or sieve elements per unit time. Mass transfer rates based on the cross-sectional area of the sieve elements are preferred, since the sieve elements are the actual conducting cells of the phloem.

In the original publications the units of velocity were in centimeters per hour, while the units of mass transfer were in grams per hour per square centimeter of phloem or sieve elements. The currently preferred units (SI units) are meters (m) or millimeters (mm) for length, seconds (s) for time, and kilogram (kg) for mass.

Velocities of Phloem Transport Average One Meter per Hour

Both velocities and mass transfer rates can be measured with radioactive tracers, and this is the technique most commonly used to determine transport velocities. In the simplest type of experiment, ^{11}C- or ^{14}C-labeled CO_2 is applied for a brief period of time to a source leaf (pulse labeling), and the arrival of label at a sink tissue or at some point along the pathway is monitored with an appropriate detector. The length of the translocation pathway divided by the time interval required for label to be first detected at the sink yields a measure of velocity, at least for the fastest-moving labeled component. A more accurate measurement of velocity is obtained by monitoring the arrival of label at two points along the pathway, since this method excludes from the time measurement the time required for fixation of labeled carbon by photosynthesis, for conversion into transport sugar, and for accumulation of sugar in the sieve elements of the source leaf. In general, velocities measured by a variety of techniques average about 1 m h^{-1} and range from 30 to 150 cm h^{-1}, with a few studies reporting speeds an order of magnitude greater than these. The variability in the measured velocities is due to species differences, differences between experimental methods, and problems inherent in the experimental methods. Despite this variability, it is clear that transport velocities in the phloem are quite high and well in excess of the rate of diffusion over long distances. Any proposed mechanism of phloem translocation must account for these high velocities.

Phloem Loading

Several transport steps are involved in the movement of photosynthetic products (photosynthate) from the mesophyll chloroplasts to the sieve elements of mature leaves:

1. In the typical case of a plant that translocates sucrose, triose phosphate formed during photosynthesis (Chapter 9) must first be transported from the chloroplast to the cytosol, where it is converted to sucrose.

2. Sucrose then moves from the mesophyll cell to the vicinity of the sieve elements in the smallest veins of the leaf (Fig. 7.14). This pathway usually involves a distance of only two or three cell diameters and is referred to as the **short-distance transport** pathway.

3. In the third step, **phloem loading**, sucrose is actively transported into the sieve elements.

Once inside the sieve elements, sucrose and other solutes are translocated away from the source, a process known as **export**. Translocation through the vascular system to the sink is referred to as **long-distance transport**.

The processes of phloem loading at the source, and unloading at the sink, are believed to produce the driving force for translocation (see page 165) and thus are of considerable basic as well as agricultural importance. Once the mechanisms are understood, it may be possible to increase crop productivity by increasing the accumulation of photosynthate by edible sink tissues, such as cereal grains.

Phloem Loading of Sugar Requires Metabolic Energy

In source leaves, sugars become more concentrated in the sieve elements and companion cells than in the mesophyll. This difference in solute concentration can be demonstrated by measuring the osmotic potential of the various cell types in the leaf, using a combination of plasmolysis and light or electron microscopy. In sugar beet, for example, the osmotic pressure of the mesophyll is approximately 1.3 MPa, while the osmotic pressures of the sieve elements and companion cells are on the order of 3.0 MPa (Geiger et al., 1973). Most of this difference in osmotic pressure is thought to be due to sugar, and specifically to sucrose, since sucrose is the major transport sugar in this species. Furthermore, ^{14}C-labeled sucrose supplied exogenously to a source leaf of sugar beet accumulates in the sieve elements and companion cells of the major veins, as does sucrose derived from $^{14}CO_2$ (Fig. 7.15). The fact that sucrose, an uncharged solute, is at a higher concentration in the sieve element–companion cell complex than in surrounding cells

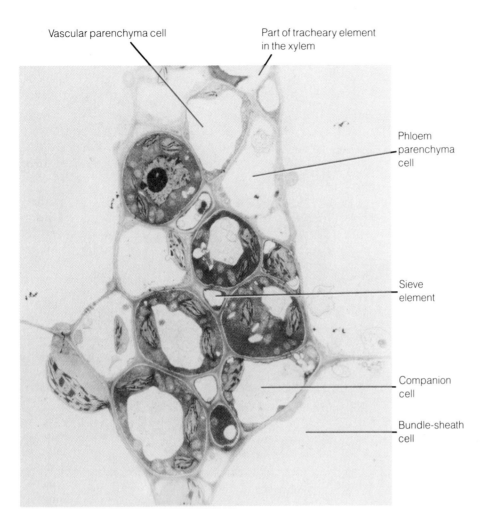

Vascular parenchyma cell

Part of tracheary element in the xylem

Phloem parenchyma cell

Sieve element

Companion cell

Bundle-sheath cell

FIGURE 7.14. Electron micrograph of a transverse section of a small vein in a source leaf of sugar beet (*Beta vulgaris*), showing the spatial relationships between the various cell types of the vein. Surrounding the compactly arranged cells of the bundle sheath layer are the photosynthetic cells (mesophyll cells) with numerous intercellular spaces (not shown in this micrograph). Photosynthetic products moving from the mesophyll to the sieve elements of the minor vein must traverse a distance equivalent to several cell diameters before being loaded. Note that the sieve elements of the small veins of sugar beet are smaller than the companion cells, the reverse of the size relationship in the large veins and main vascular system. This relationship is typical of dicots and probably reflects the importance of the companion cells in the uptake of sugars into sieve elements of small veins. (From Evert and Mierzwa, 1985.)

indicates that sucrose is transported against its chemical potential gradient and is evidence for active transport of this solute. Other data support the concept that phloem loading requires metabolic energy. For example, treating source tissues with respiratory inhibitors decreases the ATP concentration in the tissue and also inhibits the loading of exogenous sugar.

The Pathway from the Mesophyll Cells to the Sieve Elements Is at Least Partly Apoplastic

We have seen that solutes (mainly sugars) must move in source leaves from the photosynthesizing cells in the mesophyll to the veins. As illustrated in Figure 7.16, sugars might move entirely through the symplast via the plasmodesmata or, alternatively, might enter the apoplast at some point en route to the phloem. In the latter case, the sugars could be actively loaded from the apoplast into the sieve elements and companion cells by an energy-driven carrier located in the plasma membranes of these cells.

The exit of sucrose from mesophyll cells into the apoplast is at least partly controlled by the level of certain substances, such as potassium, in the apoplast. A high level of potassium in the apoplast of sugar beet source leaves increases the rate at which sucrose enters the

FIGURE 7.15. An autoradiograph of a sugar beet (*Beta vulgaris*) source leaf that was supplied with ^{14}C-labeled sucrose (26 ×). The sugar solution was applied to the upper surface of a darkened lead for 30 min. The surface of the leaf was gently rubbed with an abrasive paste to remove the cuticle and allow penetration of the solution to the interior of the leaf. The picture is a negative, so the label appears white. The label is accumulated in the small veins, sieve elements, and companion cells of the source leaf, indicating the ability of these cells to transport sucrose against its concentration gradient. (From Fondy and Geiger, 1977.)

FIGURE 7.16. Diagram of possible pathways of phloem loading in source leaves. In the totally symplastic pathway (thick arrows), sugars move from one cell to another in the plasmodesmata, all the way from the mesophyll to the sieve elements. In the partially apoplastic pathway (thin arrows), sugars enter the apoplast at some point. For simplicity, sugars are shown here entering the apoplast near the sieve element–companion cell complex, but it is also possible that they enter the apoplast over the entire mesophyll and then move to the small veins. In any case, the sugars are actively loaded into the companion cells and sieve elements from the apoplast. Sugars loaded into the companion cells are thought to move through plasmodesmata into the sieve elements. Note that this figure is only schematic. The actual structure of a source leaf of a C4 plant, such as corn, is quite different from that of a C3 plant, such as sugar beet.

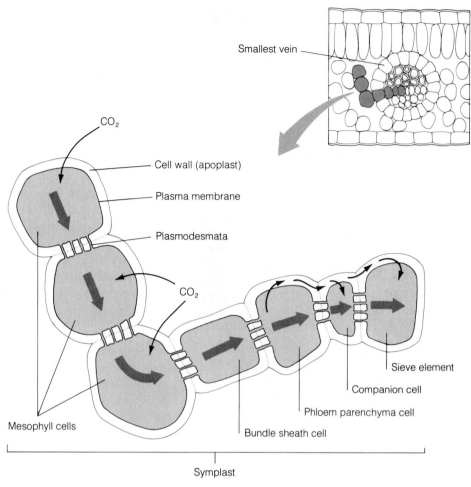

apoplast, thus coordinating a favorable nutrient supply, increased translocation to sinks, and enhanced sink growth (Doman and Geiger, 1979). The mechanism of sucrose efflux into the apoplast is unknown. However, loading of sucrose from the apoplast to the sieve elements is thought to be regulated by the osmotic pressure or, more likely, the turgor pressure of the sieve elements. According to this model, a decrease in sieve-element turgor below a certain point would lead to a compensatory increase in loading. The entry of sucrose into the sieve elements is also affected by sucrose concentration, with higher concentrations in the apoplast increasing phloem loading by the sucrose carrier.

It is important to point out that a partially apoplastic pathway does not necessarily hold true for loading in all species or even for all solutes in species like sugar beet. In cucurbit species, the distribution of plasmodesmata between various cell types and the similar osmotic pressures of the mesophyll and phloem suggest that the route from the mesophyll to the smallest vein phloem is totally symplastic in these plants.

Phloem Loading Is Specific and Selective

The composition of sieve element sap generally does not correspond to the composition of solutes in tissues surrounding the phloem. As noted above, only certain nonreducing sugars are translocated in sieve elements, and these transport sugars vary in different species. Therefore, some process in the source leaf must select for a specific transport sugar or sugars.

Selectivity could occur in loading from the apoplast, since membrane-bound carriers are specific for the transported molecule. Selectivity could also result from some other process that controls the availability of specific sugars for translocation in the phloem. For example, it could result from the preferential conversion of nontransport sugars into transport sugars, either within the mesophyll cells or within the sieve elements and companion cells. It appears that in some species the transport sugar is selected by the carrier for loading from the apoplast, whereas in other species it is selected by processes other than the loading system.

Not All Substances Transported in the Phloem Are Actively Loaded into the Sieve Elements

Many substances, such as organic acids and plant hormones, are found in the phloem sap at lower concentrations than the carbohydrates. These substances are probably not actively loaded into the sieve element–companion cell complex. They may be taken up directly by diffusion across the phospholipid bilayer of the plasma membrane of the sieve element–companion cell complex or by a passive carrier in the plasma membrane of those cells, or they may diffuse into the sieve elements via the symplast. Once in the sieve elements, they are swept along in the translocation stream by bulk flow, the motive force being generated by the active loading of only certain sugars or amino acids. Many substances not normally found in plants, such as herbicides and fungicides, can be transported in the phloem because of their ability to diffuse through membranes at an intermediate rate. In other words, they diffuse through membranes rapidly enough to allow considerable accumulation in the sieve elements, but slowly enough that they do not diffuse out completely before reaching a sink tissue. Substances that are not transported in the phloem (such as calcium) apparently cannot enter the sieve elements.

Sucrose Loading Is Driven by a Proton Gradient Generated at the Expense of ATP

Most active transport mechanisms use metabolic energy supplied in the form of ATP. As we have seen in Chapter 6, the hydrolysis of ATP can be coupled to the movement of solutes across membranes in at least two ways. In **primary active transport** the same membrane protein is thought to break down ATP and use the energy released to move the solute across the membrane barrier against a chemical potential gradient. In plant cells, the primary active transport ATPase is a proton pump. In secondary transport (cotransport; see Chapter 6), solute movement against a chemical potential gradient is driven not directly by ATP hydrolysis, but indirectly by the proton gradient established by the primary active transport ATPase. The high proton concentration in the apoplast and the spontaneous tendency toward equilibrium (equal proton concentrations and electrical neutrality in the apoplast and symplast) cause protons to diffuse back into the symplast. Specific carrier molecules couple this movement to the transport of sucrose. This type of transport is known as sucrose/proton symport or cotransport (Fig. 7.17).

A great deal of research on phloem loading has been done on relatively few species, such as sugar beet and corn. In these plants and a few others, the present data support the pathway for loading illustrated in Figure 7.16.

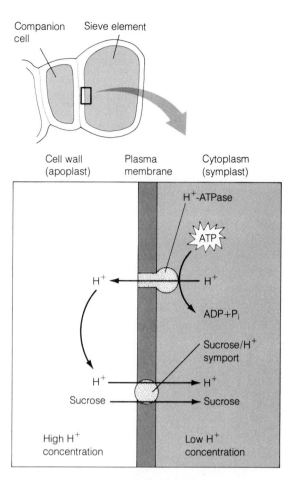

FIGURE 7.17. In the cotransport model of sucrose loading into the symplast of the sieve element–companion cell complex, the plasma membrane ATPase pumps protons out of the cell into the apoplast, establishing a high proton concentration there. The energy in this proton gradient is then used to drive sucrose transport into the symplast of the sieve element–companion cell complex through a sucrose/H^+ symporter.

In corn leaves, sugars probably diffuse in the symplast from the mesophyll to the vicinity of the minor-vein sieve elements. In this species the movement of water and dissolved solutes through the apoplast is blocked by suberized cell walls in the bundle sheath. The sugars then enter the apoplast by a largely unknown mechanism. In sugar beet leaves, sugars might initially follow a similar symplastic pathway, or they might enter the apoplast over the entire mesophyll and diffuse to the minor veins in the apoplast; the actual initial pathway is not known. In both species the sugars are then actively and selectively loaded from the apoplast into the sieve elements and the companion cells by a carrier present in their plasma membranes, giving rise to the high sugar concentrations present in these cells. The carrier mechanism appears to be a sucrose/proton symport. As we shall see, such active phloem loading plays a critical role in the mechanism of phloem translocation most widely accepted today.

Phloem Unloading and Sink-to-Source Transition

The detailed discussion above illustrates the events leading up to the export of sugars from sources. Phloem **unloading** is the process by which translocated sugars exit from the sieve elements of sink tissues.

Translocation into sink organs, such as developing roots, tubers, and reproductive structures, is termed **import**. In a sense, the events following import are simply the reverse of the steps in sources. The initial step is the unloading of sugars from the sieve elements. After unloading, the sugars are transported to cells in the sink for storage or metabolism (see page 169). For readings and references on phloem unloading, refer to the volume edited by Cronshaw et al. (1986).

Phloem Unloading and Transport into Receiver Cells Can Be Symplastic or Apoplastic

In vegetative sinks that are growing, such as roots and young leaves, unloading and transport into receiver cells are usually symplastic (Table 7.3, Fig. 7.18). Both structural observations and experimental results support this hypothesis. In other sink organs unloading is apoplastic (Fig. 7.18). In storage sinks, such as sugar beet root and the sugarcane (*Saccharum officinarum* L.) stem, sucrose is unloaded into the apoplast prior to entering the symplast of the sink. In reproductive sinks (developing seeds), an apoplastic step is necessary because there are no symplastic connections between the maternal tissues and the tissues of the embryo.

When unloading is symplastic, transport sugar moves through the plasmodesmata to the receiver cell, where it can be metabolized in the cytosol or the vacuole before entering metabolic pathways associated with growth of the tissue. When unloading is apoplastic, the transport sugar can be partially metabolized in the apoplast itself. For example, in sugarcane stem and corn kernels, sucrose is split into its components in the apoplast by an enzyme called invertase and is taken up as glucose or fructose (Table 7.3). In sugar beet root and soybean seeds, on the other hand, sucrose crosses the apoplast unchanged. Storage sinks can accumulate sucrose in the vacuole, or the absorbed sugars can be metabolized to some other solute before storage.

The information in Table 7.3 emphasizes the variety of transport mechanisms involved in unloading. The patterns discussed here are only general ones and should not be taken as hard-and-fast rules. Rather, for any particular species or organ, the operative pathway must be determined experimentally.

TABLE 7.3. Pathways of sucrose unloading and transport to cells of sink tissues.

Sink tissue	Species	Apoplastic vs. symplastic unloading	Sucrose hydrolyzed in apoplast during transport to sink cells vs. unaltered	Probable sites of energy requirement in sink cells
Vegetative tissues				
Roots	Pea, corn, tomato*	Symplastic	Not applicable	Metabolic conversion
Young leaves	Sugar beet	Symplastic	Not applicable	Metabolic conversion
Storage tissues				
Roots	Sugar beet	Apoplastic	Unaltered	Active transport at tonoplast
Stems	Sugarcane	Apoplastic	Hydrolyzed	Active transport of hexoses at cell membrane
Reproductive tissues				
Seeds	Soybean, wheat†	Apoplastic	Unaltered	Entry into apoplast and active transport into embryo (soybean)
	Corn	Apoplastic	Hydrolyzed	Active transport of hexoses into embryo

* *Lycopersicon esculentum.*

† *Triticum aestivum.*

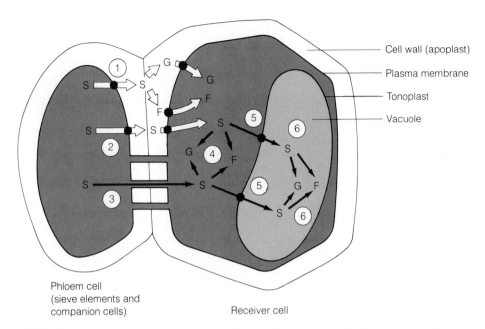

FIGURE 7.18. Schematic diagram of possible pathways of sucrose unloading in sink tissues. The phloem cell could be a sieve element or a companion cell. The receiver cell could be the cell of a growing leaf or root, of an importing storage organ, or of an embryo. The circles represent possible sites of carrier-mediated transport of sugars, either passive or active, across membranes. Sucrose (S) can enter the apoplast before being taken up into the receiver cell (1 and 2), or it can enter the receiver cell symplastically through the plasmodesmata (3). Sucrose that enters the apoplast can be split into glucose (G) and fructose (F) by a cell wall invertase before entering the receiver cell (1), or it can be taken up into the receiver cell unaltered (2). Once in the symplast of the receiver cell, sucrose can be split into glucose and fructose by a cytoplasmic invertase (4) or can enter the vacuole unaltered (5). Once in the vacuole, sucrose can be split into glucose and fructose by a vacuolar invertase (6), or it can remain unaltered. Of course, once in the cytosol or vacuole of the receiver cell, the sugars can enter pathways other than the ones shown. For example, glucose and fructose that enter the cytoplasm by route 1 may be used to resynthesize sucrose. (Adapted from Giaquinta et al., 1983.)

Transport into Sink Tissues Depends On Metabolic Activity

Numerous studies with inhibitors have shown that transport into sink tissues is energy dependent. The site or sites at which energy is required vary with the species or organ studied. In the case of apoplastic unloading, sugars must cross at least two membranes: the membrane of the sieve element–companion cell complex and the membrane of the receiver cell. When transport into the vacuole of the sink cell occurs, the tonoplast must also be traversed. Carriers are thought to function in transport across these membranes, and at least one of the membrane transport steps has been shown to be active (dependent on metabolism) in sinks where unloading is apoplastic (Table 7.3). In sinks where unloading is symplastic, no membranes are crossed during uptake into the sink cells. Unloading through the plasmodesmata is passive, since transport sugars move from a high concentration in the sieve elements to a low concentration in the sink cells. The low concentration in the sink cells is maintained by respiration and by conversion of the transport sugars to other compounds needed in growth. Metabolic energy is required directly in these growing sinks for respiration and biosynthetic reactions and only indirectly for nutrient uptake. Metabolic conversions also help to maintain the concentration gradient for sink uptake in some organs with apoplastic unloading.

The Active Transport Step During Apoplastic Unloading Depends On the Species and Organ

Storage organs with apoplastic unloading, such as sugar beet root and sugarcane stem, often accumulate sugars to high concentrations (Table 7.3). This fact is consistent with active membrane transport, since energy is generally required to move sugars into storage compartments against a concentration gradient. However, it is important to note that the actual unloading step is probably passive in both these sinks; that is, movement of sucrose out of the sieve elements into the apoplast occurs down a concentration gradient, from a high to a low concentration. A low apoplastic sucrose level is maintained in sugarcane by the cell wall invertase, which splits sucrose into glucose and fructose; in the case of sugar beet storage root, a subsequent active transport of sucrose at the tonoplast of receiver cells maintains the concentration gradient.

Developing seeds have proved to be a most interesting system in which to study unloading processes. In legumes such as soybean, the embryo can be removed from the seedcoat. Thus, unloading from the seedcoat into the apoplast can be studied without the influence of the embryo, and uptake into the embryo can also be investigated separately. Such studies have shown that in these species, at least part of the uptake into the embryo is active, carrier-mediated transport and that the entry of sucrose into the apoplast is probably carrier-mediated and may require metabolic energy.

In general, sugar/proton symport mechanisms appear to function in retrieval from the apoplast (for instance, in sucrose uptake in the soybean embryo and hexose uptake in sugarcane storage cells). Sugar transport into the vacuoles of storage cells such as those of sugarcane and sugar beet is accomplished by a **sucrose/proton antiport** (Fig. 6.13). In this case, an ATPase pumps protons into the vacuole; the antiport carrier moves sucrose into the vacuole in exchange for protons, which exit the vacuole down their electrochemical potential gradient.

The Transition from Sink to Source Is a Gradual Developmental Process in Leaves

Dicot leaves begin their development as sink organs. A transition from sink to source status begins later in development, generally when the leaf is approximately 25% expanded, and is usually complete when the leaf is 40 to 50% expanded. Export from the leaf begins at the tip or apex of the blade and progresses toward the base until the whole leaf is exporting. During the transition period, the tip exports sugar while the base imports it from the other source leaves, as shown by autoradiography (Fig. 7.19). What causes import to cease and export to begin? The maturation of leaves is accompanied by a large number of functional and anatomical changes, many of which are simply preparatory for the beginning of export. Export actually begins when phloem loading has accumulated sufficient photosynthate in the sieve elements to drive translocation out of the leaf (see page 165).

The cessation of import and the initiation of export are separate events in the development of a leaf (Turgeon, 1984). In albino leaves of tobacco, which have no chlorophyll and therefore are incapable of photosynthesis, import stops at the same developmental stage as in green leaves, even though export is not possible. Therefore some other change must occur in developing leaves

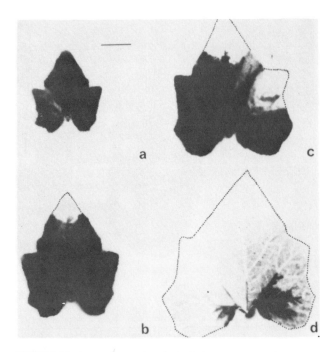

FIGURE 7.19. Autoradiographs of a leaf of summer squash (*Cucurbita pepo*), showing the transition of the leaf from sink to source status. In each case, the leaf imported ^{14}C from the source leaf on the plant for 2h. Label is visible as black accumulations. In (a) the entire leaf is a sink, importing sugar from the source leaf. In (b) to (d) the base is still a sink. As the tip of the leaf (white areas) stops importing sugar, it gains the ability to load and unload sugars, as shown by the loss of black accumulations in (c) and (d), and eventually to export them. (From Turgeon and Webb, 1973.)

that causes them to cease importing sugars from other leaves. Such a change could involve blockage of the unloading pathway at some point in the development of mature leaves. Since unloading in sink leaves appears to be symplastic in nature, plasmodesmatal closure could account for the cessation of unloading and import into transition leaves. With the unloading pathway blocked, the sieve elements could accumulate enough sugar for export to be initiated in normal leaves. Additional data support the hypothesis that the unloading pathway is blocked in mature leaves. Some translocation and import of labeled sugars into a mature leaf (from other mature leaves!) can be induced by rather extreme conditions, such as darkening of the treated leaf and elimination of any alternative sinks. However, the imported sugars are not unloaded: they remain in the veins, indicating that the symplastic pathway to the mesophyll has been blocked during the maturation of the leaf.

The Mechanism of Phloem Translocation

The mechanism of phloem translocation was a subject of research from the 1930s to the mid-1970s. Now, one theory is generally accepted as the correct explanation for translocation. This theory, called the **pressure-flow hypothesis**, accounts for most of the experimental and structural data currently available for angiosperms.

The discussion here will begin with a description of the various types of models proposed in the past to account for phloem translocation. A second section will deal with the considerable evidence in favor of the pressure-flow hypothesis in angiosperms. The discussion will end with a note of caution: pressure flow may not be a complete description of translocation in all species.

Active and Passive Mechanisms Have Been Proposed to Account For Phloem Translocation

Theories of phloem translocation can be categorized as active or passive. Both types of theories assume that energy is required for loading in the source and for sink uptake. The active theories further assume that an additional expenditure of energy is required to drive the translocation process in the sieve elements, whereas in the passive theories, energy is required only to maintain the functional integrity of the sieve elements, not to drive translocation itself.

One of the original active theories was that solutes were carried from one end of a sieve element to the other by cytoplasmic streaming or cyclosis, with transfer occurring across the sieve plate by some unknown mechanism. This theory had to be abandoned because cytoplasmic streaming has never been observed in mature, functioning sieve elements. In addition, there is no evidence that sieve elements contain actin microfilaments, which function in cytoplasmic streaming in other plant cells. Another early active model for phloem translocation suggested that P-protein might provide the motive force for solute movement by some type of contractile or peristaltic action, analogous to the action of actin microfilaments in muscle or of microtubules in cilia and flagella. However, there is no strong evidence that P-protein is similar to either actin or tubulin.

A second type of active theory involved energy-driven transport of solutes from one sieve element to another across the sieve plate. Such active transport of solutes was thought to be required because electron micrographs of phloem tissue appeared to show that functional sieve-plate pores were normally blocked by callose and P-protein. If the sieve-plate pores between sieve elements were blocked when functional, passive models of translocation based on bulk flow of solutes would be impossible. As we shall see, the apparent plugging of sieve-plate pores observed in these early electron micrographs was an artifact of the fixation method. The sieve-plate pores are actually open when functional, so it is unnecessary to invoke an active mechanism of solute transport across sieve plates.

According to the Pressure-Flow Hypothesis, Translocation in the Phloem Is Driven by a Pressure Gradient from Source to Sink

The passive theories of translocation include diffusion and the pressure-flow hypothesis. Diffusion is far too slow to account for the velocities of solute movement observed in the phloem. Translocation velocities average 1 meter per hour, while the rate of diffusion is 1 meter per *eight years*!

The pressure-flow hypothesis, on the other hand, is widely accepted as the most probable mechanism of phloem translocation. First proposed by Ernst Münch in 1930, the pressure-flow hypothesis states that the flow of solution in the sieve elements is driven by an osmotically generated pressure gradient between source and sink (Fig. 7.20). The pressure gradient is established as a consequence of phloem loading at the source and phloem unloading at the sink. That is, energy-driven phloem loading generates a high osmotic pressure in the sieve elements of the source tissue, causing a steep drop in the water potential. In response to the water potential gradient, water enters the sieve elements and causes the turgor pressure to increase. At the receiving end of the translocation pathway, phloem unloading leads to a lower osmotic pressure in the sieve elements of sink tissues. As the water potential of the phloem rises above that of the xylem, water tends to leave the phloem in response to the water potential gradient, causing a decrease in the turgor pressure of the phloem sieve elements of the sink.

If no cross walls were present in the translocation pathway—that is, if the entire pathway were a single membrane-bound compartment—the two different pressures at the source and sink would rapidly come to near equilibrium. The presence of sieve plates greatly increases the resistance along the pathway and results in the generation and maintenance of a substantial pressure gradient in the sieve elements between source and sink. The sieve-element contents are physically pushed along the translocation pathway by bulk flow, much like water flowing through a garden hose. Note that this model implies that some of the water circulates throughout the plant between the transpiration (xylem) and translocation (phloem) pathways.

From close inspection of the water potential values shown in Figure 7.20 it is apparent that water in the phloem is moving up a water potential gradient from source to sink. Such water movement does not transgress the laws of thermodynamics, however, since the water is moving by bulk flow rather than by osmosis. That is, no membranes are crossed during transport from one sieve tube to another and solutes are moving at the same rate as the water molecules. Under these conditions, the osmotic pressure, π, cannot contribute to the driving force for

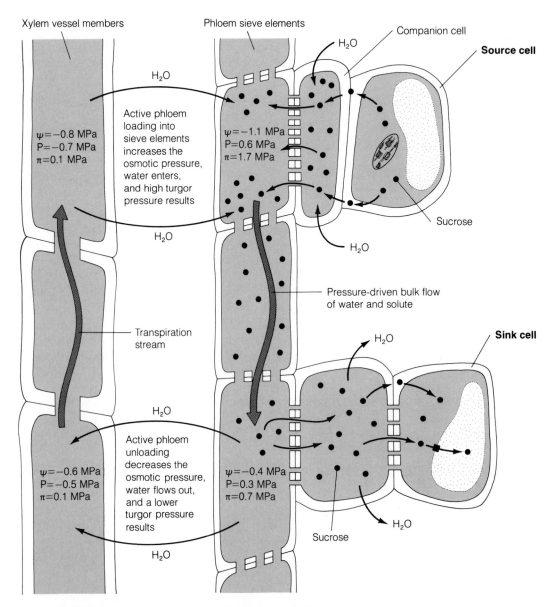

FIGURE 7.20. Schematic diagram showing the pressure-flow model. In the source, sugar is actively loaded into the sieve element–companion cell complex. Water enters the phloem cells osmotically, building up a high turgor pressure. At the sink, as sugars are unloaded, water leaves the phloem cells and a lower pressure results. Water and its dissolved solutes move by bulk flow from the area of high pressure (source) to the area of low pressure (sink). Possible values for ψ, P, and π in the xylem and phloem are illustrated. (From P. S. Nobel.)

water movement, although it still influences the water potential. Water movement in the translocation pathway is therefore driven by the pressure gradient rather than the water potential gradient. Of course, the passive, pressure-driven long-distance translocation in the sieve tubes ultimately depends on the active short-distance transport mechanisms involved in phloem loading and unloading. These active mechanisms are responsible for setting up the pressure gradient in the first place.

Predictions of the Pressure-Flow Model

Since the pressure-flow hypothesis has gained such wide acceptance, the predictions based on the model demand special attention. First of all, the sieve-plate pores must be unobstructed. If P-protein or other materials blocked the pores, the resistance to flow of the sieve-element sap would be too great. Second, true **bidirectional transport** (i.e., transport in both directions) in a single sieve element cannot occur. A mass flow of solution precludes such bidirectional movement, since a solution can flow in only one direction in a pipe! Bidirectional movement of solutes within the phloem can occur, but in different vascular bundles or in different sieve elements. Third, since the pressure-flow hypothesis is a passive theory, great expenditures of energy would not be required in order to drive translocation in the tissues along the path. Energy would be required in the path only to maintain the structure of the sieve elements and the integrity of the cell membranes. Therefore, treatments that restrict the supply of ATP, such as low temperature, anoxia, and metabolic inhibitors, should not stop translocation when applied to the path tissues. Note, however, that energy dependence of the translocation mechanism is not demonstrated if the inhibitor halts translocation by disrupting cell and membrane structure and not solely by reducing ATP supply. The cell membrane must remain intact and functional in order to retain solutes in the transport stream, and the sieve-plate pores must remain open for translocation to proceed. Fourth, the pressure-flow hypothesis demands the presence of a positive pressure gradient. Turgor pressure must be higher in sieve elements of sources than in sieve elements of sinks, and the pressure difference must be large enough to overcome the resistance of the pathway and to maintain flow at the observed velocities.

Many attempts have been made to test the predictions of the pressure-flow model. Often the results, or at least the interpretations given them, have been contradictory. Emphasis will be given here to experiments whose results can be interpreted in a relatively straightforward fashion. The following is a summary of the evidence supporting this hypothesis.

The Sieve-Plate Pores Are Essentially Open Channels Connecting One Sieve-Tube Member to Another

The ultrastructure of sieve elements is not easy to investigate because of their high internal pressure. When the phloem is excised or fixed (killed) slowly with chemical fixatives, the turgor pressure in the sieve elements is released. The contents of the cell, including P-protein, surge toward the point of pressure release, and in the case of sieve-tube members, accumulate on the sieve plates (Fig. 7.9). That is probably why many of the earlier electron micrographs show sieve plates that are obstructed. As mentioned above, if the pores of the sieve plates were normally blocked, translocation by pressure flow could not occur. Now, rapid freezing and fixation techniques are available that permit a fairly reliable picture of undisturbed sieve elements to be obtained. Electron micrographs of sieve-tube members prepared by such techniques indicate that P-protein is normally found

along the periphery of the seive-tube members (see Figs. 7.5, 7.6, and 7.7), or evenly distributed throughout the lumen of the cell. Furthermore, the pores contain P-protein in similar positions, lining the pore or in a loose network. The open condition of the pores has been observed in a number of angiosperm species, including cucurbits, sugar beet, and bean (e.g., see Fig. 7.7), consistent with the pressure-flow model.

Bidirectional Transport in a Single Sieve Element Has Not Been Demonstrated

Bidirectional transport has been investigated by applying two different tracers to two source leaves, one above the other (Eschrich, 1975). Each source receives one of the tracers, and some point in between the two sources is monitored for the presence of both tracers. Alternatively, a single tracer is applied to an internode and detected at points above and below the site of application. Detection techniques depend on the tracer used and include autoradiography and visualization of fluorescent dyes in fresh sections. In some experiments, aphids are used to collect exudate from a single sieve element, and the exudate is analyzed.

Transport in two directions has often been detected in sieve elements of different vascular bundles in stems. Transport in two directions has also been seen in adjacent sieve elements of the same bundle in petioles. This can occur in the petiole of a leaf that is undergoing the transition from sink to source and simultaneously importing and exporting assimilates through its petiole. *However, bidirectional transport in a single sieve element has never been convincingly demonstrated.* Experiments that purport to demonstrate such bidirectional transport can be interpreted in other ways.

The Rate of Translocation Is Relatively Insensitive to the Energy Supply of the Path Tissues

In plants, such as sugar beet, that are able to survive periods of low temperatures, rapidly chilling a short segment of a source-leaf petiole to approximately 1°C does not cause a sustained inhibition of mass transport out of the leaf (Fig. 7.21). Rather, there is a brief period of inhibition, after which transport returns to the control rate in one to several hours after the beginning of the cold treatment. These changes are most easily seen in time courses obtained using steady-state labeling, as in Figure 7.21. Chilling reduces the respiration rate (and ATP production) by about 90% in the petioles at a time when translocation has recovered and is proceeding normally. In fact, calculations for sugar beet indicate that for each molecule of ATP generated in a chilled sieve element–companion cell complex, approximately 600,000 molecules of sucrose are transported through the element (Coulson et al., 1972). Similar results are obtained when squash (*Cucurbita melopepo torticollis*) petioles are treated with an atmosphere consisting of 100% nitrogen in the dark; transport persists even though anaerobic metabolism is the only source of ATP throughout a significant portion of the transport pathway (Sij and Swanson, 1973). The main point here is that the energy requirement for transport through the pathway of these plants is small, consistent with the pressure-flow hypothesis.

In the case of chilling-sensitive plants, such as bean (*Phaseolus vulgaris* L.), chilling the petiole of a source leaf to 10°C does inhibit translocation out of the leaf. Treating the petiole with a metabolic inhibitor (cyanide) also inhibits translocation in this plant. However, examination of the treated tissue by electron microscopy reveals blockage of the sieve-plate pores by cellular debris in both cases (Giaquinta and Geiger, 1973, 1977). Clearly, these results do not bear on the question of whether energy is required for translocation along the pathway.

Pressure Gradients in the Sieve Elements Are Sufficient to Drive a Mass Flow of Solution

Turgor pressures in sieve elements can be determined either by calculation from the water potential and osmotic pressure ($P = \psi + \pi$) or by direct measurement. The earliest direct measurements were made with a hypodermic needle attached to a simple manometer (a thin glass capillary partially filled with fluid and sealed at the end opposite the connection to the needle). Although this technique is useful in the field, it has the disadvantage that the hypodermic needle ruptures a large number of sieve elements when it is inserted. Because of the leakage of sieve-element sap around the needle tip and pressure loss as cells lose sap into the needle, these measurements probably underestimate the true turgor (Sovonick-Dunford et al., 1981). A newer and quite elegant technique eliminates these problems by using micromanometers or pressure transducers sealed over exuding aphid stylets (Wright and Fisher, 1980). The data obtained are much more accurate because aphids pierce only a single sieve element and the cell membrane apparently seals well around the aphid stylet. The technique, however, is also technically quite difficult!

When sieve-element turgor is measured by the techniques described above, the pressure at the source is generally found to be higher than that at the sink. For example, a pressure difference of 0.11 MPa was detected between the source and sink tissues of squirting cucumber (*Ecballium elaterium*) by the hypodermic needle/manometer method (Sheikholeslam and Currier, 1977). In soybean, the observed pressure difference between source and sink has been shown to be sufficient to drive a mass flow of solution through the pathway, taking

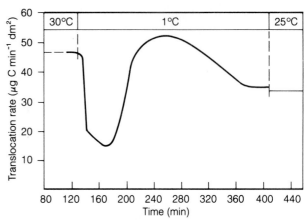

FIGURE 7.21. The response of translocation in sugar beet (*Beta vulgaris*) to loss of metabolic energy by chilling the tissue. $^{14}CO_2$ was supplied to a source leaf, and a 2-cm portion of its petiole was chilled to 1°C. Translocation was measured by monitoring the rate of arrival of ^{14}C at a sink leaf. The chilling results in a temporary inhibition and subsequent recovery of translocation through the petiole. The fact that translocation can continue when ATP production is largely inhibited by chilling indicates that the energy requirement for translocation through the pathway of this plant is small. (Data from Geiger and Sovonick, 1975.)

into account the path resistance (mainly caused by the sieve-plate pores), the path length, and the velocity of translocation (Fisher, 1978). The actual pressure difference between source and sink was calculated from the water potential and osmotic pressure to be 0.41 MPa, and the pressure difference required for translocation by pressure flow was calculated to be 0.12 to 0.46 MPa. Thus, the observed pressure difference appears to be sufficient to drive mass flow through the phloem.

All the experiments and data described above support the operation of pressure flow in angiosperm phloem. The lack of an energy requirement in the pathway and the presence of open sieve-plate pores are definitive evidence for a passive mechanism. The failure to detect bidirectional transport or motility proteins, as well as the positive data on pressure gradients, are all in accord with the pressure-flow hypothesis.

The Mechanism of Phloem Transport in Gymnosperms May Be Different from That in Angiosperms

While pressure flow is adequate to explain translocation in angiosperms, it may not be sufficient in gymnosperms. Very little physiological information on gymnosperm phloem is available, and speculation about translocation in these species is based almost entirely on ultrastructural data. As discussed previously (page 148), the sieve cells of gymnosperms are similar in many respects to sieve-tube members of angiosperms. However, the sieve areas of sieve cells are relatively unspecialized and do not appear to consist of open pores (Fig. 7.8). Rather, the pores are filled with numerous membranes that are continuous with the smooth ER adjacent to the sieve areas. Such pores are clearly inconsistent with the requirements of the pressure-flow hypothesis. Either the picture presented by these electron micrographs is an artifact of the fixation process, or translocation in gymnosperms involves some other mechanism, perhaps with the endoplasmic reticulum playing a significant role.

Assimilate Allocation and Partitioning

The photosynthetic rate determines the total amount of fixed carbon available to the leaf. However, the amount of fixed carbon available for translocation depends on subsequent metabolic events. The regulation of the diversion of fixed carbon into the various metabolic pathways is termed **allocation**.

As we have seen, once the transport sugar enters the sieve element, it and the solvent in which it is dissolved move by mass flow along a pressure gradient in the direction of a sink. The vascular bundles form a system of pipes that can direct the flow of photoassimilates to the various organs: young leaves, stems, roots, fruits, or seeds. However, the vascular system is often highly interconnected, forming an open network that allows source leaves to communicate with multiple sinks. Under these conditions, what determines the volume of flow to any given sink? The differential distribution of photoassimilates within the plant is termed **partitioning**.

Allocation Includes the Storage, Utilization, and Transport of Fixed Carbon in the Plant

The fate of fixed carbon in a source cell can be classified into three principal categories: storage, utilization, and transport.

1. *Synthesis of storage compounds.*   Starch is synthesized and stored within chloroplasts and, in most species, is the primary storage form that is mobilized for translocation during the night period. Such plants are called starch storers. In some organs of certain grasses, fructans (polymerized fructose molecules) are the storage compounds, rather than starch. (See also item 3 below.)

2. *Metabolic utilization.*   Fixed carbon can be utilized within various compartments of the photosynthesizing cell to meet the energy needs of the cell or to provide carbon skeletons for the synthesis of other compounds required by the cell.

3. *Synthesis of transport compounds.*   Fixed carbon can be incorporated into transport sugars for export to various sink tissues. A portion of the

transport sugar can also be stored temporarily in the vacuole. In most species studied, this stored sucrose is highly transitory, providing a buffer against short-term changes in sucrose synthesis. However, in some species, such as barley (*Hordeum vulgare* L.), fixed carbon is stored for use during the night period primarily as sucrose, with very little being stored as starch. Such plants are called sucrose storers.

Allocation is a key process in sink tissues as well. Once the transport sugars have been unloaded and enter the receiver cells, they can remain as such or can be transformed into various other compounds. In storage sinks, fixed carbon can be accumulated as sucrose or hexose in vacuoles or as starch in amyloplasts. In growing sinks, sugars can be utilized for respiration and for the synthesis of other molecules required for growth.

Once Synthesized, Transport Sugars Are Partitioned Among the Various Sink Tissues

The cellular processes involved in allocation largely determine the relative strength of the various sinks. That is, the greater the ability of a sink to store or metabolize imported sugars, the greater is its ability to compete for assimilates being exported by the sources. Such competition determines the distribution of transport sugars among the various sink tissues of the plant (assimilate partitioning), at least in the short term. Of course, events in sources and sinks must be synchronized, so an additional level of control lies in the interaction between areas of supply and demand. Turgor pressure in the sieve elements could be an important means of communication between sources and sinks, acting to coordinate rates of loading and unloading. Chemical messengers are also important in signaling to one organ the status of the other. These chemical messengers include plant growth regulators (hormones) and nutrients, such as sucrose, potassium, and phosphate.

There has been a surge in research on assimilate allocation and partitioning with the goal of improving yields of crop plants. In the past, efforts by plant breeders to increase yield by increasing net photosynthetic rates have generally been unsuccessful (Gifford et al., 1984). However, significant improvements in yield have resulted from increases in **harvest index**, the ratio of commercial or edible yield to total shoot yield (the latter including inedible portions of the shoot). Clearly, an understanding of partitioning should enable plant breeders and molecular biologists to select and develop varieties with improved transport to edible portions of the plant. As we will see, however, allocation and partitioning in the whole plant must be coordinated, and increased transport to edible tissues must not occur at the expense of other essential processes and structures. Thus, increasing the retention of photoassimilates that are normally "lost" by the plant will constitute a necessary, complementary approach to increasing yield. In other words, if losses due to nonessential respiration and leaching or exudation from roots can be reduced, crop yield will also be improved.

Allocation in Source Leaves Is Regulated by Key Enzymes

As shown in Figure 7.22, the quantity of sucrose available for export is influenced by a number of biochemical reactions and carrier-mediated events, including the rate of CO_2 fixation. In fact, increasing the rate of photosynthesis in a source leaf generally results in an increase in the rate of translocation from the source. The actual fraction of fixed carbon that is channeled to the so-called transport pool is determined by other processes. Control points include the allocation of triose phosphates to the regeneration of intermediates in the C3 photosynthetic carbon reduction (PCR) cycle, to starch synthesis, or to

sucrose synthesis, and the distribution of sucrose between transport and temporary storage pools. A number of enzymes operate in these pathways, and the controls are quite complex. However, one area of importance is the coordination of starch and sucrose synthesis.

The rate of starch synthesis in the chloroplast must be coordinated with sucrose synthesis in the cytosol. Triose phosphates (glyceraldehyde 3-phosphate and dihydroxyacetone phosphate) produced in the chloroplast by the C3 PCR cycle (see Chapter 9) can be used for either starch or sucrose synthesis (disregarding for the moment utilization of fixed carbon by the photosynthesizing cell itself). Since sucrose synthesis occurs in the cytosol, the triose phosphate destined for sucrose synthesis must leave the chloroplast. This is accomplished by the *phosphate translocator*, located in the chloroplast membrane. This membrane carrier exchanges triose phosphate from the chloroplast for orthophosphate from the cytosol. Whenever sucrose is synthesized, phosphate is released into the cytosol (Fig. 7.22). The entire sequence of events is summarized as follows:

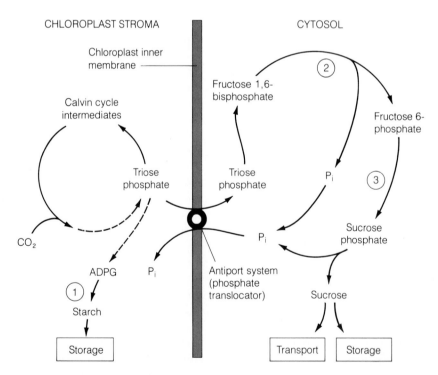

FIGURE 7.22. A simplified scheme for starch and sucrose synthesis. Triose phosphate, formed during photosynthesis in the Calvin (C3 photosynthetic carbon reduction) cycle, can either be utilized in starch formation in the chloroplast or transported into the cytosol in exchange for inorganic phosphate (P_i) via the phosphate translocator in the inner chloroplast membrane. The outer chloroplast membrane is porous to small molecules and is omitted here for clarity. In the cytosol, triose phosphate can be converted to sucrose for either storage in the vacuole or transport. Key enzymes involved are (1) starch synthetase, (2) fructose-1,6-bisphosphatase, and (3) sucrose-phosphate synthase. ADPG = adenosine diphosphate glucose. (Adapted from Preiss, 1982.)

Sucrose synthesis in cytosol → release of phosphate

→ exchange of phosphate
from cytosol for triose
phosphate from
chloroplast

→ less triose phosphate
available for starch
synthesis in chloroplast

Thus, sucrose synthesis diverts triose phosphate away from starch synthesis and storage. For example, it has been shown that when the demand for sucrose by other parts of a soybean plant is high, less carbon is stored as starch by the source leaves.

However, there is a limit to the amount of carbon that can be diverted away from starch synthesis, particularly in species that store carbon primarily as starch. Studies of allocation between starch and sucrose under different conditions suggest that a steady rate of translocation throughout the 24-hour period is a high priority for most plants. Thus, to some extent the level of starch formed in the chloroplast is predetermined. Such control is evident in studies of the effect of photoperiod on starch synthesis in soybean (Chatterton and Silvius, 1979). Short day lengths (7 h light, 17 h dark) led to increased rates of starch synthesis, while on long days (14 h light, 10 h dark) rates of starch accumulation were lower. Conversely, translocation rates were lower on short days than on long days. By synthesizing more starch during the short-day period, the plant has more stored materials available for translocation during the long dark period. The biochemical basis for such "anticipatory" controls is not known. To some extent, the control of starch synthesis has a genetic basis, as has been observed in studies of isolated cells of various species. Species that tend to accumulate more starch in the light as whole plants retain that capacity even when mesophyll cells are isolated from the remainder of the plant and investigated in cell culture (Huber, 1981).

Sink Tissues Compete for Available Translocated Assimilate

As discussed earlier (see pages 153), translocation to sink tissues depends on the position of the sink in relation to the source and on the vascular connections between source and sink. Another factor determining the pattern of transport in whole plants is competition between sinks. For example, reproductive tissues (seeds) can compete with growing vegetative tissues (young leaves and roots) for available translocate. Competition is indicated by numerous experiments in which removal of a sink tissue from a plant generally results in increased translocation to alternative, presumably competing sinks. The reverse

type of experiment involves altering the source supply while leaving sink tissues intact. When the supply of assimilates from sources to competing sinks is suddenly and drastically reduced by shading all the source leaves but one, the sink tissues now depend on a single source. In sugar beet and bean plants the rates of photosynthesis and export from the single remaining source leaf usually do not change over the short term (approximately 8 h; Fondy and Geiger, 1980). However, the roots receive less sugar from the single source, while the young leaves receive relatively more. Presumably, in these plants the young leaves are stronger sinks than the roots. A stronger sink can deplete the sugar content of the sieve elements more readily and thus steepen the pressure gradient, increasing translocation toward itself. An effect on the pressure gradient is also indicated indirectly by experiments in which transport to a sink is enhanced over the short term by making the sink water potential more negative. Soybean seedlings transplanted to vermiculite containing one-eighth as much water as controls partitioned dry matter from the cotyledons to the roots at a rate twice that of the controls (Meyer and Boyer, 1981).

Sink Strength Is a Function of Sink Size and Sink Activity

Various experiments indicate that the ability of a sink to mobilize assimilates toward itself, the **sink strength**, depends on two factors: **sink size** and **sink activity**.

$$\text{Sink strength} = \text{sink size} \times \text{sink activity}$$

Sink activity is the rate of uptake of assimilates per unit weight of sink tissue, and sink size is the total weight of the sink tissue. Altering either the size or activity of the sink results in changes in transport patterns. For example, less sugar is translocated to an ear of wheat if some of the grains are removed from the ear, lowering its sink size. This reduced translocation is partly due to an inhibition of transport velocity (Wardlaw, 1965). Changes in sink activity can be more complex, since a number of activities in sink tissues can potentially limit the rate of uptake by the sink. These activities include unloading from the sieve elements, metabolism in the cell wall, retrieval from the apoplast in some cases, and metabolic processes that use the assimilate in either growth or storage. Cooling a sink tissue inhibits activities that require metabolic energy and results in a decrease in the speed of transport toward the sink. Because cooling is so general, however, it does not permit discrimination among processes. Single gene mutations can be helpful here, as in so many other instances. In corn, a mutant that has a defective enzyme for starch synthesis in the kernels transports less material to the kernels than does its normal counterpart (Koch et al., 1982). In this mutant, a deficiency in assimilate storage leads to an inhibition of transport. These results on sink size and activity describe sink strength in a very general fashion. Much research remains to be done on the individual steps involved in unloading, so that sites of regulation and control can be identified and the mechanism of competition between sinks can be better understood.

Changes in the Source-to-Sink Ratio Bring About Long-Term Alterations in Source Metabolism

When all but one of the source leaves on a soybean plant are shaded for an extended period (e.g., 8 days), many changes occur in the single remaining source leaf, including a decrease in starch concentration and increases in photosynthetic rate, activity of ribulose bisphosphate carboxylase (Rubisco, the CO_2 fixation enzyme of photosynthesis), sucrose concentration, transport from the source, and orthophosphate concentration (Thorne and Koller, 1974). Thus, while only the distribution of assimilates among different sinks may change over the short term, over longer periods the metabolism of the source adjusts to the altered conditions. As indicated previously, sucrose and starch synthesis in the source often responds to a change in the sink demand for sucrose. When the sink demand for sucrose is high, sucrose synthesis is increased at the expense of starch synthesis, and vice versa.

Photosynthetic rate (the net amount of carbon fixed per unit leaf area per unit time) often varies over a period of several days in response to altered sink demand, increasing when sink demand increases and decreasing when sink demand decreases. It has been postulated that an accumulation of assimilates in the source leaf could account for the linkage between sink demand and photosynthetic rate. One of the following mechanisms could be operating:

1. When sink demand is low, high starch levels in the source could physically disrupt the chloroplasts, interfere with CO_2 diffusion, or block light absorption.

2. When sink demand is low, photosynthesis could be restricted by a lack of free orthophosphate in the chloroplast. Under these conditions, sucrose synthesis is usually reduced, and less phosphate is thus available for exchange with triose phosphate from the chloroplast (via the phosphate translocator). If starch synthesis, which releases orthophosphate in the chloroplast, cannot recycle phosphate fast enough, a deficiency in phosphate would result. ATP synthesis and thus CO_2 fixation would decline.

Turgor Pressure May Directly Regulate the Responses of Sources and Sinks

Turgor pressure has been implicated in the communication between source and sink. Changes in turgor could act as a signal from sinks that is rapidly transmitted to source tissues via the interconnecting system of sieve elements. For example, when unloading is rapid, turgor pressures in the sieve elements of sinks would be reduced, and this reduction would be transmitted to the sources. If loading is controlled in part by turgor in the sieve elements of the source, as proposed by a number of researchers, it would increase in response to this signal from the sinks. The opposite response would be seen when unloading is slow in the sinks. Efflux into the apoplast of sources prior to loading and retrieval from the apoplast of sinks following unloading appear to be controlled by turgor pressures of the mesophyll and sink cells. According to this model, high turgor pressure in the mesophyll cells leads to increased efflux into the apoplast, and high turgor in sink cells leads to decreased uptake into those cells. Low turgor pressures would have the opposite effects on efflux and retrieval. Compensatory changes in loading and unloading rates and in long-distance transport would follow.

There is evidence that cell turgor affects transport across plant cell membranes by modifying the activity of the proton-pumping ATPase in the membranes. In experiments with sugar beet taproot tissue, for example, incubation in a mannitol solution, which lowers the turgor pressure, stimulated proton extrusion (Wyse et al., 1986).

Plant Hormones May Act as Long-Distance Messengers Between Sources and Sinks

Shoots produce growth regulators, such as auxin (indoleacetic acid; see Chapter 16), which can be rapidly transported to the roots via the phloem, while roots produce cytokinins, which move to the shoots through the xylem. (Transport of auxin in the phloem from source to sink is different from the "polar transport" that occurs in other cell types, as will be discussed in Chapter 16.) Gibberellins and abscisic acid (see Chapters 17 and 19) are also transported throughout the plant in the vascular system. Plant hormones play at least an indirect role in regulating source-sink relationships. They affect assimilate partitioning by controlling the growth of sinks, leaf senescence, and other developmental processes. For example, when sink demand is reduced, translocation is inhibited and photosynthesis declines rapidly. Photosynthesis is inhibited because abscisic acid accumulates in the source leaf as a result of decreased phloem transport out of the source. Abscisic acid closes stomates and inhibits CO_2 fixation.

A more direct role for hormones is suggested by other experiments. For example, exogenous abscisic acid inhibits sucrose uptake (loading) in source leaves of species such as castor bean (Malek and Baker, 1978), but stimulates sucrose uptake (unloading) into sink tissues of other species, such as soybean embryos and sugar beet root tissue (Saftner and Wyse, 1984). Exogenous auxin, on the other hand, inhibits active sucrose uptake by beet roots (unloading; Saftner and Wyse, 1984), but promotes sucrose uptake by primary leaves of bean (loading; Sturgis and Rubery, 1982). These and other studies suggest that plant growth regulators play a role in controlling and coordinating translocation throughout the plant. However, more experiments are needed in this area, particularly to clarify the role of endogenous growth regulators as long-distance messengers in the whole plant. In the future, the possible mechanisms by which turgor pressure and growth regulators affect loading, unloading, and other aspects of translocation will certainly constitute an important avenue of research in phloem transport.

Summary

Phloem translocation is the movement of the products of photosynthesis from mature leaves to areas of growth and storage. The phloem also serves to redistribute water and various compounds throughout the plant body.

Some aspects of phloem translocation have been well established by extensive research over a number of years. These include:

1. *The pathway of translocation.* Sugars and other organic materials are conducted throughout the plant in the phloem and specifically in cells called sieve elements. Sieve elements display a variety of structural adaptations that make these cells well suited for their transport function.

2. *Patterns of translocation.* Phloem translocation occurs from sources (areas of photosynthate supply) to sinks (areas of metabolism or storage of photosynthate). Sources are usually mature leaves. Sinks include organs such as roots and immature leaves.

3. *Materials translocated in the phloem.* The translocated solutes are mainly carbohydrates, and sucrose is the most commonly translocated sugar. Phloem sap also contains other organic molecules, such as amino acids and plant hormones, as well as inorganic ions.

4. *Rates of movement.* Rates of movement in the phloem are quite high and well in excess of rates of

diffusion. Velocities average 1 m h^{-1}, and mass transfer rates range from 1 to 15 g h^{-1} cm^{-2} of sieve elements.

Other aspects of phloem translocation require further investigation, and most of these are being intensively studied at the present time. These include:

1. *Phloem loading and unloading.* Transport of sugars into and out of the sieve elements is called phloem loading and unloading, respectively. Loading requires metabolic energy and is specific for the transported sugar. For most species, the available data indicate that sugars must enter the apoplast of the source leaf prior to loading and that a proton gradient drives the entry of sugars into the sieve elements of the source. Transport into sink tissues also requires metabolic energy, but the transport pathway, site of metabolism of transport sugars, and site where energy is expended vary with the organ and species.

2. *Mechanism of translocation.* Pressure flow is well accepted as the most probable mechanism of phloem translocation. In this model, the bulk flow of phloem sap occurs in response to an osmotically generated pressure gradient. A variety of structural and physiological data indicate that phloem translocation occurs by pressure flow in angiosperms. The mechanism of translocation in gymnosperms requires further investigation.

3. *Assimilate allocation and partitioning.* Allocation is the regulation of the quantities of fixed carbon that are channeled into various metabolic pathways. In sources, these regulatory mechanisms determine the quantities of fixed carbon that will be stored, usually as starch; metabolized within the cells of the source; or immediately transported to sink tissues. In sinks, transport sugars are allocated to growth processes or to storage. Partitioning is the differential distribution of photoassimilates within the whole plant. Partitioning mechanisms determine the quantities of fixed carbon delivered to specific sink tissues. Phloem loading and unloading and assimilate allocation and partitioning are of great research interest because of their roles in crop productivity.

GENERAL READING

Cronshaw, J. (1981) Phloem structure and function. *Annu. Rev. Plant Physiol.* 32:465–484.
Cronshaw, J., Lucas, W. J., and Giaquinta, R. T., eds. (1986) *Phloem Transport. Proceedings of an International Conference on Phloem Transport,* Asilomar, Calif. Alan R. Liss, New York.

Evert, R. F. (1982) Sieve-tube structure in relation to function. *BioScience* 32:789–795.

Fondy, B. R., and Geiger, D. R. (1981) Regulation of export by integration of sink and source activity. *What's New in Plant Physiol.* 12:33–36.

Geiger, D. R., and Fondy, B. R. (1980) Phloem loading and unloading: Pathways and mechanisms. *What's New in Plant Physiol.* 11:25–28.

Giaquinta, R. T. (1983) Phloem loading of sucrose. *Annu. Rev. Plant Physiol.* 34:347–387.

Gifford, R. M., and Evans, L. T. (1981) Photosynthesis, carbon partitioning, and yield. *Annu. Rev. Plant Physiol.* 32:485–509.

Hendrix, J. E. (1983) Phloem function: An integrated view. *What's New in Plant Physiol.* 14:45–48.

Humphreys, T. E. (1980) Sugar proton cotransport and phloem loading. *What's New in Plant Physiol.* 11:9–12.

Lichtner, F. T. (1984) Phloem transport of xenobiotic chemicals. *What's New in Plant Physiol.* 15:29–32.

Murray, B. J., Mauk, C., and Nooden, L. D. (1982) Restricted vascular pipelines (and orthostichies) in plants. *What's New in Plant Physiol.* 13:33–36.

Pate, J. S. (1980) Transport and partitioning of nitrogenous solutes. *Annu. Rev. Plant Physiol.* 31:313–340.

Preiss, J. (1982) Regulation of the biosynthesis and degradation of starch. *Annu. Rev. Plant Physiol.* 33:431–454.

CHAPTER REFERENCES

Brentwood, B. and Cronshaw, J. (1978) *Planta* 140:111–120.

Chatterton, N. J., and Silvius, J. E. (1979) Photosynthate partitioning into starch in soybean leaves. I. Effects of photoperiod versus photosynthetic period duration. *Plant Physiol.* 64:749–753.

Christy, A. L. and Fisher, D. B. (1978) Kinetics of ^{14}C-photosynthate translocation in morning glory vines. *Plant Physio.* 61:283–290.

Coulson, C. L., Christy, A. L., Cataldo, D. A., Swanson, C. A. (1972) Carbohydrate translocation in sugar beet petioles in relation to petiolar respiration and adenosine 5'-triphosphate. *Plant Physiol.* 49:919–923.

Doman, D. C., and Geiger, D. R. (1979) Effect of exogenously supplied foliar potassium on phloem loading in *Beta vulgaris* L. *Plant Physiol.* 64:528–533.

Eschrich, W. (1975) Bidirectional transport. In: *Transport in Plants. I. Phloem Transport.* Zimmermann, M. H., and Milburn, J. A., eds. *Encyclopedia of Plant Physiology*, New Series, Vol. 1, pp. 245–255. Springer-Verlag, New York.

Evert, R. F., and Mierzwa, R. J. (1985) Pathway(s) of assimilate movement from mesophyll cells to sieve tubes in the *Beta vulgaris* leaf. In: *Phloem Transport. Proceedings of an International Conference on Phloem Transport*, Asilomar, Calif. Cronshaw, J., Lucas, W. J., and Giaquinta, R. T., eds., pp. 419–432. Alan R. Liss, New York.

Fisher, D. B. (1978) An evaluation of the Münch hypothesis for phloem transport in soybean. *Planta* 139:25–28.

Fondy, B. R., and Geiger, D. R. (1977) Sugar selectivity and other characteristics of phloem loading in *Beta vulgaris* L. *Plant Physiol.* 59:953–960.

Fondy, B. R., and Geiger, D. R. (1980) Effect of rapid changes in sink-source ratio on export and distribution of products of photosynthesis in leaves of *Beta vulgaris* L. and *Phaseolus vulgaris* L. *Plant Physiol.* 66:945–949.

Geiger, D. R. and Sovonick, S. A. (1975) Effects of temperature, anoxia and other metabolic inhibitors on translocation. In: Zimmerman, M. H. and Milburn, J. A., eds. *Transport in Plants. I. Phloem Transport.* Encyclopedia of Plant Physiology, New Series, Vol. 1., pp. 256–286.

Geiger, D. R., Giaquinta, R. T., Sovonick, S. A., and Fellows, R. J. (1973) Solute distribution in sugar beet leaves in relation to phloem loading and translocation. *Plant Physiol.* 52:585–589.

Geiger, D. R., and Swanson, C. A. (1965) Evaluation of selected parameters in a sugar beet translocation system. *Plant Physiol.* 40:942–947.

Gent, M. P. N. (1982) Effect of defoliation and depodding on long distance translocation and yield in Y-shaped soybean plants. *Crop Sci.* 22:245–250.

Giaquinta, R. (1976) Evidence for phloem loading from the apoplast. Chemical modification of membrane sulfhydryl groups. *Plant Physiol.* 57:872–875.

Giaquinta, R. T., and Geiger, D. R. (1973) Mechanism of inhibition of translocation by localized chilling. *Plant Physiol.* 51:372–377.

Giaquinta, R. T., and Geiger, D. R. (1977) Mechanism of cyanide inhibition of phloem translocation. *Plant Physiol.* 59:178–180.

Giaquinta, R. T., Lin, W., Sadler, N. L., and Franceschi, V. R. (1983) Pathway of phloem unloading of sucrose in corn roots. *Plant Physiol.* 72:362–367.

Gifford, R. M., Thorne, J. H., Hitz, W. D., and Giaquinta, R. T. (1984) Crop productivity and photoassimilate partitioning. *Science* 225:801–808.

Gougler Schmalstig, J., Geiger, D. R., and Hitz, W. D. (1986) Pathway of phloem unloading in developing leaves of sugar beet. In: *Phloem Transport.* Cronshaw, J., Lucas, W. J., and Giaquinta, R. T., eds. pp. 231–237. Alan R. Liss, New York.

Hall, S. M., and Baker, D. A. (1972) The chemical composition of *Ricinus* phloem exudate. *Planta* 106:131–140.

Hendrix, J. E. (1973) Translocation of sucrose by squash plants. *Plant Physiol.* 52:688–689.

Huber, S. C. (1980) Inter- and intra-specific variation in photosynthetic formation of starch and sucrose. *Z. Pflanzenphysiol.* 101:49–54.

Joy, K. W. (1964) Translocation in sugar beet. I. Assimilation of $^{14}CO_2$ and distribution of materials from leaves. *J. of Exper. Bot.* 15:485–494.

King, R. W., and Zeevaart, J. A. D. (1974) Enhancement of phloem exudation from cut petioles by chelating agents. *Plant Physiol.* 53:96–103.

Koch, K. E., and Avigne, W. T. (1984) Localized photosynthate deposition in citrus fruit segments relative to source-leaf position. *Plant Cell Physiol.* 25:859–866.

Koch, K. E., Tsui, C.-L., Schrader, L. E., and Nelson, O. E. (1982) Source-sink relations in maize mutants with starch-deficient endosperms. *Plant Physiol.* 70:322–325.

Layzell, D. B., and LaRue, T. A. (1982) Modeling C and N

transport to developing soybean fruits. *Plant Physiol.* 70:1290–1298.

Layzell, D. B., Pate, J. S., Atkins, C. A., and Canvin, D. T. (1981) Partitioning of carbon and nitrogen and the nutrition of root and shoot apex in a nodulated legume. *Plant Physiol.* 67:30–36.

Malek, T., and Baker, D. A. (1978) Effect of fusicoccin on proton cotransport of sugars in the phloem loading of *Ricinus communis* L. *Plant Sci. Lett.* 11:233–239.

Mason, T. G., and Lewin, C. J. (1926) On the rate of carbohydrate transport in the greater yam, *Dioscorea alata. Sci. Proc. R. Dublin Soc.* 18:203–205.

Meyer, R. F., and Boyer, J. S. (1981) Osmoregulation, solute distribution, and growth in soybean seedlings having low water potentials. *Planta* 151:482–489.

Münch, E. (1930) *Die Stoffbewegungen in der Pflanze.* Gustav Fischer, Jena.

Neuberger, D. S. and Evert, R. F. (1975) *Protoplasm* 84: 109–125.

Richardson, P. T., Baker, D. A., and Ho, L. C. (1982) The chemical composition of cucurbit vascular exudates. *J. Exp. Bot.* 33:1239–1247.

Saftner, R. A., and Wyse, R. E. (1984) Effect of plant hormones on sucrose uptake by sugar beet root tissue discs. *Plant Physiol.* 74:951–955.

Servaites, J. C., Schrader, L. E., and Jung, D. M. (1979) Energy-dependent loading of amino acids and sucrose into the phloem of soybean. *Plant Physiol.* 64:546–550.

Sheikholeslam, S. N., and Currier, H. B. (1977) Phloem pressure differences and ^{14}C-assimilate translocation in *Ecballium elaterium. Plant Physiol.* 59:376–380.

Sij, J. W., and Swanson, C. A. (1973) Effect of petiole anoxia on phloem transport in squash. *Plant Physiol.* 51:368–371.

Singh, B. K., and Pandey, R. K. (1980) Production and distribution of assimilate in chickpea (*Cicer arietinum* L.). *Aust. J. Plant Physiol.* 7:727–735.

Sovonick-Dunford, S., Lee, D. R., and Zimmermann, M. H. (1981) Direct and indirect measurements of phloem turgor pressure in white ash. *Plant Physiol.* 68:121–126.

Sturgis, J. N., and Rubery, P. H. (1982) The effects of indol-3-yl acetic acid and fusicoccin on the kinetic parameters of sucrose uptake by discs from expanded primary leaves of *Phaseolus vulgaris* L. *Plant Sci. Lett.* 24:319–326.

Thorne, J. H., and Koller, H. R. (1974) Influence of assimilate demand on photosynthesis, diffusive resistances, translocation, and carbohydrate levels of soybean leaves. *Plant Physiol.* 54:201–207.

Turgeon, R. (1984) Termination of nutrient import and development of vein loading capacity in albino tobacco leaves. *Plant Physiol.* 76:45–48.

Turgeon, R., and Webb, J. A. (1973) Leaf development and phloem transport in *Cucurbita pepo*: Transition from import to export. *Planta* 113:179–191.

Wardlaw, I. F. (1965) The velocity and pattern of assimilate translocation in wheat plants during grain development. *Aust. J. Biol. Sci.* 18:269–281.

Warmbrodt, R. D. (1985) Studies on the root of *Hordeum vulgare* L.—ultrastructure of the seminal root with special reference to the phloem. *Amer. J. Bot.* 72(3): 414–432.

Watson, B. T. (1975) The influence of low temperature on the rate of translocation in the phloem of *Salix viminalis* L. *Ann. Bot.* 39:889–900.

Wright, J. P., and Fisher, D. B. (1980) Direct measurement of sieve tube turgor pressure using severed aphid stylets. *Plant Physiol.* 65:1133–1135.

Wyse, R. E., Zamski, E., and Tomos, A. D. (1986) Turgor regulation of sucrose transport in sugar beet taproot tissue. *Plant Physiol.* 81:478–481.

Zimmerman, M. H. and Brown, C. L. (1971) *Trees: Structure and Function.* Springer-Verlag, Berlin.

Zimmermann, M. H., and Milburn, J. A., eds. (1975) *Transport in Plants.* I. *Phloem Transport. Encyclopedia of Plant Physiology*, New Series, Vol. 1. Springer-Verlag, New York.

BIOCHEMISTRY AND METABOLISM

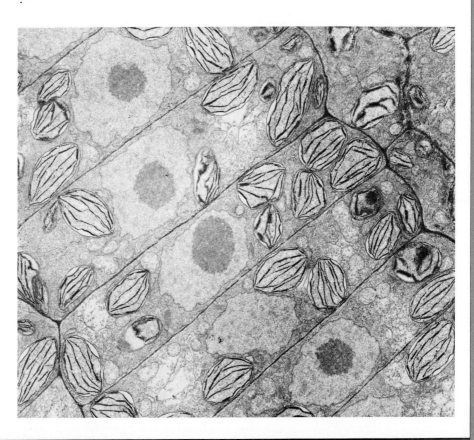

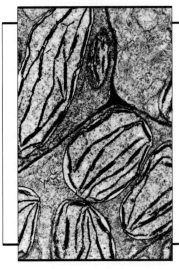

CHAPTER 8

Photosynthesis: The Light Reactions

LIFE ON EARTH ULTIMATELY depends on energy derived from the sun. Photosynthesis is the only process of biological importance that can harvest this energy. In addition, a large fraction of the planet's energy resources results from photosynthetic activity in either recent (biomass) or ancient (fossil fuels) times. This chapter introduces the basic physical principles that underlie photosynthetic energy storage and the current understanding of the structure and function of the photosynthetic apparatus.

Photosynthesis literally means "synthesis using light." As we will see in this chapter, photosynthetic organisms use solar energy to synthesize organic compounds that cannot be formed without the input of energy. Energy stored in these molecules can be used later to power cellular processes in the plant and can serve as the energy source for all forms of life.

Recently, the scientific community has debated the validity of forecasts of the effects of a nuclear war on the biosphere. Some studies predict that a nuclear war would generate huge dust clouds that would cover the sun for many months, resulting in a so-called "nuclear winter." In the absence of sunlight, both the natural vegetation and agricultural crops would die, and the interruption of the food chain would have catastrophic consequences. It has also been suggested that massive dust clouds ensuing from a collision of an asteroid with the earth caused a disruption of the food chain that led to the extinction of the dinosaurs. These issues underscore the fact that photosynthesis cannot take place without light and that all life on earth is inextricably tied to the photosynthetic process. This chapter deals with the role of light in photosynthesis.

Photosynthesis in Higher Plants

The most active photosynthetic tissue in higher plants is the mesophyll of leaves (Chapter 1). Mesophyll cells have a large number of chloroplasts, which contain the specialized light-absorbing green pigments, the **chlorophylls**. In photosynthesis, solar energy is used by the plant to oxidize water, with a concomitant release of oxygen, and to reduce CO_2 into organic compounds, primarily sugars. The complex series of reactions that culminate in the reduction of CO_2 include the **light reactions** and the **dark reactions**. The light reactions of photosynthesis take place in the specialized internal membranes of the chloroplast called thylakoids (Chapter 1). The end products of the light reactions are the high-energy compounds, ATP and NADPH, that are used for the synthesis of sugars. This synthetic phase, also called the dark reactions because they do not require light, takes place in the **stroma** of the chloroplasts, the aqueous region that surrounds the thylakoids. The light reactions of photosynthesis are the subject of this chapter; the dark reactions are discussed in Chapter 9.

In the chloroplast, light energy is harvested by two different functional units called photosystems. The absorbed light energy is used to power the transfer of electrons between a series of compounds that act as electron donors and electron acceptors. The final electron acceptor is NADP$^+$, which is reduced to NADPH. Light energy is also used to generate a proton motive force (Chapter 6) across the thylakoid membrane, and the ensuing proton gradient is used to synthesize ATP.

General Concepts and Historical Background

In this section we explore the essential concepts that provide a foundation for an understanding of the photosynthetic process. These include the nature of light, the properties of pigments, the various roles of pigments, and the overall chemical equations for photosynthesis. In addition, we discuss some of the critical experiments that have made the conceptual understanding of photosynthesis possible.

Light Has Characteristics of Both a Particle and a Wave

A triumph of physics in the early twentieth century was the realization that light has properties of both particles and waves. A wave (Fig. 8.1) is characterized by a **wavelength**, denoted by the Greek letter lambda (λ), which is the distance between successive wave crests. The **frequency**, represented by the Greek letter nu (v), is the number of wave crests that pass an observer in a given time. A simple equation relates the wavelength, the frequency, and the speed of any wave (Eq. 8.1):

$$c = \lambda v \qquad (8.1)$$

where c is the speed of the wave—in the present case, the speed of light (3.0×10^8 m s^{-1}). The light wave is a transverse (side-to-side) electromagnetic wave, in which both electric and magnetic fields oscillate perpendicular to the direction of propagation of the wave and at 90° with respect to each other.

Light is also a particle, which we call a photon. Each photon contains an amount of energy that is called a quantum (plural, quanta). The energy content of light is not continuous but rather is delivered in these discrete packets, the quanta. The energy (E) of a photon depends on the frequency of the light according to a relation known as **Planck's law** (Eq. 8.2):

$$E = hv \qquad (8.2)$$

where h is Planck's constant (6.626×10^{-34} J s).

Sunlight is like a rain of photons of different frequencies. Our eyes are sensitive to only a small range of frequencies—the visible light region of the electromagnetic spectrum (Fig. 8.2). Light of slightly higher frequencies (or shorter wavelengths) is in the ultraviolet region of the spectrum, and light of slightly lower frequencies (or longer wavelengths) is in the infrared region. The output of the sun is shown in Figure 8.3, along with the energy density that actually strikes the surface of the earth. The absorption spectrum of chlorophyll a (curve c in the figure) indicates approximately the portion of the solar output that is utilized by plants. Spectrophotometry, the technique used to measure the absorption of light by a sample, is discussed in the accompanying box.

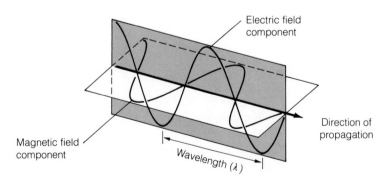

FIGURE 8.1. Light is a transverse electromagnetic wave, consisting of oscillating electric and magnetic fields that are perpendicular to each other and also perpendicular to the direction of propagation of the light. Light moves at a speed of 3×10^8 m s^{-1}. The wavelength (λ) is the distance between successive crests of the wave.

Principles of Spectrophotometry

MUCH OF WHAT we know about the photosynthetic apparatus was learned through **spectroscopy**—that is, measurements of the interaction of light and molecules. Spectrophotometry is an important branch of spectroscopy that focuses on the technique of measurement.

BEER'S LAW RELATES THE ABSORBANCE TO CONCENTRATION

An essential piece of information about any molecular species is how much of it is present. Quantitative measures of concentration are the cornerstone of much biological science. Of all the methods that have been devised for measuring concentration, by far the most widely applied is absorption spectrophotometry. In this technique, the amount of light that a sample absorbs at a particular wavelength is measured and used to determine the concentration of the sample by comparison with appropriate standards or reference data. The most useful measure of light absorption is the **absorbance** (A), also commonly called the **optical density** (OD) (Fig. 8.A). The absorbance is defined as

$$A = \log \frac{I_0}{I}$$

where I_0 is the intensity of light incident on the sample, I is the intensity of light transmitted by the sample, and log stands for the base 10 logarithm function.

The absorbance of a sample can be related to the concentration of the absorbing species through the use of **Beer's law**:

$$A = \varepsilon c l$$

where c is concentration, usually measured in moles per liter; l is the light path length, usually 1 cm; and ε is a proportionality constant known as the **molar extinction coefficient**. The units of ε are liters per mole per centimeter. The ε value is a function of both the particular compound being measured and the wavelength. Chlorophylls typically have ε values of about 100,000 liter $\mathrm{mol}^{-1}\,\mathrm{cm}^{-1}$. When more than one component of a complex mixture absorbs at a given wavelength, the absorbances due to the individual components are generally additive.

ABSORBANCE IS MEASURED WITH A SPECTROPHOTOMETER

The absorbance is measured by using an instrument called a spectrophotometer (Fig. 8.B). The essential parts of a spectrophotometer include a light source, a wavelength selection device such as a monochromator or filter, a sample chamber, a light detector, and a readout device. Modern instruments usually also include a computer, which is used for storage and analysis of the spectra. The most useful machines scan the wavelength of the light incident on the sample and produce as output spectra of absorbance versus wavelength, such as those shown in Figure 8.6.

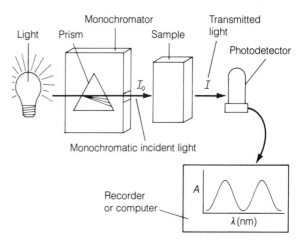

FIGURE 8.B. Schematic diagram of a spectrophotometer. The instrument consists of a light source, a monochromator containing a wavelength selection device such as a prism, a sample holder, a photodetector, and a recorder or computer. The output wavelength of the monochromator can be changed by rotating the prism; the graph of absorbance versus wavelength is called a spectrum.

FIGURE 8.A. Definition of absorbance. A monochromatic incident light beam of intensity I_0 traverses a sample contained in a cuvette of length l. Some of the light is absorbed by the chromophores in the sample, and the intensity of light that emerges is I.

ACTION SPECTRA MEASURE THE EFFECT OF ABSORBED LIGHT

The use of **action spectra** has been central to the development of our current understanding of photosynthesis. An action spectrum is a graph of the magnitude of the biological effect observed as a function of wavelength. Examples of effects measured by use of action spectra are oxygen evolution (Fig. 8.C) and hormonal growth responses due to the action of phytochrome (Chapter 20). Often, an action spectrum can identify the chromophore responsible for a particular light-induced phenomenon. Action spectra were instrumental in discovering the existence of the two photosystems in O_2-evolving photosynthetic organisms.

Some of the first action spectra were measured by T. W. Engelmann in the late 1800s (Fig. 8.D). Engelmann used a prism to disperse sunlight into a

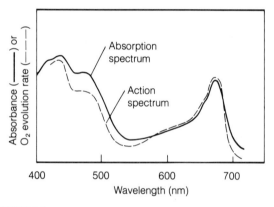

FIGURE 8.C. An action spectrum compared to an absorption spectrum. The absorption spectrum (solid line) is measured with a spectrophotometer as shown in Figure 8.B. An action spectrum is measured by plotting the effect of irradiating the sample with light as a function of wavelength—for example, plotting the rate of oxygen evolution induced by illumination with light of many different wavelengths. If the pigments that absorb the light are the same as those that cause the effect, the action spectrum and the absorption spectrum will match. In the example shown here, the action spectrum of oxygen evolution is generally similar to the total absorption spectrum of the chloroplast, indicating that light absorption by chlorophylls is responsible for photosynthetic energy storage. Discrepancies are found in the far-red region of the spectrum, where the effectiveness of the light falls off (the red drop effect) and in the region of carotenoid absorption from 450–550 nm. The red drop is a crucial piece of evidence for the cooperation of two photosystems with slightly different light-absorbing properties. The lower effectiveness of light absorbed by carotenoids reflects the low efficiency of energy transfer from carotenoid to chlorophyll.

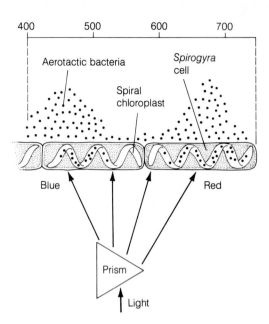

FIGURE 8.D. Schematic diagram of the action spectrum measurements by T. W. Engelmann in 1880s. Engelmann projected a spectrum of light onto the spiral chloroplast of the filamentous green alga *Spirogyra* and observed that the oxygen-seeking bacteria collected in the region of the spectrum where chlorophyll pigments absorb. This was the first action spectrum ever measured and gave the first indication of the effectiveness of light absorbed by accessory pigments in driving photosynthesis.

rainbow that was allowed to fall on an aquatic algal filament. A population of O_2-seeking bacteria were introduced into the system. The bacteria congregated in the regions of the filaments that evolved the most O_2. These were the regions illuminated by blue light and red light, which are strongly absorbed by chlorophyll. Today, action spectra can be measured in room-sized spectrographs in which the scientist actually enters a huge monochromator and places samples for irradiation in a large area of the room bathed by monochromatic light. The principle of the experiment is, however, the same as that of Engelmann's experiments.

DIFFERENCE SPECTRA MEASURE CHANGES IN ABSORBANCE

An important technique in photosynthetic studies is light-induced difference spectroscopy (Fig. 8.E). In this technique, bright light, often called **actinic** light, is used to illuminate a sample, while a dim beam of light is used to measure the absorbance of the sample at wavelengths other than that of the actinic beam.

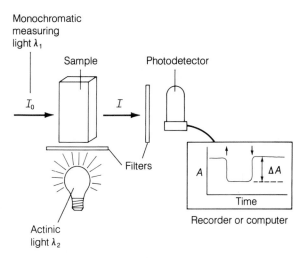

FIGURE 8.E. Principle of difference spectroscopy. Difference spectra are measured by observing as a function of time the change in the absorbance of a measuring light in a sample when an actinic light is turned on. The actinic light causes chemical changes in the sample that change its absorption spectrum. Blocking filters are necessary to prevent scattered actinic light from entering the detector. The up and down arrows signify the times when the actinic light was turned on and off, respectively. The change in absorbance induced by the actinic light, ΔA, can be either positive or negative. Usually, measurements are made at one wavelength at a time. A difference spectrum is built up by repeating the measurement at many different wavelengths.

In this way a difference spectrum is obtained, which represents the changes in the absorption spectrum of the sample induced by illumination with the actinic light. Absorption bands that disappear upon illumination appear as negative peaks, while new bands that appear upon illumination appear as positive peaks. Difference spectra give important clues to the identity of molecular species participating in the photoreactions of photosynthesis. The difference spectrum of P700 photooxidation is shown in Figure 8.F (Ke, 1973).

By use of special flash techniques, it is possible to record the difference spectrum at a given time after flash excitation. A number of difference spectra recorded at different times after flash excitation can be used to measure the kinetics of the chemical reactions that follow photon excitation of a reaction center. These techniques can have extraordinary time resolution, in some cases less than a picosecond (10^{-12} s), and have provided great insights into the earliest events in the photosynthetic energy storage process (Ke and Shuvalov, 1987).

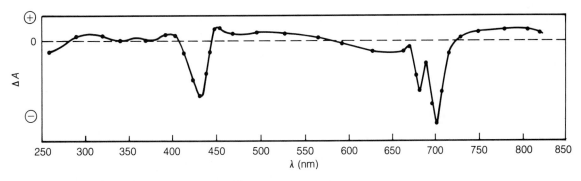

FIGURE 8.F. Light-minus-dark difference spectrum for P700 photooxidation, measured as shown in Figure 8.E. Decreases of absorption (bleaching) are observed at 700 and 430 nm, due to loss of the absorbance of P700. Increases observed around 450 nm and beyond 730 nm are due to absorption by $P700^{+}$. (Adapted from Ke, 1973.)

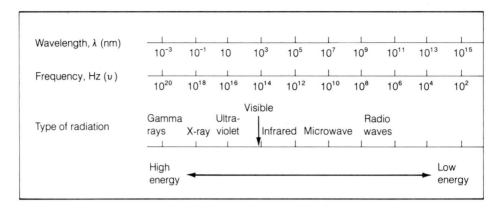

FIGURE 8.2. The electromagnetic spectrum. The wavelength (λ) and frequency (ν) are inversely related. Our eyes are sensitive to only a narrow range of wavelengths of radiation, the visible region, which extends from about 400 nm (violet) to about 700 nm (red). Short-wavelength (high-frequency) light has a high-energy content, while long-wavelength (low-frequency) light has a low-energy content.

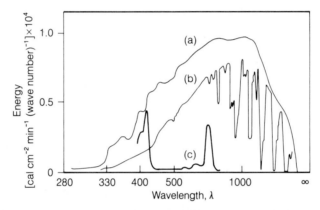

FIGURE 8.3. The solar spectrum and its relation to the absorption spectrum of chlorophyll. Curve a is the energy output of the sun as a function of wavelength. Curve b is the energy that actually strikes the surface of the earth. The sharp valleys in the infrared region beyond 700 nm are due to absorption of solar energy by molecules in the atmosphere, chiefly water vapor. Curve c is the absorption spectrum of chlorophyll a, which absorbs strongly in the blue (about 430 nm) and the red (about 660 nm). The green light in the middle of the visible region that is not absorbed is reflected into our eyes and gives plants their characteristic green color. Note that curve c has absorbance rather than energy units. Solar spectra from Calvin (1976).

When Molecules Absorb or Emit Light They Change Their Electronic State

Chlorophyll appears green to our eyes because it absorbs light in the red and blue parts of the spectrum, so that only the green light (~ 550 nm) is reflected into our eyes (Fig. 8.3). The absorption of light is represented by Equation 8.3, in which chlorophyll (chl) in its lowest-energy or ground state absorbs a photon and makes a transition to a higher-energy or excited state (chl*):

$$chl + h\nu \rightarrow chl^* \qquad (8.3)$$

The distribution of electrons in the molecule is somewhat different in the excited state than in the ground state. Figure 8.4 describes the absorption and emission of light

by chlorophyll molecules. Absorption of blue light excites the chlorophyll to a higher-energy state than absorption of red light, because the energy of photons is higher when their wavelength is shorter. In the higher excited state, chlorophyll is extremely unstable and very rapidly gives up some of its energy to the surroundings as heat and enters the lowest excited state, where it can be stable for up to several nanoseconds (10^{-9} s). In this state, the excited chlorophyll has several possible pathways for disposing of its available energy. It can re-emit a photon and thereby return to its ground state—a process known as **fluorescence**. When it does so, the wavelength of fluorescence is almost always slightly longer than the wavelength of absorption to the same electronic state, because

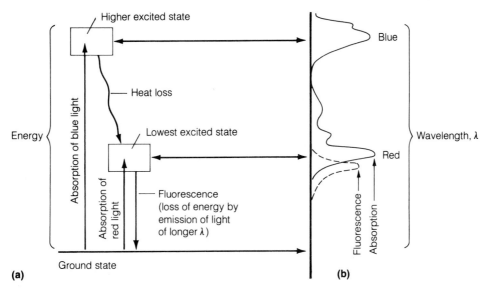

FIGURE 8.4. Light absorption and emission by chlorophyll. Part (a) shows an energy level diagram. Absorption or emission of light is indicated by vertical lines that connect the ground state with excited electronic states. The blue and red absorption bands of chlorophyll (which absorb red and blue photons, respectively) correspond to the upward vertical arrows, signifying that energy absorbed from light causes the molecule to change from the ground state to an excited state. The downward-pointing arrow indicates fluorescence, in which the molecule goes from the lowest excited state to the ground state while re-emitting energy as a photon. The spectra of absorption and fluorescence are shown in part (b). The long-wavelength absorption band of chlorophyll corresponds to light that has the energy required to cause the transition from the ground state to the first excited state. The short-wavelength absorption band corresponds to a transition to a higher excited state. (After Sauer, 1975.)

a portion of the excitation energy is converted into heat before the fluorescent photon is emitted. Conservation of energy therefore requires that the energy of the fluorescent photon be lower than that of the excitation photon, hence the shift to longer wavelength. Chlorophylls usually fluoresce in the red, and the shift in wavelength (commonly called the Stokes shift) is typically about ten nanometers.

Alternatively, the excited chlorophyll can return to its ground state by directly converting its excitation energy into heat, with no emission of a photon. A third process that deactivates excited chlorophyll is **photochemistry**, in which the energy of the excited state is used to cause chemical reactions to occur. The rate of the earliest steps in the photosynthetic energy storage process are among the fastest known chemical reactions. This extreme speed is necessary for photochemistry to compete with the other possible reactions of the excited state.

The Quantum Yield Gives Information About the Fate of the Excited State

The process with the highest rate will be the one most likely to deactivate excited chlorophyll; slower processes only occasionally win out in the race to dispose of the energy of the excited state. This concept is expressed quantitatively by way of the **quantum yield** (Clayton, 1971, 1980). The quantum yield (Φ) of a process in which molecules give up their excitation energy (or "decay") is the fraction of excited molecules that decay via that pathway. Mathematically, the quantum yield of a process, such as photochemistry, is defined by Equation 8.4.

$$\Phi_{\text{photochemistry}} = \frac{\text{yield of photochemical products}}{\text{total number of quanta absorbed}} \quad (8.4)$$

The quantum yield of the other processes is defined analogously. The value of Φ for a particular process can range from 0 (if that process is never involved in the decay of the excited state) to 1.0 (if that process always serves to deactivate the excited state). The sum of the quantum yields of all possible processes is 1.0.

In functional chloroplasts kept in dim light, the quantum yield of photochemistry is approximately 0.95, the quantum yield of fluorescence is 0.05 or lower, and the quantum yields of other processes are negligible. The vast majority of excited chlorophyll molecules therefore carry out photochemistry.

The quantum yield of formation of products of photosynthesis, such as O_2, can also be measured quite accurately. In this case the quantum yield is substantially lower than the value for photochemistry, because several photochemical events must take place before a single

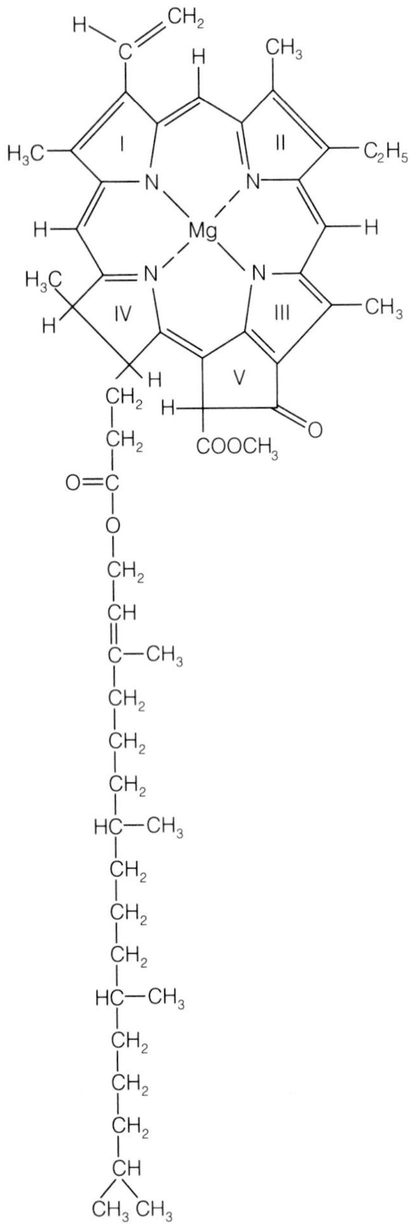

Chlorophyll a

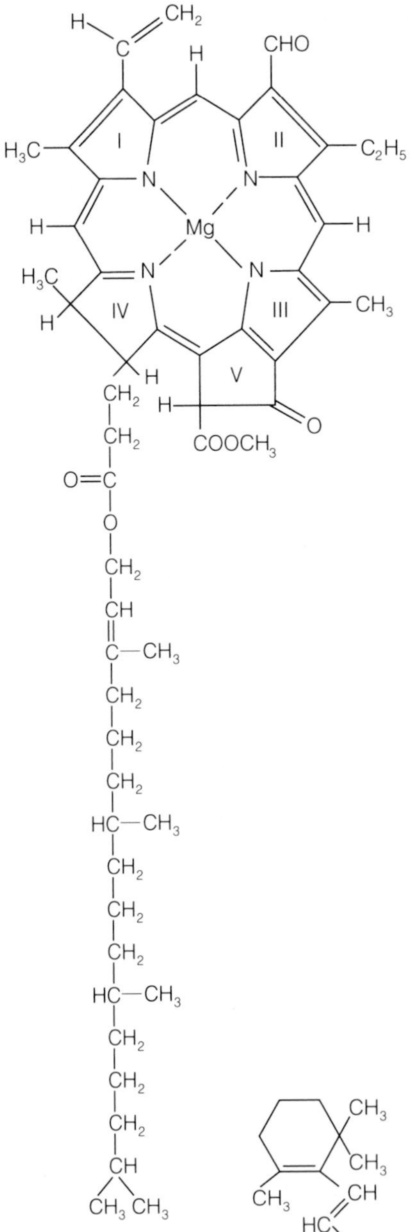

Chlorophyll b

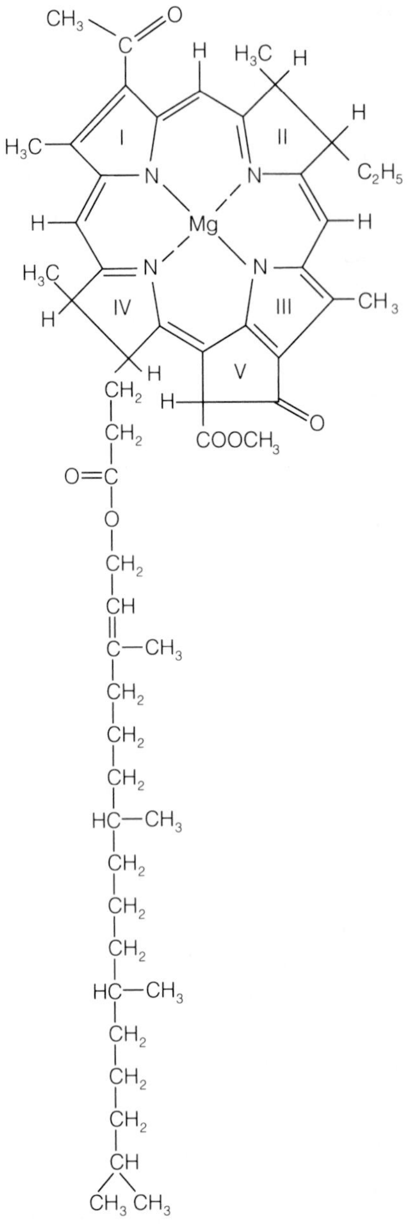

Bacteriochlorophyll a

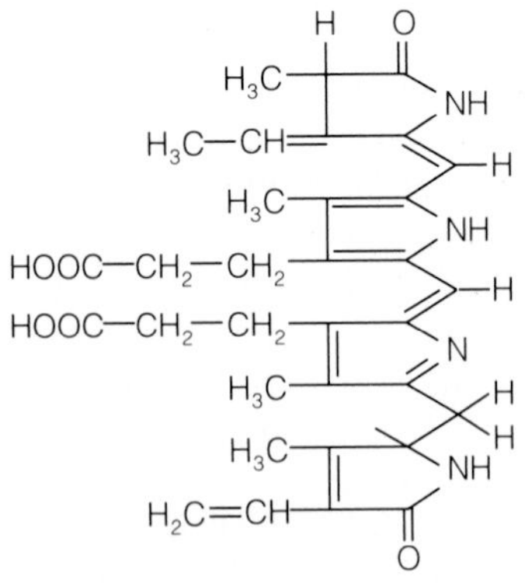

Phycoerythrobilin

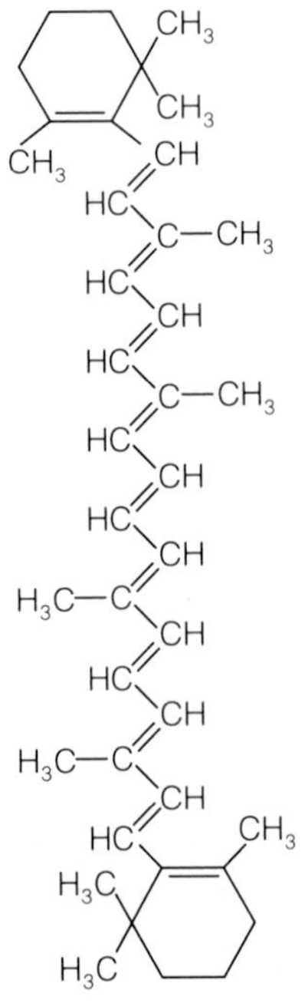

ß-Carotene

TABLE 8.1. Distribution of chlorophylls and other photosynthetic pigments

| Organism | Chlorophyll | | | | Bacteriochlorophyll | | | | | | Carotenoids | Phycobiliproteins |
	a	b	c	d	a	b	c	d	e	g		
Eukaryotes												
Mosses, ferns, seed plants	+	+	−	−							+	−
Green algae	+	+	−	−							+	−
Euglenoids	+	+	−	−							+	−
Diatoms	+	−	+	−							+	−
Dinoflagellates	+	−	+	−							+	−
Brown algae	+	−	+	−							+	−
Red algae	+	−	−	+							+	+
Prokaryotes												
Cyanobacteria	+	−	−	−							+	+
Prochlorophytes	+	+	−	−							+	−
Sulfur purple bacteria					+ or	+	−	−	−	−	+	−
Nonsulfur purple bacteria					+ or	+	−	−	−	−	+	−
Green bacteria					+	−	+ or	+ or	+	−	+	−
Heliobacteria					−	−	−	−	−	+	+	−

molecule of O_2 is formed. For O_2 production the measured maximum quantum yield is approximately 0.1, meaning that 10 quanta are absorbed for each O_2 released. The reciprocal of the quantum yield is called the **quantum requirement**. The minimum quantum requirement for O_2 evolution is therefore 10. Quantitative measurements of the absorption of light and the fate of the energy contained in the light are essential to an understanding of the photosynthetic process.

Photosynthetic Pigments Absorb the Light That Powers Photosynthesis

The energy of sunlight is first absorbed by the pigments of the plant. All pigments active in photosynthesis are found in the chloroplast. Structures and absorption spectra of several photosynthetic pigments are shown in Figures 8.5 and 8.6. Table 8.1 shows the distribution of pigments in different types of photosynthetic organisms. The **chlorophylls** and **bacteriochlorophylls** (pigments) found in certain bacteria are the typical pigments of photosynthetic organisms, but all organisms actually contain a mixture of more than one kind of pigment, each serving a specific function. All chlorophylls have a complex ring structure that is chemically related to the porphyrin-like groups found in hemoglobin and cytochromes (Fig. 8.5). In ad-

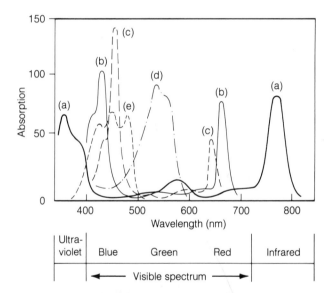

FIGURE 8.6. Absorption spectra of photosynthetic pigments. Curve a, bacteriochlorophyll *a*; curve b, chlorophyll *a*; curve c, chlorophyll *b*; curve d, phycoerythrobilin; curve e, *β*-carotene. The absorption spectra shown are for pure pigments dissolved in nonpolar solvents, except for curve d, which represents an aqueous buffer of phycoerythrin, a protein that contains a phycoerythrobilin chromophore covalently attached to the peptide chain. In many cases, the spectra of photosynthetic pigments in vivo are substantially affected by their environment in the photosynthetic membrane. (From Avers, 1985.)

FIGURE 8.5. (Left) Molecular structures of photosynthetic pigments. The chlorophylls have a porphyrin-like ring structure with a magnesium atom coordinated in the center and a long hydrophobic hydrocarbon tail that anchors them in the photosynthetic membrane. The porphyrin-like ring is the site of the electronic rearrangements that occur when the chlorophyll is excited and of the unpaired electrons when it is either oxidized or reduced. Various chlorophylls differ chiefly in the substituents around the rings and the pattern of double bonds. Carotenoids are linear polyenes that serve as both antenna pigments and photoprotective agents. Bilin pigments are open-chain tetrapyrroles found in antenna structures known as phycobilisomes that occur in cyanobacteria and red algae.

dition, a long hydrocarbon tail is almost always attached to the ring structure. The tail appears to anchor the chlorophyll to the hydrophobic portion of the membrane environment. The ring structure contains some loosely bound electrons and is the part of the molecule involved in electronic transitions and redox reactions.

Photosynthesis Takes Place in Complexes Containing Light-Gathering Antennas and Photochemical Reaction Centers

A portion of the light energy absorbed by chlorophylls and other pigments is eventually stored as chemical energy by forming chemical bonds. This conversion of energy from one form to another is a complex process that depends on cooperation between a large number of pigment molecules and a group of electron transfer proteins. The majority of the pigments serve as an antenna, collecting light and transferring the energy to the reaction center, where the chemical reactions leading to long-term energy storage take place. The basic concept of light absorption and energy transfer during photosynthesis is shown diagrammatically in Figure 8.7.

How does the plant benefit from this division of labor between antenna and reaction center pigments? Even in bright sunlight, a chlorophyll molecule absorbs only a few photons each second. If every chlorophyll had a complete reaction center associated with it, the enzymes that make up this system would be idle most of the time, only occasionally being activated by photon absorption.

However, if a large number of pigments can send energy into a common reaction center, the system is kept active a large fraction of the time.

In 1932 Robert Emerson and William Arnold performed a classic experiment that provided the first evidence for the cooperation of many chlorophyll molecules in energy conversion during photosynthesis (Emerson and Arnold, 1932). They gave very brief (10^{-5} s duration) flashes of light to a suspension of the green alga *Chlorella pyrenoidosa* and measured the amount of oxygen produced. The flashes were spaced about 0.1 s apart, a time that they had determined in earlier work was long enough for the enzymatic steps of the process to be completed before the arrival of the next flash. They varied the energy of the flashes and found that at high intensity the oxygen production did not increase when a yet more intense flash was given: the photosynthetic system was saturated with light. Figure 8.8 shows the relationship of oxygen production to flash energy measured by Emerson and Arnold. They were surprised to find that under saturating conditions, only one molecule of oxygen was produced for each 2500 chlorophyll molecules in the sample. We know now that several hundred pigments are associated with each reaction center and that each reaction center must operate four times to produce one molecule of oxygen (see Fig. 8.29a), hence the value of 2500 chlorophyll per O_2.

The reaction centers and most of the antenna complexes are integral components of the photosynthetic membrane. In eukaryotic photosynthetic organisms, these membranes are found within the chloroplast; in photosynthetic prokaryotes, the site of photosynthesis is the cell membrane or membranes derived from it.

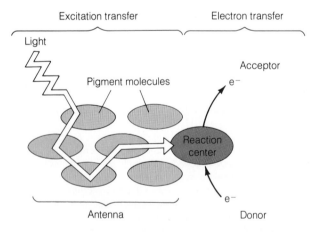

FIGURE 8.7. The basic concept of energy transfer during photosynthesis. A large number of pigments serve as an antenna, collecting light and transferring its energy to the reaction center, where chemical reactions store some of the energy by transferring electrons from a chlorophyll pigment to an electron acceptor molecule. An electron donor then rereduces the chlorophyll. The excitation transfer process in the antenna is a purely physical phenomenon and does not involve any chemical changes. The first chemical reactions following photon absorption take place in the reaction center. Subsequent reactions stabilize the unstable products of the initial chemical reaction.

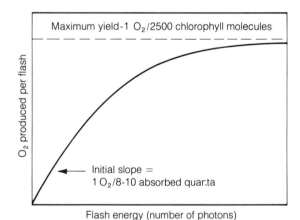

FIGURE 8.8. The relationship of oxygen production to flash energy, the first evidence for the interaction between the antenna pigments and the reaction center. This light saturation curve, obtained in 1932 by Robert Emerson and William Arnold, indicates that regardless of the amount of energy in each flash, the maximum amount of O_2 produced is 1 O_2 per 2500 chlorophyll molecules. The slope of the low-intensity part of the curve gives the quantum requirement of oxygen production.

The Chemical Reaction of Photosynthesis Is Driven by Light

Establishing the overall chemical equation of photosynthesis required several hundred years and contributions by many scientists (Rabinowitch and Govindjee, 1969). In 1771 Joseph Priestley observed that a sprig of mint growing in air in which a candle had burned out improved the air so that another candle could burn. He had discovered oxygen evolution by plants. A Dutchman, Jan Ingenhousz, documented the essential role of light in photosynthesis in 1779. Other scientists established the roles of CO_2 and H_2O and showed that organic matter, specifically carbohydrate, is a product of photosynthesis along with oxygen. By the end of the nineteenth century, the overall reaction for photosynthesis could be written as follows:

$$6CO_2 + 6H_2O \xrightarrow{\text{light, plant}} C_6H_{12}O_6 + 6O_2 \quad (8.5)$$

where $C_6H_{12}O_6$ represents a simple sugar such as glucose. As discussed in Chapter 9, glucose is not the actual product of the dark reactions. However, the energetics for the actual products are approximately the same, so the use of glucose in Equation 8.5 should be regarded a convenience but not taken literally.

It is important to realize that equilibrium for the chemical reaction shown in Equation 8.5 lies very far to the direction of the reactants (see Eq. 2.9 in Chapter 2). The equilibrium constant for Equation 8.5, calculated from tabulated free energies of formation for each of the compounds involved, is $\sim 10^{-500}$. This number is so close to zero that one can be quite confident that in the entire history of the universe there has never been a single molecule of glucose formed spontaneously from H_2O and CO_2 without external energy being provided. The energy needed to drive the photosynthetic reaction comes from light. A simpler form of Equation 8.5 is

$$CO_2 + H_2O \xrightarrow{\text{light, plant}} (CH_2O) + O_2 \quad (8.6)$$

where (CH_2O) is one-sixth of a glucose molecule. About 9 or 10 photons of light are required to drive the reaction of Equation 8.6. If red light of wavelength 680 nm is absorbed, the total energy input (Eq. 8.2) is 1760 kJ per mole of oxygen formed. This is more than enough to drive the reaction in Equation 8.6, which has a standard state free energy change of $+467$ kJ/mol. The efficiency of conversion of light energy at the optimal wavelength into chemical energy is therefore about 27%, which is actually remarkably high for an energy conversion system. Most of this stored energy is used for cellular processes, so the amount actually diverted to the formation of biomass is much less.

Photosynthesis Is a Light-Driven Redox Process

The reaction shown in Equation 8.6 is an overall reaction for photosynthesis. As such, it tells us what goes into the chloroplast and what comes out, but it tells us nothing about how the transformation takes place. The actual chemical mechanism for the reaction is complex; at least 50 intermediate steps have been identified, and there are undoubtedly some that have not yet been discovered. The first clue to the nature of the mechanism came from studies by C. B. van Niel in the 1920s utilizing non-oxygen-evolving photosynthetic bacteria such as *Chromatium vinosum*. These organisms can grow by using organic acids or reduced sulfur compounds as reductants, as represented by Equation 8.7:

$$CO_2 + 2H_2A \xrightarrow{\text{light, bacteria}} (CH_2O) + 2A + H_2O \quad (8.7)$$

where H_2A represents any one of these reduced compounds, for example, H_2S. In Equation 8.7, the reduced compound is oxidized, and CO_2 is reduced. Van Niel drew the analogy to the process shown in Equation 8.6 and correctly concluded that, in essence, photosynthesis is a redox process. In the special case of oxygen-evolving photosynthetic organisms, H_2O serves as the reductant. Although some aspects of van Niel's formulation of photosynthesis were incorrect, his conclusion about the redox nature of the process stands as the central tenet of our understanding of the subject.

In 1937 Robert Hill found that isolated chloroplast thylakoids photoreduced a variety of compounds such as iron salts (Hill, 1939). These compounds serve as oxidants in place of CO_2, as shown in Equation 8.8.

$$4Fe^{3+} + 2H_2O \rightarrow 4Fe^{2+} + O_2 + 4H^+ \quad (8.8)$$

A large number of compounds have since been shown to act as artificial electron acceptors in what has become known as the Hill reaction. Their use has been invaluable in elucidating the reactions that precede carbon reduction. We now know that during the normal functioning of the photosynthetic system, light serves to reduce nicotinamide-adenine dinucleotide phosphate (NADP), which in turn serves as the reducing agent for carbon fixation in the carbon reduction cycle (Chapter 9). ATP is also formed during the electron flow from water to NADP, and it too is used in carbon reduction. The chemical reactions in which water is oxidized to oxygen, NADP is reduced, and ATP is formed are generally known as the light reactions, while the carbon reduction reactions are called the dark reactions. This division, while somewhat arbitrary, is reasonable in that almost all the reactions up to NADP reduction take place within the thylakoids, while the carbon reduction reactions take place in the aqueous region of the chloroplast, the stroma. It is important to realize, however, that the only reactions actually driven by light are the initial photochemical reactions within the reaction center, so the terms "light reactions" and "dark reactions" should not be taken too literally. All the subsequent electron transfer reactions, usually included in the light reactions, as well as all the carbon reduction reactions, actually occur without the need for light. However, all these reactions use the energy stored in the photochemical reactions, which is why photosynthesis occurs only in the light.

Oxygen-Evolving Organisms Have Two Photosystems That Operate in Series

By the late 1950s, several experiments were puzzling the scientists who studied photosynthesis. One of these experiments measured the quantum yield of photosynthesis as a function of wavelength and revealed an effect known as the **red drop** (Fig. 8.9). The quantum yield of photosynthesis is the reciprocal of the number of absorbed photons necessary to reduce one molecule of CO_2 to carbohydrate. If the quantum yield is measured for the wavelengths at which chlorophyll absorbs light, the values found throughout most of the range are remarkably constant, indicating that any photon absorbed by chlorophyll or other pigments is as effective as any other photon in driving photosynthesis. However, at the extreme red edge of the chlorophyll absorption (>680 nm), the yield drops dramatically. This drop does not simply reflect the fact that the absorption is falling off, because the quantum yield is a measure of only light that has actually been absorbed. Thus, light with a wavelength greater than 680 nm is much less efficient than light of shorter wavelengths.

Another puzzling experimental result was the **enhancement effect**, discovered by Emerson (Emerson et al., 1957). The rate of photosynthesis was measured separately with light of two different wavelengths, and then the two beams were used simultaneously (Fig. 8.10).

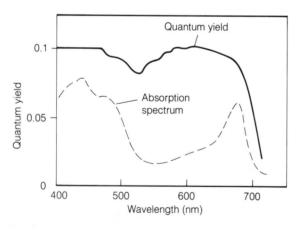

FIGURE 8.9. The "red drop" effect. The quantum yield of photosynthesis falls off drastically for "far-red" light of wavelengths >680 nm, indicating that far-red light alone is inefficient in driving photosynthesis. The slight dip near 500 nm is due to somewhat lower efficiency of photosynthesis using light absorbed by accessory pigments, carotenoids. The absorption spectrum of the chloroplast is shown by the dashed line.

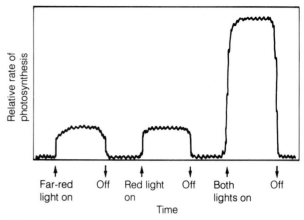

FIGURE 8.10. The "enhancement effect." The rate of photosynthesis when red and far-red light are given together is greater than the sum of the rates when they are given apart. Upward and downward arrows indicate the switching on and off of the light that drives photosynthesis. This experiment baffled scientists in the 1950s, but it provided essential evidence in favor of the concept that photosynthesis is carried out by two photochemical systems with slightly different wavelength optima.

Under certain conditions, especially when one of the wavelengths was in the far-red region (>680 nm), the photosynthetic rate obtained with both wavelengths was greater than the sum of the individual rates!

These observations were explained by experiments performed by Louis Duysens of the Netherlands (Duysens et al., 1961). Chloroplasts contain cytochromes, iron-containing proteins that function as intermediate electron carriers in photosynthesis. Duysens found that when a sample of a red alga was illuminated with long-wavelength light, the cytochrome became largely oxidized. If shorter wavelength light was also present, the effect was partially reversed (Fig. 8.11). These **antagonistic effects** were explained by Hill and Fay Bendall in 1960. They proposed a mechanism involving two photochemical events: one that tended to oxidize the cytochrome and one that tended to reduce it (Hill and Bendall, 1960). We know now that in the red region of the spectrum, one of the photoreactions, known as **photosystem I**, absorbs preferentially far-red light of wavelengths >680 nm, while the second, known as **photosystem II**, absorbs red light of 680 nm well and is driven very poorly by far-red light. This wavelength dependence explains the enhancement effect and the red drop effect. Another difference between the photosystems is that photosystem I produces a strong reductant, capable of reducing NADP, and a weak oxidant. Photosystem II produces a strong oxidant, capable of oxidizing water, and a weak reductant that rereduces the oxidant produced by photosystem I. This explains the antagonistic effect. These properties are shown schematically in Figure 8.12. The scheme of photosynthesis depicted, called the Z (for

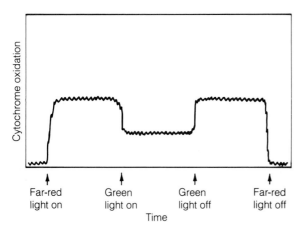

FIGURE 8.11. Antagonistic effects of light on cytochrome oxidation. Far-red light is very effective in oxidizing the cytochrome *f* in the chloroplast. If green light is also present, some of the cytochrome becomes reduced. The two wavelengths have opposite effects, hence the term *antagonistic effects*. This experiment is one of the clearest demonstrations of the existence of two photochemical systems in photosynthesis, one that reduces cytochrome and one that oxidizes it. This particular experiment was done with a red alga, in which photosystem II is driven best by green light and photosystem I is driven best by far-red light. (Adapted from Duysens et al., 1961.)

zigzag) scheme, has become the basis for understanding O_2-evolving photosynthetic organisms. It accounts for the operation of two physically and chemically distinct photosystems (I and II), each with its own antenna pigments and photochemical reaction center. The two photosystems are linked together by an electron transport chain. Non-O_2-evolving organisms, such as the purple photosynthetic bacteria of the genera *Rhodobacter* and *Rhodopseudomonas*, contain only a single photosystem and therefore do not display the effects described above. However, these simpler organisms have been very useful for detailed structural and functional studies of photosynthesis.

Structure of the Photosynthetic Apparatus

In the previous section we learned some physical principles underlying the photosynthetic process, some aspects of the functional roles of various pigments, and some of the chemical reactions carried out by photosynthetic organisms. We now turn to the architecture of the photosynthetic apparatus and the structures of its component parts (Staehelin, 1986), with the goal of describing the structures of the organelles and complexes that carry out photosynthesis.

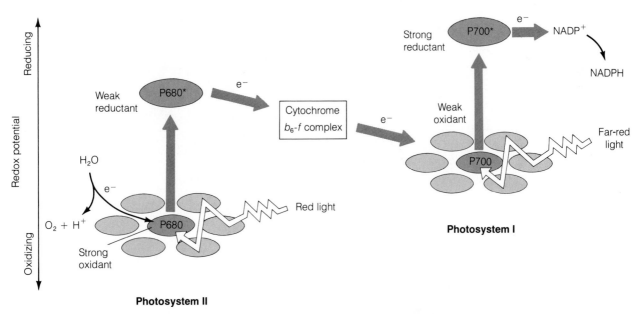

FIGURE 8.12. The Z scheme of photosynthesis. Red light absorbed by photosystem II (PSII) produces a strong oxidant and a weak reductant. Far-red light absorbed by photosystem I (PSI) produces a weak oxidant and a strong reductant. The strong oxidant generated by PSII oxidizes water, while the strong reductant produced by PSI reduces NADP. PSII produces electrons that reduce the cytochrome b_6-*f* complex, while PSI produces an oxidant that oxidizes the cytochrome b_6-*f* complex. This scheme is basic to an understanding of photosynthetic electron transport. P680 and P700 refer to the wavelengths of maximum absorption of the reaction center chlorophylls in PSII and PSI. These are discussed in more detail later in the chapter.

The Chloroplast Is the Site of Photosynthesis

In photosynthetic eukaryotes, photosynthesis takes place in the subcellular organelle known as the chloroplast. Figure 8.13 shows a transmission electron micrograph of a pea chloroplast. The most striking aspect of the structure of the chloroplast is the extensive system of internal membranes known as **thylakoids**. All the chlorophyll is contained within this membrane system, which is the site of the light reactions of photosynthesis. The carbon reduction reactions, or dark reactions, which are catalyzed by water-soluble enzymes, take place in the **stroma**, the region of the chloroplast outside the thylakoids. Most of the thylakoids appear to be very closely associated with each other. These stacked or appressed membranes are known as **grana lamellae**, while the exposed membranes in which stacking is absent are known as **stroma lamellae**. Another view of the chloroplast membrane structure, shown in Figure 8.14, was obtained by freeze-fracture electron microscopy (Chapter 1).

Two separate membranes, each composed of a lipid bilayer, surround the entire chloroplast (Fig. 8.15). This double membrane system contains a variety of metabolite transport systems (Chapter 6). The chloroplast also contains its own DNA, RNA, and ribosomes. Many of the chloroplast proteins are products of transcription and translation within the chloroplast itself, whereas others are coded for by nuclear DNA, synthesized on cytoplasmic ribosomes and then imported into the chloroplast. This remarkable division of labor, extending in many cases to different subunits of the same enzyme complex, will be discussed in detail in a later section.

Thylakoids Contain Integral Membrane Proteins

A wide variety of proteins essential to photosynthesis are embedded in the thylakoid membrane. In many cases, portions of these proteins extend into the aqueous regions on both sides of the thylakoid. These **integral membrane**

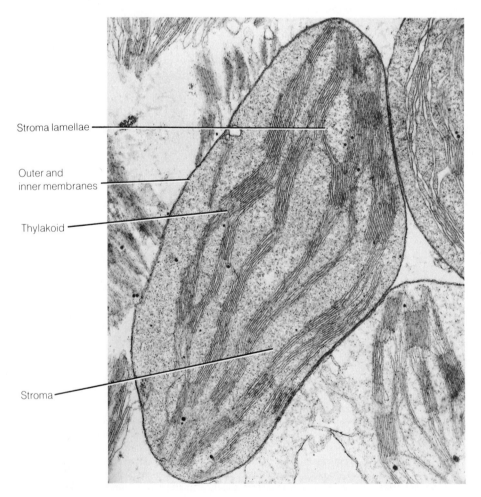

Stroma lamellae

Outer and inner membranes

Thylakoid

Stroma

FIGURE 8.13. Transmission electron micrograph of pea chloroplast, fixed in glutaraldehyde and OsO_4, embedded in plastic resin, and thin-sectioned with an ultramicrotome (18,200 ×). (Courtesy of J. Swafford, Department of Botany, Arizona State University.)

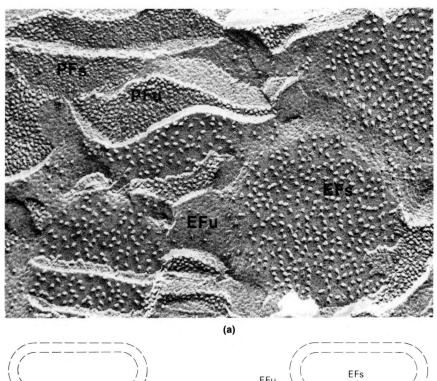

(a)

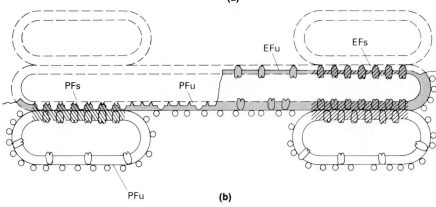

(b)

FIGURE 8.14. Freeze-fractured thylakoid membrane. (a) Electron micrograph shows four different fracture faces distinguished by different particle sizes and densities (85,000 ×). (b) Diagram showing the relative locations of the four different fracture faces seen in (a). Ef indicates an external fracture face (farthest from the cytoplasm), which may be stacked (Efs) or unstacked (Efu). Pf indicates the fracture face nearest the cytoplasm, which may exist in stacked (Pfs) or unstacked (Pfu) states. (From Avers, 1985.)

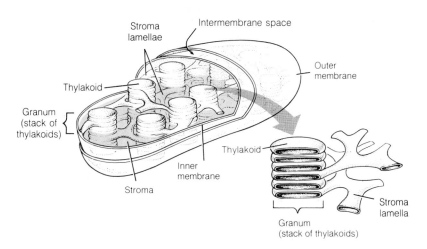

FIGURE 8.15. Schematic picture of the overall organization of the membranes in the chloroplast. The chloroplast is surrounded by the inner and outer membranes. The region of the chloroplast that is inside the inner membrane but outside the thylakoid membranes is known as the stroma. It contains the enzymes that catalyze carbon fixation. The thylakoid membranes are highly folded and appear in many pictures to be stacked like coins, although in reality they form one or a few large interconnected membrane systems, with a well-defined interior and exterior with respect to the stroma. The stacked or appressed regions are known as grana lamellae and the unstacked regions are known as stroma lamellae. The interior of the thylakoid membrane is known as the lumen. (From Becker, 1986.)

proteins contain a large proportion of hydrophobic amino acids and are therefore much more stable in a nonaqueous medium such as the hydrocarbon portion of the membrane (Fig. 1.8a). The reaction centers, the antenna pigment–protein complexes, and most of the electron transport enzymes are all integral membrane proteins and are inserted into the lipid bilayer of the membrane. The three dimensional structures of these complexes have been difficult to elucidate because membrane proteins have been resistant to crystallization, the essential first step in an x-ray structure determination. However, much progress has been made in this area in

recent years, and we can expect to understand the structures of many integral membrane proteins in the years to come. This information will be invaluable in improving our understanding of photosynthesis.

In all known cases, integral membrane proteins of the chloroplast have a unique orientation within the membrane. Many thylakoid membrane proteins have one region pointing toward the stromal side of the membrane and the other oriented toward the interior portion of the thylakoid, known as the **lumen**, as shown in Figure 8.16. This vectorial arrangement of proteins is a universal feature of all energy-conserving membranes.

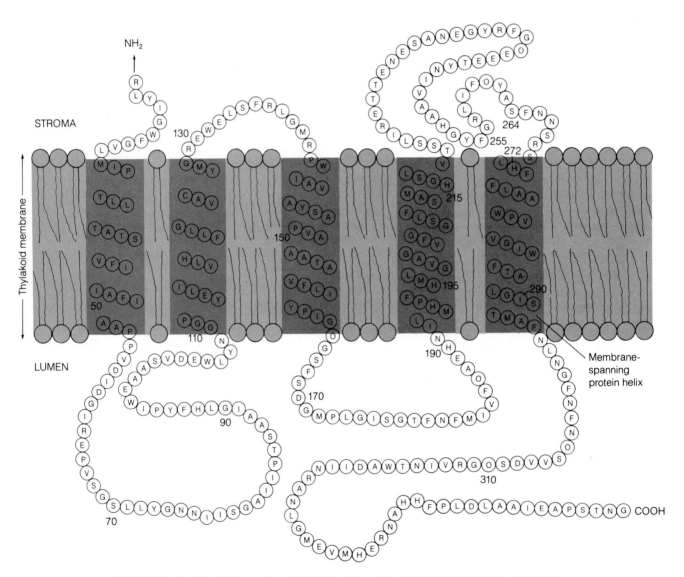

FIGURE 8.16. Predicted folding pattern of the D1 protein of the photosystem II reaction center. The hydrophobic portion of the membrane is traversed five times by the peptide chain. The protein is asymmetrically arranged in the thylakoid membrane, with the amino terminus on the stromal side of the membrane and the carboxyl terminus on the lumenal side. Specific amino acids are identified by their single-letter abbreviations. The numbers indicate the positions of amino acids within the polypeptide chain. (Adapted from Trebst, 1986.)

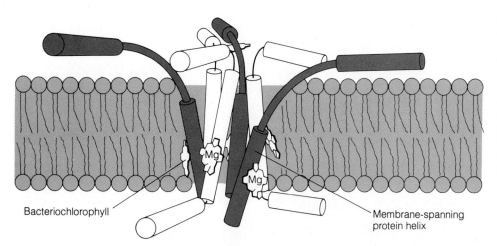

FIGURE 8.17. Proposed structure for a bacterial antenna pigment protein. The vertical tubes are transmembrane α helices of the antenna protein, while the horizontal tubes are surface helices. Magnesium atoms in the center of the bacteriochlorophyll molecules are thought to be coordinated to nitrogens in specific histidine residues of the protein. (From Hunter et al., 1989).

Bacteriochlorophyll

Membrane-spanning protein helix

The chlorophylls and accessory light-gathering pigments in the thylakoid membrane are always associated in a noncovalent but highly specific way with proteins, forming **chlorophyll proteins** (Fig. 8.17) (Thornber, 1986; Green, 1988; Hunter et al., 1989). Both antenna and reaction center chlorophylls are contained in chlorophyll proteins that are inserted into the membrane. In many cases histidine residues in the protein serve as ligands, with one of the nitrogens of the histidine imidazole ring binding to the magnesium in a chlorophyll. Several pigments are often associated with each peptide. The distances between pigments and their orientations with respect to each other appear to be relatively fixed. This precisely determined geometric arrangement optimizes the energy transfer in antenna complexes and the electron transfer process in reaction centers, while at the same time minimizing wasteful processes.

The Structures of Two Bacterial Reaction Centers Have Been Determined

In 1984 Hartmut Michel, Johann Deisenhofer, Robert Huber, and co-workers in Munich solved the three-dimensional structure of the reaction center from the purple photosynthetic bacterium *Rhodopseudomonas viridis* (Deisenhofer and Michel, 1989). This landmark achievement, for which a Nobel Prize was awarded in 1988, was the first high-resolution x-ray structure determination for an integral membrane protein and the first structure determination for a reaction center complex. The structure is shown in Figure 8.18. Of particular interest are the twofold symmetry of the structure about an axis perpendicular to the plane of the membrane, the α-helical arrangement of the transmembrane portions of the protein, and the complete absence of charged amino acid residues in the interior of the membrane. The geometric arrangement of the pigments and the quinones (electron acceptors) is shown in Figure 8.19, with the protein removed. A similar arrangement is found in the reac-

tion center of another purple photosynthetic bacterium, *Rhodobacter sphaeroides*.

Two of the bacteriochlorophyll molecules are in intimate contact with each other and are known as the **special pair**. This dimer, whose existence was predicted from magnetic resonance studies, is the photoactive portion of the complex. An electron is transferred from this dimer along the sequence of electron carriers on the right side of the complex, as pictured in Figure 8.19.

This reaction center structure is thought to be similar in many ways to that found in photosystem II from oxygen-evolving organisms, especially in the electron acceptor portion of the chain. The bacterial peptides are relatively similar in sequence to their photosystem II counterparts, implying an evolutionary relatedness.

Photosystems I and II Are Spatially Separated in the Thylakoid Membrane

In recent years it has become well established that the photosystem II reaction center, along with its antenna chlorophylls and associated electron transport proteins, is located predominantly in the stacked regions of the grana lamellae (Anderson and Andersson, 1988) (Fig. 8.20). The photosystem I reaction center and its associated antenna pigments and electron transfer proteins, as well as the coupling factor enzyme that catalyzes the formation of ATP, are found almost exclusively in the stroma lamellae and at the edges of the grana lamellae. The cytochrome b_6-f complex that connects the photosystems is evenly distributed. Thus, the two photochemical events that take place in O_2-evolving photosynthesis are spatially separated. This implies that one or more of the electron carriers that function between the photosystems diffuses from the grana region of the membrane to the stroma region, where electrons are delivered to photosystem I. The protons produced by the oxidation of water must also be able to diffuse to the stroma region, where ATP synthesis takes place. The possible functional role of this large separation (many tens of nanometers) between

FIGURE 8.18. Structure of the reaction center of the purple bacterium *Rhodopseudomonas viridis* resolved by x-ray crystallography. The 11 tubes represent transmembrane protein helices. The chlorophyll-type pigments are shown in light gray, while the heme groups of the tightly bound iron-containing cytochrome are shown in dark gray at the top of the diagram. The cytochrome subunit points into the periplasm or outside the cell, while the other side is in contact with the cytoplasm of the cell. The quinones and the cytochrome participate in electron transfer reactions with the bacteriochlorophylls. (From Jane Richardson.)

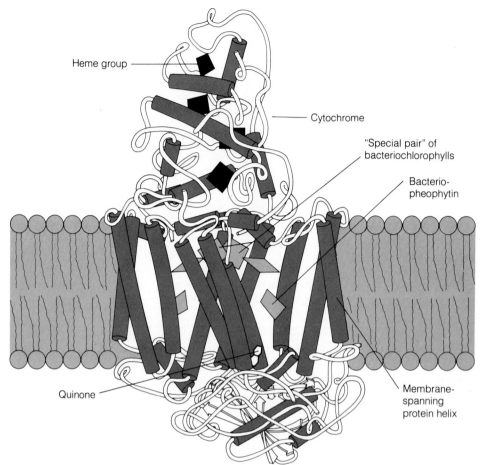

FIGURE 8.19. Geometric arrangement of pigments and other prosthetic groups in the bacterial reaction center. Bchl stands for bacteriochlorophyll, BPh stands for bacteriopheophytin (pheophytin is a chlorophyll molecule in which the magnesium atom has been replaced by two hydrogen atoms), and Q_A and Q_B are the first and second quinone acceptors. The cytochrome heme groups have been deleted from this diagram; if present, they would be at the top. The dimer of two bacteriochlorophylls at the top (P) is the "special pair" that react with light within the reaction center. The electron transfer sequence begins at the special pair and proceeds down the right side of the diagram to bacteriopheophytin. Next, the electrons are transferred to Q_A and then to Q_B. The distance between the depicted molecules and their angles within the planes of symmetry of the reaction center have been determined with precision, making it possible to analyze the path of photons and electrons in the reaction center in remarkable detail. (From Woodbury et al., 1985.)

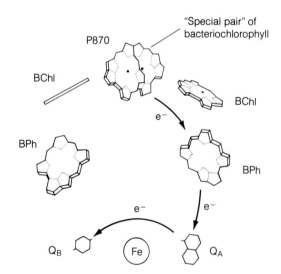

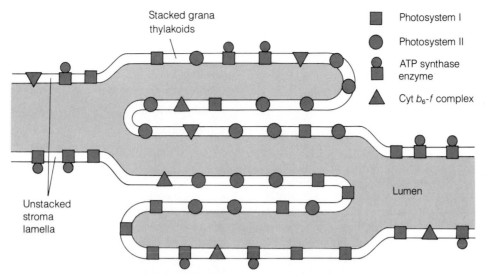

FIGURE 8.20. Organization of the protein complexes of the thylakoid membrane. Photosystem II is located predominantly in the stacked regions of the thylakoid membrane, while photosystem I and the ATP synthesizing enzyme are found in the unstacked regions, protruding into the stroma. Cytochrome b_6-f complexes are evenly distributed. This lateral separation of the two photosystems requires that electrons and protons produced by photosystem II be transported a considerable distance before they can be acted on by photosystem I and the ATP coupling enzyme.

photosystems I and II is not entirely clear, although most researchers believe that the separation is somehow involved in regulation of light energy input to the two photosystems.

The spatial separation between photosystems I and II indicates that the one-to-one stoichiometry between the two photosystems suggested by earlier investigations does not necessarily exist. Instead, photosystem II reaction centers feed reducing equivalents into a common intermediate pool of one of the electron carriers, plastoquinone, and photosystem I reaction centers remove the reducing equivalents from the common pool, rather than from any specific photosytem II. Most measurements of the relative quantities of photosystems I and II have shown that there is an excess of photosystem II. Under many conditions the PSII/PSI ratio is about 1.5:1, but it can change when plants are grown in different conditions of light (Anderson, 1986).

Organization of Light-Absorbing Antenna Systems

The antenna systems of different classes of photosynthetic organisms are remarkably varied, in contrast to the reaction centers, which appear to be similar in even distantly related organisms. The variety of antenna complexes reflects evolutionary adaptation to the diverse environments in which different organisms live, as well as the need to balance energy input to the two photosystems.

Antenna systems function to deliver energy efficiently to the reaction center(s) with which they are associated.

The size of the antenna system varies considerably in different organisms: a low of 20 to 30 bacteriochlorophylls per reaction center in some photosynthetic bacteria, 200 to 300 chlorophylls per reaction center in higher plants, and a few thousand pigments per reaction center in some types of algae and bacteria. The molecular structures that serve as antennas are also quite diverse, although they are all associated in some way with the photosynthetic membrane (Zuber, 1986; Hunter et. al. 1989).

The mechanism by which excitation energy is conveyed from the chlorophyll that absorbs the light to the reaction center is currently thought to be **resonance transfer** (also known as **Förster transfer** after the scientist who first described the phenomenon) (Borisov, 1989). In this process, photons are not simply emitted by one molecule and reabsorbed by another; rather the excitation energy is transferred from one molecule to another by a nonradiative process. A useful analogy for the resonance transfer process is transfer of energy between two tuning forks. If one tuning fork is struck and properly placed near another, the second tuning fork receives some energy from the first and begins to vibrate. The efficiency of energy transfer between the two tuning forks will depend on their distance from each other and relative orientation, as well as their pitches or vibrational frequencies, just as in resonance energy transfer in antenna complexes.

The end result of this process is that approximately 95–99% of the photons absorbed by the antenna pigments have their energy transferred to the reaction center, where it can be used for photochemistry. It is important to distinguish the process of energy transfer among pigments in the antenna from the process of electron transfer that occurs in the reaction center. The former is a purely physical phenomenon, while the latter involves chemical changes in molecules.

The Antenna Funnels Energy to the Reaction Center

The energy absorbed in antenna pigments is funneled toward the reaction center by a sequence of pigments with absorption maxima that are progressively more red-shifted (Fig. 8.21). This red shift in absorption maximum means that the energy of the excited state is somewhat lower nearer the reaction center than in the more peripheral portions of the antenna system. When excita-tation is transferred from, for example, a chlorophyll *b* molecule absorbing maximally at 650 nm to a chlorophyll *a* molecule absorbing maximally at 670 nm, the difference in energy between these two excited chlorophylls is lost to the environment as heat. For the excitation to be trans-ferred back to the chlorophyll *b*, this lost energy would have to be resupplied. The probability of reverse transfer is therefore very small simply because thermal energy is not sufficient to make up the deficit between the lower-energy and higher-energy pigments. This effect gives the energy-trapping process a degree of directionality or irreversibility and makes the delivery of excitation to

the reaction center more efficient. In essence, the system sacrifices some energy from each quantum so that nearly all of the quanta can be trapped by the reaction center.

Carotenoids Serve as Both Accessory Pigments and Photoprotective Agents

Carotenoids are found in all photosynthetic organisms, except for mutants incapable of living outside the labora-tory (Siefermann-Harms, 1985). The different types of carotenoids found in photosynthetic organisms are all linear molecules with multiple conjugated double bonds (Fig 8.5). Absorption bands in the 400–500 nm region give carotenoids their characteristic orange color. In fact, the color of carrots is due to the carotenoid β-carotene, whose structure and absorption spectrum are shown in Figures 8.5 and 8.6.

Carotenoids are usually associated intimately with both antenna and reaction center pigment proteins and are integral constituents of the membrane. The energy of light absorbed by carotenoids is rapidly transferred to chlorophylls, so carotenoids are termed **accessory pig-ments**. However, the efficiency of energy transfer from carotenoid to chlorophyll is usually less than that of energy transfer from chlorophyll to chlorophyll.

Carotenoids also play an essential role in **photopro-**

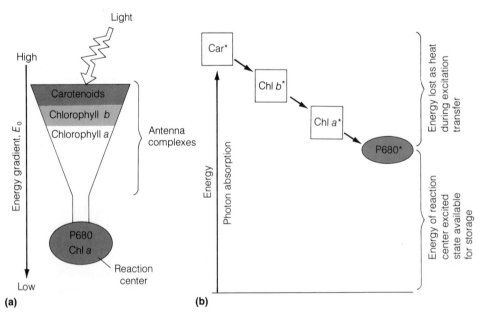

FIGURE 8.21. Funneling of excitations from the antenna system toward the reaction center. (a) The excited-state energy of pigments increases with distance from the reaction center; that is, pigments closer to the reaction center are lower in energy than those farther from the reaction center. This energy gradient ensures that excitation transfer toward the reaction center is energetically favorable and that excitation transfer back out to the peripheral portions of the antenna is energetically unfavorable. (b) Some energy is lost as heat to the environment by this process, but essentially all the excitations absorbed in the antenna complexes are delivered to the reaction center.

tection. The photosynthetic membrane can easily be damaged by the large amounts of energy absorbed by the pigments, if this energy cannot be stored by photochemistry, hence the need for a protection mechanism. The photoprotection mechanism can be thought of as a safety valve, venting excess energy before it can damage the organism. If the excited state of chlorophyll is not rapidly quenched by excitation transfer or photochemistry, a reaction takes place with molecular oxygen to form an excited state of oxygen known as **singlet oxygen**. The extremely reactive singlet oxygen species goes on to react with and damage many cellular components, especially lipids. Carotenoids exert their photoprotective action by rapidly quenching the excited state of chlorophyll. The carotenoid excited state does not have sufficient energy to form singlet oxygen, so it decays back to its ground state while losing its energy as heat.

Mutant organisms lacking carotenoids cannot live in the presence of both light and molecular oxygen, which is a rather difficult situation for an O_2-evolving photosynthetic organism. For non-O_2-evolving photosynthetic bacteria, mutants lacking carotenoids can be maintained under laboratory conditions if oxygen is excluded from the growth medium.

Thylakoid Stacking Is Involved in Energy Partitioning Between the Photosystems

The fact that photosynthesis in higher plants depends on two photochemical systems with different light-absorbing properties poses a special problem. If the rate of delivery of energy to photosystems I and II is not precisely matched and conditions are such that the rate of photosynthesis is limited by the available light (i.e., low light intensity), the rate of electron flow will be limited by the photosystem that is receiving less energy. The most efficient situation would be one in which the input of energy is the same to both photosystems. However, no single arrangement of pigments would satisfy this requirement, because at different times of the day the light intensity and spectral distribution tend to favor one photosystem or the other. The answer to this problem would be a mechanism that shifts energy from one photosystem to the other in response to different conditions, and, in fact, such a regulating mechanism has been shown to operate in different experimental conditions. Indeed, the observation that the overall quantum yield of photosynthesis is nearly independent of wavelength (Fig. 8.9) strongly suggests that such a mechanism exists.

Another benefit of the shifting of energy from one photosystem to the other is that it prevents **photoinhibition**, which occurs when excess excitation arriving at the photosystem II reaction center leads to its rapid inactivation (Powles, 1984). Controlling the distribution

of excitation energy therefore serves as both a protective mechanism and a mechanism for enhancing the efficiency of energy trapping.

Considerable progress has been made in understanding the molecular mechanism responsible for this energy redistribution (Fig. 8.22) (Staehelin and Arntzen, 1983; Allen, 1983). Thylakoid membranes contain a protein kinase that can phosphorylate a specific threonine residue on the surface of one of the membrane-bound antenna pigment proteins. This antenna complex is known as the **chlorophyll *a/b* antenna protein**, or **light harvesting complex II (LHCII)**. When LHCII is not phosphorylated it delivers more energy to photosystem II, and when it is phosphorylated it delivers more energy to photosystem I. The kinase is activated when plastoquinone, one of the electron carriers between the photosystems, accumulates in the reduced state, which occurs when photosystem II is being activated more frequently than photosystem I. The phosphorylated LHCII then migrates out of the stacked regions of the membrane into the unstacked regions, probably because of repulsive interactions with negative charges on adjacent membranes. The lateral migration of LHCII serves to shift the energy balance toward photosystem I, which is located in the stroma lamellae, and away from photosystem II, which is located in the stacked membranes of the grana (Fig. 8.20). This situation is called state 2. If the plastoquinone becomes more oxidized because of excess excitation of photosystem I, the kinase is inhibited and the level of phosphorylation of LHCII is decreased by the action of a membrane-bound phosphatase (Fig. 8.22). LHCII then moves back to the grana and the system is in state 1. The net result is very precise control of the energy distribution between the photosystems, allowing efficient use of the available energy over a wide range of environmental conditions.

Phycobilisomes and Chlorosomes Are Pigment-Protein Complexes That Are Peripherally Associated with the Photosynthetic Membrane

As described above, most antenna pigment proteins are integral membrane proteins, buried within the membrane. There are, however, a few exceptions to this general pattern. Two of the most interesting exceptions are phycobilisomes and chlorosomes, antenna proteins that are attached to the membrane, but not buried within it (Glazer, 1983; Zuber, 1986).

Phycobilisomes are antenna pigment-protein complexes found in cyanobacteria and the eukaryotic red algae (Figs. 8.23 and 8.24). They contain phycobiliproteins, in which **bilin** chromophores are covalently attached to protein. Bilins are open-chain tetrapyrrole pigments, which are chemically akin to a porphyrin that has been split open on one side (Fig. 8.5). A phycobilisome

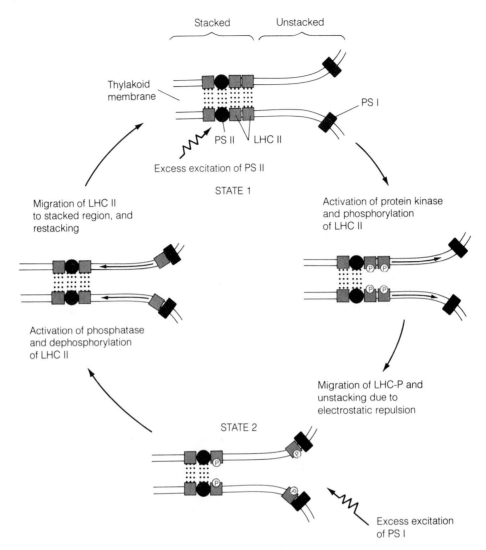

FIGURE 8.22. State transitions control the partitioning of energy between photosystems I and II. If photosystem II is preferentially excited, reduced plastoquinone accumulates. This activates a protein kinase that phosphorylates some of the photosystem II light harvesting antenna complexes (LHCII) that are present in the stacked regions of the membrane. Electrostatic repulsion between adjacent LHCII molecules displaces these negatively charged complexes into the unstacked regions of the membrane where they deliver excitation to photosystem I. This arrangement is called state 2. If PSI receives excess excitation, the plastoquinone becomes oxidized and a phosphatase is activated. This enzyme removes the phosphates from LHCII, which can then migrate back to the stacked region of the membrane and deliver excitations to PSII. This arrangement is called State 1. This mechanism ensures a more effective utilization of the light reaching the leaf.

consists of a number of phycobiliproteins organized into a large complex by interactions of the pigment proteins with one another and also with colorless proteins known as linkers, which help to hold the complex together. Excitation transfer occurs through a series of phycobiliproteins with absorption maxima at progressively longer wavelengths, in agreement with the funnel concept discussed above. The energy is eventually transferred to the membrane and finally into the reaction center. Phycobilisomes appear to transfer energy exclusively to photosystem II. Organisms that contain phycobilisomes also exhibit the state 1–state 2 phenomenon, although the

molecular mechanism of energy distribution is not yet understood in detail.

Chlorosomes are antenna complexes that are found in the non-oxygen-evolving green photosynthetic bacteria. They contain a large number of bacteriochlorophylls *c*, *d*, or *e* (formerly called *Chlorobium* chlorophylls)—up to 1500 per reaction center. The arrangement of the pigments in chlorosomes is not understood in detail. The pigments contain less protein than those in other types of antenna complexes, and their spectroscopic properties suggest that they may exist in an aggregated or oligomeric form.

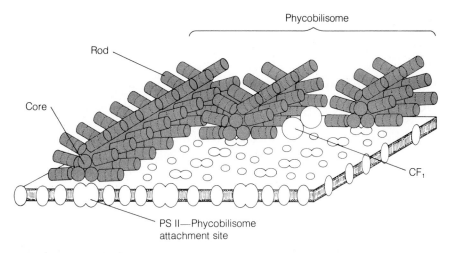

FIGURE 8.23. Structural model of the organization of phycobilisomes at the membrane. Phycobilisomes are found in cyanobacteria and eukaryotic red algae. They contain phycobiliproteins, which consist of bilin chromophores covalently linked to proteins. The biliproteins are organized into rods attached to cores. Excitations absorbed in the rods are funneled to the cores and delivered to photosystem II reaction centers in the membrane.

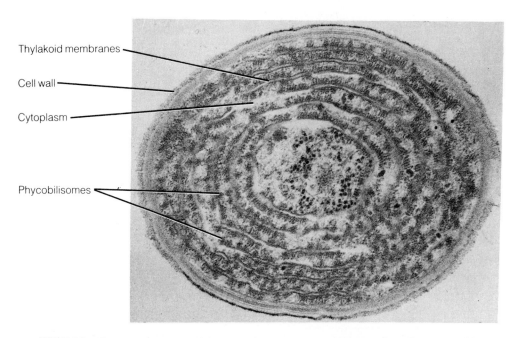

FIGURE 8.24. Electron micrograph of a section through the cyanobacterium *Synechococcus lividus*, showing the concentrically folded thylakoids with phycobilisomes attached to the cytoplasm-facing surfaces of the elongated membrane sacs ($53,500 \times$). (From Edwards et al., 1971. Photograph courtesy of E. Gantt.)

Mechanisms of Electron and Proton Transport

Some of the evidence that led to the idea of two photochemical reactions operating in series has been discussed earlier in this chapter. Here we will consider in detail the chemical reactions involved in electron transfer during photosynthesis (Glazer and Melis, 1987). Figure 8.25 shows a current version of the Z scheme, in which all the electron carriers known to function in electron flow from H_2O to NADP are arranged vertically at their midpoint redox potentials (see box on midpoint potentials and redox reactions). Components known to react with each other are connected by arrows, so the Z scheme is really a synthesis of both kinetic and thermodynamic information. The large vertical arrows represent light energy input into the system.

Midpoint Potentials and Redox Reactions

REDOX REACTIONS, MIDPOINT potentials, and their relationship to the laws of thermodynamics were discussed in Chapter 2. These concepts are useful for our discussion of electron flow from H_2O to NADPH and the interactions between the different electron carriers.

You will recall that the midpoint potential (E_m) is a measure of the tendency of a compound to take electrons from other compounds. A large positive midpoint potential means that the compound is a strong oxidant, while a large negative value means that the compound is a strong reductant (in both cases relative to the standard hydrogen electrode).

Equilibrium constants can easily be predicted from midpoint potentials, in the same way that free energies were related to equilibrium constants (Eq. 2.15). Midpoint potentials for many chemical and biochemical reactions have been measured and tabulated. The y axis on the Z scheme in Figure 8.25 shows midpoint potentials of the electron carriers, with negative values higher than positive ones. This choice makes reactions that are spontaneous (release of free energy) appear "downhill" on the graph.

Knowledge of the midpoint potentials of the various electron carriers is important in establishing the pathway of electron flow in any biochemical electron transport system, such as those found in chloroplasts or mitochondria. This measurement is

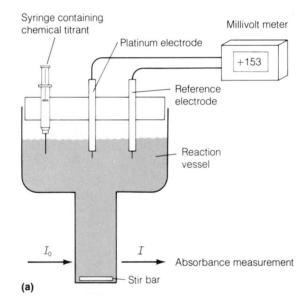

(a)

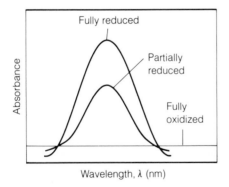

(b) Changes in absorbance as a function of redox potential

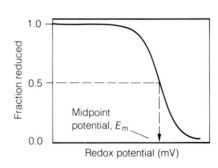

(c) Plot of redox state vs. redox potential

FIGURE 8.G

usually made by carrying out a **redox titration** (Dutton, 1978). The sample is adjusted, or poised, at a particular redox potential, usually by adding small amounts of oxidants or reductants. The redox state of the sample is measured, usually by some type of spectroscopic method. **Redox mediators,** small molecules that permit rapid equilibration between the sample and the electrodes of the measurement system, must be included to ensure that the system is truly at equilibrium when the measurement is made. A large number of measurements are made at a variety of redox potentials. Then the fraction of the carrier that is oxidized (or reduced) is plotted against the redox potential of the medium, and the redox potential at which the sample is half reduced and half oxidized is the midpoint potential.

Redox experiments are straightforward. The sample is stirred in a special cell that contains platinum and reference electrodes, and chemical oxidants and reductants are added to adjust the redox potential, which is read with a voltmeter (Fig. 8.Ga). The extent of the redox reaction is measured by following some property of the sample, usually absorbance, at each potential. In the example illustrated here, the reduced form of the compound has an absorption that decreases as the compound is oxidized (Fig. 8.Gb). The fraction of reduced form at each potential is plotted against redox potential, and the midpoint potential (E_m) is determined as the potential at which the compound is half oxidized and half reduced (Fig. 8.Gc).

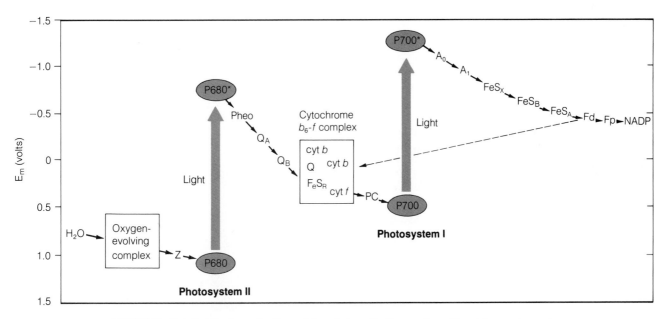

FIGURE 8.25. Detailed Z scheme for O_2-evolving photosynthetic organisms. The redox carriers are placed at their accepted midpoint redox potentials (at pH 7). The vertical arrows signify photon absorption by the reaction center chlorophylls: P680 for PSII and P700 for PSI. The excited PSII reaction center chlorophyll, P680*, transfers an electron to pheophytin (Pheo). On the oxidizing side of PSII (to the left of the arrow joining P680 with P680*), P680$^+$ is rereduced by Z, the immediate donor to PSII. Z is a tyrosine side chain on the reaction center protein D1. Electrons are extracted from water by the oxygen-evolving complex and rereduce Z$^+$. On the reducing side of PSII (to the right of the arrow joining P680 with P680*), the pheophytin transfers electrons to the plastoquinone acceptors Q_A and Q_B. The cytochrome b_6-f complex transfers electrons to plastocyanin (PC), which in turn reduces P700$^+$. The b_6-f complex contains a Rieske iron-sulfur protein (FeS$_R$), two b-type cytochromes (cyt b), and cytochrome f (cyt f). Electron flow within this complex is shown in Figure 8.32. The acceptor of electrons from P700* (A_0) is thought to be a chlorophyll, and the next acceptor (A_1) may be a quinone. A series of membrane-bound iron-sulfur proteins (FeS$_X$, FeS$_B$, and FeS$_A$) transfer electrons to soluble ferredoxin (Fd). The flavoprotein ferredoxin-NADP reductase (Fp) serves to reduce NADP, which is used in the Calvin cycle to reduce CO_2 (Chapter 9). The dashed line indicates cyclic electron flow around PSI. (Adapted from Blankenship and Prince, 1985.)

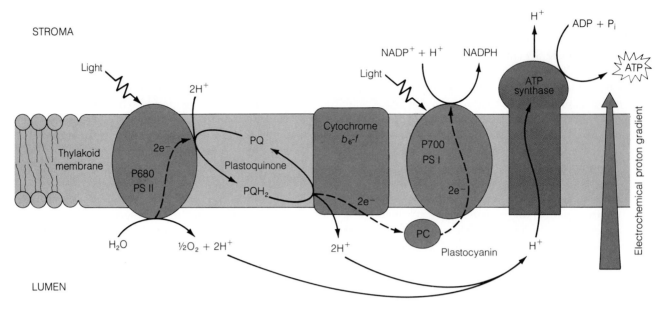

FIGURE 8.26. Electron and proton transfer in the thylakoid membrane is carried out vectorially by four protein complexes. Water oxidation and proton release takes place in the lumen by PSII. Photosystem I reduces NADP$^+$ to NADPH in the stroma. Protons are also transported into the lumen by the action of the cytochrome b_6-f complex and contribute to the electrochemical proton gradient. These protons must then diffuse to the ATP synthase enzyme, where their diffusion down the electrochemical energy gradient is used to synthesize ATP in the stroma. Dashed lines represent electron transfer; solid lines represent proton movement.

Electron and Proton Transport Is Carried Out by Four Thylakoid Protein Complexes

Almost all the chemical processes that make up the light reactions of photosynthesis are carried out by four major protein complexes: photosystem II, the cytochrome b_6-f complex, photosystem I, and the ATP synthase. These integral membrane complexes are vectorially oriented in the thylakoid membrane so that water is oxidized to O_2 in the thylakoid lumen, NADP$^+$ is reduced to NADPH on the stromal side of the membrane, and ATP is released into the stroma by H$^+$ moving from the lumen to the stroma (Fig. 8.26).

Energy Storage Takes Place When an Excited Chlorophyll Reduces an Electron Acceptor Molecule

As discussed earlier in the chapter, the function of light is to excite a specialized chlorophyll present in the reaction center from the ground electronic state to an excited electronic state, either by direct absorption or, more likely, via energy transfer from an antenna pigment. This excitation process can be envisioned as the promotion of an electron from the highest-energy filled orbital to the lowest-energy unfilled orbital (Fig. 8.27). The electron in the upper orbital is only loosely bound to the chlorophyll

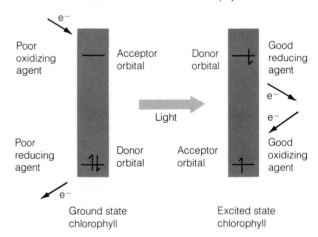

Redox properties of ground and excited states of reaction center chlorophyll

FIGURE 8.27. Orbital occupation diagram for the ground and excited states of reaction center chlorophyll. In the ground state the molecule is a poor reducing agent (loses electrons from a low-energy orbital) and a poor oxidizing agent (accepts electrons only into a high-energy orbital). In the excited state the situation is reversed, and an electron can be lost from the high-energy orbital, making the molecule an extremely powerful reducing agent. This is the reason for the extremely negative excited state redox potential shown by P680* and P700* in Figure 8.25. The excited state can also act as a strong oxidant by accepting an electron into the lower-energy orbital, although this is not a significant pathway in reaction centers. (Adapted from Blankenship and Prince, 1985.)

and is easily lost if a molecule that can accept the electron is nearby. The excited state of the reaction center chlorophyll is therefore an extremely strong reducing agent, strong enough to reduce molecules that do not readily accept an electron (Blankenship and Prince, 1985). It is not possible to carry out a standard redox titration (see box on midpoint potentials) on the excited chlorophyll because of its short lifetime. However, values for the redox potential of the excited state can be estimated from the energy of excitation and the redox potential of the ground state (Fig. 8.25).

The first reaction that converts electronic energy into chemical energy—that is, the primary photochemical event—is the transfer of an electron from the excited state of a chlorophyll in the reaction center to an acceptor molecule. The energy of light is used to convert the chlorophyll from a weak reducing agent to a very strong one. An equivalent way to view this process is that the absorbed photon causes an intramolecular electron rearrangement in the reaction center chlorophyll, followed by an intermolecular electron transfer process in which part of the energy in the photon is captured in the form of redox energy.

Immediately after the photochemical event, the reaction center chlorophyll reverts to an oxidized state and the nearby electron acceptor molecule is reduced. The system is now poised at a critical juncture. The lower-energy orbital of the positively charged oxidized reaction center chlorophyll shown in Figure 8.27 has a vacancy and can accept an electron. If the acceptor molecule donates its electron back to the reaction center chlorophyll, the system will be returned to the same state that existed prior to the light excitation, and all the absorbed energy will be converted into heat. However, this wasteful reversal process does not appear to occur to any substantial degree in functioning reaction centers. Instead, the acceptor transfers its extra electron to a secondary acceptor, and so on down the electron transport chain. The oxidized reaction center chlorophyll that had donated an electron is rereduced by a secondary donor, which is in turn reduced by a tertiary donor. The ultimate electron donor is H_2O, and the ultimate electron acceptor is $NADP^+$ (see Figs. 8.12 and 8.25).

The essence of photosynthetic energy storage is thus the transfer of an electron from an excited chlorophyll to an acceptor molecule, followed by separation of the positive and negative charges by a very rapid series of chemical reactions (Kirmaier and Holten, 1987). These secondary reactions separate the charges to opposite sides of the thylakoid membrane in approximately 200 picoseconds. With the charges thus separated, the reversal reaction is many orders of magnitude slower, and the energy has effectively been captured.

The quantum yield for production of stable products in purified reaction centers from photosynthetic bacteria has been measured as 1.0; that is, every photon produces stable products and essentially no reversal reactions occur. Although these types of measurements have not been made on purified reaction centers from higher plants, the measured quantum requirements for O_2 production under optimal conditions (low-intensity light) indicate that the values for the primary photochemical events are very close to 1.0. This follows from the fact that four electrons must be transferred to form one oxygen molecule (see next section) and that the two photosystems must be activated for the overall process to be completed. The theoretical minimum quantum requirement is therefore 8, compared to the best measured values of 9–10 (Fig. 8.8). Of course, this represents the maximum attainable efficiency. Under less than ideal conditions, such as high light intensity, the actual quantum requirement is much higher.

The structure of the reaction center appears to be extremely fine-tuned for maximal rates of productive reactions and minimal rates of energy-wasting reactions. Exactly how this is accomplished is not well understood and is a very active research area. The x-ray crystal structure of the bacterial reaction center described earlier is an important advance toward understanding photosynthetic energy conversion. Knowledge of the structure of a biological system is an essential step toward understanding its function.

The Reaction Center Chlorophylls of Photosystems I and II Have Maximal Absorbances at 700 and 680 nm, Respectively

The reaction center chlorophyll is transiently in an oxidized state after losing an electron and before it is rereduced by its electron donor. In the oxidized state, the strong light absorbance in the red region of the spectrum that is characteristic of chlorophylls is lost, or **bleached**. It is therefore possible to monitor the redox state of these chlorophylls by time-resolved optical absorbance measurements in which this bleaching is monitored directly (see box on principles of spectrophotometry). Using such techniques, Bessel Kok found that the reaction center chlorophyll of photosystem I absorbs maximally at 700 nm in its reduced state. Accordingly, it is named **P700** (the P stands for pigment). H. T. Witt and co-workers found the analogous optical transient of photosystem II at 680 nm, so its reaction center chlorophyll is known as **P680**. Earlier, Louis Duysens had identified the reaction center bacteriochlorophyll from purple photosynthetic bacteria as P870. The x-ray structure of the bacterial reaction center described earlier clearly indicates that P870 is a closely coupled pair or dimer of bacteriochlorophylls, rather than a single molecule. Most evidence suggests that P700 is also a dimer. On the other hand, the aggregation state of P680 is yet to be characterized.

In the oxidized state, the reaction center chlorophyll(s) contains an unpaired electron. Molecules with unpaired electrons often can be detected by a magnetic resonance technique known as **electron spin resonance (ESR)**. ESR studies, along with the spectroscopic measurements described above, have led to the discovery of a large number of intermediate electron carriers in the photosynthetic electron transport system.

The Photosystem II Reaction Center Complex Oxidizes Water and Reduces Plastoquinone

Photosystem II is contained in a multisubunit protein complex (Fig. 8.28). The core of the reaction center is made of two membrane proteins with molecular masses of 32 and 34 kilodaltons (kDa) known as D1 and D2. P680, pheophytin, and plastoquinone are bound to the membrane proteins. D1 and D2 have some sequence similarity to the L and M peptides of purple bacteria. The current understanding of the structure and function of photosystem II owes much to the detailed information available on the bacterial system (Michel and Deisenhofer, 1988). Other proteins serve as antenna complexes (43 and 47 kDa) or are involved in oxygen evolution (33, 23, and 18 kDa). Some, such as cytochrome *b*-559, have no known function but appear to be important because no stable photosystem II complex is assembled if the genes that code for the polypeptides are deleted.

Water Is Oxidized to Oxygen by Photosystem II

The chemical reaction by which water is oxidized is given by Equation 8.9 (Babcock, 1987; Rutherford, 1989; Govindjee and Coleman, 1990).

$$2H_2O \rightarrow O_2 + 4H^+ + 4e^- \tag{8.9}$$

This equation indicates that four electrons are removed from two water molecules, generating an oxygen molecule and four hydrogen ions. (For an introduction to oxidation-reduction reactions, see Chapter 2 and box on midpoint potentials and redox reactions.) Little is known about the actual chemical mechanism of photosynthetic water oxidation, although there is a great deal of indirect evidence about the process. If a sample of dark-adapted photosynthetic membranes is exposed to a sequence of very brief intense flashes, a characteristic pattern of oxygen production is observed (Fig. 8.29a): little or no oxygen is produced on the first two flashes, and maximal oxygen is released on the third flash and every fourth flash thereafter, until eventually the yield per flash damps to a constant value. This remarkable result was first observed by Pierre Joliot in France in the 1960s. A model (Fig. 8.29b) explaining these observations, proposed by Kok and co-workers, has been widely accepted (Kok et al., 1970). The model consists of a series of five states, known as S_0 to S_4, which represent successively more oxidized forms of the water-oxidizing enzyme system, or

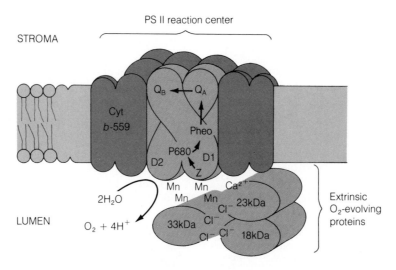

FIGURE 8.28. Schematic model of the PSII reaction center and the oxygen-evolving complex. P680 is associated with the D1 and D2 proteins, which are analogous to the L and M proteins of the bacterial reaction center. Electrons are transferred from P680 to pheophytin (Pheo) and then to Q_A and Q_B. P680$^+$ is rereduced by Z, a tyrosine in the D1 protein. The oxygen evolution complex involves Mn, Ca^{2+}, and Cl^- as cofactors. Three extrinsic polypeptides are involved in regulation of oxygen evolution. Cytochrome *b*-559 is intimately associated with the PSII reaction center, although its function is unknown. (Adapted from Murata and Miyao, 1985, and Barber, 1987.)

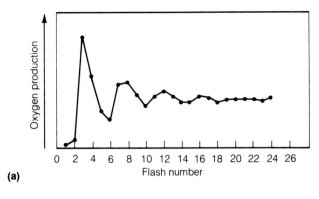

(a)

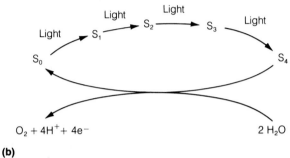

(b)

FIGURE 8.29. The pattern of oxygen evolution in flashing light and the S state mechanism for oxygen evolution. (a) A series of short flashes is given to a sample of dark-adapted chloroplasts. The maximum production of oxygen occurs on the third flash, secondary maxima occur on every fourth flash, and the yield damps to a steady-state value after about 20 flashes. The Kok S state model (b) was advanced to explain the observations of part (a). The oxygen-evolving system can exist in five states, S_0 to S_4. Successive photons advance the system from S_0 to S_1, S_1 to S_2, etc. until state S_4 is reached. S_4 is unstable and reacts with two water molecules to produce O_2.

oxygen-evolving complex. The sequential flashes of light advance the system from one S state to the next, until state S_4 is reached. State S_4 produces O_2 without further light input and returns the system to S_0. Occasionally, a center does not advance to the next S state upon flash excitation, and less frequently, a center is activated twice by a single flash. These "misses" and "double hits" cause the synchrony achieved by dark adaptation to be lost and the oxygen yield to damp to a constant value. In this scheme, a complex has the same probability of being in any of the states S_0 to S_3 (S_4 is unstable and occurs only transiently). States S_2 and S_3 decay in the dark, but only as far back as S_1, which is assumed to be stable in the dark. Therefore, after dark adaptation approximately three-fourths of the oxygen-evolving complexes appear to be in state S_1 and one-fourth in state S_0. Hence, the maximum yield of O_2 is observed after the third of a series of flashes given to dark-adapted chloroplasts.

This **S state mechanism** explains the observed pattern of O_2 release, but not the chemical nature of the S states or whether any partially oxidized intermediates, such as H_2O_2, are formed. Additional information was obtained by measuring the pattern of proton release (see Eq. 8.9) as the S states are advanced with flashes. For the advancement from S_0 to S_1, S_1 to S_2, etc., the numbers of protons released were found to be 1, 0, 1, and 2. Clearly, the release of protons is not fully coupled to O_2 release. There is some evidence that the protons released before O_2 is produced do not come directly from partially oxidized water but may come from ionizable protein groups such as certain amino acids.

The protons produced by water oxidation are released into the lumen of the thylakoid, not directly into the stromal compartment (Figs. 8.15 and 8.26). They are released into the lumen because of the vectorial nature of the membrane and the fact that the oxygen-evolving complex is localized on the interior surface of the thylakoid. These protons are eventually released from the lumen to the stroma through the process of ATP synthesis. In this way, the electrochemical potential formed by the release of protons during water oxidation contributes to ATP formation.

It has been known for many years that manganese (Mn) is an essential cofactor in the water-oxidizing process (Chapter 5), and a classical hypothesis in photosynthesis research postulates that the S states represent successively oxidized states of an Mn-containing enzyme. This hypothesis has received strong support from a variety of experiments, most notably ESR studies in which the S_2 state has been linked to an ESR spectrum characteristic of Mn (Dismukes, 1986). Analytical experiments indicate that four Mn atoms are associated with each oxygen-evolving complex. Other experiments have shown that Cl^- and Ca^{2+} ions are essential for O_2 evolution, although few details are known about their role.

Considerable progress has been made on the structural aspects of the oxygen-evolving system. Several proteins, including three peripheral membrane polypeptides of molecular mass 18, 23, and 33 kDa, have been implicated in the water-oxidizing process and its control. A tentative model of these proteins and their relation to the thylakoid membrane and the PSII reaction center is shown in Figure 8.28 (Murata and Miyao, 1985).

There is at least one electron carrier, generally identified by the letter Z, that functions between the oxygen-evolving complex and P680. To function in this region, it must have an extremely positive midpoint redox potential, that is, a very strong tendency to retain its electrons. This species has recently been identified as a radical formed from a tyrosine residue in the PSII reaction center proteins.

The Acceptor Region of Photosystem II Contains Pheophytin and Two Quinones as Electron Carriers

Spectral and ESR evidence indicates that pheophytin acts as an early acceptor in photosystem II, followed by a complex of two plastoquinones in close proximity to an iron atom. **Pheophytin** is a chlorophyll in which the central Mg atom has been replaced by two H atoms. This chemical change gives the pheophytin chemical and spectral properties slightly different from those of chlorophyll. The precise arrangement of the carriers in the electron acceptor complex is not yet known, but is probably very similar to that of the reaction center of purple bacteria (Fig. 8.19).

The plastoquinones bound to the reaction center operate as a **two-electron gate**. An electron is transferred from pheophytin to the first quinone of the complex and is then transferred rapidly to the second quinone, where it remains. During this time, oxidized P680 regains an electron from Z (see Fig. 8.25), thereby returning to a photochemically competent state. A second electron is transferred from pheophytin to the first quinone and then to the second quinone. Two protons are picked up from the surrounding medium, producing a fully reduced hydroquinone (Fig. 8.30). The hydroquinone then dissociates from the complex and enters the hydrocarbon portion of the membrane, where it in turn transfers its electrons to the cytochrome b_6-f complex.

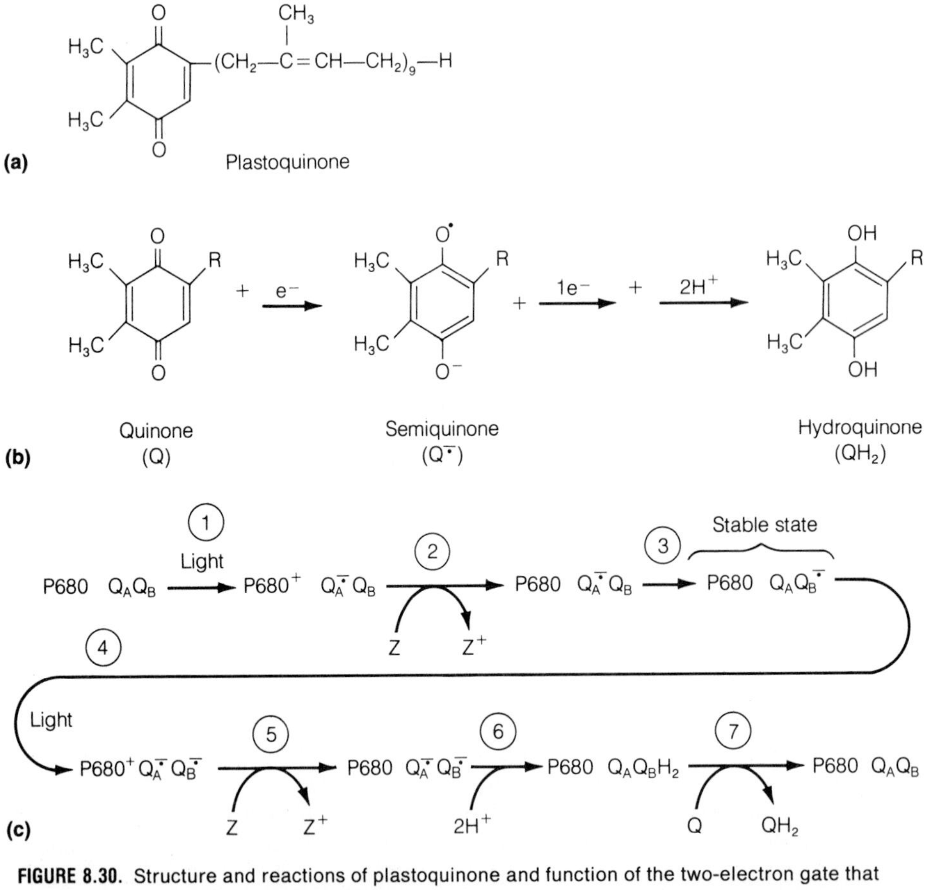

FIGURE 8.30. Structure and reactions of plastoquinone and function of the two-electron gate that operates in photosystem II. (a) The plastoquinone consists of a quinoid head and a long nonpolar tail that anchors it in the membrane. (b) Redox reactions of plastoquinone. The fully oxidized quinone (Q), anionic semiquinone (Q$^-$), and reduced hydroquinone (QH$_2$) forms are shown; R represents the side chain. (c) Function of the two-electron gate of photosystem II. The complex between the reaction center and plastoquinones (Q$_A$ and Q$_B$) couples the one-electron photochemistry of the reaction center to the two-electron chemistry of quinones. Reaction 1: Photon absorption induces electron transfer from P680 to Q$_A$ (intermediate reactions are omitted for simplicity). Reaction 2: P680$^+$ is rereduced by Z, the electron donor to PSII. Reaction 3: Q$_A$ transfers an electron to Q$_B$, producing a stable state that lasts until another photon excites P680. Reaction 4: Photon absorption induces a second electron transfer event from P680 to Q$_A$. Reaction 5: Z rereduces P680$^+$. Reaction 6: Q$_A$ transfers an electron to Q$_B$, which simultaneously binds two protons, forming Q$_B$H$_2$. Reaction 7: Q$_B$H$_2$ is released from the PSII reaction center and is replaced by an oxidized quinone, which becomes the new Q$_B$.

Electron Flow Through the Cytochrome b_6-f Complex Results in Proton Accumulation in the Thylakoid Lumen

The **cytochrome b_6-f complex** is a large multisubunit protein with several prosthetic groups (Hauska et al., 1983; O'Keefe, 1988). It contains two b-type hemes and one c-type heme (cytochrome f). In c-type cytochromes the heme is covalently attached to the peptide, whereas in b-type cytochromes the chemically similar protoheme group is not covalently attached (Fig. 8.31). In addition, the complex contains a **Rieske iron-sulfur protein** (named for the scientist who discovered it) in which two iron atoms are bridged by two sulfur atoms.

The mechanism of electron and proton flow through the cytochrome b_6-f complex is not yet fully understood, but several possible mechanisms have been postulated. In one of them, known as the **Q cycle**, the oxidized Rieske protein (FeS$_R$) accepts an electron from reduced plastoquinone (QH$_2$) and transfers it to cytochrome f (cyt f) (Fig. 8.32). Cytochrome f then transfers an electron to the blue-colored copper protein **plastocyanin** (PC), which in in turn reduces oxidized P700 of PSI. Simultaneously with the electron transfer to FeS$_R$, the plastoquinone transfers its other electron to one of the b-type hemes (cyt b), releasing its protons to the lumenal side of the membrane. The b-type heme transfers its electron through the second b-type heme to an oxidized plastoquinone molecule, reducing it to the semiquinone form. Another similar sequence of electron flow fully reduces the plastoquinone, which picks up protons from the stromal side of the membrane and becomes plastohydroquinone. The net result is that two electrons are transferred to P700, two plastohydroquinones are oxidized to the quinone form, and one oxidized plastoquinone is reduced to the hydroquinone form. In addition, four protons are transferred

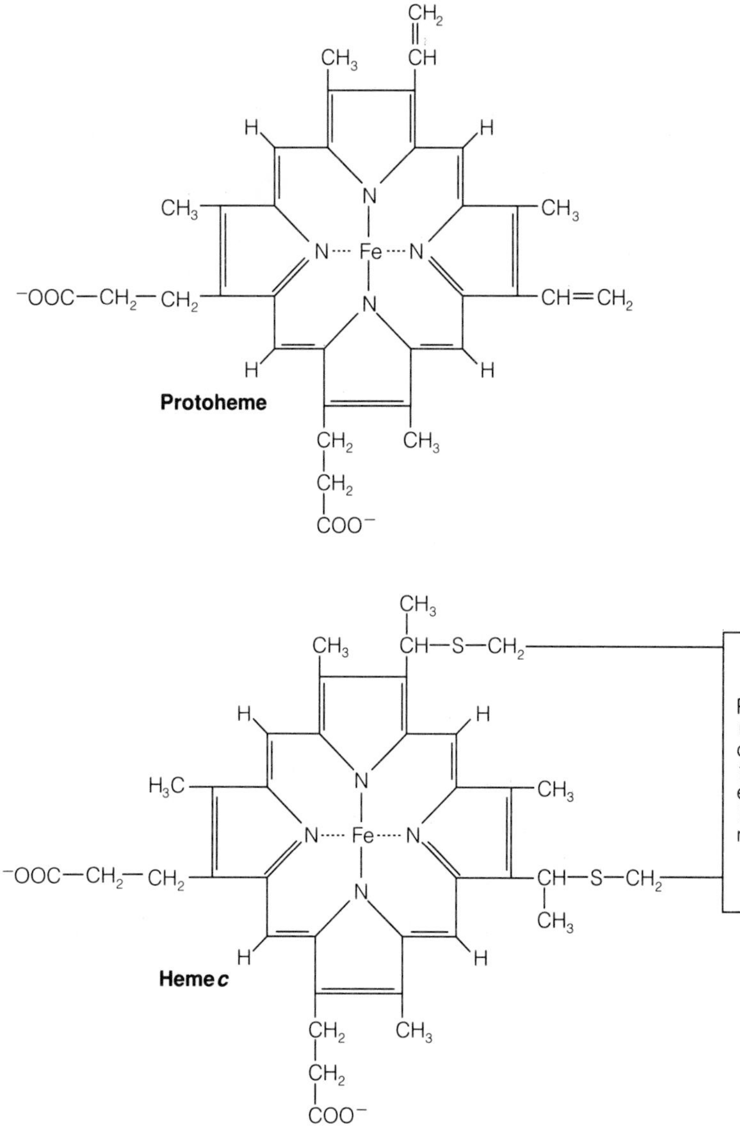

FIGURE 8.31. Structure of prosthetic groups of b- and c-type cytochromes. The protoheme group (also called protoporphyrin IX) is found in b-type cytochromes, the heme c group in c-type cytochromes. The heme c group is covalently attached to the protein by thioether linkages with two cysteine residues in the protein, while the protoheme group is not covalently attached to the protein. The Fe atom is in the 2+ oxidation state in reduced cytochromes and in the 3+ oxidation state in oxidized cytochromes.

(a) First QH$_2$ oxidized

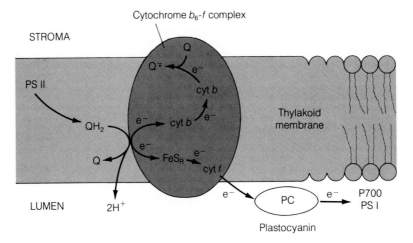

(b) Second QH$_2$ oxidized

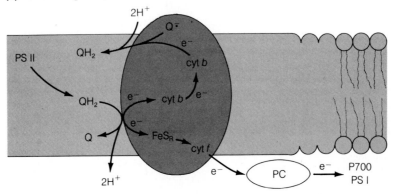

FIGURE 8.32. Mechanism of electron and proton transfer in the cytochrome b_6-f complex. This complex contains two b-type cytochromes, a c-type cytochrome (historically called cytochrome f), a Rieske Fe-S protein (FeS$_R$), and two quinone oxidation-reduction sites. (a) A QH$_2$ molecule produced by the action of PSII (see Fig. 8.30) is oxidized near the lumenal side of the complex, transferring its two electrons to the Rieske Fe-S protein and one of the b-type cytochromes and simultaneously expelling two protons to the lumen. The electron transferred to FeS$_R$ is passed to cyt f and then to plastocyanin (PC), which reduces P700 of PSI. The reduced b-type cytochrome transfers an electron to the other b-type cytochrome, which reduces a quinone to the semiquinone state (Fig. 8.30). (b) A second QH$_2$ is oxidized, with one electron going from FeS$_R$ to PC and finally to P700. The second electron goes through the two b-type cytochromes and reduces the semiquinone to the hydroquinone, at the same time picking up two protons from the stroma. Overall, four protons are transported across the membrane for every two electrons delivered to P700.

from the stromal to the lumenal side of the membrane. In this way, electron flow connecting the acceptor side of the photosystem II reaction center to the donor side of the photosystem I reaction center also gives rise to an electrochemical potential across the membrane, due in part to H$^+$ concentration differences on the two sides of the membrane. This electrochemical potential is used to power the synthesis of ATP. The cyclic electron flow through the cyt b and plastoquinone increases the number of protons pumped per electron beyond what could be achieved in a strictly linear sequence.

Many types of bacteria as well as mitochondria contain a **cytochrome b-c_1 complex** similar to the cyt b_6-f

complex (Prince, 1985). The reaction sequence described above is fairly well established in these b-c_1 complexes, but the precise sequence in the cytochrome b_6-f complex in chloroplasts remains to be characterized.

Plastoquinone and Plastocyanin Are Candidates for Diffusible Intermediates

The lateral heterogeneity of the photosynthetic membrane (Fig. 8.20) requires that at least one component be capable of moving along or within the membrane, in order to deliver electrons produced by photosystem II to photosystem I. The cytochrome b_6-f complex is equally distributed between the grana and stroma regions of the membrane. However, its large size makes it unlikely that it is the mobile carrier. Instead, plastoquinone or plastocyanin or possibly both are thought to serve as mobile carriers to connect the two photosystems.

Plastocyanin is a small (10.5 kDa) copper-containing protein that transfers electrons between the cyt b_6-f complex and P700. This protein is found in the lumenal space. In certain green algae and cyanobacteria, a c-type cytochrome is sometimes found instead of plastocyanin; which protein is synthesized depends on the amount of copper available to the organism.

Some cytochrome b_6-f complex is found in the stroma region of the membrane, where photosystem I is located. Under certain conditions **cyclic electron flow** from the reducing side of photosystem I, through the b_6-f complex and back to P700, is known to occur. This cyclic electron flow pumps protons through the action of the cytochrome b_6-f complex but does not serve to oxidize water or reduce NADP$^+$. In fact, cyclic electron flow appears to play a minor role in the overall pattern of photosynthetic electron flow.

The Photosystem I Reaction Center Reduces NADP$^+$

The photosystem I reaction center is contained in a multiprotein complex. P700 and about 140 core antenna chlorophylls are bound to at least two peptides with molecular masses in the 66–70 kDa range (Golbeck, 1987; Knaff, 1988). Photosystem I reaction center complexes have been isolated from several organisms and found to contain the 66–70 kDa peptides, along with a variable number of smaller peptides in the 4–25 kDa range. Some of these proteins serve as binding sites for the soluble carriers plastocyanin and ferredoxin. The functions of these other peptides are not well understood. The 8-kDa protein contains some of the bound iron-sulfur centers that serve as early electron acceptors in photosystem I.

In their reduced form, the electron carriers that function in the acceptor region of photosystem I are all extremely strong reducing agents. These reduced forms are very unstable and thus difficult to identify. Recent evidence indicates that one of these early acceptors is a chlorophyll molecule, and another appears to be a quinone species.

Additional electron acceptors include a series of three membrane-associated iron-sulfur proteins, or **bound ferredoxins**, also known as **Fe-S centers X, A, and B**. Fe-S center X is thought to be part of the P700 binding protein, while centers A and B reside on the 8-kDa protein that is part of the PSI reaction center complex. Electrons are transferred through centers A and B to ferredoxin, a water-soluble iron-sulfur protein. The soluble flavoprotein **ferredoxin-NADP reductase** serves to reduce NADP$^+$ to NADPH (Fig. 8.33), thus completing the sequence of noncyclic electron transport that begins with the oxidation of water.

The photosystem I reaction center appears to have some functional similarity to the reaction center found in the anaerobic green sulfur bacteria. These bacteria contain low-potential Fe-S centers as early electron acceptors and are capable of NAD reduction similar to the NADP$^+$ reduction function of photosystem I.

Some Herbicides Kill Plants by Blocking Photosynthetic Electron Flow

A major class of herbicides—about half of the commercially important compounds—act by interrupting photosynthetic electron flow (Ashton and Crafts, 1981). The chemical structure of two of these compounds are shown in Figure 8.34a. The precise sites of action of many of these agents have been found to lie either at the reducing side of photosystem I (for example, paraquat) or in the quinone acceptor complex in the electron transport chain between the two photosystems (for example, DCMU) (Fig. 8.34a).

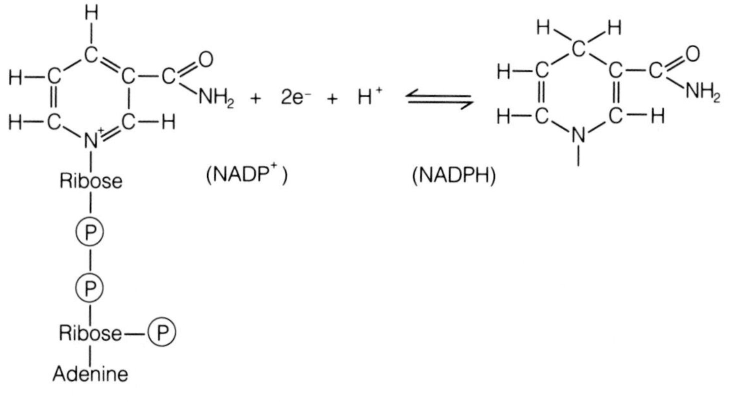

FIGURE 8.33. The reduction of NADP$^+$ to NADPH is the last step in the series of redox reactions that starts with the oxidation of water. Photosystem I reduces NADP$^+$ to NADPH by addition of two electrons and a proton to the nicotinamide portion of the molecule.

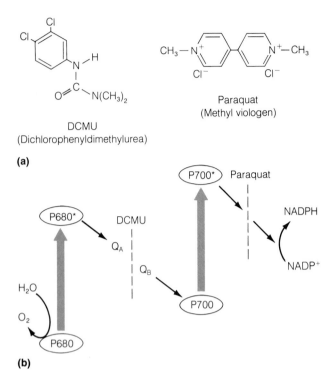

FIGURE 8.34. The use of herbicides to kill unwanted plants is widespread in modern agriculture. Many different classes of herbicides have been developed, and they act by blocking amino acid, carotenoid, or lipid biosynthesis or by disrupting cell division. Understanding the mode of action of herbicides has been an important tool in research on plant metabolism and has facilitated their application under different agricultural practices (Ashton and Crafts, 1981). (a) Chemical structures of two herbicides that block photosynthetic electron flow. DCMU is also known as diuron. Paraquat has acquired public notoriety because of its use on marijuana crops. (b) Sites of action of herbicides. Many herbicides, such as DCMU, act by blocking electron flow at the quinone acceptors of photosystem II, by competing for the binding site of plastoquinone normally occupied by Q_B. Other herbicides, such as paraquat, act by accepting electrons from the early acceptors of photosystem I and then reacting with oxygen to form superoxide, O_2^-, a species that is very damaging to chloroplast components, especially lipids.

Paraquat acts by intercepting electrons between the bound ferredoxin acceptors and NADP and then reducing oxygen to **superoxide** (O_2^-). Superoxide is a free radical that reacts nonspecifically with a wide range of molecules in the chloroplast, leading to rapid loss of chloroplast activity. Lipid molecules in cell membranes are especially sensitive.

The herbicides that act on the quinone acceptor complex compete with plastoquinone for the Q_B binding site. If herbicide is present, it displaces the oxidized form of plastoquinone and occupies the specific binding site for the quinone acceptor, which is thought to lie on the D1 herbicide-binding protein. The herbicide is not able to accept electrons, so the electron is unable to leave Q_A, the first quinone acceptor. Thus, the herbicide binding effectively blocks electron flow and inhibits photosyn-

thesis. Many herbicides that act in this manner also inhibit electron flow in photosynthetic bacteria that have quinone-type electron acceptor complexes (Trebst, 1986).

In recent years, herbicide-resistant biotypes of common weeds have appeared in areas where a single type of herbicide has been used continuously for several years. These biotypes can be orders of magnitude more resistant to certain classes of herbicides than is the nonresistant plant. In several cases the resistance factor has been traced to a single amino acid substitution in the D1 protein. This change, presumably in the quinone (and herbicide) binding region of the peptide, lowers the binding affinity of the herbicide, making it much less effective.

The possibility of fine-tuning the herbicide sensitivity of crop plants by making subtle changes in the proteins of the photosystem II reaction center has created a great deal of interest in the agricultural chemical industry. Through biotechnology, it is now possible to make a crop plant resistant to a particular herbicide, which could then be applied to control undesirable plants that are not resistant. The success of this approach will depend on whether undesirable side effects of the herbicide-resistant mutations can be controlled and how rapidly the weeds acquire resistance through natural selection. Thus, it is a race between science and evolutionary change (Dover and Croft, 1986).

A Chemiosmotic Mechanism Converts the Energy Stored in Chemical and Membrane Potentials to ATP

In addition to the energy stored as redox equivalents (NADPH) by the light reactions, a portion of the photon's energy is captured as the high-energy phosphate bond in ATP (Arnon, 1984). Electron flow must occur for this **photophosphorylation** to take place, although under some conditions electron flow may take place without accompanying phosphorylation. In the latter case, electron flow is said to be uncoupled from phosphorylation. It is now widely accepted that photophosphorylation works via the **chemiosmotic** mechanism, first proposed in the 1960s by Peter Mitchell (Mitchell, 1979). The same general mechanism is applicable to phosphorylation during aerobic respiration in bacteria and mitochondria (Chapter 11) and also underlies many ion and metabolite transfers across membranes (Chapter 6). The chemiosmotic mechanism appears to be a unifying aspect of membrane processes in all forms of life.

The basic principle of chemiosmosis is that ion concentration differences and electrical potential differences across membranes are a source of free energy that can be utilized by the cell. As described by the second law of thermodynamics (Chapter 2), any nonuniform distribution of matter or energy represents a source of energy. **Chemical potential** differences of any molecular

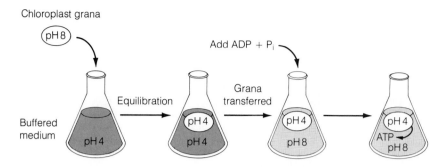

FIGURE 8.35. Summary of the experiment carried out by Jagendorf and co-workers. Isolated chloroplast grana previously at pH 8 were transferred to an acid medium at pH 4 and equilibrated. The grana were then transferred to a medium of pH 8, and ADP and P_i were added shortly afterward. This manipulation generates a pH gradient (pH 4 inside the grana and pH 8 outside) that in the intact cell is produced by electron flow and proton translocation. In these conditions, ATP was synthesized without any need for light. This experiment provided evidence in favor of Mitchell's chemiosmotic theory, which stated that pH gradients and electrical potentials across a membrane are a source of energy that can be converted into the high-energy phosphate bond of ATP.

species whose concentrations are not the same on opposite sides of a membrane provide such a source of energy.

The asymmetric nature of the photosynthetic membrane and the fact that proton flow from one side of the membrane to the other accompanies electron flow were discussed earlier. The direction of proton translocation is such that the stroma becomes more alkaline and the lumen becomes more acidic when electron transport occurs (Figs. 8.26 and 8.32).

Some of the early evidence in support of a chemiosmotic mechanism driving photophosphorylation was obtained in an elegant experiment carried out by Andre Jagendorf and co-workers (Fig. 8.35) (Jagendorf, 1967). They suspended chloroplasts in a pH 4 buffer, which permeated the membrane and caused the interior, as well as the exterior, of the thylakoid to equilibrate at this acidic pH. They then rapidly injected the suspension into a pH 8 buffer solution, thereby creating a pH difference of 4 units across the thylakoid membrane, with the inside acidic relative to the outside. They found that large amounts of ATP were formed from ADP and P_i by this process, without any light input or electron transport. This result supports the predictions of the chemiosmotic hypothesis.

The ATP is synthesized by a large (400-kDa) enzyme complex known as the **coupling factor**, **ATPase** (after the reverse reaction of ATP hydrolysis), **ATP synthase**, or **CF_0-CF_1** (McCarty, 1985; McCarty and Hammes, 1987). This enzyme consists of two parts: a hydrophobic membrane-bound portion called CF_0 and a portion that sticks out into the stroma called CF_1 (see Fig. 8.36). CF_0 appears to be a channel across the membrane through which protons can pass, while CF_1 is the portion of the complex that actually synthesizes ATP.

In Chapter 6, we discussed ATPases in the plasma membrane and their role in chemiosmosis and ion trans-

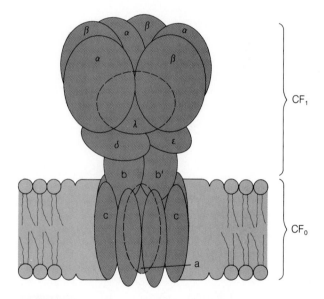

FIGURE 8.36. Schematic model of the ATP synthase, or coupling factor enzyme. This enzyme consists of a large multisubunit complex, CF_1, attached on the stromal side of the membrane to an integral membrane portion, known as CF_0. Protons enter the channel through the membrane formed by CF_0 and are expelled by CF_1. The catalytic sites for ATP synthesis are found on CF_1. CF_1 consists of five different polypeptides, with a stoichiometry of α_3, β_3, γ, δ, ε. CF_0 contains probably four different polypeptides, with a stoichiometry of a, b, b', c_{10}. The ATP synthase has a fundamental role in the harnessing of light energy and its storage in the high-energy chemical bond of ATP. (Adapted from Curtis, 1988.)

port at the cellular level. In that context, the ATPase utilizes ATP made available by photophosphorylation in the chloroplast and oxidative phosphorylation in the mitochondrion (Fig. 8.37). Here we are concerned with chemiosmosis and proton gradients used to make ATP in the chloroplast.

(a) Purple bacteria

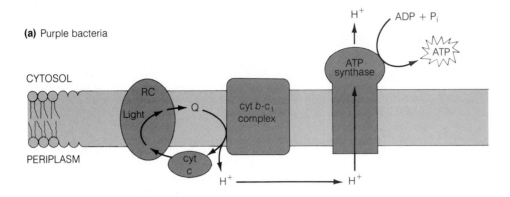

(b) Chloroplasts

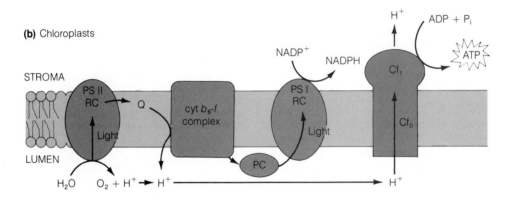

(c) Mitochondria

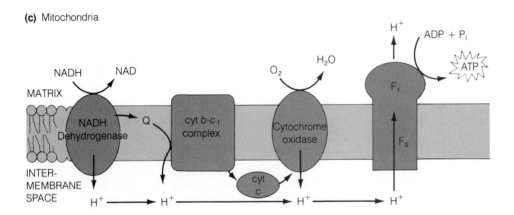

FIGURE 8.37. Similarities of photosynthetic and respiratory electron flow. In all three types of energy-conserving systems, electron flow is coupled to proton translocation, creating a transmembrane proton motive force (Δp). The energy in the proton motive force is then used for synthesis of ATP by the coupling factor enzyme complex (CF_0-CF_1), also called ATP synthase. (a) Purple photosynthetic bacteria carry out cyclic electron flow, generating a proton gradient by the action of the cytochrome b-c_1 complex. (b) Chloroplasts carry out noncyclic electron flow, oxidizing water and reducing NADP. Protons are produced by the oxidation of water and by the action of the cytochrome b_6-f complex. (c) Mitochondria oxidize NADH to NAD and reduce oxygen to water. Protons are pumped by the enzyme NADH dehydrogenase, the cytochrome b-c_1 complex, and cytochrome oxidase. The coupling factor enzymes in the three systems are very similar in structure.

Mitchell proposed that the total energy available for ATP synthesis, which he called the **proton motive force** (Δp), is the sum of a proton chemical potential and a transmembrane electrical potential (Lowe and Jones, 1984). These two components of the proton motive force from the outside of the membrane to the inside are given by

$$\Delta p = \Delta E_m - 59(pH_i - pH_o) \qquad (8.10)$$

In Equation 8.10, ΔE_m (or $\Delta \psi$) is the transmembrane potential and $pH_i - pH_o$ (or ΔpH) is the pH difference across the membrane. The constant of proportionality (at 25°C) is 59 mV/pH unit, so that a transmembrane pH difference of one unit is equivalent to a membrane potential of 59 mV.

Many experiments have established that the two components of the proton motive force are interchangeable, so that a large ΔE_m, a large ΔpH, or intermediate amounts of both are all equally effective in forming ATP (Hangarter and Good, 1988). Under conditions of steady-state electron transport in chloroplasts, the membrane electrical potential is quite small, so Δp is due almost entirely to ΔpH. The stoichiometry of protons translocated to ATP synthesized has been found to be $3H^+/ATP$.

The lateral heterogeneity of the thylakoid membranes (Fig. 8.20) places constraints on the formation of ATP in addition to the need for mobile electron carriers discussed earlier. ATP synthase is found only in the stroma lamellae and at the edges of the grana stacks. Protons pumped across the membrane by the cytochrome b_6-f complex or protons produced by water oxidation in the middle of the grana must move laterally up to several tens of nanometers to reach ATP synthase. Until recently, this was thought to occur by simple diffusion of protons in the aqueous portions of the lumen, a pathway known as **delocalized coupling**. However, a growing body of evidence indicates that these protons do not move by simple diffusion but instead travel to the ATP synthase via a different, more specific pathway that is not yet understood. This **localized coupling model** is the focus of much current research (Dilley et al., 1987).

The Entire Chloroplast Genome Has Been Sequenced

The complete chloroplast genome has now been sequenced in several organisms (Ohyama et al., 1988). Chloroplast DNA is circular, with a size that ranges from 120 to 160 kilobases. It almost always contains a duplicated region known as an **inverted repeat**, which is flanked by a large and a small single-copy region (Whitfield and Bottomley, 1983). The chloroplast genome contains coding sequences for approximately 120 proteins. Many of these DNA sequences code for pro-

teins that are yet to be characterized. It is still uncertain whether all these genes are transcribed into mRNA and translated into protein, but it seems likely that there are protein constituents of the chloroplast that remain to be identified.

Chloroplast Genes Exhibit Non-Mendelian Patterns of Inheritance

Chloroplasts and mitochondria reproduce by division rather than by **de novo** synthesis. This is not surprising because they contain genetic information that is not present in the nucleus. During cell division, chloroplasts partition between the two daughter cells. In most sexual plants, however, chloroplasts are contributed to the zygote only by the maternal parent. In these plants, the normal Mendelian pattern of inheritance does not apply to chloroplast-encoded genes, because the offspring receive chloroplasts only from one parent. This gives rise to **non-Mendelian** or **maternal inheritance**. There are numerous examples of traits that are inherited in this way; one is the herbicide-resistance trait discussed earlier.

Many Chloroplast Proteins Are Imported from the Cytoplasm

Proteins found in chloroplasts are coded for by either chloroplast DNA or nuclear DNA. The chloroplast-encoded proteins are synthesized on chloroplast ribosomes, while the nuclear-encoded ones are synthesized on cytoplasmic ribosomes and then transported into the chloroplast. Many nuclear genes contain introns—that is, base sequences that do not code for protein—and are located between the base sequences that are translated into proteins. The mRNA is processed to remove the introns, and the proteins are then synthesized in the cytoplasm. The genes needed for chloroplast function are distributed in the nucleus and in the chloroplast genome without any evident pattern, but both sets are essential for the viability of the chloroplast. Chloroplast genes, on the other hand, do not appear necessary for other cellular functions, and cells in which chloroplasts are missing appear normal in all respects except photosynthesis. Control of the expression of the nuclear genes that code for chloroplast proteins is complex, involving photoactive regulators—both phytochrome (see Chapter 20) and an unidentified receptor for blue light (Chapter 6)—as well as other factors (Tobin and Silverthorne, 1985).

The transport of chloroplast proteins that are synthesized in the cytoplasm shows well-regulated properties. For example, the enzyme Rubisco (Chapter 9) that functions in carbon fixation has two types of subunits, a chloroplast-encoded large subunit and a nuclear-encoded small subunit. Small subunits of Rubisco are synthesized in the cytoplasm and transported into the chloroplast, where assembly of the enzyme takes place (Schmidt and Mishkind, 1986). In this and other known cases, the nuclear-encoded chloroplast proteins are synthesized as precursor proteins containing an N-terminal amino acid sequence known as a **transit peptide**. In some manner, this sequence directs the precursor protein to the chloroplast, facilitates its passage through both the outer and inner envelope membranes, and is then clipped off. Plastocyanin is a water-soluble protein that is coded for in the nucleus but functions in the lumen of the chloroplast. It therefore must cross three membranes and yet end up as a hydrophilic protein! In the case of plastocyanin, the transit peptide is very large and may be processed in more than one step.

Summary

Photosynthesis is the process of storage of solar energy carried out by plants and certain bacteria. Photosynthetic antenna pigments absorb light and transfer the energy to a specialized chlorophyll-protein complex known as a reaction center. The reaction center initiates a complex series of chemical oxidation-reduction reactions that capture energy in the form of chemical bonds.

Plants and some photosynthetic prokaryotes have two-reaction centers, photosystems I and II, that are connected in tandem. These organisms carry out noncyclic electron transport by oxidizing water to molecular oxygen and reducing $NADP^+$ to NADPH. A portion of the energy of the photon is also stored as chemical potential energy in the form of a pH difference and an electrical potential across the photosynthetic membrane. This energy is converted into the high-energy phosphate bonds of ATP by action of an enzyme complex known as the coupling factor. NADPH and ATP provide the energy for carbon reduction.

Photosynthesis takes place in the thylakoid membranes of the chloroplast. Photosystems I and II are spatially separated, with photosystem I found exclusively in the nonstacked stroma membranes and photosystem II found largely in the stacked grana membranes. Reversible protein phosphorylation of antenna complexes serves to regulate the distribution of energy between the two photosystems.

Chloroplasts contain DNA and code for and synthesize some of the proteins that are essential for photosynthesis. Additional proteins are coded for by nuclear DNA, synthesized in the cytoplasm, and imported into the chloroplast, where they are assembled into the thylakoid membrane.

GENERAL READING

Amesz, J., ed. (1987) *Photosynthesis* (New Comprehensive Biochemistry, Vol. 15). Elsevier, Amsterdam.

Anderson, J. M. (1986) Photoregulation of the composition, function and structure of thylakoid membranes. *Annu. Rev. Plant Physiol.* 37:93–136.

Ashton, F. M., and Crafts, A. S. (1981) *Mode of Action of Herbicides*, 2d ed. Wiley, New York.

Babcock, G. T. (1987) The photosynthetic oxygen-evolving process. In: *Photosynthesis* (New Comprehensive Biochemistry, Vol. 15), Amesz, J., ed. pp. 125–158. Elsevier, Amsterdam.

Barber, J., ed. (1987) *The Light Reactions* (Topics in Photosynthesis, Vol. 8). Elsevier, Amsterdam.

Clayton, R. K. (1971) *Light and Living Matter: A Guide to the Study of Photobiology*. McGraw-Hill, New York.

Clayton, R. K. (1980) *Photosynthesis: Physical Mechanisms and Chemical Patterns*. Cambridge University Press, Cambridge, England.

Dilley, R. A., Theg, S. M., and Beard, W. A. (1987) Membrane-proton interactions in chloroplast bioenergetics: Localized proton domains. *Annu. Rev. Plant Physiol.* 38:347–389.

Glazer, A. N., and Melis, A. (1987) Photochemical reaction centers: Structure, organization, and function. *Annu. Rev. Plant Physiol.* 38:11–45.

Govindjee, ed. (1982) *Photosynthesis: Energy Conversion by Plants and Bacteria*. Academic Press, New York.

Govindjee, ed. (1988) *Molecular Biology of Photosynthesis*. Kluwer, Dordrecht, Netherlands.

Haehnel, W. (1984) Photosynthetic electron transport in higher plants. *Annu. Rev. Plant Physiol.* 35:659–693.

Hall, D. O., and Rao, K. K. (1987) *Photosynthesis*, 4th ed. Edward Arnold, London.

Hipkins, M. F., and Baker, N. R. (1986) *Photosynthesis: Energy Transduction: A Practical Approach*. IRL Press, Oxford.

Ke, B., and Shuvalov, V. A. (1987) Picosecond absorption spectroscopy in photosynthesis and primary electron transfer processes. In: *The Light Reactions* (Topics in Photosynthesis, Vol. 8), Barber, J., ed., pp. 31–93. Elsevier, Amsterdam.

Powles, S. B. (1984) Photoinhibition of photosynthesis induced by visible light. *Annu. Rev. Plant Physiol.* 35:15–44.

Rabinowitch, E. and Govindjee (1969) *Photosynthesis*. Wiley, New York.

Schmidt, G. W., and Mishkind, M. L. (1986) The transport of proteins into chloroplasts. *Annu. Rev. Biochem.* 55:879–912.

Staehelin, L. A. (1986) Chloroplast structure and supramolecular organization of photosynthetic membranes. In: *Photosynthesis. III. Photosynthetic Membranes and Light Harvesting Systems* (Encyclopedia of Plant Physiology, New Series, Vol. 19), Staehelin, L. A., and Arntzen, C. J., eds. pp. 1–84. Springer-Verlag, Berlin.

Staehelin, L. A., and Arntzen, C. J., eds. (1986) *Photosynthesis. III. Photosynthetic Membranes and Light Harvesting Systems* (Encyclopedia of Plant Physiology, New Series, Vol. 19). Springer-Verlag, Berlin.

Stevens, E., Bryant, D., and Shannon, J., eds. (1988) *Light-Energy Transduction in Photosynthesis: Higher Plants and Bacterial Models*. Am. Soc. Plant Physiol., Rockville, Md.

Thornber, J. P. (1986) Biochemical characterization and structure of pigment-proteins of photosynthetic organism. In: *Photosynthesis. III. Photosynthetic Membranes and Light Harvesting Systems* (Encyclopedia of Plant Physiology, New Series, Vol. 19), Staehelin, L. A., and Arntzen, C. J., eds. pp. 98–142. Springer-Verlag, Berlin.

Tobin, E. M., and Silverthorne, J. (1985) Light regulation of gene expression in higher plants. *Annu. Rev. Plant Physiol.* 36:569–593.

Whitfield, P. R., and Bottomley, W. (1983) Organization and structure of chloroplast genes. *Annu. Rev. Plant Physiol.* 34:279–310.

Woodbury, N. W., Becker, M., Middendorf, D., and Parson, W. W. (1985) Picosecond kinetics of the initial photochemical electron-transfer reaction in bacterial photosynthetic reaction centers. *Biochemistry* 24:7516–7521.

CHAPTER REFERENCES

Allen, J. F. (1983) Protein phosphorylation—carburetor of photosynthesis? *Trends Biochem. Sci.* 8:369–373.

Anderson, J. M., and Andersson, B. (1988) The dynamic photosynthetic membrane and regulation of solar energy conversion. *Trends Biochem. Sci.* 13:351–355.

Arnon, D. I. (1984) The discovery of photosynthetic phosphorylation. *Trends Biochem. Sci.* 9:258–262.

Avers, C. G. (1985) *Molecular Cell Biology.* Addison-Wesley, Reading, Mass.

Becker, W. M. (1986) *The World of the Cell.* Benjamin/Cummings, Menlo Park, Calif.

Blankenship, R. E., and Prince, R. C. (1985) Excited-state redox potentials and the Z scheme of photosynthesis. *Trends Biochem. Sci.* 10:382–383.

Borisov, A. Y. (1989) Transfer of excitation energy in photosynthesis: Some thoughts. *Photosynth. Res.* 20:35–58.

Deisenhofer, J., and Michel, H. (1989) The photosynthetic reaction center from the purple bacterium *Rhodopseudomonas viridis. Science* 245:1463–1473.

Dismukes, G. C. (1986) The metal centers of the photosynthetic oxygen-evolving complex. *Photochem. Photobiol.* 43:99–115.

Dover, M. J., and Croft, B. A. (1986) Pesticide resistance and public policy. *BioScience* 36:78–85.

Dutton, P. L. (1978) Redox potentiometry: Determination of midpoint potentials of oxidation-reduction components of biological electron-transfer systems. *Methods Enzymol.* 54:411–435.

Duysens, L. N. M., Amesz, J., and Kamp, B. M. (1961) Two photochemical systems in photosynthesis. *Nature* 190:510–511.

Emerson, R., and Arnold, W. (1932) The photochemical reaction in photosynthesis. *J. Gen. Physiol.* 16:191–205.

Emerson, R., Chalmers, R., and Cederstrand, C. (1957) Some factors influencing the long-wave limit of photosynthesis. *Proc. Natl. Acad. Sci. USA* 43:133–143.

Glazer, A. N. (1983) Comparative biochemistry of photosynthetic light-harvesting systems. *Annu. Rev. Biochem.* 52:125–157.

Golbeck, J. H. (1987) Structure, function and organization of the photosystem I reaction center complex. *Biochim. Biophys. Acta* 895:167–204.

Govindjee and Coleman, W. J. (1990) How plants make oxygen. *Scientific American* 262:50–58.

Green, B. R. (1988) The chlorophyll-protein complexes of higher plant photosynthetic membranes: Or, just what green band is that? *Photosynth. Res.* 15:3–32.

Hangarter, R. P., and Good, N. E. (1988) Active transport, ion movements and pH changes. II. Changes of pH and ATP synthesis. *Photosynth. Res.* 19:237–250.

Hauska, G., Hurt, E., Gabellini, N., and Lockau, W. (1983) Comparative aspects of quinol-cytochrome *c*/plastocyanin oxidoreductases. *Biochim. Biophys. Acta* 726:97–133.

Hill, R. (1939) Oxygen produced by isolated chloroplasts. *Proc. R. Soc. London Ser. B* 127:192–210.

Hill, R., and Bendall, F. (1960) Function of the two cytochrome components in chloroplasts: A working hypothesis. *Nature* 186:136–137.

Hunter, C. N., van Grondelle, R., and Olsen, J. D. (1989) Photosynthetic antenna proteins: 100 ps before photochemistry starts. *Trends Biochem. Sci.* 14:72–76.

Jagendorf, A. T. (1967) Acid-base transitions and phosphorylation by chloroplasts. *Fed. Proc. Fed. Am. Soc. Exp. Biol.* 26:1361–1369.

Ke, B. (1973) The primary electron acceptor of photosystem I *Biochem. Biophys.* Acta 301:1–33.

Kirmaier, C., and Holten, D. (1987) Primary photochemistry of reaction centers from the photosynthetic purple bacteria. *Photosynth. Res.* 13:225–260.

Knaff, D. B. (1988) The photosystem I reaction centre. *Trends Biochem. Sci.* 13:460–461.

Kok, B., Forbush, B., and McGloin, M. (1970) Cooperation of charges in photosynthetic O_2 evolution. I. A linear four step mechanism. *Photochem. Photobiol.* 11:457–475.

Lowe, A. G., and Jones, M. N. (1984) Proton motive force—what price Δp? *Trends Biochem. Sci.* 9:11–12.

McCarty, R. E. (1985) H^+-ATPases in oxidative and photosynthetic phosphorylation. *BioScience* 35:27–30.

McCarty, R. E., and Hammes, G. G. (1987) Molecular architecture of chloroplast coupling factor 1. *Trends Biochem. Sci.* 12:234–237.

Melis, A. (1984) Light regulation of photosynthetic membrane structure, organization and function. *J. Cell. Biochem.* 24:271–285.

Michel, H., and Deisenhofer, J. (1988) Relevance of the photosynthetic reaction center from purple bacteria to the structure of photosystem II. *Biochemistry* 27:1–7.

Mitchell, P. (1979) Keilin's respiratory chain concept and its chemiosmotic consequences. *Science* 206:1148–1159.

Murata, N., and Miyao, M. (1985) Extrinsic membrane proteins in the photosynthetic oxygen-evolving complex. *Trends Biochem. Sci.* 10:122–124.

Nanba, O., and Satoh, K. (1987) Isolation of a photosystem II reaction center consisting of D-1 and D-2 polypeptides and cytochrome *b*-559. *Proc. Natl. Acad. Sci. USA* 84:109–112.

Ohyama, K., Kohchi, T., Sano, T., and Yamada, Y. (1988) Newly identified groups of genes in chloroplasts. *Trends Biochem. Sci.* 13:19–22.

O'Keefe, D. P. (1988) Structure and function of the chloroplast cytochrome *bf* complex. *Photosynth. Res.* 17:189–216.

Olson, J. M., and Pierson, B. K. (1987) Evolution of reaction centers in photosynthetic prokaryotes. *Int. Rev. Cytol.* 108:209–248.

Prince, R. C. (1985) Redox-driven proton gradients. *BioScience* 35:22–26.

Rutherford, A. W. (1989) Photosystem II, the water-splitting enzyme. *Trends Biochem. Sci.* 14:227–232.

Sauer, K. (1975) Primary events and the trapping of energy. In: *Bioenergetics of Photosynthesis*, Govindjee, ed., pp. 116–181. Academic Press, New York.

Siefermann-Harms, D. (1985) Carotenoids in photosynthesis. I. Location in photosynthetic membranes and light-harvesting function. *Biochim. Biophys. Acta* 811:325–335.

Staehelin, L. A., and Arntzen, C. J. (1983) Regulation of chloroplast membrane function: Protein phosphorylation changes the spatial organization of membrane components. *J. Cell Biol.* 97:1327–1337.

Trebst, A. (1986) The topology of the plastoquinone and herbicide binding peptides of photosystem II in the thylakoid membrane. *Z. Naturforsch. Teil C* 41:240–245.

Wood, P. M. (1985) What is the Nernst equation? *Trends Biochem. Sci.* 10:106–108.

Zuber, H. (1986) Structure of light-harvesting antenna complexes of photosynthetic bacteria, cyanobacteria and red algae. *Trends Biochem. Sci.* 11:414–419.

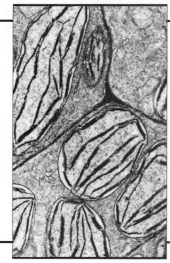

CHAPTER 9

Photosynthesis: Carbon Metabolism

W E DISCUSSED IN CHAPTER 5 that plants require mineral nutrients and light to grow and complete their life cycle. Because of interactions between living organisms and their environment and among living organisms themselves, mineral nutrients cycle through the biosphere. These cycles involve complex interactions, and each cycle is critical in its own right. However, in terms of magnitude, the flux of matter passing through the carbon cycle dwarfs that of all other cycles. Recent estimates indicate that the amount of CO_2 converted to biomass reaches about 200 billion tons annually. Some 40% of this mass is attributed to the activities of marine phytoplankton. Most of this carbon is incorporated into organic compounds by the operation of a single metabolic device, the **C3 photosynthetic carbon reduction cycle**.

In the previous chapter we have seen how, during the **light reactions of photosynthesis**, the photochemical oxidation of water to molecular oxygen is coupled to the generation of reduced pyridine nucleotide (NADPH) and ATP. The reactions associated with the reduction of CO_2 to carbohydrate are coupled to the consumption of NADPH and ATP. These reactions are referred to as the **dark reactions of photosynthesis**, since light is not directly involved. The only requirement for the synthetic carbon reduction cycle is that it be coupled to the capacity to generate ATP and NADPH. In photosynthetic organisms this capacity is provided by light (Fig. 9.1). But in chemosynthetic organisms (such as the hydrogen bacteria) ATP and NADPH are generated by the oxidation of some inorganic substrate (in this case, the oxidation of hydrogen to water). These organisms reduce CO_2 to carbohydrate in the dark using the same metabolic device as do photosynthetic organisms: the C3 carbon reduction cycle.

The C3 Photosynthetic Carbon Reduction Cycle

All photosynthetic eukaryotes, from the most primitive alga to the most advanced angiosperm, reduce CO_2 to carbohydrate via the same basic mechanism, the C3 photosynthetic carbon reduction (PCR) cycle. The PCR cycle is sometimes referred to as the **Calvin cycle** in honor of its discoverer, the American biochemist Melvin Calvin. Other metabolic pathways associated with the photosynthetic fixation of

219

Photosynthesis
(as performed by algae and higher plants)

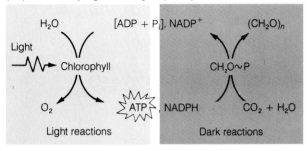

Chemosynthesis
(as performed, for example, by hydrogen bacteria)

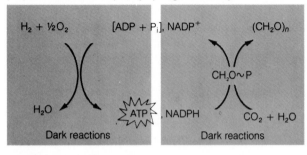

FIGURE 9.1. The reduction of CO_2 to carbohydrate by autotrophic organisms does not directly require light and is therefore referred to as the **dark reaction**. In photosynthetic organisms, light is required to generate ATP and NADPH. These are in turn consumed by the dark reactions, which reduce CO_2 to carbohydrate. In chemosynthetic organisms, which derive their energy from the oxidation of inorganic material, the generation of ATP and NADPH also occurs as a dark reaction.

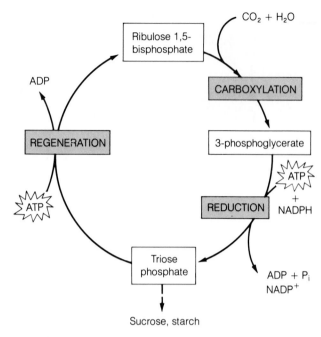

FIGURE 9.2. The C3 photosynthetic carbon reduction cycle (PCR cycle) proceeds in three stages: (1) carboxylation, during which CO_2 is covalently linked to a carbon skeleton; (2) reduction, during which carbohydrate is formed at the expense of the photochemically derived ATP and reducing equivalents, NADPH; and (3) regeneration, during which the CO_2-acceptor molecule, ribulose 1,5-bisphosphate is re-formed.

CO_2, such as the C4 photosynthetic carbon assimilation (PCA) cycle and the C2 photorespiratory carbon oxidation (PCO) cycle, are either auxiliary to or dependent on the basic PCR cycle.

The C3 PCR Cycle Includes Carboxylation, Reduction, and Regeneration Steps

In the C3 PCR cycle, carbon dioxide from the atmosphere and water are enzymatically combined with a five-carbon acceptor molecule to generate two molecules of a three-carbon intermediate. These intermediates are reduced to carbohydrate using the photochemically generated ATP and NADPH. The cycle is completed by regeneration of the five-carbon acceptor. The C3 PCR cycle proceeds in three stages (Fig. 9.2):

1. Carboxylation of the CO_2 acceptor, ribulose 1,5-bisphosphate, to form two molecules of 3-phosphoglycerate, the first stable intermediate of the PCR cycle.

2. Reduction of this carboxylic acid to a carbohydrate in the form of glyceraldehyde 3-phosphate.

3. Regeneration of the CO_2 acceptor, ribulose 1,5-bisphosphate, from glyceraldehyde 3-phosphate.

The Carboxylation of Ribulose Bisphosphate. CO_2 enters the PCR cycle by reacting with ribulose 1,5-bis-phosphate to yield two molecules of 3-phosphoglycerate (Table 9.1, reaction 1), a reaction that is catalyzed by the chloroplast enzyme ribulose bisphosphate carboxylase/oxygenase, referred to by the acronym **Rubisco** (Andrews and Lorimer, 1987).

Rubisco, the enzyme responsible for fixing 200 billion tons of CO_2 annually, is without doubt the world's most abundant enzyme. To accomplish its gargantuan task, an estimated 10^7 tons of enzyme (20 kg for every man, woman, and child!) are required globally. Rubisco accounts for some 40% of the total soluble protein of most leaves. The concentration of Rubisco active sites within the chloroplast stroma is calculated to be about 4 mM, or about 500 times greater than the concentration of one of its substrates, CO_2 (see box on carbon dioxide).

Rubisco (molecular mass 560 kDa) is composed of eight large (L) subunits ($\sim$56 kDa), each of which has an active site, and eight small (S) subunits ($\sim$14 kDa),

Carbon Dioxide: Some Important Physicochemical Properties

AN UNDERSTANDING OF the mechanism of CO_2 fixation requires knowledge of the physical and chemical properties of CO_2, particularly those related to its interaction with water. The amount of any gas dissolved in water is proportional to its partial pressure (P_{gas}) above the solution (Henry's law) and its Bunsen absorption coefficient (α). The Bunsen absorption coefficient is the volume of gas absorbed by one volume of water at a pressure of one atmosphere and is temperature-dependent, decreasing as the temperature rises. The solubility of a gas therefore decreases with increasing temperature. Thus, for a given temperature

$$[\text{gas}] \, (\mu M) = P_{gas} \cdot \alpha \cdot 10^6 / V_0$$

where V_0 is the normal volume of an ideal gas at standard temperature and pressure ($V_0 = 22.4$ L mol^{-1}).

The partial pressure of a gas is obtained by multiplying the mole fraction of the gas by the total pressure. The mole fraction of a gas is obtained by dividing its partial volume by the total volume of all the gases present. Thus, the mole fractions of CO_2 and O_2 in air are 0.0345% and 20.95%, respectively. At sea level, atmospheric pressure is about 1 bar, so the partial pressures of CO_2 and O_2 are 345 μbar and 209.5 mbar, respectively. From these values and those of α, the corresponding solution concentrations of CO_2 and O_2 can be computed using the above equation. The table presents values for these concentrations at different temperatures.

These values place considerable constraints on the carboxylation process. As a carboxylase, Rubisco must be capable of operating efficiently even at the rather low concentrations of CO_2 available to it. Rubisco also functions as an oxygenase in photorespiration (page 229), so, the solution concentration of O_2 is also important. Because of the different temperature dependences of the Bunsen absorption coefficients α (CO_2) and α (O_2), the concentrations of these two gases vary with temperature so that the $[CO_2]/[O_2]$ ratio decreases as the temperature increases. This effect is important biologically, because as the temperature increases the ratio of carboxylation to oxygenation catalyzed by Rubisco decreases and the ratio of photorespiration to photosynthesis increases.

Temperature (°C)	α (CO_2)	$[CO_2]$ (μM in solution)	α (O_2)	$[O_2]$ (μM in solution)	$\dfrac{[CO_2]}{[O_2]}$
5	1.424	21.93	0.0429	401.2	0.0515
15	1.019	15.69	0.0342	319.8	0.0462
25	0.759	11.68	0.0283	264.6	0.0416
35	0.592	9.11	0.0244	228.2	0.0376

whose function is unknown. The gene for the large subunit is encoded by the chloroplast genome and this subunit is synthesized by the chloroplast ribosomes. Genes for the small subunit are encoded by the nuclear genome. The small subunit is synthesized by cytosolic ribosomes, transported into the chloroplast, and assembled there with large subunits to form L_8S_8 Rubisco molecules.

The carboxylase reaction brings about the incorporation of CO_2 into the carboxyl group of the "upper" molecule of 3-phosphoglycerate (Fig. 9.3). Two properties of the carboxylase reaction are especially important. First, the negative free energy change (Chapter 2) associated with the carboxylation of ribulose bisphosphate is large ($\Delta G^{\circ\prime} = -12.4$ kcal mol^{-1}). As a consequence, the carboxylation reaction proceeds to completion even at the low concentrations of CO_2 that are available to it. In other words, the equilibrium constant (Chapter 2) strongly favors the forward reaction. Second, the affinity of Rubisco for CO_2 is sufficiently high to ensure rapid carboxylation at the low concentrations of CO_2 found in photosynthetic cells (see box on carbon dioxide).

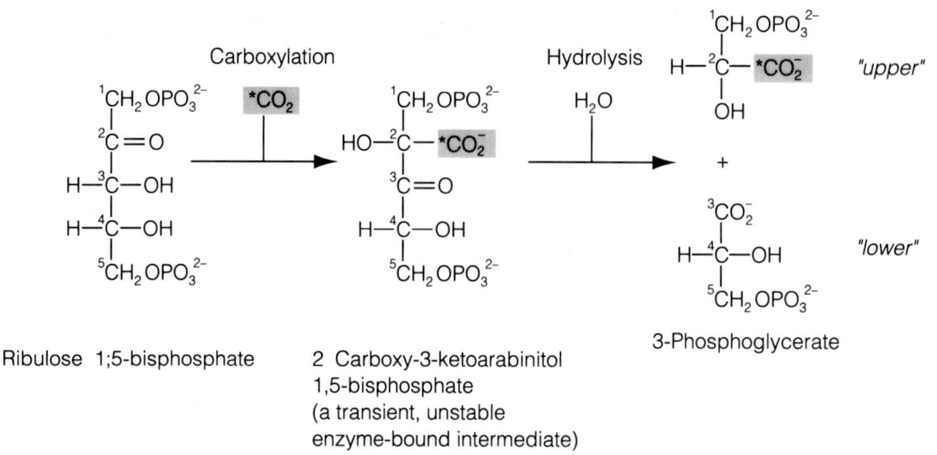

FIGURE 9.3. The carboxylation of ribulose 1,5-bisphosphate, catalyzed by Rubisco, proceeds in two stages, carboxylation and hydrolysis. Carboxylation involves the addition of CO_2 to carbon 2 (C-2) of ribulose 1,5-bisphosphate to form the unstable, enzyme-bound intermediate, 2-carboxy-3-ketoarabinitol 1,5-bisphosphate, which undergoes hydrolysis to yield two molecules of the stable product, 3-phosphoglycerate. The two molecules of 3-phosphoglycerate, "upper" and "lower," are distinguished by the fact that the "upper" molecule contains the newly incorporated carbon dioxide, designated as $*CO_2$.

The Reduction Step of the C3 PCR Cycle. In this stage, the 3-phosphoglycerate formed as a result of the carboxylation (Table 9.1, reaction 1) of ribulose bisphosphate is first phosphorylated to 1,3-bisphosphoglycerate (Table 9.1, reaction 2) by the ATP generated in the light reactions and is then reduced to glyceraldehyde 3-phosphate (Table 9.1, reaction 3), using the NADPH generated by the light reactions. The chloroplast enzyme NADP: glyceraldehyde-3-phosphate dehydrogenase catalyzes this step.

The Regeneration of Ribulose 1,5-Bisphosphate. Continued fixation of CO_2 requires that the CO_2 acceptor, ribulose 1,5-bisphosphate, be constantly regenerated. To avoid depleting the cycle of intermediates, 3 molecules of ribulose 1,5-bisphosphate are formed by reshuffling the carbons from 5 molecules of triose phosphate (Fig. 9.4).

One molecule of glyceraldehyde 3-phosphate is converted to dihydroxyacetone 3-phosphate in a reaction involving proton transfer between C-2 and C-1 (Table 9.1, reaction 4). Dihydroxyacetone 3-phosphate then undergoes aldol condensation with a molecule of glyceraldehyde 3-phosphate (Table 9.1, reaction 5) to give fructose 1,6-bisphosphate. This product is hydrolyzed to fructose 6-phosphate (Table 9.1, reaction 6), which then reacts with transketolase, an enzyme that requires thiamine pyrophosphate as a cofactor. A two-carbon unit (C-1 and C-2 of fructose 6-phosphate) is transferred via the cofactor to a third molecule of glyceraldehyde 3-phosphate to give erythrose 4-phosphate (from C-3 to C-6 of fructose) and xylulose 5-phosphate (Table 9.1, reaction 7). Erythrose 4-phosphate then combines with a fourth molecule of triose phosphate (as dihydroxyacetone 3-phosphate, Table 9.1, reaction 8) to yield the seven-carbon sugar sedoheptulose 1,7-bisphosphate, which is further hydrolyzed to give sedoheptulose 7-phosphate (Table 9.1, reaction 9). Sedoheptulose 7-phosphate donates a two-carbon unit to the fifth molecule of glyceraldehyde 3-phosphate and produces ribose 5-phosphate (from C-3 to C-7 of sedoheptulose) and xylulose 5-phosphate as products (Table 9.1, reaction 10). The two molecules of xylulose 5-phosphate (Table 9.1, reactions 7 and 10) are epimerized to give ribulose 5-phosphate (Table 9.1, reaction 11). The third molecule of ribulose 5-phosphate is formed by the isomerization of ribose 5-phosphate (Table 9.1, reaction 12). Finally, ribulose 5-phosphate is phosphorylated with ATP, thus regenerating the CO_2 acceptor ribulose 1,5-bisphosphate (Table 9.1, reaction 13).

FIGURE 9.4. (Right) The stoichiometry of the C3 photosynthetic carbon reduction cycle. The carboxylation of three molecules of ribulose 1,5-bisphosphate leads to *net* synthesis of one molecule of glyceraldehyde 3-phosphate and regeneration of the three molecules of starting material. The reactions are more fully described in Table 9.1; here 1-1, 1-2, etc. refer to reactions 1, 2, etc. in that table. Note that this process starts and ends with three molecules of ribulose 1,5-bisphosphate, reflecting the cyclic nature of the PCR cycle.

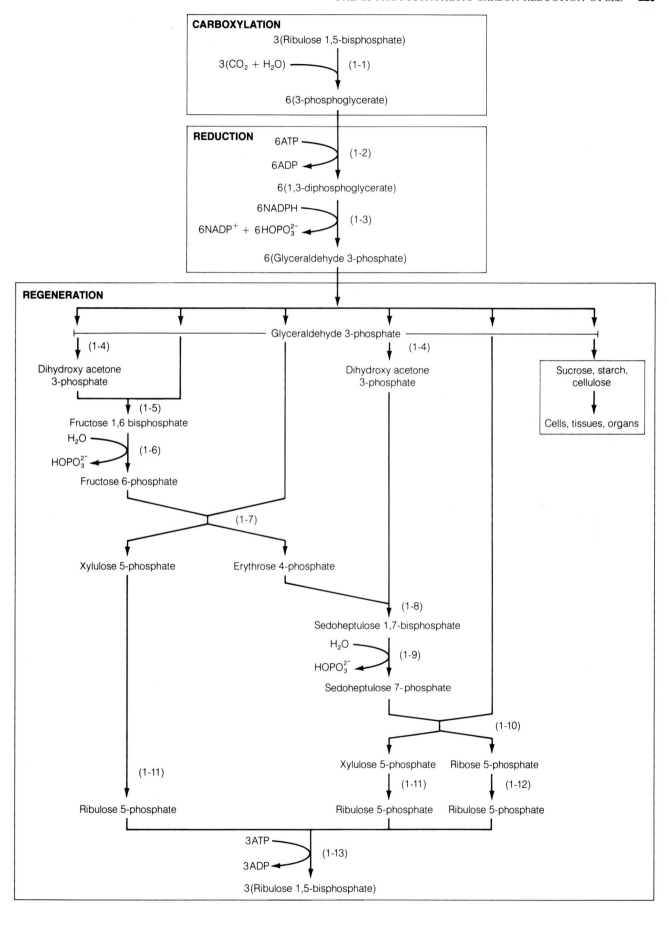

TABLE 9.1. Reactions of the C3 photosynthetic carbon reduction cycle

1. *Ribulose-1,5-bisphosphate carboxylase*
   ribulose 1,5-bisphosphate + CO_2 + H_2O → 2(3-phosphoglycerate) + $2H^+$

2. *3-Phosphoglycerate kinase*
   3-phosphoglycerate + ATP → 1,3-bisphosphoglycerate + ADP

3. *NADP:glyceraldehyde-3-phosphate dehydrogenase*
   1,3-bisphosphoglycerate + NADPH + H^+ → glyceraldehyde 3-phosphate + $NADP^+$ + $HOPO_3^{2-}$

4. *Triose-phosphate isomerase*
   glyceraldehyde 3-phosphate → dihydroxyacetone 3-phosphate

5. *Aldolase*
   glyceraldehyde 3-phosphate + dihydroxyacetone 3-phosphate → fructose 1,6-bisphosphate

6. *Fructose-1,6-bisphosphate phosphatase*
   fructose 1,6-bisphosphate + H_2O → fructose 6-phosphate + $HOPO_3^{2-}$

7. *Transketolase*
   fructose 6-phosphate + glyceraldehyde 3-phosphate → erythrose 4-phosphate + xylulose 5-phosphate

continued

TABLE 9.1. (continued)

8. *Aldolase*

erythrose 4-phosphate + dihydroxyacetone 3-phosphate → sedoheptulose 1,7-bisphosphate

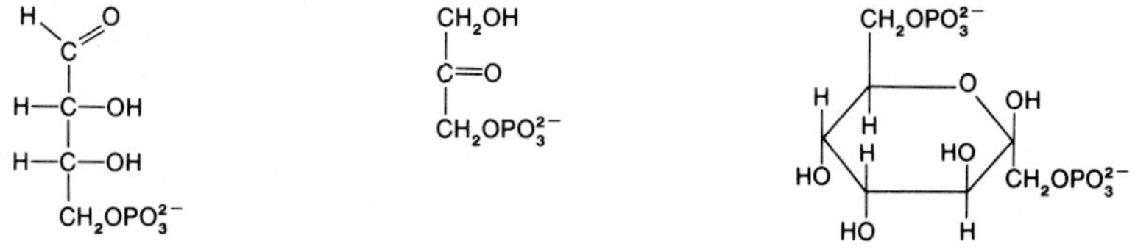

9. *Sedoheptulose-1,7-bisphosphate phosphatase*

sedoheptulose 1,7-bisphosphate + H_2O → sedoheptulose 7-phosphate + $HOPO_3^{2-}$

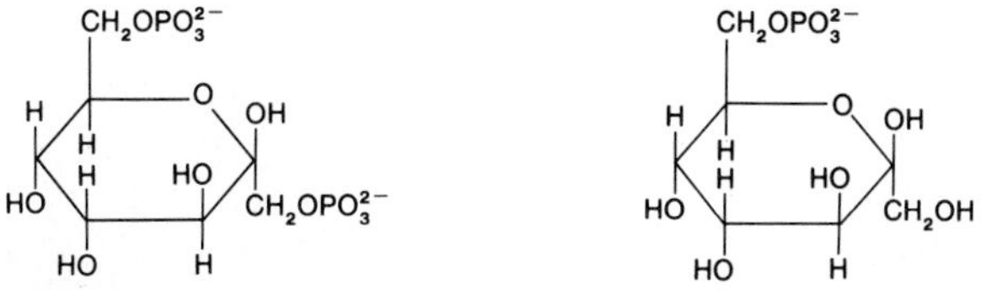

10. *Transketolase*

sedoheptulose 7-phosphate + glyceraldehyde 3-phosphate → ribose 5-phosphate + xylulose 5-phosphate

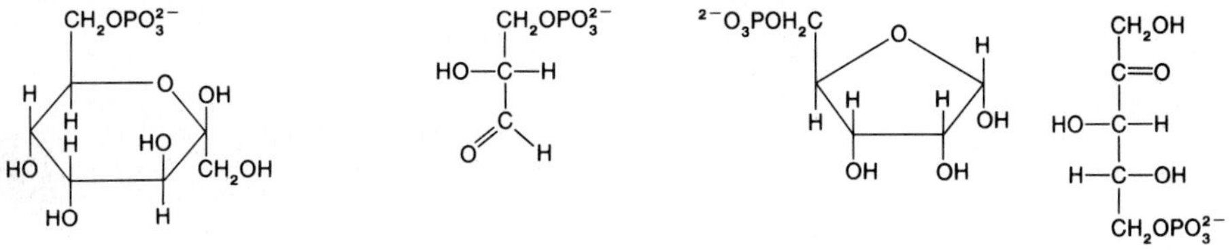

11. *Ribulose-5-phosphate epimerase*

xylulose 5-phosphate → ribulose 5-phosphate

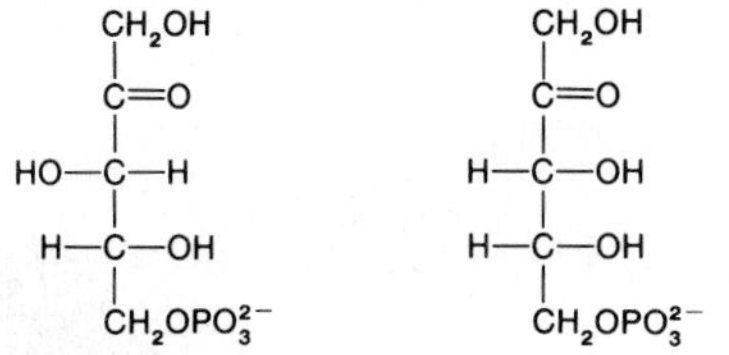

12. *Ribose-5-phosphate isomerase*

ribose 5-phosphate → ribulose 5-phosphate

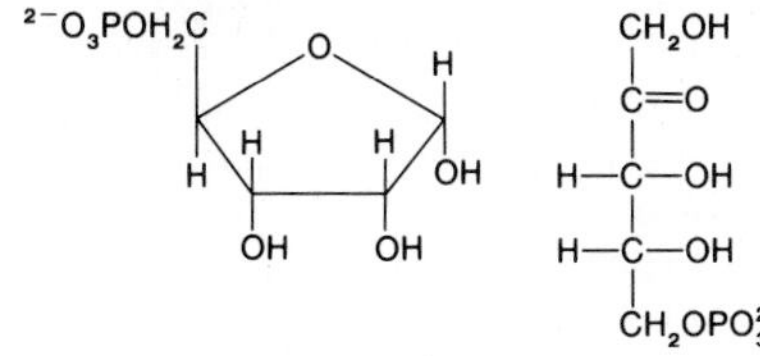

13. *Ribulose-5-phosphate kinase*

ribulose 5-phosphate + ATP → ribulose 1,5-bisphosphate + ADP + H^+

The PCR Cycle Was Elucidated Using Radioactive Carbon Compounds

The elucidation of the PCR cycle was the result of a series of elegant experiments by Melvin Calvin and his colleagues in the 1950s. Uniform suspensions of the unicellular eukaryotic green alga *Chlorella* were used. The cells were first exposed to constant conditions of light and CO_2 to establish steady-state photosynthesis. Then radioactive $^{14}CO_2$ was added for a brief period to label the various intermediates of the cycle. The cells were then killed and their enzymes inactivated by dropping them into boiling alcohol. The ^{14}C-labeled compounds were separated from one another and identified by their positions on two-dimensional paper chromatograms (Fig. 9.5). In this manner, 3-phosphoglycerate and the various sugar phosphates were identified as intermediates in the fixation process. By exposing the cells to $^{14}CO_2$ for progressively shorter periods of time, Calvin was able to identify 3-phosphoglycerate as the first, stable intermediate. It followed that the other labeled sugar phosphates must be derived from a subsequent reduction of 3-phosphoglycerate (Bassham, 1965).

To deduce the path of carbon, it was necessary to determine the distribution of ^{14}C in each of the labeled sugars. Following a brief exposure to $^{14}CO_2$, each of the intermediates was isolated and chemically degraded so that the amount of ^{14}C in each carbon atom could be determined. The results showed that 3-phosphoglycerate was predominantly labeled in the carboxyl group. This suggested that the initial CO_2 acceptor was a two-carbon compound and prompted a long and futile search for such a compound. The subsequent discovery that pentose monophosphates and a pentose bisphosphate (ribulose) participated in the cycle raised the possibility that the initial CO_2 acceptor was a five-carbon compound, which, after reacting with CO_2, generated two molecules of a 3-phosphoglycerate. This conceptual breakthrough rapidly led to the identification of ribulose 1,5-bisphosphate as the CO_2 acceptor and to the formulation of the complete cycle. Once the cycle was understood, the ^{14}C-labeling pattern of the other intermediates could also be explained.

To demonstrate conclusively that a metabolic pathway exists, it is necessary to prove that the postulated enzymes catalyze the proposed reactions in the test tube (in vitro). Ideally, the rates of these in vitro reactions should be equal to or in excess of those observed in the intact cell (in vivo). Keep in mind, however, that this evidence can be used only to support a proposed pathway. Failure to demonstrate a reaction in vitro does not prove that the reaction does not occur—absence of evidence is not evidence of absence! All of the reactions of the PCR cycle have been demonstrated in vitro. With one exception, the in vitro rates of enzymatic activity are well in excess of the maximal observed rates of photosynthesis. That exception is Rubisco, which, when assayed under

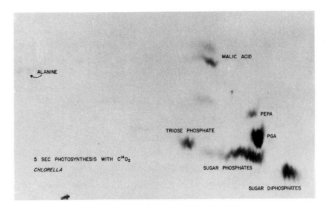

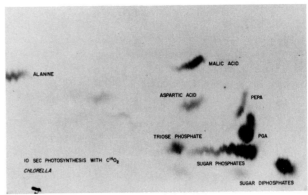

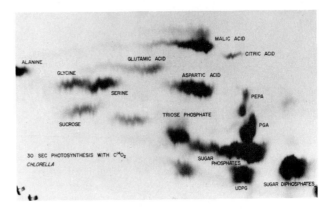

FIGURE 9.5. Labeling of carbon compounds in the alga *Chlorella* after exposure to $^{14}CO_2$. The time intervals shown in the figure indicate the length of exposure to the radiolabel. At the indicated time intervals, the reaction was terminated, and the labeled compounds in the cell homogenates were separated with paper chromatography. The heavy labeling of 3-phosphoglyceric acid (PGA) after the shortest exposure indicates that it is the first stable intermediate of the PCR cycle. (From Bassham, 1965.)

the levels of CO_2 and O_2 in air, has barely enough in vitro activity to account for the observed rate of photosynthesis in air. This suggests that Rubisco may be one of the rate-determining factors in photosynthesis (Lorimer, 1981).

During Its Operation, the C3 PCR Cycle Regenerates Its Own Biochemical Components

An important property of the PCR cycle is that it can increase its rate of operation by increasing the concentrations of its intermediates; that is, the cycle is **autocatalytic**. In addition, the PCR cycle has the metabolically unique feature that it produces more substrate (ribulose 1,5-bisphosphate, or RuBP) than it consumes:

$$5RuBP + 5CO_2 + 9H_2O + 16ATP + 10NADPH \rightarrow$$

$$6RuBP + 14HOPO_3^{2-} + 6H^+ + 16ADP + 10NADP^+$$

The importance of this property is shown by experiments in which previously darkened leaves or isolated chloroplasts are illuminated. In such experiments, CO_2 fixation commences only after a lag, called the induction period, and photosynthetic rates increase with time in the first few minutes after the onset of illumination. The increase in the rate of photosynthesis during the induction period is partly attributable to the increase in the concentration of intermediates of the PCR cycle.

Another important feature of the PCR cycle is its overall stoichiometry. Let us analyze it by following the reactions involved in fixing six molecules of CO_2 (see also Fig. 9.4):

$$6(\text{ribulose 1,5-bisP}) + 6CO_2 + 6H_2O \rightarrow$$
$$12(\text{3-P-glycerate}) + 12H^+$$

$$12(\text{3-P-glycerate}) + 12ATP \rightarrow$$
$$12(\text{1,3-bisP-glycerate}) + 12ADP$$

$$12(\text{1,3-bisP-glycerate}) + 12(NADPH + H^+) \rightarrow$$
$$12(\text{triose-P}) + 12NADP^+ + 12P_i$$

$$6(\text{triose-P}) \rightarrow 3(\text{fructose 1,6-bisP})$$

$$3(\text{fructose 1,6-bisP}) + 3H_2O \rightarrow 3(\text{fructose 6-P}) + 3P_i$$

$$2(\text{fructose 6-P}) + 2(\text{triose-P}) \rightarrow$$
$$2(\text{xylulose 5-P}) + 2(\text{erythrose 4-P})$$

$$2(\text{erythrose 4-P}) + 2(\text{triose-P}) \rightarrow$$
$$2(\text{sedoheptulose 1,7-bisP})$$

$$2(\text{sedoheptulose 1,7-bisP}) + 2H_2O \rightarrow$$
$$2(\text{sedoheptulose 7-P}) + 2P_i$$

$$2(\text{sedoheptulose 7-P}) + 2(\text{triose-P}) \rightarrow$$
$$2(\text{xylulose 5-P}) + 2(\text{ribose 5-P})$$

$$4(\text{xylulose 5-P}) \rightarrow 4(\text{ribulose 5-P})$$

$$2(\text{ribose 5-P}) \rightarrow 2(\text{ribulose 5-P})$$

$$6(\text{ribulose 5-P}) + 6ATP \rightarrow 6(\text{ribulose 1,5-bisP}) + 6ADP$$

Net: $6CO_2 + 11H_2O + 12NADPH + 18ATP \rightarrow$
$\text{fructose 6-P} + 12NADP^+ + 6H^+ + 18ADP + 17P_i$

Our calculation shows that in order to synthesize the equivalent of 1 molecule of hexose sugar, 6 molecules of CO_2 are fixed at the expense of 18 ATP and 12 NADPH. In other words, the PCR cycle consumes 2 molecules of NADPH and 3 molecules of ATP for every CO_2 fixed. We can compute the maximal overall thermody-

namic efficiency of photosynthesis if we know the energy content of the light, the minimum quantum requirement (moles of quanta absorbed per mole of CO_2 fixed), and the energy stored in a mole of carbohydrate (hexose). Red light at 680 nm contains 175 kJ (42 kcal) per quantum mole of photons. The minimum quantum requirement is usually calculated to be 8 photons per CO_2 fixed, although experimentally the number obtained is 9 to 10 (see Chapter 8). Therefore, the minimum light energy needed to reduce 6 moles of CO_2 to a mole of hexose is $6 \times 8 \times 175 = 8400$ kJ (2016 kcal). However, a mole of hexose, such as fructose, yields only 2804 kJ (673 kcal) when totally oxidized. Comparing 8400 and 2804 kJ, we can see that the maximum overall thermodynamic efficiency of photosynthesis is about 33%. But most of the light energy is lost in the generation of ATP and NADPH by the light reactions (Chapter 8) rather than during the operation of the PCR cycle. We can calculate the efficiency of the PCR cycle more directly by computing the free energy changes associated with the hydrolysis of ATP and the oxidation of NADPH, which are 29 and 217 kJ (7 and 52 kcal) per mole, respectively. We saw in the list of reactions above that the synthesis of one fructose 6-phosphate from 6 CO_2 used 12 NADPH and 18 ATP. Therefore the PCR consumes $12 \times 217 + 18 \times 29 = 3126$ kJ (750 kcal) in the form of ATP and ADP, resulting in a thermodynamic efficiency close to 90%.

The C3 PCR Cycle Is Regulated by Light-Dependent Enzyme Activation, by Ionic Changes in the Stroma, and by Transport Processes in the Chloroplast Envelope

The high energy efficiency of the PCR cycle indicates that there must be some form of regulation ensuring that all intermediates in the cycle are present at adequate concentrations. The amount of each enzyme present in the chloroplast stroma is regulated at the genetic level by mechanisms that control the expression of the nuclear and chloroplast genomes (Whitfeld and Bottomley, 1983). In addition, short-term regulation of the PCR cycle is achieved by a variety of mechanisms.

Light-dependent ionic movements regulate several PCR cycle enzymes. Although the reactions of the cycle are "dark" reactions, the presence or absence of light induces reversible changes in the chloroplast stroma that influence the activity of various enzymes. For example, upon illumination, protons are pumped from the stroma into the lumen of the thylakoids in exchange for Mg^{2+} (Heldt, 1979). Consequently, the stromal Mg^{2+} concentration increases, and the pH rises from about 7 to about 8. The reverse is thought to occur upon darkening. Since several PCR cycle enzymes (e.g., Rubisco, fructose-1,6-bisphosphate phosphatase, phosphoribulokinase) are

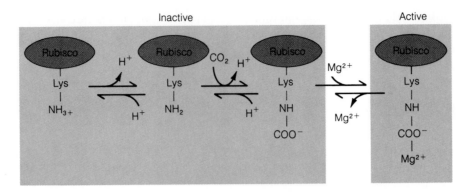

FIGURE 9.6. The activation of Rubisco involves the formation of a carbamate-Mg^{2+} complex on the ε-amino group of a lysine within the active site of the enzyme. Two protons are released, so activation is stimulated by pH increases as well as increases in the Mg^{2+} concentrations. Increases in stromal pH and Mg^{2+} concentrations are thought to be a result of illumination. Note that the CO_2 involved in the carbamate-Mg^{2+} reaction is not the same as the CO_2 involved in the carboxylation of ribulose 1,5-bisphosphate.

more active at pH 8 than at pH 7, these ionic movements increase enzyme activity.

Light-dependent ionic movements are also thought to promote the activation of Rubisco by carbamate formation (Fig. 9.6). The catalytically active species of Rubisco is formed when activator CO_2 (a different molecule from the substrate CO_2 that becomes fixed) reacts with the uncharged ε-NH_2 group of lysine 201 within the active site of the enzyme. The resulting carbamate (a new anionic site) then binds Mg^{2+} to form the activated complex. Since two protons are released during these reactions, activation is promoted by increases in pH as well as by increases in the concentration of Mg^{2+}. The extent of activation is thought to be regulated at least in part by light-dependent changes in the stroma, the concentration of Mg^{2+}, and pH.

Light also stimulates the activity of several PCR cycle enzymes via a covalent thiol-based oxidation-reduction system. These enzymes (notably fructose-1,6-bisphosphate phosphatase and phosphoribulokinase) contain two adjacent cysteinyl residues. In the light these residues exist in the reduced, sulfhydryl state (—SH HS—), in which the enzymes are active. In the dark, however, the residues are oxidized to the disulfide state (—S—S—), in which the enzymes are inactive. Buchanan (1980) and his colleagues have worked out the details of this mechanism (Fig. 9.7). In the light photosystem I reduces the iron-containing protein ferredoxin. The enzyme ferredoxin:thioredoxin reductase catalyzes the ferredoxin-dependent reduction of a disulfide bond in the sulfur-containing protein thioredoxin. Finally, the reduced thioredoxin activates the target enzyme by reduction of the critical disulfide bond. Inactivation of the target enzymes (by oxidation) upon darkening appears to occur by reversal of this pathway.

Compartmentalization also plays a role in regulation of the PCR cycle by controlling the rate at which triose

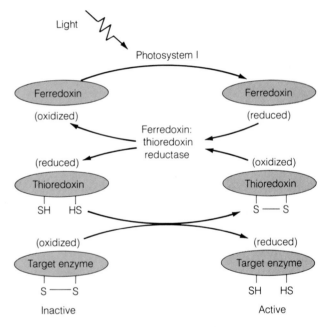

FIGURE 9.7. The ferredoxin-thioredoxin system activates (reduces) specific enzymes (e.g., fructose-1,6-bisphosphate phosphatase, phosphoribulokinase) in the light. In the dark, the enzymes are oxidized and thereby inactivated.

phosphates are withdrawn from the chloroplast in exchange for orthophosphate via the phosphate translocator in the chloroplast envelope membrane. To ensure continued operation of the PCR cycle, at least five-sixths of the triose phosphate must be recycled (Fig. 9.4); at most one-sixth can be exported for sucrose synthesis in the cytosol or diverted to starch synthesis within the chloroplast. The regulation of this aspect of photosynthetic carbon metabolism will be discussed when sucrose and starch synthesis is considered in more detail (see page 247.

The C2 Photorespiratory Carbon Oxidation (PCO) Cycle

One of the most interesting properties of Rubisco is that it catalyzes not only the carboxylation of ribulose 1,5-bisphosphate but also its oxygenation (Table 9.2, reaction 1). That oxygenation is the primary reaction in a process known as **photorespiration**. Because photosynthesis and photorespiration work in diametrically opposite directions, photorespiration results in loss of CO_2 from cells that are simultaneously fixing CO_2 by the PCR cycle.

Photosynthetic CO_2 Fixation and Photorespiratory Oxygenation of Ribulose 1,5-Bisphosphate Are Competing Reactions at the Same Active Site of Rubisco

The oxygenation of ribulose 1,5-bisphosphate results in the incorporation of one atom of molecular O_2 into one of the products of the reaction, phosphoglycolate. The ability to catalyze oxygenation of ribulose 1,5-bisphosphate is a property of all Rubiscos, regardless of taxonomic origin. Even Rubiscos from anaerobic, autotrophic bacteria catalyze the oxygenase reaction! As alternative substrates for the Rubisco, CO_2 and O_2 compete for reaction with ribulose 1,5-bisphosphate, since carboxylation and oxygenation occur within the same active site of the enzyme. Offered equal concentrations of CO_2 and O_2 in a test tube, angiosperm Rubiscos fix CO_2 about 80 times faster than they oxygenate. However, an aqueous solution in equilibrium with air at 25°C has a CO_2/O_2 ratio of 0.0416 (see box on carbon dioxide). Thus, in air carboxylation outruns oxygenation by a scant 3:1.

The C2 PCO cycle acts as a scavenger operation. The phosphoglycolate formed in the chloroplast by oxygenation of ribulose 1,5-bisphosphate is rapidly hydrolyzed to glycolate by a specific phosphatase (Table 9.2, reaction 2). Subsequent metabolism of the glycolate (Fig. 9.8) involves the cooperation of two other organelles, peroxisomes and mitochondria (Chapter 1). Glycolate leaves the chloroplast via a specific transporter protein in the envelope membrane and diffuses to a peroxisome. There it is oxidized by an oxidase (Table 9.2, reaction 3) to glyoxylate and hydrogen peroxide. The peroxide is destroyed by the action of catalase (Table 9.2, reaction 4), and the glyoxylate undergoes transamination (Table 9.2, reaction 5). The amino donor for this transamination is probably glutamate, and the product is the amino acid glycine. The glycine leaves the peroxisome and enters a mitochondrion, where two molecules of glycine are converted to one of serine (Table 9.2, reactions 6 and 7). Glycine is thus the immediate source of the photorespiratory CO_2. Serine leaves the mitochondrion and enters a peroxisome, where it is

converted first by transamination (Table 9.2, reaction 8) to hydroxypyruvate and then by reduction to glycerate (Table 9.2, reaction 9). Finally, glycerate reenters the chloroplast, where it is phosphorylated (Table 9.2, reaction 10) to yield 3-phosphoglycerate. Thus, two molecules of phosphoglycolate (four carbon atoms), lost from the C3 PCR cycle by the oxygenation of ribulose bisphosphate (Table 9.2, reaction 1), are converted into one molecule of 3-phosphoglycerate (three carbon atoms) and one of CO_2. In other words, 75% of the carbon is recovered by the C2 PCO cycle and returned to the C3 PCR cycle (Lorimer, 1981).

For many years the mechanism of (phospho)glycolate synthesis in leaves was an enigma. Studies by Tolbert, Zelitch, and others showed that glycolate is involved in photorespiration and characterized its metabolism. The discovery that Rubisco also catalyzes the oxygenation of ribulose bisphosphate was the key to understanding the relationship between the PCO cycle (Fig. 9.8) and the PCR cycle (Fig. 9.2).

The evidence in support of the PCO cycle is as strong as that in support of the PCR cycle (Ogren, 1984). After a short period of labeling in $^{14}CO_2$, C-1 and C-2 of glycolate are equally labeled. So, too, are C-1 and C-2 of ribulose bisphosphate. In vivo, one of the carboxyl oxygen atoms of glycolate is derived from molecular O_2. In vitro, Rubisco catalyzes the incorporation of one atom of molecular O_2 into the carboxyl group of phosphoglycolate. In mutant plants lacking phosphoglycolate phosphatase (Table 9.2, reaction 2), radioactivity from $^{14}CO_2$ accumulates in phosphoglycolate under conditions favoring photorespiration (high concentrations of O_2, low concentrations of CO_2), but not under nonphotorespiratory conditions (low O_2, high CO_2). Finally, glycolate synthesis in vivo is dependent on O_2 and is inhibited by CO_2 in a competitive manner. In vitro, the carboxylation of ribulose 1,5-bisphosphate is inhibited by O_2 (competitively with respect to CO_2), whereas oxygenation is inhibited by CO_2 (competitively with respect to O_2). All these observations strongly indicate that oxygenation of ribulose bisphosphate is the primary event in the PCO cycle.

Studies with specifically labeled intermediates such as $[1\text{-}^{14}C]$glycolate, $[2\text{-}^{14}C]$glycolate, or $(3\text{-}^{14}C)$serine have shown that glycolate is metabolized via the PCO cycle. When the labeled intermediates were supplied ("fed") to illuminated leaves or algae, the ^{14}C moved quickly from glycolate through the other PCO cycle intermediates into the sugar phosphates of the PCR cycle. For example, feeding of $[2\text{-}^{14}C]$glycolate resulted in the formation of $[2,3\text{-}^{14}C]$serine, $[2,3\text{-}^{14}C]$glycerate, and hexoses labeled predominantly in carbons 1, 2, 5, and 6. Feeding of $[1\text{-}^{14}C]$glycolate resulted in the formation of $^{14}CO_2$, $[1\text{-}^{14}C]$serine, $[1\text{-}^{14}C]$glycerate, and hexose labeled predominantly in C-3 and C-4. These labeling patterns are entirely consistent with the formulation of the PCO

TABLE 9.2. Reactions of the C2 photorespiratory carbon oxidation cycle

1. *Ribulose-1,5-bisphosphate oxygenase (chloroplast)*
   ribulose 1,5-bisphosphate + O_2 → 2-phosphoglycolate + 3-phosphoglycerate + $2H^+$

2. *Phosphoglycolate phosphatase (chloroplast)*
   phosphoglycolate + H_2O → glycolate + $HOPO_3^{2+}$

3. *Glycolate oxidase (peroxisome)*
   glycolate + O_2 → glyoxylate + H_2O_2

4. *Catalase (peroxisome)*
   $2H_2O_2$ → $2H_2O$ + O_2

5. *Glyoxylate:glutamate aminotransferase (peroxisome)*
   glyoxylate + glutamate → glycine + α-ketoglutarate

6. *Glycine decarboxylase (mitochondrion)*
   glycine + NAD^+ + H_4-folate → NADH + H^+ + CO_2 + NH_3 + methylene H_4-folate

7. *Serine hydroxymethyltransferase (mitochondrion)*
   methylene H_4-folate + H_2O + glycine → serine + H_4-folate

8. *Serine aminotransferase (peroxisome)*
   serine + α-ketoglutarate → hydroxypyruvate + glutamate

9. *Hydroxypyruvate reductase (peroxisome)*
   hydroxypyruvate + NADH + H^+ → glycerate + NAD^+

10. *Glycerate kinase (chloroplast)*
    glycerate + ATP → 3-phosphoglycerate + ADP + H^+

cycle shown in Figure 9.8. Similarly, ^{18}O from molecular $^{18}O_2$ has been traced as it progressively labels the PCO cycle intermediates from glycolate through 3-phosphoglycerate.

As for the PCR cycle, enzymes have been shown to catalyze all the proposed reactions of the PCO cycle in vitro at rates higher than the in vivo rate of photorespiration. These enzymes are localized in the chloroplasts (Rubisco, phosphoglycolate phosphatase, glycerate kinase), the peroxisomes (glycolate oxidase, catalase, hydroxypyruvate reductase, and the two aminotransferases), and the mitochondria (glycine decarboxylase and serine hydroxymethyltransferase) (Table 9.2) (Hatch and Osmond, 1976). The transport of PCO cycle intermediates between these organelles is inferred from the locations of these enzymes (Tolbert, 1981).

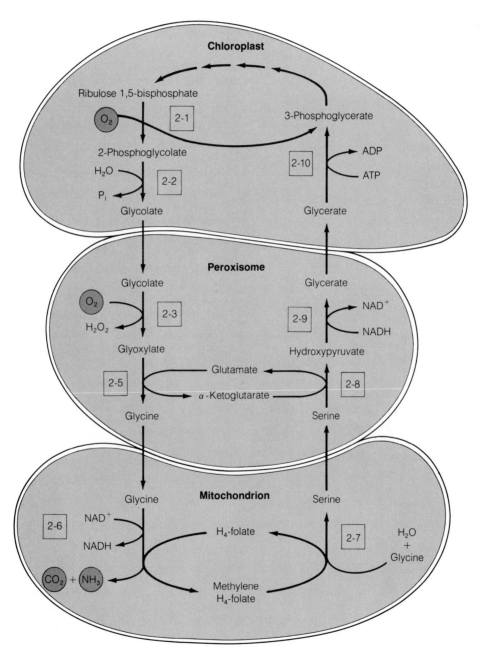

FIGURE 9.8. The operation of the C2 photorespiratory carbon oxidation cycle involves cooperative interactions between three organelles: chloroplasts, mitochondria, and peroxisomes. Oxygen consumption occurs in each organelle. Shuttling of metabolites between these organelles is inferred from the localization of the various enzymes. The transport of metabolites across the organellar membranes is not well understood, but it is thought that transport of glycolate out the chloroplast and of glycerate back into the chloroplast involves a transporter protein in the chloroplast membrane. The reactions are more fully described in Table 9.2; here 2-1, 2-2, etc. refer to reactions 1, 2, etc. in that table.

Competition Between the Carboxylation and Oxygenation Reactions in Vivo Decreases the Thermodynamic Efficiency of Photosynthesis

Since photorespiration is concurrent with photosynthesis, it is difficult to measure the rate of photorespiration in intact cells. Two molecules of phosphoglycolate, a total of four carbon atoms, are needed to make one molecule of 3-phosphoglycerate, three carbon atoms, and one molecule of CO_2; so theoretically, the fraction of the carbon entering the PCO cycle that is released as CO_2 is 0.25. Steady-state measurements with sunflower leaves in air at 25°C yielded values for the rate of CO_2 release by the PCO that were 20–25% of the rate of net photosynthesis. This indicates that the rate of "true" photosynthesis is 120–125% of the rate of net photosynthesis. This "true" rate can be equated with the in vivo rate of carboxylation of ribulose bisphosphate. Stoichiometry requires that the in vivo rate of oxygenation of ribulose bisphosphate is twice the rate of photorespiratory CO_2 release; so oxygenation in vivo must proceed in air at 25°C at 40–50% of the rate of net photosynthesis. On the basis of these physiological measurements, the ratio of carboxylation to oxygenation in air at 25°C is computed to be between 2.5 and 3. This value is reassuring, since it is close to that predicted from the in vitro kinetic properties of Rubisco and the concentrations of the substrates CO_2 and O_2.

The energetic consequences of the simultaneous carboxylation and oxygenation by Rubisco are significant (Table 9.3). Carboxylation of ribulose bisphosphate and operation of the PCR cycle uses 3 ATP and 2 NADPH (521 kJ, or 125 kcal) for every CO_2 fixed. Oxygenation of ribulose bisphosphate and operation of the PCO cycle consumes 2 ATP and 2.5 NAD(P)H (600 kJ, or 144 kcal) for each ribulose bisphosphate oxygenated.

In air, with a carboxylation/oxygenation ratio of 3:1, considerably more energy in the form of ATP and NADPH must be supplied to offset the photorespiratory losses. Three ribulose bisphosphates carboxylated plus one ribulose bisphosphate oxygenated requires an input of $3 \times 521 + 600 = 2163$ kJ (519 kcal) for net fixation of 2.5 CO_2. Thus, in air the oxygenation of ribulose bisphosphate and operation of the PCO cycle increase the energy requirement (ATP + NADPH) from 521 kJ (125 kcal) per mole of CO_2 fixed (p. 227) to 867 kJ (208 kcal) per mole of CO_2 fixed. The thermodynamic efficiency is reduced from 90% ($467/521 = 0.90$) to 54% ($467/867 = 0.54$). This decreased efficiency is reflected by increases in the measured quantum requirement for CO_2 fixation under photorespiratory conditions (normal air with high O_2 and low CO_2) as opposed to nonphotorespiratory conditions (low O_2 and high CO_2).

The Relationship Between Carboxylation and Oxygenation in the Intact Leaf Depends On the Kinetic Properties of Rubisco, Prevailing Temperatures, and the Concentrations of CO_2 and O_2

Photosynthetic carbon metabolism is presently understood as the integrated balance of two mutually opposing and interlocking cycles (Fig. 9.9). The PCR cycle is capable of independent operation, but the PCO cycle is ultimately "parasitic," depending on the PCR cycle for the supply of ribulose bisphosphate. The balance between the two cycles is determined by the kinetic properties of Rubisco and the concentrations of the substrates CO_2 and O_2. Aside from the concentrations of CO_2 and O_2, the major environmental influence on the balance between the PCR and PCO cycles is temperature. As the temperature increases, the concentration of CO_2 in a solution in equilibrium with air decreases (see box on carbon dioxide) more than the concentration of O_2. As a result, the concentration ratio $[CO_2]/[O_2]$ decreases as the temperature rises. This temperature effect on the relative solubilities of CO_2 and O_2 results in increased oxygenation (photorespiration) relative to carboxylation (photosynthesis) as the temperature increases. Compounding this effect, the kinetic properties of Rubisco are influenced by temperature increases, which also enhance oxygenation relative to carboxylation (Ku and Edwards, 1978). Overall, then, increasing temperatures progressively tilt the balance between the PCR and PCO cycles toward the latter (Chapter 10).

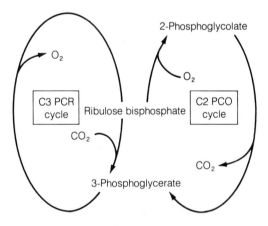

FIGURE 9.9. The flow of carbon in the leaf is shaped by the balance between two mutually opposing cycles. Operation of the C3 PCR cycle results in consumption of CO_2, release of O_2, and a gain in dry matter (carbon gain). In contrast, operation of the C2 PCO cycle results in release of CO_2, consumption of O_2, and a loss of dry matter (carbon loss). In normal air (see box on carbon dioxide) the two cycles operate simultaneously, carboxylating and oxygenating ribulose bisphosphate in a ratio of about 3:1. The C3 PCR cycle is capable of independent operation, whereas the C2 PCO cycle requires continued operation of the C3 PCR cycle to regenerate its starting material, ribulose bisphosphate.

TABLE 9.3. Energetic consequences of carboxylating or oxygenating ribulose bisphosphate*

Energetic consequences of carboxylating ribulose 1,5-bisphosphate

$3(\text{ribulose } 1,5\text{-P}_2) + 3CO_2 + 3H_2O$	$\rightarrow$ 6(3-P-glycerate)
6(3-P-glycerate) + 6ATP + 6(NAD(P)H + H⁺)	$\rightarrow$ 6(triose-P) + 6ADP + 6P$_i$ + 6NAD(P)⁺
5(triose-P)	$\rightarrow$ 3(ribulose 5-P) + 2P$_i$
3(ribulose 5-P) + 3ATP	$\rightarrow$ $3(\text{ribulose } 1,5\text{-P}_2) + 3ADP$

Net: $3CO_2 + 3H_2O + 9ATP + 6(NAD(P)H + H^+)$ → triose-P + 9ADP + 6(NAD(P)⁺) + 8P$_i$

> Cost of fixing one $CO_2 = 3ATP + 2(NAD(P)H + H^+)$

Energetic consequences of oxygenating ribulose 1,5-bisphosphate

10(ribulose 5-P) + 10ATP	$\rightarrow$ $10 \text{ ribulose } 1,5\text{-P}_2 + 10ADP$
$10(\text{ribulose } 1,5\text{-P}_2) + 10O_2$	$\rightarrow$ 10(2-P-glycolate) + 10(3-P-glycerate)
10(2-P-glycolate) + 10 glutamate	$\rightarrow$ 10P$_i$ + 10 glycine + 10 α-ketoglutarate
10 glycine + 15ADP + 15P$_i$	$\rightarrow$ 15ATP + 5NH$_3$ + 5CO$_2$ + 5 serine
5 serine + 5 α-ketoglutarate	$\rightarrow$ 5 glutamate + 5 hydroxypyruvate
5 hydroxypyruvate + 5(NAD(P)H + H⁺)	$\rightarrow$ 5 glycerate + 5NAD(P)⁺
5 glycerate + 5ATP	$\rightarrow$ 5(3-P-glycerate) + 5ADP
5NH$_3$ + 5ATP + 5 glutamate	$\rightarrow$ 5 glutamine + 5ADP + 5P$_i$
5 glutamine + 5 α-ketoglutarate + 5(NAD(P)H + H⁺)	$\rightarrow$ 10 glutamate
15(3-P-glycerate) + 15ATP + 15(NAD(P)H + H⁺)	$\rightarrow$ 15-triose-P + 15ADP + 15P$_i$
15 triose-*P*	$\rightarrow$ 9 ribulose 5-P + 6P$_i$

Net: ribulose 5-P + O$_2$ + 2ATP + 2.5(NAD(P)H + H⁺) → 0.9(ribulose 5-P) + 0.5CO$_2$ + 2ADP + 2.5NAD(P)⁺ + 2P$_i$

> Cost of fixing one $O_2 = 2ATP + 2.5(NAD(P)H + H^+)$

* The notation NAD(P)H indicates that the involved reactions can utilize either NADPH or NADH.

The Biological Function of Photorespiration Remains an Intriguing Question in Biology

Although it is true that the PCO cycle recovers 75% of the carbon originally lost from the PCR cycle as phosphoglycolate, why is phosphoglycolate formed in the first place? One possible explanation is that phosphoglycolate formation is a consequence of the chemistry of the carboxylation reaction, which would involve an intermediate that can react with both CO_2 and O_2. Such a reaction would have had little consequence in early evolutionary times if the CO_2/O_2 ratio in air was higher than it is today. In this view, photorespiration is a response to the low CO_2/O_2 ratios prevalent in the present-day atmosphere and has no functional role. Another possible explanation is that photorespiration is necessary under conditions of high light intensity and low CO_2 concentration (e.g., when stomata are closed because of water stress) to dissipate excess ATP and reducing power from the light reactions, thus preventing damage to the photosynthetic apparatus. To date, there is no conclusive evidence for a role of photorespiration in the carbon metabolism of the leaf.

Many plants do not photorespire. This is not because their Rubiscos have different properties; rather, it is a consequence of interesting mechanisms that concentrate CO_2 in the Rubisco environment. In the presence of sufficiently high CO_2 concentrations, the oxygenation reaction is suppressed.

Three mechanisms for concentrating CO_2 at the site of carboxylation have been recognized. One is a **CO_2 pump**, and it is confined to aquatic plants. It has been extensively studied with unicellular cyanobacterial and algal cells. The second and third systems of CO_2 concentration are found in angiosperms and involve a different type of carbon assimilation called the C4 photosynthetic carbon assimilation (PCA) cycle. Plants that have the PCA cycle also have the PCR cycle, but the PCA cycle is separated from the PCR cycle either spatially, by a specialized leaf anatomy (the classical C4 photosynthesis), or temporally, by a metabolic device known as **Crassulacean acid metabolism (CAM)**.

CO₂ Concentrating Mechanisms I: Algal and Cyanobacterial CO₂/HCO₃⁻ Pumps

When algal and cyanobacterial cells are grown in air enriched with 5% CO_2, they display the symptoms typical of photorespiration (O_2 inhibition of photosynthesis at low concentration of CO_2). However, if the cells are transferred to air containing 0.03% CO_2 they rapidly develop the ability to concentrate inorganic carbon (CO_2 plus HCO_3^-) within themselves. Under these conditions, the cells lose the ability to photorespire. The accumulation of inorganic carbon is accomplished by

CO_2/HCO_3^- pump(s) at the plasma membrane (Chapter 6) that are driven by energy derived from the light reactions. It is not always clear which species of inorganic carbon (CO_2 and/or HCO_3^-) is transported, but substantial differences in the concentration of inorganic carbon inside and outside the cells have been measured (up to 10^3-fold in some cyanobacterial cells) (Ogawa and Kaplan, 1987). The proteins functioning as CO_2/HCO_3^- pumps are not present in cells growing in high concentrations of CO_2 but are induced upon exposure to low concentrations of CO_2. The metabolic consequence of this CO_2 enrichment is suppression of the oxygenation of ribulose bisphosphate and hence also of photorespiration.

CO₂ Concentrating Mechanisms II: The C4 Photosynthetic Carbon Assimilation (PCA) Cycle

There are differences in leaf anatomy between plants that have a C4 PCA cycle (called C4 plants) and those that photosynthesize solely via the PCR and PCO cycles (called C3 plants). A cross section of a typical C3 leaf reveals essentially one type of photosynthetic, chloroplast-containing cell, the **mesophyll**. In constrast, a typical C4 leaf has two distinct chloroplast-containing cell types, the mesophyll and the **bundle sheath** (or *Kranz*, German for "wreath") cells (Fig. 9.10). There is considerable anatomic variation in the arrangement of the bundle sheath cells with respect to the mesophyll and vascular tissue. However, the operation of the C4 PCA cycle requires the cooperative effort of both cell types, and no mesophyll cell of a C4 plant is more than two or three cells distant from the nearest bundle sheath cell (Fig. 9.10a). An extensive network of plasmodesmata connects mesophyll and bundle sheath cells, providing a pathway for the flow of metabolites between the cells (Fig. 9.10d).

The C4 PCA Cycle Increases the Concentration of CO₂ in the Bundle Sheath Cells

The basic C4 PCA cycle (Fig. 9.11) consists of four stages: (1) assimilation of CO_2, involving carboxylation of phosphoenolpyruvate in the mesophyll cells to form the C4 acids (malate and/or aspartate); (2) transport of C4 acids to the bundle sheath cells; (3) decarboxylation of the C4 acids within the bundle sheath cells and generation of CO_2, which is reduced to carbohydrate via the C3 PCR cycle; and (4) transport of the C3 acid formed by the decarboxylation (pyruvate or alanine) back to the mesophyll cell and regeneration of the CO_2 acceptor, phosphoenolpyruvate. Thus, the PCA cycle effectively shuttles CO_2 from the atmosphere into the

bundle sheath cells. This transport process generates a much higher concentration of CO_2 in the bundle sheath cells than would occur in equilibrium with the external atmosphere. This elevated concentration of CO_2 at the site of carboxylation of ribulose bisphosphate results in suppression of ribulose bisphosphate oxygenation (photorespiration).

The PCA cycle was first discovered in tropical grasses (e.g., sugarcane and maize) and is now known to occur in 16 plant families. It occurs in both monocotyledonous and dicotyledonous plants and is particularly prominent in species of the Gramineae, Chenopodiaceae, and Cyperaceae (Edwards and Walker, 1983). This distribution seems to indicate that the PCA cycle evolved independently on several occasions, a conclusion supported by the fact that there are three variations on the basic PCA pathway.

The three variations of the PCA cycle are shown in Table 9.4 and Figure 9.12. They differ principally in the C4 acid transported into the bundle sheath cells (malate and aspartate) and in the manner of decarboxylation, and they are named after the enzymes that catalyze their decarboxylation reactions: NADP-dependent malic enzyme (NADP-ME), found in chloroplasts; NAD-dependent malic enzyme (NAD-ME), found in mitochondria; and phosphoenolpyruvate (PEP) carboxykinase (PEP-CK). The primary carboxylation reaction, common to all three variants, occurs in the cytosol of the mesophyll cells and is catalyzed by phosphoenolpyruvate carboxylase, using HCO_3^-, rather than CO_2, as a substrate (Table 9.5, reaction 1). The fate of the oxaloacetate produced in this reaction depends on the PCA variant (Gutierrez et al., 1974). In the NADP-ME species, oxaloacetate is rapidly reduced to malate in the mesophyll choroplasts by NADPH (Table 9.5, reaction 2). In the NAD-ME and PEP-CK species, oxaloacetate undergoes transamination (Table 9.5, reaction 3) in the cytosol, with glutamate as the amino donor. The C4 acids are then transported to the bundle sheath cells, where they are further metabolized in reactions specific to the PCA variant. In the NADP-ME species, the malate enters the chloroplast of the bundle sheath cell and there undergoes oxidative decarboxylation (Table 9.5, reaction 4). In the NAD-ME and PEP-CK species, the aspartate transported into the bundle sheath cells is first reconverted to oxaloacetate by transamination (Table 9.5, reaction 3) in the mitochondrion (NAD-ME species) or the cytosol (PEP-CK species). The oxaloacetate in the mitochondrion of the NAD-ME species is then reduced (Table 9.5, reaction 2) and decarboxylated (Table 9.5, reaction 4) by NAD-malic enzyme. The oxaloacetate in the cytosol of the PEP-CK species is decarboxylated by PEP carboxykinase (Table 9.5, reaction 5). In all three PCA variants, the CO_2 released within the bundle sheath cells is reduced to carbohydrate by the PCR cycle.

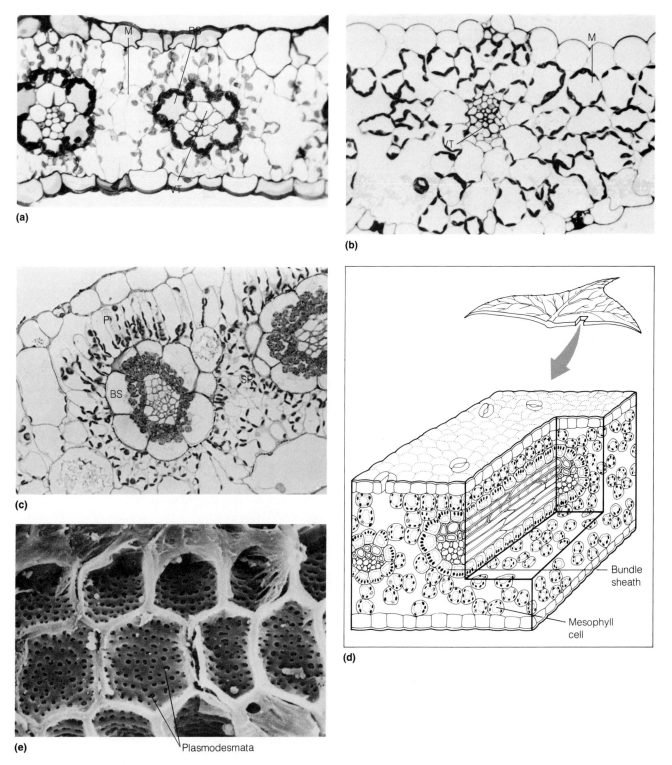

FIGURE 9.10. Cross sections of leaves showing the anatomic differences between C3 and C4 plants: (a) a C4 monocot, *Zea mays* (350 ×); (b) a C3 monocot, *Avena sativa* (380 ×); (c) a C4 dicot *Gomphrena* (740 ×). (d) A tridimensional model of a C4 leaf. The bundle sheath cells are large in the C4 leaves (a and c), and no mesophyll cell is more than two or three cells away from the nearest bundle sheath cell. These anatomic features are absent in the C3 leaf (b). (a, b, and c from Edwards and Walker, 1983; photographs by S. E. Frederick and E. H. Newcomb; d from Lüttge and Higinbotham, 1979.) (e) A scanning electron micrograph of a C4 leaf from *Triodia irritans* (1450 ×), showing the plasmodesmata in the bundle sheath cell walls through which transport of metabolites of the C4 PCA cycle is thought to occur. (From Craig and Goodchild, 1977.)

TABLE 9.4. Three variants of the C4 photosynthetic carbon assimilation cycle

Principal C4 acid transported to the bundle sheath cells	Decarboxylating enzyme	Variant name	Principal C3 acid returned to mesophyll cells	Examples
Malate	NADP-dependent malic enzyme (chloroplast)	NADP-ME	Pyruvate	Maize, crabgrass sugarcane, sorghum
Aspartate	NAD-dependent malic enzyme (mitochondria)	NAD-ME	Alanine	Millet, pigweed (*Panicum miliaceum*)
Aspartate	Phosphoenolpyruvate carboxykinase (cytoplasm)	PEP-CK	Alanine/pyruvate	Guinea grass (*Panicum maximum*)

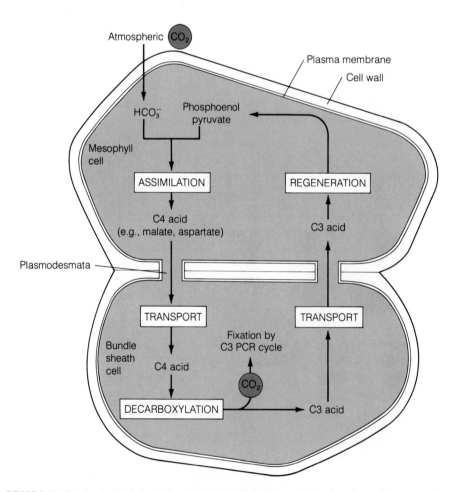

FIGURE 9.11. The basic C4 photosynthetic carbon assimilation cycle involves two cell types and proceeds in four stages: *assimilation* of CO_2 and its incorporation into a four-carbon acid in the mesophyll cells; *transport* of the four-carbon acid from the mesophyll cells to the bundle sheath cells; *decarboxylation* of the four-carbon acid, generating a high concentration of CO_2 in the bundle sheath cells; and *transport* of a three-carbon acid back to the mesophyll cells, where the original CO_2 acceptor, phosphoenolpyruvate, is regenerated. The C4 PCA cycle effectively concentrates CO_2 in the bundle sheath cells, to which Rubisco and the other enzymes of the C3 PCR cycle are restricted. This high concentration of CO_2 in the bundle sheath cells effectively suppresses oxygenation via the C2 PCO cycle.

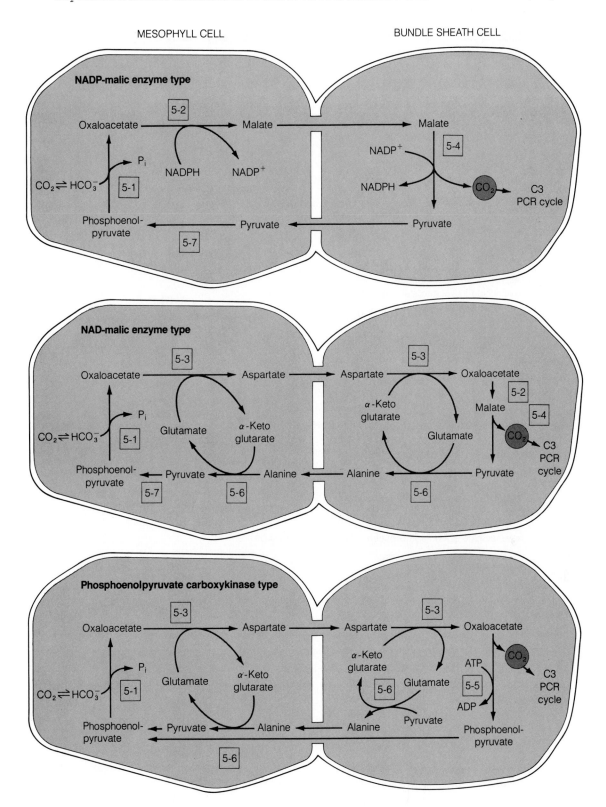

FIGURE 9.12. Three variants of the C4 photosynthetic carbon assimilation cycle. The variants differ principally in (1) the nature of the four-carbon acid (malate or aspartate) transported into the bundle sheath cell and of the three-carbon-acid (pyruvate or alanine) returned to the mesophyll cell and (2) the nature of the enzyme catalyzing the decarboxylation step in the bundle sheath cell. The three variants are named after the decarboxylation mechanism. The reactions (5-1, 5-2, etc.) are more fully described in Table 9.5. Within a mesophyll cell or a bundle sheath cell, different reactions can occur in different compartments, such as the cytosol, the mitochondrion, or the chloroplast.

TABLE 9.5. Reactions of the C4 photosynthetic carbon assimilation cycle

1. *Phosphoenolpyruvate carboxylase*
   phosphoenolpyruvate + HCO_3^- → oxaloacetate + $HOPO_3^{2-}$

$$
\begin{array}{ll}
CH_2 & COO^- \\
\| & | \\
C\!-\!O\!-\!PO_3^{2-} & CH_2 \\
| & | \\
COO^- & C\!=\!O \\
 & | \\
 & COO^-
\end{array}
$$

2. *NADP malate dehydrogenase*
   oxaloacetate + NADPH + H^+ → malate + $NADP^+$

$$
\begin{array}{ll}
COO^- & COO^- \\
| & | \\
CH_2 & CH_2 \\
| & | \\
C\!=\!O & H\!-\!C\!-\!OH \\
| & | \\
COO^- & COO^-
\end{array}
$$

3. *Aspartate aminotransferase*
   oxaloacetate + glutamate → aspartate + α-ketoglutarate

$$
\begin{array}{llll}
COO^- & COO^- & COO^- & COO^- \\
| & | & | & | \\
CH_2 & CH_2 & CH_2 & CH_2 \\
| & | & | & | \\
C\!=\!O & CH_2 & H\!-\!C\!-\!NH_2 & CH_2 \\
| & | & | & | \\
COO^- & H\!-\!C\!-\!NH_2 & COO^- & C\!=\!O \\
 & | & & | \\
 & COO^- & & COO^-
\end{array}
$$

4. *NADP malic enzyme*
   malate + $NADP^+$ → pyruvate + CO_2 + NADPH + H^+

$$
\begin{array}{ll}
COO^- & CH_3 \\
| & | \\
CH_2 & C\!=\!O \\
| & | \\
H\!-\!C\!-\!OH & COO^- \\
| & \\
COO^- &
\end{array}
$$

5. *Phosphoenolpyruvate carboxykinase*
   oxaloacetate + ATP → phosphoenolpyruvate + CO_2 + ADP

$$
\begin{array}{ll}
COO^- & CH_2 \\
| & \| \\
CH_2 & C\!-\!O\!-\!PO_3^{2-} \\
| & | \\
C\!=\!O & COO^- \\
| & \\
COO^- &
\end{array}
$$

6. *Alanine aminotransferase*
   pyruvate + glutamate ⇌ alanine + α-ketoglutarate

$$
\begin{array}{llll}
CH_3 & COO^- & CH_3 & COO^- \\
| & | & | & | \\
C\!=\!O & CH_2 & H\!-\!C\!-\!NH_2 & CH_2 \\
| & | & | & | \\
COO^- & CH_2 & COO^- & CH_2 \\
 & | & & | \\
 & H\!-\!C\!-\!NH_2 & & C\!=\!O \\
 & | & & | \\
 & COO^- & & COO^-
\end{array}
$$

7. *Pyruvate,orthophosphate dikinase*
   pyruvate + $HOPO_3^{2-}$ + ATP → phosphoenolpyruvate + AMP + $H_2P_2O_7^{2-}$

$$
\begin{array}{ll}
CH_3 & CH_2 \\
| & \| \\
C\!=\!O & C\!-\!O\!-\!PO_3^{2-} \\
| & | \\
COO^- & COO^-
\end{array}
$$

Which species of C3 acid is returned to the mesophyll also depends on the PCA variant. In the NADP-ME species, the C3 acid transported back to the mesophyll is pyruvate. In the other two variants it is probably alanine. In all three variants, a four-carbon dicarboxylic acid is transported into the bundle sheath cell in exchange for a three-carbon monocarboxylic acid. The transport of these metabolites is probably accompanied by compensatory movements of protons and other ionic species in order to maintain pH and charge balance. In the mesophyll cytosol of the NAD-ME and PEP-CK species, alanine appears to be converted to pyruvate via transamination (Table 9.5, reaction 6). The final step of the PCA cycle, which is common to all three variants, involves the conversion of pyruvate to phosphoenolpyruvate (Table 9.5, reaction 7) within the mesophyll chloroplast.

Malate and Aspartate Are the First Stable Products of Carboxylation by the C4 PCA Cycle

The discovery of the C4 PCA cycle can be traced to studies involving $^{14}CO_2$ labeling of sugarcane by Hugo Kortschack and colleagues in Hawaii and of maize by Y. Karpilov and co-workers in the Soviet Union. When the leaves were exposed for a few seconds to $^{14}CO_2$ in the light, 70–80% of the label was found in the C4 acids malate and aspartate—a result very different from the pattern expected for leaves that photosynthesize solely via the C3 PCR cycle. In pursuing these initial observations, Hal Hatch and Roger Slack in Australia elucidated what is now known as the C4 PCA cycle (Hatch and Slack, 1966). Using sugarcane leaves, they established that the C4 acids malate and aspartate were the first stable, detectable intermediates of photosynthesis in leaves of C4 plants and that carbon atom 4 of malate subsequently became carbon atom 1 of 3-phosphoglycerate. Clearly, the primary carboxylation in these leaves was not catalyzed by Rubisco but rather by PEP carboxylase (O'Leary, 1982).

The manner in which carbon was transferred from carbon atom 4 of the C4 acids to carbon atom 1 of 3-phosphoglycerate became clear when the involvement of two cell types (mesophyll and bundle sheath cells) was fully appreciated. It then became evident that the enzymes involved are divided between the two cell types: PEP carboxylase and pyruvate,orthophosphate dikinase are restricted to the mesophyll cells, while the decarboxylases and the enzymes of the complete PCR cycle are confined to the bundle sheath cells. With this knowledge, Hatch and Slack were able to formulate the basic model of the PCA cycle (Fig. 9.11). Subsequent work by Hatch, Edwards, and others using isolated leaf cells and organelles derived from those cells uncovered the three variants of the PCA cycle (Edwards and Walker, 1983). In most cases, enzymes from appropriate cells and organelles have been shown to catalyze the proposed reactions in vitro, at rates higher than the observed in vivo rates of photosynthesis by C4 species.

The CO_2 concentration in the bundle sheath cells can be calculated from measurements of the CO_2 released in the cells and the cell volume. Estimates of up to 60 μM CO_2 have been obtained (Edwards and Walker, 1983). This value is 8- to 10-fold higher than the estimated concentration of CO_2 in the leaves of C3 plants photosynthesizing in normal air. Thus, the PCA cycle truly increases the concentration of CO_2 at the site of carboxylation of ribulose bisphosphate.

The Energy Cost of Concentrating CO_2 in the Bundle Sheath Cells Lowers the Efficiency of Photosynthesis

The net effect of the C4 PCA cycle is to convert the dilute solution of CO_2 found in the mesophyll cells into a concentrated CO_2 solution in the bundle sheath cells. Thermodynamics (Chapter 2) tells us that work must be done to establish and maintain a concentration gradient; and this principle also holds true for operation of the C4 PCA cycle. From a summation of the reactions involved (Table 9.6), we can calculate the energy cost to the plant. The calculation shows that the CO_2- concentrating mechanism is coupled to the hydrolysis of two ATP per CO_2 transported. Thus, the total energy requirement for fixing CO_2 by the combined C4 PCA and C3 PCR cycles is

TABLE 9.6. Energetics of the C4 photosynthetic carbon assimilation cycle

phosphoenolpyruvate + H_2O + NADPH + CO_2 (mesophyll)	→ malate + $NADP^+$ + $HOPO_3^{2-}$
malate + $NADP^+$	→ pyruvate + NADPH + CO_2 (bundle sheath cell)
pyruvate + $HOPO_3^{2-}$ + ATP	→ phosphoenolpyruvate + AMP + $2H^+$ + $P_2O_7^{4-}$
$P_2O_7^{4-}$ + H_2O	→ $2HOPO_3^{2-}$
AMP + ATP	→ 2ADP
Net: CO_2 (mesophyll) + 2ATP + $2H_2O$	→ CO_2 (bundle sheath cell) + 2ADP + $2HOPO_3^{2-}$ + $2H^+$

Cost of concentrating CO_2 within the bundle sheath cell = 2ATP per CO_2

five ATP plus two NADPH per CO_2 fixed. Because of this higher energy demand, C4 leaves photosynthesizing under nonphotorespiratory conditions (high CO_2 and low O_2) require more quanta of light per CO_2 fixed than C3 leaves. In normal air, the quantum requirement of C3 leaves changes with factors that affect the balance between photosynthesis and photorespiration, such as temperature (see Fig. 10.14), while the quantum requirement of C4 plants remains relatively constant.

Light Regulates the Activity of Key Enzymes of the C4 PCA Cycle

The operation of the C4 PCA cycle requires several levels of metabolic regulation. Shuttling of key metabolites between mesophyll and bundle sheath cells is probably carefully regulated, but that aspect of C4 metabolism is poorly understood. On the other hand, the effect of light on enzyme activation has been intensively studied. For example, the activities of phosphoenolpyruvate carboxylase (Table 9.5, reaction 1), NADP malate dehydrogenase (Table 9.5, reaction 2), and pyruvate,orthophosphate dikinase (Table 9.5, reaction 7) are regulated in response to variations in the light intensity. NADP malate dehydrogenase is subject to regulation via the thioredoxin system (Fig. 9.7); so it is

reduced (activated) upon illumination of the leaves and oxidized (inactivated) upon darkening.

Pyruvate,orthophosphate dikinase is rapidly inactivated when the light intensity is decreased and is activated by increased illumination (Burnell and Hatch, 1985) (Fig. 9.13). Inactivation results from ADP-dependent phosphorylation of a threonine residue on the enzyme. Activation is accomplished by phosphorolytic cleavage of this threonyl phosphate group. Both reactions, phosphorylation and pyrophosphorolysis, appear to be catalyzed by a single protein—pyruvate,orthophosphate dikinase-regulatory protein. The catalytic action of pyruvate,orthophosphate dikinase includes the formation of a phosphorylated histidine residue. In the normal course of catalysis, this histidinyl phosphate is transferred to pyruvate to generate phosphoenolpyruvate. In the dark, however, when there is insufficient pyruvate, the active, histidinyl phosphorylated enzyme becomes the substrate for an ADP-dependent phosphorylation, catalyzed by the regulatory protein, that inactivates the enzyme. The same regulatory protein also catalyzes the pyrophosphorolytic reactivation of pyruvate,orthophosphate dikinase. It is clear from Figure 9.13 that, besides the level of pyruvate, the levels of orthophosphate and of the adenylates (AMP, ADP, and ATP) will determine how much of the available enzyme is in the catalytically active form.

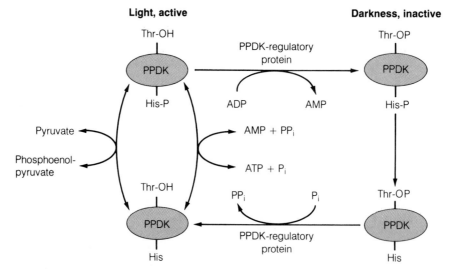

FIGURE 9.13. The regulation of the enzyme pyruvate,orthophosphate dikinase (PPDK) involves inactivation through the phosphorylation of a threonine residue, a reaction that is catalyzed by the enzyme PPDK-regulatory protein. The same regulatory protein catalyzes the phosphorolysis of the phosphothreonine residue, regenerating active PPDK. Illumination of the leaves appears to favor the formation of active PPDK, whereas darkness leads to the inactivation of PPDK. The histidine residue of PPDK is also phosphorylated and dephosphorylated in the cycle, but those reactions are not well understood.

In Hot, Dry Climates, the Operation of the C4 PCA Cycle Reduces Photorespiration and Water Loss

We saw earlier (p. 232) that one of the principal environmental factors influencing photorespiration in C3 plants is temperature. At higher temperatures, the oxygenase activity of Rubisco is increasingly favored over carboxylation, with a concomitant stimulation of photorespiration and loss of photosynthetic efficiency. In addition, higher temperatures lower the relative humidity of the air and increase the humidity gradient between the leaf and the air. As a consequence, the rate of water loss may increase. Combined, these effects can result in a drastic reduction of the water-use efficiency (Chapter 4) of C3 plants as the temperature rises.

In C4 plants, two features of the C4 PCA cycle overcome the deleterious effects of increasing temperature. First, the affinity of phosphoenolpyruvate carboxylase for its substrate, HCO_3^-, is so high that it is effectively saturated by the equivalent of air levels of CO_2. This enables C4 plants to reduce the stomatal aperture (and thereby conserve water) while fixing CO_2 at rates equal to or greater than those of C3 plants. Second, by concentrating CO_2 in the bundle sheath cells, photorespiration is suppressed. These two features enable C4 plants to photosynthesize efficiently at higher temperatures than C3 plants. The preponderance of C4 plants in drier, hotter climates may be an expression of these advantages of the C4 PCA cycle at high temperatures.

CO$_2$ Concentrating Mechanisms III: Crassulacean Acid Metabolism (CAM)

A third mechanism for concentrating CO_2 at the site of Rubisco is found in CAM plants. CAM is an acronym for **Crassulacean acid metabolism**, but the mechanism is not restricted to the family Crassulaceae. Instead, like the C4 PCA cycle, it is found in many angiosperm families. CAM plants are especially adapted to arid environments—cacti are fine examples. The CAM mechanism enables plants to maximize their water-use efficiency. Typically, a CAM plant loses 50–100 g of water for every gram of CO_2 gained, compared with values of 250–300 and 400–500 g for C4 and C3 plants, respectively (Chapter 4). Thus, in desert environments, CAM plants have a competitive advantage.

The CAM mechanism is similar in many respects to the C4 PCA cycle, but differs from it in two important features. In C4 plants the formation of the C4 acids is spatially, but not temporally, separated from the decarboxylation of the C4 acids and refixation of the resulting CO_2 by the C3 PCR cycle. A specialized anatomy is needed to effect this spatial separation. In CAM plants, on the other hand, the formation of the C4 acids is temporally, but not spatially, separated from decarboxylation and refixation. CAM plants lack the specialized leaf anatomy typical of C4 plants.

CAM Plants Open Their Stomata at Night and Keep Them Closed During the Day

CAM plants achieve their high water-use efficiency by opening their stomata during the cool, desert nights and by closing them during the hot, dry day. This minimizes water loss, but since H_2O and CO_2 share the same diffusion pathway, CO_2 must be assimilated at night. The CO_2 assimilation is accomplished by carboxylation of phosphoenolpyruvate to oxaloacetate, which is then reduced to malate. The phosphoenolpyruvate originates from the breakdown of starch and other sugars by the glycolytic pathway (Fig. 9.14). The C4 acid accumulates as malic acid in the large vacuoles that are a typical, but not obligatory, anatomic feature of the leaf cells of CAM plants. The accumulation of substantial amounts of malic acid, equivalent to the amount of CO_2 assimilated at night, has long been recognized as a dark acidification of the leaf (Bonner and Bonner, 1948).

With the onset of day, the stomata close, preventing loss of water and further acquisition of CO_2. The leaf cells become deacidified as the reserves of vacuolar malic acid are consumed. Decarboxylation is achieved by the action of NADP malic enzyme on malate or of phosphoenolpyruvate carboxykinase on oxaloacetate. Because the stomata are closed, the internally released CO_2 cannot escape from the leaf and instead is reduced to carbohydrate by operation of the C3 PCR cycle. The elevated internal concentration of CO_2 effectively suppresses the photorespiratory oxygenation of ribulose bisphosphate and favors carboxylation. The C3 acid remaining after decarboxylation is thought to be converted first to triose phosphate and then to starch or sucrose, thus recovering the original starting material.

CAM Metabolism Is Regulated by Different Forms of the Enzyme Phosphoenolpyruvate Carboxylase

The mechanism of CAM outlined above requires that phosphoenolpyruvate carboxylase and the decarboxylases, both of which are located in the cytosol, function at different times. To avoid a futile cycle of carboxylation-decarboxylation, the carboxylase must be "switched on" at night and "switched off" during the day. Conversely, the decarboxylases must be activated during the day and inactivated at night. Recently, it became evident that CAM phosphoenolpyruvate carboxylase exists in two forms. The molecular species found at night is insensitive to malate, whereas the one found during

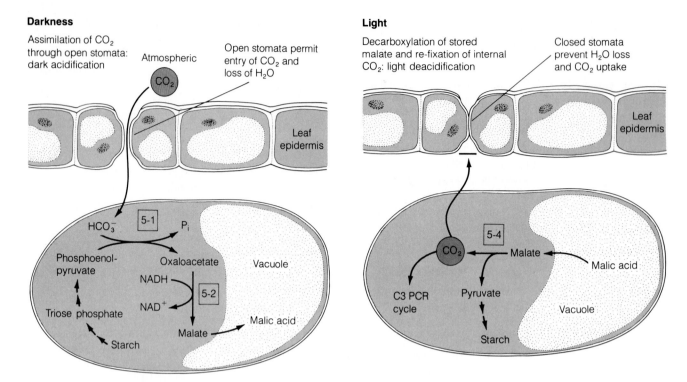

FIGURE 9.14. Crassulacean acid metabolism (CAM) involves the temporal separation of CO_2 assimilation at night and decarboxylation and refixation of the internally released CO_2 during day. CAM is primarily an adaptation to minimize the loss of water that occurs whenever stomata are opened to permit the entry of CO_2. In CAM plants, the stomata are opened in the cool of the night. CO_2 is assimilated as malic acid, which is stored in the vacuole. As malic acid accumulates, the leaf vacuoles undergo dark acidification. Upon illumination, the stomata close, and the leaf deacidifies. The malic acid is recovered from the vacuole and undergoes decarboxylation. The CO_2 that is released is prevented from escaping by stomatal closure and is fixed via the C3 PCR cycle, using photochemically generated ATP and NADPH.

the day is inhibited by low concentrations of malate. Conversion between the two forms is achieved by phosphorylation-dephosphorylation (Brulfert et al., 1986) (Fig. 9.15). The diurnal regulation of the decarboxylases is not understood.

In addition to short-term (day-night) regulation, some CAM plants show longer-term regulation and are able to adjust their degree of "CAMness" according to environmental conditions. For example, the ice plant, *Mesembryanthemum crystallinum*, fixes CO_2 more or less exclusively during the day via the C3 PCR cycle when it is well watered. Under such conditions, there is little diurnal variation in leaf acidity or assimilation of CO_2 at night. However, when these plants are water-stressed, they develop the CAM mode of photosynthesis characterized by a diurnal cycle of CO_2 assimilation and acidification at night and CO_2 reduction and deacidification during the day. Several species whose CAM metabolism is induced by different types of stress have been shown to be facultative CAM plants. This form of regulation indicates that the expression of the CAM genes is susceptible to environmental control.

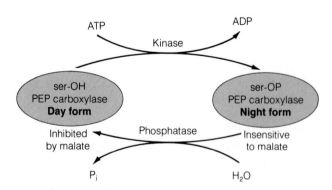

FIGURE 9.15. Diurnal regulation of CAM phosphoenolpyruvate (PEP) carboxylase appears to be achieved by phosphorylation of a serine residue of the day form of the enzyme, which is sensitive to inhibition by malate, to yield the phosphorylated night form, which is relatively insensitive to malate. Dephosphorylation of the night form reverses the processes. Consequently, PEP carboxylase is active at night during the period of malate formation but inactive during the day, when malate decarboxylation occurs.

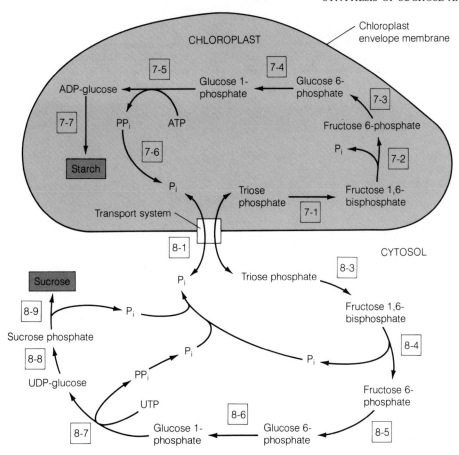

FIGURE 9.16. The syntheses of starch and sucrose are competing processes that occur in different cellular compartments. Starch synthesis and deposition is localized in the chloroplast, (Figure 9.17). The reactions of starch synthesis (7-1, 7-2, etc.) are more fully described in Table 9.7. Sucrose, the principal carbohydrate transported throughout the plant, is synthesized in the cytosol. The reactions of sucrose synthesis (8-1, 8-2, etc.) are more fully described in Table 9.8. The partitioning of triose phosphate between starch synthesis in the chloroplast and sucrose synthesis in the cytosol is largely regulated by the orthophosphate (P_i) concentrations in these cellular compartments. When the cytosolic P_i concentration is high, chloroplast triose phosphate is exported to the cytosol in exchange for P_i, and sucrose synthesis occurs. When the cytosolic P_i concentration is low, triose phosphate is retained within the chloroplast, and starch synthesis occurs.

Synthesis of Sucrose and Starch

Both starch and sucrose are synthesized from excess triose phosphate generated by the C3 PCR cycle (Beck and Ziegler, 1989). The pathways of starch and sucrose synthesis are shown in Figure 9.16, and the component reactions are shown in Tables 9.7 and 9.8. Starch synthesis, occuring in the chloroplast, and sucrose synthesis, occuring in the cytosol, are competing processes. Sucrose is the principal form of carbohydrate translocated throughout the plant. In rapidly growing plants, sucrose, rather than starch, is the major end-product of photosynthesis.

Starch Is Synthesized and Stored in the Chloroplast While Sucrose Synthesis Takes Place in the Cytosol

Electron micrographs (Fig. 9.17) showing massive deposits of starch within the chloroplast leave no doubt about the site of starch synthesis. Evidence that the cytosol is the site of sucrose synthesis is based on cell fractionation studies in which the organelles are separated from the contents of the cytosol. The enzymes sucrose-phosphate synthetase (Table 9.8, reaction 8) and sucrose-phosphate phosphatase (Table 9.8, reaction 9) are found only in the cytosolic fraction.

The pathways to starch and sucrose have a number of steps in common (e.g., those involving fructose-1,6-bisphosphate phosphatase and hexosephosphate-isomerase). However, these enzymes have isozymes (different forms of the enzymes that catalyze the same reaction) that have different properties and are unique to the appropriate cellular compartments. For example, the chloroplast fructose-1,6-bisphosphate phosphatase is regulated by the thioredoxin system and is insensitive to fructose 2,6-bisphosphate and AMP. Conversely, the cytosolic form of the enzyme is regulated by fructose 2,6-bisphosphate.

TABLE 9.7. Reactions of starch synthesis from triose phosphate in chloroplasts

1. *Fructose-1,6-bisphosphate aldolase*
   dihydroxyacetone 3-phosphate + glyceraldehyde 3-phosphate → fructose 1,6-bisphosphate

2. *Fructose-1,6-bisphosphate phosphatase (chloroplast)*
   fructose 1,6-bisphosphate + H_2O → fructose 6-phosphate + $HOPO_3^{2-}$

3. *Hexosephosphate isomerase*
   fructose 6-phosphate → glucose 6-phosphate

4. *Phosphoglucomutase*
   glucose 6-phosphate → glucose 1-phosphate

5. *ADPglucose pyrophosphorylase*
   glucose 1-phosphate + ATP → ADP-glucose + $H_2P_2O_7^{2-}$

6. *Pyrophosphorylase*
   $H_2P_2O_7^{2-} + H_2O$ → $2HOPO_3^{2-} + 2H^+$

7. *Starch synthetase*
   ADP-glucose + $(1,4-\alpha\text{-}D\text{-glucosyl})_n$ → ADP + $(1,4-\alpha\text{-}D\text{-glucosyl})_{n+1}$

Starch

TABLE 9.8. Reactions of sucrose synthesis from triose phosphate in cytoplasm

1. *Phosphate/triose phosphate translocator (chloroplast envelope membrane)*
   triose phosphate (chloroplast) + $HOPO_3^{2-}$ (cytoplasm) → triose phosphate (cytoplasm) + $HOPO_3^{2-}$ (chloroplast)

2. *Triose-phosphate isomerase (cytoplasm)*
   dihydroxyacetone 3-phosphate → glyceraldehyde 3-phosphate

3. *Fructose-1,6-bisphosphate aldolase (cytoplasm)*
   dihydroxyacetone 3-phosphate + glyceraldehyde 3-phosphate → fructose 1,6-bisphosphate

4. *Fructose-1,6-bisphosphate phosphatase (cytoplasm)*
   fructose 1,6-bisphosphate + H_2O → fructose 6-phosphate + $HOPO_3^{2-}$

4a. *PP_i-linked phosphofructokinase (cytoplasm)*
   fructose 6-phosphate + $H_2P_2O_7^{2-}$ → fructose 1,6-bisphosphate + $HOPO_3^{2-}$

5. *Hexosephosphate isomerase (cytoplasm)*
   fructose 6-phosphate   →   glucose 6-phosphate

6. *Phosphoglucomutase (cytoplasm)*
   glucose 6-phosphate → glucose 1-phosphate

7. *UDPglucose pyrophosphorylase (cytoplasm)*
   glucose 1-phosphate   +   UTP   →   UDP-glucose + $H_2P_2O_7^{2-}$

continued on p. 246

TABLE 9.8. (continued)

8. *Sucrose-phosphate synthase (cytoplasm)*
   UDP-glucose   +   fructose 6-phosphate   →   UDP + sucrose 6-phosphate

9. *Sucrose-phosphate phosphatase (cytoplasm)*
   sucrose 6-phosphate + H_2O → sucrose + $HOPO_3^{2-}$

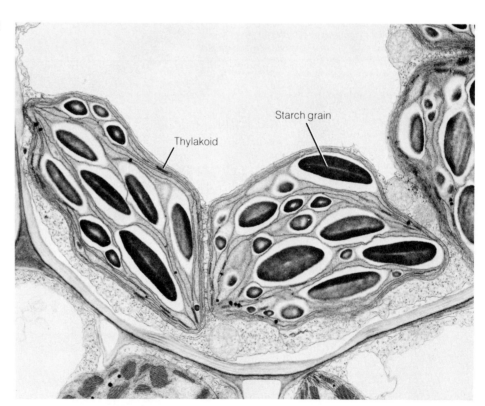

FIGURE 9.17. Electron micrograph of a bundle sheath cell from corn, showing the starch grains in the chloroplasts (16,500 ×). (Photo by S. E. Frederick, courtesy of E. H. Newcomb.)

Thylakoid

Starch grain

The Syntheses of Sucrose and Starch Are Competing Reactions Regulated by Key Metabolites

Excess triose phosphate can be used for either starch synthesis in the chloroplast or sucrose synthesis in the cytosol. Thus, starch and sucrose syntheses are competing processes, and conditions that promote one process inhibit the other. The key components of the system that regulate its partitioning are the relative concentrations of orthophosphate and triose phosphate in the cytosol and the chloroplast and the concentration of fructose 2,6-bisphosphate in the cytosol. The two compartments communicate with one another via the phosphate/triose phosphate translocator (Table 9.8, reaction 1), a transport protein in the chloroplast envelope membrane that catalyzes the movement of orthophosphate and triose phosphate in opposite directions between the compartments. A low cytosolic orthophosphate concentration limits the export of triose phosphate through the translocator. Instead, the triose phosphate is used for the synthesis of starch. The chloroplast enzyme ADPglucose pyrophosphorylase (Table 9.7, reaction 5), a component of the pathway leading to starch, is stimulated by the triose phosphate 3-phosphoglycerate and inhibited by orthophosphate. A high concentration ratio of 3-phosphoglycerate to orthophosphate is typically found in chloroplasts actively synthesizing starch. Conversely, an abundance of orthophosphate in the cytosol inhibits starch synthesis within the chloroplast and promotes the export of triose phosphate into the cytosol, where it is converted to sucrose.

Cytosolic sucrose synthesis is strongly regulated by fructose 2,6-bisphosphate. This recently discovered compound is found in the cytosol in minute concentrations, and it exerts a regulatory effect on the cytosolic interconversion of fructose 1,6-bisphosphate and fructose 6-phosphate (Huber, 1986). Increased cytosolic fructose 2,6-bisphosphate is associated with decreased rates of sucrose synthesis, because fructose 2,6-bisphosphate is a powerful inhibitor of cytosolic fructose 1,6-bisphosphate phosphatase (Table 9.8, reaction 4) and an activator of the pyrophosphate-dependent phosphofructokinase (Table 9.8, reaction 4a). But what, in turn, controls the cytosolic concentration of fructose 2,6-bisphosphate? Fructose 2,6-bisphosphate is synthesized from fructose 6-phosphate by a special fructose-6-phosphate 2-kinase (not to be confused with fructose-6-phosphate 1-kinase!) and is degraded by fructose-2,6-bisphosphate phosphatase (not to be confused with fructose-1,6-bisphosphate phosphatase!). Orthophosphate both stimulates fructose-6-phosphate 2-kinase and inhibits fructose-2,6-bisphosphate phosphatase, whereas triose phosphate inhibits the 2-kinase (Fig. 9.18). Consequently, a low cytosolic triose phosphate/orthophosphate ratio promotes the formation of fructose 2,6-bisphosphate, which in turn inhibits the hydrolysis of cytosolic fructose 1,6-bisphosphate and hence slows the rate of sucrose synthesis. A high cytosolic triose phosphate/orthophosphate ratio has the opposite effect.

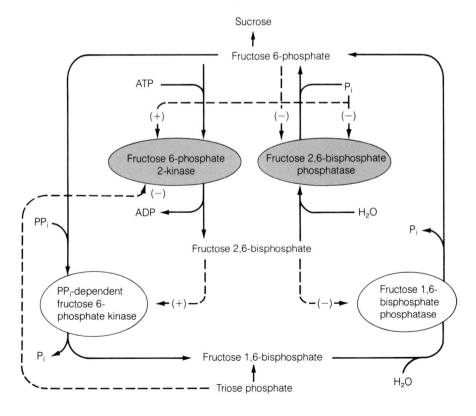

FIGURE 9.18. The regulation of cytosolic fructose 6-phosphate/ fructose 1,6-bisphosphate metabolism by fructose 2,6-bisphosphate, orthophosphate (P_i), and triose phosphate. The interconversion (solid lines) of the various metabolites is catalyzed by the indicated enzymes. The influence of the metabolites on the activities of these enzymes is shown with dashed lines indicating (+) for stimulation and (−) for inhibition. For example, fructose 2,6-bisphosphate stimulates the activity of the PP_i-dependent fructose-6-phosphate kinase but inhibits the activity of fructose-1,6-bisphosphate phosphatase.

Summary

The reduction of CO_2 to carbohydrate is coupled to the consumption of NADPH and ATP synthesized during the light reactions of photosynthesis. All photosynthetic eukaryotes reduce CO_2 via the photosynthetic carbon reduction (PCR) cycle. In the PCR cycle, CO_2 from the atmosphere and water are combined with ribulose 1,5-bisphosphate to form two molecules of 3-phosphoglycerate. The continued operation of the cycle is ensured by the regeneration of ribulose 1,5-bisphosphate. The PCR cycle consumes two molecules of NADPH and three molecules of ATP for every CO_2 fixed and has a thermodynamic efficiency of about 90%.

Rubisco, the enzyme that catalyzes the carboxylation ribulose 1,5-bisphosphate, also acts as an oxygenase. In that process, called photorespiration, ribulose 1,5-bisphosphate combines with oxygen and yields phosphoglycolate and 3-phosphoglycerate. The carboxylation and oxygenation reactions take place at the same active site of Rubisco and decrease the efficiency of photosynthesis in the intact leaf.

The dissipative effects of photorespiration are avoided in some plants by mechanisms that concentrate CO_2 at the carboxylation sites in the chloroplast. These mechanisms include C4 and CAM metabolism and active accumulation of inorganic carbon by algae and cyanobacteria.

The carbohydrates synthesized by the PCR cycle are metabolized into sugars and starch. Sucrose is synthesized in the cytosol, whereas starch metabolism takes place in the chloroplast. The balance between these two metabolic pathways is regulated by the relative concentrations of orthophosphate, triose phosphate, and fructose 2,6-bisphosphate.

GENERAL READING

Andrews, T. J., and Lorimer, G. H. (1987) Rubisco: Structure, mechanisms, and prospects for improvement. In: *The Biochemistry of Plants*, Vol. 10, Hatch, M. D., and Boardman, N. K., eds, pp. 131–218. Academic Press, San Diego.

Bassham, J. A. (1965) Photosynthesis: The path of carbon. In: *Plant Biochemistry*, 2d ed., Bonner, J., and Varner, J. E., eds., pp. 875–902. Academic Press, New York.

Beck, E., and Ziegler, P. (1989) Biosynthesis and degradation of starch in higher plants. *Annu. Rev. Plant Physiol. Plant Mol. Biol.* 40:95–117.

Buchanan, B. B. (1980) Role of light in the regulation of chloroplast enzymes. *Annu. Rev. Plant Physiol.* 31:341–374.

Burnell, J. N., and Hatch, M. D. (1985) Light-dark modulation of leaf pyruvate, P_i dikinase. *Trends Biochem. Sci.* 10:288–291.

Edwards, G., and Walker, D. (1983) C_3, C_4: *Mechanisms, and Cellular and Environmental Regulation, of Photosynthesis*. University of California Press, Berkeley.

Hatch, M. D., and Boardman, N. K., eds. (1987) *The Biochemistry of Plants*, Vol. 10: *Photosynthesis*. Academic Press, San Diego. (This volume contains several chapters on carbon metabolism and photosynthesis.)

Hatch, M. D., and Osmond, C. B. (1976) Compartmentation and transport in C_4 photosynthesis. In: *Transport in Plants* (Encyclopedia of Plant Physiology, New Series, Vol. 3), Stocking, C. R., and Heber, U., eds., pp. 144–184. Springer-Verlag, Berlin.

Heldt, H. W. (1979) Light-dependent changes of stromal H^+ and Mg^{+2} concentrations controlling CO_2 fixation. In: *Photosynthesis II* (Encyclopedia of Plant Physiology, New Series, Vol. 6), Gibbs, M., and Latzko, E., eds, pp. 202–207. Springer-Verlag, Berlin.

Huber, S. C. (1986) Fructose 2,6-bisphosphate as a regulatory metabolite in plants. *Annu. Rev. Plant Physiol.* 37:233–246.

Lorimer, G. H. (1981) The carboxylation and oxygenation of ribulose 1,5-bisphosphate: The primary events in photosynthesis and photorespiration. *Annu. Rev. Plant Physiol.* 32:349–383.

Lüttge, U., and Higinbotham, N. (1979) *Transport in Plants*. Springer-Verlag, New York.

Ogren, W. L. (1984) Photorespiration: Pathways, regulation and modification. *Annu. Rev. Plant Physiol.* 35:415–442.

O'Leary, M. H. (1982) Phosphoenolpyruvate carboxylase: An enzymologist's view. *Annu. Rev. Plant Physiol.* 33:297–315.

Tolbert, N. E. (1981) Metabolic pathways in peroxisomes and glyoxysomes. *Annu. Rev. Biochem.* 50:133–157.

Whitfeld, P. R., and Bottomley, W. (1983) Organization and structure of chloroplast genes. *Annu. Rev. Plant Physiol.* 34:279–310.

CHAPTER REFERENCES

Bonner, W., and Bonner, J. (1948) The role of carbon dioxide in acid formation by succulent plants. *Am. J. Bot.* 35:113–117.

Brulfert, J., Vidal, J., Le Marechal, P., Gadal, P., and Queiroz, O. (1986) Phosphorylation-dephosphorylation process as a probable mechanism for the diurnal regulatory changes of phosphoenolpyruvate carboxylase in CAM plants. *Biophys. Res. Commun.* 136:151–159.

Craig, S., and Goodchild, D. J. (1977) Leaf ultrastructure of *Triodia irritans*: A C4 grass possessing an unusual arrangement of photosynthetic tissues. *Aust. J. Bot.* 25:277–290.

Frederick, S. E., and Newcomb, E. H. (1969) Cytochemical localization of catalase in leaf microbodies (peroxisomes). *J. Cell Biol.* 43:343–353.

Gutierrez, M., Gracen, V. E., and Edwards, G. E. (1974) Biochemical and cytological relationships in C_4 plants. *Planta* 119:279–300.

Hatch, M. D., and Slack, C. R. (1966) Photosynthesis by sugarcane leaves. A new carboxylation reaction and the pathway of sugar formation. *Biochem. J.* 101:103–111.

Ku, S. B., and Edwards, G. E. (1978) Oxygen inhibition of photosynthesis. III. Temperature dependence of quantum yield and its relation to O_2/CO_2 solubility ratio. *Planta* 140:1–6.

Ogawa, T., and Kaplan, A. (1987) The stoichiometry between CO_2 and H^+ fluxes involved in the transport of inorganic carbon in cyanobacteria. *Plant Physiol.* 83:888–891.

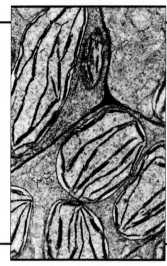

CHAPTER 10

Photosynthesis: Physiological and Ecological Considerations

CONVERSION OF SOLAR ENERGY into the organic compounds required by all living organisms is a complex process that includes electron transport and photophosphorylation (covered in Chapter 8) and photosynthetic carbon metabolism (discussed in Chapter 9). The emphasis that has been placed on some specific aspects of this large number of photochemical and biochemical reactions should not overshadow the fact that, under natural conditions, photosynthesis is an integral process taking place in intact organisms that are continuously responding to internal and external changes. This chapter addresses the photosynthetic responses of the intact leaf and their environmental correlations. Additional photosynthetic responses to different types of stress are included in Chapter 14.

The impact of the environment on photosynthesis is of interest to both plant physiologists and agronomists. From a physiological standpoint, we wish to understand how photosynthesis responds to environmental factors such as light, ambient CO_2 concentrations, and temperature. The environmental dependence of photosynthetic processes is important to agronomists because plant productivity, and hence crop yields, is strongly dependent on prevailing photosynthetic rates in a dynamic environment.

In studying the environmental dependence of photosynthesis, a central question arises: How many environmental factors can limit photosynthesis at one time? The British plant physiologist F. F. Blackman postulated in 1905 that, under any particular conditions, the rate of photosynthesis will be limited by the slowest step, the so-called **limiting factor**. The implication is that at any given time, photosynthesis can be limited by either light or CO_2 concentration, but not by both factors. This hypothesis has had a marked influence on the approach used by plant physiologists to study photosynthesis—that is, varying one factor and keeping all other environmental conditions constant (see, for example, Fig. 10.6). However, the hypothesis has some limitations, as can be understood by an analogy with a car. If a car breaks down, the malfunction can usually be traced to a single factor, such as a gas pump or the car battery. On the other hand, the car will run more *efficiently* after a tune up, in which several of its components are adjusted for optimal performance and a single factor is not usually critical. In the intact leaf, three major metabolic steps have been identified as important for optimal photosynthetic performance: Rubisco activity, the regeneration of ribulose bisphosphate (RuBP), and the metabolism of the triose phosphates. The first two steps are the most prevalent under natural conditions. Table 10.1 provides some examples of how light and CO_2 can affect these key metabolic steps

TABLE 10.1 Some characteristics of limitations to the rate of photosynthesis

Limiting factor	Conditions that lead to this limitation		Response of photosynthesis under this limitation to		
	CO_2	Light	CO_2	O_2	Light
Rubisco activity	Low	High	Strong	Strong	Absent
RuBP regeneration	High	Low	Moderate	Moderate	Strong

and how ensuing limitations in these steps are expressed in the photosynthetic responses to light or to ambient CO_2 and O_2 concentrations. In the following sections, the effects of different environmental factors on photosynthesis are discussed in detail.

Light and Photosynthesis in the Intact Leaf

Roughly 1.3 kW m^{-2} of radiant energy from the sun reaches the earth. Recall that in light energy calculations in Chapter 8 we used joules as units, whereas here we have used watts (W). Joules are used as a measure of total energy, whereas watts per square meter are units of *irradiance*, the flux of radiant energy per unit time. Since 1 W equals 1 J s^{-1}, 1.3 kW m^{-2} equals 1.3 kJ s^{-1} m^{-2}. Another unit used in light measurements is the mole of light per unit surface per unit time, mol m^{-2} s^{-1} (for instance, see Fig. 10.5). One mole of light equals 6.02×10^{23} quanta (i.e., Avogadro's number of quanta). Many studies in plant physiology have used the unit einstein (or microeinstein), where 1 einstein (E) equals 1 mole of light. However, the einstein is not a unit in the SI system (Systeme Internationale), and moles or micromoles should be used instead.

When a photobiological process, such as photosynthesis, depends on the number of quanta absorbed rather than on the absorbed energy, a response to light is more accurately expressed in moles rather than in watts or joules. **Energy fluxes**, expressed in watts or joules, can be converted to **photon flux densities**, expressed in moles, when the precise wavelength composition of the light under consideration is known. Plant physiologists also use **photon fluence rates** (in units of mol m^{-2} s^{-1}) to express the number of quanta impinging on a leaf surface. Instruments that measure photon flux densities in moles are commercially available and are often called *quantum sensors*. The sensor is made of a chemical that generates a voltage proportional to the absorbed number of quanta. This voltage is amplified and read in a panel that is calibrated in moles per square meter per second. The photon flux density of full sunlight in the wave band

from 400 to 700 nm is about 2000 μmol m^{-2} s^{-1}, though higher values can be measured at high altitudes.

Only about 5% of the solar energy reaching the earth can be converted into carbohydrates by a photosynthesizing leaf (Fig. 10.1). This is because a major fraction of the incident light is of a wavelength either too short or too long to be absorbed by the photosynthetic pigments (Fig. 8.3). In addition, some of the absorbed light energy is lost as heat (Chapter 8), used in the biochemical processes required to run the photosynthetic carbon reduction (PCR) cycle (Chapter 9), or used in the general metabolism of the leaf (Chapter 11).

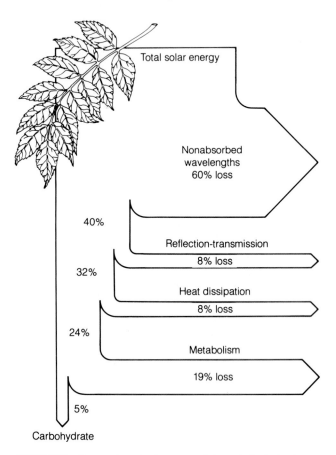

FIGURE 10.1. Conversion of solar energy into carbohydrates by a leaf. Of the total incident energy, only 5% is converted into carbohydrates.

The Architecture and Composition of a Leaf Maximize Light Absorption

The light energy in the wave band 400–700 nm is often referred to as **photosynthetically active radiation**, or PAR. Of the PAR reaching a leaf, 10–15% will not be absorbed by the photosynthetic pigments but will be either reflected at the leaf surface or transmitted throughout the leaf (Fig. 10.2). Because chlorophyll absorbs very strongly in the blue and the red (Fig. 8.3), the transmitted light and reflected light are enriched in green, hence the green color of the vegetation. The transmitted light is very important to leaves that grow in the shade of other leaves. Leaf hairs, salt glands, and other epidermal characteristics can influence the reflection of light by leaves (Ehleringer et al., 1976). Some pubescent leaves absorb about 40% less light than leaves that have the same chlorophyll content but lack hairs in their epidermis.

The anatomy of the leaf is highly specialized for light absorption. The top layers of photosynthetic cells below the epidermis are called **palisade cells** and are shaped like pillars that stand in parallel columns one to three layers thick (Fig. 10.3). In the case of leaves with more than one layer of palisade cells, we may wonder how efficient it is for a plant to invest energy in the development of more than one cell layer when the high chlorophyll content of the first layer would appear to allow little transmission of the incident light to the leaf interior. In fact, it is efficient to have several layers of palisade cells. More light than

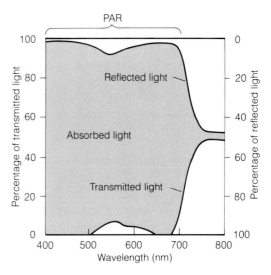

FIGURE 10.2. Optical properties of a bean leaf. The figure shows the percentage of light absorbed, reflected, and transmitted, as a function of wavelength. The transmitted and reflected green light in the wave band 500–600 nm gives leaves their green color. Note that most of the light above 700 nm is not absorbed by the leaf. (From Smith, 1986.)

might be expected goes through the first layer of palisade cells because of the sieve and light guide effects (Terashima and Saeki, 1983).

The **sieve effect** is due to the fact that the chlorophyll is not uniformly distributed in the cells but is confined to

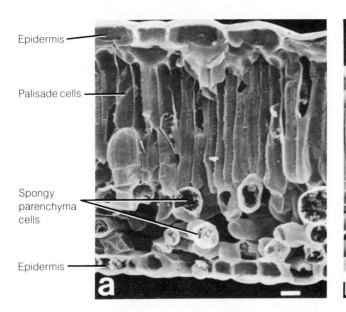
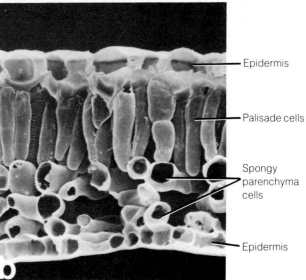

FIGURE 10.3. Anatomical features of leaves grown in different light environments. These are scanning electron micrographs of (a) a leaf grown in sunlight and (b) a leaf from the same tree but grown in the shaded environment within the canopy. The palisade (column-like) cells are much longer in leaves grown in sunlight. Layers of spongy parenchyma cells can be seen below the palisade cells. Bar = 10 μm. (From McCain et al., 1988.)

the chloroplasts. This clustering of the chlorophyll results in shading between the chorophyll molecules, so that the total absorption of light by a given amount of chlorophyll in a chloroplast is less than that of the same amount in a solution. The sieve effect is greatest for the strongly absorbed blue and red wavelengths (Sharkey, 1985).

The **light guide** effect refers to channeling of some of the incident light through the intercellular spaces between the palisade cells, in a manner analogous to transmission of light by an optical fiber. The arrangement of the palisade cells in columns facilitates this type of transmission of some of the light into the leaf interior.

Below the palisade layers is the **spongy mesophyll**. The mesophyll cells are very irregular in shape (Fig. 10.3), and as a result there are large air spaces between them. This architecture results in many air-water interfaces that cause a high degree of light scattering, which increases the probability that light escaping from one cell is absorbed by another cell. The marked light scattering in the spongy mesophyll causes the lower surface of many leaves to appear lighter in color than the upper surface. The contrasting properties of the palisade and spongy mesophyll tissues—with the former allowing light to pass through and the latter trapping as much light as possible—results in more uniform light absorption throughout the leaf.

Chloroplast Rearrangement and Leaf Movement Can Change the Amount of Light Absorbed by the Leaf

In addition to their anatomical properties, leaves maximize light absorption by changing the spatial relationship between the leaf and the incident sunlight. This is achieved by movement of chloroplasts within the mesophyll cells and by movement of the leaf blade with respect to the sun rays.

Chloroplast movements can substantially change the amount of light absorbed by a photosynthesizing cell. When the photon flux densities reaching a leaf are below the values that saturate the light response of photosynthesis, chloroplasts gather at the cell surfaces parallel to the plane of the leaf and perpendicular to the incident light (Fig. 10.4a). When photon flux densities are too high, the chloroplasts move to the cell surfaces that are parallel to the incident light. Light absorption is maximal when the surfaces occupied by chloroplasts are parallel to the plane of the leaf and minimal when they are parallel to the incident light. In the latter, damage to the photosynthetic apparatus by excessive irradiance can, to an extent, be prevented. Chloroplasts move along cables of actin microfilaments in the cytoplasm. The light signal triggering those movements is perceived in the cytoplasm and not in the chloroplast proper (Blatt and Briggs, 1980).

The architectural properties of the leaf that facilitate light absorption are most efficient when the leaf surface is perpendicular to the incident light. For stationary leaves, that condition prevails for only a short time of the day. Many plants can substantially increase the time during which the leaf blades are perpendicular to the sun rays by **solar tracking** (Fig. 10.4b). A leaf tracking the sun has its blade in the nearly vertical position at sunset, facing the

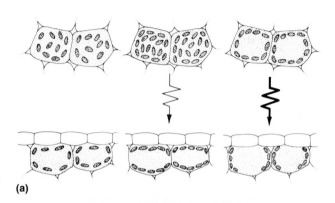

(a)

(b)

FIGURE 10.4. (a) Chloroplast distribution in photosynthesizing cells of the duckweed, *Lemna*. The upper sequence depicts a surface view and the lower sequence a cross section of the cells. Chloroplast distribution is shown in the dark (left), at low photon flux densities (center), and at high photon flux densities (right). A distribution parallel to the leaf surface maximizes light absorption. Under high photon flux densities, the chloroplasts are found along the walls parallel to the incident light (arrows). (From Haupt, 1986.) (b) Leaf movement toward the incident light. Seedlings of *Lavatera cretica* L., an annual weed that grows along roadsides, are shown with their leaves oriented toward the incident light. The direction of the light beam is indicated by the arrows. The movement is generated by asymmetric swelling of a pulvinus, found at the junction between the lamina and the petiole. In natural conditions, the leaves track the sun's trajectory in the sky. (From Schwartz and Koller, 1980.)

western horizon where the sun is setting. During the night the leaf takes a horizontal position, and at sunrise it faces east, to the location where the sun came up the day before. During the day, the blade moves at a velocity that matches the movement of the sun in the sky (Schwartz and Koller, 1980). Alfalfa, cotton, legumes including soybean and bean, and some wild species of the Malvaceae are examples of the numerous species capable of solar tracking. In many cases, leaf orientation is controlled by a specialized organ called the **pulvinus**, found at the junction between the blade and the petiole. The pulvinus has motor cells that change their osmotic potential and generate mechanical forces that cause the blade displacement. In other plants, leaf orientation is controlled by small mechanical changes along the length of the petiole and by movements of the younger parts of the stem. Both phytochrome (Chapter 20) and a blue light-sensitive pigment (Schwartz and Koller, 1980) have been implicated as photoreceptors in solar tracking.

In some species, particularly those living in deserts, the leaves move *away* from the sun, thus minimizing heat absorption and water loss (Ehleringer and Forseth, 1980). On the basis of a term often used for leaf movements, **heliotropism**, these leaves are called paraheliotropic, whereas leaves that maximize light interception by solar tracking are called diaheliotropic. In some species, a plant can display diaheliotropic movements when well watered and paraheliotropic movements when it experiences water stress.

Plant Growth and Development Often Reflect Competition for Sunlight

Sunlight is a resource, and usually plants compete for sunlight. Held upright by stems and trunks, leaves form a canopy that absorbs light and influences photosynthetic rates and growth beneath them. Leaves that are shaded by other leaves have much lower photosynthetic rates. Some plants have very thick leaves that transmit little, if any, light. Some plants, like the dandelion (*Taraxum* sp.), have a rosette growth habit in which leaves grow radially very close to each other and to the stem, thus precluding the growth of any leaves below them. Stems and trunks allow plants to capture light above the ground and to outgrow nonstemmed plants. Trees represent an outstanding adaptation for light interception. The elaborate branching structure of trees results in a vast increase in the interception of sunlight. Very little PAR penetrates the canopy of many forests (Fig. 10.5).

Leaves grown under low light intensities are usually damaged when exposed to high photon flux densities, a phenomenon called **photoinhibition** (Powles, 1984). Any leaf can undergo photoinhibition when exposed to unusually high photon flux densities, but photoinhibition is most common in plants growing in the shade that are

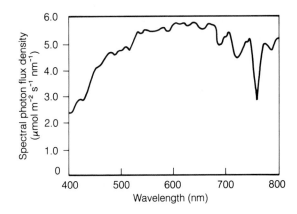

(a)

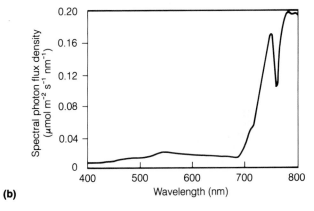

(b)

FIGURE 10.5. The spectral distribution of sunlight at the top of the canopy (a) and under the canopy (b). The total photon flux density was (a) 1.9 mmol m^{-2} s^{-1} and (b) 17.7 μmol m^{-2} s^{-1}. Most of the sunlight in the blue and in the red has been absorbed by the leaves in the canopy. (From Smith, 1986.)

suddenly exposed to sunlight. Photoinhibition lowers photosynthetic rates and impairs electron transport and photophosphorylation.

Another feature of the habitat of leaves growing in the shade is **sunflecks**, patches of full sunlight that pass through discontinuities of the canopy and travel along the shaded leaves as the sun moves. In a dense forest, sunflecks can change the photon flux density impinging on a leaf in the forest floor over 10-fold within seconds! For some of these leaves, a sunfleck contains nearly 50% of the total light energy available during the day, but this critical energy is available for only a few minutes in a very high dose. Sunflecks also play a role in the carbon metabolism of lower leaves in dense crops that are shaded by the upper leaves of the plant. Rapid responses of both the photosynthetic apparatus and the stomata to sunflecks have been of substantial interest to plant physiologists and ecologists (Pearcy, 1988).

Since leaves absorb little of the far-red light, habitats below canopies are enriched in far-red (wavelengths > 700 nm) light (Fig. 10.5), and that light environment changes the properties of the light-sensitive growth regulator phytochrome (Chapter 20). As a result, plants in such environments may be tall and spindly. When a canopy tree dies, a light opening is created, and the more abundant PAR facilitates faster growth of stunted plants below the canopy.

The Photosynthetic Response to Light in the Intact Leaf Reflects Basic Properties of the Photosynthetic Apparatus

Measurement of CO_2 fixation in intact leaves at increasing photon flux densities allows the construction of useful dose-response curves (Fig. 10.6). In the dark, CO_2 fixation is negative, and there is net CO_2 evolution from the leaf because of respiration (Chapter 11). As the photon flux density increases, net CO_2 evolution decreases, and after it reaches zero the plant shifts to net CO_2 fixation. The photon flux density at which the net CO_2 exchange in the leaf is zero is called the *light compensation point*. At this point, the amount of CO_2 evolved by mitochondrial respiration is balanced by the amount of CO_2 fixed by photosynthesis. The photon flux densities at which different leaves reach the light compensation point vary with species and developmental conditions. One of the more interesting differences is found between plants grown in full sunlight and those grown in the shade (Fig. 10.6). Light compensation points of sun plants are in the range 10–20 μmol m^{-2} s^{-1}, whereas corresponding values for shade plants are 1–5 μmol m^{-2} s^{-1}. The values for shade plants are lower because respiration rates in shade plants are very low, so little net photosynthesis suffices to bring the rates of CO_2 evolution to zero. Low respiratory rates seem to represent a basic adaptation allowing shade plants to survive in light-limited environments.

The linear portion of the light-response curves of photosynthesis represent the part of the process in which photosynthesis is strictly light-limited. Under this condition, each increment in light elicits a proportional increase in photosynthetic rate (Fig. 10.6), resulting in a linear relationship between photosynthetic rates and photon flux densities. This portion of the curve can be used to calculate the quantum yield of photosynthesis in the intact leaf. Recall from Chapter 8 that the quantum yield of photochemistry is about 0.95 and the quantum yield of oxygen evolution by isolated chloroplasts is about 0.1. (Quantum yields can vary between zero, when a process is not involved in the use of light energy, and 1, when all the light absorbed is used to drive the process under consideration.) In the intact leaf, measured quantum yields vary between 0.04 and 0.1. Healthy leaves from many species of C3 plants kept under low O_2 concentrations that inhibit photorespiration usually show a quantum yield of 0.1 (Bjorkman and Demmig, 1987). In normal air, the quantum yield of C3 plants is lower, typically 0.05 (Berry and Downton, 1982). Quantum

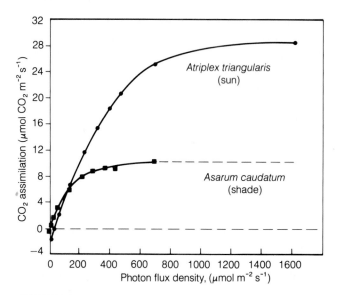

FIGURE 10.6. Dose-response curves of photosynthetic carbon fixation as a function of photon flux densities. *Atriplex triangularis* is a sun plant, and *Asarum caudatum* is a shade plant. Typically, shade plants have a low light compensation point and have lower maximal photosynthetic rates than sun plants. (From Harvey, 1979.)

yields vary with temperature and CO_2 concentration, because of the effect of these factors on the ratio of the carboxylase and oxygenase reactions of Rubisco (Chapter 9). Below 30°C, quantum yields of C3 plants are generally higher than those of C4 plants; above 30°C the situation is usually reversed (see Fig. 10.14). Despite their different growth habitats, sun and shade plants show similar quantum yields.

At higher photon flux densities, the photosynthetic response to light starts to level off (Fig. 10.6) and reaches *saturation*. Once saturation is reached, further increases in photon flux densities no longer affect photosynthetic rates, indicating that other factors, such as Rubisco activity or the metabolism of triose phosphates, have become limiting. Typical light saturation rates for shade plants are substantially lower than those for sun plants (Fig. 10.6). These saturation levels usually reflect the maximal photon flux densities to which the leaf is exposed in its growing environment.

Leaves Must Dissipate Vast Quantities of Heat

Absorption of the energy of sunlight by a leaf results in heating. The heat load on a leaf exposed to full sunlight is very high. A leaf with an effective thickness of water of 300 μm would warm up by 100°C every minute if all available solar energy were absorbed and no heat loss occurred! However, this enormous heat load is dissipated by emission of long-wave radiation, by sensible (or perceptible) heat loss, and by evaporative (or latent) heat loss (Fig. 10.7). Air circulation around the leaf removes heat from the leaf surfaces if the temperature of the leaf is warmer than that of the air; this is called **sensible heat loss**. On the other hand, **evaporative heat loss** occurs because evaporation of water requires energy. Thus, as water evaporates from a leaf it withdraws heat from the leaf and cools it. (The same principle is involved in the use of an evaporative cooler to cool a room.)

Sensible heat loss and evaporative heat loss are the most important processes in the regulation of leaf temperature, and the ratio of the two is called the **Bowen ratio** (Campbell, 1977):

$$\text{Bowen ratio} = \frac{\text{sensible heat loss}}{\text{evaporative heat loss}}$$

In well-watered crops, transpiration, and hence water evaporation from the leaf, is high, and the Bowen ratio is low. In these conditions, one can calculate the water loss from the crop by computing the amount of water that, when vaporized, will dissipate the net radiant energy being absorbed by the crop. Very low Bowen ratios can also be measured in a lawn on a relatively still day. In these conditions there is no sensible heat loss because the air around the leaf is at the same temperature as the leaf; the Bowen ratio therefore approaches zero, and the water loss of the lawn is determined primarily by the solar energy input.

In other cases, such as cotton leaves in the afternoon,

water loss from stomatal transpiration cools the leaf below the air temperature by evaporative heat loss. In that case, there is sensible heat gain rather than heat loss, and the Bowen ratio becomes negative. In contrast, the Bowen ratio can be very high in leaves from desert plants. In some cacti, stomata are tightly closed, precluding any evaporative cooling; all the heat is dissipated by sensible heat loss, and the Bowen ratio is infinite. Plants with very high Bowen ratios conserve water but have to endure very high leaf temperatures in order to maintain a sufficient temperature gradient between the leaf and the air. Slow growth is usually correlated with these adaptations.

Plants, Leaves, and Cells Adapt to Their Light Environment

Some plants are sufficiently plastic to adapt to a range of light regimes, growing as sun plants in sunny areas and as shade plants in shaded habitats (Bjorkman, 1981). Figure 10.8 illustrates some changes in the photosynthetic properties of leaves grown at high and low photon flux densities. But some shaded habitats receive less than 1% of the PAR available in an exposed habitat. Such a dynamic range cannot be covered by any leaf, so leaves that are adapted to very sunny or very shaded environments are unable to survive in the other type of habitat.

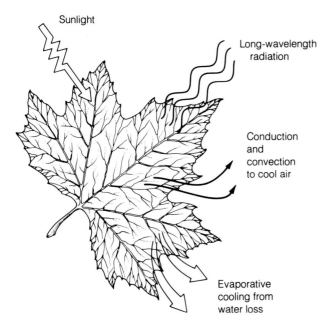

FIGURE 10.7. The absorption of energy from sunlight by the leaf. The imposed heat load must be dissipated in order to avoid damaging heating of the leaf. The heat load is dissipated by emission of long-wavelength radiation, by sensible heat loss to the air surrounding the leaf, and by the evaporative cooling caused by transpiration.

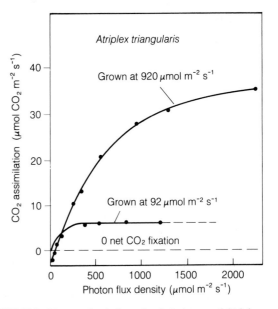

FIGURE 10.8. Changes in photosynthesis in leaves of *Atriplex triangularis* as a function of photon flux densities. The lower curve represents a leaf grown at 92 μmol m^{-2} s^{-1}, the upper curve a leaf grown at 10 times that photon flux density. In the leaf grown at the lower light levels, photosynthesis saturates at a substantially lower photon flux density, indicating that the photosynthetic properties of a leaf depend on its growing conditions. (From Bjorkman, 1981.)

Sun and shade leaves have some contrasting characteristics. For example, shade leaves have more total chlorophyll per reaction center, have a higher chlorophyll *b* to chlorophyll *a* ratio, and are usually thinner than sun leaves. Sun leaves have higher concentrations of soluble protein and Rubisco. Contrasting morphological characteristics can also be found in leaves of the same plant that are exposed to different light regimes. Figure 10.3 shows some differences between a leaf grown in the crown of a tree and exposed to bright sunlight and another leaf from the same tree shaded by the upper leaves. Adaptations to the light environment also occur within a single leaf. Cells in the upper surface of the leaf, which are exposed to the highest prevailing photon flux densities, have characteristics of cells from leaves grown in full sunlight, whereas cells in the lower surface of the leaf have characteristics of cells found in shade leaves (Terashima and Inoue, 1984).

These morphological and biochemical modifications are associated with physiological specializations. The higher photosynthetic rates of sun leaves are closely correlated with high respiration rates and a high light compensation point (Fig. 10.6). Shade plants show adaptations to light intensity and quality, such as a 3:1 ratio of photosystem II to photosystem I reaction centers, compared with the 2:1 ratio found in sun plants (Glazer and Melis, 1987). These properties appear to enhance light absorption and energy transfer in the shaded environments.

Carbon Dioxide and Photosynthesis in Leaves

Carbon dioxide is a trace gas in the atmosphere, constituting about 0.035% of the atmosphere, or 350 parts per million (ppm). The partial pressure of CO_2 varies with atmospheric pressure and is approximately 3.5×10^{-5} MPa (350 μbar) at sea level (see box on working with gases). Water vapor usually accounts for up to 2% of the atmosphere and O_2 for about 20%. The bulk of the atmosphere, nearly 80%, is nitrogen.

The carbon dioxide content of the atmosphere has become a matter of concern because of the greenhouse effect. The glass in a greenhouse roof transmits visible light, which is absorbed by the plants and other surfaces inside the greenhouse. The absorbed energy warms up these surfaces and is partially re-emitted as long-wavelength radiation. Because glass transmits long-wavelength radiation very poorly, this radiation cannot leave the greenhouse through the glass roof, and the greenhouse heats up. Certain gases in the atmosphere, particularly CO_2 and methane, play the same role as the glass roof in a greenhouse. Hence the warming up of the earth's climate because of the trapping of long-wavelength radiation by the atmosphere is called the

greenhouse effect. Because carbon dioxide accounts for about one-half of the greenhouse effect, there is concern about CO_2 levels in the air.

The CO_2 concentration of the atmosphere is increasing about 1 ppm per year (Fig. 10.9), primarily because of the burning of fossil fuels. Since 1958, when systematic measurements of CO_2 began at Mauna Loa, Hawaii, atmospheric CO_2 concentrations have increased by over 10% (Bacastow et al., 1985). The consequences of this increase are the subject of intense scrutiny by scientists and government agencies, particularly because of predictions that ensuing temperature increases could have profound effects on the world's climate. Plant physiologists are also interested in the effect of higher atmospheric CO_2 concentrations on leaf photosynthesis. The growth of most plants is limited by CO_2, and some plants grow twice as fast when the current levels of atmospheric CO_2 are doubled under laboratory conditions. Most plants grow 30 to 60% faster in an atmosphere containing twice the present levels of CO_2, and some species do not respond to a CO_2 doubling. Carbon dioxide enrichment of greenhouses is used commercially to increase the productivity of many crops, such as tomatoes, lettuce, cucumbers, and roses.

The Supply of CO_2 for Photosynthesis Depends On the Diffusion of CO_2 from the Atmosphere to the Chloroplast

For photosynthesis to occur, carbon dioxide has to diffuse from the atmosphere surrounding the leaf to the carboxylation site of Rubisco. Since diffusion rates depend on concentration gradients (Chapter 6), appropriate gradients are needed to ensure adequate diffusion of CO_2 from the leaf surface to the chloroplast. The cuticle that

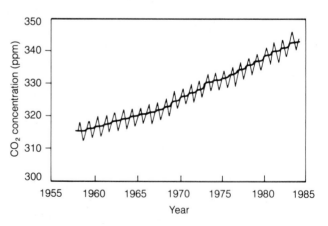

FIGURE 10.9. Concentration of atmospheric CO_2 measured at Mauna Loa, Hawaii, since 1958. Each year, the highest concentration is observed in May, just before the northern hemisphere growing season, and the lowest concentration is observed in October. In 1985, the May-to-October decrease was about 6 ppm. Superimposed on the yearly variation is a trend toward overall higher CO_2 concentrations. (From Bacastow et al., 1985.)

Working with Gases

THE CONCENTRATION OF oxygen in the atmosphere is independent of altitude. Why, then, is it more difficult to breath at high altitudes? The answer is that the binding of oxygen to hemoglobin in the blood depends on the partial pressure of oxygen, not its concentration, and the partial pressure of a gas does vary with atmospheric pressure. Many biological reactions involving gases, such as the responses of Rubisco to CO_2, depend on the partial pressure of the gas, hence the importance of measuring partial pressures.

The measure of concentration of a gas with which we are most familiar is mole fraction, not partial pressure. The **mole fraction** of a gas in a mixture is the number of moles of the gas of interest in a mole of the mixture. For example, the atmosphere has 21% oxygen and 350 ppm CO_2. Both of these measures are mole fractions: they tell us the number of moles of O_2 or CO_2 present in a mole of air. The partial volume of a gas divided by the volume of the mixture, the number of moles divided by the total number of moles, and the partial pressure divided by the total pressure are all expressions of mole fractions and are unitless and interchangeable. **Partial pressure** is obtained by multiplying mole fraction by total pressure. For example, at sea level, where the atmospheric pressure is about 0.1 MPa (1 bar), the partial pressure of oxygen is 2.1×10^{-2} MPa (210 mbar), and the partial pressure of CO_2 is 3.5×10^{-5} MPa (350 μbar). At the top of a 5000-foot mountain, where the atmospheric pressure is about 8.5×10^{-2} MPa (850 mbar), the partial pressure of oxygen is 1.8×10^{-2} MPa (180 mbar) and that of CO_2 is 2.95×10^{-5} MPa (295 μbar). This is why it is harder for people to breath and for leaves to photosynthesize at higher altitudes.

---

covers the leaf is nearly impermeable to CO_2, so the main port of entry of CO_2 into the leaf is the stomatal pore. From the pore, CO_2 diffuses into the substomatal cavity and into the intercellular air spaces between the mesophyll cells (see Fig. 4.9). This portion of the diffusion path of CO_2 into the chloroplast is a gaseous phase. The remainder of the diffusion path to the chloroplast is a liquid phase, which begins at the water layer wetting the cell walls of the mesophyll cells and continues through the cell membranes, the cytosol, and the chloroplast proper (see box on carbon dioxide in Chapter 9 for the properties of CO_2 in solution). Each portion of this diffusion pathway imposes a resistance to CO_2 diffusion, so the supply of CO_2 for photosynthesis can be described as a series of resistances. An evaluation of the magnitude of each resistance is helpful for understanding CO_2 limitations on photosynthesis.

Gas Phase Diffusion. Recall from Chapter 4 that the water transpired by the leaf evaporates from the wet cell walls into the intercellular air spaces and exits the leaf through the stomatal pores. The same path is traveled in the reverse direction by CO_2 on its way to the chloroplast. The sharing of this pathway by CO_2 and water presents the plant with a functional dilemma. In air of high relative humidity, the diffusion gradient driving water loss is about 50 times bigger than the gradient driving CO_2 uptake (Chapter 4). In drier air, this gradient can be significantly larger. Therefore, a decrease in stomatal resistance facilitating higher CO_2 uptake unavoidably entails a substantial water loss. Plant physiologists have used different approaches to understand how plants cope with this dilemma, including mathematical modeling of optimal strategies that maximize CO_2 uptake while minimizing water loss (Cowan and Farquhar, 1977). Another approach involves characterization of the physiological processes that regulate CO_2 uptake and water loss in the leaf, some of which are described below.

The gas phase of CO_2 diffusion into the leaf can be divided into three components—the boundary layer, the stomata, and the intercellular spaces of the leaf—each of which imposes a resistance to CO_2 diffusion. These components were discussed in connection with water loss in Chapter 4. The boundary layer is a layer of relatively unstirred air at the leaf surface, and its resistance to diffusion is called the boundary layer resistance. The magnitude of the boundary layer resistance decreases with leaf size and with wind speed. The boundary layer resistance to water and CO_2 diffusion is physically related to the boundary layer resistance for sensible heat loss. Smaller leaves have a lower boundary layer resistance to CO_2 and water diffusion and to sensible heat loss. Leaves of desert plants are usually small, which facilitates sensible heat loss. The large leaves often found in the humid tropics can have large boundary layer resistances, but these leaves can dissipate the radiation heat load by evaporative cooling because of the high transpiration rates made possible by the abundant water supply in these habitats.

After diffusing through the boundary layer, CO_2 enters the leaf through the stomatal pores, which impose the next resistance in the diffusion pathway, the stomatal resistance. Under most conditions in nature, in which the air around a leaf is seldom completely still, the boundary layer resistance is much smaller than the stomatal resistance, and the main limitation to CO_2 diffusion is imposed by the stomatal resistance. There is also a resistance to CO_2 diffusion in the air spaces separating the substomatal cavity from the walls of the mesophyll cells, called the intercellular air space resistance. This resistance is also usually small—a drop of 5×10^{-7} MPa (5 μbar) or less in partial pressure of CO_2, compared with the 3.5×10^{-5} MPa (350 μbar) outside the leaf.

Plants can regulate the magnitude of the stomatal resistance by changing the apertures of the stomatal pores. Since the stomatal pores usually impose the largest resistance to CO_2 uptake and water loss in the diffusion pathway, this regulation provides the plant with an effective way of controlling the gas exchange between the leaf and the environment. In experimental measurements of gas exchange from leaves, the boundary layer resistance and the intercellular air space resistance are usually ignored, and the stomatal resistance is used as the single parameter describing the gas phase resistance to CO_2.

To evaluate the effect of changes in CO_2 supply on photosynthesis, it is necessary to measure photosynthetic rates as a function of prevailing partial pressures of CO_2 (C_i) in the intercellular air spaces of the leaf. It is not technically feasible to measure C_i in the intercellular spaces of the leaf directly. However, since the CO_2 diffusing into the leaf travels by the same pathway as water diffusing out, C_i can be calculated from measured evaporation rates and stomatal resistance.

Evaporation (E) can be defined using Fick's law (Chapter 3):

$$E = \frac{(e_i - e_a)}{Pr} \qquad (10.1)$$

where E is the evaporation rate, e_i and e_a are the partial pressures of water inside the leaf and in the ambient air, P is the atmospheric pressure, and r is the stomatal resistance. The ratio $(e_i - e_a)/P$ is the difference of water vapor between the inside and the outside of the leaf (in mole fraction; see box on working with gases).

The concept of a resistance to gas diffusion is intuitively obvious but has the disadvantage that evaporation is inversely related to r. However, one can define conductance to water vapor (g) as the inverse of r. In contrast to r, g is directly related to E, and it is the parameter usually used in gas exchange analysis. Equation 10.1 can be rearranged to

$$E = \frac{(e_i - e_a)g}{P} \qquad (10.2)$$

and Equation 10.2 can be used to solve for g:

$$g = \frac{EP}{e_i - e_a} \qquad (10.3)$$

The rate of evaporation from a leaf can be determined with devices that measure the amount of water leaving the leaf, and the vapor pressure of water in the ambient air can also be measured. The air in the intercellular spaces of the leaf is assumed to be at 100% humidity, and the vapor pressure of water at 100% relative humidity is a function of temperature. Therefore, e_i is determined by measuring the leaf temperature.

Since CO_2 diffuses along the same pathway as water, we can write an equation defining photosynthetic CO_2 assimilation (A) based on Equation 10.2:

$$A = \frac{C_a - C_i}{1.6Pr} = \frac{(C_a - C_i)g}{1.6P} \qquad (10.4)$$

where C_a and C_i are the partial pressure of CO_2 in the ambient air and in the intercellular spaces of the leaf, respectively. The CO_2 molecule is larger than H_2O and therefore has a smaller diffusion coefficient; the difference between the two diffusion coefficients has been empirically determined to be 1.6 (Jarvis, 1971). Use of this correction factor in Equation 10.4 allows us to convert the measured resistance to water vapor diffusion into a resistance to CO_2 diffusion.

From Equation 10.4 we can calculate the partial pressure of CO_2 in the intercellular spaces of the leaf:

$$C_i = C_a - \frac{1.6AP}{g} \qquad (10.5)$$

Since all the variables on the right side of Equation 10.5 can be either measured or calculated, this equation is the basis for the calculation of C_i in experiments in which A is measured as a function of C_i (von Caemmerer and Farquhar, 1981). Changes in photosynthesis as a function of ambient and intercellular CO_2 concentrations in representative C3 and C4 plants are shown in Figure 10.10. By expressing the photosynthetic rates as a function of C_i, stomatal effects on these rates mediated by changes in CO_2 concentrations inside the leaf are no longer reflected in the response, and the intrinsic limitation imposed by the CO_2 supply can be evaluated. The remarkable differences between the photosynthetic responses of C3 and C4 plants to CO_2 became apparent in this type of analysis. In C4 plants, photosynthetic rates saturate at C_i values of about 2×10^{-5} MPa (200 μbar) (Fig. 10.10), reflecting the effective CO_2 concentrating mechanisms operating in plants with C4 metabolism (Chapter 9). In C3 plants, on the other hand, increasing C_i levels continue to stimulate photosynthesis over a much broader range. Also markedly different in C3 and C4 plants is the CO_2 **compensation point**, the CO_2 concentration at which CO_2 fixation by photosynthesis balances CO_2 loss by respiration and net CO_2 exchange is

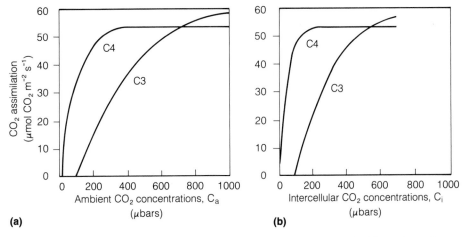

FIGURE 10.10. Changes in photosynthesis as a function of ambient intercellular CO_2 concentrations in *Tidestromia oblongifolia*, a C4 plant, and *Larrea divaricata*, a C3 plant. Photosynthetic rates are plotted against (a) ambient CO_2 concentrations and (b) calculated intercellular CO_2 concentrations inside the leaf (Eq. 10.5). The CO_2 concentration at which CO_2 assimilation is zero defines the CO_2 compensation point. 100 μbar = 1×10^{-5} MPa. (From Berry and Downton, 1982.)

zero. This concept is analogous to that of the light compensation point discussed earlier in the chapter: the CO_2 compensation point reflects the balance between photosynthesis and respiration as a function of CO_2 concentration, and the light compensation point reflecting such a balance as a function of photon flux density. Plants with C4 metabolism have a CO_2 compensation point of zero or nearly zero (Fig. 10.10), reflecting their very low levels of photorespiration (Chapter 9). This difference between C3 and C4 plants is no longer evident when the experiments are conducted at low oxygen concentrations.

Liquid Phase Diffusion. The resistance to CO_2 diffusion of the liquid phase, also called mesophyll resistance, encompasses diffusion from the intercellular leaf spaces to the carboxylation sites in the chloroplast. It has been calculated that the liquid phase resistance to CO_2 diffusion is approximately one-tenth of the combined boundary layer resistance and stomatal resistance when the stomata are fully open (Evans et al., 1986). This low resistance can be partially attributed to the large surface area of mesophyll cells exposed to the intercellular air spaces, which can be as much as 10–30 times the projected leaf area (Longstreth and Nobel, 1980). The liquid phase resistance, however, can be larger in thick leaves or in leaves from trees.

There Is a Cost in Water for Every Mole of CO_2 Taken Up by the Leaf

Now that we have discussed CO_2 diffusion into the leaf and its relation to water loss, we can review the concept of water-use efficiency introduced in Chapter 4. Water-

use efficiency can be quantified by combining Equations 10.2 and 10.4:

$$\frac{A}{E} = \frac{C_a - C_i}{(e_i - e_a)1.6} \tag{10.6}$$

Let us assume a leaf temperature of 30°C and an atmosphere with a relative humidity of 50%. For a C3 plant, $C_a - C_i$ is typically 1×10^{-5} MPa (100 μbar). At a leaf temperature of 30°C and a relative humidity of 50%, $e_i - e_a = 2.1 \times 10^{-3}$ MPa (21 mbar). From Equation 10.6, we compute that $A/E = 0.003$ mol CO_2/mol H_2O, or a transpiration ratio of 336. If water availability is reduced, stomata close and evaporation decreases, leading to improved water use efficiency. Total CO_2 assimilation also decreases, but the plant conserves water and increases its chances of survival. Agronomists can use rainfall amounts to evaluate the suitability of different crops for a particular region in terms of the water-use efficiency of each crop. Areas that have insufficient rainfall for a crop with very low water-use efficiency can be allocated to crops requiring less water.

Both Stomatal and Nonstomatal Factors Can Limit Photosynthesis

When a plant experiences adverse environmental conditions such as water stress, stomata close and intercellular CO_2 concentrations decrease. The drop in C_i increases the CO_2 limitation on photosynthesis, so photosynthetic rates decrease. This interaction poses a dilemma for the analysis of photosynthetic responses to environmental changes: Are the observed decreases in photosynthetic rates caused solely by stomatal closure and the decrease in C_i, or is there a direct, nonstomatal response of photosynthesis to the perturbation? One experimental approach to this problem is to increase the ambient CO_2 concentration to the extent necessary to increase the C_i level of a leaf to 3.5×10^{-5} MPa (350 μbar). Since that C_i level is the same as the normal ambient CO_2 concentration, the experiment simulates a condition in which the stomatal resistance is zero. In this way the changes in photosynthetic rates can be partitioned into stomatal and nonstomatal components (Farquhar and Sharkey, 1982). Experiments of this type have shown that stomatal limitations in C3 plants range between 10 and 40%.

Photosynthetic Carbon Fixation in Guard Cells Could Serve as a Signal Between the Stomata and the Mesophyll

The characterization of stomatal and nonstomatal limitations on photosynthesis under different environmental conditions has attracted substantial interest. Some recent studies are designed to investigate the mechanism that regulates the coupling of stomatal movements and photosynthetic rates in mesophyll cells.

Several classical hypotheses on stomatal function have included a role of the guard cell chloroplast in stomatal movements (Chapter 6). In one of the earliest experiments with radioisotopes, the Canadian plant physiologists M. Shaw and G. A. Maclachlan showed in 1954 that guard cells incorporate $^{14}CO_2$, which they interpreted as an expression of photosynthetic carbon fixation. However, evidence for photosynthetic activity in guard cells has remained controversial because of methodological difficulties, including contamination from mesophyll chloroplasts, and the need to distinguish between CO_2 fixation by the PCR cycle and that catalyzed by phosphoenolpyruvate carboxylase (Chapter 9) in the cytosol. Recent progress in work with guard cell protoplasts (Chapter 6) is providing new information related to this important question. Guard cell protoplasts illuminated with red light and exposed to $^{14}CO_2$ were shown to have radioactive 3-phosphoglyceric acid (PGA) within 5 seconds of the initiation of the experiment (Gotow et al., 1988). In time-course experiments, the percentage of radioactivity in PGA decreased with time, while that in sugar monophosphates increased (Fig. 10.11). These findings satisfy the criteria used in work on mesophyll photosynthesis (Chapter 9) to identify PGA as a primary carboxylation product of the PCR cycle. Other investigators have reported that the activity of Rubisco and other enzymes of the PCR cycle increases when guard cell chloroplasts are illuminated (Shimazaki et al., 1989).

The functional role of the PCR cycle in the guard cell chloroplast is under investigation. Competition for energy supply between photosynthetic carbon fixation in the chloroplast and the proton-pumping ATPase at the

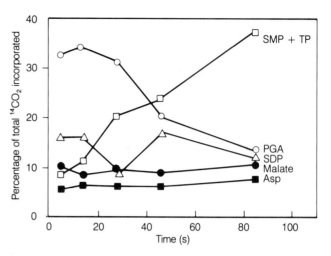

FIGURE 10.11. Percentage of total radioactivity in different fractions extracted from guard cell protoplasts of *Vicia faba*, plotted against time. At time zero, illuminated guard cell protoplasts were exposed to $^{14}CO_2$. Radioactivity was incorporated into the organic acids malate and aspartate (Asp), 3-phosphoglyceric acid (PGA), sugar monophosphates and triose phosphates (SMP + TP), and sugar diphosphates (SDP). With time, the percentage of radioactivity in PGA decreased and that in sugar monophosphates increased, indicating that PGA was the primary carboxylation product and that it participates in the synthesis of sugar monophosphates. These observations indicate that guard cell chloroplasts have the photosynthetic carbon reduction pathway. (From Gotow, et. al, 1988.)

plasma membrane (Chapter 6) could play a regulatory role in stomatal movements. Photosynthetic carbon fixation in guard cells could also serve as a coupling signal between the stomata and the mesophyll.

CO_2 Concentrating Mechanisms Affect the Photosynthetic Responses of the Intact Leaf

In plants with CO_2 concentrating mechanisms, which include C4 and CAM plants, the CO_2 concentrations at the carboxylation sites are often saturating. This physiological feature has several implications. For instance, we would not expect a C4 plant such as corn to increase its photosynthetic performance as CO_2 concentrations in the atmosphere increase.

Plants with C4 metabolism need less Rubisco to achieve a given rate of photosynthesis and therefore require less nitrogen to grow. In addition, the CO_2 concentrating mechanism allows the leaf to maintain high photosynthetic rates at lower C_i values, which require lower rates of stomatal conductance for a given rate of photosynthesis. Thus, C4 plants can use water and nitrogen more efficiently than C3 plants. On the other hand, the additional energy cost of the concentrating mechanism (Chapter 9) makes C4 plants less efficient in their utilization of light. This is probably one of the reasons why most shade-adapted plants are C3 plants.

Many cacti and other succulent plants with CAM metabolism have their stomata open at night and closed during the day (Fig. 10.12). The CO_2 taken up during the night is fixed into malate (Chapter 9). Since air temperatures are much lower at night than during the day, $e_i - e_a$ is smaller and a significant amount of water is saved. The main constraint on CAM metabolism is that the capacity to store malic acid is limited, which restricts the maximal rates of CO_2 uptake. However, many CAM plants can fix CO_2 via the C3 photosynthetic reduction cycle at the end of the day, when the temperature gradients are less extreme (Osmond, 1978).

Cladodes (the leaf-like structures) of *Opuntia* sp. can survive after detachment from the plant for several months without water. Their stomata are closed all the time, and the CO_2 released in respiration is refixed into malate. This process, which has been called CAM idling, allows the plant to survive for prolonged periods of time while losing remarkably little water (Ting and Gibbs, 1982).

Temperature Responses of Photosynthesis

Plants can photosynthesize in habitats having a surprisingly broad range of temperatures. In the lower temperature range, plants growing in alpine areas are capable of net CO_2 uptake at temperatures close to 0°C; at the other extreme, plants living in Death Valley, California, have optimal rates of photosynthesis at temperatures approaching 50°C.

When photosynthetic rates are plotted as a function of temperature, the curves have a characteristic bell shape (see upper curve in Fig. 10.13). The ascending arm of the curve represents a temperature-dependent stimulation of photosynthesis up to an optimum; the descending arm is associated with deleterious effects, some of which are reversible while others are not.

Temperature affects all the biochemical reactions of photosynthesis, so it is not surprising that the responses to temperature are complex. We can gain insight into the underlying mechanisms by comparing photosynthetic

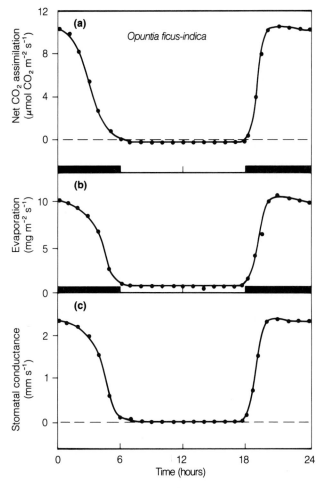

FIGURE 10.12. Photosynthetic carbon fixation, evaporation, and stomatal conductance of a cactus, *Opuntia ficus-indica*, over a 24-hour period. The whole plant was kept in a gas exchange chamber in the laboratory. The dark period is indicated by shaded bars. The units used here to express evaporation and stomatal conductance (mg m^{-2} s^{-1} and mm s^{-1}, respectively) are no longer used for these types of measurements (see Jarvis, 1971). In contrast to plants with C3 or C4 metabolism, CAM plants open their stomata and fix CO_2 at night. (From Gibson and Nobel, 1986.)

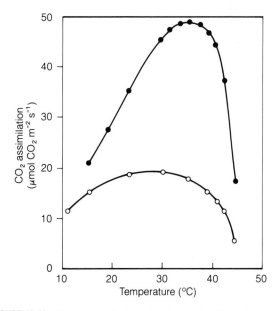

FIGURE 10.13. Changes in photosynthesis as a function of temperature at normal atmospheric CO_2 concentrations (open circles) and at concentrations that saturate photosynthetic CO_2 assimilation (filled circles). Photosynthesis has a strong temperature dependence at saturating CO_2 concentrations. (Redrawn from Berry and Bjorkman, 1980.)

rates in air at normal and at high CO_2 concentrations (Fig. 10.13). At high CO_2, there is an ample supply of CO_2 at the carboxylation sites, and the rate of photosynthesis is limited primarily by biochemical reactions connected with electron transfer (Chapter 8). In these conditions, temperature changes have large effects on fixation rates. At ambient CO_2 concentrations, photosynthesis is limited by the activity of Rubisco, and the response reflects two conflicting processes: an increase in carboxylation rate with temperature and a decrease in the affinity of Rubisco for CO_2 as the temperature rises (Chapter 9). These opposing effects damp the temperature response of photosynthesis at ambient CO_2 concentrations (Fig. 10.13).

Respiration rates increase as a function of temperature, and the interaction between photorespiration and photosynthesis becomes apparent in temperature responses. Figure 10.14 shows changes in quantum yield as function of temperature in a C3 and a C4 plant. In the C4 plant, the quantum yield remains constant with temperature, reflecting typical low rates of photorespiration. In the C3 plant, the quantum yield decreases sharply with temperature, reflecting a stimulation of photorespiration by temperature and an ensuing higher energy demand per net CO_2 fixed.

At low temperatures, photosynthesis is often limited by phosphate availability at the chloroplast (Sivak and

Walker, 1986; Sage and Sharkey, 1987). When triose phosphates are exported from the chloroplast to the cytosol, an equimolar amount of inorganic phosphate is taken up via translocators in the chloroplast membrane. If the rate of triose phosphate utilization in the cytosol decreases, phosphate uptake into the chloroplast is inhibited and photosynthesis becomes phosphate-limited. Starch synthesis and sucrose synthesis decrease rapidly with temperature, reducing the demand for triose phosphates and causing the phosphate limitation observed at low temperatures.

The highest photosynthetic rates seen in temperature responses (Fig. 10.13) represent the so-called optimal temperature response. When these temperatures are exceeded, photosynthetic rates decrease again. It has been argued that this optimal temperature is the point at which the capacities of the various steps of photosynthesis are optimally balanced (Berry and Downton, 1982), with some of the steps becoming limiting as the temperature decreases or increases. Optimal temperatures have strong genetic and physiological components. Plants of different species growing in habitats with different temperatures have different optimal temperatures for photosynthesis, and plants of the same species, grown at different temperatures and then tested for their photosynthetic responses, show temperature optima that correlate with the temperature at which they were grown. Plants growing at low temperatures maintain higher photosynthetic rates at low temperatures than plants grown at high temperatures. These changes in photosynthetic properties in response to temperature play an important role in plant adaptations to different environments.

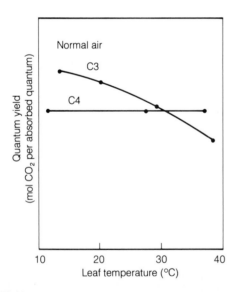

FIGURE 10.14. The quantum yield of photosynthetic carbon fixation in a C3 and a C4 plant as a function of leaf temperature. In normal air, photorespiration increases with temperature in C3 plants and the energy cost of net CO_2 fixation increases accordingly. This higher energy cost is expressed in lower quantum yields at higher temperatures. Because of the CO_2 concentrating mechanisms of C4 plants, photorespiration is low in these plants, and the quantum yield does not show a temperature dependence. Note, however, that at lower temperatures the quantum yield of C3 plants is higher than that of C4 plants. (From Berry and Downton, 1982.)

Summary

Photosynthetic activity in the intact leaf is an integral process that depends on many biochemical reactions. Different environmental factors can limit photosynthetic rates. Leaf architecture is highly specialized for light absorption, and the properties of palisade and mesophyll cells ensure uniform light absorption throughout the leaf. In addition to the structural features of the leaf, chloroplast movements within the cells and solar tracking by the leaf blade help to maximize light absorption. Light transmitted by upper leaves is absorbed by leaves growing beneath them. Many properties of the photosynthetic apparatus change as a function of the available light, including the light compensation point, which is higher in sun leaves than in shade leaves. The linear portion of the photosynthetic response to light provides a measure of the quantum yield of photosynthesis in the intact leaf. In temperate areas, quantum yields of C3 plants are generally higher than those of C4 plants.

Sunlight imposes a substantial heat load on the leaf, which is dissipated back into the air by sensible heat loss or by evaporative cooling. Increasing CO_2 concentrations in the atmosphere are increasing the heat load on the biosphere. This process could cause damaging changes in the world's climate but it could also reduce the CO_2 limitations on photosynthesis. At high photon flux densities, photosynthesis in most plants is CO_2 limited, but the limitation is substantially lower in C4 and CAM plants because of their CO_2 concentrating mechanisms. Diffusion of CO_2 into the leaf is constrained by a series of resistances. The largest resistance is usually that imposed by the stomata, so modulation of stomatal apertures provides the plant with an effective means of controlling water loss and CO_2 uptake. Both stomatal and nonstomatal factors affect CO_2 limitations on photosynthesis.

Temperature responses of photosynthesis reflect the temperature sensitivity of the biochemical reactions of photosynthesis and are most pronounced at high CO_2 concentrations. Because of the role of photorespiration, the quantum yield is strongly dependent on temperature in C3 plants but is nearly independent of temperature in C4 plants. Leaves growing in cold climates can maintain higher photosynthetic rates at low temperatures than leaves growing in warmer climates. Leaves grown at high temperatures perform better at high temperatures. Functional changes in the photosynthetic apparatus in response to prevailing temperatures in their environment have an important effect on the capacity of plants to live in diverse habitats.

GENERAL READING

Berry, J., and Bjorkman, O. (1980) Photosynthetic response and adaptation to temperature in higher plants. *Annu. Rev. Plant Physiol.* 31:491–543.

Berry, J. A., and Downton, J. S. (1982) Environmental regulation of photosynthesis. In: *Photosynthesis, Development, Carbon Metabolism and Plant Productivity*, Vol. II, pp. 263–343, Govindjee, ed. Academic Press, New York.

Bjorkman, O. (1981) Responses to different quantum flux densities. In: *Encyclopedia of Plant Physiology*, New Series, Vol. 12A, pp. 57–107, Lange, O. L., Nobel, P. S., Osmond, C. B., and Zeigler, H., eds. Springer-Verlag, Berlin.

Campbell, G. S. (1977) *An Introduction to Environmental Biophysics*. Springer-Verlag, New York.

Cowan, I. R., and Farquhar, G. D. (1977) Stomatal function in relation to leaf metabolism and environment. In: *Integration of Activity in the Higher Plant*, pp. 471–505, Jennings, D. H., ed. Harvard University Press, Cambridge, Mass.

Farquhar, G. D., and Sharkey, T. D. (1982) Stomatal conductance and photosynthesis. *Annu. Rev. Plant Physiol.* 33:317–345.

Gibson, A. C., and Nobel, P. S. (1986) *The Cactus Primer*. Harvard University Press, Cambridge, Mass.

Glazer, A. N., and Melis, A. (1987) Photochemical reaction centers: Structure, organization, and function. *Annu. Rev. Plant Physiol.* 38:11–45.

Osmond, C. B. (1978) Crassulacean acid metabolism: A curiosity in context. *Annu. Rev. Plant Physiol.* 29:379–414.

Powles, S. B. (1984) Photoinhibition of photosynthesis induced by visible light. *Annu. Rev. Plant Physiol.* 35:15–44.

Sharkey, T. D. (1985) Photosynthesis in intact leaves of C3 plants: Physics, physiology and rate limitations. *Bot. Rev.* 51:53–105.

Ting, I. P., and Gibbs, M., eds. (1982) *Crassulacean Acid Metabolism*. Waverly Press, Baltimore.

CHAPTER REFERENCES

Bacastow, R. B., Keeling, C. D., and Whorf, T. P. (1985) Seasonal amplitude increase in atmospheric CO_2 concentration at Mauna Loa, Hawaii, 1959–1982. *J. Geophys. Res.* 90:10529–10540.

Bjorkman, O., and Demmig, B. (1987) Photon yield of O_2 evolution and chlorophyll fluorescence characteristics at 77 K among vascular plants of diverse origins. *Planta* 170:489–504.

Bjorkman, O., and Powles, S. B. (1984) Inhibition of photosynthetic reactions under water stress: Interaction with light level. *Planta* 161:490–504.

Blatt, M. R., and Briggs, W. R. (1980) Blue light-induced cortical fiber reticulation concomitant with chloroplast aggregation in the alga *Vaucheria sessilis*. *Planta* 147:355–362.

Ehleringer, J., Bjorkman, O., and Mooney, H. A. (1976) Leaf pubescence: Effects on absorptance and photosynthesis in a desert shrub. *Science* 192:376–377.

Ehleringer, J., and Forseth, I. (1980) Solar tracking by plants. *Science* 210:1094–1098.

Evans, J. R., Sharkey, T. D., Berry, J. A., and Farquhar, G. D. (1986) Carbon isotope discrimination measured concurrently with gas exchange to investigate CO_2 diffusion in leaves of higher plants. *Aust. J. Plant Physiol.* 13:281–292.

Gotow, K., Taylor, S., and Zeiger, E. (1988) Photosynthetic carbon fixation in guard cell protoplasts of *Vicia faba* L.: Evidence from radiolabel experiments. *Plant Physiol.* 86:700–705.

Harvey, G. W. (1979) Photosynthetic performance of isolated leaf cells from sun and shade plants. *Carnegie Inst. Washington Yearb.* 79:161–164.

Haupt, W. (1986) Photomovement. In: *Photomorphogenesis in Plants*, pp. 415–441, Kendrick, R. E., and Kronenberg, G. H. M., eds. Martinus Nijhoff, Dordrecht, Netherlands.

Jarvis, P. G. (1971) The estimation of resistances to carbon dioxide transfer. In: *Plant Photosynthetic Production. Manual of Methods*, pp. 566–631, Seztak, Z., Catsky, J., and Jarvis, P. G., eds. Junk, The Hague.

Kaiser, W. M. (1982) Correlation between changes in photosynthetic activity and changes in total protoplast volume in leaf tissue from hygro-, meso- and xerophytes under osmotic stress. *Planta* 154:538–545.

Longstreth, D. J., and Nobel, P. S. (1980) Nutrient influences on leaf photosynthesis: Effects of nitrogen, phosphorus and potassium for *Gossypium hirsutum* L. Plant Physiol. 65:541–543.

McCain, D. C., Croxdale, J., and Markley, J. L. (1988) Water is allocated differently to chloroplasts in sun and shade leaves. *Plant Physiol.* 86:16–18.

Pearcy, R. W. (1988) Photosynthetic utilization of lightflecks by understory plants. *Aust. J. Plant Physiol.* 15:223–238.

Sage, R. F., and Sharkey, T. D. (1987) The effect of temperature on the occurrence of O_2 and CO_2 insensitive photosynthesis in field grown plants. *Plant Physiol.* 84:658–664.

Schwartz, A., and Koller, D. (1980) Role of the cotyledons in the phototropic response of *Lavatera cretica* seedlings. *Plant Physiol.* 66:82–87.

Shimazaki, K., Terada, J., Tanaka, K., and Kondo, N. (1989) Calvin-Benson cycle enzymes in guard-cell protoplasts from *Vicia faba* L.: Implications for the greater utilization of phosphoglycerate/dihydroxyacetone phosphate shuttle between chloroplasts and the cytosol. *Plant Physiol.* 90:1057–1064.

Sivak, M. N., and Walker, D. A. (1986) Photosynthesis *in vivo* can be limited by phosphate supply. *New Phytol.* 102:499–512.

Smith, H. (1986) The perception of light quality. In: *Photomorphogenesis in Plants*, pp. 187–217, Kendrick, R. E., and Kronenberg, G. H. M., eds. Martinus Nijhoff, Dordrecht, Netherlands.

Terashima, I., and Inoue, Y. (1984) Comparative photosynthetic properties of palisade tissue chloroplasts and spongy tissue chloroplasts of *Camellia japonica* L.: Functional adjustment of the photosynthetic apparatus to light environment within a leaf. *Plant Cell Physiol.* 25:555–563.

Terashima, I., and Saeki, T. (1983) Light environment within a leaf. I. Optical properties of paradermal sections of *Camellia* leaves with special reference to differences in the optical properties of palisade and spongy tissues. *Plant Cell Physiol.* 24:1493–1501.

von Caemmerer, S., and Farquhar, G. D. (1981) Some relationships between the biochemistry of photosynthesis and the gas exchange of leaves. *Planta* 153:376–387.

Younis, H. M., Boyer, J. S., and Govindjee (1979) Conformation and activity of chloroplast coupling factor exposed to low chemical potential of water in cells. *Biochim. Biophys. Acta* 548:328–340.

CHAPTER 11

Respiration and Lipid Metabolism

WHEREAS PHOTOSYNTHESIS PROVIDES THE carbohydrate substrate upon which the plant (and all life) depends, glycolysis and respiration are the processes whereby the energy stored in carbohydrates is released in a controlled manner. In this chapter we will review the basic features of these two vital interconnecting pathways, taking care to note the features that are peculiar to plants. Since many plants store their photosynthate in a more highly reduced form as lipids, a phenomenon common in seeds, we will also consider pathways of lipid biosynthesis and degradation in relation to the energy metabolism of seeds.

Respiration

Aerobic respiration is common to all eukaryotic organisms, and in its broad outlines, the respiratory process in higher plants is similar to that found in animals and lower eukaryotes. However, a number of specific aspects of plant respiration distinguish it from its animal counterpart. **Aerobic respiration** refers to the biological process by which reduced organic compounds are mobilized and subsequently oxidized in a controlled manner. During respiration, free energy is released and incorporated into a form (ATP) that can be readily utilized for the maintenance and development of the plant. From a chemical standpoint, respiration is most commonly expressed in terms of the oxidation of the six-carbon sugar glucose, as outlined below.

$$C_6H_{12}O_6 + 6O_2 + 6H_2O \rightarrow 6CO_2 + 12H_2O$$

This equation is, of course, the reverse of that used to describe the photosynthetic process (Chapters 8 and 9), and it represents a coupled redox reaction in which glucose is completely oxidized to CO_2 while oxygen serves as the ultimate electron acceptor, being reduced to water. The standard free energy change for the reaction as written involves the release of roughly 2880 kJ (686 kcal) per mole (180 g) of glucose oxidized, and it is the controlled release of this free energy and its coupling to the synthesis of ATP that serves as the primary, though by no means the only, role of respiratory metabolism. While glucose is most commonly cited as the substrate for respiration, it should be recognized that in a functioning plant cell the reduced carbon is actually derived from such sources as the glucose polymer starch, the disaccharide sucrose, fructose-containing polymers (fructosans), and other sugars, as well as lipids (primarily triacylglycerols), organic acids, and, on occasion, proteins.

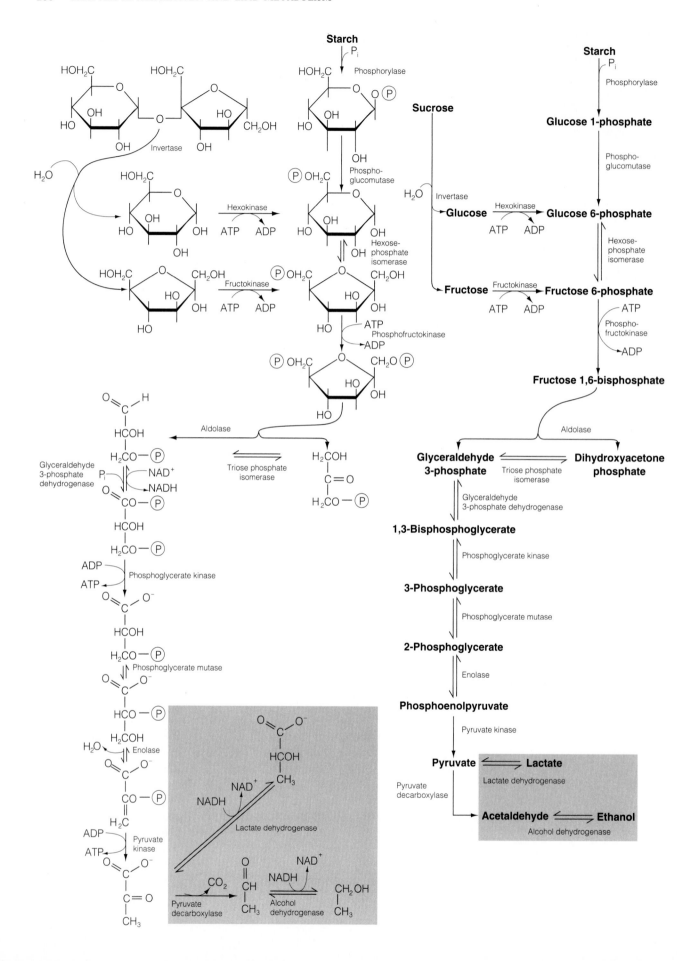

Considering the large amount of free energy released in the oxidation of glucose, it is not surprising that respiration is a multistep process in which glucose is oxidized through a series of reactions. These reactions can be subdivided into three stages: (1) glycolysis, (2) the tricarboxylic acid (TCA) cycle, and (3) the electron transport chain. **Glycolysis** is carried out by a group of soluble enzymes located in the cytosol. Chemically, glucose undergoes a limited amount of oxidation to produce two molecules of pyruvate (a three-carbon compound), a little ATP, and stored reducing power in the form of a reduced pyridine nucleotide, NADH.

The TCA cycle and the electron transport chain are both located within the confines of the membrane-bounded organelle known as the mitochondrion. The **TCA cycle** brings about the complete oxidation of pyruvate to CO_2 and, in so doing, generates a considerable amount of reducing power (about 10 NADH equivalents per glucose). With one exception, these reactions involve a series of soluble enzymes located in the internal aqueous compartment, or matrix, of the mitochondrion (see Fig. 11.2). The **electron transport chain** consists of a collection of electron transport proteins bound to the inner of the two mitochondrial membranes. This system transfers electrons from NADH (and related species), produced during glycolysis and the TCA cycle, to oxygen. The electron transfer releases a large amount of free energy, much of which is conserved through the conversion of ADP and P_i to ATP. This final stage completes the oxidation of glucose. We can now formulate a more complete picture of respiration as related to its role in cellular energy metabolism:

$$C_6H_{12}O_6 + 6O_2 + 6H_2O + 32ADP + 32P_i \rightarrow$$

$$6CO_2 + 12H_2O + 32ATP$$

As will become apparent, not all the carbon that enters the respiratory pathway ends up as CO_2. Many important metabolic intermediates appear in the reactions of the glycolytic and TCA cycle pathways, allowing these pathways to serve as starting points for many other cellular pathways in addition to their central role in energy metabolism (see Fig. 11.10). Examples of metabolites derived from respiratory intermediates include several amino acids, the pentoses used in cell wall and nucleotide biosynthesis, precursors of porphyrin biosynthesis, and the primary compound (acetyl CoA) needed to synthesize fatty acids and isoprenoid compounds (carotenoids, gibberellins, and abscisic acid, among others).

In Glycolysis, Glucose Is Converted into Pyruvate, and the Energy Released Is Stored in NADH and ATP

Glycolysis is a process that occurs in all living organisms (prokaryotes and eukaryotes). From an evolutionary standpoint, it is the oldest of the three stages of respiration. No oxygen is required to convert glucose to pyruvate, and glycolytic metabolism can become the primary mode of energy production in plant tissues where oxygen levels are low—for example, in roots in a flooded soil. As noted, glycolysis involves a series of soluble enzymes located in the cytosol, and until recently it was not associated with any particular organelle or subcellular organizational network. However, there is now evidence that in animal cells the enzymes of the glycolytic pathway do not exist independently in the cytosol but are associated in a supramolecular complex that may be bound to the surface of the outer mitochondrial membrane. This complex is thought to facilitate the conversion of substrates to products in the multistep glycolytic process.

The principal reactions associated with the classical glycolytic and fermentative pathways are shown in Figure 11.1. The two monosaccharides, glucose and fructose, can readily enter the pathway. In vivo, these two sugars are most commonly derived from the breakdown of either sucrose or starch. However, in higher plants, starch synthesis and catabolism take place in plastids, so carbon obtained from starch probably enters the glycolytic sequence in the cytosol at the level of the three-carbon sugars glyceraldehyde 3-phosphate and dihydroxyacetone phosphate (Chapter 9). Because sucrose is the major translocated sugar in higher plants and is therefore the form of carbon that most nonphotosynthetic tissues import, sucrose (*not* glucose) should be thought of as the true "sugar substrate" for plant respiration. All the enzymes shown in Figure 11.1 have been measured in plant tissues at levels sufficient to support the respiration rates observed in intact tissue.

In the initial series of glycolytic reactions, the entering hexose (glucose or fructose) is phosphorylated twice and then split, producing two molecules of the three-carbon sugar, glyceraldehyde 3-phosphate. This requires the input of two molecules of ATP per glucose and involves two of the three essentially irreversible (i.e., $K_{eq} > 500$) reactions of the glycolytic pathway catalyzed by hexokinase and phosphofructokinase. The phosphofructokinase reaction serves as one of the control

FIGURE 11.1. (left) The reactions of glycolysis and anaerobic fermentation. For each molecule of glucose that is metabolized, two molecules of glyceraldehyde 3-phosphate are formed, which ultimately produce two molecules of pyruvate. NAD^+ is reduced to NADH by glyceraldehyde-3-phosphate dehydrogenase, and the resulting NADH is reoxidized during fermentative metabolism (indicated by the shaded areas) by either lactate dehydrogenase or alcohol dehydrogenase. ATP input is required at the level of the hexo- and phosphofructokinase reactions, while ATP is synthesized by the phosphoglycerate and pyruvate kinases. The double arrows ($\rightleftharpoons$) denote the reversible reactions of glycolysis and the single arrows ($\rightarrow$) denote irreversible reactions.

points of glycolysis in both plants and animals. Recent work on plants has indicated that this part of the pathway is even more complex than shown in Figure 11.1, and this will be discussed in a later section. With the appearance of glyceraldehyde 3-phosphate, the glycolytic pathway can begin to extract usable energy. The enzyme glyceraldehyde-3-phosphate dehydrogenase catalyzes the oxidation of the aldehyde to a carboxylic acid, releasing sufficent free energy to allow the concomitant reduction of NAD^+ to NADH and the phosphorylation (using inorganic phosphate) of glyceraldehyde 3-phosphate to produce 1,3-bisphosphoglycerate.

NAD^+ (nicotinamide-adenine dinucleotide) is an organic cofactor associated with many enzymes that catalyze cellular redox reactions. NAD^+ is the oxidized form of the cofactor, and it undergoes a reversible two-electron reaction yielding NADH ($NAD^+ + 2e^- + H^+ \rightleftharpoons NADH$). The standard reduction potential for this redox couple is about -320 mV, which, biologically speaking, makes it a relatively strong reductant (i.e., electron donor). NADH is thus a good species in which to store free energy released during the stepwise oxidations of glycolysis and the TCA cycle. The subsequent oxidation of NADH by oxygen via the electron transport chain then releases sufficient free energy (220 kJ mol^{-1} or 52 kcal mol^{-1}) to drive the synthesis of ATP. A related compound, NADP (nicotinamide-adenine dinucleotide phosphate), performs a redox function in photosynthesis (Chapter 9) and the oxidative pentose phosphate pathway to be discussed later.

The phosphorylated carboxylic acid on C-1 of 1,3-bisphosphoglycerate represents a mixed acid anhydride that has a large free energy of hydrolysis (-54.5 kJ mol^{-1} or -11.8 kcal mol^{-1}). This makes 1,3-bisphosphoglycerate a strong donor of phosphate groups. In the next step of glycolysis, catalyzed by phosphoglycerate kinase, the phosphate of carbon 1 is transferred to a molecule of ADP to give ATP and 3-phosphoglycerate. For each glucose entering the pathway, two ATPs are generated by this reaction: one for each 1,3-bisphosphoglycerate. This type of ATP synthesis is referred to as a **substrate-level phosphorylation** because it involves the direct transfer of a phosphate moiety from a substrate molecule to ATP. It is distinct from the ATP synthesis by oxidative phosphorylation that is utilized by the electron transport chain in the final stage of respiration.

In the last series of glycolytic reactions, the phosphate on 3-phosphoglycerate is transferred to carbon 2 and a molecule of water is removed to give the compound phosphoenolpyruvate (PEP). The phosphate group on PEP also has a high free energy of hydrolysis (-62 kJ mol^{-1} or -14.8 kcal mol^{-1}) by virtue of its locking the parent compound, pyruvate, into an energetically unfavored enol ($-C=C-OH$) configuration, and this makes PEP an extremely good phosphate donor

as well. In the final step of glycolysis, the enzyme pyruvate kinase catalyzes a second substrate-level phosphorylation to yield ATP and pyruvate. This final step, which is the third irreversible step in glycolysis, yields two additional molecules of ATP for each glucose that enters the pathway.

In the Absence of O_2, Fermentation Allows the Regeneration of NAD^+ Needed for Glycolysis

In the absence of oxygen, the TCA cycle and electron transport chain cannot function. This creates a problem for the continued operation of glycolysis because the cell's supply of NAD^+ is limited, and once all the NAD^+ becomes tied up in the reduced state (NADH) the glyceraldehyde-3-phosphate dehydrogenase reaction will not be able to take place. To overcome this problem, plants and other organisms can further metabolize pyruvate by carrying out one or more forms of **fermentative metabolism** (Fig. 11.1). In lactic acid fermentation (common to mammalian muscle as well as plants), the enzyme lactate dehydrogenase uses NADH to reduce pyruvate to lactate, thus regenerating NAD^+. In alcoholic fermentation (also common in plants but more widely known in brewer's yeast), the two enzymes pyruvate decarboxylase and alcohol dehydrogenase act on pyruvate, ultimately producing ethanol and CO_2 and oxidizing NADH in the process.

Under some circumstances, plant tissues may be subjected to low (hypoxic) or zero (anoxic) concentrations of ambient oxygen, which forces them to carry out fermentative metabolism. The best-known example of this involves flooded or waterlogged soils in which the diffusion of oxygen to the roots is sufficiently reduced to cause root tissues to become hypoxic. In corn, the initial response involves lactic acid fermentation, but subsequent induction of the enzyme alcohol dehydrogenase promotes a switch to ethanolic fermentation (Roberts et al., 1984). Ethanol is thought to be a less deleterious end product of fermentation than lactate, because accumulation of lactate promotes acidification of the cytosol. There are numerous other examples of plants functioning under near-anaerobic conditions by carrying out some form of fermentation.

Plants Can Also Use the Glycolytic Pathway in the Reverse Direction to Synthesize Glucose

The reactions described above occur in all organisms that carry out glycolysis. In addition, organisms can utilize this pathway operating in the opposite direction to synthesize glucose from other organic molecules. The operation of the glycolytic pathway to synthesize glucose is known as **gluconeogenesis**. Gluconeogenesis operates in higher plants in the germination of seeds, such as castor bean and sunflower, that store a significant quantity of their carbon reserves in the form of fat. During germination, much of the fat is converted via gluconeogenesis to sucrose, which is then used to support the growing seedling. Because the reaction catalyzed by phosphofructokinase is not readily reversible, an additional enzyme, fructose bisphosphatase, acts at this step to convert fructose 1,6-bisphosphate to fructose 6-phosphate during gluconeogenesis. Fructose bisphosphatase is found in higher plants. Further, these two enzymes appear, in both plants and animals, to represent a major control point of carbon flux through the glycolytic and gluconeogenic pathways. However, in higher plants the interconversion of fructose 6-phosphate and fructose 1,6-bisphosphate is made more complex by the presence in plant tissues of an additional enzyme, a pyrophosphate (PP_i)-dependent phosphofructokinase (pyrophosphate–fructose-6-phosphate 1-phosphotransferase), which catalyzes the reversible interconversion of fructose 6-phosphate and fructose 1,6-bisphosphate:

$$\text{Fructose 6-P} + PP_i \rightleftharpoons \text{fructose 1,6-P}_2 + P_i$$

This enzyme is found in the cytosol of many plant tissues at levels that can be considerably higher than those of phosphofructokinase. Since the PP_i-phosphofructokinase reaction is readily reversible, this enzyme could operate during either glycolytic or gluconeogenic carbon fluxes, although its operation in glycolysis would require a source of pyrophosphate (Black et al., 1985). More important, this enzyme, like phosphofructokinase and fructose bisphosphatase, appears to be under metabolic control, leading to the conclusion that, under some circumstances, the operation of the glycolytic (or gluconeogenic) pathway in higher plants may not conform exactly to that commonly associated with other organisms (Fig. 11.1).

Anaerobic Fermentation Liberates Only a Fraction of the Energy Available in Each Molecule of Sugar

Before leaving glycolysis, some consideration should be given to the efficiency of fermentation. The standard free energy change ($\Delta G^{\circ\prime}$) for the complete oxidation of one mole of glucose is $-2880 \text{ kJ mol}^{-1}$ ($-686 \text{ kcal mol}^{-1}$). The $\Delta G^{\circ\prime}$ for the synthesis of ATP is $-31.8 \text{ kJ mol}^{-1}$ ($-7.6 \text{ kcal mol}^{-1}$). However, under the nonstandard conditions that normally exist in plant cells, the synthesis of ATP requires an input of free energy of approximately $-50.2 \text{ kJ mol}^{-1}$ ($-12 \text{ kcal mol}^{-1}$). Given the net synthesis of two molecules of ATP for each glucose that is converted to lactate (or ethanol), the efficiency of anaerobic fermentation—that is, the energy stored as ATP relative to the energy potentially available in a molecule of glucose—is only about 4%. Most of the energy available in the glucose remains in the reduced by-product of fermentation, lactate (or ethanol). During aerobic respiration, the pyruvate produced by glycolysis is transported into the mitochondria, where it is further oxidized, resulting in much more efficient conversion of the free energy originally available in the glucose.

Mitochondria Are Semiautonomous Organelles Surrounded by a Double Membrane

The next two stages of respiration take place within a double membrane-bound organelle, the **mitochondrion**. Plant mitochondria were originally identified by light microscopy as particles that stained with the dye Janus green B. With the advent of electron microscopy, mitochondrial morphology became clearly defined (Fig. 11.2). Plant mitochondria generally appear as spherical or rodlike entities ranging from 0.5 to 1.0 μm in diameter and up to 3 μm in length (Douce, 1985). Highly reticulate networks representing a single complex mitochondrion are commonly observed in animal cells and have been reported in unicellular algae, but their occurrence is not common in higher plant tissues. The number of mitochondria observed per plant cell is usually directly related to the metabolic activity of the tissue, reflecting the mitochondrial role in energy metabolism.

The ultrastructural features of plant mitochondria are similar to those of mitochondria in nonplant tissues (Fig. 11.2). Plant mitochondria have two membranes, a smooth **outer membrane** that completely surrounds a highly invaginated **inner membrane**. The aqueous phase contained within the inner membrane is referred to as the mitochondrial **matrix**, while the region between the two mitochondrial membranes is known as the **intermembrane space**. Invaginations of the inner membrane give rise to structures known as **cristae**.

Intact mitochondria are osmotically active; that is, they take up water and swell when placed in a hypoosmotic medium. Most inorganic ions and charged organic molecules are not able to diffuse freely into the matrix space. The site of the osmotic barrier is the inner membrane, the outer membrane being permeable to solutes having molecular masses less than approximately 10,000 Da (i.e., most cellular metabolites and ions). The lipid fraction of both membranes is made up primarily of phospholipids, 80% of which are either phosphatidylcholine or phosphatidylethanolamine.

Mitochondria, like chloroplasts, are semiautonomous organelles, since they contain ribosomes, RNA, and DNA, which encodes a limited number (15–25) of mitochondrial proteins. Plant mitochondria are thus able to carry out the various steps of protein synthesis and to transmit genetic information from one genera-

FIGURE 11.2. (a) Three-dimensional representation of a mitochondrion showing the invaginations of the inner membrane that give rise to the cristae and the location of the matrix and intermembrane spaces. The outer membrane is permeable to molecules smaller than 10,000 Da; the inner membrane serves as the major osmotic barrier of the organelle. (b) Electron micrograph of corn mesocotyl mitochondria (31,600 ×). (Courtesy of Carl Braun, North Carolina State University.)

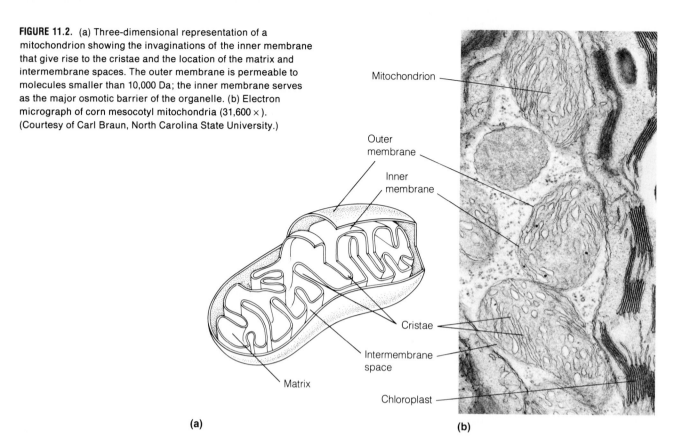

(a)   (b)

tion to the next. The phenomenon of cytoplasmic male sterility, which is commonly used by plant breeders to develop hybrid lines, in most instances seems to involve a mitochondrially linked trait (Hanson and Conde, 1985).

Pyruvate Is Oxidized in the Mitochondrion via the TCA Cycle

The TCA cycle is also known as the **Krebs cycle** in honor of Hans A. Krebs, who first proposed a cyclic series of reactions in the aerobic breakdown of pyruvate, and as the **citric acid cycle**, because of the importance of citrate as an early intermediate (Fig. 11.3). This cycle represents the second stage in respiration and takes place in the mitochondrial matrix. Its operation necessitates transport of the pyruvate generated in the cytosol during glycolysis through the rather impermeable inner mitochondrial membrane. This transport involves a pyruvate (monocarboxylate) translocator that catalyzes an electroneutral exchange of pyruvate and OH^- (see Figs. 11.4 and 11.7) across the inner membrane.

Once inside the mitochondrial matrix, pyruvate is oxidatively decarboxylated by the enzyme pyruvate dehydrogenase to produce NADH (from NAD^+), CO_2, and acetic acid, which is linked via a thioester bond to a sulfur-containing cofactor, coenzyme A (CoA), to form acetyl CoA (Fig. 11.3). Pyruvate dehydrogenase exists as

a complex of several enzymatic activities that catalyze the overall reaction in three separate steps. Following this reaction, the enzyme citrate synthase combines acetyl CoA with a four-carbon dicarboxylic acid, oxaloacetic acid (OAA), to give a six-carbon tricarboxylic acid, citrate. Following isomerization of citrate to isocitrate by the enzyme aconitase, the next two reactions involve successive oxidative decarboxylations, each of which produces one NADH and releases a molecule of CO_2. At this point, three molecules of CO_2 have been produced for each pyruvate that enters the mitochondrion, so the complete oxidation of glucose has actually been accomplished.

The remainder of the TCA cycle involves the conversion of succinyl CoA to OAA to allow continued operation of the cycle. Initially, the large amount of free energy available in the thioester bond of succinyl CoA is conserved through the synthesis of ATP from ADP and P_i via a substrate-level phosphorylation catalyzed by succinyl-CoA synthetase. (In the citrate synthase step, the free energy available in acetyl CoA was used to form a carbon-carbon bond in citrate.) The resulting succinate is oxidized to fumarate by succinate dehydrogenase, which is the only membrane-associated enzyme of the TCA cycle and is also considered to be a component of complex II of the electron transport chain (see p. 272). The electrons removed from succinate end up not on

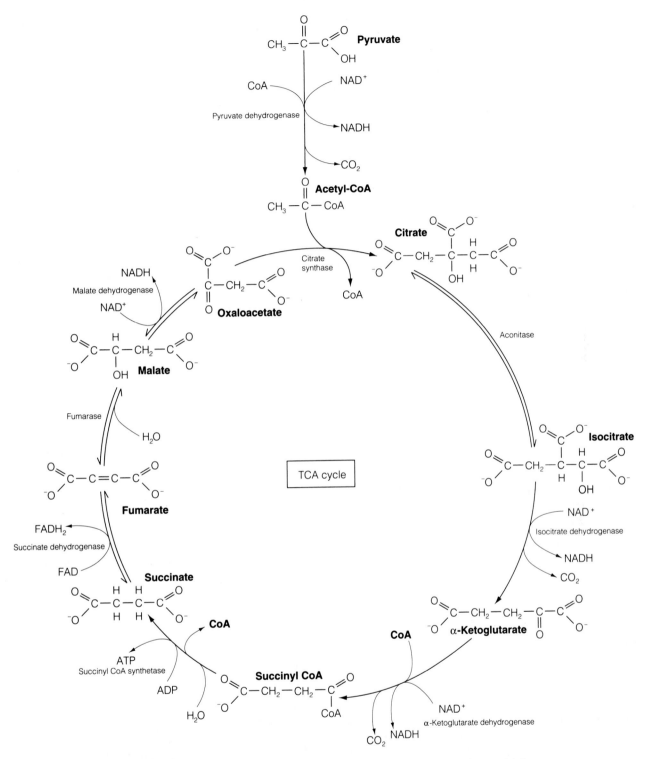

FIGURE 11.3. The reactions and enzymes of the TCA cycle. In the cycle, pyruvate is completely oxidized to three molecules of CO_2. The electrons generated during these oxidations are used to reduce four molecules of NAD^+ to NADH and one molecule of FAD to $FADH_2$. One molecule of ATP is also synthesized by a substrate-level phosphorylation during the succinyl-CoA synthetase reaction.

NAD^+ but on another electron-transferring cofactor, FAD (flavin-adenine dinucleotide). FAD is covalently bound to the active site of succinate dehydrogenase and undergoes a reversible two-electron reduction to produce $FADH_2$ ($FAD + 2e^- + 2H^+ \rightleftharpoons FADH_2$). In the final two reactions of the TCA cycle, fumarate is hydrated to produce malate, which is subsequently oxidized by malate dehydrogenase to regenerate OAA and produce another molecule of NADH. The OAA produced is now able to react with another acetyl CoA and continue the cycle.

The stepwise oxidation of pyruvate in the TCA cycle gives rise to three molecules of CO_2, and much of the free energy released during these oxidations is stored in the form of reduced NAD^+ (four NADH) and FAD (one $FADH_2$). In addition, one molecule of ATP is produced by a substrate-level phosphorylation. All the enzyme activities associated with the TCA cycle have been measured in mitochondria of higher plants, and it is likely that the enzymes are associated together in a multi-enzyme complex.

The TCA Cycle of Plant Cells Has Some Unique Features

The TCA cycle reactions outlined in Figure 11.3 are not all indentical to those carried out by animal mitochondria. For example, the succinyl-CoA synthetase step produces ATP in plants and GTP in animals. In animal mitochondria, ATP is formed in a second enzyme reaction that involves the transfer of a phosphate from GTP to ADP.

Another feature of the TCA cycle that is generally unique to plants is the significant NAD^+ malic enzyme activity, which has been observed in all plant mitochondria measured to date. This enzyme catalyzes the oxidative decarboxylation of malate.

$$\text{Malate} + NAD^+ \rightarrow \text{pyruvate} + CO_2 + NADH$$

The presence of NAD^+ malic enzyme enables plant mitochondria to operate an alternative pathway for the metabolism of PEP derived from glycolysis. Malate can be synthesized from PEP in the cytosol via the enzymes PEP carboxylase and malate dehydrogenase (Fig. 11.4). The malate is then readily transported into the mitochondrial matrix by a dicarboxylate transporter on the inner membrane that catalyzes the electroneutral exchange of malate and P_i (see Fig. 11.7). In the matrix, NAD^+ malic enzyme promotes the metabolism of malate by oxidizing it to pyruvate, which is then oxidized via the TCA cycle as outlined previously. The presence of NAD^+ malic enzyme thus allows the complete oxidation of organic acids (e.g., malate, citrate, α-oxoglutarate) in the absence of the normal TCA cycle substrate, pyruvate (Wiskich and Dry, 1985). The presence of an alternative pathway for the oxidation of malate is consistent with the observation that many plants, in addition to those that carry out Crassulacean acid metabolism (Chapter 9), store significant levels of malate in their central vacuole.

The Electron Transport Chain of the Mitochondrion Catalyzes an Electron Flow from NADH to O_2

For each molecule of glucose oxidized through glycolysis and the TCA cycle pathways, two molecules of NADH are generated in the cytosol while eight molecules of NADH plus two molecules of $FADH_2$ (on succinate dehydrogenase) appear in the mitochondrial matrix. As discussed previously, these reduced compounds must be reoxidized or the entire respiratory process will come to a halt. The electron transport chain catalyzes an electron flow from NADH (and $FADH_2$) to oxygen, the final electron acceptor of the respiratory process. For the oxidation of NADH, the overall two-electron transfer is represented as

$$NADH + H^+ + \tfrac{1}{2}O_2 \rightarrow NAD^+ + H_2O$$

From the midpoint reduction potentials for the $NADH/NAD^+$ couple, -320 mV, and the $H_2O/\frac{1}{2}O_2$ couple, $+810$ mV, the standard free energy released during this overall reaction ($= -nF \Delta E^{\circ\prime}$) is about 220 kJ mol^{-1} (52 kcal mol^{-1}) (per two electrons). Because the $FADH_2/FAD$ reduction potential (-45 mV) is somewhat higher than that of $NADH/NAD^+$, only 167.5 kJ mol^{-1} (40 kcal mol^{-1}) is released for each two electrons generated during the oxidation of succinate. The role of the electron transport chain is to bring about the oxidation of NADH (and $FADH_2$) and, in the process, utilize some of the free energy released to drive the synthesis of ATP.

The electron transport chain of higher plants contains approximately the same general set of electron carriers found in nonplant mitochondria (Fig. 11.5). The individual electron transport proteins are organized into a series of four multiprotein complexes (I–IV), each localized on the inner mitochondrial membrane (Fig. 11.6). Electrons from the NADH generated in the mitochondrial matrix during the TCA cycle are oxidized by complex I (an NADH dehydrogenase), which in turn transfers these electrons to ubiquinone, a p-benzoquinone species that exists as a pool within the inner membrane and is not tightly associated with any of the membrane proteins. Chemically and functionally, ubiquinone is similar to plastoquinone in the photosynthetic electron transport chain (Chapter 8). The electron carriers in complex I include a tightly bound FMN (flavin mononucleotide, chemically similar to FAD) and three or four iron-sulfur proteins. The iron-sulfur proteins can be distinguished on the basis of their redox potentials and their electron spin resonance (ESR) spectra (Moore and Rich, 1985).

The TCA cycle enzyme succinate dehydrogenase is part of complex II, so that electrons derived from the oxidation of succinate are transferred via the $FADH_2$ and a group of three iron-sulfur proteins into the ubiquinone pool. Complex III acts as a ubiquinone:cytochrome-c oxidoreductase, oxidizing reduced ubiquinone (ubiquinol) and transferring the electrons through an iron-sulfur center, two b-type cytochromes, and a membrane-bound cytochrome c_1 to cytochrome c. The electron flow through this part of the electron transport chain is complex and does not proceed in the linear fashion suggested by Figure 11.5 (see Moore and Rich, 1985, for a further explanation). Cytochrome c is the only protein in the electron transport chain that is not an integral membrane protein, and it serves as a mobile carrier

to transfer electrons between complexes III and IV (cytochrome-c oxidase). Cytochrome oxidase contains two a-type cytochromes and two atoms of copper, which bring about the four-electron reduction of O_2 to two molecules of H_2O.

The organization of these complexes on the inner membrane is quite specific (Fig. 11.6). The oxidation of NADH and succinate takes place on the matrix or M side of the membrane, as does the reduction of oxygen. Cytochrome c is located in the intermembrane space, on the cytoplasmic or C side of the membrane. It appears that the complex III iron-sulfur center and cytochrome c_1, though both membrane-bound, are oriented facing the C side of the inner membrane, while the two cytochromes b of complex III and cytochromes a and a_3 of cytochrome

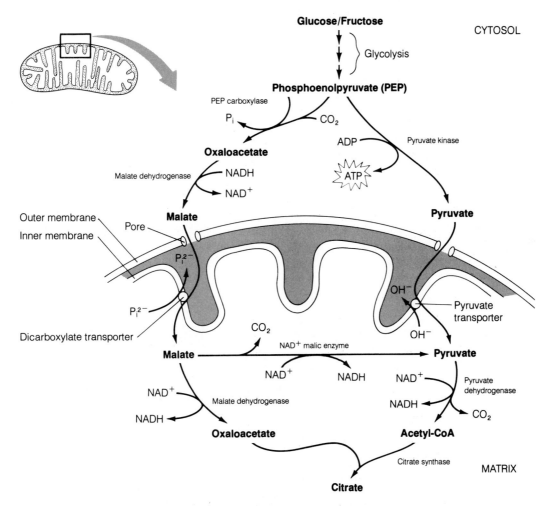

FIGURE 11.4. Pathways for the movement of malate and pyruvate into the TCA cycle from glycolysis in higher plants. Pyruvate generated in the cytosol during glycolysis is transported across the inner mitochondrial membrane in an electroneutral exchange with OH^- involving a pyruvate-specific transporter. Because plant mitochondria contain NAD^+ malic enzyme, they are also capable of completely oxidizing malate (in the absence of added pyruvate). This pathway involves the conversion of PEP, generated in the cytosol during glycolysis, to malate and the movement of malate into the mitochondria through the dicarboxylate transporter. The latter transporter catalyzes the electroneutral exchange of malate (and other dicarboxylates) and P_i^{2-} across the inner mitochondrial membrane. See Figures 11.1 and 11.3 for the structures of compounds.

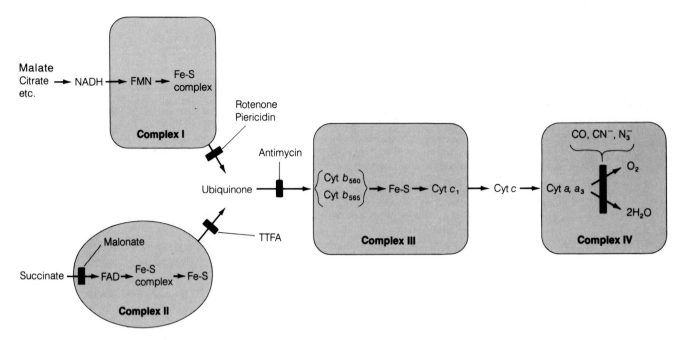

FIGURE 11.5. A linear representation of the cyanide-sensitive (main) pathway of plant mitochondrial electron transport, showing complexes I–IV and the sites of action of several inhibitors. Complex I, NADH:ubiquinone oxidoreductase; complex II, succinate:ubiquinone oxidoreductase; complex III, ubiquinone:cytochrome-c oxidoreductase; complex IV, cytochrome-c oxidase. Specific inhibitors include rotenone and piericidin, which inhibit electron transfer between the complex I iron-sulfur (Fe-S) complex and ubiquinone pool; malonate, which acts as a competitive inhibitor of succinate in complex II; TTFA (thenoyltrifluoroacetone), which blocks electron transfer from complex II into the ubiquinone pool; antimycin, which inhibits electron transfer in the cytochrome b region of complex III; and CO, cyanide (CN^-), and azide (N_3^-), all of which are competitive inhibitors of oxygen at cytochrome-c oxidase (complex IV).

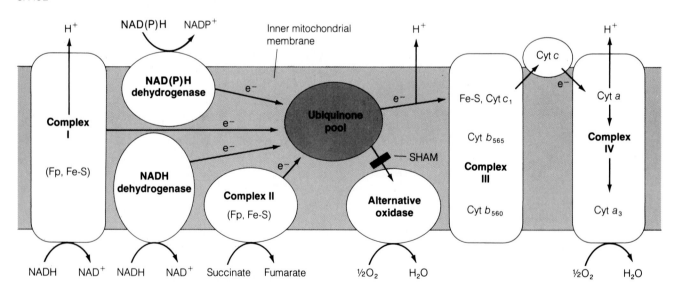

FIGURE 11.6. A two-dimensional representation of the electron carriers on the plant inner mitochondrial membrane (complexes I–IV, cytochrome c, two additional NADH dehydrogenases, and the cyanide-resistant [alternative] oxidase), the three sites of energy conservation (vertical arrows indicating H^+ extrusion), and the site of action of the specific inhibitor of the alternative pathway, salicylhydroxamic acid (SHAM). (Adapted from Siedow and Berthold 1986.)

oxidase are localized toward opposite sides of the inner membrane (Fig. 11.6). This sort of organization necessitates transmembrane movement of electrons as they traverse the electron transport chain, and its functional significance will be dealt with more fully when we consider the mechanism of ATP synthesis later in this chapter. Finally, these individual electron transfer complexes apparently are not permanently associated into larger "supercomplexes"; rather, electron transfer occurs during random collisions between reacting species (e.g., between complex I and ubiquinone) as they diffuse laterally within the fluid phospholipid bilayer.

Some of the Electron Carriers of Plant Mitochondria Are Absent from Animal Mitochondria

In addition to the collection of electron carriers described above, plant mitochondria contain some components not commonly found in other mitochondria (Fig. 11.6). These include an NADH dehydrogenase complex that faces the intermembrane space or C side of the inner membrane and apparently facilitates the oxidation of cytoplasmic NADH and, possibly, NADPH (Moller and Lin, 1986). Electrons from this *external* NADH dehydrogenase enter the main electron transport chain at the level of the ubiquinone pool. A second unusual feature of plant mitochondria is the presence of two pathways for oxidizing matrix NADH. Electron flow through the classical complex I described above is sensitive to inhibition by a number of compounds, including rotenone and piericidin. Isolated plant mitochondria consistently display the presence of a rotenone-resistant pathway for the oxidation of NADH derived from TCA cycle substrates. Although much has been learned about the properties of this rotenone-resistant bypass, its exact role in plant metabolism is still a matter of speculation (Palmer and Ward, 1985). Finally, most, if not all, plant mitochondria have varying levels of an "alternative" pathway for the reduction of oxygen. This pathway involves an oxidase that, unlike cytochrome-*c* oxidase, is insensitive to inhibition by cyanide, azide, or carbon monoxide. The nature and physiological significance of this cyanide-resistant pathway will be considered more fully below.

ATP Synthesis in the Mitochondrion Is Coupled to Electron Transport

The transfer of electrons to oxygen via complexes I–IV is coupled to the synthesis of ATP, with the number of ATPs synthesized depending on the nature of the electron donor (Table 11.1). With isolated mitochondria, electrons derived from internal (matrix) NADH give ADP:O ratios (number of ATPs synthesized per two electrons transferred to oxygen) of 2.4–2.7. Succinate and externally

added NADH each give values in the range 1.6–1.8, while ascorbic acid, which serves as an artificial electron donor to cytochrome *c*, gives values of 0.8–0.9. Results such as these (for both plant and animal mitochondria) have led to the general concept that three sites of energy conservation exist along the electron transport chain, at complexes I, III, and IV, respectively. Because of the observed connection between mitochondrial electron flow and ATP synthesis, this conversion of ADP and P_i to ATP became known as **oxidative phosphorylation** long before any of its mechanistic features were understood. However, in recent years a remarkable amount of progress has been achieved in understanding the mechanism of this energy transformation.

The currently accepted mechanism of mitochondrial ATP synthesis is based on the **chemiosmotic hypothesis**, proposed in the early 1960s by Peter Mitchell as a general mechanism of energy conservation across biological membranes (Nicholls, 1982). In this theory, the asymmetric orientation of electron carriers within the mitochondrial inner membrane allows for the transfer of protons (H^+) across the inner membrane during electron flow (Moore and Rich, 1985). Numerous studies have confirmed that mitochondrial electron transport is associated with a net transfer of protons from the mitochondrial matrix to the intermembrane space (Fig. 11.7). The free energy associated with the formation of this proton electrochemical gradient ($\Delta\bar{\mu}_{H^+}$, also referred to as a **proton motive force** when expressed in units of volts) is made up of an electrical (membrane) potential component, ΔE_m, and a chemical potential component, ΔpH. The ΔE_m results from the asymmetric distribution of a charged species (H^+) across the membrane, and the ΔpH is due to the proton concentration difference across the membrane. Both contribute to the proton motive force in plant mitochondria, although ΔE_m is consistently found to be of greater magnitude.

The input of free energy required to generate $\Delta\bar{\mu}_{H^+}$ comes from the free energy released during electron transport. Exactly how electron transport is coupled to proton translocation is not well understood in all cases. Mitchell originally proposed a series of alternating hydrogen ($H^+ + e^-$) and electron (e^-) carriers suitably situated within the membrane to give rise to a series of redox loops. This idea holds, in somewhat modified form, for proton translocation at complex III but seems unlikely to explain the coupling of proton and electron transfer at complexes I and IV.

Once generated, the $\Delta\bar{\mu}_{H^+}$ is reasonably stable because of the low permeability (conductance) of the inner membrane to protons, and the free energy stored in the proton electrochemical gradient can be utilized to carry out chemical work (i.e., ATP synthesis). The $\Delta\bar{\mu}_{H^+}$ is coupled to the synthesis of ATP by an additional protein complex associated with the inner membrane, the F_oF_1-**ATP synthase** (also called complex V). This complex

FIGURE 11.7. A proton electrochemical gradient ($\Delta\bar{\mu}_{H^+}$) is established across the inner mitochondrial membrane (acidic outside) during electron transport as outlined in the text. The membrane potential component (ΔE_m) of the gradient drives the electrogenic exchange of ADP in the cytosol for ATP in the mitochondria via the adenine nucleotide transporter, and the ΔpH drives the electroneutral uptake of P_i through the phosphate transporter. The free energy stored in the proton gradient is then coupled to the synthesis of ATP from ADP and P_i via the many F_oF_1-ATP synthase complexes that line the inner membrane. The F_o component forms a channel for the movement of protons across the inner membrane, while the F_1 complex provides the catalytic site for the condensation of ADP and P_i to ATP. Uncouplers provide a pathway for movement of protons across the inner membrane, preventing the buildup of the proton electrochemical gradient and inhibiting ATP synthesis but not electron transfer. (Modified from Douce, 1985.)

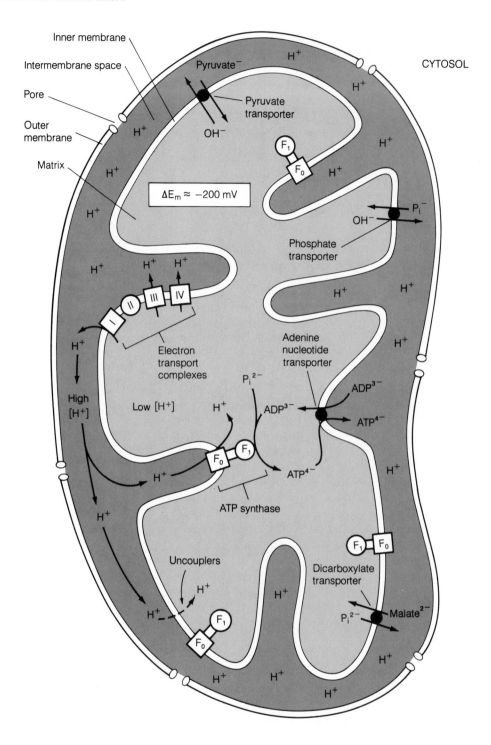

consists of two major components, F_1 and F_o (Fig. 11.7). F_1 is a peripheral membrane protein complex that is composed of at least five different subunits and contains the catalytic site for converting ADP and P_i to ATP (or hydrolyzing ATP to ADP and P_i when the ATPase activity is being measured). This complex is attached to the matrix side of the inner membrane. F_o is an integral membrane protein complex that consists of at least three different polypeptides that form the channel through which protons are able to cross the inner membrane. The passage of protons through the channel activates the

ATP synthase to synthesize ATP and simultaneously dissipates the $\Delta\bar{\mu}_{H^+}$. The structure and function of the mitochondrial ATP synthase are similar to those of the CF_o-CF_1 ATPase in photosynthetic photophosphorylation (Chapter 8).

Acceptance of the general features of a chemiosmotic mechanism of ATP synthesis has several implications. First, the true site of ATP formation on the mitochondrial inner membrane is the ATP synthase complex, not complex I, III, or IV. The latter serve as sites of *energy conservation* whereby electron transport is coupled to the generation of a $\Delta\bar{\mu}_{H^+}$. In fact, electron transport is not obligatorily linked to ATP synthesis if an alternative method of generating a proton gradient across the inner membrane exists. This can be demonstrated experimentally when electron transport is inhibited and ATP synthesis is made to proceed through an artificially generated pH gradient (Nicholls, 1982). Second, it is not necessary for an integral number of ATPs to be produced at each of the three sites of energy conservation. The overall ADP:O ratio will simply be a function of the number of protons translocated per two electrons times the number of ATPs synthesized per proton translocated. This need not be an integral value, so the ADP:O values of 2.4–2.7 measured for internal NADH oxidation (Table 11.1) are probably correct and do not have to be rounded off to 3.0 when attempting to calculate the total number of ATPs generated during aerobic respiration.

Finally, the chemiosmotic theory readily explains the mechanism of action of **uncouplers**, a wide range of chemically unrelated compounds (including dinitrophenol, FCCP [*p*-trifluoromethoxycarbonylcyanide phenylhydrazone], and many detergents) that inhibit mitochondrial ATP synthesis but are often observed to stimulate the rate of electron transport. These compounds all make the inner membrane leaky to protons, which prevents the buildup of a sufficiently large $\Delta\bar{\mu}_{H^+}$ to drive ATP synthesis. The stimulation of electron transport upon addition of uncouplers is related to an effect seen with isolated mitochondria. Experimentally, lower rates of electron flow (measured as the rate of oxygen uptake in the presence of an appropriate TCA cycle substrate, Fig. 11.8) are observed in the absence of ADP (referred to as *state 4*) than upon addition of ADP (*state 3*). ADP provides a substrate that stimulates the dissipation of the $\Delta\bar{\mu}_{H^+}$ through the F_oF_1-ATPase during ATP synthesis. Once all the ADP has been converted to ATP, the $\Delta\bar{\mu}_{H^+}$ builds up again and exerts a restraint on the rate of electron flow. The ratio of the state 3 and state 4 rates is referred to as the respiratory control ratio (RCR) and, although some caveats apply (Douce, 1985), it and the ADP:O ratio are used as characteristic measures of the quality of a mitochondrial preparation.

The proton electrochemical gradient also plays a role in the movement of the substrates and products of the TCA cycle and oxidative phosphorylation into and out of mitochondria. Whereas mitochondria synthesize ATP in the mitochondrial matrix, most of the ATP is used outside the organelle, so an efficient mechanism is needed for moving ADP into and ATP out of the mitochondrion. This mechanism involves another inner membrane protein, the ADP/ATP (adenine nucleotide) transporter (Fig. 11.7), which catalyzes an exchange of ADP and ATP across the inner membrane. The electrical potential gradient (ΔE_m) generated during electron transfer (negative inside) is such that there will be a net movement of the more negatively charged ATP^{4-} out of the mitochondria in exchange for ADP^{3-}. Likewise, the uptake of P_i involves an active phosphate transporter that uses the ΔpH component of the proton motive force to catalyze the electroneutral exchange of P_i^- (in) for OH^- (out). As long as a ΔpH is maintained across the inner membrane, the P_i content within the matrix will remain high. Similar reasoning applies to the uptake of pyruvate, which is driven by the electroneutral exchange of pyruvate for OH^-, leading to continued uptake of pyruvate from the cytoplasm (Fig. 11.7). Most of the proton electrochemical potential (~ -200 mV) across the mitochondrial inner membrane is due to the ΔE_m.

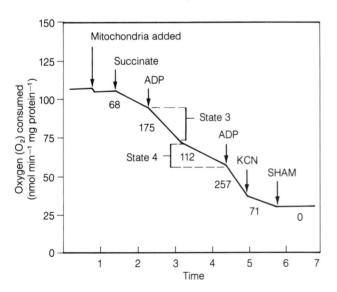

FIGURE 11.8. Regulation of respiratory rate by ADP in isolated mitochondria during succinate oxidation. The vertical arrows indicate successive additions of purified mitochondria, 10 mM succinate, 150 μM ADP, 450 μM ADP, 0.5 mM KCN, and 1.0 mM SHAM. Addition of succinate initiates mitochondrial electron transfer, which is measured with an oxygen electrode as the rate of oxygen reduction (to H_2O). Addition of ADP stimulates electron transfer (state 3) by facilitating the dissipation of the proton electrochemical gradient. Once all the ADP has been converted to ATP, electron transfer reverts to a lower (state 4) rate. Addition of cyanide (CN^-) inhibits electron flow through the main pathway and diverts it to the cyanide-resistant pathway, which is subsequently inhibited by the addition of SHAM. The numbers below the traces refer to the rate of oxygen uptake expressed as O_2 consumed. (Courtesy of Steven J. Steginck.)

Aerobic Respiration Yields 32 Molecules of ATP per Molecule of Glucose

The complete oxidation of glucose leads to the net formation of four molecules of ATP by substrate-level phosphorylation (two during glycolysis and two in the TCA cycle), two molecules of NADH in the cytosol, and eight molecules of NADH and two molecules of $FADH_2$ (via succinate dehydrogenase) in the mitochondrial matrix. Based on measured ADP:O values (Table 11.1), a total of approximately 28 molecules of ATP will be

TABLE 11.1. ADP:O Ratios in isolated plant mitochondria

Substrate	ADP:O ratio
Malate	2.4–2.7
Succinate	1.6–1.8
$NADH_{ext}$	1.6–1.8
Ascorbate	0.8–0.9

generated per glucose by oxidative phosphorylation. This results in a total of 32 ATPs per glucose. (Note the difference between this number and the value of 36 commonly seen in textbooks; the latter value is based on rounded-off integral values for the number of ATPs synthesized during the oxidation of each molecule of matrix or external NADH and succinate rather than the actual measured values given in Table 11.1.) Using 50.2 kJ mol^{-1} (12 kcal mol^{-1}) for the actual free energy of formation of ATP in vivo, we find that about 1606 kJ mol^{-1} (384 kcal mol^{-1}) of free energy is conserved per mole of glucose oxidized during aerobic respiration. This represents about 56% of the total free energy available; the rest is lost as heat. This is a marked improvement over the 3–5% efficiency associated with fermentative metabolism.

Plants Have a Cyanide-Resistant Respiration Pathway Not Found in Animals

If cyanide (1 mM) is added to actively respiring animal tissues, cytochrome oxidase is inhibited and the respiration rate quickly drops to less than 1% of its initial level. However, most plant tissues display a level of **cyanide-resistant respiration** that represents 10–25%, but in some tissues can be up to 100%, of the uninhibited control rate. The enzyme responsible for this oxygen uptake has been identified as a cyanide-resistant oxidase of the plant mitochondrial electron transport chain (Siedow and Berthold, 1986). Although the detailed mechanism of the unusual oxidase associated with this **alternative pathway** has yet to be characterized, current evidence indicates that electrons feed off the main electron transport chain into the alternative pathway at the level of the ubiquinone pool. The cyanide-resistant terminal oxidase associated with the alternative pathway catalyzes a four-electron reduction of oxygen to water and is specifically inhibited by several compounds, most notably salicylhydroxamic acid (SHAM) and n-propyl gallate. Because electrons branch to the alternative pathway from the ubiquinone pool, two sites of energy conservation (at complexes III and IV) are bypassed, and there is no evidence for an energy conservation site on the alternative pathway between ubiquinone and oxygen. Therefore, the free energy that would normally be stored as ATP is lost as heat when electrons are shunted through the alternative pathway.

How can such an energetically wasteful process as the cyanide-resistant pathway contribute to plant metabolism? One example of its utility appears during floral development in certain members of the Araceae (skunk cabbage, *Symplocarpus foetidus*). Just prior to pollination, tissue within a clublike organ on the developing floral apex (the spadix) undergoes high rates of respiration through the alternative pathway, causing the spadix to heat up by as much as 14°C above ambient temperature and to volatilize odiferous compounds that attract pollinating insects. Salicylic acid, a compound related to aspirin, has been identified as the chemical signal responsible for initiating this thermogenic event (Raskin et al., 1989). (See box on salicylic acid in Chapter 13.)

In most plants, however, both the respiratory rates and the level of cyanide-resistant respiration are too low to generate significant levels of heat. It has been suggested that the alternative pathway might serve as an energy overflow, oxidizing respiratory substrates that accumulate in excess of those needed for growth, storage, or synthesis of ATP (Lambers, 1985). In this view, electrons "spill" into the alternative pathway only after the capacity of the main pathway is saturated. This occurs experimentally when cyanide is added, but it could take place in vivo if a state 4 situation developed and the respiration rate exceeded the cell's demands for ATP. There is evidence from research that the alternative pathway serves an energy overflow function in higher plants (Lambers, 1985).

Respiration Is Regulated by Energy Demand and the Concentration of Key Metabolites

The control of carbohydrate metabolism in plants is poorly understood, but it is clear that it operates at numerous levels, starting with the rate at which reduced carbon is synthesized and/or imported into a plant tissue. In addition, because one of the major functions of respiration is to generate ATP, the rate of ATP utilization by a cell or tissue will be of primary importance in controlling the level of ADP present, and this will markedly influence the rates of glycolysis in the cytosol and oxidative phosphorylation in the mitochondria through mechanisms described below.

Control points appear to exist at all three stages of respiration, and only a brief overview of some obvious features will be given here (see Douce, 1985, for a more detailed discussion). In vivo, glycolysis appears to be regulated at the level of the enzymes phosphofructokinase and pyruvate kinase, just as it is in animal respiration. Pyruvate kinase is markedly inhibited by its product, ATP. As a result, high cytosolic levels of the substrate ADP promote the pyruvate kinase reaction, whereas high levels of ATP in the cytosol inhibit this reaction. Further, PEP, which builds up in the cytosol in response to ATP inhibition of pyruvate kinase, is a potent inhibitor of the ATP-dependent phosphofructokinase activity. This inhibitory effect of PEP on phosphofructokinase is enhanced in the presence of ATP but strongly attenuated by inorganic phosphate. These regulatory controls allow the flow of carbon through glycolysis to respond, to a first approximation, to the cell's energy demands by monitoring changes in the cytoplasmic [ATP]/[ADP] ratio.

However, the regulation of glycolysis is more complex than outlined above. For example, the ability of plant mitochondria to metabolize malate in addition to pyruvate (Fig. 11.4) makes it possible for plant cells to bypass the regulatory features of the pyruvate kinase reaction. With phosphofructokinase, the situation is made more complex by the fact that plants can contain appreciable levels of the PP_i-dependent phosphofructokinase activity as well as the metabolic effector fructose 2,6-bisphosphate. Fructose 2,6-bisphosphate (fructose 2,6-P_2) plays an important regulatory role in glycolysis in animals, stimulating phosphofructokinase activity and inhibiting that of fructose 1,6-bisphosphatase. Fructose 2,6-P_2 is found in the cytosol of plant tissues, as are the enzymes fructose 6-phosphate 2-kinase and fructose 2,6-bisphosphatase, which act to synthesize and break down fructose 2,6-P_2, respectively. Although the role of fructose 2,6-P_2 in the regulation of plant carbohydrate metabolism is not fully understood, several points have been established (Huber, 1986). Fructose 2,6-P_2 is present at varying levels in the cytoplasm. It markedly inhibits the activity of fructose 1,6-bisphosphatase, stimulates the activity of PP_i-dependent phosphofructokinase, and has little or no effect on the activity of the ATP-dependent phosphofructokinase. Further information is needed before we will fully understand the regulation at this point of the glycolytic/gluconeogenic pathways. However, it is clear that the PP_i-dependent phosphofructokinase and fructose 2,6-P_2 play a role in plant carbohydrate metabolism.

The primary control of plant mitochondrial oxidations seems to take place at the level of cellular adenine nucleotides. Exactly what feature is most important is less certain. Both the **adenylate energy charge**—$([ATP] + \frac{1}{2}[ADP])/([ATP] + [ADP] + [AMP])$—and the **[ATP]/[ADP] ratio** have been cited as regulatory indicators of the cell's energy status. From the limited number of studies of the mechanisms of regulatory control in plant mitochondria, it appears that the absolute concentration of ADP available (i.e., in the cytosol) is the most important factor in the regulation of mitochondrial respiratory rates. According to this view, as the cell's demand for ATP in the cytosol decreases relative to the rate of synthesis of ATP in the mitochondria, less ADP will be available, and the electron transfer chain will operate in more of a state 4 situation (Fig. 11.8). The resulting slowdown of electron transport could be communicated to TCA cycle enzymes through increases in (1) the matrix NADH:NAD^+ ratio, which affects several TCA cycle dehydrogenase activities, and (2) the matrix concentration of ATP, which has an allosteric influence on the pyruvate dehydrogenase and citrate synthase reactions (Wiskich and Dry, 1985). Decreased electron flow through the main electron transport chain would also lead to use of the cyanide-resistant pathway as the steady-state level of reduction of the ubiquinone pool increased in response to decreased cellular energy demands (consistent with its role in energy overflow). Clearly, much more needs to be learned about the various controls that modulate plant respiration rates.

The Pentose Phosphate Pathway Oxidizes Glucose to Ribulose 5-Phosphate and Reduces NADPH

The glycolytic pathway is not the only route available for oxidation of glucose in the cytosol of plant cells. The **oxidative pentose phosphate pathway** (also known as the hexose monophosphate shunt) can also accomplish this (Fig. 11.9). The first two reactions of this series represent the oxidative events of the pathway, converting the six-carbon glucose 6-phosphate to a five-carbon sugar, ribulose 5-phosphate, with loss of a CO_2 and generation of two molecules of NADPH (not NADH). The remaining reactions of the pathway bring about the conversion of ribulose 5-phosphate to the glycolytic intermediates glyceraldehyde 3-phosphate and fructose 6-phosphate. Studies of the release of $^{14}CO_2$ from isotopically labeled glucose indicate that glycolysis is the more dominant pathway, accounting for 80–95% of the total carbon flux in most plant tissues. However, the pentose phosphate pathway does contribute, and developmental studies indicate that its contribution increases as plant cells go from a meristematic to a more differentiated state (ap Rees, 1980).

The pentose phosphate pathway plays several roles in plant metabolism. The products of the two oxidative steps are NADPH, and it is thought that this NADPH is used to drive reductive steps associated with various biosynthetic reactions that occur in the cytosol. Such a role has commonly been accepted for the operation of this pathway in animal tissues. However, because the NADH dehydrogenase facing the cytosol on the mitochondrial inner membrane is also capable of oxidizing NADPH, it is possible that in plant cells some of the reducing power generated by the pentose phosphate pathway contributes to cellular energy metabolism—that is, electrons from NADPH may end up reducing O_2 and generating ATP. The pathway also produces ribose 5-phosphate, a precursor of the ribose and deoxyribose needed in the synthesis of RNA and DNA. Another intermediate on this pathway, the four-carbon erythrose 4-phosphate, combines with PEP in the initial reaction in the production of plant phenolic compounds, including the aromatic amino acids and the precursors of lignin, flavonoids, and phytoalexins (Chapter 13).

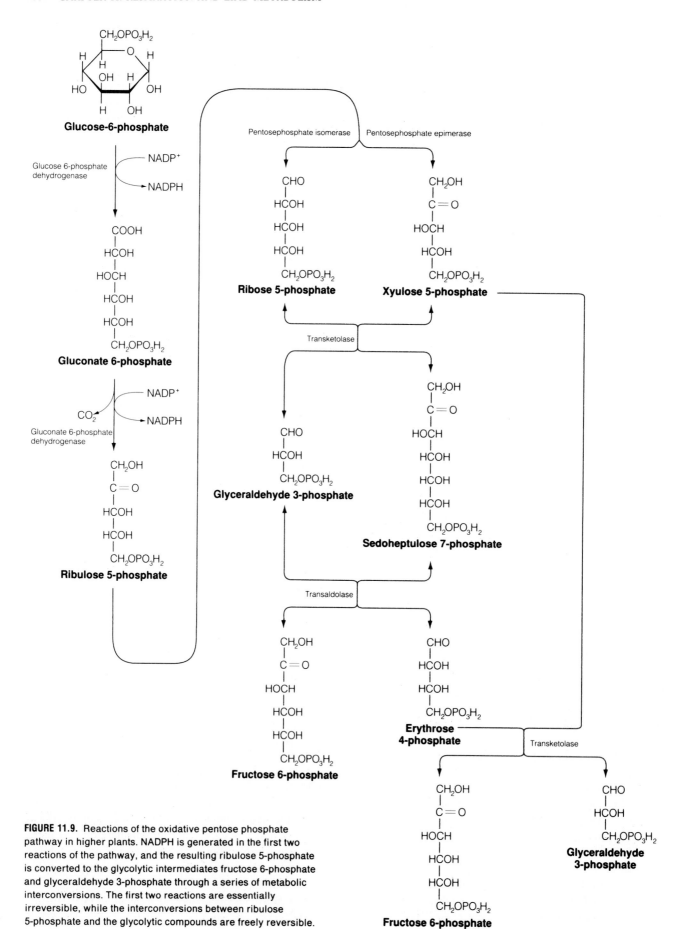

FIGURE 11.9. Reactions of the oxidative pentose phosphate pathway in higher plants. NADPH is generated in the first two reactions of the pathway, and the resulting ribulose 5-phosphate is converted to the glycolytic intermediates fructose 6-phosphate and glyceraldehyde 3-phosphate through a series of metabolic interconversions. The first two reactions are essentially irreversible, while the interconversions between ribulose 5-phosphate and the glycolytic compounds are freely reversible.

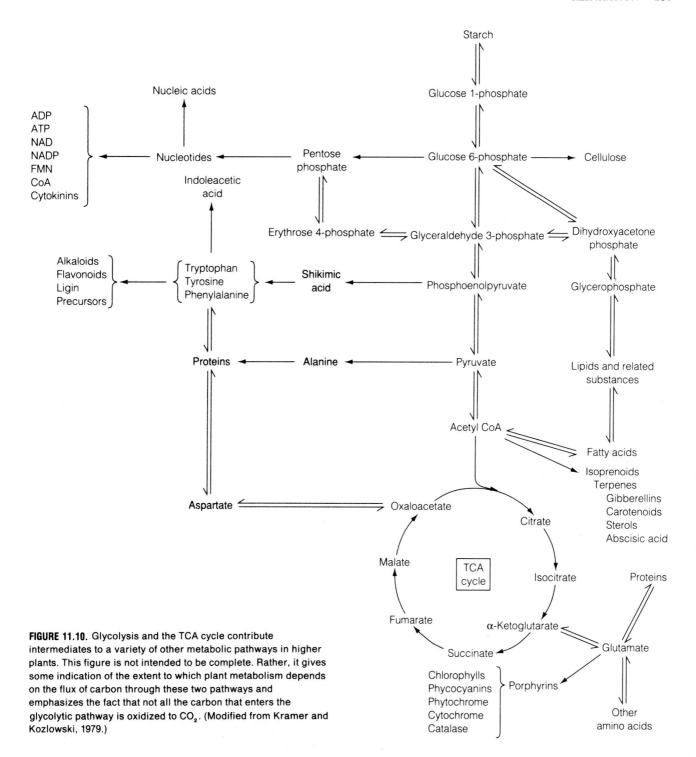

FIGURE 11.10. Glycolysis and the TCA cycle contribute intermediates to a variety of other metabolic pathways in higher plants. This figure is not intended to be complete. Rather, it gives some indication of the extent to which plant metabolism depends on the flux of carbon through these two pathways and emphasizes the fact that not all the carbon that enters the glycolytic pathway is oxidized to CO_2. (Modified from Kramer and Kozlowski, 1979.)

The pentose phosphate pathway is controlled by the initial reaction of the pathway catalyzed by glucose-6-phosphate dehydrogenase, whose activity is markedly inhibited as the NADPH:NADP ratio increases. However, measurements of the enzyme activities of the oxidative pentose phosphate pathway in green tissues are complicated by the fact that many of the same activities are also associated with chloroplast enzymes that catalyze the reactions of the reductive pentose phosphate pathway or Calvin cycle (Chapter 9). Both of the dehydrogenase

activities of the pentose phosphate pathway appear in chloroplasts and other plastids. This has led to speculation that the oxidative pentose phosphate pathway might function in chloroplasts under some conditions, notably in the dark. Little operation of the oxidative pathway is likely to occur in the chloroplast in the light, because mass action will drive the nonoxidative interconversions of the pathway in the direction of pentose synthesis. Moreover, glucose-6-phosphate dehydrogenase will be inhibited in the light by the high NADPH:NADP ratio in the chloroplast as well as by a reductive inactivation involving the ferredoxin : thioredoxin system (Chapter 9).

Respiration Is Tightly Coupled to Other Metabolic Pathways in the Cell

Although much of this chapter has focused on the role of respiration in energy metabolism, it should be pointed out that glycolysis and the TCA cycle are both linked to a number of important metabolic pathways, which will be covered in greater detail in Chapter 13. Glycolysis and the TCA cycle are central to the production of a wide variety of plant metabolites, including amino acids, lipids and related compounds, isoprenoids, and porphyrins (Fig. 11.10).

Not shown in Figure 11.10 are two additional reactions associated with plant mitochondria. First, two enzyme activities found in the matrix of mitochondria from chlorophyll-containing tissues, glycine decarboxylase and serine hydroxymethyltransferase, catalyze the CO_2-releasing step in photorespiration (Chapter 9). Second, plant mitochondria provide some of the reactions associated with the breakdown of stored fats in developing seeds, which is considered more extensively later in this chapter.

Whole-Plant Respiration

Many factors can affect the respiration rates of a plant as well as the individual organs of the plant. These include the species and growth habit of the plant, the type and age of the specific organ, and environmental variables such as the external oxygen concentration, temperature, and plant water status (see Chapter 14). Whole-plant respiration rates, particularly when considered on a fresh-weight basis, are generally lower than respiration rates reported for animal tissues. This is due in large part to the considerable fraction of a mature plant cell that is metabolically inactive in respiration (i.e., the central vacuole and the cell wall). Nonetheless, respiration rates in some plant tissues equal the rates observed in actively respiring animal tissues (Seymour et al., 1983), so the plant respiratory process is not inherently slower than respiration in animals.

Even though plants generally have low respiration rates, the contribution of respiration to the overall carbon economy of the plant can be substantial. A survey of several herbaceous species indicated that 30–60% of the daily gain in photosynthetic carbon was lost to dark respiration, although these values tended to decrease with age of the plants (Lambers, 1985). Young trees lose roughly a third of their daily photosynthate to respiration, and this loss can double in older trees as the ratio of photosynthetic to nonphotosynthetic tissue decreases. In tropical areas, 70–80% of the daily photosyn-

thetic gain can be lost to respiration owing to the high nighttime respiration rates associated with elevated night temperatures.

Different Tissues and Organs Respire at Different Rates

A useful rule of thumb is that the greater the overall metabolic activity of a given tissue, the higher is its repiration rate. Developing buds usually show very high rates of respiration (on a dry-weight basis), and respiration rates of vegetative tissues usually decrease from the growing tip to more differentiated regions. In mature vegetative tissues, stems generally have the lowest respiration rates, and leaf and root respiration varies with the plant species and the conditions under which the plants are growing.

When a plant tissue has reached maturity, its respiration rate will either remain roughly constant or decrease slowly as the tissue ages and ultimately undergoes senescence. An exception to this behavior is the marked rise in respiration known as the **climacteric** that accompanies the onset of ripening in many fruits (avocado, apple, banana) and senescence in detached leaves and flowers. Both ripening and the climacteric respiratory rise are triggered by the endogenous production of ethylene, and both processes can be stimulated to occur prematurely by exogenous application of ethylene (Chapter 19). Although ripening in these fruits is associated with increased rates of protein synthesis, which require cellular energy, the exact role of the respiratory rise is not clear. Some fruits (citrus, pineapple, grapes) do not show a respiratory climacteric, and these fruits are also not generally stimulated to ripen by addition of ethylene. Many studies have shown that fruits that undergo a climacteric respiratory rise display a high level of the cyanide-resistant pathway prior to ripening, but apparently not during the respiratory burst itself (Laties, 1982).

Environmental Factors Can Alter Respiration Rates

The environmental factors that can affect plant respiration clearly include oxygen because of its role as a substrate in the overall process. At 25°C, the equilibrium concentration of O_2 in an air-saturated (20% O_2) aqueous solution is about 250 μM. The K_m for oxygen in the cytochrome oxidase reaction is difficult to measure exactly but is well below 1 μM, so there should be no apparent dependence of the respiration rate on external O_2 concentrations (see Chapter 2 for discussion of K_m). In fact, little effect on measured respiration rates is observed until the atmospheric oxygen concentration drops below 5% for whole tissues or below 2–3% for tissue slices. Isolated mitochondria are unaffected by oxygen depletion approaching 0%. That there is any effect of oxygen

Does Respiration Reduce Crop Yields?

PLANT RESPIRATION CAN consume an appreciable amount of the carbon fixed each day during photosynthesis over and above the losses due to photorespiration (Chapter 9). To what extent can changes in a plant's respiratory metabolism affect crop yields? Attempts to establish a quantitative relation between respiratory energy metabolism and the various processes going on in the cell have led to a splitting of respiration into two components (Lambers, 1985). **Growth respiration** involves the processing of reduced carbon to bring about the growth of new plant matter. **Maintenance respiration** is the component of respiration needed to keep existing, mature cells in a viable state. Utilization of energy by the latter is not well understood, but estimates indicate that it can represent more than 50% of the total respiratory flux. On top of all this is the specter of the cyanide-resistant alternative pathway, with its potential for utilizing considerable amounts of the cell's reduced carbon to no apparent useful end. Estimates of the alternative pathway in wheat roots alone suggest a loss of carbon equivalent to 6% of the final grain yield.

Although numerous questions remain regarding the issues cited above, there are several examples of empirical relations between plant respiration rates and crop yield. In the forage crop perennial ryegrass (*Lolium perenne*), yield increases of 10–20% were correlated with a 20% decrease in the leaf respiration rate (Wilson and Jones, 1982). Similar correlations have been found for other plants, including corn and tall fescue (Lambers, 1985). Although there is a potential for increasing crop yield through reduction of respiration rates, a better understanding of the sites and mechanisms that control plant respiration is needed before such changes can be exploited in a logical fashion by plant physiologists, geneticists, and molecular biologists. Much remains to be established regarding the general applicability of such observations and the conditions under which slower-respiring lines might actually be at a disadvantage.

concentration at all is due to limitations on the diffusion of oxygen through the aqueous phase between the atmosphere and the mitochondria. The diffusion coefficient of oxygen dissolved in an aqueous medium is 10^4 less than that of oxygen diffusing through air. The fact that atmospheric oxygen tensions as high as 5% can reduce the observed respiration rate in whole tissues indicates that the diffusive movement of oxygen represents a true limitation on plant respiration. Further, the existence of a diffusion-limited, aqueous pathway of oxygen movement implies that intercellular air spaces are important in facilitating oxygen movement within plant tissues. If there were no gaseous diffusion pathway throughout the plant, the cellular respiration rates of many plants would be limited by an insufficient supply of oxygen.

This diffusion limitation is even more significant when the medium in which the tissue is growing is a liquid. When plants are grown hydroponically it is necessary to aerate the solutions vigorously to keep oxygen levels high in the vicinity of the roots, and the problem of oxygen supply also arises with plants growing in predominantly wet or flooded soils (see Chapter 14). Some plants, particularly trees, are restricted in geographic distribution by the need to maintain a supply of oxygen to their roots. For instance, dogwood and tulip tree poplar can survive only in well-drained, aerated soils because their roots are unable to tolerate more than a limited exposure to a flooded condition. Many other plants, however, can adapt to growth in flooded soils. Herbaceous species such as rice and sunflower often rely on a network of intercellular air spaces running from the leaves to the roots to provide a continuous, gaseous pathway for the movement of oxygen to the flooded roots. The problem can be even more acute for trees that have very deep roots growing in wet soils. Such roots must survive on anaerobic (fermentative) metabolism and/or develop structures that facilitate the movement of oxygen to the roots. Good examples of the latter are outgrowths of the roots, called **pneumatophores**, which protrude out of the water and provide a gaseous pathway for oxygen diffusion into the roots. Pneumatophores are found in *Avicennia* and *Rhizophora* trees that grow in mangrove swamps under continuously flooded conditions. While superficially similar in appearance to mangrove pneumatophores, the "knees" of cypress (*Taxodium*) do not perform such a function.

Although lowering the external oxygen concentration ultimately reduces the rate of aerobic respiration, it does not necessarily decrease the rate of carbohydrate utilization. In 1861, Louis Pasteur discovered that yeast cells consume less glucose in the presence of air than under anaerobic conditions. This response, known as the **Pasteur effect**, occurs in higher plant tissues and manifests itself as an increased rate of fermentative metabolism under anaerobic conditions. In aerobic respiration, cytosolic levels of ATP are high, producing an inhibitory effect on the regulatory enzymes of the glycolytic pathway, ATP-dependent phosphofructokinase and pyruvate kinase. In tissues maintained under anaerobic conditions, the cytosolic ATP level and its attendant inhibitory effects are decreased, leading to higher rates of carbohydrate metabolism.

The effect of atmospheric oxygen (and temperature) on respiration is utilized in fruit storage, where it is common to maintain fruits in the cold under conditions of 2–3% oxygen and 3–5% CO_2. The reduced temperature serves to lower the respiration rate (see below), as does the reduced oxygen. Low levels of oxygen are used instead of anoxic conditions to avoid lowering tissue oxygen tensions enough to stimulate fermentative metabolism. Carbon dioxide has little direct effect on the respiration rate even at 3–5%, well in excess of the 0.035% normally found in the atmosphere. High concentrations of CO_2 are used in fruit storage to block the effect of the hormone ethylene on fruit ripening (Chapter 19).

In the physiological temperature range, respiration is temperature-dependent. The increase in respiration rate for every 10°C increase in temperature, commonly referred to as the Q_{10}, is slightly greater than two, although in some plants an unexplained decrease in the Q_{10} has been observed at temperatures above 30°C. High night temperatures are thought to account for the large respiratory component in tropical plants, and lowered temperatures are utilized to retard postharvest respiration rates during the storage of fruits and vegetables. However, complications may arise from this practice. For instance, when potato tubers are stored at temperatures above 10°C, respiration and ancillary metabolic activities are sufficient to allow sprouting to proceed. Below 5°C, respiration rates and sprouting are reduced in most tissues, but the breakdown of stored starch and its conversion to sucrose impart an unwanted sweetness to the tubers. As a compromise, potatoes are stored at 7–9°C, which prevents the breakdown of starch while minimizing respiration and germination.

Other factors that can affect plant respiration include inorganic ions and injury. Addition of ions to plants previously grown in distilled water stimulates respiration—a phenomenon called salt respiration. The simplest explanation for this effect is that respiratory-linked ATP is required to support the enhanced ion uptake, but other factors also seem to be involved. Physical injury to higher plant tissues often stimulates oxygen uptake because of increases in respiration (mitochondria-linked oxygen uptake) and in nonmito-chondrial activities (e.g., lipoxygenase, polyphenol oxidase, peroxidase).

Lipid Metabolism

Starch and sucrose are not the only substrates available for energy production in the plant. Fats and oils are important storage forms of reduced carbon in many seeds, including those of agriculturally important species such as soybean, sunflower, peanut, and cotton. Oils often serve a major storage function in nondomesticated plants that produce small seeds. Some fruits, such as olives and avocados, also store fats and oils.

Fats and oils belong to the general class of *lipids*, a structurally diverse group of hydrophobic compounds that are soluble in organic solvents and highly insoluble in water. They represent a more reduced form of carbon than carbohydrates, so the complete oxidation of one gram of fat or oil (~40 kJ or 9.3 kcal) can produce considerably more ATP than the oxidation of one gram of starch (~15.9 kJ or 3.79 kcal). Conversely, the biosynthesis of fats, oils, and related molecules such as the phospholipids of membranes (see Chapter 1) requires a correspondingly large investment of metabolic energy. Other lipids are important for plant structure and function but are not used for energy storage. These include waxes, which make up the protective cuticle that reduces water loss from exposed plant tissues, and terpenoids (also known as isoprenoids), a family of compounds derived from the condensation of successive five-carbon isoprene units. Important terpenoids include the carotenoids involved in photosynthesis and the sterols, which are present in many plant membranes. The metabolism of terpenoids will be covered in greater detail in Chapter 13.

Fats and Oils Are Triglycerides and Are Stored in Spherosomes

Fats and oils exist mainly in the form of triglycerides, or triacylglycerols, in which fatty acid molecules are linked by ester bonds to the three hydroxyl groups of glycerol (Fig. 11.11). The fatty acids in plants are usually straight-chain carboxylic acids having an even number of carbon atoms. The carbon chain lengths can be as short as 12 and as long as 20 but are more commonly 16 and 18. Oils are liquid at room temperature, primarily due to the presence of a number of unsaturated bonds in their component fatty acids, whereas fats are solid, having a higher proportion of saturated fatty acids. The major fatty acids appearing in plant lipids are shown in Table 11.2. The percent composition of fatty acids in plant lipids varies with the species. For example, peanut oil is about 9% palmitate, 59% oleate, and 21% linoleic acid, while cottonseed oil is 20% palmitate, 30% oleate, and 45% linoleic acid.

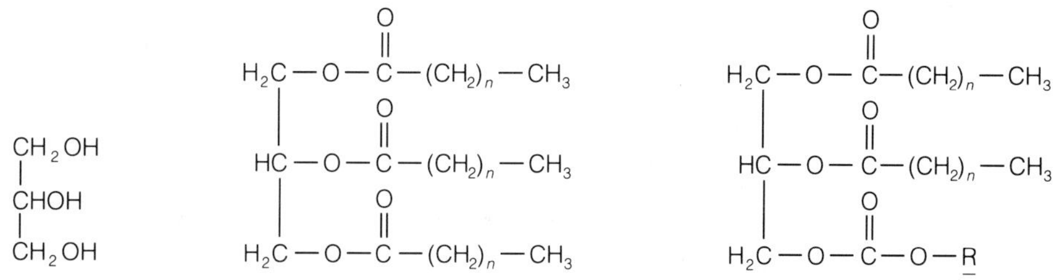

Glycerol

$$CH_2OH$$
$$CHOH$$
$$CH_2OH$$

Triglyceride

$$H_2C-O-\overset{\overset{O}{\|}}{C}-(CH_2)_n-CH_3$$
$$HC-O-\overset{\overset{O}{\|}}{C}-(CH_2)_n-CH_3$$
$$H_2C-O-\overset{\overset{O}{\|}}{C}-(CH_2)_n-CH_3$$

Diglycerides

$$H_2C-O-\overset{\overset{O}{\|}}{C}-(CH_2)_n-CH_3$$
$$HC-O-\overset{\overset{O}{\|}}{C}-(CH_2)_n-CH_3$$
$$H_2C-O-\overset{\overset{O}{\|}}{C}-O-\underline{R}$$

$R = -H$ **Diacylglycerol (DAG)**

$\quad = -HPO_3^{-2}$ **Phosphatidic acid (PA)**

$\quad = -PO_3-CH_2-CH_2-\overset{+}{N}(CH_3)_3$ **Phosphatidylcholine (lecithin)**

$\quad = -PO_3-CH_2-CH_2-NH_2$ **Phosphatidylethanolamine**

$\quad = -$ galactose(s) **Galactolipids**

FIGURE 11.11. The structural features of triacylglycerols and amphipathic lipids in higher plants. The carbon chain lengths of the fatty acids, which always have even numbers of carbons, range from 12–20 but are typically 16–18. Thus, the value of *n* is usually 14–16.

TABLE 11.2. Common fatty acids in higher plant tissues

Name	*Structure*
Saturated fatty acids	
Lauric acid (12:0)*	$CH_3(CH_2)_{10}CO_2H$
Myristic acid (14:0)	$CH_3(CH_2)_{12}CO_2H$
Palmitic acid (16:0)	$CH_3(CH_2)_{14}CO_2H$
Stearic acid (18:0)	$CH_3(CH_2)_{16}CO_2H$
Unsaturated fatty acids	
Oleic acid (18:1)	$CH_3(CH_2)_7CH{=}CH(CH_2)_7CO_2H$
Linoleic acid (18:2)	$CH_3(CH_2)_4CH{=}CH-CH_2-CH{=}CH(CH_2)_7CO_2H$
Linolenic acid (18:3)	$CH_3CH_2CH{=}CH-CH_2-CH{=}CH-CH_2-CH{=}CH-(CH_2)_7CO_2H$
Arachidonic acid (20:4)	$CH_3(CH_2)_3(CH_2-CH{=}CH)_4(CH_2)_3CO_2H$

* Each fatty acid has a numerical abbreviation. The number before the colon represents the total number of carbons, while the number after the colon is the number of double bonds.

Triacylglycerols in most seeds are stored in the cytoplasm of either cotyledon or endosperm cells in organelles known as **spherosomes** (also called lipid bodies and oleosomes). Spherosomes have an unusual membrane barrier that separates the triglycerides from the aqueous cytoplasm. A single layer of phospholipids (i.e., a half bilayer) surrounds the spherosome with the hydrophilic ends of the phospholipids exposed to the cytosol and the hydrophobic acyl side chains facing the triacylglycerol interior. Several proteins are also located within the phospholipid half bilayer. This unique structure appears to result from the pattern of triglyceride biosynthesis. Triglyceride synthesis is completed by enzymes in the endoplasmic reticulum (ER), and the resulting fats accumulate between the two monolayers of the ER membrane bilayer. The bilayer swells apart as more fats are put into the growing structure, and ultimately a mature spherosome buds off from the ER (Wanner et al., 1981).

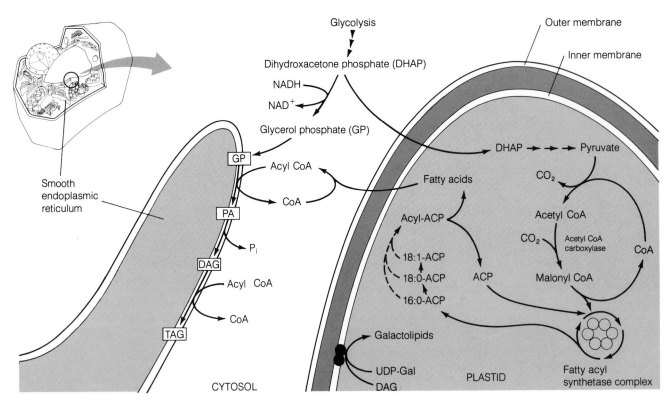

FIGURE 11.12. Biosynthesis of fatty acids and triacylglycerol (TAG) in higher plants. Fatty acid biosynthesis involves a complex series of enzymes localized in the chloroplast, which leads to the synthesis of saturated C_{16} and C_{18} acids and the monounsaturated C_{18}, oleic acid. The subsequent formation of triacylglycerols occurs in the cytosol on the membranes of the endoplasmic reticulum, as does additional unsaturation of fatty acids. PA, Phosphatidic acid; DAG, diacylglycerol; TAG, triacylglycerol; ACP, acyl carrier protein. (Adapted from Douce, 1985.)

Triacylglycerol Biosynthesis Is Energetically Expensive and Takes Place in Several Cell Organelles

Triacylglycerol biosynthesis is a multicompartment process, and the biochemical pathway in plants is similar to that in animals. Saturated fatty acids are synthesized by the repeated addition of two-carbon units derived from acetyl CoA. However, the acetyl CoA is first carboxylated to malonyl CoA, which serves as the two-carbon donor (Fig. 11.12). The reaction series involves several enzymes, including acetyl-CoA carboxylase and a fatty acyl synthetase complex. Throughout the biosynthetic process, the growing fatty acid remains covalently attached to a low-molecular-weight protein known as an **acyl carrier protein** (ACP). The biosynthesis of fatty acids is not cheap energetically; one molecule of ATP and two molecules of NADPH are required for the addition of each two-carbon unit.

In animals, fatty acid biosynthesis takes place in the cytosol, but in plants it is localized in chloroplasts in leaves and in proplastids in nongreen tissue. The initial product of this series of reactions is the C_{16} saturated fatty acid palmitate (16:0, Table 11.2). A two-carbon

elongation to stearate (18:0) also takes place in the plastid (by a mechanism that does not involve the fatty acid synthetase complex), as does the initial desaturation of stearate to oleate (18:1). Further modifications of the fatty acids (e.g., desaturation of oleate to linoleate) take place in the ER following transfer of the fatty acids from the plastid to the ER as a fatty acyl CoA intermediate (Fig. 11.12). The glycerol moiety of triacylglycerols is derived from the reduction of dihydroxyacetone phosphate to glycerol 3-phosphate in the cytosol. Three fatty acyl CoAs are subsequently esterified to the glycerol by enzymes in the ER membrane, leading to the buildup of fats in the growing spherosome.

Phospholipid Biosynthesis Occurs in the ER and Mitochondrial Membranes

The lipids that make up plant membranes are generally **amphipathic** molecules composed of a hydrophobic diacylglycerol moiety connected to one of several types of hydrophilic head groups. The nature of the head groups varies from one membrane to the next. In mitochondria and most other plant membranes, phospholipids predominate, whereas in chloroplasts the thylakoid mem-

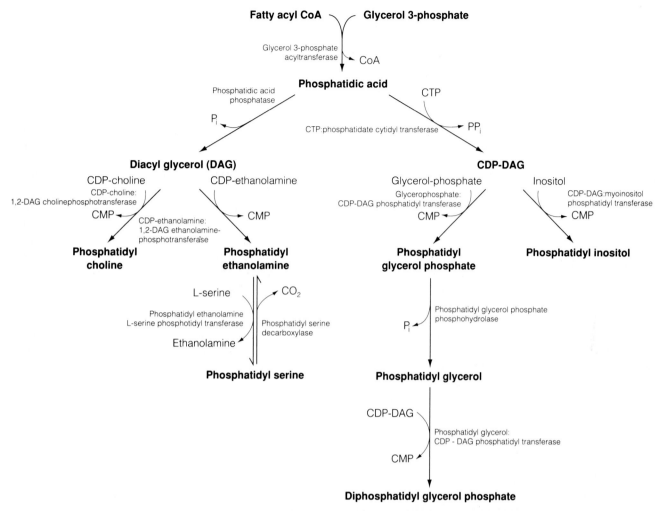

FIGURE 11.13. Phosphatidic acid is synthesized from fatty acyl CoA and glycerol 3-phosphate in the endoplasmic reticulum membrane as shown in Figure 11.12. Subsequent metabolism involves either the hydrolysis of phosphatidic acid to diacylglycerol (DAG) or reaction with CTP to give CDP-diacylglycerol. Diacylglycerol leads to the formation of phosphatidylcholine and phosphatidylethanolamine by reacting with the appropriate CDP adduct, and CDP-diacylglycerol reacts with inositol to produce phosphatidylinositol. Diphosphatidylglycerol (cardiolipin) is also synthesized from CDP-diacylglycerol, and phosphatidylserine is produced from phosphatidylethanolamine. (Adapted from Moore, 1982.)

branes are largely composed of galactolipids (Chapter 8). The biosynthesis of membrane lipids is complex, involves several different cell compartments, and is not completely understood in most cases (Moore, 1982). Here we will give a brief overview of the facts that seem most well established.

Two routes of phospholipid biosynthesis are recognized within the ER, both of which share a common pathway initially. These are presented in Figure 11.13. At one time, it was thought that the addition of head groups in phospholipid biosynthesis took place exclusively in the ER. Now there is evidence that some phospholipid biosynthesis can also occur in other cell compartments. Most notably, mitochondria contain the enzymes necessary for converting diacylglycerol to phosphatidylcholine

and phosphatidylethanolamine, and plant mitochondria are capable of synthesizing phosphatidic acid, CDP-diacylglycerol, and phosphatidylglycerol. Further, mitochondria appear to be the sole site of synthesis of a unique mitochondrial inner membrane lipid, cardiolipin (diphosphatidylglycerol). The extent to which mitochondrial membranes import lipids from the ER during organelle biogenesis is not known, nor is the mechanism of exchange between the ER and the outer and inner mitochondrial membranes.

Phospholipids in the Golgi complex, plasma membrane, and nuclear envelope (all components of the endomembrane system that includes the ER) are probably derived from phospholipids initially synthesized in the ER and transferred to the appropriate membrane by vesiculation of specific regions of the ER (Sabatini et al., 1982). This ER-vesiculation model also explains most of the observed features of microbody (peroxisome, glyoxysome) membrane biogenesis (Trelease, 1984).

The synthesis of diacylglycerol and galactolipids for the chloroplast thylakoid membranes has been localized in spinach to a group of enzymes within the inner membrane of the chloroplast envelope (Fig. 11.12) (Heemskerk and Wintermans, 1987). Synthesis of galactolipids is carried out by an enzyme that transfers the galactosyl moiety from UDP-galactose to diacylglycerol to form monogalactosyldiacylglycerol. Formation of digalactosyldiacylglycerol involves reaction with a second molecule of UDP-galactose.

In Germinating Seeds, Lipids Are Converted into Carbohydrates

Metabolism of stored fats by oil-containing seeds involves the conversion of lipids to carbohydrates following germination by a process involving several different cellular sites (Huang et al., 1983). Plants are not able to transport fats from the endosperm to the root and shoot tissues of the germinating seedling, so they must convert stored lipids to a more mobile form of carbon, generally sucrose.

The process of converting lipid to sugar in oil seeds is triggered by germination and begins with the hydrolysis of triglycerides stored in the spherosomes to free fatty acids, followed by oxidative breakdown of the fatty acids to produce acetyl CoA (Fig. 11.14). The latter reaction takes place in a type of microbody called **glyoxysomes**, organelles bounded by a single membrane that are found in the oil-rich storage tissues of seeds. The discovery of glyoxysomes in castor beans and much of our understanding of glyoxysomal function are due to the work of Harry Beevers and his colleagues at Purdue University and the University of California, Santa Cruz (see Beevers, 1990). Acetyl CoA is metabolized in the glyoxysome via the glyoxylate cycle (Fig. 11.14) to produce succinate, which is transported from the glyoxysome to the mitochondria, where it is further metabolized to oxaloacetate. The process ends in the cytosol with the conversion of oxaloacetate to glucose via gluconeogenesis. Although some of this carbon is diverted to other metabolic reactions in certain oil seeds, in castor bean the process is so efficient that each gram of lipid metabolized results in the formation of one gram of carbohydrate, equivalent to 40% recovery of free energy in the form of carbon bonds.

The initial step in the conversion of lipids to carbohydrate involves the breakdown of triglycerides stored in the spherosomes by the enzyme lipase, which, at least in castor bean, is located on the half membrane that serves as the spherosome outer boundary. The lipase hydrolyzes triglycerides to three molecules of fatty acid and glycerol. Corn and cotton also contain a lipase activity in the sphe-

rosome, but peanut, soybean, and cucumber show lipase activity in the glyoxysome instead. During the breakdown of lipids, there is generally a close physical association between spherosomes and glyoxysomes.

Following hydrolysis of the triglycerides, the resulting fatty acids enter the glyoxysome, where they are activated to fatty acyl CoA by the enzyme fatty-acyl-CoA synthase. Fatty acyl CoA is the initial substrate for the **β-oxidation** series of reactions in which C_n fatty acids are sequentially broken down to $n/2$ molecules of acetyl CoA (Fig. 11.14). This reaction sequence is similar to that associated with the breakdown of fatty acids in animal tissues and involves the reduction of $\frac{1}{2}O_2$ to H_2O and the formation of one NADH for each acetyl CoA produced. In mammalian tissues, the four enzyme activities associated with β-oxidation appear in the mitochondrion, whereas in plant seed storage tissues they are exclusively localized in the glyoxysome. Interestingly, in plant vegetative tissues (e.g., mung bean hypocotyl, potato tuber), the β-oxidation reactions are localized in a related organelle, the peroxisome, rather than in the mitochondria (Gerhardt, 1983), suggesting that the limited breakdown of lipids that does occur in plant nonstorage tissues takes place in peroxisomes and not mitochondria.

The acetyl CoA produced by β-oxidation is further metabolized in the glyoxysomes through a series of reactions that make up the **glyoxylate cycle** (Fig. 11.14). Initially, the acetyl CoA reacts with oxaloacetate to give citrate, which in turn undergoes isomerization to isocitrate. Both of the enzymes involved here (citrate synthase and aconitase) catalyze the same reactions as in the TCA cycle in the mitochondria. The next two reactions are unique to the glyoxylate pathway. First, isocitrate (C_6) is cleaved by the enzyme isocitrate lyase to give succinate (C_4) and glyoxylate (C_2). Next, malate synthase combines a second molecule of acetyl CoA with glyoxylate to produce malate. According to the classical

FIGURE 11.14. (right) (a) Carbon flow during fatty acid breakdown and gluconeogenesis during the germination of oil-storing seeds. Triacylglycerols in the spherosome are hydrolyzed to yield fatty acids, which are metabolized to acetyl CoA in the glyoxysome. The latter conversion involves a series of reactions known as the β-oxidation pathway. In this pathway, fatty-acyl-CoA synthase first converts the free fatty acid to its fatty acyl CoA adduct, and the fatty acyl CoA is then degraded in two-carbon units through a reaction sequence involving a series of four enzymes to release one molecule of acetyl CoA. Every two molecules of acetyl CoA produced are further metabolized through the glyoxylate cycle to generate one succinate, which moves into the mitochondrion and is converted to malate. The resulting malate is transported into the cytosol and oxidized to oxaloacitic acid, which is converted to phosphoenolpyruvate by the enzyme PEP carboxykinase. The resulting PEP is then metabolized to produce glucose via the gluconeogenic pathway. Refer to Figures 11.1 and 11.3 for structures. (b) Electron micrograph of a cell from the oil-storing cotyledon of a cucumber seedling showing glyoxygomes, mitochondria, and spherosomes (8600 ×). (From Trelease et. al., 1971).

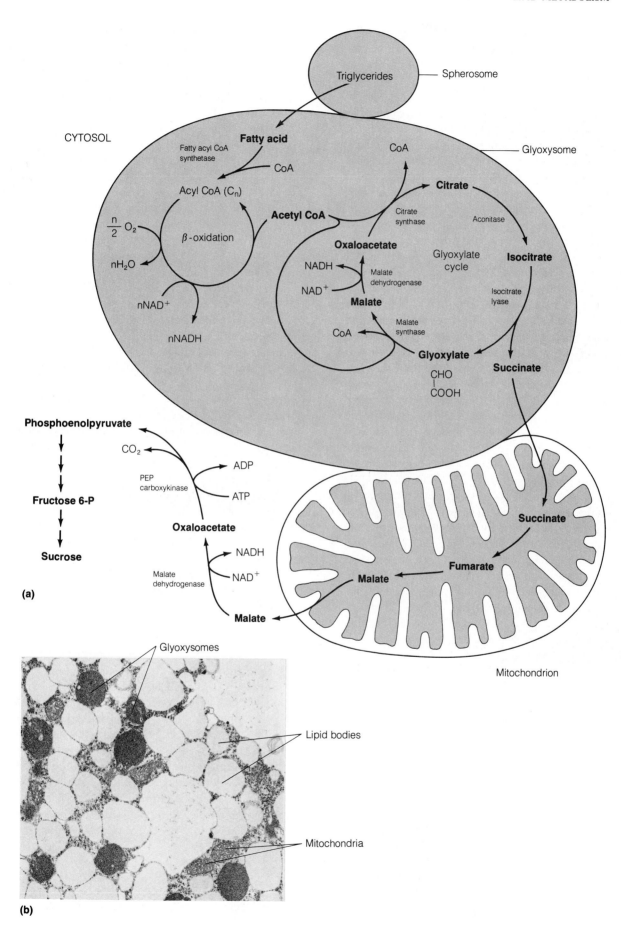

(a)

(b)

view of the operation of the glyoxylate cycle, the malate is then oxidized by malate dehydrogenase to oxaloacetate, which can combine with another acetyl CoA to continue the cycle (Fig. 11.14).

The function of the glyoxylate cycle is, in essence, to convert two molecules of acetyl CoA to succinate. The succinate then moves to the mitochondria, where it is converted to malate by the normal TCA cycle reactions. The resulting malate can be moved out of the mitochondria in exchange for the incoming succinate via the dicarboxylate transporter (Fig. 11.4). The malate is then oxidized to oxaloacetate by cytosolic malate dehydrogenase, and the resulting oxaloacetate is converted to carbohydrate. This conversion requires circumventing the irreversibility of the pyruvate kinase reaction (see Fig. 11.1) and is facilitated by the enzyme PEP carboxykinase, which utilizes the phosphorylating ability of ATP to convert oxaloacetate to PEP and CO_2 (Fig. 11.14). From PEP, gluconeogenesis can proceed to the production of glucose, as outlined on page 268. Sucrose is the ultimate product of this process, being the primary form in which reduced carbon is translocated from the cotyledons to the growing seedling tissues.

Although the pathway for the mobilization of triglycerides has been best characterized in castor bean (Ricinus communis), it seems to be similar in the storage tissues of other oil seeds. However, not all seeds carry out a quantitative conversion of fat to sugar. In castor bean, the endosperm degenerates after the fat and protein reserves are fully utilized. In many oil seeds, such as sunflower (Helianthus annus), cotton (Gossipium hirsutum), and members of the squash family (Cucurbits), the cotyledons differentiate into actively photosynthesizing organs after the food reserves are used up. In these tissues only part of the stored lipid is converted to exported carbohydrate. Much of the lipid-derived carbon remains in the cotyledons, where it contributes to the synthesis of chloroplasts and other cellular structures. As the greening process takes place, there is a transition in the microbody population of these cells that shows fewer of the characteristics of glyoxysomes and more of those of leaf-type peroxisomes. Such a transition is in keeping with the decreased requirement for the breakdown of stored lipids and the increased need to metabolize the products of photorespiration as the tissue goes from a heterotrophic to a more autotrophic mode of metabolism.

Summary

Higher plant respiration couples the complete oxidation of reduced cellular carbon generated during photosynthesis to the synthesis of ATP. Respiration takes place in three stages: glycolysis, the tricarboxylic acid cycle, and the electron transport chain. In glycolysis, carbohydrate is converted in the cytosol to pyruvate with the synthesis of a small amount of ATP via substrate-level phosphorylation. Pyruvate is subsequently oxidized within the mitochondrial matrix through the TCA cycle, generating a large number of reducing equivalents in the form of NADH and $FADH_2$. In the third stage, NADH and $FADH_2$ are oxidized by the electron transport chain, which is associated with the inner mitochondrial membrane. The free energy released during electron transport is used to synthesize a large amount of ATP in a process known as oxidative phosphorylation. ATP synthesis is accomplished by the generation of a proton gradient across the inner mitochondrial membrane during electron transfer. The F_oF_1-ATP synthase complex then couples the break-down of the proton gradient to the conversion of ADP and P_i to ATP.

The regulation of substrate oxidation during respiration involves control points at glycolysis, the TCA cycle, and the electron transport chain. Aerobic respiration in higher plants has several unique features, including the presence of a cyanide-resistant oxidative pathway. Carbohydrates can also be oxidized via the pentose phosphate pathway, in which the reducing power generated produces NADPH for biosynthetic purposes. Numerous glycolytic and TCA cycle reactions also provide the starting points for different biosynthetic pathways.

Many factors can affect the respiration rate observed at the whole-plant level. These include the nature and age of the plant tissue and such environmental factors as the external oxygen concentration and temperature. Because respiration rates contribute to the overall net carbon balance of a plant, variations in whole-plant respiration rates can affect final agronomic yields.

Lipids play a major role in higher plants; amphipathic lipids serve as the primary nonprotein components of plant membranes, while triglycerides (fats and oils) are an efficient storage form of reduced carbon, particularly in seeds. Triglycerides are stored in subcellular organelles known as spherosomes. Synthesis of the long-chain fatty acids that make up triglycerides takes place in the plastids of higher plants, and the conversion of the fatty acids to triglycerides occurs on the membranes of the endoplasmic reticulum. During germination in seeds that store triglycerides, the stored lipids are metabolized to carbohydrates in a series of reactions that involve a metabolic sequence known as the glyoxylate cycle. This cycle takes place in organelles called glyoxysomes. The reduced carbon generated during lipid breakdown in the glyoxysomes is ultimately converted to carbohydrate in the cytosol by the process known as gluconeogenesis.

GENERAL READING

Beevers, H. (1990) Forty years on. *Annu. Rev. Plant Physiol. Plant Mol. Biol.* (in press).

Douce, R. (1985) *Mitochondria in Higher Plants: Structure, Function, and Biogenesis.* Academic Press, Orlando, FL.

Huang, A. H. C., Trelease, R. N., and Moore, T. S., Jr. (1983) *Plant Peroxisomes.* Academic Press, New York.

Nicholls, D. G. (1982) *Bioenergetics: An Introduction to the Chemiosmotic Theory.* Academic Press, London.

CHAPTER REFERENCES

ap Rees, T. (1980) Assessment of the contributions of metabolic pathways to plant respiration. In: *The Biochemistry of Plants*, Vol. 2, pp. 1–29, Davies, D. D., ed. Academic Press, New York.

Black, C. C., Smyth, D. A., and Wu, M.-Y. (1985) Pyrophosphate-dependent glycolysis and regulation by frustose 2,6-bisphosphate in plants. In: *Nitrogen Fixation and CO_2 Metabolism*, pp. 361–370, Ludden, P. W., and Burris, J. E., eds. Elsevier, New York.

Gerhardt, B. (1983) Localization of β-oxidation enzymes in peroxisomes isolated from nonfatty plant tissues. *Planta* 159:238–246.

Hanson, M. R., and Conde, M. F. (1985) Functioning and variation of cytoplasmic genomes: Lessons from cytoplasmic-nuclear interactions affecting male fertility in plants. *Int. Rev. Cytol.* 94:213–267.

Heemskerk, J. W. M., and Wintermans, J. F. G. M. (1987) Role of the chloroplast in the leaf acyl-lipid synthesis. *Physiol. Plant.* 70:558–568.

Huber, S. C. (1986) Fructose 2,6-bisphosphate as a regulatory metabolite in plants. *Annu. Rev. Plant Physiol.* 37:233–246.

Lambers, H. (1985) Respiration in intact plants and tissues. Its regulation and dependence on environmental factors, metabolism and invaded organisms. In: *Higher Plant Cell Respiration* (Encyclopedia of Plant Physiology, New Series, Vol. 18), pp. 418–473, Douce, R., and Day, D. A., eds. Springer-Verlag, Berlin.

Laties, G. G. (1982) The cyanide-resistant, alternative path in higher plant respiration. *Annu. Rev. Plant Physiol.* 33:519–555.

Møller, I. M., and Lin, W. (1986) Membrane-bound NAD(P)H dehydrogenases in higher plant cells. *Annu. Rev. Plant Physiol.* 37:309–334.

Moore, A. L., and Rich, P. R. (1985) Organization of the respiratory chain and oxidative phosphorylation. In: *Higher Plant Cell Respiration* (Encyclopedia of Plant Physiology, New Series, Vol. 18), pp. 134–172, Douce, R., and Day, D. A., eds. Springer-Verlag, Berlin.

Moore, T. S., Jr. (1982) Phospholipid biosynthesis. *Annu. Rev. Plant Physiol.* 33:235–259.

Palmer, J. M., and Ward, J. A. (1985) The oxidation of NADH by plant mitochondria. In: *Higher Plant Cell Respiration* (Encyclopedia of Plant Physiology, New Series, Vol. 18), pp. 173–201, Douce, R., and Day, D. A., eds. Springer-Verlag, Berlin.

Raskin, I., Turner, I. M., and Melander, W. R. (1989) Regulation of heat production in the inflorescences of an *Arum* lily by endogenous salicyclic acid. *Proc. Natl. Acad. Sci. USA* 86:2214–2218.

Roberts, J. K. M., Callis, J., Wemmer, D., Walbot, V., and Jardetzky, O. (1984) Mechanism of cytoplasmic pH regulation in hypoxic maize root tips and its role in survival under hypoxia. *Proc. Natl. Acad. Sci. USA* 81:3379–3383.

Sabatini, D. D., Kreibich, G., Morimoto, T., and Adesnik, M. (1982) Mechanisms for the incorporation of proteins into membranes and organelles. *J. Cell Biol.* 92:1–22.

Seymour, R. S., Bartholomew, G. A., and Barnhart, M. C. (1983) Respiration and heat production by the inflorescence of *Philodendron selloum* Koch. *Planta* 157:336–343.

Siedow, J. N., and Berthold, D. A. (1986) The alternative oxidase: A cyanide-resistant respiratory pathway in higher plants. *Physiol. Plant.* 66:569–573.

Trelease, R. N. (1984) Biogenesis of glyoxysomes. *Annu. Rev. Plant Physiol.* 35:321–347.

Trelease, R. N., Gruber, P. J., Becker, W. M., and Newcomb, E. H. (1971) *Plant Physiology* 48:461.

Wanner, G., Formanek, H., and Theimer, R. R. (1981) The ontogeny of lipid bodies (spherosomes) in plant cells. *Planta* 151:109–123.

Wilson, D., and Jones, J. G. (1982) Effect of selection for dark respiration rate of mature leaves on crop yields of *Lolium perenne* cv. S23. *Ann. Bot.* 49:313–320.

Wiskich, J. T., and Dry, I. B. (1985) The tricarboxylic acid cycle in plant mitochondria: Its operation and regulation. In: *Higher Plant Cell Respiration* (Encyclopedia of Plant Physiology, New Series, Vol. 18), pp. 281–313, Douce, R., and Day, D. A. Springer-Verlag, Berlin.

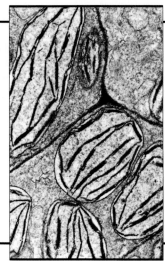

CHAPTER 12

Assimilation of Mineral Nutrients

HIGHER PLANTS ARE AUTOTROPHIC organisms that can synthesize all of their molecular components from inorganic nutrients obtained from the local environment. For many mineral nutrients, this involves uptake by the plant (Chapter 5) and incorporation into the carbon skeletons that are essential for growth and development. This conversion of the mineral nutrients into carbohydrates, lipids, and amino acids is termed **nutrient assimilation**.

For some nutrients, assimilation requires complex biochemical reactions. For example, nitrate (NO_3^-) is first reduced to ammonia (NH_3) and then introduced into intermediary metabolism through a series of enzyme-mediated steps. Other nutrients, especially the macronutrient and micronutrient cations (Chapter 5), are assimilated by forming complexes with organic compounds, such as Ca^{2+}-pectate associated with the cell wall. In leguminous plants that can assimilate nitrogen from the atmosphere, this process involves an intricate interaction between the host plant and the nitrogen-fixing bacteria with which it lives in a symbiotic relationship.

In this chapter, the basic reactions by which nutrients (sulfur, nitrogen, phosphate, oxygen, and cations) are assimilated will be discussed with an emphasis on the assimilation of nitrogen.

Nitrogen Assimilation

Nitrogen is a key element in many of the compounds present in plant cells. It is found in the nucleoside phosphates and amino acids that form the building blocks of nucleic acids and proteins, respectively. Availability of nitrogen for crop plants is an important limiting factor in agricultural production, and the importance of nitrogen is demonstrated by the fact that only oxygen, carbon, and hydrogen are more abundant in higher plant cells.

Nitrogen Occurs in Several Forms That Are Interconverted in the Nitrogen Cycle

Nitrogen is present in many forms in the biosphere. The atmosphere contains vast quantities of molecular nitrogen (N_2), about 78% of the atmosphere by volume. For the most part, this large reservoir of nitrogen is not directly available to plants. Acquisition of nitrogen from the atmosphere requires the breaking of an exceptionally stable triple covalent bond between two nitrogen atoms ($N\equiv N$), and higher plants do not have the capacity to carry out this reaction directly. On the other hand, nitrogen present in the form of nitrate or ammonia is readily assimilated by higher plants (Chapter 5). The consumption of plants by animals allows nitrogen to move further up the food chain, and the nitrogen is returned to the soil through the death and decomposition of plants and animals. These processes are a part of the nitrogen cycle (Fig. 12.1a).

The conversion of molecular nitrogen into other forms such as ammonia or nitrate is known as **nitrogen fixation**, and it can occur as a result of both natural and industrial processes. Under the conditions of elevated temperature (about 200°C) and high pressure (about 200 atmospheres), molecular nitrogen will combine with hydrogen to form ammonia. The extreme conditions are required to overcome the high activation energy of the reaction. This reaction, called the Haber process, is a starting point in the production of a variety of industrial and agricultural products. Worldwide industrial nitrogen fixation amounts to about 50 million tons annually, equivalent to about 12% of the global nitrogen removed from the atmospheric pool each year (Fig. 12.1b) (Marschner, 1986).

Natural processes account for a larger amount of nitrogen fixation (Delwiche, 1983). Of the total amount fixed, about 10% occurs as a result of lightning. Lightning causes the formation of hydroxyl free radicals, free hydrogen atoms, and free oxygen atoms from water vapor and oxygen in the atmosphere, and these reactive species attack molecular nitrogen to form nitric acid (HNO_3), which is carried down to the soil by rain. A much smaller amount of nitrogen fixation occurs in the stratosphere as a result of photochemical reactions involving gaseous nitric oxide and ozone. The nitric acid produced in this manner eventually finds its way into the lower atmosphere, from which it is removed by rain.

The remaining 90% of the nitrogen made available by natural processes is fixed through the action of microorganisms in a process termed **biological nitrogen fixation** (Fig. 12.1b). Biological nitrogen fixation is carried out by free-living bacteria, including cyanobacteria, and by bacteria that are in symbiotic association with plants (Burris, 1976; Marschner, 1986). These organisms contain enzyme systems that can catalyze chemical reactions involving molecular nitrogen. From an agricultural standpoint, biological nitrogen fixation is an important source of soil nitrogen; on a worldwide basis, the demands for nitrogen in crop production cannot be met through the use of chemical fertilizers (Schubert and Wolk, 1982).

The reactions of the nitrogen cycle that follow nitrogen fixation involve interconversions of fixed forms of nitrogen and the return of nitrogen to the atmosphere. Organic forms of nitrogen such as amino acids that are returned to the soil in wastes and by the decay of dead plants and animals are converted to ammonia in a process called **ammonification**. Bacteria and fungi in the soil transform the amino nitrogen into free ammonia, which can react with other components in the soil to form ammonium salts. This released ammonia can also be oxidized first to nitrite (NO_2^-) and then to nitrate (NO_3^-) by the process of **nitrification**. Oxidation of ammonia to nitrite is carried out by soil bacteria of the *Nitroso* group (*Nitrosomonas*), while the further oxidation from nitrite to nitrate is carried out by bacteria of the *Nitro* group (*Nitrobacter*) (Beevers, 1976). Nitrogen present as nitrate can be readily used by plants (Chapter 5), or it can be returned to the atmosphere through **denitrification**, a process in which nitrate is reduced to nitrogen. Many bacteria use denitrification as a source of respiratory energy (Chapter 11) when oxygen is absent and either NO_3^-, NO_2^-, or N_2O is present (Beevers, 1976). Denitrification is an important biological process because, in its absence, nitrate would rapidly accumulate in the oceans. Nevertheless, when the amount of nitrogen fixed annually (125 million tons) is compared with amount of nitrogen lost to the atmosphere by denitrification (93 million tons), a net gain is observed (Delwiche, 1983). This gain presumably represents fixed nitrogen that is slowly accumulating in soil, lakes, and the ocean.

Biological Nitrogen Fixation Is Carried Out by Both Free-Living and Symbiotic Bacteria

Nitrogen-fixing bacteria occur either as free-living microorganisms or in symbiotic association with higher plants, primarily those of the family Leguminosae (Burris, 1976). In the symbiotic relationship, nitrogen fixed by the bacterium is acquired by the host plant in exchange for other nutrients. The wide range of microorganisms known to carry out nitrogen fixation are listed in Table 12.1.

The nitrogenase enzymes involved in nitrogen fixation are irreversibly inactivated by oxygen, so the nitrogen-fixing process must occur under anaerobic conditions. Thus, each of the organisms listed in Table 12.1 either functions under natural anaerobic conditions or can create an internal anaerobic environment in the presence of oxygen. In the filamentous cyanobacteria, this is achieved in specialized cells called **heterocysts** (Fig. 12.2). Heterocysts are thick-walled cells that differentiate when heterocystous cyanobacteria are deprived of NH_4^+. Photosystem II, the oxygen-producing photosystem of photosynthesis (Chapter 8), is absent from heterocysts, so oxygen is not generated (Burris, 1976). Heterocysts appear to represent an adaptation for nitrogen fixation, since they are widespread among cyanobacteria that are capable of fixing nitrogen aerobically (Beevers, 1976). Cyanobacteria that do not have heterocysts can carry out nitrogen fixation only

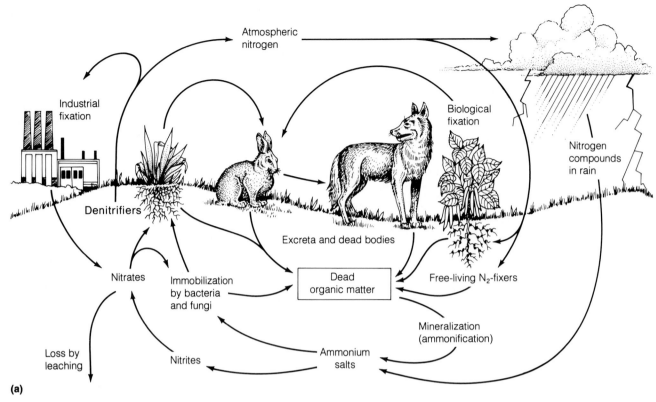

(a)

FIGURE 12.1. Nitrogen cycles through the atmosphere as it changes from a gaseous form to reduced ions before being incorporated into organic compounds in living organisms. (a) Some of the steps involved in the nitrogen cycle. (b) Estimated amounts of nitrogen transferred from one global pool to another expressed in millions of metric tons per year. $h\nu$ represents photochemical reactions. (From Burns and Hardy, 1975.)

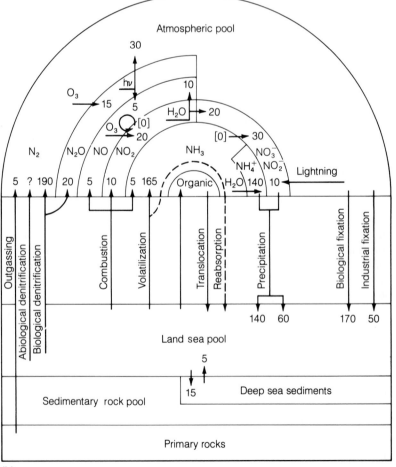

(b)

TABLE 12.1. Some examples of organisms that can carry out nitrogen fixation

Symbiotic nitrogen fixation

1. Leguminous plants, *Parasponia*: *Rhizobium*, *Bradyrhizobium*
2. Alder (tree), *Ceonothus* (shrub), *Casuarina* (tree): *Frankia* (actinomycete)
3. Tropical grasses: *Azospirillum* (actinomycete)
4. *Azolla* water fern: *Anabaena*

Nonsymbiotic nitrogen fixation

1. Cyanobacteria (blue-green algae): *Nostoc*, *Anabaena*, *Calothrix*, etc.
2. Other bacteria
   a. Aerobic: *Azotobacter*, *Beijerinckia*, *Derxia*, etc.
   b. Facultative: *Bacillus*, *Klebsiella*, etc.
   c. Anaerobic:
      i. Nonphotosynthetic: *Clostridium*, *Methanococcus* (archaebacterium), etc.
      ii. Photosynthetic: *Rhodospirillum*, *Chromatium*, etc.

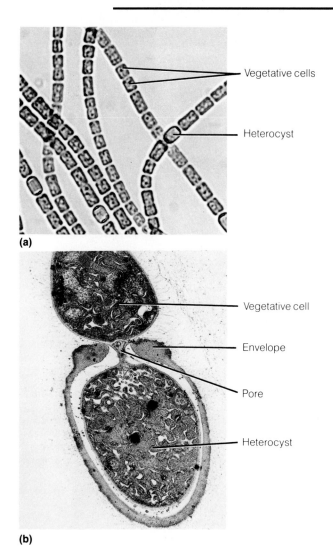

(a)

(b)

FIGURE 12.2. Filaments of the nitrogen-fixing cyanobacterium *Anabena cylindrica*. (a) Vegetative cells and heterocysts (1000 ×). Nitrogen fixation takes place in these specialized cells. (b) An electron micrograph of a vegetative cell and a heterocyst of *A. cylindrica* (8700 ×). The two types of cells communicate through a pore that provides protoplasmic continuity between them. The heterocyst is surrounded by a thick envelope. (From Stewart, 1977.)

under anaerobic conditions such as those that occur in flooded fields. In Asian countries, nitrogen-fixing cyanobacteria of both the heterocyst and nonheterocyst types are a major means of maintaining an adequate nitrogen supply to the soil in rice fields (Brady, 1979). The microorganisms fix nitrogen when the fields are flooded and decompose and release it when the fields dry out. An important contributor to the available nitrogen of flooded rice fields is the tiny water fern *Azolla*, which maintains a symbiotic association with the nitrogen-fixing cyanobacterium *Anabaena*. The *Azolla-Anabaena* complex can fix as much as 1.2 kg of atmospheric nitrogen per acre per day.

Free-living bacteria that are capable of fixing nitrogen can either be aerobic, facultative, or anaerobic. Aerobic nitrogen-fixing bacteria such as *Azotobacter* are thought to maintain reduced oxygen conditions (**microaerobic** conditions) through their high levels of respiration (Burris, 1976). Others, like *Gloeotheca*, evolve O_2 photosynthetically during the day and fix nitrogen during the night. Facultative organisms, which are able to grow under both aerobic and anaerobic conditions, generally fix nitrogen only under anaerobic conditions. For anaerobic nitrogen-fixing bacteria, oxygen does not pose a problem; these anaerobic organisms can be either photosynthetic, as in the case of *Rhodospirillum*, or nonphotosynthetic, as in the case of *Clostridium*.

In symbiotic nitrogen fixation, such as the association of nitrogen-fixing bacteria with the roots of legumes, the amount of free oxygen is reduced by the high respiration rates of the symbionts and by the presence of an oxygen-binding heme protein, *leghemoglobin* (Appleby, 1984; Quispel, 1983). This protein is present in the cytosol of infected nodule cells at rather high concentrations (700 μM in soybean nodules) and gives the nodules their pink color. The globin portion of leghemoglobin is produced by the host plant in response to infection by the bacteria (Marschner, 1986), and the heme portion is produced by the bacterial symbiont (Appleby,

1984). The rate of nitrogen fixation is closely correlated with the concentration of leghemoglobin (Werner et al., 1981). The protein has a very high affinity for oxygen; half-saturation takes place at 10–20 nM (10^{-9} M) oxygen, compared with 126 nM for the β chain of human hemoglobin. It is thought that leghemoglobin buffers the oxygen concentration within the root nodule, allowing respiration without inhibition of the nitrogenase (Dixon and Wheeler, 1986). In addition, leghemoglobin acts as an oxygen carrier, facilitating oxygen diffusion across a thin, unstirred layer surrounding the respiring symbiotic bacterial cells (Ludwig and de Vries, 1986). Leghemoglobin thus functions at the cellular level in a manner analogous to hemoglobin in animal circulatory systems.

The First Step in the Establishment of Symbiotic Nitrogen Fixation Is the Attachment of *Rhizobium* to the Root

The symbiotic association of nitrogen-fixing bacteria with plant roots generally occurs in multicellular structures called nodules. The best-characterized symbiotic association involving nodules is the one occurring on the roots of leguminous plants (Fig. 12.3a). These nodules are produced by the host plant root upon infection by nitrogen-fixing gram-negative bacteria of the genus *Rhizobium*. Members of the family Leguminosae (Fabaceae) are the only plants capable of forming nitrogen-fixing nodules with *Rhizobium*, with the single exception of the nonlegume genus *Parasponia*, a tropical tree. The ability to fix nitrogen in association with *Rhizobium* may account for the fact that the legumes are the third-largest angiosperm family, with a wide distribution from the tropics to the arctic zones (Long, 1989).

Infectious *Rhizobium* species exhibit some specificity for different leguminous host plants (Table 12.2). Such host selection implies that the *Rhizobium* species can recognize its plant host and vice versa. *Rhizobium* typically occurs free-living in the soil. The first stage in the formation of the symbiotic relationship is the migration of the bacteria toward the *rhizosphere*, the area immediately surrounding the roots. This is a chemotactic response and is probably mediated by chemical attractants, especially flavonoids, produced by the roots

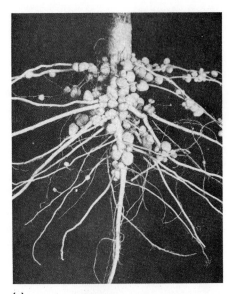

(a)

(b)

FIGURE 12.3. Root nodules on soybean (a) and alder tree (b). The nodules in soybean are a result of infection by *Rhizobium japonicum*; those in alder (*Alnus*) are caused by *Frankia*, a bacterium that belongs to the actinomycetes. The associations between actinomycetes and their hosts are called actinorhizal associations. All known actinorhizal associations occur in shrubs or trees. (Part (a) from The Nitragen Company; (b) from H. J. Evans.)

TABLE 12.2. Host preference among *Rhizobium* species

Rhizobium species	Preferred host genus
R. leguminosarum	*Pisum* (pea), *Vicia* (broad bean), *Lens* (lentil), *Cicer* (chick pea)
R. trifolii	*Trifolium* (clover)
R. phaseoli	*Phaseolus* (kidney bean)
R. meliloti	*Medicago* (alfalfa), *Melilotus* (sweet clover)
R. japonicum	*Glycine* (soybean)

Adapted from Burris (1976), Table IV, p. 900.

The Genes Involved in Symbiotic Nitrogen Fixation

THE ABILITY OF rhizobia to become endosymbiotic nitrogen-fixing organelles in legume root cells supports the endosymbiotic origin of mitochondria and chloroplasts. Indeed, in the bacteroid we may be witnessing an intermediate stage in the evolution of a permanent nitrogen-fixing, self-replicating plant organelle.

To carry out its dual existence as a free-living organism and an organelle, *Rhizobium* must be able to alter its genetic program. Many of the genes necessary for nodulation and nitrogen fixation are present on a large circular plasmid, the Sym (for symbiosis) plasmid. "Curing" rhizobia of their Sym plasmid by incubation at high temperature renders them unable to infect roots or induce nodulation. Reintroduction of the plasmid restores the original infectious phenotype.

The Sym plasmid genes involved in host-specific infection and nodule formation are referred to as *nod* genes. Our understanding of the *nod* genes is limited, and the biochemical functions of *nod* gene products have not yet been elucidated. However, a number of *nod* gene mutants of *Rhizobium* have been described that are blocked at various stages of infection and nodulation. This has allowed mapping of *nod* genes to two distinct regions on the Sym plasmid corresponding to "common" genes, which are interchangeable with plasmid genes from bacteria covering a wide host range, and "host-specific" genes, which function only in a particular plant host.

The genes involved in nitrogen fixation, which become active during the latter stages of nodule development, are called either *nif* genes or *fix* genes and are often clustered. Mutations in either type result in bacteria that can induce nodulation but cannot carry out nitrogen fixation. *Nif* genes are present in both free-living and symbiotic nitrogen-fixers. They include the structural genes for nitrogenase as well as a number of regulatory genes. The *nifH* gene encodes the nitrogenase Fe protein subunit, while *nifD* and *nifK* encode the α and β subunits, respectively, of the MoFe protein of nitrogenase. *Fix* genes seem to be present only in symbiotic nitrogen-fixers and may occur either within or outside the *nif* gene cluster. One possible function of the *fix* genes is heme biosynthesis. The precursor for heme is δ-aminolevulinic acid (ALA). Mutations in the *Rhizobium* gene for ALA synthase cause the production of white nodules that are unable to fix nitrogen. Apparently, heme production by the plant host cell is insufficient to support massive leghemoglobin synthesis.

In addition to leghemoglobin, nodules contain a number of proteins of plant origin that are nodule-specific. Such proteins are called *nodulins*, and their genes are termed *Nod* genes. It has been estimated that legume nodules contain 20–30 nodulins, many of which have been identified. Nodulins are expressed at specific stages of nodule development and include leghemoglobin and the enzymes involving in ureide metabolism.

An important aspect of symbiotic nitrogen fixation is the interactions between the plant *Nod* genes and the bacterial *nod*, *nif*, and *fix* genes. Just how the complex processes of infection and nodule development are coordinated remains a fascinating topic for future research. Such studies are not only of theoretical interest but also of practical importance, for they may one day enable plant scientists to broaden the range of *Rhizobium* to include nonleguminous crops such as cereals. This might greatly increase the productivity of nitrogen-poor soils as well as reduce the dependence of agriculture on energy-expensive nitrogen fertilizers.

(Long, 1989). Rhizobia have evolved to use flavonoid compounds released by legume roots as positive regulators of the genes involved in nodulation (*nod* genes). In soybean, the rhizobia are thought to secrete factors that stimulate subepidermal cell divisions in the root (Fig. 12.4). The dividing cells in the root evidently potentiate the root hairs above them to become target sites for infection.

Once in the rhizosphere, the bacteria attach to the root surface, particularly to the root hairs. The mechanism of attachment is still poorly understood. According to an early idea, legume roots secrete sugar-binding proteins called lectins, which interact with the surface polysaccharides of *Rhizobium* cells and facilitate binding to the root cell walls (Halverson and Stacey, 1986). Lectins may explain the association between clover and *Rhizobium trifolii*, but in the case of the pea-*Rhizobium leguminosarum* association a different, calcium-dependent protein has been implicated (Smit et al., 1987). It is now thought that multiple mechanisms of attachment are probably involved, some general and some host-specific (Kijne et al., 1988).

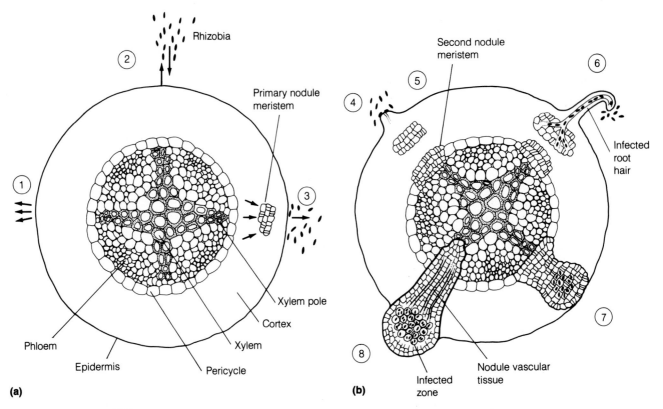

FIGURE 12.4. Stages in the initiation and development of a soybean root nodule. (a) Events involved in the initiation of the nodule: (1) the root excretes substances; (2) these substances attract rhizobia and stimulate them to produce cell-division factors; and (3) cells in the root cortex divide to form the primary nodule meristem. (b) Stages of infection and nodule formation: (4) bacteria attach to the root hair; (5) cells in the pericycle near the xylem poles are stimulated to divide; (6) the infection thread forms and extends inward as the primary nodule meristem and the pericycle continue to divide; (7) the two masses of dividing cells fuse into a single clump while the infection thread continues to grow; and (8) the nodule elongates and differentiates, including the vascular connection to the root stele. Bacteroids are released into the cells in the center. (From Rolfe and Gresshoff, 1988.)

Infection of Legume Roots by *Rhizobium* Is Facilitated by the Formation of an Infection Thread

The various stages in the infection process are shown in Figure 12.5.

The rhizobia that bind to emerging root hairs induce a pronounced curling of these cells, trapping colonies of rhizobia within the coils (Fig. 12.5b). Curling of root hairs is thought to be stimulated by bacterial production of the plant hormone auxin. The root hair cell wall becomes degraded in these regions, allowing the bacterial cells direct access to the outer surface of the plant plasma membrane.

The next step involves the formation of a remarkable structure called the infection thread (Fig. 12.5c). The **infection thread** begins as an internal tubular extension of the plasma membrane caused by the fusion of Golgi-derived membrane vesicles at the site of infection. It grows at the tip, much as a stalagmite grows upward from the floor of a cave as the result of the dripping of mineral-rich water; that is, secretory vesicles "drip" onto the end of the tube. As the membrane tube extends inward, a thin layer of cellulosic cell wall is deposited. The wall material lines the inner surface of the infection thread and is physically connected with the host cell wall. This makes sense when we consider that the inner surface of the infection thread membrane is topologically equivalent to the outer surface of the host cell plasma membrane. Electron micrographs of infection threads are shown in Figure 12.6.

The growth of the infection thread is closely correlated with the proliferation of enclosed rhizobia, which still technically occupy an extracellular compartment. Branching of the infection thread (Fig. 12.5f) enables the bacteria to spread to many cells. At the same time, cortical cells near the epidermis undergo rapid divisions and give rise to the body of the nodule. When the infection thread reaches the base of the root hair cell (Fig. 12.5d) it fuses with the plasma membrane, releasing free bacteria into solubilized pockets of the apoplast (Bauer, 1981). The infection thread membrane fuses with the host cell plasma

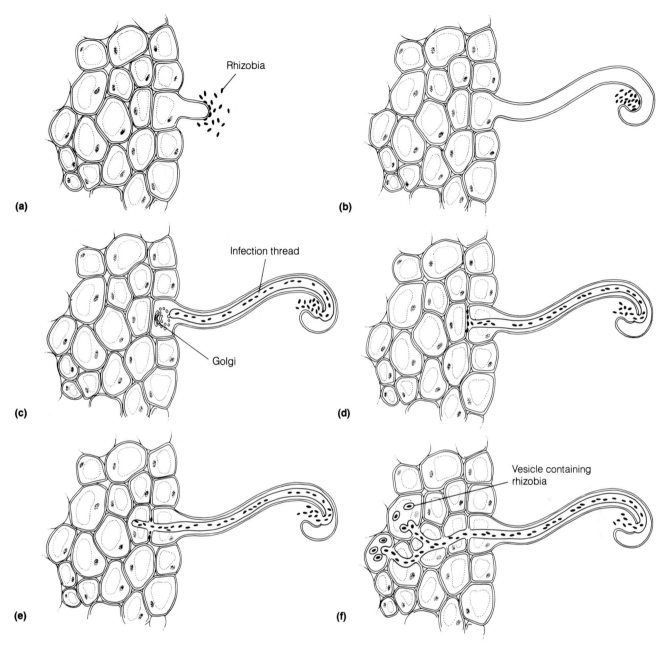

(a)

Rhizobia

(b)

(c)

Infection thread

Golgi

(d)

(e)

(f)

Vesicle containing
rhizobia

membrane. The bacteria in the apoplast apparently degrade the cell wall to gain access to the plasma membrane of the subepidermal cell, and the infection process repeats itself (Fig. 12.5e). In this way, infection threads spread from cell to cell, allowing the bacteria to penetrate deep within the developing nodule without ever entering the cytoplasm of the host (Sprent, 1989).

In the final step in the process, the *Rhizobium* cells are internalized into the cytoplasm of the host cells by endocytosis. This occurs when the infection thread reaches specialized cells within the nodule. The tip of the infection thread vesiculates, releasing *Rhizobium* cells packaged in a membrane derived from the host cell plasma membrane (Fig. 12.5f). At first the bacteria continue to divide, and the surrounding membrane increases in surface area to accommodate this growth by fusion with

FIGURE 12.5. The infection process. (a) Rhizobia bind to emerging root hair. (b) In response to factors produced by the bacteria, the root hair exhibits abnormal curling growth, and rhizobia cells proliferate within the coils. (c) Localized root hair wall degradation leads to infection and formation of the infection thread from Golgi secretory vesicles. (d) The infection thread reaches the end of the cell, and its membrane fuses with the plasma membrane of the root hair cell. (e) Rhizobia are released into the apoplast and penetrate the compound middle lamella to the subepidermal cell plasma membrane, leading to the initiation of a new infection thread, which forms an open channel with the first. The two infection threads are actually discontinuous because the infection thread membrane is absent from the apoplast. (f) The infection thread extends and branches until it reaches target cells, where vesicles containing bacterial cells are released into the cytosol.

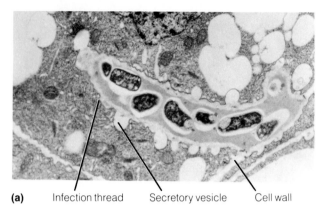

(a)   Infection thread   Secretory vesicle   Cell wall

FIGURE 12.6. Electron micrographs of infection threads.
(a) Growth of the infection thread containing bacteria. Note the fusion of vesicles (v) with the infection thread (it). The plasma membrane and cell wall of the root cell are continuous with the infection thread. Each bacterium is surrounded by a halo of capsular polysaccharides. The bacteria continue to divide as the thread grows (1cm = 1μm). (b) Scanning electron micrograph of an infection thread which is in the early stages of branching near the tip, indicated by the arrow (3750 ×). (a and b courtesy of Dr. Ann Hirsch, UCLA.) (c) Electron micrograph of bacteroids surrounded by a plant-derived peribacteroid membrane in an infected soybean root nodule cell (2.5cm = 1μm) (From E. H. Newcomb and S. R. Tandon).

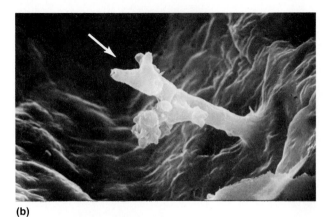

(b)

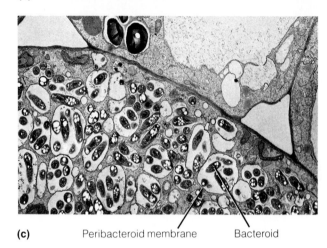

(c)   Peribacteroid membrane   Bacteroid

smaller vesicles. Very soon, however, the bacteria stop dividing; they enlarge and differentiate into nitrogen-fixing endosymbiotic organelles called **bacteroids**. This differentiation is thought to be triggered by a signal from the plant. The membrane surrounding the bacteroids is called the **peribacteroid membrane** (Fig. 12.6c).

In response to signals from the bacteria, the host cells undergo marked morphological and biochemical changes, including the production of leghemoglobin. The nodule as a whole develops such features as a vascular system, which facilitates the exchange of fixed nitrogen produced by the bacteroids for nutrients contributed by the plant, and a layer of cells to exclude O_2 from the root nodule interior. In some temperate legumes (e.g., peas), the nodules are elongated and cylindrical due to the presence of a nodule meristem. The nodules of tropical legumes, such as soybeans and peanuts, lack a persistent meristem and are spherical in shape (Rolfe and Gresshoff, 1988).

Plants other than legumes can also develop symbiotic relationships with nitrogen-fixing organisms. With some plants, such as the deciduous tree alder (*Alnus*), this involves the formation of root nodules by the actinomycete, *Frankia*. The nodules in *Alnus* occur in clusters, which are formed by repeated branching (Fig. 12.3b). Grasses can also develop symbiotic relationships with nitrogen-fixing organisms, but these associations do not involve the production of root nodules (Smith et al., 1976). Instead, the nitrogen-fixing bacteria (actinomycetes) seem

to be anchored in the root surfaces, mainly around the elongation zone and the root hairs. Bacteria such as *Azospirillum* are being intensively studied for possible application in the cultivation of corn and other grasses (von Bulow and Döbereiner, 1975). The potential nitrogen-fixation properties of these bacteria could be exploited to reduce the use of nitrogen fertilizer (Okon, 1985).

The Biochemical Process of Nitrogen Fixation Is Carried Out by the Nitrogenase Enzyme Complex

When organisms carry out nitrogen fixation, the end product, as in industrial nitrogen fixation, is ammonia. The ammonia is subsequently incorporated into organic compounds such as glutamine or glutamate and is utilized by the microorganism, or by the host plant in the case of a symbiotic association. The overall reaction for the fixation of molecular nitrogen into ammonia is

$$N_2 + 8e^- + 8H^+ + 16ATP \rightarrow$$
$$2NH_3 + H_2 + 16ADP + 16P_i \quad (12.1)$$

Note that the reduction of N_2 to $2NH_3$, a six-electron transfer, appears to be coupled to the evolution of H_2

from the reduction of two protons. The entire reaction is catalyzed by the enzyme nitrogenase.

Nitrogenase is an enzyme complex that can be separated into two components, the MoFe protein and the Fe protein, neither of which has catalytic activity by itself. The Fe protein is the smaller of the two components and has two identical subunits of 30,000 to 72,000 Da, depending on the organism. It contains one 4Fe-4S iron-sulfur cluster that participates in the redox reactions involved in the conversion of N_2 to NH_3. The Fe protein is extremely sensitive to oxygen and is irreversibly inactivated by O_2 with a half-life of 30–45 s (Dixon and Wheeler, 1986). The MoFe protein has four subunits with a total molecular mass of 180,000 to 235,000 Da, depending on the species. It has two molybdenum atoms per molecule in two Mo-Fe-S clusters and a variable number of Fe-S clusters. The MoFe protein is also inactivated by oxygen, with a half-life of 10 min.

In the overall reaction catalyzed by nitrogenase (Fig. 12.7), the nonheme iron protein ferredoxin (Chapter 8) serves as an electron donor to the Fe protein, which in turn reduces the MoFe protein. ATP hydrolysis is involved in this step. The MoFe protein reduces the N_2 substrate.

Nitrogenase can reduce a number of substrates (Table 12.3), although under natural conditions it reacts only with N_2 and H^+. One of the reactions catalyzed by the nitrogenase, the reduction of acetylene, is used in estimating nitrogenase activity by gas chromotography (Dilworth, 1966). Direct measurements of N_2 fixation requires expensive instrumentation that is not readily available, but the reduction of acetylene to ethylene can easily be measured. Acetylene reduction involves two electrons, as opposed to the eight electrons required for the reduction of N_2 and $2H^+$ (Eq. 12.1). Therefore, four

TABLE 12.3. Reactions catalyzed by nitrogenase

$N_2 \rightarrow NH_3$	Denitrogen fixation
$N_2O \rightarrow H_2 + H_2O$	Nitrous oxide reduction
$N_3^- \rightarrow N_2 + NH_3$	Azide reduction
$C_2H_2 \rightarrow C_2H_4$	Acetylene reduction
$2H^+ \rightarrow H_2$	H_2 production
$ATP \rightarrow ADP + P_i$	ATP hydrolytic activity

Adapted from Burris (1976), Table V, p. 901.

ethylene molecules correspond to the reduction of one N_2 molecule. The acetylene reaction has some limitations, however, such as ethylene production by some micro-organisms. For these reasons, the acetylene method is valuable for comparative studies, whereas direct measurements of N_2 reduction by mass spectroscopy are required for precise quantitation of nitrogen fixation.

The energetics of nitrogen fixation is complex. Production of NH_3 from N_2 and H_2 is an exergonic reaction with a $\Delta G^{\circ\prime}$ of -27 kJ mol^{-1} (Dixon and Wheeler, 1986). However, industrial production of NH_3 from N_2 and H_2 is a very energy-intensive process because of the activation energy required to break the triple bond in N_2. It is also clear that the enzymatic reduction of N_2 by nitrogenase requires a large investment of ATP (Eq. 12.1), although the specific free energy changes of the reactions involved remain to be established. The energy cost of nitrogen fixation can also be evaluated on the basis of carbohydrate utilization by an intact plant (Heytler et al., 1984). Calculations show that 12 g of organic carbon are consumed per gram of N_2 fixed. Based on Equation 12.1, the $\Delta G^{\circ\prime}$ for the overall reaction for biological nitrogen fixation has been estimated as -203 kJ mol^{-1}. Since the overall reaction is highly

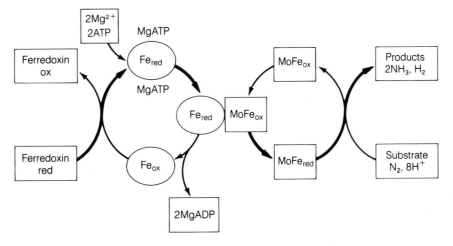

FIGURE 12.7. The reaction catalyzed by nitrogenase. Ferredoxin reduces the Fe protein, while binding of ATP takes place. The Fe protein reduces the MoFe protein, and the MoFe protein reduces the nitrogen. The flow of electrons is shown by the darker arrows. (From Dixon and Wheeler, 1986.)

exergonic, ammonium production is limited by the rather slow turnover rate of the nitrogenase complex (Ludwig and deVries, 1986).

Under natural conditions, the reduction of H^+ to H_2 gas can be substantial and can compete with N_2 for electrons from nitrogenase. In *Rhizobium*, 30 to 60% of the energy supplied to nitrogenase may be lost as H_2 (Schubert et al., 1978). Thus, this reaction would limit the efficiency of nitrogen fixation by these microorganisms. However, some *Rhizobium* strains contain an enzyme, hydrogenase, that can split H_2 and recycle electrons for N_2 reduction, thus increasing the efficiency of nitrogen fixation (Marschner, 1986).

The Nitrate Assimilation Pathway

Nitrate is the principal source of nitrogen that is available to higher plants under normal field conditions; hence, the nitrate assimilation pathway is the major point of entry of inorganic nitrogen into organic compounds (Hewitt et al., 1976). In most temperate soils, ammonia is rapidly converted to nitrate through nitrification by bacteria. Only in poorly aerated soils with inadequate drainage, where the bacteria that normally carry out nitrification cannot grow well, does ammonia accumulate to a significant extent. Although nitrate is generally the major form of nitrogen available to plants, some plants directly utilize ammonia under certain conditions, such as when growing in acid soils in which nitrification by microorganism is inhibited (Rice and Pancholy, 1972).

Nitrate Is Taken Up by an Inducible Transport System

Before entering the assimilation pathway, nitrate must be taken up by the cell. In plants not previously exposed to nitrate, nitrate stimulates its own uptake after a lag period. This fact and the inhibition of nitrate uptake by protein and RNA synthesis inhibitors suggest that nitrate induces its own specific transport system (Rao and Rains, 1976). Nitrate uptake is inhibited by cyanide, anaerobic conditions, and dinitrophenol, an uncoupler of oxidative phosphorylation, suggesting that metabolic energy is required. Efforts to purify a plasma membrane nitrate carrier are under way.

Once inside the cell, excess nitrate may be taken up by the vacuole. Several types of channels on the tonoplast potentially could facilitate vacuolar nitrate accumulation. Accumulation of high concentrations of nitrate in vacuoles of fodder or agricultural crops, such as spinach, is economically important because of its toxicity to livestock and humans. In the liver, nitrate is reduced to nitrite, which is very reactive and combines strongly with hemoglobin, rendering it unable to bind oxygen.

Nitrate Absorbed by Plants Is Reduced to Nitrite by Nitrate Reductase

Following its uptake from the soil by plant roots (Chapter 5), nitrate is reduced to ammonia prior to assimilation into organic compounds. The first step in this process is the reduction of nitrate to nitrite (Beevers, 1976). This reaction, which occurs in the cytosol, can be written as

$$AH_2 + NO_3^- \rightarrow A + NO_2^- + H_2O$$

for the redox couple

$$NO_3^- + 2H^+ + 2e^- \rightarrow NO_2^- + H_2O$$

and is carried out by the enzyme **nitrate reductase** (Beevers and Hageman, 1983). The electron donor of nitrate reductases in higher plants may be either NADH or NADPH, but the most common form of the enzyme utilizes NADH according to the equation:

$$NO_3^- + NAD(P)H + H^+ \xrightarrow{2e^-}$$
$$NO_2^- + NAD(P)^+ + H_2O \quad (12.2)$$

The NADH nitrate reductases of higher plants appear to be homodimers, that is, composed of two identical subunits. The molecular mass of the holoenzyme ranges from 200,000 to 270,000 Da, depending on the species, while the subunit molecular mass ranges from 100,000 to 120,000 Da. All eukaryotic nitrate reductases contain three prosthetic groups: FAD, heme, and a molybdenum complex. It is thought that one FAD, one heme, and one Mo are present per subunit. Molybdenum is bound to the enzyme via a complex with a small organic molecule called a **pterin**, which acts as a chelator of the metal (Wahl et al., 1984). Nitrate reductase is the principal molybdenum-containing protein in vegetative tissues (Hewitt, 1983). Under conditions of molybdenum deficiency, the activity of nitrate reductase in cells is much reduced.

The sequences of nitrate reductase mRNA from barley, *Arabidopsis*, and tobacco have now been determined from clones isolated from cDNA libraries. By comparing the deduced amino acid sequences with those

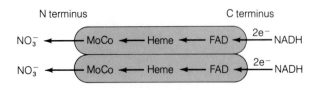

FIGURE 12.8. A model of the nitrate reductase dimer of higher plants, illustrating the three binding domains: molybdenum complex (MoCo), heme, and FAD. NADH binds at the FAD-binding region of each subunit and initiates a two-electron transfer from the carboxyl (C) terminus, through each of the electron transfer components, to the amino (N) terminus. Nitrate is reduced at the molybdenum complex near the amino terminus.

of other proteins that bind FAD, heme, or molybdenum, Crawford et al. (1988) were able to assign three functional domains to the subunit (Fig. 12.8). A sequence found at the C-terminal end of the protein was similar to the FAD-binding domain of cytochrome-b_5 reductase, and another sequence near the middle of the protein was similar to the heme-binding domain of cytochrome b_5. Finally, two short sequences near the N terminus were similar to the molybdenum-binding site of the enzyme, sulfite oxidase. Based on these sequence similarities, the model shown in Figure 12.8 has been proposed.

Nitrate Reductase Is an Inducible Enzyme

The synthesis of nitrate reductase is regulated by its substrate; that is, nitrate induces the de novo synthesis of the enzyme (Hewitt et al., 1976; Beevers and Hageman, 1983). The steady-state level of nitrate reductase is determined by the rate of degradation (turnover) as well as by the rate of synthesis. A newly synthesized nitrate reductase protein has a half-life of only a few hours in the cell. Thus, when nitrate is withdrawn, the level of nitrate reductase rapidly decreases.

In all plants studied so far, the induction of nitrate reductase by nitrate is preceded by an increase in nitrate reductase mRNA. In barley seedlings, nitrate reductase mRNA was detected approximately 40 min after addition of nitrate (Kleinhofs et al., 1989), and maximum levels were attained by about 2 h. Slightly different patterns were observed in shoots and roots (Fig. 12.9). The pattern of mRNA accumulation contrasts with the gradual linear increase in nitrate reductase activity (Fig. 12.9).

Although nitrate is the primary inducer of nitrate reductase, light is also required for the maximum effect. Somers et al. (1983) showed that barley seedlings treated with nitrate in the dark accumulated very low levels of

nitrate reductase. Exposure of the seedlings to white light resulted in a strong promotion of nitrate reductase activity. The effect of white light can be seen at the level of mRNA. However, light alone is insufficient to induce nitrate reductase in most cases; the role of light is to enhance the effect of nitrate.

Nitrite Is Reduced to Ammonium by Nitrite Reductase

Nitrate is reduced in the cytosol to nitrite by nitrate reductase. Rather than accumulating to toxic levels in the cytosol, the nitrite is rapidly transported into plastids—chloroplasts in green leaves and proplastids in nongreen tissues. Each type of plastid has its own assimilatory **nitrite reductase**. Chloroplasts contain a ferredoxin-nitrite reductase, while proplastids contain an NAD(P)H-nitrite reductase activity. Both enzymes catalyze the six-electron reduction of nitrite to ammonium with the overall reaction:

$$NO_2^- + 6e^- + 8H^+ \rightarrow NH_4^+ + 2H_2O \qquad (12.3)$$

Ferredoxin (Fd) is the electron donor in photosynthetic tissues, as shown in the following reaction:

$$NO_2^- + 6Fd_{red} + 8H^+ \xrightarrow{6e^-}$$
$$NH_4^+ + 6Fd_{ox} + 2H_2O \quad (12.4)$$

where the subscripts red and ox stand for reduced and oxidized, respectively. In contrast, nongreen tissues utilize NADH or NADPH supplied by respiration as the electron donor for nitrite reduction:

$$NO_2^- + 3NAD(P)H + 5H^+ \xrightarrow{6e^-}$$
$$NH_4^+ + 3NAD(P)^+ + 2H_2O \quad (12.5)$$

A "ferredoxin-like" protein may also be involved in

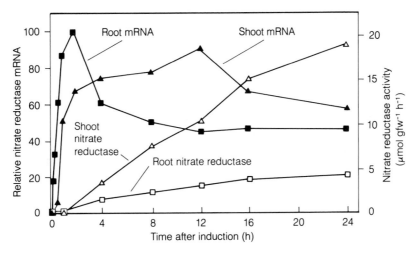

FIGURE 12.9. The pattern of nitrate reductase and nitrate reductase mRNA induction in shoots and roots of barley. The induction of mRNA precedes the appearance of the enzyme in both cases, although the precise timing is slightly different in the two tissues. (From Kleinhofs et al., 1989.)

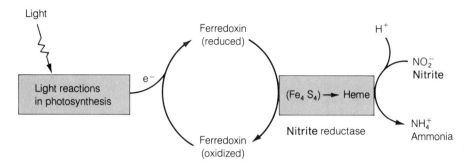

FIGURE 12.10. Model for coupling photosynthetic electron flow, via ferredoxin, to the reduction of nitrite by nitrite reductase. The enzyme contains two prosthetic groups, Fe_4S_4 and heme, which participate in the reduction of nitrite to ammonium.

transferring electrons to the enzyme in proplastids (Oaks and Hirel, 1985).

The best-characterized nitrite reductase is the photosynthetic enzyme of spinach. It consists of a single polypeptide of 60,000 Da and contains two prosthetic groups: an iron-sulfur cluster (Fe_4S_4) and a specialized heme (Wilkerson et al., 1983). Kinetic experiments suggest that nitrite is bound to the coupled Fe_4S_4-heme pair on the enzyme and is reduced directly to ammonium, without accumulation of nitrogen compounds of intermediate redox states. Thus, the pathway of electron flow from ferredoxin can be represented as shown in Figure 12.10.

Although localized in plastids, nitrite reductase, like nitrate reductase, is nuclear-encoded. The nitrite reductase of wheat leaves is synthesized on cytosolic ribosomes as a 64,000-Da precursor, which is processed to 60,500 Da during transit across the chloroplast membranes (Small and Gray, 1984).

The Anatomical Site of Nitrate Metabolism Differs According to Growth Conditions, Plant Age, and Plant Species

In most plant species both the roots and the shoot have the capacity to carry out nitrate metabolism. However, the degree to which nitrate reduction takes place in the roots or the leaves depends on a number of factors, including the level of nitrate supplied to the roots, plant species, and plant age. In many plants, when the amount of nitrate supplied to the roots is low, nitrate reduction takes place primarily in the roots. When the supply of nitrate is increased, a greater proportion of nitrate metabolism will occur in the leaves because root nitrate reduction becomes limiting under these conditions (Marschner, 1986). However, species differ in the amount of nitrate metabolized in the roots, even under similar conditions of nitrate supply. When the xylem exudate from the stems of various plants grown with their roots exposed solely to nitrate was examined, the proportion of nitrate to other nitrogen compounds varied from species

to species (Fig. 12.11) (Pate, 1973). In plants such as the cockleburr (*Xanthium* sp.), nitrate metabolism appears to be restricted to the shoots, whereas the roots of the white lupine (*Lupinus* sp.) have the capacity to metabolize nitrate into organic forms. In all cases in which nitrate is metabolized by the roots, nitrogen is translocated in an organic form, primarily as amino acids (see next section), and not as free ammonia. This is undoubtedly due to the toxicity of the latter compound.

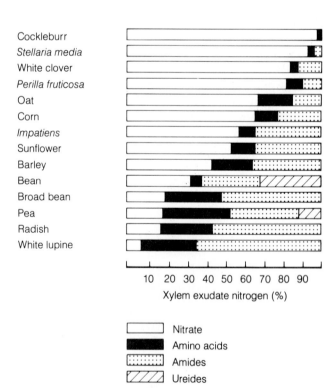

FIGURE 12.11. Relative amounts of nitrate and other nitrogen compounds in the xylem exudate of various plant species. The plants were grown with their roots exposed to nitrate solutions, and xylem sap was collected by severing the stem. Note the presence of ureides, specialized nitrogen compounds, in bean and pea (see text). (Adapted from Pate, 1973.)

Ammonia Is Rapidly Incorporated into Organic Compounds

Ammonia and ammonium ions do not accumulate in plant cells but are instead rapidly incorporated into organic compounds. Structural diagrams of a number of the compounds involved in ammonia metabolism and their pathways are shown in Figure 12.12a. Both ammonia and ammonium ions (generally grouped together as ammonia-nitrogen) are toxic to plant cells if present in high enough concentrations. Only certain plants such as *Begonia* are able to accumulate ammonia-nitrogen, and this occurs within the large central vacuole.

The assimilation of ammonia-nitrogen into carbon compounds can proceed through two possible pathways. The first pathway involves the reductive amination of α-ketoglutarate to produce glutamate:

$$\alpha\text{-Ketoglutarate} + NADH + NH_3 \rightarrow$$
$$glutamate + NAD^+ + H_2O. \quad (12.6)$$

This reaction is catalyzed by glutamate dehydrogenase (GDH), which is found in chloroplasts as well as in mitochondria. In the mitochondria, the enzyme would have available an abundant supply of NADH and α-ketoglutarate continually produced by the tricarboxylic acid (TCA) cycle. The second pathway for assimilation of ammonia involves a reaction with glutamate to form its amide, glutamine.

$$Glutamate + NH_3 + ATP \rightarrow glutamine + ADP + P_i$$
$$(12.7)$$

This reaction is catalyzed by glutamine synthetase (GS) and requires input of energy in the form of ATP. Glutamine can then be converted to glutamate by the following reaction:

$$Glutamine + \alpha\text{-ketoglutarate} \rightarrow 2 \text{ glutamate} \quad (12.8)$$

which is catalyzed by glutamate synthase (GOGAT[1]). Glutamine synthetase is localized in the chloroplast and cytosol in the leaves and in the cytosol in cells of the roots (Oaks and Hirel, 1985). Glutamate synthase is localized in the chloroplasts in the leaves (Miflin and Lea, 1980) and in plastids in roots (Emes and Fowler, 1979). Three different forms (isoenzymes) of glutamate synthase may exist: a photosynthetic form using ferredoxin as electron donor and two nonphotosynthetic forms using either NADH or NADPH (Oaks and Hirel, 1985).

Both of these possible pathways result in the same final product, glutamate, but which is the more important pathway in terms of the assimilation of ammonia-nitrogen? Studies of the individual enzymes have revealed that glutamate dehydrogenase, the enzyme of the first pathway, has a relatively high K_m value for ammonia

(Beevers, 1976). The K_m value corresponds to the concentration of ammonia at which the enzyme operates at one-half the maximal rate. This K_m value is within the concentration levels considered to be toxic for plant cells. Therefore, this enzyme would not contribute to the metabolism of NH_3 at the much lower concentrations of this compound expected in the cell. Glutamine synthetase has a much lower K_m value for ammonia. This difference in the range of ammonia concentrations at which these two enzymes operate has led most investigators to consider the second pathway the more logical one for the incorporation of ammonia-nitrogen. Even stronger evidence for the GS/GOGAT pathway of ammonia incorporation has come from studies using ^{15}N-labeled precursors and specific inhibitors of these enzymes. In kinetic experiments in which $^{15}NO_3^-$ or $^{15}NH_4^+$ was used as the precursor, labeled nitrogen appeared first in the amide group of glutamine, followed by other amino acids such as glutamate, alanine, serine, glycine, aspartate, and asparagine (Oaks and Hirel, 1985; Goodwin and Mercer, 1983). When specific inhibitors of glutamine synthetase (methionine sulfoximine) or glutamate synthase (azaserine) were used with ^{15}N-labeled precursors, the label was not incorporated into amino acids (Oaks and Hirel, 1985, and references therein).

Nitrogen Is Incorporated into Other Amino Acids Through Transamination Reactions

Once assimilated into glutamate, the nitrogen can be further incorporated into other amino acids through transamination reactions. These reactions are carried out by enzymes known as aminotransferases. An example of a transamination reaction is

$$Glutamate + oxaloacetate \rightarrow$$
$$aspartate + \alpha\text{-ketoglutarate} \quad (12.9)$$

in which the amino group of glutamate is transferred to the carboxyl atom of the α-keto acid. All transamination reactions require pyridoxal phosphate (vitamin B_6) as a cofactor. Aminotransferases are found in the cytosol, chloroplasts, and microbodies (Beevers, 1976). The enzymes localized in the chloroplasts may have a significant role in amino acid biosynthesis, since rapid incorporation of radioactive carbon-14 into glutamate, aspartate, alanine, serine, and glycine occurs when plant leaves are exposed to labeled carbon dioxide (Roberts et al., 1970).

The Major Export Forms of Nitrogen in Nitrogen-Fixing Plants Are Amides and Ureides

Fixed nitrogen is converted to organic form in the root nodules before being transported to the shoot via the xylem. Based on the composition of the xylem sap, nitrogen-

---

1. GOGAT is the acronym for glutamine-oxoglutarate amidotransferase. (Oxoglutarate is an alternative name for α-ketoglutarate.)

(a) Major Ammonia Assimilation Pathways

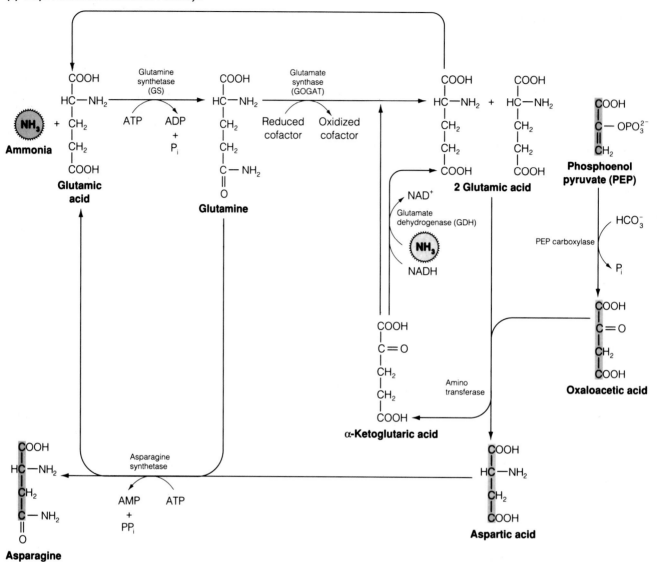

(b) Ureides

Allantoic acid   **Allantoin**   **Citrulline**

FIGURE 12.12. Structure and pathways of compounds involved in ammonia metabolism. (a) Ammonia can be assimilated either by combination with glutamate to form the amide glutamine (glutamine synthetase) or by reductive amination of α-ketoglutarate to produce glutamate (glutamate dehydrogenase). Two glutamates are produced from glutamine and α-ketoglutarate (glutamate synthase). A reduced cofactor is required for the reaction: ferredoxin in green leaves and NADPH in root tissues. Since the formation of glutamine is an uphill reaction requiring ATP, glutamate production helps to drive the reaction forward. Further reactions yield aspartate and the amide asparagine. (b) Structures of the major ureide compounds found in nitrogen-fixing plants.

fixing legumes can be divided into amide exporters or ureide exporters. Temperate-region legumes, such as pea (*Pisum*), clover (*Trifolium*), broad bean (*Vicia*), and lentil (*Lens*), tend to be exporters of amides, principally asparagine and glutamine. Legumes of tropical origin, such as soybean (*Glycine*), kidney bean (*Phaseolus*), peanut (*Arachis*), and southern pea (*Vigna*), export nitrogen in the form of **ureides**. The three major ureides are allantoin, allantoic acid, and citrulline (Fig. 12.12b). Allantoin is synthesized in peroxisomes from uric acid, and allantoic acid is synthesized from allantoin in the endoplasmic reticulum. The site of citrulline synthesis from the amino acid ornithine has not yet been determined. All three compounds are ultimately released into the xylem and transported to the shoot, where they are rapidly metabolized. During catabolism, ammonium is released and enters the assimilation pathway.

Sulfur Assimilation

The sulfur present in the carbon constituents of higher plant cells comes mostly from sulfate acquired from the soil by plant roots. The bulk of the soil sulfate comes from the weathering of parent rock material; however, with industrialization, an additional source of sulfate is atmospheric pollution. The burning of fossil fuels releases several gaseous forms of sulfur, including sulfur dioxide (SO_2) and hydrogen sulfide (H_2S), which find their way to the soil in rain. When dissolved in water, sulfur dioxide is hydrolyzed to become H_2SO_4, a strong acid, which is the major source of acid rain. Plants can also metabolize sulfur dioxide taken up in the gaseous form through the stomata. However, prolonged exposure (>8 h) to high atmospheric concentrations (>0.3 ppm) causes extensive tissue damage because of the formation of sulfuric acid. The major sulfur-containing organic compounds in higher plant cells are the amino acids cysteine and methionine.

Uptake of sulfate into roots is an active process thought to be mediated by specific carrier proteins. Once inside the cell, sulfate is either converted to organic compounds via the assimilation pathway or transported as free sulfate into the vacuole through anion channels in the tonoplast. Vacuolar sulfate has been shown to make up 99% of the total sulfur in the duckweed, *Lemna minor* (Rennenberg, 1984). The vacuole is thus an important site of accumulation of excess sulfate in certain species.

Sulfate Must Be Reduced Prior to Assimilation into Carbon Compounds

Like nitrate, the sulfate acquired by plant cells must be reduced before it can be incorporated into organic compounds (Wilson and Reuveny, 1976; Schiff, 1983). This reduction process involves the formation of several in-termediate sulfur compounds prior to incorporation of sulfur into amino acids. In higher plants, the enzymes responsible for the assimilatory reduction of sulfate are located in the chloroplasts of leaf cells and in the proplastids of root cortical cells (Frankhauser and Brunold, 1978). In plants that utilize the C4 pathway of photosynthesis, the bundle sheath chloroplasts are the main sites of sulfate reduction (Schmutz and Brunold, 1982).

The Activation of Sulfate. Among the intermediate sulfur compounds are two compounds, adenosine 5'-phosphosulfate (APS) and 3'-phosphoadenosine 5'-phosphosulfate (PAPS), that represent the activated forms of sulfur. Only activated forms of sulfate can be reduced and converted into other organic compounds (Fig. 12.13). Sulfate activation involves adenosine 5'-triphosphate (ATP) and sulfate in the following successive reactions:

$$\text{ATP} + \text{sulfate} \rightarrow \text{APS} + \text{PP}_i$$

$$\Delta G^{\circ\prime} = 45 \text{ kJ mol}^{-1} \quad (12.10)$$

$$\text{APS} + \text{ATP} \rightarrow \text{PAPS} + \text{ADP}$$

$$\Delta G^{\circ\prime} = -25 \text{ kJ mol}^{-1} \quad (12.11)$$

which are catalyzed by the enzymes ATP sulfurylase and APS kinase, respectively (Wilson and Reuveny, 1976; Schiff, 1983). Although the first reaction, as written, is not energetically favorable for product formation, its coupling to the favorable reaction involving PAPS formation drives sulfate activation. An additional driving force for sulfate activation is the subsequent hydrolysis of pyrophosphate (PP_i) to 2 moles of inorganic phosphate.

$$\text{PP}_i + \text{H}_2\text{O} \rightarrow 2\text{P}_i \quad \Delta G^{\circ\prime} = -33.5 \text{ kJ mol}^{-1} \quad (12.12)$$

This energetically favorable reaction is catalyzed by an enzyme known as an inorganic pyrophosphatase.

From PAPS, sulfate may be directly assimilated into organic compounds such as sulfolipids, which contain sulfur bound as a sulfate ester. However, assimilation of sulfur into amino acids and proteins requires reduction to the level of sulfide.

Reduction of Sulfate to Sulfide. The next step in the assimilation of sulfur is the actual reduction of the activated form to the level of sulfide. Eight electrons are required, and the oxidation number of sulfur changes from $+6$ to -2. However, the exact mechanism of APS reduction is poorly understood. Because of the energetic constraints discussed in the previous section, APS should not accumulate significantly in plant cells. Therefore, for some time it was assumed that PAPS was the activated form of sulfate that was reduced to sulfide (Wilson and Reuveny, 1976). However, other studies of sulfur reduction have suggested that PAPS is first converted back

FIGURE 12.13. The structure and synthesis of the two activated sulfur compounds, APS and PAPS. The enzyme ATP sulfurylase cleaves pyrophosphate from ATP and replaces it with sulfate. APS is used in the synthesis of sulfur-containing amino acids. PAPS is formed from APS by the enzyme APS kinase.

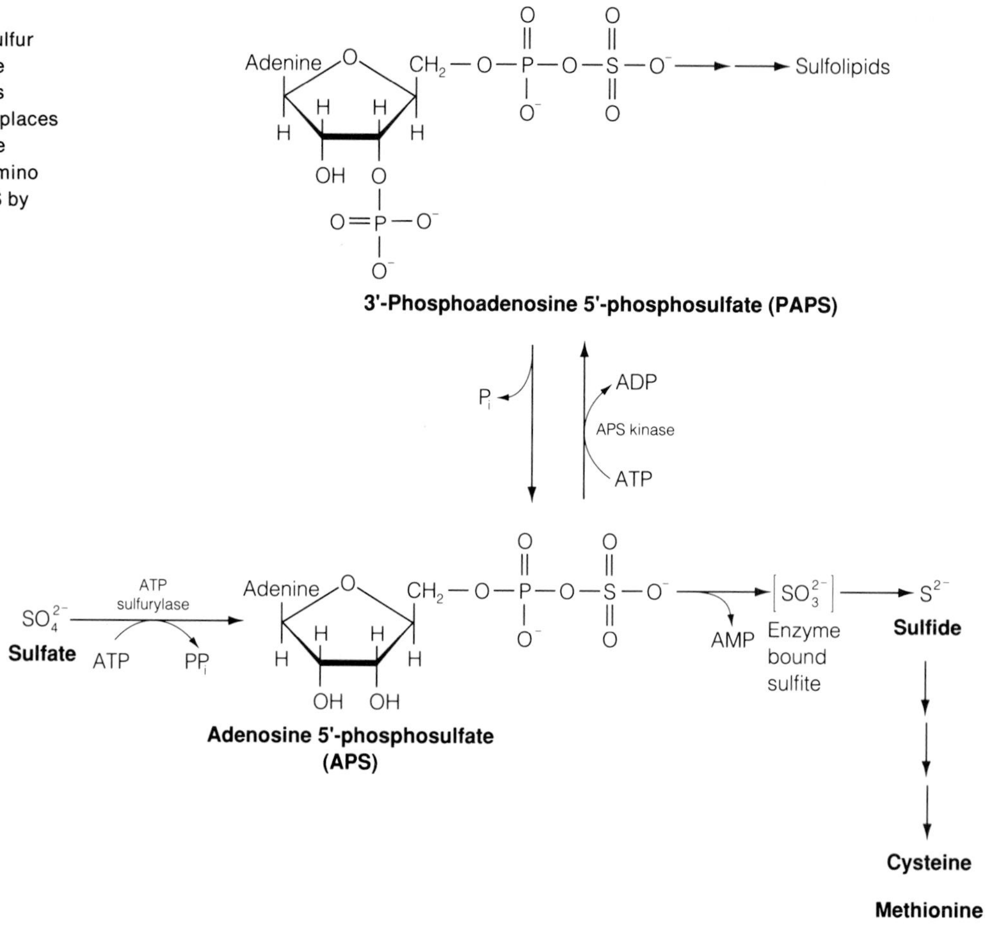

Sulfolipids

3'-Phosphoadenosine 5'-phosphosulfate (PAPS)

P_i → ADP

APS kinase

ATP

SO_4^{2-}

Sulfate ATP PP_i

ATP sulfurylase

AMP

$\left[SO_3^{2-}\right]$ → S^{2-}

Enzyme bound sulfite

Sulfide

Adenosine 5'-phosphosulfate (APS)

Cysteine

Methionine

to APS, which then serves as the substrate for reduction (Hodson and Schiff, 1971). This reaction involves the hydrolysis of the 3'-phosphate groups of PAPS by a magnesium-dependent 3'-phosphonucleotidase. Therefore, in reductive sulfur metabolism the formation of PAPS may be primarily a means of carrying out the energetically unfavorable reaction of APS formation at the expense of ATP hydrolysis.

In the reduction of APS to sulfide, reduced ferredoxin acts as an electron donor for the reaction:

$$APS + 8Fd_{red} + 5H^+ \rightarrow$$
$$sulfide + AMP + 8Fd_{ox} + 3H_2O \quad (12.13)$$

There is evidence that this overall reaction may involve the production of sulfite, SO_3^-, as a protein-bound intermediate (Wilson and Reuveny, 1976; Goodwin and Mercer, 1983). Although the reduction of APS to sulfide can take place in the roots, the rate is generally several-fold higher in the leaves and is light-stimulated (Marshner, 1986). This stimulation by light could be due to the enhanced production of reduced ferredoxin by the activity of the photosystems. In nonphotosynthetic cells, NAD(P)H is the likely electron donor.

Incorporation of Sulfide into Cysteine and Methionine. The sulfide produced by the reduction of APS does not accumulate in plant cells but is instead rapidly incorporated into the sulfur-containing amino acids (cysteine and methionine). The formation of cysteine involves the addition of sulfide to the three-carbon acceptor O-acetylserine in a reaction catalyzed by O-acetylserine sulfhydrylase:

$$O\text{-Acetylserine} + sulfide \rightarrow cysteine + acetate \quad (12.14)$$

The O-acetylserine is an activated form of serine produced by a reaction with coenzyme A catalyzed by serine transacetylase:

$$Serine + acetyl\ CoA \rightarrow O\text{-acetylserine} + CoA \quad (12.15)$$

Methionine can be formed by two possible pathways, transsulfuration or direct sulfhydration, involving the series of reactions shown in Figure 12.14. There is evidence that in higher plants transsulfuration is the primary pathway for methionine biosynthesis from sulfide (Wilson and Reuveny, 1976). Following the synthesis of cysteine and methionine, sulfur can be incorporated into proteins and a variety of other compounds such as

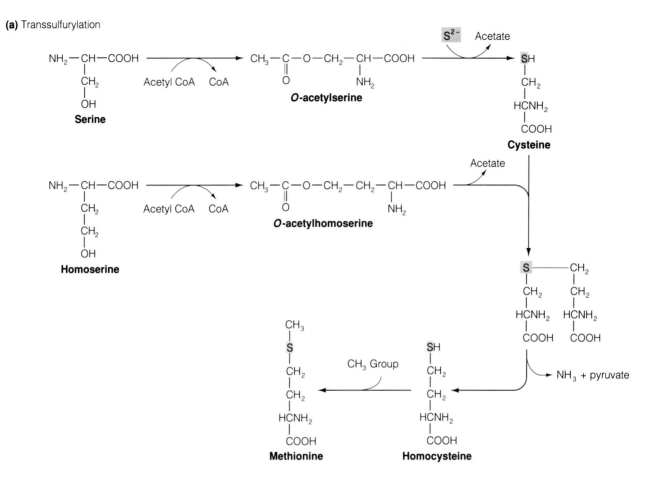

FIGURE 12.14. The two pathways of methionine biosynthesis: (a) transsulfuration; (b) direct sulfurylation. The transsulfuration pathway is probably the major route of methionine production.

acetyl coenzyme A and *S*-adenosylmethionine. The latter compound is important in synthetic reactions involving the transfer of methyl groups, as in lignin synthesis (see Chapter 13).

Phosphate Assimilation

Phosphate in the soil solution is readily taken up by plant roots and incorporated into a variety of organic compounds including sugar phosphates, phospholipids, and nucleotides. The main entry point of phosphate into assimilatory pathways occurs during the formation of ATP, the energy "currency" of the cell. The overall reaction for this process involves the addition of inorganic phosphate to the second phosphate group in adenosine diphosphate to form a phosphate ester bond. In mitochondria, the energy for ATP synthesis is derived from the oxidation of NADH produced during the breakdown of sugars in respiration, a process called oxidative phosphorylation (Chapter 11). The formation of ATP is also driven by light energy in the process of **photophosphorylation** that occurs in the chloroplasts (Chapter 8). In addition to these reactions in mitochondria and chloroplasts, phosphate is assimilated by reactions in the cytosol. In glycolysis, inorganic phosphate is incorporated into 1,3-bisphosphoglyceric acid, forming a high-energy acyl phosphate group. This phosphate can be donated to ADP to form ATP in a **substrate-level phosphorylation** reaction (Fig. 12.15). Once incorporated in ATP, the phosphate group can be transferred in reactions to form the various phosphorylated compounds found in higher plant cells.

Cation Assimilation

Assimilation of cations taken up by plant cells involves the formation of complexes with organic compounds without any chemical transformation of the nutrient. That is, the cation becomes bound to the carbon compound without the formation of covalent bonds. This occurs for macronutrient cations such as potassium, magnesium, and calcium as well as for micronutrient cations such as copper, iron, manganese, cobalt, sodium, and zinc. The assimilation of iron will be discussed in more detail in the next section.

Cation Complexes with Carbon Compounds Involve the Formation of Coordination or Electrostatic Bonds

In the formation of a coordination complex, oxygen or nitrogen atoms of a carbon compound donate unshared electrons to form a bond with the cation nutrient. As a result, the positive charge on the cation is neutralized. Coordination complexes typically occur between polyvalent cations and organic molecules, such as copper and tartaric acid (Fig. 12.16a) and magnesium and chlorophyll *a*, in which the magnesium is bound by coordination bonds with the nitrogen group atoms in the porphyrin ring (Fig. 12.16b). The nutrients that are assimilated with the formation of coordinate complexes include copper, zinc, iron, and magnesium. Calcium can also form coordinate complexes with cell wall polygalacturonic acid (Fig. 12.16c).

Electrostatic bonds are formed because of the attraction of a positively charged cation for a negatively charged group on an organic compound. In electrostatic bonds, unlike coordinate bonds, the cation retains its positive charge. An important negatively charged group is the ionized form of a carboxylic acid. For example, monovalent cations such as potassium can form electrostatic bonds with the carboxylic groups of many organic acids, as illustrated in Figure 12.17a. The incorporation of calcium into the pectic components of the cell wall involves coordinate bonds as well as electrostatic bonds with the carboxylic groups of polygalacturonic acid (Fig. 12.17b). In general, assimilation of magnesium and calcium that is not accounted for by the formation of coordination complexes takes place through the forma-

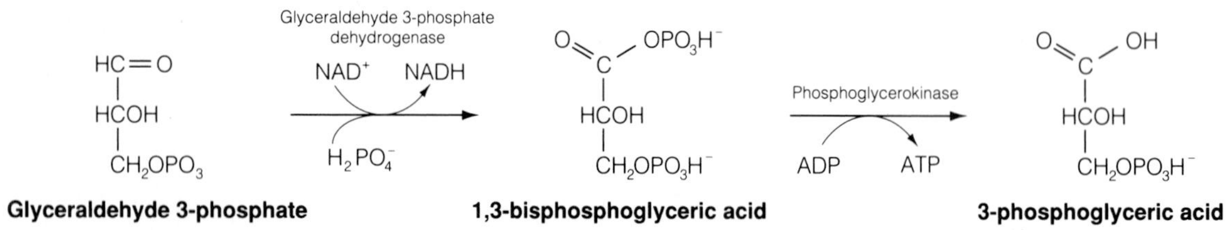

FIGURE 12.15. Substrate-level phosphorylation with the production of ATP. This reaction, which occurs in glycolysis, involves the incorporation of inorganic phosphate into 1,3-bisphosphoglyceric acid (first reaction) and the breakdown of this compound to produce ATP from ADP.

FIGURE 12.16. Examples of coordination complexes. (a) Copper ions share electrons with the hydroxyl oxygens of tartaric acid. (b) Magnesium ions share electrons with nitrogen atoms in chlorophyll *a*. (c) The "eggbox" model of the interaction of polygalacturonic acid and calcium ions. On the left, calcium ions are held in the spaces between two polygalacturonic acid chains, indicated by the kinked, horizontal lines. The arrow shows a blow-up of a single calcium ion forming a coordination complex with the hydroxyl oxygens of the galacturonic acid residues. Much of the calcium in the cell wall is thought to be bound in this fashion (From Rees, 1977.)

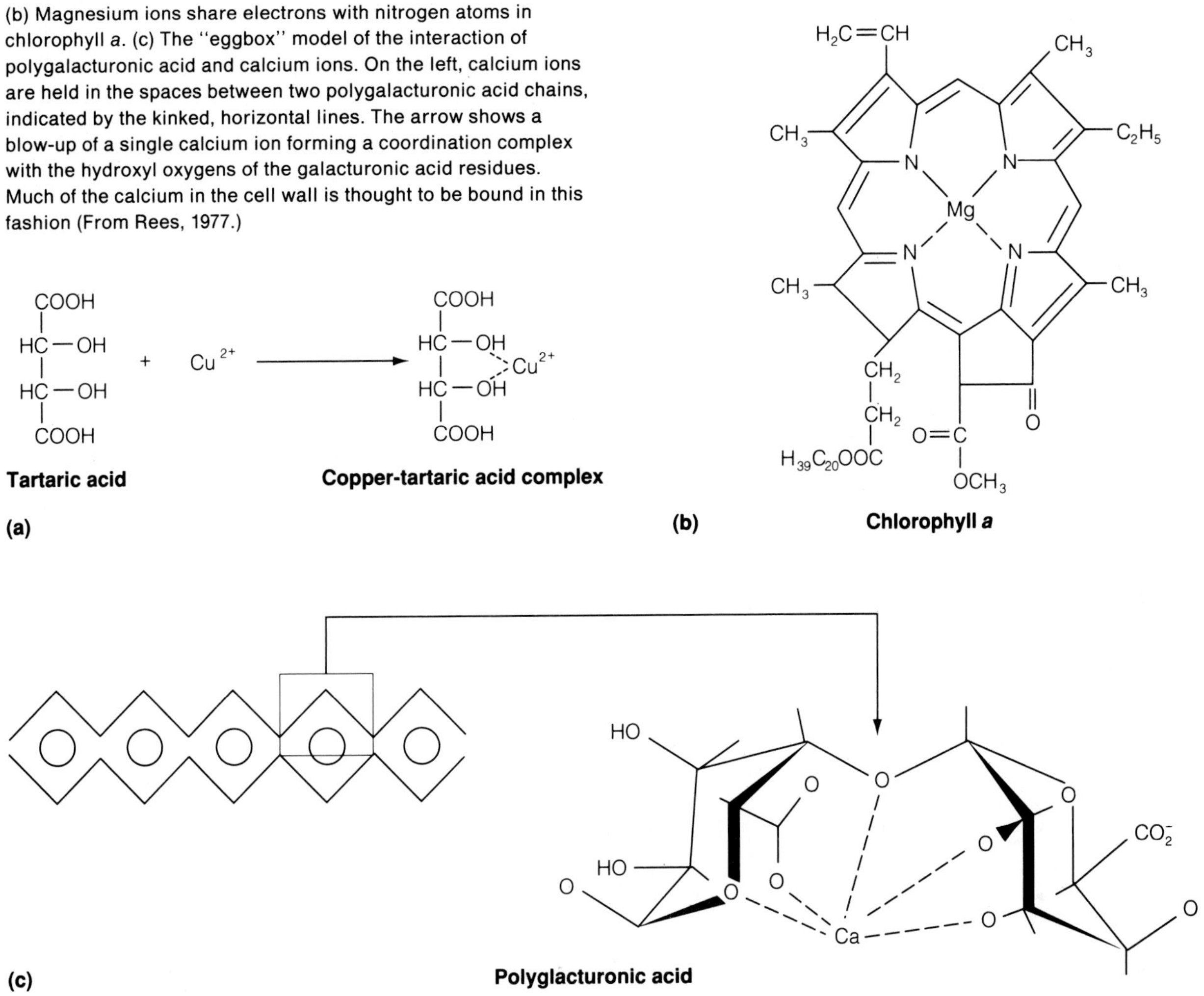

Tartaric acid   **Copper-tartaric acid complex**

(a)

(b)   **Chlorophyll *a***

(c)   **Polyglacturonic acid**

tion of electrostatic bonds with amino acids, phospholipids, and other negatively charged molecules. It should be noted that much of the potassium that is accumulated by plant cells remains as the free ion in the cytosol and the vacuole and functions in osmotic regulation and enzyme activation.

Iron Assimilation Involves Redox Reactions and Complex Formation

Iron is important in plant cells as a catalyst in enzyme-mediated redox reactions and in chlorophyll biosynthesis (Bienfait and Van der Mark, 1983). Iron is acquired by plant roots from the soil, where it is present primarily in the ferric form (Fe^{3+}) as oxides such as $Fe(OH)_2^+$, $Fe(OH)_3$, and $Fe(OH)_4^-$ (Olsen et al., 1981). This form of iron is extremely insoluble, as indicated by the K_{sp} (2×10^{-39}), which describes the equilibrium between the solid and dissolved species. It is therefore difficult for

plant roots to obtain iron from the soil. The process by which plant roots can solubilize and acquire iron has been studied in detail for dicots. In response to iron stress, these plants: (1) acidify the medium surrounding the root, (2) chemically reduce soluble ferric chelates (Fe coordination complexes), (3) take up ferrous iron (Fe^{2+}), and (4) accumulate organic acids such as citrate (Bienfait and Van der Mark, 1983). Acidification and reduction are key events in the solubilization of iron, since acidification of the soil surrounding the roots greatly increases the solubility of the ferric form of iron (Olsen et al., 1981) and the reduced ferrous iron (Fe^{2+}) is much more soluble than the oxidized ferric form (Fe^{3+}). There has been a controversy as to whether iron reduction takes place as a result of the release of "reducing substances" such as caffeic acid into the soil from root cells (Olsen et al., 1981) or whether reduction occurs on the membrane surface of root cells by the

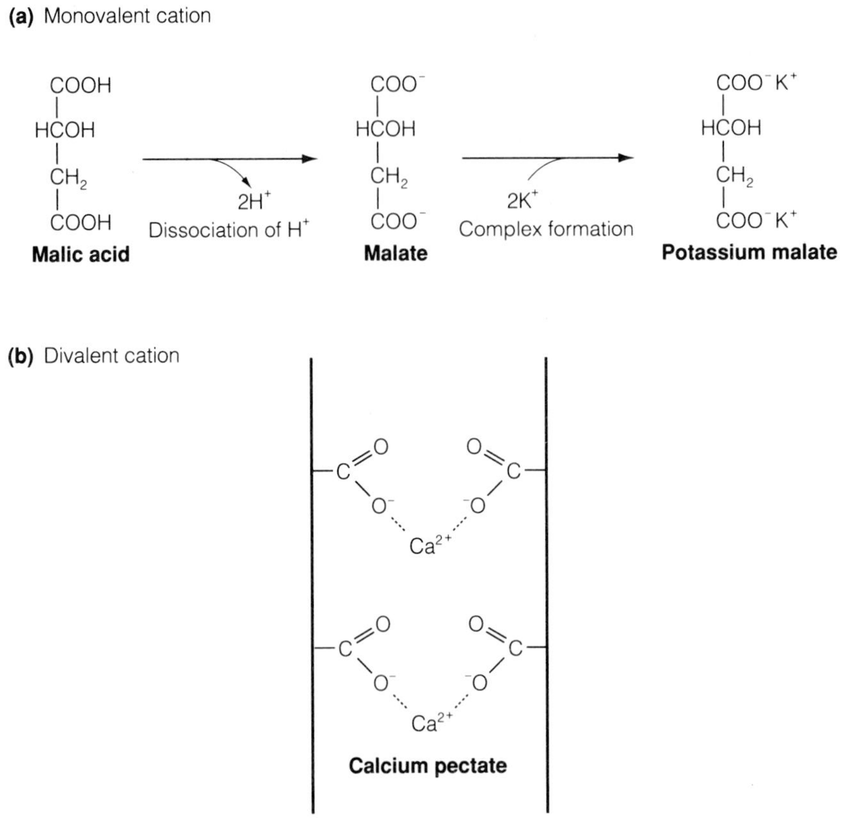

(a) Monovalent cation

(b) Divalent cation

FIGURE 12.17. Examples of the formation of electrostatic (ionic) complexes. (a) Monovalent cation, potassium malate. (b) Divalent cation, calcium pectate. Divalent cations can form cross-links between parallel strands containing negatively charged carboxyl groups. Calcium cross-links may play a structural role in the cell walls of some algae and higher plants.

action of a redox electron transport chain (Bienfait and Van der Mark, 1983). Under conditions of iron deficiency, the activity of a trans-plasma membrane electron transport system greatly increases in bean roots and certain other species and may represent an adaptation to low Fe^{3+} concentrations in the soil (Crane et al., 1985). In any case, reduction appears to be a prerequisite for the uptake of iron into plant cells (Chaney and Brown, 1972). Most evidence suggests that the iron is then reoxidized in the cells of the root and transported to the leaves as an electrostatic complex with citrate (Olsen et al., 1981).

When iron-deficient plants receive iron, they may often take up massive amounts by an augmentation of the uptake system. Large amounts of "free" iron in plant cells can pose problems because iron can interact with oxygen to form superoxide anions (O_2^-), which can damage membranes by degrading unsaturated lipid components (Halliwell, 1974; Trelstad et al., 1981). Plant cells can, however, provide safe storage for iron in the leaves through the use of an iron storage protein called **phytoferritin** (Bienfait and Van der Mark, 1983). Phytoferritin is made up of 24 identical protein subunits that

form a hollow sphere with a molecular mass of about 480,000 Da. Iron incorporation into phytoferritin is thought to occur by oxidation of ferrous iron inside the phytoferritin protein and formation of an iron core (Bienfait and Van der Mark, 1983). The core consists of about 5400 to 6200 iron atoms present as a ferric oxide-phosphate complex. How iron is released from phytoferritin is uncertain, but it is generally thought to involve a breakdown of the protein complex. It is also thought that the level of free iron in plant cells may regulate the de novo biosynthesis of phytoferritin (Zähringer et al., 1976).

When iron acts as a redox component, it is often present as a complex within a porphyrin group in the active site of an enzyme. This is the case, for example, in the cytochromes that mediate electron transport in the mitochondrial inner membrane. Therefore, an important assimilatory reaction for iron involves its insertion into the porphyrin system. This reaction is called the ferrochelase reaction, and it is catalyzed by the enzyme ferrochelase (Fig. 12.18) (Jones, 1983). Ferrochelase is associated with chloroplasts and mitochondria and most

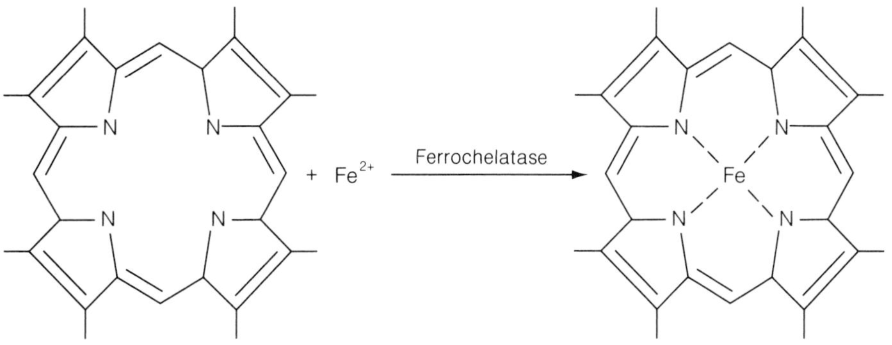

FIGURE 12.18. The ferrochelase reaction. The enzyme ferrochelase catalyzes the insertion of iron into the porphyrin ring to form a coordination complex.

likely functions in active metalloporphyrin synthesis for electron transport systems in these organelles. In addition, iron-sulfur proteins of the electron transport chain (Chapter 9) contain Fe covalently attached to the cysteine sulfur atoms of the apoprotein, as well as Fe atoms bound to other Fe atoms by means of sulfur bridges.

Oxygen Assimilation

Although respiration accounts for the bulk of the oxygen (O_2) utilized by plant cells, a small proportion of oxygen may be directly assimilated into organic compounds in the process of **oxygen fixation**. Oxygen fixation involves direct addition of molecular oxygen to an organic compound in reactions carried out by enzymes known as oxygenases (Metzler, 1977). Other than oxygen fixation, the major pathway for the assimilation of oxygen into organic compounds involves reactions in which the oxygen from water is incorporated. For many years, it was assumed that only the latter pathway occurs in living organisms.

Oxygenases are classified as **dioxygenases** or **monooxygenases** according to the number of atoms of oxygen that are transferred to an organic compound in the enzyme-catalyzed reaction. Examples of dioxygenases in plant cells are lipoxygenase, which catalyzes the addition of two atoms of oxygen to unsaturated fatty acids, and prolyl hydroxylase, the enzyme that converts proline to the rare amino acid hydroxyproline (Fig. 12.19). Hydroxyproline is an important component of cell wall protein, **extensin** (see Chapter 1). Prolyl hydroxylase is considered a dioxygenase because one oxygen atom is incorporated into proline to form hydroxyproline, while a second oxygen is used to convert α-ketoglutarate to succinate. Ferrous iron and ascorbate are also needed as cofactors but do not participate directly in the redox

reaction (Fig. 12.19b). The synthesis of hydroxyproline from proline differs from the synthesis of all other amino acids in that the reaction occurs *after* the proline has been incorporated into protein and is therefore a post-translational modification reaction. Prolyl hydroxylase is localized in the endoplasmic reticulum, suggesting that most, if not all, proteins containing hydroxyproline are found in the secretory pathway.

Monooxygenases add one of the atoms in molecular oxygen to an organic compound; the other oxygen atom is converted into water. Monooxygenases are sometimes referred to as *mixed-function oxidases* because of their ability to catalyze simultaneously both the oxygenation reaction and the oxidase reaction (reduction of oxygen to water). The monooxygenase reaction also requires a reduced substrate (NADH or NADPH) as an electron donor, according to the following equation:

$$A + O_2 + BH_2 \rightarrow AO + H_2O + B$$

where A represents an organic compound and B represents the electron donor. An important monooxygenase in plant cells is the family of heme proteins collectively called cytochrome P-450, which catalyzes the hydroxylation of cinnamic acid to *p*-coumaric acid. In monooxygenases, the oxygen is first activated by combining with the iron atom of the heme group. NADPH serves as the electron donor, as shown in Figure 12-19c. The mixed-function oxidase system is localized on the endoplasmic reticulum and is capable of oxidizing a variety of substrates, including mono- and diterpenes and fatty acids.

Another reaction in which oxygen is directly incorporated into an organic compound occurs during photorespiration (see Chapter 9) and involves the oxygenase activity of ribulose-1,5-bisphosphate carboxylase/oxygenase (Rubisco), the enzyme of CO_2 fixation (Ogren, 1984). The first stable product that contains oxygen originating from molecular oxygen is 2-phosphoglycolate.

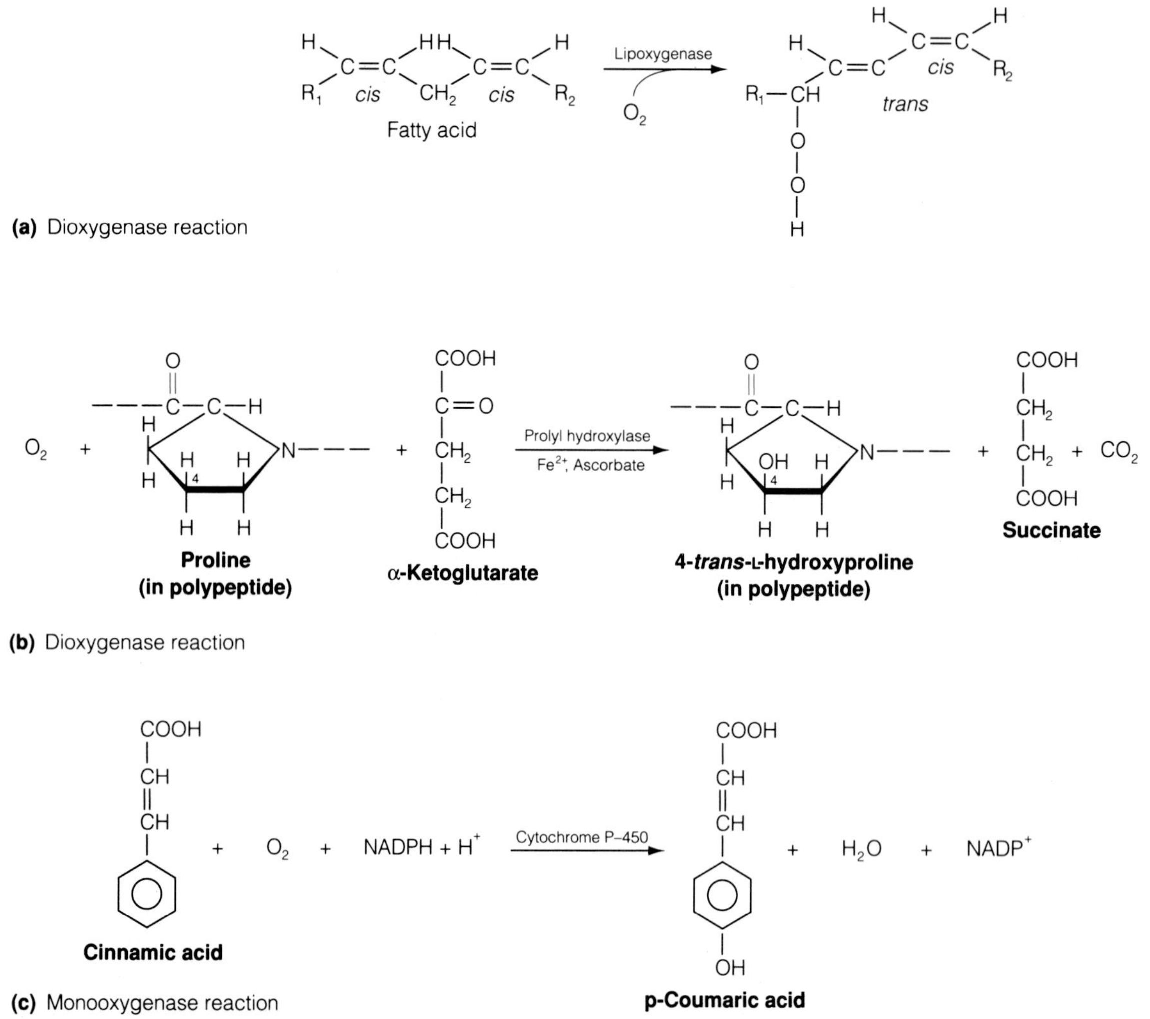

(a) Dioxygenase reaction

(b) Dioxygenase reaction

Proline
(in polypeptide)

α-Ketoglutarate

4-*trans*-ʟ-hydroxyproline
(in polypeptide)

Succinate

(c) Monooxygenase reaction

Cinnamic acid

p-Coumaric acid

FIGURE 12.19 Examples of the two types of oxygenase reactions in cells of higher plants. (a) The dioxygenase lipoxygenase catalyzes the addition of two atoms of oxygen to the conjugated portion of an unsaturated fatty acid to form a hydroperoxide with a pair of *cis-trans* conjugated double bonds. The hydroxy peroxy fatty acid may then be enzymatically converted to hydroxy fatty acids and other metabolites. Such changes can contribute to membrane deterioration during senescence. (b) The dioxygenase prolyl hydroxylase catalyzes the addition of one oxygen to peptide proline and one oxygen to α-ketoglutarate. (c) The monooxygenase cytochrome P-450 uses one oxygen to hydroxylate cinnamic acid (and other substrates) and the other oxygen to produce water. NAD(P)H serves as the electron donor for monooxygenase reactions.

Summary

Nutrient assimilation is the process by which nutrients acquired by plants are incorporated into the carbon constituents necessary for growth and development. For nitrogen, this is one step in a series of processes that constitute the nitrogen cycle, which encompasses the various states of nitrogen in the biosphere and their interconversions. The principal source of nitrogen available to most plants is nitrate (NO_3^-), which must be reduced to ammonia (NH_3) prior to incorporation into organic compounds. Under some conditions, ammonia is present in significant amounts in the soil and can be used directly by plants. A number of plants, primarily legumes, form a symbiotic relationship with nitrogen-fixing bacteria and utilize nitrogen (N_2) obtained from the atmosphere. These bacteria contain an enzyme complex, nitrogenase, that can reduce atmospheric nitrogen to ammonia. Nitrogen-fixing prokaryotic microorganisms also exist that do not form symbiotic relationships with higher plants but benefit plants by enriching the soil nitrogen content.

Like nitrate, sulfate (SO_4^{2-}) must be reduced in the process of assimilation. Reduction of sulfate involves the formation of two activated forms of sulfate called adenosine 5'-phosphosulfate (APS) and 3'-phosphoadenosine 5'-phosphosulfate (PAPS). Sulfide (H_2S), the end product of sulfate reduction, does not accumulate in plant cells but is instead rapidly incorporated into the amino acids cysteine and methionine.

Phosphate (PO_4^{3-}) is present in a variety of compounds found in plant cells, including sugar phosphates, lipids, nucleic acids, and free nucleotides. The initial product of its assimilation is ATP, which is produced by substrate-level phosphorylations in the cytosol, oxidative phosphorylation in the mitochondria, and photophosphorylation in the chloroplasts.

Whereas nitrogen, sulfur, and phosphorus assimilation requires the formation of covalent bonds with carbon compounds, many macro- and micronutrient cations simply form complexes. These complexes may be of the electrostatic or coordination type. Iron assimilation involves not only complex formation but also several steps of reduction and oxidation. To be taken up, Fe^{3+} must be reduced to Fe^{2+}, most likely by a plasma membrane electron transport system. Iron is transported within the plant as Fe^{3+} complexed with citrate. In order to store large amounts of iron, plant cells synthesize phytoferritin, an iron storage protein. An important function of iron in plant cells is to act as a redox component in the active site of enzymes, often as an iron-porphyrin complex. The insertion of iron into a porphyrin group occurs in the ferrochelase reaction.

In addition to utilization in respiration, molecular oxygen can be assimilated in the process of oxygen fixation, the direct addition of oxygen to organic compounds. This process is catalyzed by enzymes known as oxygenases, which are classified as monooxygenases or dioxygenases.

GENERAL READING

Beevers, L. (1976) *Nitrogen Metabolism in Plants*. Elsevier, London.

Clarkson, D. T., and Hanson, J. B. (1980) The mineral nutrition of higher plants. *Annu. Rev. Plant Physiol.* 31:239–298.

Delwiche, C. C. (1983) Cycling of elements in the biosphere. In: *Encyclopedia of Plant Physiology*, New Series, Vol. 15A, pp. 212–238, Läuchli, A., and Bieleski, R. L., eds. Springer-Verlag, Berlin.

Dixon, R. O. D., and Wheeler, C. T. (1986) *Nitrogen Fixation in Plants*. Chapman and Hall, New York.

Goodwin, T. W., and Mercer, E. I. (1983) *Introduction to Plant Biochemistry*. Pergamon, Oxford, England.

Hardy, R. W. F., ed. (1977) *A Treatise on Dinitrogen Fixation*. Wiley, New York.

Hewitt, E. J. (1983) A perspective on mineral nutrition. In: *Metals and Micronutrients: Uptake and Utilization by Plants*, pp. 277–323, Robb, D. A., and Pierpoint, W. S., eds. Academic Press, New York.

Läuchli, A., and Bieleski, R. L., eds. (1983) *Inorganic Plant Nutrition* (Encyclopedia of Plant Physiology, New Series, Vol. 15A,B). Springer-Verlag, Berlin.

Marschner, H. (1986) *Mineral Nutrition of Higher Plants*. Academic Press, London.

Oaks, A., and Hirel, B. (1985) Nitrogen metabolism in roots. *Annu. Rev. Plant Physiol.* 36:345–365.

Olsen, R. A., Clark, R. B., and Bennett, J. H. (1981) The enhancement of soil fertility by plant roots. *Am. Sci.* 69:378–384.

Rao, S., ed. (1988) *Biological Nitrogen Fixation. Recent Developments*. Gordon and Breach, New York.

Ullrich, W. R., Aparicio, P. J., Syrett, P. J., and Castillo, F., eds. (1987) *Inorganic Nitrogen Metabolism*. Springer-Verlag, Berlin.

CHAPTER REFERENCES

Appleby, C. A. (1984) Leghemoglobin and rhizobium respiration. *Annu. Rev. Plant Physiol.* 35:443–478.

Bauer, W. D. (1981) Infection of legumes by rhizobia. *Annu. Rev. Plant Physiol.* 32:407–449.

Beevers, L., and Hageman, R. H. (1983) Uptake and reduction of nitrate: Bacteria and higher plants. In: *Inorganic Plant Nutrition* (Encyclopedia of Plant Physiology, New Series, Vol. 15A), pp. 351–375, Läuchli, A., and Bielski, R. L., eds. Springer-Verlag, Berlin.

Bienfait, H. F., and Van der Mark, F. (1983) Phytoferritin and its role in iron metabolism. In: *Metals and Micronutrients: Uptake and Utilization by Plants*, pp. 111–123, Robb, D. A., and Pierpoint, W. S., eds. Academic Press, New York.

Brady, N. C. (1979) *Nitrogen and Rice*. International Rice Research Institute, Manila.

Burns, R. C., and Hardy, R. W. F. (1975) *Nitrogen Fixation in Bacteria and Higher Plants*. Springer-Verlag, New York.

Burris, R. H. (1976) Nitrogen fixation. In: *Plant Biochemistry*, pp. 887–908, Bonner, J., and Varner, J. E., eds. Academic Press, New York.

Chaney, R. L., Brown, J. C. and Tiffin, L. O. (1972) Obligatory reduction of ferric chelates in iron uptake by soybeans. *Plant Physiol.* 50:208–213.

Crane, F. L., Sun, I. L., Clark, M. G., Grebing, C., and Löw, H. (1985) Transplasma-membrane redox systems in growth and development. *Biochim. Biophys. Acta* 811:233–264.

Crawford, N. M., Smith, M., Bellissimo, D., and Davis, R. W. (1988) Sequence and nitrate regulation of the *Arabidopsis thaliana* mRNA encoding nitrate reductase, a metalloflavoprotein with three functional domains. *Proc. Natl. Acad. Sci. USA* 85:5006–5010.

Dilworth, M. J. (1966) Acetylene reduction by nitrogen-fixing preparations from *Clostridium pasteurianum*. *Biochim. Biophys. Acta* 127:285–294.

Emes, M. J., and Fowler, M. W. (1979) The intracellular location of the enzymes of nitrate assimilation in the apices of seedling pea roots. *Planta* 144:249–253.

Etzler, M. E. (1985) Plant lectins: Molecular and biological aspects. *Annu. Rev. Plant Physiol.* 36:209–234.

Frankhauser, H., and Brunold, C. (1978) Localization of adenosine 5′-phosphosulfate sulfotransferase in spinach leaves. *Planta* 143:285–289.

Halliwell, B. (1974) Superoxide dismutase, catalase, and glutathione peroxidase: Solutions to the problems of living with oxygen. *New Phytol.* 73:1075–1086.

Halverson, L., and Stacey, G. (1986) Signal exchange in plant-microbe interactions. *Microbiol. Rev.* 50:193–225.

Hewitt, E. J., Hucklesby, D. P., and Notton, B. A. (1976) Nitrate metabolism. In: *Plant Biochemistry*, pp. 633–681, Bonner, J., and Varner, J. E., eds. Academic Press, New York.

Heytler, P. G., Reddy, G. S., and Hardy, R. W. F. (1984) *In vivo* energetics of symbiotic nitrogen fixation in soybeans. In: *Nitrogen Fixation and CO$_2$ Metabolism*, pp. 283–292, Ludden, P. W., and Burris, J. E., eds. Elsevier, New York.

Hodson, R. C., and Schiff, J. A. (1971) Studies of sulfate utilization by algae. 9. Fractionation of a cell-free system from *Chlorella* into two activities necessary for the reduction of adenosine 3′-phosphate 5′-phosphosulfate to acid volatile radioactivity. *Plant Physiol.* 47:300–305.

Jones, O. T. G. (1983) Ferrochelatase. In: *Metals and Micronutrients: Uptake and Utilization by Plants*, pp. 125–144, Robb. D. A., and Pierpoint, W. S., eds. Academic Press, New York.

Kijne, J. W., Smit, G., Diaz, C. L., and Lugtenberg, B. J. J. (1988) Lectin-enhanced accumulation of manganese-limited *Rhizobium leguminosarum* cells on pea root hair tips. *J. Bacteriol.* 170:2994–3000.

Kleinhofs, A., Warner, R. L., and Melzer, J. M. (1989) Genetics and molecular biology of higher plant nitrate reductases. In: *Recent Advances in Phytochemistry*, Vol. 23, *Plant Nitrogen Metabolism*, pp. 117–155, Poulton, J. E., Romeo, J. T., and Conn, E., eds. Plenum, New York.

Long, S. (1989) Rhizobium-legume nodulation: Life together in the underground. *Cell* 56:203–214.

Ludwig, R. A., and de Vries, G. E. (1986) Biochemical physiology of *Rhizobium* dinitrogen fixation. In: *Nitrogen Fixation*, Vol. 4, *Molecular Biology*, pp. 50–69, Broughton, W. J., and Puhler, S., eds. Clarendon Press, Oxford.

Metzler, D. E. (1977) *Biochemistry*, pp. 615–625. Academic Press, New York.

Miflin, B. J., and Lea, P. J. (1980) Ammonia assimilation. In: *The Biochemistry of Plants: Amino Acids and Derivatives*, Vol. 5, pp. 169–202, Miflin, B. J., ed. Academic Press, New York.

Notton, B. A. (1983) Micronutrients and nitrate reductase. In: *Metals and Micronutrients: Uptake and Utilization by Plants*, pp. 219–239, Robb, D. A., and Pierpoint, W. S., eds. Academic Press, New York.

Ogren, W. L. (1984) Photorespiration: Pathways, regulation, and modification. *Annu. Rev. Plant Physiol.* 35:415–442.

Okon, Y. (1985) The physiology of *Azospirillum* in relation to its utilization as inoculum for promoting growth of plants.

In: *Nitrogen Fixation and CO$_2$ Metabolism*, pp. 165–174, Ludden, P. W., and Burris, J. E., eds. Elsevier, New York.

Pate, J. S. (1973) Uptake, assimilation and transport of nitrogen compounds by plants. *Soil Biol. Biochem.* 5:109–119.

Postgate, J. R. (1982) *The Fundamentals of Nitrogen Fixation.* Cambridge University Press, Cambridge, England.

Quispel, A. (1983) Dinitrogen-fixing symbioses with legumes, non-legume angiosperms and associative symbioses. In: *Encyclopedia of Plant Physiology*, New Series, Vol. 15A, pp. 286–329, Läuchli, A., and Bieleski, R. L., eds. Springer-Verlag, Berlin.

Rao, K. P., and Rains, D. W. (1976) Nitrate absorption by barley. 1. Kinetics and energetics. *Plant Physiol.* 57:55–58.

Rees, D. A. (1977) *Polysaccharide Shapes*, p. 52, Chapman and Hall, London.

Rennenberg, H. (1984) The fate of excess sulfur in higher plants. *Annu. Rev. Plant. Physiol.* 35:121–153.

Rice, E. L., and Pancholy, S. K. (1972) Inhibition of nitrification by climax ecosystems. *Am. J. Bot.* 59:1033–1040.

Roberts, G. R., Keys, A. J., and Whittingham, C. P. (1970) The transport of photosynthetic products from the chloroplast of tobacco leaves. *J. Exp. Bot.* 21:683–692.

Rolfe, B. G., and Gresshoff, P. M. (1988) Genetic analysis of legume nodule initiation. *Annu. Rev. Plant Physiol. Plant Mol. Biol.* 39:297–319.

Schiff, J. A. (1983) Reduction and other metabolic reactions of sulfate. In: *Inorganic Plant Nutrition* (Encyclopedia of Plant Physiology, New Series, Vol. 15A), pp. 401–421, Läuchli, A., and Bieleski, R. L., eds. Springer-Verlag, Berlin.

Schmutz, D., and Brunold, C. (1982) Regulation of sulfate assimilation in plants. XIII. Assimilatory sulfate reduction during ontogenesis of primary leaves of *Phaseolis vulgaris*. *Plant Physiol.* 70:524–527.

Schubert, K. R., Jennings, N. T., and Evans, H. J. (1978) Hydrogen reactions of nodulated leguminous plants. *Plant Physiol.* 61:398–401.

Small, I. S., and Gray, J. C. (1984) Synthesis of wheat leaf nitrite reductase *de novo* following induction with nitrate and light. *Eur. J. Biochem.* 145:291–297.

Smit, G., Kijne, J. W., and Lugtenberg, B. J. J. (1987) Involvement of both cellulose fibrils and Ca^{2+}-dependent adhesion in the attachment of *Rhizobium leguminosarum* to pea root hair tips. *J. Bacteriol.* 169:4294–4301.

Smith, R. I., Bouton, J. H., Schank, S. C. Quesenberry, K. H., Tyler, M. E., Milam, J. R., Gaskins, M. H. and Littell, R. C. (1976) Nitrogen fixation in grasses inoculated with *Spirillum lipoferum*. *Science* 193:1003–1005.

Somers, D. A., Kuo, T.-M., Kleinhofs, A., Warner, R. L., and Oaks, A. (1983) Synthesis and degradation of barley nitrate reductase. *Plant Physiol.* 72:949–952.

Sprent, J. I. (1989) Which steps are essential for the formation of functional legume nodules? *New Phytol.* 111:129–153.

Stewart, W. D. P. (1977) Blue-green algae. In: *A Treatise on Dinitrogen Fixation*, Vol. III, pp. 63–123, Hardy, R. W. F., and Silver, W. S., eds. Wiley, New York.

Trelstad, R. L., Lawley, K. R., and Holmes, L. B. (1981) Nonenzymatic hydroxylations of proline and lysine by reduced oxygen derivatives. *Nature* 289:310–312.

Von Bulow, J. F. W., and Döbereiner, J. (1975) Potential for nitrogen fixation in maize genotypes in Brazil. *Proc. Natl. Acad. Sci. USA* 72:2389–2393.

Wahl, R. C., Hageman, R. V., and Rajagopalan, K. V. (1984) The relationship of Mo, molybdopterin, and the cyanolyzable sulphur in the Mo cofactor. *Arch. Biochem. Biophys.* 230:264–273.

Werner, D., Wilcockson, J., Stripf, R., Morschel, E., and Papen, H. (1981) Limitations of symbiotic and associated nitrogen fixation by developmental stages in the systems *Rhizobium japonicum* with *Glycine max* and *Azospirillum brasilense* with grasses, e.g., *Triticum aestivum*. In: *Biology of Inorganic Nitrogen and Sulfur*, pp. 299–308, Bothe, H., and Trebst, A., eds. Springer-Verlag, Berlin.

Wilkerson, J. O., Janick, P. A., and Siegel, L. M. (1983) Siroheme-Fe$_4$S$_4$ interaction in spinach nitrite reductase (NiR). *Fed. Proc. Fed. Am. Soc. Exp. Biol.* 42:2060.

Wilson, L. G., and Reuveny, Z. (1976) Sulfate reduction. In *Plant Biochemistry*, pp. 599–632, Bonner, J., and Varner, J. E., eds. Academic Press, New York.

Zähringer, J., Baliga, B. S., and Munro, N. H. (1976) Novel mechanism for translational control in regulation of ferritin synthesis by iron. *Proc. Natl. Acad. Sci. USA* 73:857–861.

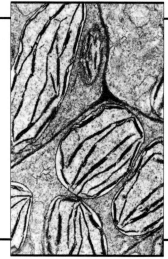

CHAPTER 13

Surface Protection and Secondary Defense Compounds

I N NATURAL HABITATS, PLANTS are surrounded by a great number of potential predators and pathogens. Nearly all ecosystems contain a wide variety of bacteria, fungi, nematodes, mites, insects, mammals, and other herbivorous animals. By their nature, plants cannot avoid these enemies simply by moving away, but they protect themselves in other ways. The cuticle and periderm, in addition to retarding water loss (Chapter 4), provide barriers to bacterial and fungal entry. Thorns, stinging hairs, and tough, leathery leaves help deter plant-feeding animals. Toxic substances belonging to a group of plant chemical compounds called secondary products kill or repel herbivores and pathogenic microbes.

In this chapter we will examine the metabolism of a variety of secondary products and their physiological and ecological roles. First, we will consider the structures and functions of several protective coverings of plant surfaces.

Cutin, Suberin, and Waxes

All plant parts exposed to the atmosphere are coated with layers of lipid material that reduce water loss and help to block the entry of pathogenic fungi and bacteria. The principal constituents of these coatings are cutin, suberin, and waxes. Cutin is found on most aboveground parts, while suberin is present on underground parts, woody stems, and healed wounds. Waxes are associated with both cutin and suberin.

Cutin, Suberin, and Waxes Are Made Up of Saturated Long-Chain Carbon Compounds

Cutin is a macromolecule, a polymer consisting of many long-chain hydroxy fatty acids that are attached to each other by ester linkages forming a rigid three-dimensional network. The fatty acids are saturated straight-chain compounds, typically with 16 or 18 carbon atoms and a hydroxyl group at the opposite end of the chain from the carboxylic acid function (Fig. 13.1).

Cutin is the main constituent of the **cuticle**, a multilayered structure that coats the outer cell walls of the epidermis on the aerial parts of all herbaceous plants (Fig. 13.2). The cuticle is composed of a top coating of wax, a thick middle layer

Examples of hydroxy fatty acids that polymerize to make cutin:

$$HOCH_2(CH_2)_{14}COOH$$

$$HOCH_2(CH_2)_8CH(CH_2)_5COOH$$
$$\mid$$
$$OH$$

Examples of hydroxy fatty acids that polymerize to make suberin:

$$HOCH_2(CH_2)_{14}COOH$$

$$HOOC(CH_2)_{14}COOH \quad \text{(a dicarboxylic acid)}$$

Examples of common wax components:

Straight-chain alkanes	$CH_3(CH_2)_{27}CH_3$
	$CH_3(CH_2)_{29}CH_3$
Fatty acid ester	$CH_3(CH_2)_{22}\overset{\displaystyle O}{\overset{\displaystyle \|}{C}}-O(CH_2)_{25}CH_3$
Long-chain fatty acid	$CH_3(CH_2)_{22}COOH$
Long-chain alcohol	$CH_3(CH_2)_{24}CH_2OH$

FIGURE 13.1. Constituents of cutin, suberin, and waxes.

containing cutin embedded in wax, and a lower layer formed of cutin and wax blended with the cell wall substances pectin and cellulose.

Waxes are not macromolecules but complex mixtures of free long-chain lipids that are extremely hydrophobic. The most common components of wax are straight-chain alkanes of 21–37 carbon atoms and fatty acid esters (Fig. 13.1). Long-chain aldehydes, ketones, free fatty acids, and free alcohols are also found. The waxes of the cuticle are synthesized by epidermal cells. They leave the epidermal cells as droplets that pass through pores in the cell wall. The top coating of cuticle wax often crystallizes in an intricate pattern of rods, tubes, or plates (Fig. 13.3).

Suberin, a polymer similar to cutin, contains long-chain acids, hydroxy acids, and alcohols. Although its structure is poorly understood, suberin appears to differ from cutin in having dicarboxylic acids (Fig. 13.1), more long-chain components, and a significant proportion of

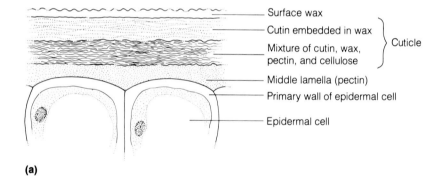

Surface wax
Cutin embedded in wax
Mixture of cutin, wax, pectin, and cellulose } Cuticle
Middle lamella (pectin)
Primary wall of epidermal cell
Epidermal cell

(a)

FIGURE 13.2. (a) Structure of the plant cuticle, the protective covering on the epidermis of leaves and young stems. (Adapted from Kolattukudy, 1980.) (b) Electron micrograph of oat coleoptile epidermal cell (×7000). Note the presence of the cuticle layers shown in the schematic drawing, except for surface waxes, which are not visible. (From N. D. Hallam.)

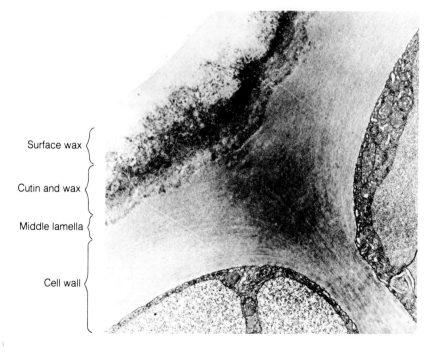

Surface wax
Cutin and wax
Middle lamella
Cell wall

(b)

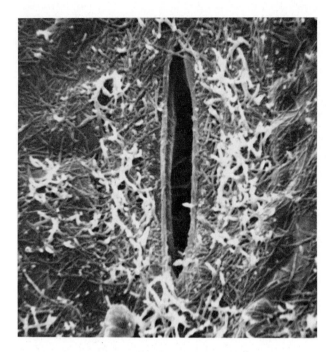

FIGURE 13.3. Surface wax deposits, which form the top layer of the cuticle, are seen in this scanning electron micrograph of the lower surface of a wheat leaf (1950 ×). Here, the wax has crystallized into long rods. In the center of the micrograph is a large stomate. (From Troughton and Donaldson, 1972.)

phenolic compounds. Suberin is a cell wall constituent found in many locations throughout the plant. We have already noted its presence in the Casparian strip of the root endodermis, which forms a barrier between the apoplast of the cortex and the stele (Chapter 4). It is a principal component of the outer cell walls of all underground organs and is associated with the cork cells of the **periderm**, the tissue that forms the outer bark of stems and roots during secondary growth of woody plants. Suberin also forms at sites of leaf abscission and in areas damaged by disease or wounding.

Cutin, Waxes, and Suberin Help Reduce Transpiration and Pathogen Invasion

Cutin, suberin, and their associated waxes form barriers between the plant and its environment that function to keep water in and pathogens out. The cuticle is very effective at limiting water loss from aerial parts of the plant but does not block transpiration completely, since even with the stomata closed some water loss occurs. The thickness of the cuticle varies with environmental conditions. Plant species native to arid areas typically have very thick cuticles, and plants from moist habitats often develop thick cuticles when grown under dry conditions.

The cuticle and suberized tissue are both important in excluding fungi and bacteria, although they do not appear to be as important in pathogen resistance as some of the other defenses we will discuss later in this chapter. Because of its waxy nature, the cuticle repels water and

may therefore slow fungal germination on plant surfaces, since most fungal spores require moist conditions for germination. After germination, some fungi produce cutinase, an enzyme that hydrolyzes cutin and thus facilitates entry of fungal hyphae into the plant (Lin and Kolattukudy, 1980).

Secondary Plant Products

Plants produce a large, diverse array of organic compounds that do not appear to have any direct function in growth and development. These substances are known as **secondary products**, secondary compounds, or secondary metabolites. Unlike primary metabolites, such as chlorophyll, amino acids, nucleotides, or simple carbohydrates, secondary products have no generally recognized roles in the processes of assimilation, respiration, transport, and differentiation discussed elsewhere in this book. Secondary products also differ from primary metabolites in having a restricted distribution in the plant kingdom. That is, particular secondary products are typically found in only one plant species or a taxonomically related group of species, whereas the basic primary metabolites are found throughout the plant kingdom.

The Principal Function of Secondary Plant Products Is to Defend Plants Against Herbivore and Pathogen Attack

For many years, the adaptive significance of most plant secondary products was unknown. These compounds were thought to be simply functionless end products of metabolism or metabolic wastes. Their study was pioneered by organic chemists of the nineteenth and early twentieth centuries who were interested in these substances, which they called "natural products," because of their importance as drugs, poisons, flavors, and industrial materials.

Beginning in the late 1960s, many secondary products were shown to have important ecological functions in plants (Harborne, 1982). Chief among these functions is protection against herbivory and infection by microbes. Secondary products have also been shown to serve as attractants for pollinators and fruit-dispersing animals and as agents of plant-plant competition. In the remainder of this chapter we will discuss the major types of plant secondary products, their syntheses, and what is known about their functions in the plant, particularly their roles in defense.

Plant Defenses Evolved Because Herbivores and Pathogens Reduce Evolutionary Fitness

We can begin by asking how plants came to have defenses in the first place. According to evolutionary biologists, plant defenses must have arisen through the well-known phenomena of heritable mutations, natural selection, and evolutionary change. Random mutations in basic metabolic pathways led to the appearance of new compounds that happened to be toxic to herbivores or pathogenic microbes. As long as the cost of producing these compounds was not detrimental, they gave the plants that possessed them greater evolutionary fitness than undefended plants. Thus, the defended plants left more descendants than undefended ones and passed their defensive traits on to the next generation. For the defended plants to predominate, herbivory or microbial infection must have significantly reduced the evolutionary fitness of plants. Experiments designed to measure the effects of plant defoliation have shown that even minor amounts of herbivory can drastically reduce growth, survivorship, and reproductive capability (Krischik and Denno, 1983). Fungal or bacterial attack has been demonstrated to have similar dramatic effects. Thus, as long as metabolic mutations occurred, the evolution of antiherbivorous and antimicrobial defenses was inevitable.

Conversely, the very defense compounds that increase the evolutionary fitness of plants by warding off fungi, bacteria, and herbivores also make them unde-

sirable as food for humans. Agriculturally important crop plants have been artificially selected for producing relatively low levels of these compounds. This, of course, makes them all the more susceptible to insects and disease.

There Are Three Principal Groups of Secondary Products Based on Biosynthetic Criteria

Plant secondary products can be divided into three groups according to their mode of biosynthesis: terpenes, phenolics, and nitrogen-containing compounds. Figure 13.4 shows in simplified form the pathways involved in the biosynthesis of secondary products and their interconnections with primary metabolism. **Terpenes** are lipids synthesized from acetyl CoA via the mevalonic acid pathway (Chapter 17). **Phenolic compounds** are aromatic substances formed via the shikimic acid pathway or the malonic acid pathway in various ways. The nitrogen-containing secondary products, such as **alkaloids**, are biosynthesized primarily from amino acids.

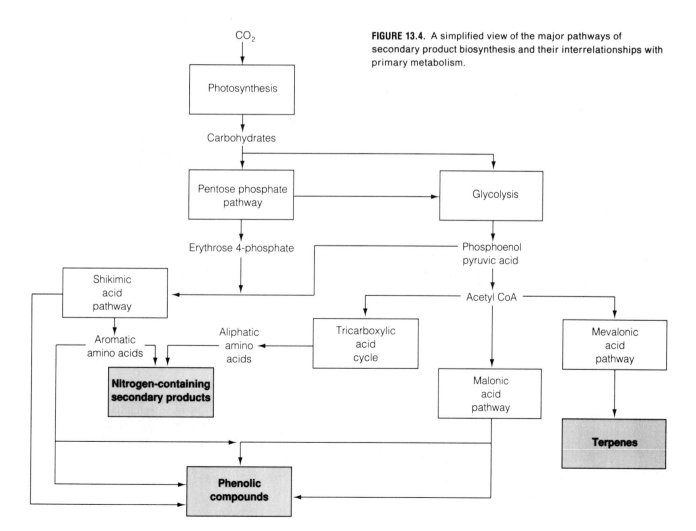

FIGURE 13.4. A simplified view of the major pathways of secondary product biosynthesis and their interrelationships with primary metabolism.

Terpenes

The terpenes or terpenoids comprise the largest class of secondary products. The diverse substances of this class are generally insoluble in water and are united by their common biosynthetic origin.

Terpenes Are Formed by the Fusion of Five-Carbon Units

All terpenes are derived from the union of 5-carbon elements that have the branched carbon skeleton of isopentane:

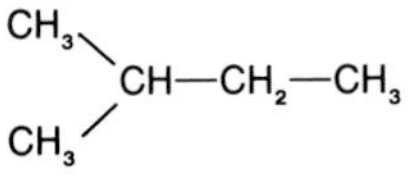

The basic structural elements of terpenes are sometimes called **isoprene units** because terpenes can decompose at high temperatures to give isoprene:

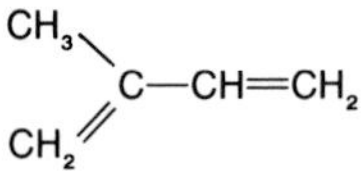

Thus, all terpenes are occasionally referred to as isoprenoids.

Terpenes are classified by the number of 5-carbon units they contain, although because of extensive metabolic modifications it is sometimes difficult to pick out the original five-carbon residues. Ten-carbon terpenes, which contain two C_5 units, are called **monoterpenes**; 15-carbon terpenes (three C_5 units) are **sesquiterpenes**; and 20-carbon terpenes (four C_5 units) are **diterpenes**. Larger terpenes include **triterpenes** (30 carbons), **tetraterpenes** (40 carbons), and **polyterpenoids** [$(C_5)_n$ carbons where $n > 20$].

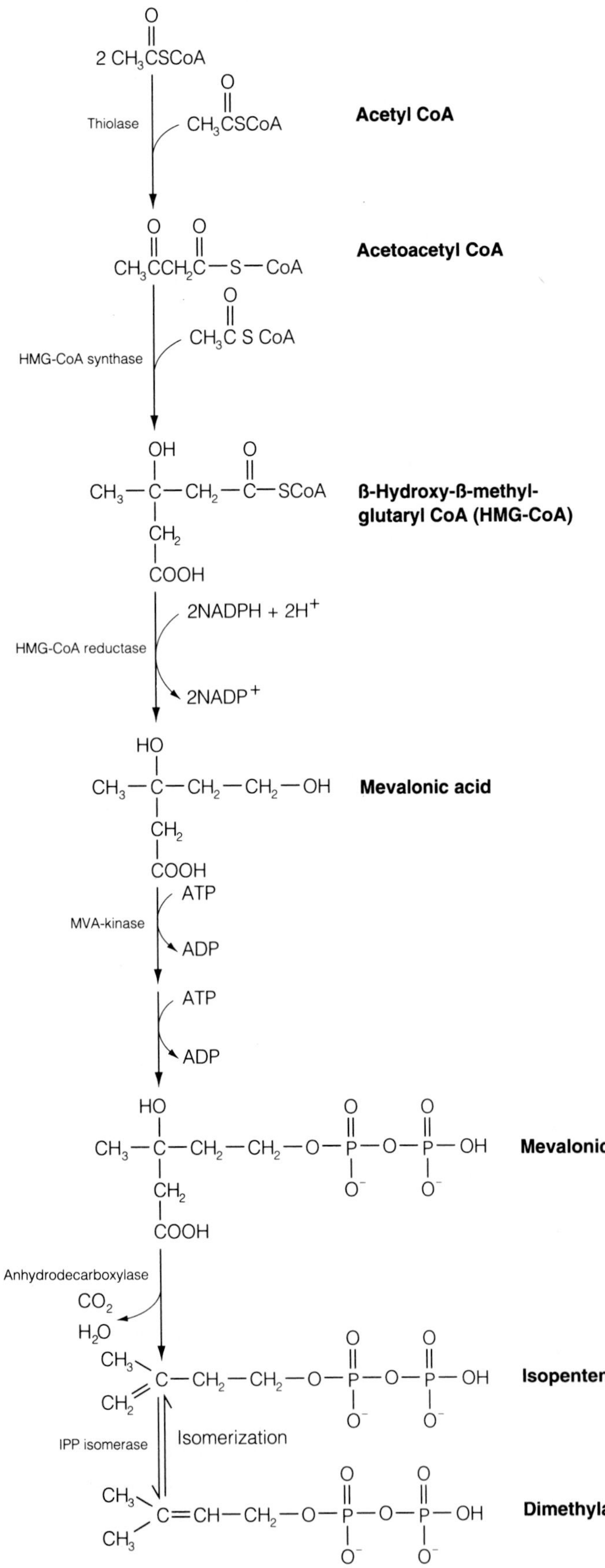

Pathway diagram (left column):

- 2 CH_3CSCoA — with Thiolase (and Acetyl CoA: CH_3CSCoA)
- CH_3CCH_2C—S—CoA — **Acetoacetyl CoA** (with HMG-CoA synthase, CH_3CSCoA)
- CH_3—C—CH_2—C—SCoA, with CH_2, COOH — **ß-Hydroxy-ß-methyl-glutaryl CoA (HMG-CoA)**; HMG-CoA reductase, 2NADPH + 2H⁺ → 2NADP⁺
- CH_3—C—CH_2—CH_2—OH, with HO, CH_2, COOH — **Mevalonic acid**; MVA-kinase, ATP → ADP, ATP → ADP
- CH_3—C—CH_2—CH_2—O—P—O—P—OH, with HO, CH_2, COOH — **Mevalonic acid-5PP (MVA-PP)**; Anhydrodecarboxylase, CO_2, H_2O
- **IsopentenylPP (IPP)**; IPP isomerase, Isomerization
- **DimethylallylPP**

FIGURE 13.5. The mevalonic acid pathway, in which the basic five-carbon units of terpenes are synthesized from acetyl CoA. Each five-carbon unit requires three molecules of acetyl CoA. PP = pyrophosphate.

The Five-Carbon Building Blocks of Terpenes Are Synthesized by the Mevalonic Acid Pathway

The biosynthesis of terpenes proceeds from acetyl CoA via the **mevalonic acid pathway** (Fig. 13.5). In this sequence, three molecules of acetyl CoA are joined together stepwise to form the key 6-carbon intermediate mevalonic acid. Mevalonic acid is pyrophosphorylated and then decarboxylated and dehydrated to yield isopentenyl pyrophosphate (isopentenylPP). IsopentenylPP and its isomer dimethyllylPP are the activated 5-carbon building blocks of terpenes that combine to form larger molecules. IsopentenylPP and dimethyllylPP react to give

geranylPP, the 10-carbon precursor of nearly all the monoterpenes (Fig. 13.6). GeranylPP can link to another molecule of isopentenylPP to give farnesylPP (C_{15}), the precursor of nearly all the sesquiterpenes.

Many terpenes have a well-characterized function in growth or development and so can be considered primary rather than secondary metabolites. For example, a sesquiterpene is a precursor of abscisic acid (Chapter 19), and the diterpene, *ent*-kaurene, is an intermediate in the synthesis of gibberellic acids (Chapter 17). Both abscisic acid and gibberellic acid are important plant hormones. Steroids are triterpene derivatives that are essential components of cell membranes, which they stabilize by interacting with phospholipids. The red, orange, and yellow carotenoids

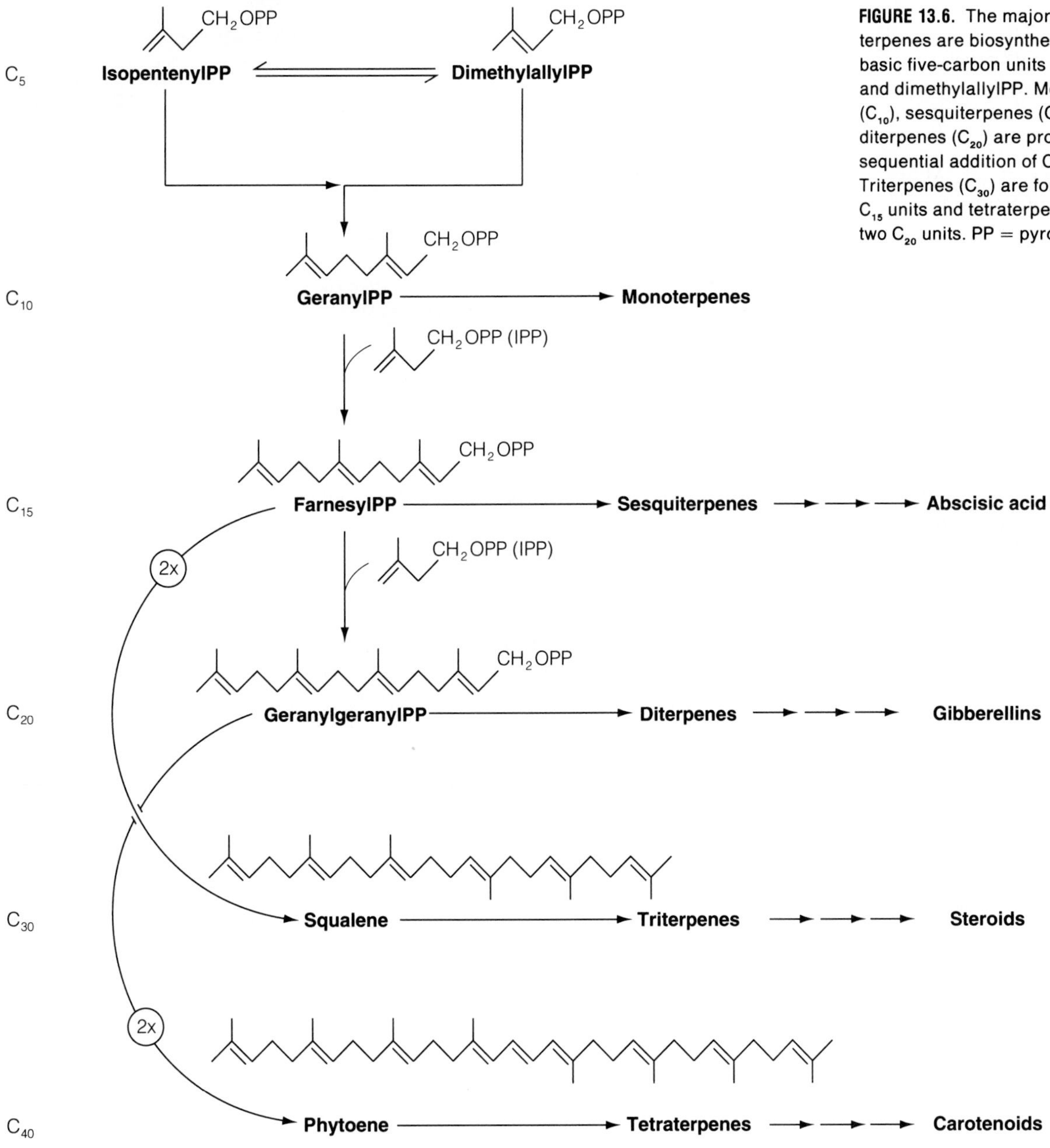

FIGURE 13.6. The major subclasses of terpenes are biosynthesized from the basic five-carbon units isopentenylPP and dimethylallylPP. Monoterpenes (C_{10}), sesquiterpenes (C_{15}), and diterpenes (C_{20}) are produced by the sequential addition of C_5 units. Triterpenes (C_{30}) are formed from two C_{15} units and tetraterpenes (C_{40}) from two C_{20} units. PP = pyrophosphate.

are tetraterpenes that function as accessory pigments in photosynthesis and protect photosynthetic tissues from photooxidation (Chapter 8). Long-chain polyterpene alcohols appear to be carriers of sugars in cell wall and glycoprotein synthesis. Terpene-derived side chains, such as the phytol side chain of chlorophyll, help anchor certain molecules in membranes. Thus, a number of terpenes have important primary roles in plants. However, the bulk of plant terpenes are secondary metabolites involved in defense.

Terpenes Serve as Antiherbivore Defense Compounds in Many Plants

Terpenes are toxins that deter a large number of plant-feeding insects and mammals and thus appear to play important defensive roles in the plant kingdom. To illustrate various aspects of these roles, we will discuss examples drawn from each of the major subclasses of terpenes.

Monoterpenes (C$_{10}$). Many monoterpenes and their derivatives are important agents of insect toxicity. For example, the monoterpene esters called **pyrethroids** that occur in the leaves and flowers of *Chrysanthemum* species show very striking insecticidal activity. Both natural and synthetic pyrethroids are popular commercial insecticides because of their low persistence in the environment and their negligible toxicity to mammals.

In conifers such as pine and fir, monoterpenes accumulate in resin ducts in the needles, twigs, and trunk. The principal monoterpenes of conifer resin are α-pinene, β-pinene, limonene, and myrcene (Fig. 13.7). These compounds are toxic to a number of insects, including bark beetles, which are serious pests of conifer species throughout the world. Many conifers respond to bark beetle infestation by producing additional quantities of monoterpenes (Fig. 13.8). The synthesis of new defensive compounds following herbivore attack is an important process analogous to the synthesis of antimicrobial compounds called phytoalexins after pathogen attack, as discussed later in this chapter.

Certain plants contain mixtures of volatile monoterpenes and sesquiterpenes, called **essential oils**, which

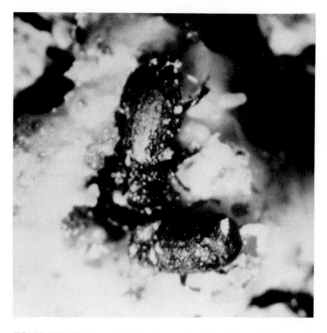

FIGURE 13.8. Photograph showing two bark beetles and a lump of newly synthesized resin on the trunk of a lodgepole pine (*Pinus contorta*) under attack by the mountain pine beetle, a species of bark beetle. Lodgepole pine produces a monoterpene-containing resin in response to bark beetle infestation. The monoterpenes are toxic to adult beetles and larvae (Raffa and Berryman, 1982). (Photograph by Ross Miller.)

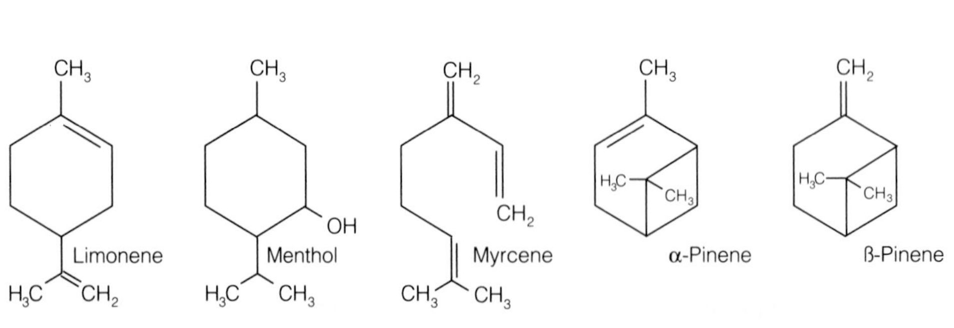

lend a characteristic odor to their foliage. Peppermint, lemon, basil, and sage are examples of plants containing essential oils. The chief monoterpene constituent of peppermint oil is menthol, and that of lemon oil is limonene (Fig. 13.7). Essential oils have well-known insect-repellent properties. They are frequently found in glandular hairs that project outward from the epidermis and serve to "advertise" the toxicity of the plant, repelling potential herbivores even before they take a trial bite (Croteau and Johnson, 1984). In the glandular hairs, the terpenes are stored in a modified extracellular space between the cuticle and the cell wall (Fig. 13.9). Essential oils can be extracted from plants by steam distillation and are important commercially in flavoring foods and making perfumes.

FIGURE 13.7. Structures of some common monoterpenes. These are common constituents of the resin of conifers and are toxic defenses against insects and other organisms that infest these plants.

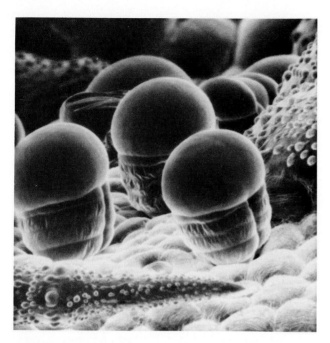

FIGURE 13.9. Monoterpenes and sesquiterpenes are commonly found in glandular hairs on the plant surface. This scanning electron micrograph shows several glandular hairs on a young leaf of Maximilian's sunflower (*Helianthus maximiliani*). Terpenes are thought to be synthesized in the cells of the hair and are stored in the rounded cap at the top. This cap is actually an extracellular space that is formed when the cuticle of the uppermost cells pulls away from the cell wall. The width of the field is 159 μm. (Photograph by G. Kreitner.)

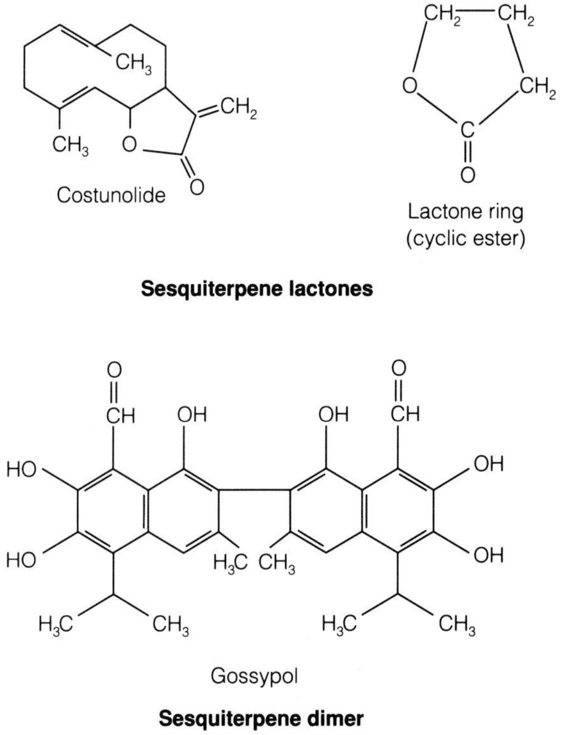

Sesquiterpenes (C_{15}). Among the many sesquiterpenes known to be antiherbivore agents are the **sesquiterpene lactones** found in the glandular hairs of members of the composite family, such as sunflower and sagebrush. These compounds are characterized by a five-membered lactone ring, a cyclic ester (Fig. 13.10). Experiments with sesquiterpene lactones have shown that they serve as strong feeding repellents to many herbivorous insects and mammals (Picman, 1986). Like many other mammalian feeding deterrents, sesquiterpene lactones taste bitter to humans. It is thought by ecologists that bitterness is not necessarily an intrinsic property of defensive plant compounds, but that through evolution, herbivores have learned to associate unpleasant tastes with certain defensive substances and thus avoid feeding on plants containing those substances.

Another defensive compound of sesquiterpene origin is **gossypol**, an aromatic sesquiterpene dimer from cotton (Fig. 13.10). Gossypol, which is found in subepidermal pigment glands, is responsible for significant insect resistance in certain varieties of cotton. Gossypol and other sesquiterpenes are also important defenses against fungal and bacterial pathogens, as we shall discuss in more detail on page 336. Recent studies in China have shown that gossypol has considerable potential as a male contraceptive in humans.

Diterpenes (C_{20}). Many diterpenes have been shown to be herbivore toxins and feeding deterrents. Plant resins, including those of pines and certain tropical leguminous trees, often contain significant amounts of diterpenes such as abietic acid (Fig. 13.11). When resin canals in the trunk are pierced by insect feeding, the outflow of resin

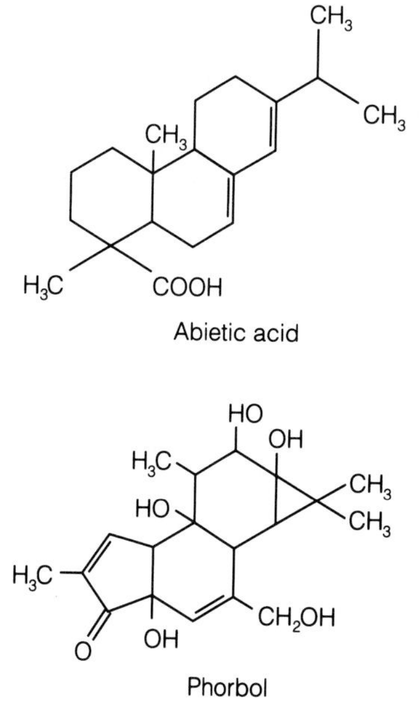

Costunolide

Lactone ring
(cyclic ester)

Sesquiterpene lactones

Gossypol

Sesquiterpene dimer

FIGURE 13.10. Structures of sesquiterpenes. Costunolide is a sesquiterpene lactone. Gossypol is a sesquiterpene dimer, a compound made up of two identical sesquiterpene units.

Abietic acid

Phorbol

FIGURE 13.11. Structures of diterpenes. Abietic acid is found in the resins of pines. Phorbol esters are protective compounds produced by species of the Euphorbiaceae, the spurge family.

may physically block feeding and serve as a chemical deterrent to continued predation. On exposure to air, the resin hardens and seals the wound. Plants in the spurge family (the Euphorbiaceae) produce diterpene esters of phorbol (Fig. 13.11) and other compounds that are severe skin irritants and internal toxins to mammals. Currently, phorbol-type diterpenes are of great interest as model tumor promoters in studies of carcinogenesis in animals.

Triterpenes (C$_{30}$). This category includes a variety of structurally diverse compounds including steroids, many of which have been modified to have fewer than 30 carbon atoms. As mentioned previously, a number of steroid alcohols (sterols) are important components of plant cell membranes. Other steroids are defensive secondary products. For example, the **phytoecdysones** are a group of plant steroids with the same basic structure as insect molting hormones (Fig. 13.12). Ingestion of phytoecdysones by insects disrupts their molting cycle, interfering with the shedding of the old exoskeleton and the production of a new one, and thereby kills them.

Another group of triterpene antiherbivore compounds are the **limonoids**, which are well known as bitter substances in citrus fruit. Perhaps the most powerful deterrent to insect feeding known is azadirachtin (Fig. 13.12), a complex limonoid from the neem tree (*Azadirachta indica*) of Africa and Asia. Azadirachtin is a feeding deterrent to some insects at doses as low as 0.1 parts per million (Butterworth and Morgan, 1971). Its potential use to limit insect damage to crops is being widely investigated.

Triterpenes that are active against vertebrate herbivores include cardenolides and saponins. **Cardenolides** are glycosides (compounds containing an attached sugar

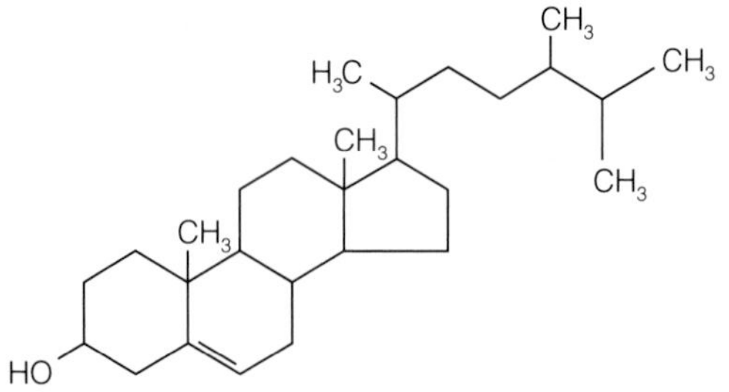

Sitosterol, a plant sterol

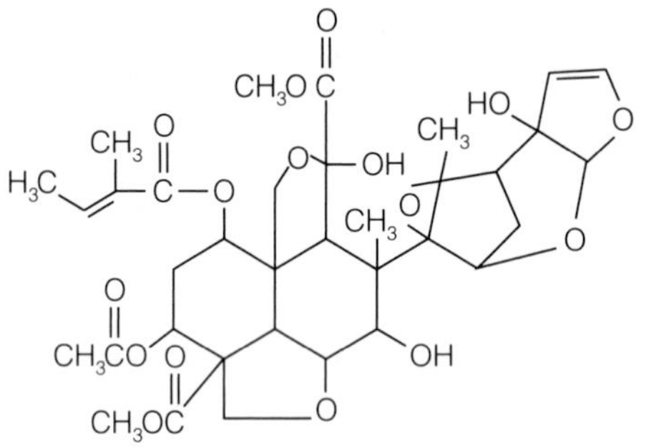

Azadirachtin, a limonoid

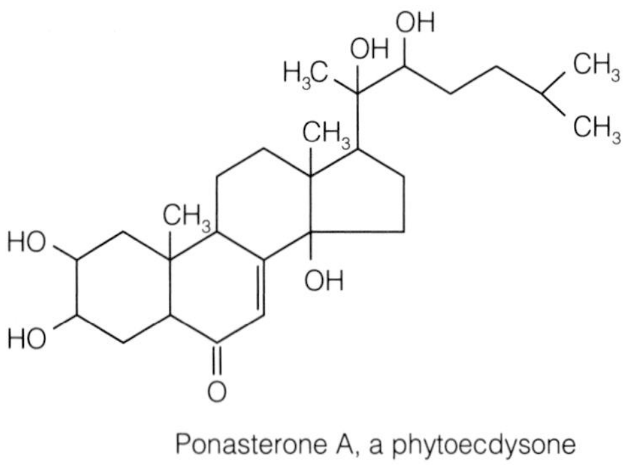

Ponasterone A, a phytoecdysone

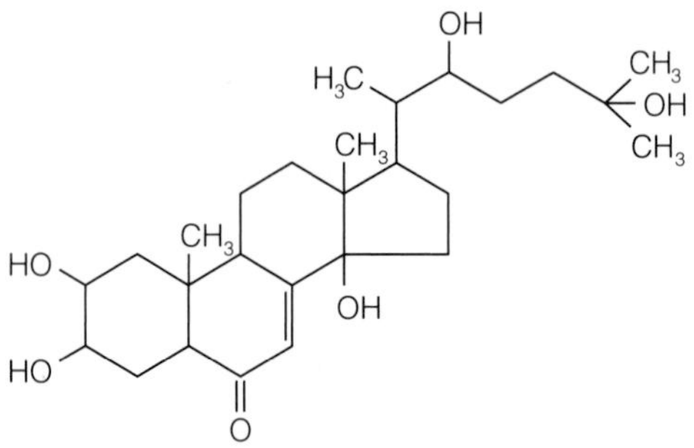

α-Ecdysone, an insect molting hormone

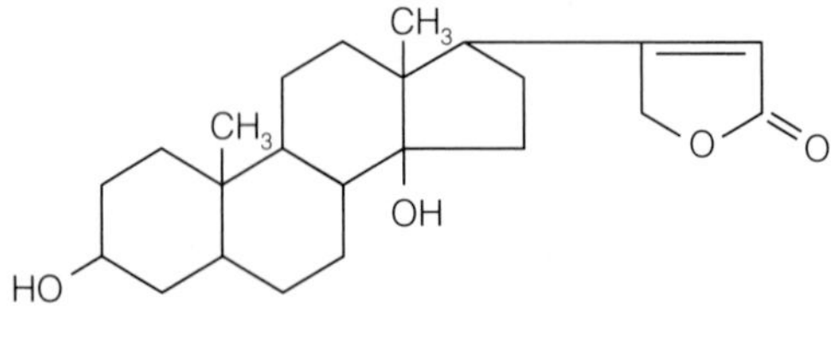

Digitoxigenin, the aglycone of digitoxin, a cardenolide

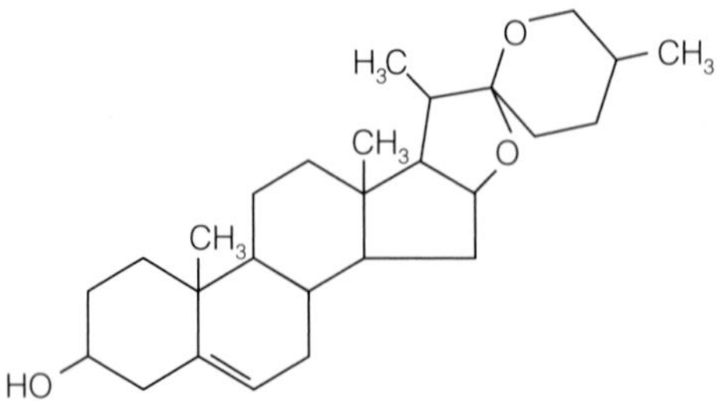

Yamogenin, a saponin

FIGURE 13.12. Structures of various triterpenes.

or sugars) that are bitter-tasting and extremely toxic to higher animals. In humans, they have dramatic effects on the heart muscle through their influence on Na$^+$/K$^+$-activated ATPases. In carefully regulated doses, they slow and strengthen the heartbeat. Cardenolides extracted from species of foxglove (*Digitalis*) are prescribed to millions of patients for the treatment of heart disease. Digitoxigenin (Fig. 13.12) is the aglycone (sugarless) triterpene portion of the naturally occurring digitanides, which contain one molecule of the rare sugar D-digitoxose and one molecule of acetyldigitoxose.

Saponins are steroid and triterpene glycosides so named because of their soaplike properties. The presence of both lipid-soluble (the triterpene) and water-soluble (the sugar) elements in one molecule gives saponins detergent properties, and they form a soapy lather when shaken with water. The toxicity of saponins is presumably a result of their ability to disrupt membranes and cause hemolysis of red blood cells. Saponins in certain varieties of alfalfa are poisonous to livestock. Saponins from the yam *Dioscorea* (e.g., yamogenin) are widely used as starting materials in the synthesis of progesterone-like compounds for birth control pills (Fig. 13.12).

Polyterpenes (C$_s$)$_n$. A number of high-molecular-weight polyterpenes occur in plants. Rubber, the best-known of these, is a polymer containing 1500 to 15,000 isopentenyl units in which all the carbon-carbon double bonds have a cis configuration (Fig. 13.13). Rubber is found in a number of plants, but the most commercially important one is the rubber tree, *Hevea brasiliensis*. Gutta rubber has its double bonds in the trans configuration and is produced by the balata plant. In the rubber tree, rubber occurs as small particles suspended in a milky fluid called latex, which is found in long vessels known as laticifers. Laticifers are found in many plant species and may contain other secondary products such as triterpenes instead of rubber. The function of laticifers in the plant has been debated since their discovery by De Bary in 1877. Since rubber is not readily mobilized, its role as a storage compound is doubtful. The most probable primary function of laticifers appears to be protection, both as a mechanism for wound healing and as a defense against herbivores and microorganisms.

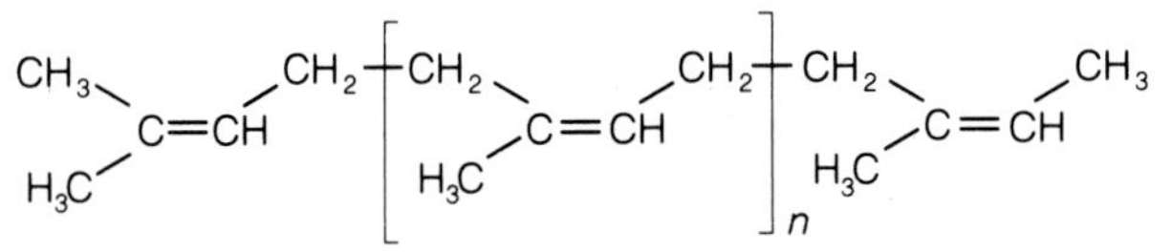

Rubber

FIGURE 13.13. Structure of rubber, a terpene polymer containing thousands of C$_s$ units. All the double bonds in rubber have a cis configuration. Other plant polyterpenes include gutta, which has similar properties except that the double bonds are all trans.

Some Herbivores Can Circumvent the Toxic Effects of Terpenes and Other Secondary Plant Products

Plant scientists have sometimes been hesistant to accept the defensive functions of terpenes and other secondary products because herbivores are often seen to eat plant parts containing high levels of these compounds without ill effect and may even be attracted to certain secondary products rather than repelled. These observations can be explained by the evolution of counteradaptations to secondary products in certain herbivore species that permit them to ingest a considerable amount of "defended" plant material without being poisoned. Some herbivores, for instance, efficiently detoxify lipophilic secondary products by converting them to water-soluble derivatives that can be excreted. Other herbivores have physiological modifications that make them no longer susceptible to the toxic effects of particular products. Still others simply feed preferentially on plant parts with low concentrations of secondary products.

Once herbivore species have developed the ability to ingest toxic compounds with impunity, they may come to specialize on plants containing those toxins, since such plants should be relatively unavailable to other herbivores. In these cases, it is often advantageous for adapted herbivores to use the distinctive secondary chemistry of their host plant species as an aid to locating that species in nature. For example, bark beetles have evolved the ability to metabolize the major monoterpenes of certain conifer species and have become specialist feeders on these trees. They are also attracted to various volatile monoterpenes of conifers (and to some metabolites of their own terpene detoxification systems) and use these compounds as cues for finding their host trees (Wood, 1982).

The ability of specialist feeders such as bark beetles to detoxify host secondary products at least partially means that a plant's secondary compounds may provide it with only partial protection against certain herbivores. Other types of defense, such as tough leaves or low nutrient concentrations, could be important deterrents to specialist herbivores. These findings do not invalidate the generalization that secondary products are protective. Secondary products still protect the plant against nonspecialist herbivores and sufficient quantities may deter specialists, since toxicity often depends on the quantity ingested.

Some herbivores not only use secondary products to find their hosts but also are adapted to store ingested secondary products for protection against their own predators. The classical example of this phenomenon is the monarch butterfly, which is a specialist feeder on milkweeds (*Asclepias*). Milkweeds contain cardenolides, which are toxic to most herbivores. Investigations by

Lincoln Brower at the University of Florida and James Seiber at the University of California, Davis have shown that, while feeding on milkweeds, monarch caterpillars accumulate cardenolides in their bodies without any observable ill effects. Both the caterpillars and the resulting brightly colored adult butterflies are toxic to predators such as birds.

Phenolic Compounds

Plants produce a large variety of secondary products containing a phenol group, a hydroxyl function on an aromatic ring:

These substances are classified as **phenolic compounds**. Plant phenolics are a chemically heterogeneous group: Some are soluble only in organic solvents, some are water-soluble carboxylic acids and glycosides, while others are large insoluble polymers.

In keeping with their chemical diversity, phenolics play a variety of roles in the plant. Many have some role in defense against herbivores and pathogens. Others function in mechanical support, in attracting pollinators and fruit dispersers, or in reducing the growth of nearby competing plants. After a brief account of phenolic biosynthesis, we will discuss several principal groups of phenolic compounds and what is known about their roles in the plant.

Most Plant Phenolics Are Synthesized from Phenylalanine, a Product of the Shikimic Acid Pathway

Plant phenolics are biosynthesized by several different routes and thus constitute a heterogeneous group from a metabolic point of view. Two basic pathways are involved: the shikimic acid pathway and the malonic acid pathway (Fig. 13.14). The shikimic acid pathway participates in the biosynthesis of most plant phenolics. The malonic acid pathway, although an important source of phenolic secondary products in fungi and bacteria, is of less significance in higher plants.

The **shikimic acid pathway** begins from simple carbohydrates and proceeds to the aromatic amino acids (Fig. 13.15). One of the pathway intermediates is shikimic acid, which has given its name to this whole sequence of reactions. The well-known herbicide glyphosate (available commercially as Round-up®) kills plants by blocking a step in this pathway (see Chapter 2, page 46). The shikimic acid pathway is present in plants, fungi, and bacteria but is not found in animals. Animals have no way to synthesize the three aromatic amino acids, phenylalanine, tyrosine, and tryptophan, which are therefore essential nutrients in animal diets.

Most classes of secondary phenolic compounds in plants are derived from phenylalanine and tyrosine, and in most plant species the key step in these syntheses is the conversion of phenylalanine to cinnamic acid by the elimination of an ammonia molecule (Fig. 13.16). This reaction is catalyzed by **phenylalanine ammonia-lyase**

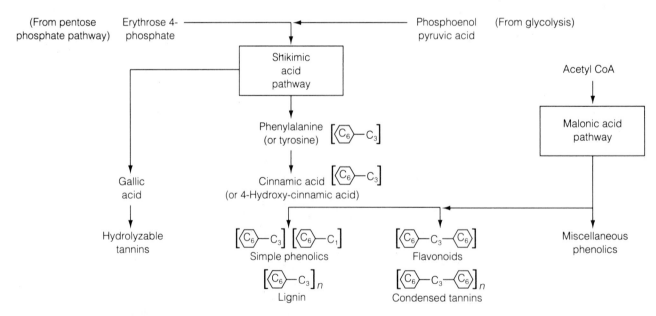

FIGURE 13.14. Plant phenolics are biosynthesized in several different ways. In higher plants, most secondary phenolics are derived at least in part from phenylalanine, a product of the shikimic acid pathway. Formulas in brackets indicate the basic arrangement of carbon skeletons: C_6 indicates a benzene ring, and C_3 is a three-carbon chain.

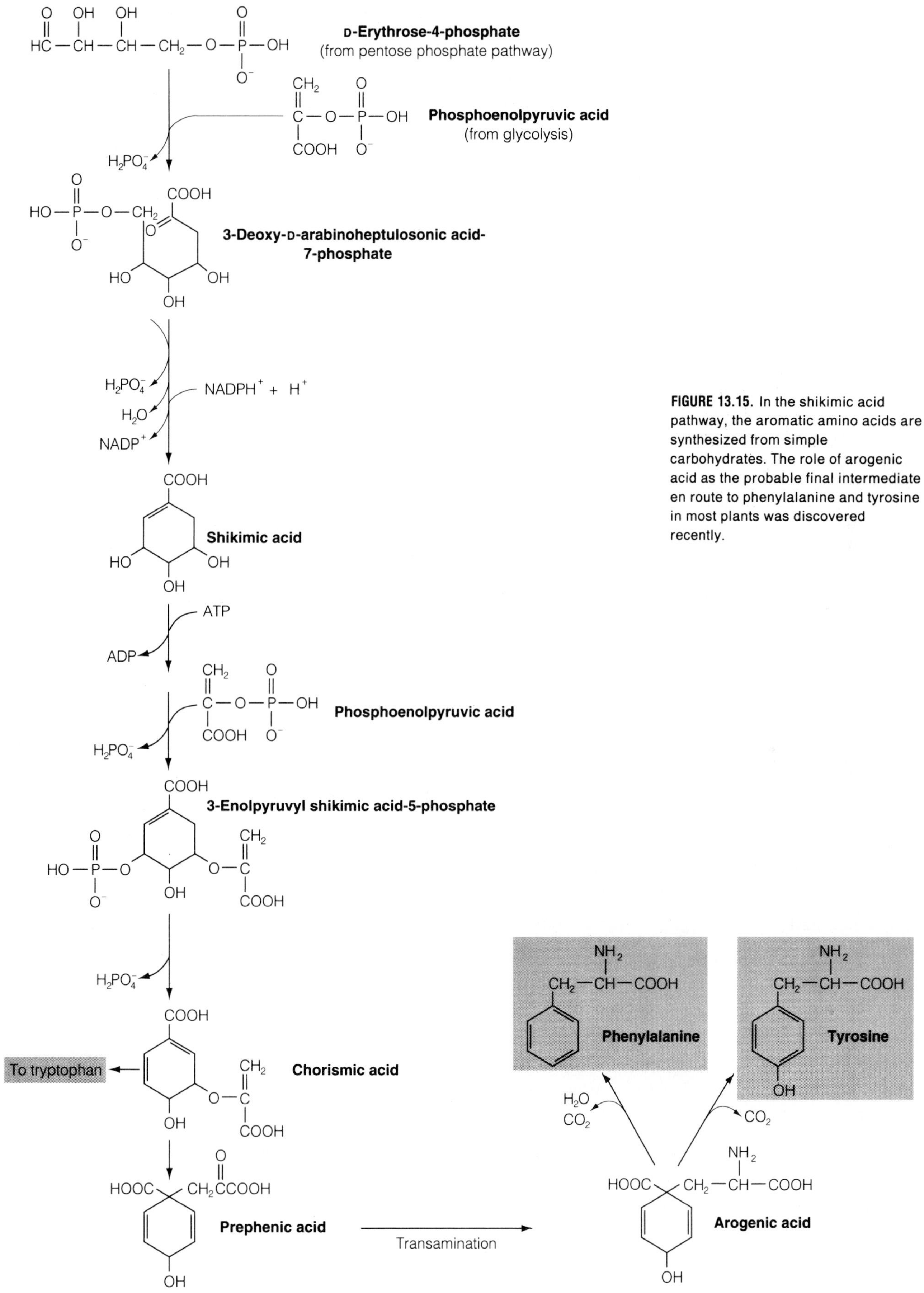

FIGURE 13.15. In the shikimic acid pathway, the aromatic amino acids are synthesized from simple carbohydrates. The role of arogenic acid as the probable final intermediate en route to phenylalanine and tyrosine in most plants was discovered recently.

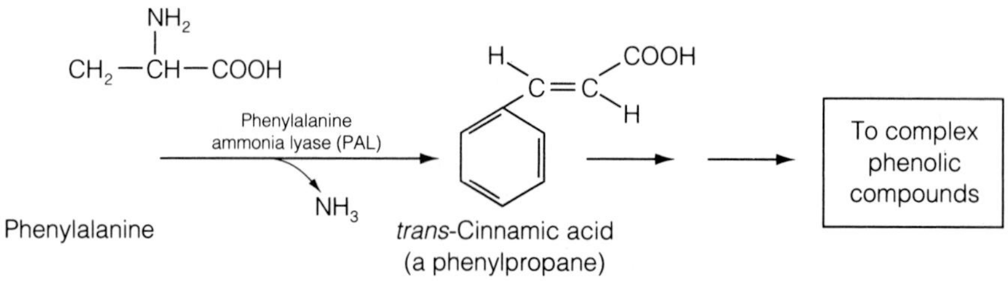

FIGURE 13.16. The deamination of phenylalanine to *trans*-cinnamic acid is an important step in phenolic biosynthesis in most species of plants. *trans*-Cinnamic acid, a phenylpropane, is an important building block of more complex phenolic compounds. This reaction is catalyzed by phenylalanine ammonia-lyase (PAL).

(PAL), an important regulatory enzyme of secondary metabolism. In a few plants, particularly grasses, the key reaction in phenolic formation appears to be the analogous conversion of tyrosine to 4-hydroxycinnamic acid.

The activity of PAL in plants is under the control of various external and internal factors, such as hormones, nutrient levels, light (through its effect on phytochrome), fungal infection, and wounding. Fungal invasion, for example, triggers the transcription of messenger RNA that codes for PAL, thus increasing the synthesis of PAL in the plant and stimulating the synthesis of phenolic compounds (Kuhn et al., 1984).

The product of PAL is *trans*-cinnamic acid, a simple C_9 plant phenolic compound known as a phenylpropane because it contains a benzene ring $\left(\langle C_6 \rangle\right)$ and a three-carbon side chain. Phenylpropanes are important building blocks of the more complex phenolic compounds discussed in the next section.

Simple Phenolics Take Part in Plant-Herbivore, Plant-Fungus, and Plant-Plant Interactions

Simple phenolic compounds are widespread in vascular plants and appear to function in different capacities. They include (1) simple phenylpropanes such as *trans*-cinnamic acid, which have a basic $\langle C_6 \rangle$—C_3 carbon skeleton; (2) phenylpropane lactones (cyclic esters) called coumarins, also with a $\langle C_6 \rangle$—C_3 skeleton; and (3) benzoic acid derivatives having a $\langle C_6 \rangle$ C_1 skeleton, formed from phenylpropanes by cleavage of a two-carbon fragment from the side chain (Fig. 13.17). As with many other secondary products, plants can elaborate on the basic carbon skeleton of simple phenolic compounds to make more complex products.

Many simple phenolic compounds have important roles in plants as defenses against insect herbivores and fungi. Of special interest is the phototoxicity of cer-

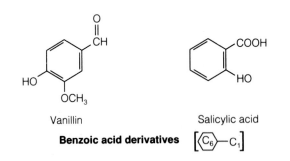

FIGURE 13.17. Some simple phenolic compounds that can act as inhibitors of plant growth. Caffeic acid and ferulic acid are present in certain soils in significant amounts. Psoralen is a furocoumarin that exhibits phototoxicity. Salicylic acid, whose acetate ester is aspirin, is abundant in many plants. The ancient headache remedy, chewing the bark of willow trees, is based on the high content of salicylic acid in this tissue. Salicylic acid was also recently identified as the thermogenic substance responsible for heat production in the inflorescences of *Arum* lilies (see box essay).

tain coumarins called furanocoumarins, which have an extra furan ring (Fig. 13.17). These compounds are not toxic until they are activated by light. Sunlight in the ultraviolet A (UV-A) region (320–400 nm) causes some furanocoumarins to become activated to a high-energy electronic state. Activated furanocoumarins can insert themselves into the double helix of DNA and bind to the pyrimidine bases cytosine and thymine, thus blocking transcription and repair and leading eventually to cell death. Phototoxic furanocoumarins are especially abundant in members of the Umbelliferae family, including celery, parsnips, and parsley. In celery, the level of these compounds can increase about 100-fold if the plant is stressed or diseased. Celery pickers, and even shoppers, have been known to develop skin rashes from handling stressed or diseased celery (Ames, 1983). Some insects have adapted to survive on plants containing phototoxic coumarins by living in rolled-up leaves, which screen out the activating wavelengths (Berenbaum, 1978).

Simple Phenolics That Escape into the Environment May Affect the Growth of Other Plants

From leaves, roots, and decaying litter, plants release a variety of primary and secondary products into the environment. Investigation of the effects of these compounds on neighboring plants is the study of **allelopathy**. If a plant can reduce the growth of nearby plants by releasing chemicals into the environment, it may increase its access to light, water, and nutrients and thus its evolutionary fitness. Generally speaking, the term allelopathy has come to be applied to the harmful effects of plants on their neighbors, although a precise definition also includes beneficial effects.

Simple $\langle C_6 \rangle$—C_1 and $\langle C_6 \rangle$—C_3 phenolic compounds are frequently cited as having allelopathic activity. Compounds such as caffeic acid and ferulic acid (Fig. 13.17) occur in soil in appreciable amounts and have been shown in laboratory experiments to inhibit the germination and growth of many plants (Rice, 1987).

In spite of results such as these, the importance of allelopathy in natural ecosystems is still controversial. Many scientists doubt that allelopathy is a significant factor in plant-plant interactions because good evidence for this phenomenon has been hard to obtain. It is easy to show that extracts or purified compounds from one plant can inhibit the growth of other plants in laboratory experiments, but it has proved very difficult to demonstrate that substances are present in the soil in sufficient concentration to cause growth inhibition. Furthermore, organic substances in the soil are often bound to soil particles and may be degraded by microbes (Dao, 1987).

Still, allelopathy is currently of great interest because of its potential agricultural applications. Reductions in crop yields caused by weeds or residues from the previous crop may in some cases be a result of allelopathy. An exciting future prospect is the development of crop plants genetically engineered to be allelopathic to weeds.

Lignin Is a Complex Phenolic Macromolecule with Both Primary and Secondary Roles

After cellulose, the most abundant organic substance in plants is **lignin**, a highly branched polymer of $\langle C_6 \rangle$—C_3 phenylpropane groups. The precise structure of lignin is not known because it is difficult to extract lignin from plants, where it is covalently bound to cellulose and other polysaccharides of the cell wall.

Lignin is generally formed from three different phenylpropane alcohols, coniferyl, coumaryl, and sinapyl alcohols (Fig. 13.18), which are synthesized from phenylalanine via various cinnamic acid derivatives. The phenylpropane alcohols are joined into a polymer through the action of enzymes called peroxidases. Peroxidases catalyze the oxidation of phenylpropane alcohols, generating free radical intermediates that combine nonenzymatically in a random fashion to give lignin. There are often multiple C—C and C—O—C bonds to each phenylpropane alcohol unit in lignin, resulting in a complex structure that branches in three dimensions. Therefore, unlike polymers such as starch, rubber, or cellulose, the units of lignin are

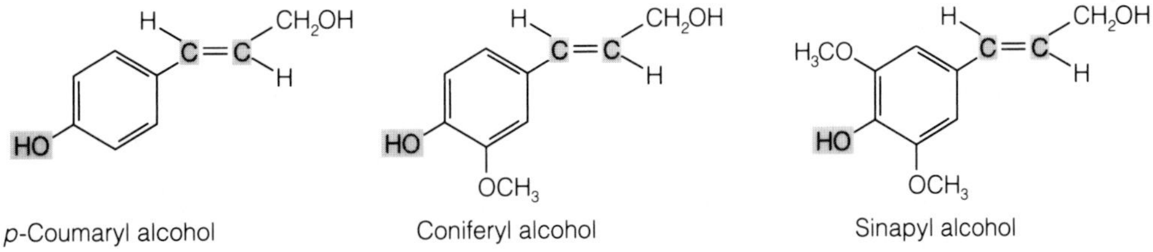

p-Coumaryl alcohol   Coniferyl alcohol   Sinapyl alcohol

FIGURE 13.18. The three phenylpropane alcohols $\left(\langle C_6 \rangle\text{—}C_3\right)$ that join to form lignin. The shaded atoms represent the sites at which linkage between the monomers occurs most frequently (see Fig. 13.19). Less frequent linkages are via the CH_2OH groups.

Salicylic Acid and Thermogenesis in *Arum* Lilies

SINCE THE TIME of Lamarck it has been known that the inflorescences of certain species of *Arum* lilies generate heat during blooming. Heat production occurs in the central column of the inflorescence called the spadix (Fig. 13.A). Prior to blooming, the spadix, which bears male and female flowers, is surrounded by a large modified bract called a spathe. Shortly after the opening of the spathe, there is a dramatic increase in the rate of respiration of the upper spadix (comparable to that of flying hummingbirds), and the temperature of the upper spadix increases by as much as 14°C. This condition lasts for about 7 hours. This extraordinary production of heat by a plant tissue apparently serves to volatize certain amines and indoles, thereby giving off a putrid odor that attracts insect pollinators. A hormone-like substance, termed "calorigen," was long thought to be responsible for the induction of the heat response (Meeuse, 1975).

Recently, Ilya Raskin and his colleagues at the Central Research and Development at du Pont de Nemours and Co. in Wilmington, Delaware, have identified salicylic acid as the agent responsible for heat production in the *Arum* spadix (Raskin et al. 1989). When salicylic acid was applied to the inflorescence exogenously, it elicited heat production by the spadix. The spadix also responded to aspirin (acetylsalicylic acid), although not quite so strongly as to salicylic acid. On the day preceding blooming, the levels of endogenous salicylic acid in the spadix increased 100-fold to 1 μg per gram fresh weight and returned to control levels at the end of the thermogenic period. Thus, salicylic acid appears to be calorigen, the heat-inducing hormone-like substance predicted by earlier workers. The increase in the respiration rate appears to be due mainly to an increase in the alternative or cyanide-resistant electron transport pathway (see Chapter 11). Studies are under way to determine whether salicylic acid may alter respiratory rates, thereby increasing heat production, in other plant species as well. In the meantime, we are left to ponder the significance of the fact that aspirin, which lowers the body temperature of humans, seems to increase the temperature of at least one plant!

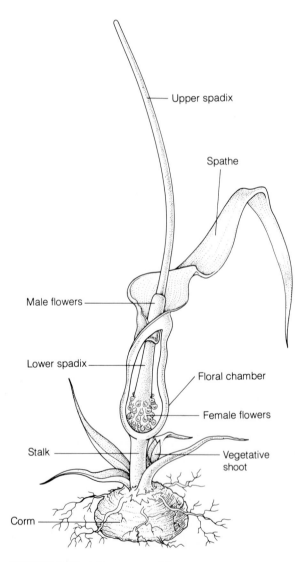

FIGURE 13.A. The inflorescence of *Sauromatum guttatum* (voodoo lily).

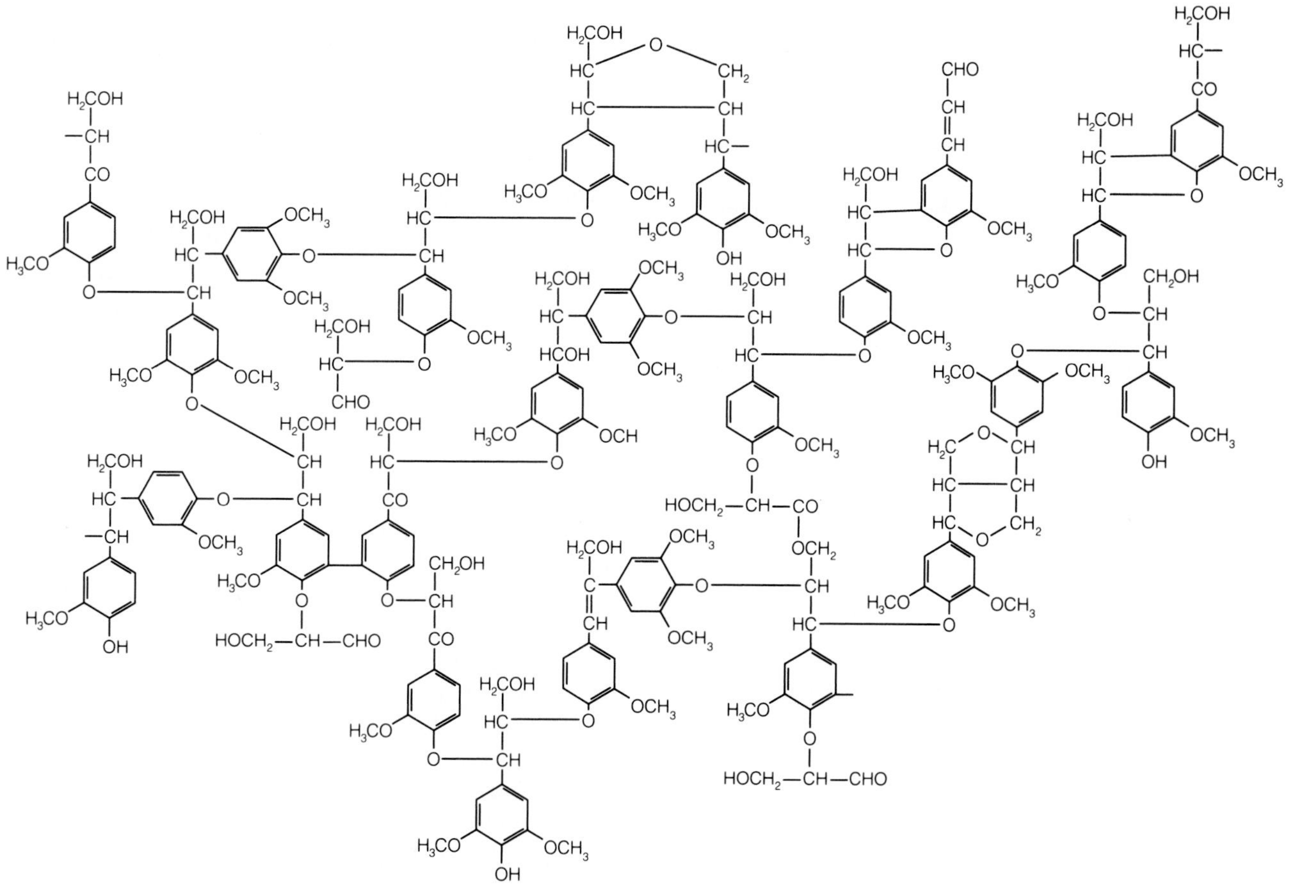

FIGURE 13.19. Partial structure of a hypothetical lignin molecule, $\left(\langle C_6\rangle - C_3\right)_n$, from European beech (*Fagus sylvatica*). The phenylpropane alcohol units that make up lignin are linked in an unorganized, nonrepeating way. The lignin of beech contains units derived from coniferyl alcohol, sinapyl alcohol, and p-coumaryl alcohol in the approximate ratios 100:70:7 and is typical of angiosperm lignin. Gymnosperm lignin contains relatively fewer sinapyl alcohol units. (Adapted from Nimz, 1974.)

not linked in an organized, repeating way, and each lignin molecule may be unique. A partial structure of a hypothetical lignin molecule is shown in Figure 13.19.

Lignin is found in the cell walls of various types of supporting and conducting tissue, notably the tracheids and vessel elements of the xylem. It is deposited chiefly in the thickened secondary wall but can also occur in the primary wall and middle lamella in close contact with the celluloses and hemicelluloses already present. The mechanical rigidity of lignin strengthens stems and vascular tissue, allowing upward growth and permitting water and minerals to be conducted through the xylem under negative pressure without collapse of the tissue. Because lignin is such a key component of water-transport tissue, the ability to make lignin must have been one of the most important adaptations permitting primitive plants to colonize dry land.

Besides providing mechanical support, lignin has significant protective functions in plants. Its physical toughness deters feeding by animals, and its chemical durability makes it relatively indigestible to herbivores. By bonding to cellulose and protein, lignin also reduces the digestibility of these substances. Lignification blocks the growth of pathogens and is a frequent response to infection or wounding.

The Flavonoids Form a Large Group of Phenolic Compounds Whose Structures Are Formed by the Action of Two Different Biosynthetic Pathways

One of the largest classes of plant phenolics is the **flavonoids**. The basic carbon skeleton of a flavonoid contains 15 carbons in a $\langle C_6\rangle - C_3 - \langle C_6\rangle$ arrangement, with two aromatic rings connected by a 3-carbon bridge (Fig. 13.20). This structure results from two separate biosynthetic pathways. The bridge and one aromatic ring (ring B) are a phenylpropane unit synthesized from phenylalanine produced in the shikimic acid pathway. The 6 carbons of the other aromatic ring (ring A) originate from three acetate units via the malonic acid pathway.

Flavonoids are classified into different groups based

primarily on the degree of oxidation of the three-carbon bridge (Fig. 13.20). We will discuss four of these groups: the anthocyanins, the flavones, the flavonols, and the isoflavonoids. The basic flavonoid carbon skeleton may have numerous substituents. Hydroxyl groups are usually present at positions 5, 7, and 4′ but may also be found at other positions. Sugars are very common as well, and, in fact, the majority of flavonoids exist naturally as glycosides. Whereas both hydroxyl groups and sugars increase the water solubility of flavonoids, other substituents such as methyl ethers or modified isopentyl units make flavonoids lipophilic. Different types of flavonoids perform very different functions in the plant, such as pigmentation and defense.

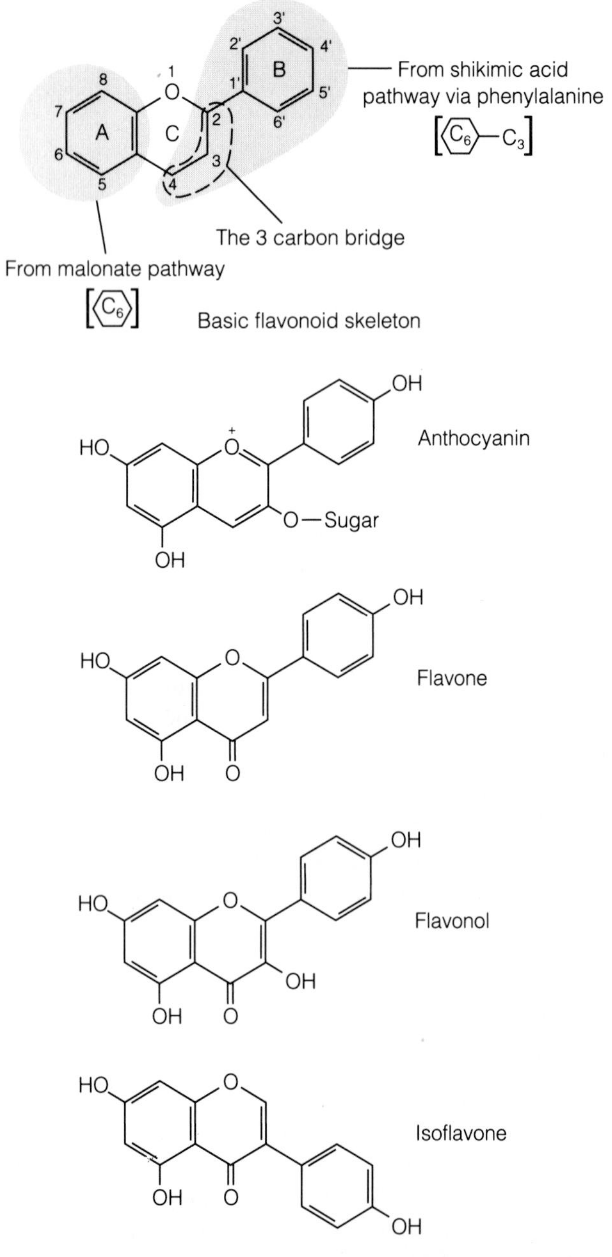

Anthocyanins Are Colored Flavonoids Found in Flowers and Fruit That Help Attract Animals for Pollination and Seed Dispersal

Interactions between plants and animals are not limited to antagonistic relationships, such as predator-prey interactions. There are mutualistic associations too. In return for the reward of ingesting nectar or fruit pulp, animals perform extremely important services for plants as carriers of pollen and seeds. Secondary products are involved in these plant-animal interactions, helping to attract animals to flowers and fruit by providing visual and olfactory signals.

The colored pigments of plants are of two principal types: carotenoids and flavonoids. Carotenoids, as we have already seen, are yellow, orange, and red terpenoid compounds that also serve as accessory pigments in photosynthesis. Flavonoids are phenolic compounds that include a wide range of colored substances. The most widespread group of pigmented flavonoids are the **anthocyanins**, which are responsible for most of the red, pink, purple, and blue colors observed in plant parts. By coloring flowers and fruits, the anthocyanins are vitally important in attracting animals for pollination and seed dispersal.

Anthocyanins are glycosides that have sugars at position 3 (Fig. 13.20) and sometimes elsewhere. Without their sugars, anthocyanins are known as **anthocyanidins**. The most common anthocyanidins and their colors are shown in Figure 13.21. Anthocyanin color is influenced

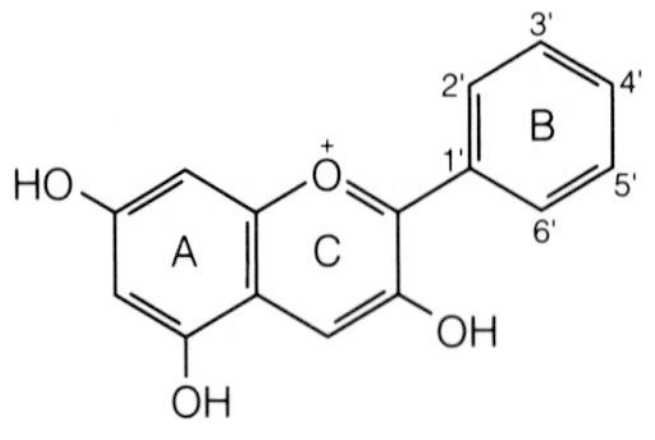

FIGURE 13.20. Carbon skeletons of major flavonoid types. Flavonoids are biosynthesized from products of the shikimic acid and malonate pathways. Positions on the flavonoid ring system are numbered as shown.

Anthocyanidin	Substituents	Color
Pelargonidin	4′—OH	Orange-red
Cyanidin	3′—OH, 4′—OH	Purplish red
Delphinidin	3′—OH, 4′—OH, 5′—OH	Bluish purple
Peonidin	3′—OCH_3, 4′—OH	Rosy red
Petunidin	3′—OCH_3, 4′—OH, 5′—OCH_3,	Purple

FIGURE 13.21. Common plant anthocyanidins and their colors. Note that these structures differ only in the substituents attached to ring B. An increase in the number of hydroxyl groups shifts absorption to a longer wavelength and gives a bluer color. Replacement of a hydroxyl group with a methoxyl group (OCH_3) shifts absorption to a slightly shorter wavelength and gives a redder color.

A Bee's-Eye View of a Golden-Eye's "Bull's-Eye"

ULTRAVIOLET ABSORBING PIGMENTS are invisible to humans, since the human eye cannot perceive ultraviolet light. Hence, patterns formed by such pigments in flowers cannot be detected by the human eye. In contrast, honeybees, which can perceive ultraviolet light, are able to detect such pigments. Plants have evolved to exploit the visual range of pollinating insects in order to attract them to their flowers. Shown in Figure 13.B are photographs of a golden-eye (*Viguiera*) as seen by humans (left) and as it might appear to honeybees (right). Special lighting was used to simulate the spectral sensitiv-

ity of the honeybee visual system. To humans, the golden-eye has yellow rays and a brown central disk; to bees, the tips of the rays appear "light yellow," the inner portion of the rays "dark yellow," and the central disk "black." Ultraviolet-absorbing flavonols are found in the inner parts of the rays but not in the tips. The distribution of flavonols in the rays and the sensitivity of insects to part of the UV spectrum contribute to the "bull's-eye" pattern seen by honeybees, which presumably helps them to locate pollen and nectar. Bees and other insects perceive not only the presence of UV-absorbing pigments but also the whole floral color pattern, which is a result of carotenoids and other flavonoid pigments as well as flavonols (McCrea and Levy, 1983).

FIGURE 13.B.

by many factors, including the number of hydroxyl and methoxyl groups in ring B of the anthocyanidin, the presence of chelating metals such as iron and aluminum, the presence of flavone or flavonol copigments, and the pH of the cell vacuole in which these compounds are stored.

Considering the variety of factors affecting anthocyanin coloration and the possible presence of carotenoids, it is not surprising that so many different shades of flower and fruit color are found in nature. The evolution of flower color may have been governed by selection pressures for different sorts of pollinators, which often

have different color preferences (see box on the visual system of bees). Color, of course, is just one type of signal used to attract pollinators to flowers. Volatile chemicals, particularly monoterpenes, frequently provide attractive scents.

Flavonoids Are Thought to Serve as Defenses and UV Protectants as Well as Attractants

Two other major groups of flavonoids found in flowers are **flavones** and **flavonols** (Fig. 13.20). These flavonoids generally absorb light at shorter wavelengths than anthocyanins and so are not visible to the human eye. However, insects such as bees, which see farther into the ultraviolet range of the spectrum than we do, could respond to flavones and flavonols as attractant cues. By using ultraviolet photography, investigators such as Morris Levy and his co-workers at Purdue University have shown that flavonols in a flower often form symmetric patterns of stripes, spots, or concentric circles called nectar guides (see box, Fig. 13.A) (McCrea and Levy, 1983). These patterns may be conspicuous to insects and are thought to help point out the location of pollen and nectar.

Flavones and flavonols are not restricted to flowers but are also present in the leaves of all green plants. It has been suggested that these two classes of flavonoids protect cells from excessive UV radiation because they absorb light strongly in the UV region while letting the visible (photosynthetically active) wavelengths pass through uninterrupted (Caldwell et al., 1983). Exposure of plants to increased UV light has been demonstrated to increase the synthesis of flavones and flavonols.

New functions of flavonoids are being discovered. For example, flavones and flavonols secreted into the soil by legume roots regulate gene expression in nodulating nitrogen-fixing bacteria. This phenomenon was described in Chapter 12. As we shall see in Chapter 16, the flavonol quercitin and the flavone apigenin have recently been implicated as endogenous regulators of polar auxin transport, again illustrating the point that secondary compounds may play important physiological as well as ecological roles.

Isoflavonoids Act as Antifungal and Antibacterial Defenses Called Phytoalexins, Which Are Synthesized Immediately Following Pathogen Infection

The **isoflavonoids** are a group of flavonoids in which the position of one aromatic ring (ring B) is shifted (Fig. 13.20). Isoflavonoids are found mostly in legumes and have several different functions. Some, such as the rotenoids, have strong insecticidal activities, whereas others have antiestrogenic effects and so cause infertility in mammals.

In the past few years, isoflavonoids have become best known for their role as **phytoalexins**, antimicrobial compounds that accumulate in high concentrations following bacterial or fungal infection and help limit the spread of the invading pathogen. Phytoalexins have several interesting characteristics:

1. Prior to infection, they are generally undetectable in the plant.

2. They are synthesized very rapidly, within hours following microbial attack.

3. Their formation is restricted to a local region around the infection site.

4. They are toxic to a broad spectrum of fungal and bacterial pathogens of plants.

Phytoalexin production appears to be a common mechanism of resistance to pathogenic microbes in a wide range of plants. However, different plant families employ different types of secondary products as phytoalexins. For example, isoflavonoids are common phytoalexins in the legume family, whereas in plants of the Solanaceae, such as potato, tobacco, and tomato, various sesquiterpenes are produced as phytoalexins (Fig. 13.22).

To respond effectively to fungal or bacterial invasion, plants must quickly recognize the presence of the pathogen and then rapidly initiate phytoalexin production. Currently, much scientific investigation is being focused on these two steps. Researchers, such as Peter Albersheim of the University of Georgia and Christopher Lamb of the Salk Institute in San Diego, have shown that polysaccharide fragments from fungal cell walls are sometimes involved in the recognition of fungal pathogens (Darvill and Albersheim, 1984). These polysaccharide fragments, which probably arise from the action of hydrolytic defense enzymes secreted by the plant, trigger the synthesis of phytoalexins. In general, substances that stimulate the synthesis of phytoalexins are referred to as **elicitors**. In addition to polysaccharide fragments, other fungal molecules including glycoproteins, peptides, and fatty acids have been found to serve as elicitors of phytoalexins in various plant species. Interestingly, fragments of pectin from the plant's own cell wall, which may act as a general signal for the presence of a wound, can also elicit phytoalexin synthesis in some species.

Now that the chemical nature of elicitors is becoming clear, researchers are beginning to study the mechanisms by which elicitors induce phytoalexin production. In several species, elicitors stimulate the transcription of messenger RNA that codes for enzymes involved in phytoalexin biosynthesis. For example, in soybean cells, which produce the isoflavonoid phytoalexin glyceollin I (Fig. 13.22), fungal elicitors trigger a dramatic increase in the rate of transcription of mRNA for phenylalanine ammonia-lyase, the chief regulatory enzyme in phenolic biosynthesis (page 330). This occurs within 1 hour after treatment of soybean cells with elicitor. Next, the various enzymes required for glyceollin I biosynthesis gradually appear at three to four times the levels in uninfected plants, and the accumulation of glyceollin I begins. Within 8 hours after exposure to elicitor, soybean cells contain sufficient concentrations of phytoalexins to inhibit pathogen growth (Ebel and Grisebach, 1988). Thus, plants do not appear to store any of the enzymatic machinery required for phytoalexin synthesis. Instead, they begin transcribing the appropriate mRNAs for the enzymes involved soon after microbial invasion.

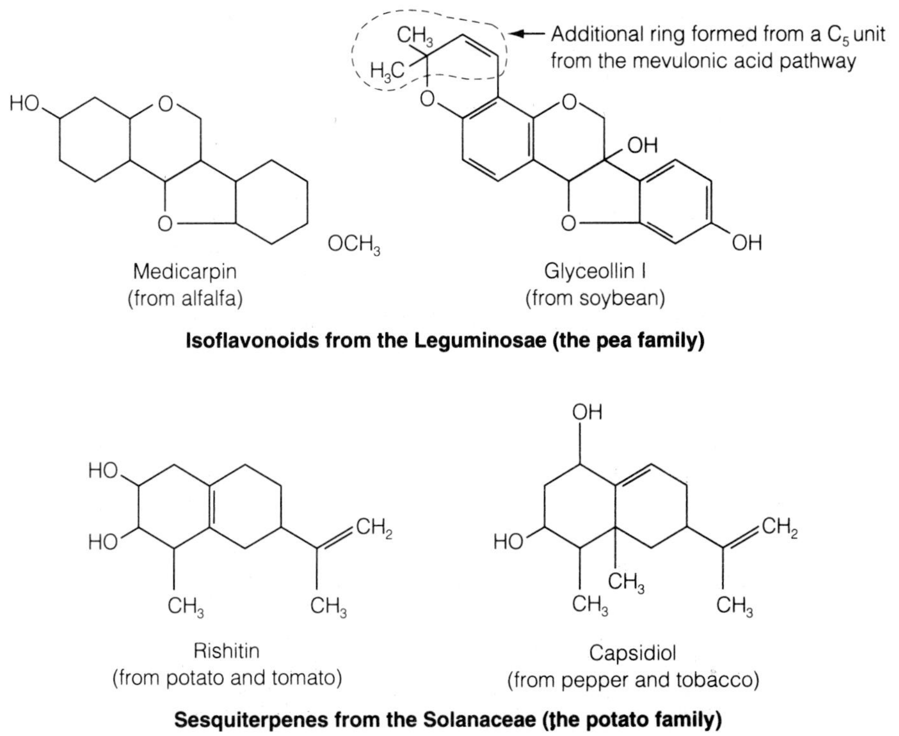

FIGURE 13.22. Structures of some phytoalexins with antimicrobial properties that are rapidly synthesized following microbial infection.

Medicarpin (from alfalfa)

Glyceollin I (from soybean)

Isoflavonoids from the Leguminosae (the pea family)

Rishitin (from potato and tomato)

Capsidiol (from pepper and tobacco)

Sesquiterpenes from the Solanaceae (the potato family)

We have seen that phytoalexins have a direct role in preventing fungal and bacterial disease in many plants. However, certain species that produce phytoalexins are still susceptible to infection. Apparently, susceptible plants sometimes do not synthesize sufficiently high concentrations of phytoalexins rapidly enough at the site of infection to stop the growth of a particular pathogen. In other cases, pathogens can specifically detoxify certain phytoalexins.

Plants possess other types of antimicrobial chemical defenses that seem to work in concert with phytoalexins. For example, many species respond to fungal or bacterial invasion by synthesizing lignin or callose. These polymers are thought to serve as barriers, walling off the fungus from the rest of the plant and physically blocking its spread. Other species produce hydrolytic enzymes, such as chitinase, which break down fungal cell walls. In addition to defenses that are induced by infection, many plants contain preformed antimicrobial secondary products that are always present, whether or not an infection is in progress.

Some researchers have suggested that phytoalexins are general plant stress metabolites because their synthesis may be induced by factors other than fungal or bacterial attack. Stresses such as wounding, freezing, high levels of UV light, and the application of fungicides or salts of various heavy metals can trigger the production of phytoalexins. Since phytoalexins are toxic to a broad range of herbivores as well as pathogens, they may serve plants as general defenses in times of stress.

Tannins Are Polymeric Phenolic Compounds That Function as Feeding Deterrents to Herbivores

Besides lignin, a second type of plant phenolic polymer with defensive properties is **tannin**. The term tannin was first used to describe compounds that could convert raw animal hides into leather in the process known as tanning. Tannins bind the collagen proteins of animal hides, increasing their resistance to heat, water, and microbes.

There are two categories of tannins, condensed and hydrolyzable. **Condensed tannins** are

$$\left(\left\langle C_6 \right\rangle - C_3 - \left\langle C_6 \right\rangle \right)_n$$

compounds formed by the linkage of flavonoid units (Fig. 13.23a). They are frequent constituents of woody plants. Condensed tannins can often be hydrolyzed to anthocyanidins by treatment with strong acids and so are called proanthocyanidins by some authors. **Hydrolyzable tannins** are heterogeneous polymers containing phenolic acids, especially gallic acid, and simple sugars (Fig. 13.23b). They are smaller than condensed tannins and may be hydrolyzed more easily; only dilute acid is needed. Most tannins have molecular masses between 600 and 3000.

Tannins are general toxins that significantly reduce the growth and survivorship of many herbivores when added to their diets. In addition, tannins act as feeding repellents to a great diversity of animals. In humans,

FIGURE 13.23. Structures of some tannins formed from phenolic acids or flavonoid units. The general structure of a condensed tannin is shown in (a), where *n* is usually 1–10. There may also be a third OH group on ring B. The hydrolyzable tannin (b) is from the sumac (*Rhus semialata*) and consists of glucose and eight molecules of gallic acid.

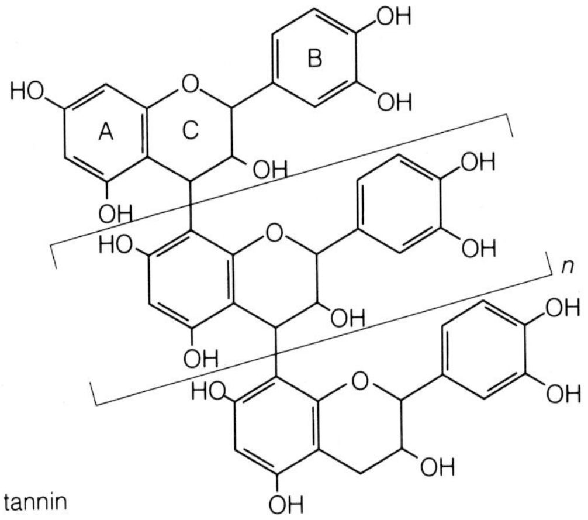

(a) Condensed tannin

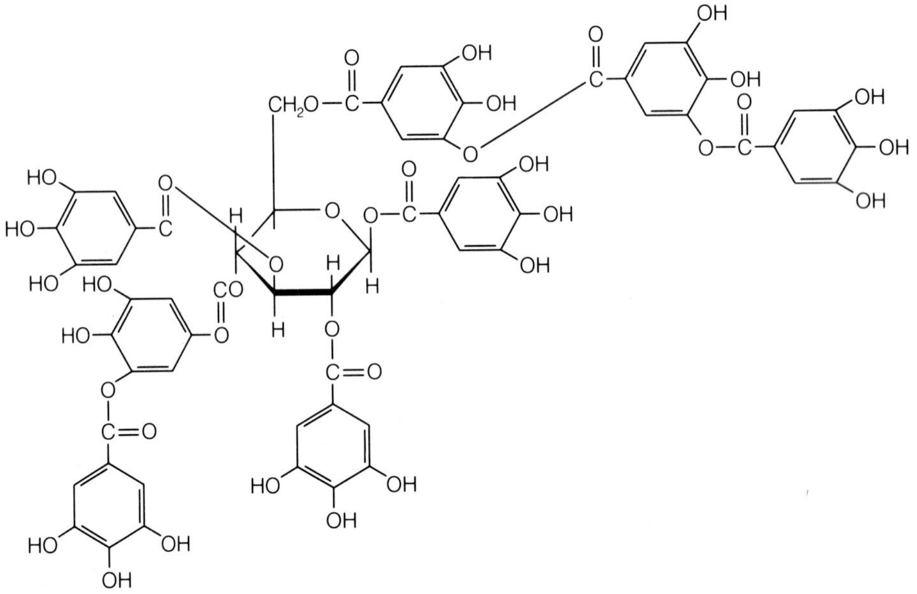

(b) Hydrolyzable tannin

tannins cause a sharp, unpleasant, astringent sensation in the mouth due to their binding of salivary proteins. Mammals, such as cattle, deer, and apes, characteristically avoid plants or parts of plants with high tannin contents (Oates et al., 1980). Unripe fruits, for instance, frequently have very high tannin levels, which may be concentrated in the "skin." Interestingly, humans often desire a certain level of astringency in tannin-containing foods such as apples, blackberries, tea, and red wine.

The defensive properties of tannins have been attributed to their ability to bind proteins. It has long been thought that plant tannins complex proteins in the guts of herbivores by forming hydrogen bonds between phenolic hydroxyl groups and sites on the protein, thus inactivating the herbivore's digestive enzymes and creating complex aggregates of tannins and plant proteins that are difficult to digest. However, the validity of this theory has been called into question by more recent experimental results. A number of animals investigated possess digestive adaptations, such as the detergent-like compounds

found in the guts of some insects, that appear to prevent tannin binding to proteins in their digestive systems (Martin et al., 1985). Therefore, it is possible that tannins do not act simply as digestibility-reducing substances but may exert their toxicity in other ways.

Plant tannins also serve as defenses against microorganisms. For example, the nonliving heartwood of many trees contains high concentrations of tannins that help prevent fungal and bacterial decay.

Nitrogen-Containing Compounds

A large variety of plant secondary products have nitrogen in their structures. Included in this category are such well-known antiherbivore defenses as alkaloids and cyanogenic glycosides, which are of considerable interest because of their toxicity to humans and their medicinal properties. Most nitrogenous secondary products are biosynthesized from common amino acids.

Alkaloids Are Nitrogen-Containing Heterocyclic Compounds That Have Marked Physiological Effects on Animals

Alkaloids are organic nitrogen-containing compounds found in 20–30% of vascular plants. The nitrogen atom in these substances is usually part of a **heterocyclic ring,** a ring containing both nitrogen and carbon atoms. Most alkaloids are alkaline, as their name would suggest. At pH values commonly found in the cytosol (where the pH is approximately 7) or the vacuole (pH usually 5–6), the nitrogen atom is protonated; hence, alkaloids are positively charged and are generally water-soluble.

Alkaloids are usually synthesized from one of a few common amino acids, in particular, aspartic acid, lysine, ornithine, tyrosine, and tryptophan. However, much of the carbon skeleton of some alkaloids is derived from the mevalonic acid pathway. The major alkaloid types and their amino acid precursors are given in Table 13.1. The biosynthetic pathway leading to nicotine is of importance not only because of the harmful effects of nicotine as a component of tobacco smoke but also because the

TABLE 13.1. Major types of alkaloids, their amino acid precursors, and well-known examples of each type.

Alkaloid Class	Structure	Biosynthetic Precursor	Examples
Pyrrolidine		Ornithine	Nicotine
Tropane		Ornithine	Atropine, cocaine
Piperidine		Lysine (or acetate)	Coniine
Pyrrolizidine		Ornithine	Retrorsine
Quinolizidine		Lysine	Lupinine
Isoquinoline		Tyrosine	Codeine, morphine
Indole		Tryptophan	Psilocybin, reserpine, strychnine

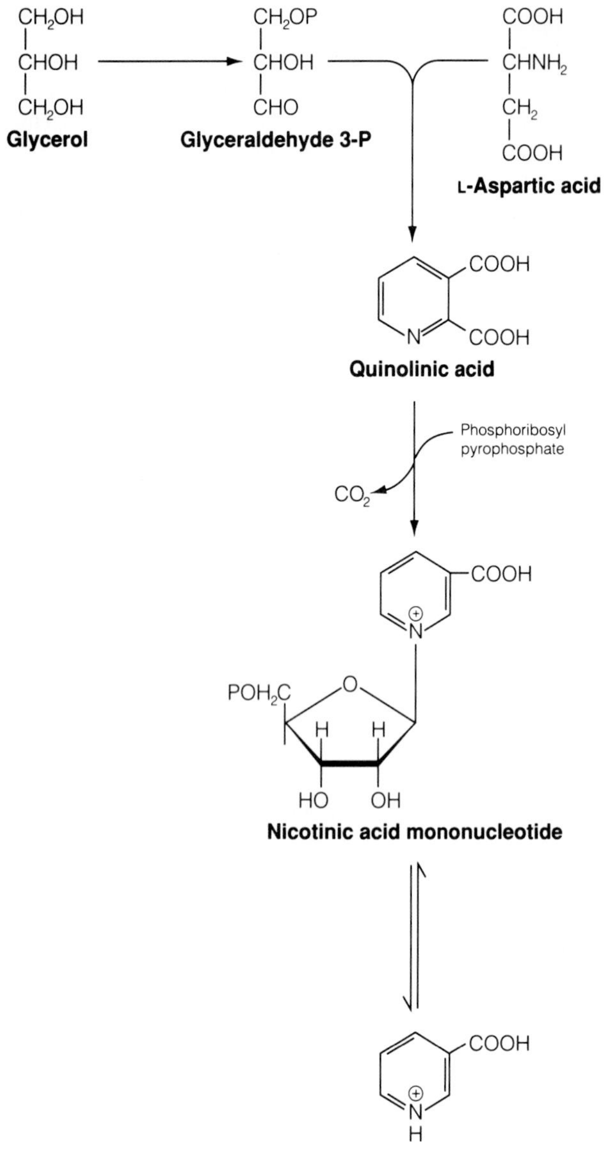

FIGURE 13.24. The biosynthesis of nicotinic acid, a precursor of the alkaloid nicotine. Nicotinic acid is also a component of NAD and NADP, important participants in biological oxidation-reduction reactions. The pathway shown is thought to operate in both plants and bacteria.

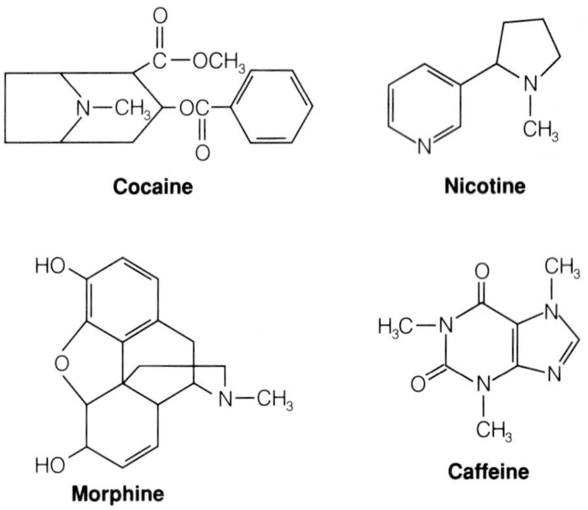

FIGURE 13.25. Examples of alkaloids, a diverse group of secondary products that contain nitrogen, often in a heterocyclic ring. Note that caffeine is a purine-type alkaloid similar to the nucleic acid bases adenine and guanine.

B vitamin nicotinic acid (niacin) is a precursor of this alkaloid. Nicotinic acid is a constituent of NAD and NADP, which serve as electron carriers in metabolism. In plants and bacteria, nicotinic acid is probably synthesized from aspartic acid and glyceraldehyde 3-phosphate (Fig. 13.24), whereas animals and higher fungi utilize tryptophan as a precursor (Robinson, 1981).

The role of alkaloids in plants has been a subject of speculation for at least 100 years. Alkaloids were once thought to be nitrogenous wastes (analogous to urea and uric acid in animals), nitrogen storage compounds, or growth regulators, but there is little evidence to support any of these functions. Many alkaloids are toxic to herbivores, so they may function as defenses against predators, especially mammals.

Large numbers of livestock deaths are caused by the ingestion of alkaloid-containing plants. In the United States, significant percentages of all grazing livestock are poisoned each year by feeding on alkaloid-containing plants such as lupines (*Lupinus*), larkspur (*Delphinium*), and groundsel (*Senecio*). This may be due to the fact that domestic animals, unlike wild animals, have not been subjected to natural selection for the avoidance of toxic plants. Indeed, some livestock even prefer alkaloid-containing plants to less harmful forage.

All alkaloids, when taken in sufficient quantity, are also toxic to humans. For example, strychnine, atropine, and coniine (from poison hemlock) are classical alkaloid poisoning agents. However, at lower doses many are useful pharmacologically. Morphine, codeine, atropine, and ephedrine are just a few of the plant alkaloids currently used in medicine. Other alkaloids, including cocaine, nicotine, and caffeine (Fig. 13.25), enjoy widespread nonmedical use as stimulants or sedatives.

On a cellular level, the mode of action of alkaloids is quite variable. Many bind to nerve receptors and affect neurotransmission; others affect membrane transport, protein synthesis, or miscellaneous enzyme activities.

Alkaloids are not solely defensive substances. A group of red- and yellow-colored alkaloids called betalains are flower and fruit pigments found in taxonomically related plant families, including cactuses and beets. Plants of these families that contain betalain pigments do not contain any anthocyanins (Kimler et al., 1970).

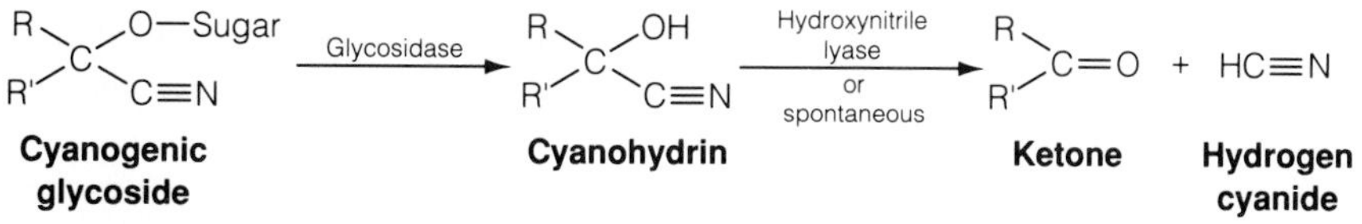

FIGURE 13.26. The enzyme-catalyzed hydrolysis of cyanogenic glycosides to hydrogen cyanide. R and R′ represent various alkyl or aryl substituents.

Cyanogenic Glycosides Release Hydrogen Cyanide When the Plant Tissue Containing Them Is Damaged by Herbivores

Various nitrogenous protective compounds other than alkaloids are found in plants. Two groups of these substances, cyanogenic glycosides and glucosinolates, are not in themselves toxic but are readily broken down to give off volatile poisons when the plant is crushed. Cyanogenic glycosides release the well-known poisonous gas hydrogen cyanide (HCN).

The breakdown of cyanogenic glycosides in plants is a two-step enzymatic process. Species that make cyanogenic glycosides also make the enzymes necessary to hydrolyze the sugar and liberate HCN. In the first step, the sugar is cleaved by a **glycosidase**, an enzyme that separates sugars from other molecules to which they are linked (Fig. 13.26). The resulting hydrolysis product, called an α-hydroxynitrile, can decompose spontaneously at a low rate to liberate HCN, or this second step can be accelerated by the enzyme **hydroxynitrile lyase**.

Cyanogenic glycosides are not normally broken down in the intact plant because the glycoside and the degradative enzymes are spatially separated. In sorghum the cyanogenic glycoside dhurrin is present in the vacuoles of epidermal cells, whereas the hydrolytic and lytic enzymes are found in the mesophyll (Kojima et al., 1979). Under ordinary conditions, this compartmentation prevents decomposition of the glycoside. However, when the leaf is damaged, as during herbivore feeding, the cell contents of different tissues mix, and HCN is formed. Cyanogenic glycosides are widely distributed in the plant kingdom and are frequently encountered in legumes, grasses, and species of the rose family.

Considerable evidence indicates that cyanogenic glycosides have a protective function in certain plants. HCN is a fast-acting toxin that blocks cellular respiration by binding to the iron-containing heme group of cytochrome oxidase and other respiratory enzymes. The presence of cyanogenic glycosides deters feeding by insects and other herbivores, such as snails and slugs.

Studies of bird's-foot trefoil (*Lotus corniculatus*), which produces cyanogenic glycosides, showed that in some natural populations not every plant is cyanogenic and, not surprisingly, herbivores significantly preferred to feed on the noncyanogenic individuals (Jones et al., 1978). The persistence of the noncyanogenic form in such populations is not easy to explain. Perhaps the natural selection pressure exerted by herbivores is not sufficiently high to remove all of the noncyanogenic plants. When herbivores are scarce, the noncyanogenic form may well have higher evolutionary fitness because it does not divert any of its energy and nitrogen to the production of unneeded cyanogenic glycosides.

Glucosinolates or Mustard Oil Glycosides Also Release Volatile Toxins When the Plant Is Damaged

A second class of plant glycosides that breaks down to release volatile defensive substances consists of the **glucosinolates** or mustard oil glycosides. Found principally in crucifers and related plant families, glucosinolates give off the compounds responsible for the smell and taste of such vegetables as cabbage, broccoli, and radishes. The release of these mustard-smelling volatiles from glucosinolates is catalyzed by a hydrolytic enzyme, called a **thioglucosidase**, that cleaves glucose from its bond with the sulfur atom (Fig. 13.27). The resulting **aglycone**, the nonsugar portion of the molecule, rearranges to give pungent and chemically reactive products including isothiocyanates, thiocyanates, and nitriles, depending on

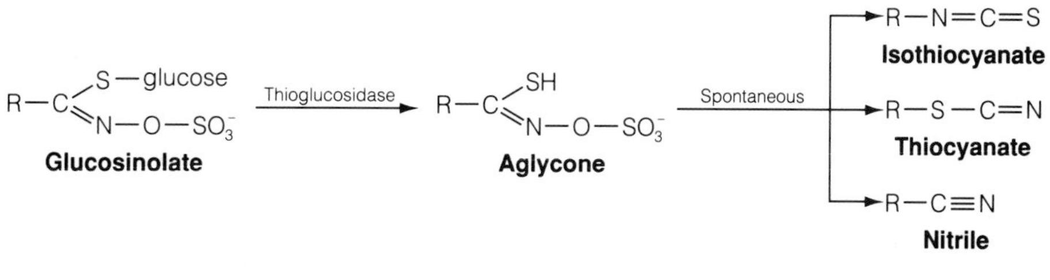

FIGURE 13.27. The hydrolysis of glucosinolates to mustard-smelling volatiles. R represents various alkyl or aryl substituents.

the conditions of hydrolysis. These products function in defense as herbivore toxins and feeding repellents. Like cyanogenic glycosides, glucosinolates are stored in the intact plant separately from the enzymes that hydrolyze them and are brought into contact with these enzymes only when the plant is crushed.

As with other secondary products, certain animals are adapted to feed on plants containing glucosinolates without ill effects. For adapted herbivores, such as the cabbage aphid, glucosinolates often serve as attractants and stimulate feeding and egg-laying. Cabbage aphids will not feed on any plant or synthetic diet that does not contain some glucosinolates (van Emden, 1972).

Certain Plants Contain Unusual Amino Acids That Are Not Constituents of Proteins but Function as Antiherbivore Defenses

Plants and animals incorporate the same 22 amino acids into their proteins. However, many plants also contain unusual amino acids, called nonprotein amino acids, that are not incorporated into proteins but are present instead in the free form and act as protective substances. Nonprotein amino acids are often very similar to common protein amino acids. Canavanine, for example, is a close analog of arginine, and azetidine-2-carboxylic acid has a structure very much like that of proline (Fig. 13.28).

Nonprotein amino acids exert their toxicity in various ways. Some block the synthesis or uptake of protein amino acids; others can be mistakenly incorporated into proteins. The basis of canavanine's toxicity to herbivores has been well studied in Gerald Rosenthal's laboratory at the University of Kentucky. After ingestion, canavanine is recognized by the herbivore's enzyme that normally binds arginine to the arginine transfer RNA molecule. In this way, canavanine becomes bound to a tRNA with the arginine anticodon and is incorporated into proteins in place of arginine. The usual result is a protein that is

nonfunctional either because its tertiary structure is disrupted or because of other changes that affect its catalytic properties. Canavanine, whose isoelectric point (pI) is 8.2, is less basic than arginine (pI = 10.8) and may alter an enzyme's ability to bind substrates or catalyze chemical reactions.

Plants that synthesize nonprotein amino acids are not susceptible to the toxicity of these compounds. The jack bean (*Canavalia ensiformis*), which synthesizes large amounts of canavanine in its seeds, has protein-synthesizing machinery that can discriminate between canavanine and arginine and does not incorporate canavanine into its own proteins. Some insects that specialize on plants containing nonprotein amino acids have similar biochemical adaptations.

Certain Plant Proteins Selectively Inhibit the Protein-Digesting Enzymes of Herbivores

Defensive proteins that block the action of an herbivore's proteolytic enzymes are found in legumes, tomatoes, and other plants. These substances, known as **proteinase inhibitors,** enter the herbivore's digestive tract and hinder protein digestion by binding tightly and specifically to the active site of protein-hydrolyzing enzymes such as trypsin and chymotrypsin.

In tomatoes, the synthesis of proteinase inhibitors is induced by herbivory. Feeding by insects leads to rapid accumulation of proteinase inhibitors throughout the plant, even in undamaged areas far from the initial feeding site. Research in Clarence A. Ryan's laboratory at Washington State University has shown that wounded tomato leaves produce a hormone-like substance that is transported out of the leaf and throughout the plant via the vascular system. This substance triggers the biosynthesis of proteinase inhibitors that discourage further predation (Ryan, 1983).

Earlier in the chapter we discussed another example of the production of defensive compounds in response to

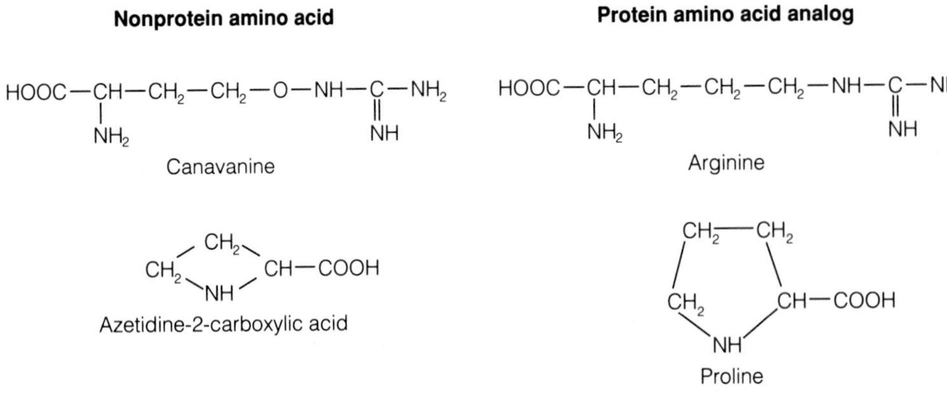

FIGURE 13.28. Nonprotein amino acids and their protein amino acid analogs. The nonprotein amino acids are not incorporated into proteins but are found free in the cell.

insect damage: the synthesis of resin monoterpenes in conifers infested by bark beetles. Such induced defenses, which are produced only after initial herbivore damage, should require a smaller investment of plant resources than defenses that are always present.

Many Plant Secondary Products Can Be Catabolized Back to Primary Metabolic Intermediates

It was once assumed that all secondary products are metabolically inactive wastes or storage products that accumulate unchanged during the life of the plant. This view now appears to be largely incorrect, except in the case of polymers such as tannins, lignin, or rubber. According to more recent research, various secondary products are degraded to simple primary metabolic intermediates at some time after their formation and thus may function in part as reserves of energy, nitrogen, or fixed carbon.

Evidence for the catabolism of secondary products comes from two types of investigations. First, many plants show dramatic seasonal, monthly, or even daily oscillations in the amounts of secondary products present. Second, when radioactively labeled precursors of secondary products are fed to plants, the radioactivity incorporated into secondary products is usually lost after several hours or days. Therefore, like most primary metabolites, many secondary products appear to be dynamic metabolites in plants.

Several nitrogen-containing secondary products are found in high concentrations in seeds, where they represent a significant portion of the total nitrogen content of the seed. Immediately following germination, these compounds are catabolized. Apparently they act as reserves for the growth of the young seedling when their role in protecting the seed from predation ends. Jack bean seeds, for example, can contain as much as 8% canavanine by weight, which represents over one-third of the total nitrogen in the seed. After germination, all this canavanine is translocated into the young plant and catabolized (Rosenthal and Rhodes, 1984).

The Distribution of Defensive Secondary Products Within Plants

The extent of qualitative and quantitative variation in the secondary products found in plant species is enormous and somewhat bewildering. Variability is apparent at several levels of organization. First, different species of plants usually have different complements of secondary products. Second, a single plant frequently contains a mixture of secondary substances, with varying amounts in different organs, tissues, and cells. Third, at the level of

the individual organ or tissue, the amounts and types of secondary products often vary with age or environmental conditions. Finally, there is a variable distribution at the subcellular level. A number of theories have been proposed to explain the patterns of secondary product distribution observed in plants. One of these theories is that plants need to store their secondary metabolites in such a way that they do not accidentally poison themselves.

Plants Are Not Usually Poisoned by Their Own Toxic Secondary Metabolites

Many aspects of the physiology and biochemistry of plant and animal cells are fundamentally similar. Because of this, a large number of plant defensive compounds are potentially as toxic to the plant that produces them as they are to herbivores. For instance, tannins can readily bind to plant proteins, many terpenes inhibit plant growth, and the release of HCN in a plant blocks cellular respiration. Therefore, plants are potentially susceptible to being poisoned by their own secondary products. They avoid this fate in several different ways.

Some secondary products, including cyanogenic glycosides and glucosinolates, are stored as inactive precursors separate from their activating enzymes. In other cases, plants that synthesize particular defensive compounds have modified enzymes, receptors, or other cellular components that are insensitive to the toxic effects of these compounds. For example, plants that produce the nonprotein amino acid canavanine have an altered arginine aminoacyl-tRNA synthetase capable of discriminating between arginine and canavanine.

Perhaps the most common way in which plants avoid the toxic effects of secondary products is to store these products away from sensitive metabolic processes. Many types of water-soluble secondary products, such as phenolics, alkaloids, and most other nitrogen-containing compounds, are deposited in the vacuole. It has been suggested that one of the purposes of glycosylation (the addition of sugars) is to make toxic compounds sufficiently water-soluble that they can be safely sequestered in the vacuole. Lipid-like secondary products, including many terpenes and nonpolar phenolics, accumulate in various extracellular sites, such as epidermal waxes, glandular hairs, resin canals, and laticifers.

Summary

Plants produce a great variety of chemical compounds, called secondary products, that apparently are not involved directly in growth or development. Scientists have only recently begun to learn about the roles of these substances in plants. Secondary products appear to function mainly as defenses against predators and pathogens.

There are three major groups of secondary products: terpenes, phenolic compounds, and nitrogen-containing compounds. Terpenes are composed of five-carbon units synthesized via the mevalonic acid pathway. Many plant terpenes are toxins and feeding deterrents to herbivores. Phenolic compounds, which are synthesized primarily from products of the shikimic acid pathway, have several important roles in plants. Tannins, lignin, flavonoids, and some simple phenolic compounds serve as defenses against herbivores and pathogens. Lignin also mechanically strengthens cell walls, and many flavonoid pigments are important attractants for pollinators and seed dispersers. Some simple phenolic compounds affect the growth of neighboring plants and thus have allelopathic activity. Nitrogen-containing secondary products, such as alkaloids, cyanogenic glycosides, glucosinolates, nonprotein amino acids, and proteinase inhibitors, protect plants from a variety of herbivorous animals. These compounds are synthesized principally from common amino acids.

Although not usually considered secondary products, cutin, suberin, and waxes form protective coverings on plants that limit transpiration and help block fungal attack.

For millions of years, plants have produced defenses against herbivory and microbial attack. Well-defended plants have tended to leave more survivors than poorly defended plants, and so the capacity to produce effective defensive products has become widely established in the plant kingdom. In response, many species of herbivores and microbes have evolved the ability to feed on or infect plants containing secondary products without being adversely affected, and this herbivore and pathogen pressure has in turn selected for new defensive products in plants. Along with their synthesis of defensive products, plants have had to develop adaptations to prevent themselves from being poisoned by their own defenses.

The study of plant secondary products has many practical applications. Many of these substances that are physiologically active against herbivores or pathogens are now used as insecticides, fungicides, or pharmaceuticals. By breeding increased levels of secondary products directly into crop plants, it may be possible to reduce the need for certain costly and potentially harmful pesticides. On the other hand, the possible toxicity effects of high levels of naturally occurring defense compounds to humans must also be considered.

GENERAL READING

Ames, B. N. (1983) Dietary carcinogens and anticarcinogens. *Science* 221:1256–1264.

Croteau, R. B., and Johnson, M. A. (1984) Biosynthesis of terpenoids in glandular trichomes. In: *Biology and Chemistry of Plant Trichomes*, Rodriguez, E., Healey, P. L., and Mehta, I., eds., pp. 133–185. Plenum Press, New York.

Darvill, A. G., and Albersheim, P. (1984) Phytoalexins and their elicitors—a defense against microbial infection in plants. *Annu. Rev. Plant Physiol.* 35:243–275.

Ebel, J., and Grisebach, H. (1988) Defense strategies of soybean against the fungus *Phytophthora megasperma* f.sp. *glycinea*: A molecular analysis. *Trends Biochem. Sci.* 13:23–27.

Harborne, J. B. (1982) *Introduction to Ecological Biochemistry*, 2d ed. Academic Press, New York.

Kolattukudy, P. E. (1980) Cutin, suberin and waxes. In: *The Biochemistry of Plants*, Vol. 4, *Lipids: Structure and Function*, Stumpf, P. K., ed., pp. 571–645. Academic Press, New York.

Meeuse, B. J. D. (1975) Thermogenic respiration in Aroids. *Annu. Rev. Plant Physiol.* 26:117–126.

Rice, E. L. (1987) Allelopathy: An overview. In: *Allelochemicals: Role in Agriculture and Forestry*, Waller, G. R., ed., pp. 8–22. ACS Symposium Series No. 330, American Chemical Society, Washington, D.C.

Robinson, T. (1981) *The Biochemistry of Alkaloids. Molecular Biology, Biochemistry and Biophysics*, 2d ed., Vol. 3, Kleinzeller, A., Springer, G. F., and Wittmann, H. G., eds. Springer-Verlag, Berlin.

Wood, D. L. (1982) The role of pheromones, kairomones and allomones in the host selection and colonization behavior of bark beetles. *Annu. Rev. Entomol.* 27:411–446.

CHAPTER REFERENCES

Berenbaum, M. (1978) Toxicity of a furanocoumarin to armyworms: A case of biosynthetic escape from insect herbivores. *Science* 201:532–534.

Butterworth, J. H., and Morgan, E. D. (1971) Investigation of the locust feeding inhibition of the seeds of the neem tree, *Azadirachta indica*. *J. Insect Physiol.* 17:969–977.

Caldwell, M. M., Robberecht, R., and Flint, S. D. (1983) Internal filters: Prospects for UV-acclimation in higher plants. *Physiol. Plant.* 58:445–450.

Dao, T. H. (1987) Sorption and mineralization of plant phenolic acids in soil. In: *Allelochemicals: Role in Agriculture and Forestry*, Waller, R. R., ed., pp. 358–370. ACS Symposium Series No. 330, American Chemical Society, Washington, D.C.

Jones, D. A., Keymer, R. J., and Ellis, W. M. (1978) Cyanogenesis in plants and animal feeding. In: *Biochemical Aspects of Plant and Animal Coevolution*, Harborne, J. B., ed., pp. 21–34. Academic Press, London.

Kimler, L., Mears, J., Mabry, T. J., and Rösler, H. (1970) On the question of the mutual exclusiveness of betalains and anthocyanins. *Taxon.* 19:875–878.

Kojima, M., Poulton, J. E., Thayer, S. S., and Conn, E. E. (1979) Tissue distributions of dhurrin and of enzymes involved in its metabolism in leaves of *Sorghum bicolor*. *Plant Physiol.* 63:1022–1028.

Krischik, V. A., and Denno, R. F. (1983) Individual, population and geographic patterns in plant defense. In: *Variable Plants and Herbivores in Natural and Managed Systems*, Denno, R. F., and McClure, M. S., eds., pp. 463–512. Academic Press, New York.

Kuhn, D. N., Chappell, J., Boudet, A., and Hahlbrock, K. (1984) Induction of phenylalanine ammonia-lyase and 4-coumarate:CoA ligase mRNAs in cultured plant cells by UV light or fungal elicitor. *Proc. Natl. Acad. Sci. USA* 81:1102–1106.

Lin, T. S., and Kolattukudy, P. E. (1980) Isolation and characterization of a cuticular polyester (cutin) hydrolyzing enzyme from phytopathogenic fungi. *Physiol. Plant Pathol.* 17:1–15.

Martin, M. M., Rockholm, D. C., and Martin, J. S. (1985) Effects of surfactants, pH and certain cations on precipitation of proteins by tannins. *J. Chem. Ecol.* 11:485–494.

McCrea, K. D., and Levy, M. (1983) Photographic visualization of floral colors as perceived by honeybee pollinators. *Am. J. Bot.* 70:369–375.

Nimz, H. (1974) Beech lignin—proposal of a constitutional scheme. *Angew. Chem. Int. Ed.* 13:313–321.

Oates, J. F., Waterman, P. G., and Choo, G. M. (1980) Food selection by the South Indian leaf-monkey, *Presbytis johnii*, in relation to leaf chemistry, *Oecologia* 45:45–56.

O'Brien, T. P. (1967) Observations on the fine structure of the oat coleoptile. I. The epidermal cells of the extreme apex. *Protoplasma* 63:385–416.

Picman, A. K. (1986) Biological activities of sesquiterpene lactones. *Biochem. Syst. Ecol.* 14:255–281.

Raffa, K. F., and Berryman, A. A. (1982) Physiological differences between lodgepole pines resistant and susceptible to the mountain pine beetle and associated microorganisms. *Environ. Entomol.* 11:486–492.

Raskin, I., Turner, I. M., and Melander, W. R. (1989) Regulation of heat production in the inflorescences of an *Arum* lily by endogenous salicylic acid. *Proc. Natl. Acad. Sci. USA* 86:2214–2218.

Rosenthal, G. A., and Rhodes, D. (1984) L-Canavanine transport and utilization in developing jack bean, *Canavalia ensiformis* (L.) DC. (Leguminosae). *Plant Physiol.* 76:541–544.

Ryan, C. A. (1983) Insect-induced chemical signals regulating natural plant protection responses. In: *Variable Plants and Herbivores in Natural and Managed Systems*, Denno, R. F., and McClure, M. S., eds., pp. 43–60. Academic Press, New York.

Troughton, J., and Donaldson, L. A. (1972) *Probing Plant Structure*. McGraw-Hill, New York.

van Emden, H. F. (1972) Aphids as phytochemists. In: *Phytochemical Ecology*, Harborne, J. B., ed., pp. 25–43. Academic Press, London.

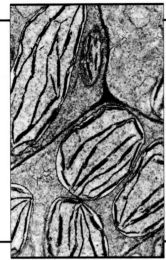

CHAPTER 14

Stress Physiology

I T IS USEFUL TO begin a discussion of stress with some definitions, since the concept is often used imprecisely and the terminology can be confusing. **Stress** is usually defined as an external factor that exerts a disadvantageous influence on the plant. In most cases, stress is measured in relation to growth (biomass accumulation) or to the primary assimilation processes (CO_2 and mineral uptake), which are related to overall growth. However, it is helpful to keep in mind that special situations may require a different evaluation. For example, plants given large amounts of water and fertilizer can grow so tall that they are blown over in windstorms (an effect known as lodging) and become unharvestable. In this situation, lower water and fertilizer levels that are suboptimal for biomass production are beneficial because they result in the highest harvestable yields. One could therefore argue that in this case the higher water and fertilizer levels, not the lower ones, represent "stress."

Because stress is defined solely in terms of plant responses (sometimes called **strain** in conformance with engineering terminology), the concept of stress is intimately associated with that of **stress resistance**, which is the plant's fitness for the unfavorable environment. An environment that is stressful for one plant may not be so for another. Consider, for example, the pea (*Pisum sativum*) and the soybean (*Glycine max*), which grow best at about 20°C and 30°C, respectively. As the temperature increases, the pea shows signs of heat stress much sooner than the soybean. Thus, the soybean has greater heat stress resistance. If resistance increases as a result of exposure to prior stress, the plant is said to be **acclimated** (or **hardened**). Acclimation must be distinguished from adaptation, which is a genetically determined level of resistance acquired over generations by selection.

Under both natural and agricultural conditions, plants are constantly exposed to stress. Some environmental factors (such as air temperature) can become stressful in just a few minutes, whereas others may take days to weeks (soil water) or even months (some mineral nutrients) to become stressful. It has been estimated that, because of physicochemical stress, the yield of field-grown crops in the United States is only 22% of the genetic potential yield (Boyer, 1982). The physiological processes underlying adaptation and acclimation, as well as the mechanisms of stress injury, are thus of immense practical importance.

In this chapter we examine ways in which plants respond to water deficit, chilling and freezing, heat, salinity, oxygen deficiency in the root zone, and air pollution. The focus is mostly on the physiology of stress resistance. While it is convenient to examine each of these factors separately, many are interrelated. For example, water deficit is often associated with root-zone salinity and/or with leaf heat stress. We will also see that plants often display **cross-resistance**, or resistance to one stress induced by acclimation to another. This behavior implies that mechanisms of resistance to several stresses share many common features.

Water Deficit and Drought Resistance

Drought resistance mechanisms have been divided into several types. First we can distinguish between **desiccation postponement** (ability to maintain tissue hydration) and **desiccation tolerance** (ability to function while dehydrated), which are sometimes referred to as drought tolerance at high and low water potentials, respectively. The older literature often refers to "drought avoidance" as opposed to "drought tolerance," but these terms are misnomers because drought is a meteorological condition that is tolerated by all plants that survive it and avoided by none. A third category, **drought escape**, comprises plants that complete their life cycles during the wet season, before the onset of drought. These are the only true "drought avoiders."

Among the desiccation postponers are *water savers* and *water spenders*. The water savers use water conservatively, preserving some in the soil for use late in the life cycle, whereas the water spenders aggressively consume water, often using prodigious quantities. The mesquite tree (*Prosopis* species) is a well-known example. This extremely deep-rooted species has ravaged semiarid rangelands in the southwestern United States and, because of its profligate water use, has prevented the reestablishment of grasses with agronomic value.

Drought Resistance Strategies Vary with Climatic or Soil Conditions

There are regions of the world, such as portions of the Sahara desert of northern Africa, that receive on average 5 mm of rainfall or less per year. Of course, this extreme aridity is atypical of most of the world's land area, particularly arable land that is used for crops or pastures. Nonetheless, most of the world's agriculture is subject to drought problems. Arid and semiarid zones are defined as areas in which plant transpiration totals 50% or less of the transpiration that would occur with unlimited water availability. In these areas water is the major factor limiting plant growth, and stress is usually alleviated by irrigation whenever possible. Unfortunately, water deficit is not limited to such regions, as even in wetter climatic zones the irregular distribution of rainfall leads to periods in which water availability limits growth (Boyer, 1982). The year-to-year variability of yields in most rain-fed agricultural areas illustrates the severity of the problem (Table 14.1).

TABLE 14.1. Yields of corn and soybean crops in the United States.

Year	Crop yield (percentage of 10-year average)		
	Corn	Soybeans	
1979	104	106	
1980	87	88	Severe drought
1981	104	100	
1982	108	104	
1983	77	87	Severe drought
1984	101	93	
1985	112	113	
1986	113	110	
1987	114	111	
1988	80	89	Severe drought

Adapted from U.S. Department of Agriculture (1989).

The water-limited productivity of plants will depend on the total amount of water available and the water-use efficiency of the plant (Chapters 4 and 10). A plant that is capable of acquiring more water or that has higher water-use efficiency will have greater stress resistance to drought. Some plants possess adaptations, such as the C4 and CAM modes of metabolism, that allow them to exploit more arid environments. In addition, plants possess acclimation mechanisms that are activated in response to water stress.

When water deficit develops slowly enough to allow changes in developmental processes, water stress has several effects on growth. Of particular importance is a specific limitation to leaf expansion. Although leaf area is important because photosynthesis is usually proportional to it, rapid leaf expansion can adversely affect water availability. If precipitation occurs only during the winter and spring, and summers are dry, young plants in the springtime will have ample soil moisture, and stress will increase rapidly in summer. Accelerated early growth can lead to large leaf areas, rapid early water depletion, and too little residual soil moisture for the plant to complete its life cycle. In this situation, only plants that retain some water for reproduction late in the season or that complete the life cycle very quickly before the onset of drought (drought escape) will produce seeds for the next generation. Either strategy will allow some reproductive success.

The situation is different if significant but erratic summer rainfall is the prevalent pattern. In this case a plant with large leaf area, or one capable of developing large leaf area very quickly, is better suited to take advantage of occasional wet summers. One acclimation strategy in these conditions is a capacity for both vegetative growth and flowering over an extended period. Such plants are said to be **indeterminate** in their growth habit, in contrast to **determinate** plants, which develop preset numbers of leaves and flower only over very short periods. Sorghum (*Sorghum bicolor*) is more drought-resistant than corn (*Zea mays*) in part because sorghum readily develops secondary shoots that can continue to grow and flower after the main shoot is fully mature. Corn does not form secondary shoots, so if fruiting fails because of drought the corn plant has no opportunity later to recoup the losses should rain occur. In two of the three drought years shown in Table 14.1 the soybean yields were less affected than the corn yields, partly because the indeterminacy of the soybeans allowed them to take advantage of late-occurring rains to a greater extent than the corn. For maximum reproductive success in an erratic environment, then, a plant must be able to respond rapidly to both stress and relief of stress.

Decreased Leaf Area Is an Early Response to Water Deficit

The earliest responses to stress appear to be mediated by biophysical events rather than by changes in chemical reactions due to dehydration. As the water content of the plant decreases, the cells shrink and the cell walls relax (Chapter 3). This decrease in cell volume results in lower hydrostatic pressure, or turgor. As water loss progresses and the cells contract further, the solutes in the cells become more concentrated. The plasma membrane becomes thicker and more compressed, as it covers a smaller area than before. Because turgor loss is the earliest significant biophysical effect of water stress, we might expect turgor-dependent activities to be the most sensitive to water deficits.

Cell expansion is a turgor-dependent process (Chapter 3) and is extremely sensitive to water deficit. Cell expansion is described by the relationship

$$GR = m(P - Y) \qquad (14.1)$$

in which GR is growth rate, P is turgor, Y is the **yield threshold** (the pressure below which the cell wall resists plastic, or nonreversible, deformation), and m is **wall extensibility** (the responsiveness of the wall to pressure). Equation 14.1 shows that a decrease in turgor causes a decrease in growth. Furthermore, not only is growth slowed when stress reduces P but also P must decline only to Y, not to zero, to eliminate expansion completely—another reason for the high sensitivity of growth to stress. In normal conditions, Y is often found to be only 0.1 to 0.2 MPa less than P, so changes in growth rate take place over a very narrow range of turgor (and of cell water content). The relationship between turgor, cell expansion, and growth under nonlimiting water conditions is discussed in detail in Chapter 16.

Inhibition of cell expansion results in a slowing of leaf expansion early in the development of water deficits. The smaller leaf area transpires less water, effectively conserving a limited supply in the soil for use over a longer

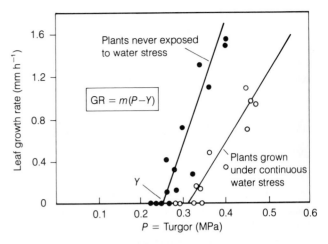

FIGURE 14.1. Dependence of leaf expansion on leaf turgor. Sunflower (*Helianthus annuus*) plants were grown either with ample water or with limited soil water to produce mild water stress. After rewatering, plants of both treatment groups were stressed by withholding water and leaf growth rates (*GR*) and turgor (*P*) were periodically measured. Both decreased extensibility (*m*) and increased threshold turgor for growth (*Y*) limit the leaf's capacity to grow after exposure to stress. (Adapted from Matthews et al., 1984.)

period. Leaf area limitation can be considered a first line of defense against drought.

Because leaf expansion depends mostly on cell expansion, the principles underlying the two processes are similar. In intact leaves, water stress not only decreases turgor, but also decreases m and increases Y. In unstressed plants, extensibility is normally greatest when the cell wall solution is slightly acid. Stress decreases m because it inhibits the transfer of H^+ ions from the cell into the wall space, and the required pH decrease does not occur. The effects of stress on Y are less well understood but presumably involve complex structural changes in the wall (Chapter 1). These changes are very important because, unlike direct effects of turgor, they are relatively slow to be reversed after relief of stress. Water-deficient plants tend to become rehydrated at night, and as a result substantial leaf growth occurs at night. Nonetheless, because of the changes in m and Y, the growth rate is still less than that of unstressed plants at the same turgor (Fig. 14.1).

Water stress not only limits the size of individual leaves, it also limits the number of leaves on an indeterminate plant because it decreases both the number and the growth rate of branches. The process of stem growth has been studied less than that of leaf expansion, but it is probably affected by the same forces that limit leaf growth during stress.

Water Deficit Stimulates Leaf Abscission

The total leaf area of a plant does not remain constant after all the leaves have matured. If plants become water-

FIGURE 14.2. Young cotton (*Gossypium hirsutum*) plants undergo leaf abscission in response to water stress. Plants at the left were watered throughout the experiment, whereas those in the middle and at the right were subjected to moderate stress and severe stress, respectively, before rewatering. Only a tuft of leaves at the top of the stem is left on the severely stressed plants. (Photograph courtesy of B. L. McMichael.)

stressed after a substantial leaf area has developed, leaves will senesce and eventually fall off (Fig. 14.2). This *leaf area adjustment* is an important long-term change that improves the plant's fitness for a water-limited environment. Indeed, many drought-deciduous desert plants depend on an extreme version of this response to stress, dropping all their leaves during drought and sprouting new ones after a rain. This cycle can occur two or more times in a single season. The abscission process during water stress results largely from enhanced synthesis of and responsiveness to the endogenous plant hormone ethylene (Chapter 19).

Water Deficit Enhances Root Extension into Deeper, Moist Soil

Mild water deficits also affect the development of the root system. Root-shoot relations appear to be governed by a functional balance between water uptake by the root and photosynthesis by the shoot. Although root-shoot relations depend on complex developmental and nutritional processes, the concept of functional balance can be simply stated: A shoot will grow until it is so large that root water uptake becomes limiting to further growth; conversely, roots will grow until their demand for photosynthate from the shoot equals the supply. This functional balance is shifted if water supply decreases. When water uptake is curtailed, the process of leaf expansion is affected very early, but photosynthetic activity is much less affected. Inhibition of leaf expansion reduces consumption of carbon and energy, and a greater proportion of the plant's assimilates can be distributed to the root system, where they can support further growth. At the same time, the

root apices in dry soil lose turgor, and the drying soil itself presents an increasingly rigid structure. All these factors lead to root growth into the soil zones that remain moist. As water deficits progress, the upper layers of the soil usually dry first. Thus, it is common to see a shallow root system when all soil layers are wetted but a loss of shallow roots and proliferation of deep roots as water is depleted. Root growth into wet soil can be considered a second line of defense against drought.

Enhancement of root growth into moist soil zones during stress depends on the allocation of assimilates to the growing root tips. As alternative sinks for assimilates, fruits usually predominate over roots, and assimilates are directed to the fruits and away from the roots. For this reason, the enhanced water uptake resulting from root growth is less pronounced in reproductive plants than in vegetative plants. Competition for assimilates between roots and fruits is one explanation for the fact that plants are generally more sensitive to water stress during reproduction.

Plant breeders have attempted to increase the capacity for root growth in breeding programs aimed at improved drought resistance of agricultural crops. Most crops show considerable genetic variability in root growth rate, and under the right circumstances breeding for increased root growth can be effective. For this approach to be successful, however, the soil must contain water that would otherwise not be absorbed, which is often the case after a wet winter that drives water below the normal root zone. On the other hand, if the winter is dry, the cost to the plant of building and maintaining the extra roots is not offset by gains in water uptake, and aboveground yields suffer.

Stomata Close During Water Deficit in Response to Abscisic Acid

The preceding sections discussed charges in plant development during slow, long-term dehydration. When the onset of stress is more rapid or the plant has reached its full leaf area before initiation of stress, there are other responses that protect the plant against immediate desiccation. Under these conditions, the stomata close to reduce evaporation from the existing leaf area. Stomatal closure can be considered a third line of defense against drought.

Uptake and loss of water in guard cells changes their turgor and modulates stomatal opening and closing (Chapter 4). Because guard cells are exposed to the atmosphere, they can lose water directly by evaporation, a process called **hydropassive closure** of stomata. Hydropassive closure is likely to take place in air of low humidity, when direct water loss from the guard cells occurs too rapidly to be balanced by water movement into the guard cells from adjacent epidermal cells. A second mechanism of stomatal closure, called **hydroactive closure**, occurs when the whole leaf is dehydrated and depends on metabolic processes in the guard cells. The mechanism of hydroactive closure is essentially a reversal of that of stomatal opening described in Chapter 6. A reduction in the solute content of the guard cells results in water loss and decreased turgor, leading to stomatal closing.

The process of solute loss from guard cells is triggered by decreasing water status in the rest of the leaf, and there is much evidence that **abscisic acid** (ABA) (Chapter 19) plays an important role in this process. Abscisic acid is continuously synthesized at a low rate in mesophyll cells and tends to accumulate mostly in the chloroplasts. When the mesophyll becomes mildly dehydrated, two things happen. First, some of the ABA stored in the mesophyll cells is released to the apoplast (the cell wall space exterior to the plasma membrane), making it possible for the transpiration stream to carry some of it to the guard cells. Second, the rate of net synthesis of ABA is increased. Stomatal closure is initiated by the redistribution of stored ABA from the mesophyll chloroplast into the apoplast (Cornish and Zeevaart, 1985). Increased ABA synthesis occurs after closure has begun and appears to enhance or prolong the initial closing effect of the stored ABA.

The process of ABA redistribution depends on pH gradients within the leaf, the weak-acid properties of the ABA molecule, and the permeability properties of cell membranes (Fig. 14.3). In an unstressed photosynthesizing

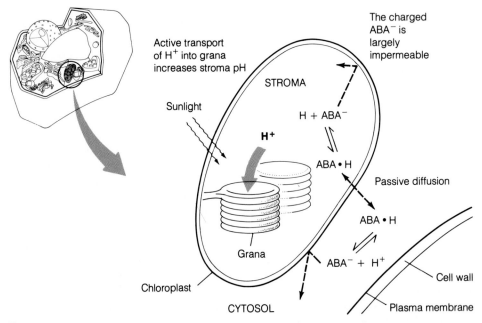

FIGURE 14.3. Accumulation of ABA by chloroplasts in the light. Dashed lines indicate passive movement of substances by diffusion; the wide arrow indicates movement by active processes. Light stimulates proton uptake into the grana, making the stroma more alkaline. The increased alkalinity causes the weak acid ABA · H to dissociate into H^+ and the ABA^- anion. The concentration of ABA · H in the stroma is lowered below the concentration in the cytosol, and the concentration difference drives passive diffusion of ABA · H across the chloroplast membrane. At the same time, the concentration of ABA^- in the stroma increases, but the chloroplast membrane is impermeable to the anion and it remains trapped. This process continues until the ABA · H concentrations are equal in the stroma and the cytosol. As long as the stroma remains more alkaline, though, the total ABA concentration (ABA · H + ABA^-) in the stroma greatly exceeds the concentration in the cytosol.

Gene Action During Water Deficit

WATER STRESS CAUSES many changes in metabolism and development that can improve fitness for a water-deficient environment. The hormone abscisic acid seems to be particularly important in at least one major drought response, namely, stomatal closure. To what extent is ABA involved in all other drought responses? To approach this question, Elizabeth Bray, of the University of California at Riverside, turned to a plant used extensively in plant physiological studies, the *flacca* mutant of tomato (*Lycopersicon esculentum*). Plants having the *flacca* mutation wilt readily in response to drought, and their tendency to wilt correlates with a substantially lower ABA content compared with that of wild-type (normal) plants. In leaves of the wild-type tomato, water stress initiated or noticeably increased the synthesis of 21 polypeptides. In *flacca* leaves, water stress caused some similar changes, but not all 21 polypeptides were synthesized and there were smaller amounts of those that were produced. Treatment of the *flacca* plants with ABA restored the polypeptide pattern to that seen in the wild type (Bray, 1988). The mRNA population was similarly affected. These results indicate that at the molecular level, many (but by no means all) drought-initiated events are responses to increases in ABA concentration.

The relatively small number of new gene products synthesized during water stress or in response to experimentally applied ABA indicates that it should be possible to isolate these genes and identify their functions, gaining insight into plant acclimation to water deficits. Other studies with different plants and tissues lend support to this expectation. For example, in hypocotyls of drought-stressed soybean (*Glycine max*) seedlings, a 28-kDa polypeptide appears in the cell walls very early during the inhibition of growth. Since this protein is largely confined to the walls of cells whose growth is inhibited, it may have a role in growth regulation (Bozarth et al., 1987).

---

leaf, the pH of the chloroplast stroma is normally higher than that of the cytosol. This pH differential leads to a large accumulation of ABA in chloroplasts. One effect of dehydration is to lower the chloroplast pH, allowing the release of some ABA. Coupled to this process is an increase in the pH of the apoplast. These concerted changes during water deficit cause the net transfer of ABA from plastid to apoplast (Hartung et al., 1988).

Stomatal responses to leaf dehydration can vary widely both within and across species. Stomata of some dehydration-postponing species, such as cowpea (*Vigna unguiculata*) and cassava (*Manihot esculenta*), are unusually responsive to decreasing water availability, and stomatal conductance and transpiration decrease so much that leaf water potential (ψ; see Chapters 3 and 4) may remain nearly constant during drought. In cotton (*Gossypium hirsutum*), factors such as nitrogen supply affect ABA accumulation or ABA redistribution, or both, and thus greatly alter stomatal responses to water stress (Radin and Hendrix, 1988).

Messengers from the root system may affect the stomatal responses to water stress. The evidence for root messengers is of two types. First, stomatal conductance is often much more closely related to soil water status than to leaf water status, and the only plant part that can be *directly* affected by soil water status is the root system. In fact, dehydrating only part of the root system can cause stomatal closure even though the well-watered portion of the root system may still deliver ample water to the tops. When corn (*Zea mays*) plants were grown with roots trained into two separate pots and water was withheld from one of the pots (with the other continuing to receive water), stomata partially closed, and the leaf water potential increased, just as in the dehydration postponers described above (Blackman and Davies, 1985). These results show that stomata can respond to conditions sensed in the roots. Second, roots are known to produce and export abscisic acid to the xylem sap. When dayflower (*Commelina communis*) plants were grown with their root systems divided between two pots and water was then withheld from one pot, there was a large increase in ABA concentration in roots in the dry pot (Zhang et al., 1987). Stomata closed in response to the treatment, without any change in leaf water potential. When the epidermis was stripped from the leaf, its ABA content was found to be closely related to the degree of stomatal closure. The ABA from the roots was presumably delivered to the leaf via the transpiration stream. These results imply that stomata respond to two different types of signals during soil drying: (1) an "early warning system" involving root messengers, indicating that some roots are drying; and (2) ABA translocation within the leaf resulting from desiccation of the leaves themselves.

Water Deficit Limits Photosynthesis Within the Chloroplast

The photosynthetic rate of the leaf (expressed per unit leaf area) is seldom as responsive to mild water stress as is leaf expansion (Fig. 14.4). The reason is that photosynthesis is much less sensitive to turgor than is leaf expansion. On the other hand, there is evidence that the Mg^{2+} concentration in chloroplasts may influence photosynthesis during water stress through its role in coupling electron transport to ATP production (Chapter 8). In isolated chloroplasts, photosynthesis is very sensitive to increasing Mg^{2+} concentration, and a similar process could occur during cell shrinkage induced by water stress. When sunflower (*Helianthus annuus*) plants were grown with different Mg^{2+} levels in the nutrient solution (Rao et al., 1987), the plants with the lower tissue Mg^{2+} concentrations maintained higher photosynthetic rates as leaves became dehydrated (Fig. 14.5).

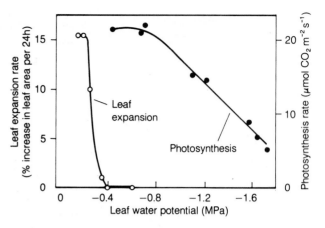

FIGURE 14.4. Effects of water stress on photosynthesis and leaf expansion of sunflower (*Helianthus annuus*) plants. This species is typical of many plants in which leaf expansion is much more sensitive than photosynthetic rate to dehydration. (Adapted from Boyer, 1970.)

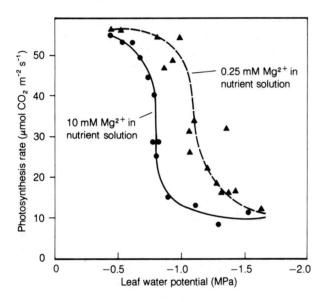

0.25 mM Mg²⁺
in nutrient solution

10 mM Mg²⁺
in nutrient solution

FIGURE 14.5. Photosynthetic rates (measured at CO_2 and light saturation) in water-stressed sunflower (*Helianthus annuus*) plants. The plants were grown with levels of Mg^{2+} in the nutrient solution chosen to produce different tissue Mg^{2+} concentrations but no change in growth rate, as shown in the photograph. As water stress progressed, photosynthesis was affected much earlier in the plants with higher Mg^{2+} levels. (Adapted from Rao et al., 1987. Photograph courtesy of J. S. Boyer.)

Water stress usually affects both stomatal conductance and photosynthetic activity in the leaf. Upon stomatal closure during early stages of water stress, water-use efficiency may increase (more CO_2 taken up per unit of water transpired) because stomatal closure inhibits transpiration more than decreasing intercellular CO_2 concentrations and dehydration of mesophyll cells inhibit photosynthesis. However, as stress becomes severe, water-use efficiency usually decreases and the inhibition of mesophyll metabolism becomes stronger. The relative effect of the stress on stomatal conductance and photosynthesis can be evaluated by putting stressed leaves into air containing high concentrations of CO_2. Stomatal limitations to photosynthesis can be overcome by a higher external concentration of CO_2, but any direct effect of water stress on mesophyll metabolism will not be removed by exposure to high CO_2 concentrations.

Another critical question regarding effects of water stress is whether the stress process directly affects translocation. Phloem transport is coupled to both photosynthesis, which provides the substrates, and metabolism in the sinks. Water stress decreases photosynthesis and the consumption of assimilates in the expanding leaves. As a consequence, water stress indirectly decreases the amount of photosynthate exported from leaves. Because phloem transport is turgor-dependent (Chapter 7), it has been argued that if water potential decreases in the phloem during stress, the decrease in turgor should also inhibit assimilate movement. However, experiments have shown that translocation is unaffected until late in the stress period, when other processes such as photosynthesis have already been strongly inhibited (Fig. 14.6). This relative insensitivity of translocation to stress allows plants to mobilize and use reserves where they are needed (for

example, in seed growth), even when stress is extremely severe. The ability to continue translocating carbon is a key factor in virtually all aspects of plant resistance to drought.

Osmotic Adjustment of Cells Helps Maintain Plant Water Balance

As soil dries, its matric potential becomes more negative. Plants can continue to absorb water only as long as their water potential (ψ) is below that of the source of water. **Osmotic adjustment**, or accumulation of solutes by cells, is a process by which water potential can be decreased without an accompanying decrease in turgor. The change in tissue water potential results simply from changes in osmotic pressure (π), the osmotic component of ψ (Eq. 3.8). Osmotic adjustment should not be confused with the increase in solute concentration that occurs during cell dehydration and shrinkage. Instead, it is a net increase in solute content per cell independent of the volume changes that take place because of loss of water. The increase in π is typically fairly small, between 0.2 and 0.8 MPa, except in plants adapted to extremely dry conditions. Most of the adjustment can usually be accounted for by increases in concentration of a variety of common solutes, including sugars, organic acids, and ions (especially K^+).

Enzymes extracted from the cytosol of plant cells have been shown to be severely inhibited by high concentrations of ions. The accumulation of ions during osmotic adjustment appears to occur mainly within the vacuoles, where the ions are kept out of contact with enzymes in the cytosol or subcellular organelles. Because of this compartmentation of ions, some other solutes must accumulate in the cytoplasm to maintain water potential equilibrium within the cell. These other solutes, called **compatible solutes** (or compatible osmolytes), are organic compounds that do not interfere with enzyme functions. Proline is a commonly accumulated compatible solute; others are a sugar alcohol, sorbitol, and a quaternary amine, glycine betaine. Synthesis of compatible solutes is also important in the adjustment of plants to increased salinity in the rooting zone, discussed later in this chapter.

Osmotic adjustment develops slowly in response to tissue dehydration. It is therefore important, when studying osmotic adjustment, to allow for a slow onset of stress in order to provide enough time for the substantial shifts in solute synthesis and transport patterns to take place. However, over the course of several days, other changes (growth, photosynthesis, etc.) are also taking place. For this reason, it is not clear whether osmotic adjustment is an independent and direct response to water deficit or a result of some other factor such as decreased growth rate. Nonetheless, it is clear that leaves capable of osmotic

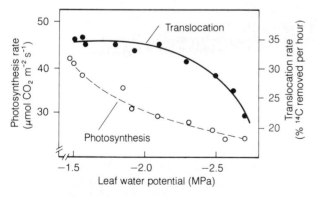

FIGURE 14.6. Relative effects of water stress on photosynthesis and translocation in sorghum (*Sorghum bicolor*). Plants were exposed to $^{14}CO_2$ for a short interval. The radioactivity fixed in the leaf was a measure of photosynthesis, and the loss of radioactivity after removal of the $^{14}CO_2$ was a measure of the rate of assimilate translocation. Although photosynthesis was affected by stress, translocation was unaffected until stress was very severe. (Adapted from Sung and Krieg, 1979.)

adjustment can maintain turgor at lower water potentials than nonadjusted leaves. Maintenance of turgor enables the continuation of cell elongation and facilitates higher stomatal conductances at lower water potentials (Turner and Jones, 1980). In this sense, osmotic adjustment is an acclimation that enhances true dehydration tolerance.

How much extra water can be acquired by the plant because of osmotic adjustment in the leaf cells? Most of the extractable soil water is held in large pores from which it is readily removed by roots. As the soil dries, this water is used first, leaving behind the small amount of water more tightly held in small pores. Osmotic adjustment enables the plant to extract more of this tightly held water, but the increase in available water is small. Thus, the cost of osmotic adjustment in the leaf is offset by rapidly diminishing returns in terms of extra water to the plant, as can be seen by comparing the water relations of adjusting and nonadjusting species. Figure 14.7 shows such a comparison for pot-grown sugar beet (*Beta*

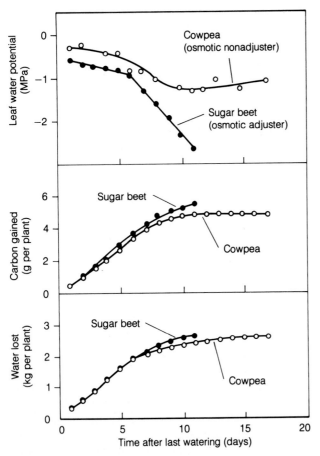

FIGURE 14.7. Water loss and carbon gain by sugar beet (*Beta vulgaris*), an osmotically adjusting species, and cowpea (*Vigna unguiculata*), a nonadjusting species, grown in pots and subjected to water stress. Although the leaf water potential reached much lower values in sugar beet because of its osmotic adjustment, the total water loss and carbon gain during stress were little affected. (Adapted from McCree and Richardson, 1987.)

vulgaris), an osmotically adjusting species, and cowpea (*Vigna unguiculata*), a nonadjusting species that instead conserves water by stomatal closure during stress. On any given day after the last watering, the sugar beet leaves maintained a lower water potential than the cowpea leaves; however, photosynthesis and transpiration during stress were only slightly greater in the sugar beet. The major difference between the two plants was the leaf water potential at which they were operating. These results show that osmotic adjustment promoted true dehydration tolerance but did not have a major effect on productivity (McCree and Richardson, 1987).

Osmotic adjustment also occurs in roots, although the process has not been so extensively studied as in leaves. The absolute magnitude of the adjustment is less in roots than in leaves, but as a percentage of the original tissue π, it can be greater in roots than in leaves. Again, these changes in many cases only slightly increase water extraction from the previously explored soil. However, osmotic adjustment can occur in the root meristems, enhancing turgor and maintaining root growth, and is an important component of the changes in root growth patterns during water depletion in the field (p. 349).

Is osmotic adjustment a valuable mechanism of stress acclimation? Do solutes accumulate in one tissue simply because of stress-related inhibition of growth elsewhere? These questions might be difficult to answer. Attempts to increase osmotic adjustment in leaves, either genetically (by breeding and selection) or physiologically (by inducing adjustment with controlled water deficits), resulted in plants that grew more slowly. Thus, the use of osmotic adjustment to improve agricultural performance is yet to be realized.

Water Deficit Alters Energy Dissipation from Leaves

You will recall from Chapter 10 that evaporative cooling lowers leaf temperature. This cooling effect can be remarkable: In Death Valley, California, one of the hottest places in the world, leaf temperatures were measured 8°C below air temperature on plants with access to ample water. In warm dry climates, leaves from irrigated crop plants can also maintain a temperature difference. In such climates, an experienced farmer can decide whether plants need water simply by touching the leaves, because a rapidly transpiring leaf is distinctly cool to the touch. When water stress limits transpiration, the leaf heats up unless some other process offsets the loss of cooling. Because of these interactions, water stress and heat stress are closely interrelated (see discussion of heat stress later in this chapter).

Maintaining a leaf much cooler than the air requires evaporation of vast quantities of water. Thus, adaptations that cool leaves by other means are very effective in conserving water. When transpiration slows and the leaf temperature increases, some of the extra energy in the leaf is dissipated as sensible heat loss (Chapter 10). Many arid-zone plants have very small leaves, which minimize the resistance of the boundary layer to transfer of heat from the leaf to the air. Small leaves tend to remain close to air temperature even when transpiration is greatly slowed. Large leaves have thicker boundary layers and dissipate less thermal energy (per unit leaf area) by direct transfer. This limitation can be compensated by leaf movements that provide additional protection against heating during water stress. Leaves that orient themselves away from the sun are called **paraheliotropic**, whereas those that gain energy by orienting themselves normal (perpendicular) to the sunlight are **diaheliotropic** (Chapter 10). Figure 14.8 shows the strong effect of water stress on leaf position in soybeans.

Other factors that can alter the interception of radiation include wilting, which changes the angle of the leaf, and leaf rolling in grasses, which minimizes the profile of tissue exposed to the sun. Leaf rolling is facilitated by specialized **bulliform cells** that act as "pivots" for the rest of the leaf. Absorption of energy can also be decreased by hairs on the leaf surface or by layers of reflective epicuticular wax outside the cuticle. Leaves of some plants have a gray-white appearance because densely packed hairs reflect a large amount of light. This hairiness, or *pubescence*, keeps leaves cooler by reflecting radiation, but it also reflects the visible wavelengths active in photosynthesis and thus decreases carbon gain. Because of this problem, attempts to breed pubescence into crops to improve their water-use efficiencies have been generally unsuccessful.

Water Deficit Increases Resistances to Liquid-Phase Water Flow

When a soil dries, its resistance to the flow of water increases very sharply, particularly near the *permanent wilting point*, usually observed when the soil water potential reaches −1.5 MPa. At the permanent wilting point, water delivery to the roots is too slow to allow overnight rehydration of plants that have wilted during the day. However, the soil is not the only source of increased resistance to flow. In fact, the resistance within the plant has been found to be larger than the resistance within the soil over a wide range of water deficits (Blizzard and Boyer, 1980).

Several factors may contribute to the increased plant resistance to water flow during drying. As plant cells lose water, they shrink. When root shrinkage during the day becomes pronounced, the root surface can move away from the soil particles that hold the water, and the delicate root hairs that actually penetrate the particles may be damaged as they are pulled away. Also, as root extension slows during soil drying, the outer layer of the cortex (the hypodermis) often becomes more extensively covered

(a)

(b)

(c)

FIGURE 14.8. Orientation of leaflets of field-grown soybean (*Glycine max*) plants in the normal, unstressed, position (a); during mild water stress (b); and during severe water stress (c). The large leaf movements induced by mild stress are quite different from wilting, which occurs during severe stress. Note that during mild stress the terminal leaflet has been raised, whereas the two lateral leaflets have been lowered; each is almost vertical. (Photographs courtesy of D. M. Oosterhuis.)

with **suberin**, a water-impermeable lipid, increasing the resistance to water flow. Another important factor is **cavitation**, or breakage of water columns under tension. As we saw in Chapter 4, transpiration from leaves "pulls" water through the plant by creating a tension on the water column. The cohesive forces required to support large tensions are present only in very narrow columns in which the water adheres to the walls.

Cavitation begins in most plants at quite moderate water potentials (-1 to -2 MPa), with the largest vessels cavitating first. Thus, in ring-porous trees such as oak (*Quercus*), the large-diameter vessels that are laid down in the spring function as a low-resistance pathway early in the growing season, when ample water is available. These vessels cease functioning during the summer, leaving the small-diameter vessels produced during the stress period to carry the transpiration stream. This shift has long-lasting consequences: even if the plant is rewatered, the original low-resistance pathway remains nonfunctional.

Water Deficit Increases Wax Deposition on the Leaf Surface

A common effect of development during water stress is the production of a thicker cuticle that reduces water loss from the epidermis (cuticular transpiration). A thicker cuticle also blocks CO_2 uptake, but this does not affect leaf photosynthesis because the epidermal cells underneath the cuticle are nonphotosynthetic. Cuticular transpiration, however, is only 5 to 10% of the total leaf transpiration, so it becomes significant only if stress is extremely severe or if the cuticle has been damaged (e.g., by wind-driven sand).

Water Deficits May Induce Crassulacean Acid Metabolism

Crassulacean acid metabolism (CAM), discussed in Chapters 9 and 10, is a plant adaptation in which stomata open at night and close during the day. The leaf-to-air vapor pressure difference driving transpiration is much reduced at night, when leaf and air are both cool. As a result, the water-use efficiencies of CAM plants are among the highest measured in all higher plants. A CAM plant may gain 1 g of dry matter for only 125 g of water used—three to five times greater than the ratio for a typical C3 plant.

The phenomenon of CAM is characteristic of succulent plants such as cacti. A few succulent species display **facultative CAM**, switching to CAM when subjected to water deficits or saline conditions (Hanscom and Ting, 1978). This switch in metabolism is rather complex and remarkable, involving accumulation of phosphoenolpyruvate carboxylase (PEPC) (Fig. 14.9), changes in carboxylation and decarboxylation patterns, transport of

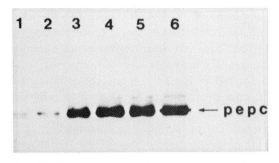

FIGURE 14.9. Increases in phosphoenolpyruvate carboxylase (PEPC) content in the ice plant, *Mesembryanthemum crystallinum*, during the shift from C3 metabolism to CAM induced by salt stress. The numbers at the top represent days after the time at which the plants were salt-stressed by addition of 500 mM NaCl to the irrigation water. The PEPC protein was revealed in the gels by use of antibodies and a stain (Bohnert et al., 1989).

large quantities of malate into and out of the vacuoles, and reversal of the periodicity of stomatal movements. Although the environmental triggers for these charges have been identified, the metabolic factors that regulate them are little understood at present.

Chilling and Freezing

Chilling injury occurs in sensitive species at temperatures that are too low for normal growth but not low enough for ice to form. Typically, chilling injury occurs in species of tropical or subtropical origin. Among crops, maize, *Phaseolus* beans, rice, tomato, cucumber, sweet potato, and cotton are sensitive. *Passiflora, Coleus*, and *Gloxinia* are examples among the ornamentals. When plants growing at relatively warm temperatures ($25–35°C$) are cooled to $10–15°C$, growth is slowed, discolorations or lesions appears on leaves, and the foliage develops a water-soaked appearance. If roots are chilled, the plants may wilt.

Species that are considered to be generally sensitive to chilling show appreciable variation in resistance. Genetic adaptation to the colder temperatures associated with high altitude improves chill resistance (Fig. 14.10). In addition, resistance often increases if plants are first hardened by exposure to cool, but noninjurious, temperatures. Chilling damage thus can be minimized if exposure is slow and gradual. Sudden exposure to temperatures near $0°C$, or cold shock, is certain to produce injury.

Membrane Properties Change upon Chilling Injury

Leaves from chill-injured plants show inhibition of photosynthesis and carbohydrate translocation, slower respiration, inhibition of protein synthesis, and increased

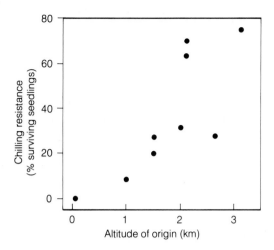

FIGURE 14.10. Survival of different varieties of tomato (*Lycopersicon*) seedlings collected from different altitudes in South America. Seedlings were chilled for 7 days at 0°C and then kept for 7 days in a warm growth room, after which the numbers of survivors were counted. Seedlings collected from high altitudes showed greater resistance to chilling (cold shock) than those collected from lower altitudes. (From Patterson et al., 1978.)

degradation of existing proteins. All of these responses probably depend on a common primary mechanism involving loss of membrane properties during chilling. For instance, solutes leak from the leaves of chill-sensitive *Passiflora maliformis* floated on water at 0°C but not from those of chill-resistant *Passiflora caerulea* (passion flower). Loss of solutes to the water reflects damage to the plasma membrane and possibly also the tonoplast. In turn, inhibition of photosynthesis and of respiration reflects injury to chloroplast and mitochondrial membranes.

Why are the membranes of chill-sensitive plants affected by chilling? Plant membranes consist of a lipid bilayer interspersed with proteins (see Chapter 1). The physical properties of the lipids greatly influence the activities of the integral membrane proteins, including channel-forming proteins that regulate transport of ions and other solutes (Chapter 6) and enzymes on which metabolism depends. In chill-sensitive plants the lipids in the bilayer have a high percentage of saturated fatty acid chains, and this type of membrane tends to solidify into a semicrystalline state at a temperature well above 0°C.[1] As the membranes become less fluid, their protein components can no longer function normally. Thus, solute transport into and out of cells is impaired, energy transduction (Chapters 8 and 11) is inhibited, and enzyme-dependent metabolism is affected.

Membrane lipids from chill-resistant plants often have a greater proportion of unsaturated fatty acids than those from chill-sensitive ones (Table 14.2). Furthermore, during the process of acclimation to cool temperatures, the proportion of unsaturated lipids increases (Williams et al., 1988). This process lowers the temperature at which the membrane lipids begin a gradual phase change from fluid to semicrystalline and thereby provides some protection against damage from chilling. In these hardened plants, the modified membranes are better able to maintain fluidity and function at lower temperatures.

---

1. Saturated lipids, which have no double bonds, solidify at a higher temperature than unsaturated lipids. Butter, which is a saturated lipid, is a solid at room temperature, whereas oil from seeds is more unsaturated (has more double bonds) and remains liquid to much lower temperatures.

TABLE 14.2. Fatty acid composition of mitochondria isolated from chill-resistant and chill-sensitive species.

Major fatty acids*	Chill-resistant species			Chill-sensitive species		
	Cauliflower bud	Turnip root	Pea shoot	Bean shoot	Sweet potato	Maize shoot
	Percentage by weight of total fatty acid content					
Palmitic (16:0)	21.3	19.0	12.8	24.0	24.9	28.3
Stearic (18:0)	1.9	1.1	2.9	2.2	2.6	1.6
Oleic (18:1)	7.0	12.2	3.1	3.8	0.6	4.6
Linoleic (18:2)	16.4	20.6	61.9	43.6	50.8	54.6
Linolenic (18:3)	49.4	44.9	13.2	24.3	10.6	6.8
Ratio of unsaturated to saturated fatty acids	3.2	3.9	3.8	2.8	1.7	2.1

*Shown in parentheses are the number of carbon atoms in the fatty acid chain and the number of double bonds.

Adapted from Lyons et al. (1964).

Freezing Kills Cells Because of the Formation of Large Intracellular Ice Crystals

The ability to tolerate freezing temperatures under natural conditions varies greatly among chill-resistant plants. Seeds, other partially dehydrated tissues, and fungal spores can be kept indefinitely long at temperatures near absolute zero (0 K), so very low temperatures are not intrinsically harmful. Fully hydrated, vegetative cells can also retain viability if they are cooled very quickly to avoid formation of large, slowly growing ice crystals that would puncture and destroy cellular fine structure within the living protoplast. Ice crystals that form during very rapid freeezing are too small to cause mechanical damage to subcellular structures. Warming back to normal temperatures must also be extremely rapid to prevent the conversion of ice crystals to water vapor by sublimation or the growth of crystals to a damaging size, which take place at intermediate temperatures (-100 to $-10°C$). Under natural conditions, however, cooling of intact, multicellular plant organs is never fast enough to prevent the formation of large injurious ice crystals in fully hydrated cells.

When tissue is cooled under natural conditions, ice usually forms first within the cell walls. This ice formation is not lethal, and the tissue fully recovers if warmed. However, when plants are exposed to freezing temperatures for an extended period, the crystals in the walls continue to grow and extend into the protoplast, causing lethal damage. Freeze-resistant species somehow limit the growth of crystals to the cell walls and intercellular spaces. The box on ice formation in higher plant cells discusses the process of freezing in more detail.

Acclimation to Freezing Involves ABA and Changes in Gene Expression

In seedlings of alfalfa (*Medicago* species), tolerance to freezing at $-10°C$ is greatly improved by previous cold exposure ($4°C$) or by treatment with exogenous ABA (Chapter 19) without cold exposure. These treatments cause changes in the pattern of newly synthesized proteins that can be resolved on two-dimensional gels. Some of the changes are unique to the particular treatment (cold or ABA), but some of the newly synthesized proteins induced by cold appear to be the same as those induced by ABA. Furthermore, examination of one gene in detail showed that the same mRNA was induced after treatment with either cold or exogenous ABA (Mohapatra et al., 1988). Similar changes in the pattern of protein synthesis and freezing resistance upon exposure to cold or treatment with ABA have been found in cells of bromegrass (*Bromus inermis*) in tissue culture. The identity and role of the proteins whose induction is associated with cold resistance remain to be characterized.

Typically, several days of exposure to cool temper-atures are required for freezing resistance to be fully induced. In potato, resistance is fully induced after 15 days of exposure to cold. On the other hand, when rewarmed, plants lose their acclimation in as little as 24 h and become susceptible to freezing once again. The need for cool temperatures to induce acclimation to even lower temperatures and the rapid loss of acclimation on warming explain the susceptibility of plants in the southern United States (and similar climatic zones with highly variable winters) to extremes of temperature in the winter months, when air temperature can drop from $20-25°C$ to below $0°C$ in a few hours.

Some Woody Plants Can Acclimate to Very Low Temperatures

When in a dormant state, some woody plants are extremely resistant to low temperatures. Resistance is in part determined by previous cold acclimation, but genetics plays an important role in determining the degree of acclimation. Clones of red osier dogwood (*Cornus stolonifera*) obtained from different climatic regions showed different degrees of resistance to low temperature in the fall and early winter when grown together in the same environment (Weiser, 1970). Likewise, native species of *Prunus* from northern cooler climates in North America are hardier after acclimation than those from milder climates. When the species were tested together in the laboratory, those with a northern geographic distribution showed greater ability to avoid intracellular ice formation, indicating distinct genetic differences (Burke and Stushnoff, 1979).

Under natural conditions, acclimation in woody species takes place in two distinct stages (Weiser, 1970). In stage 1, hardening is induced in the early autumn by exposure to short days and nonfreezing chilling temperatures, both of which combine to stop growth. A diffusible factor that promotes acclimation (probably ABA) moves in the phloem from leaves to overwintering stems and may be responsible for the changes. During this period, woody species also withdraw water from the xylem vessels, thereby avoiding splitting of the stem by expansion of water during later freezing. Cells in stage 1 of acclimation can survive temperatures well below $0°C$, but they are not fully hardened. In stage 2, direct exposure to freezing is the stimulus; no factor that is known to be translocated within the plant can substitute for this exposure and confer hardening. When fully hardened, the cells can tolerate exposure to temperatures of -50 to $-100°C$.

Deep Supercooling and Dehydration Are Involved in Resistance to Intracellular Freezing

In many species of the hardwood forests of southeastern Canada and the eastern United States, acclimation in-

Ice Formation in Higher Plant Cells

DURING THE INITIAL phase of loss of heat from tissue, the temperature drops below the true freezing point of the cytosol and the vacuole without a phase change from liquid to solid (Fig. 14.A), and the liquid phase in the cytosol and vacuole is said to be **supercooled**. As the temperature drops further, ice forms in the cell walls, and there is a release of heat energy from the latent heat of fusion of water. (Transformation of water from liquid to solid releases about 0.33 kJ (80 calories) per gram.) At this stage, the temperature of the tissue reflects the balance between heat gain from ice formation and heat loss to the environment. As a result, when ice first forms, the temperature rises rapidly (point B in Fig. 14.A), and it remains at that level until all the water in the cell walls is frozen (point C in Fig. 14.A). When that point is reached, release of heat stops, and the temperature begins to fall again. The release of heat energy during ice formation is the basis for the common practice of spraying crops with water during frost: as long as the water continues to freeze extracellularly, it releases heat that prevents intracellular freezing.

Ice formation within the cell walls and intercellular spaces of freeze-sensitive cells is not lethal, but upon extended exposure to freezing temperatures water vapor moves from the unfrozen protoplasts to the cell wall, causing ice crystals to grow within the wall. This slow dehydration concentrates solutes within the protoplast, depressing the freezing point by 1–2°C. As the temperature continues to drop, a second phase of release of heat of fusion of water is detectable (points along D–E in Fig. 14.A). This phase reflects a series of small freezing events: each "spike" of heat release represents the freezing of cell protoplasts and coincides with loss of viability.

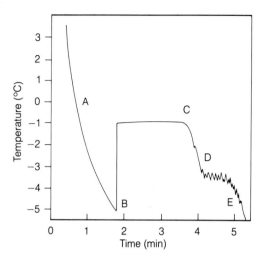

FIGURE 14.A. Temperature of cucumber (*Cucumis sativus*) fruit parenchyma cells during freezing. The temperature was recorded with an electronic device, a thermistor, inserted into a 5 × 20 mm cylinder of tissue and immersed in a coolant at −5.8°C. A–B, Supercooling; B–C, release of heat during freezing in cell walls and intercellular spaces; C–D, supercooling; D–E, small heat spikes released during intracellular freezing of individual protoplasts. (From Brown et al., 1974.)

The formation of ice crystals in cell walls or in the protoplasm requires the presence of structures called **ice nucleation points** on which crystals can be initiated and grow. In some species, acclimation confers an ability to suppress ice nucleation in the protoplast, allowing **deep supercooling** to many degrees below the freezing point without ice formation. However, deep supercooling of intracellular water has a lower limit of about −40°C, the spontaneous ice nucleation temperature of water. (The homogeneous ice nucleation temperature of pure water droplets is −38.1°C, but the presence of solutes lowers this temperature.) At or below −40°C, ice crystals form without nucleation points, and intracellular freezing and cell death are unavoidable.

volves development of the ability to suppress ice crystal formation, even at temperatures far below the theoretical freezing point (see box on ice formation). This **deep supercooling** is seen in species such as oak, elm, maple, beech, ash, walnut, hickory, rose, rhododendron, apple, pear, peach, and plum (Burke and Stushnoff, 1979). Deep supercooling also takes place in the Rocky Mountains of Colorado in stem and leaf tissue of tree species such as Engelmann spruce (*Picea engelmannii*) and subalpine fir

(*Abies lasiocarpa*), but their freezing resistance is quickly weakened once growth resumes in the spring (Becwar et al., 1981). Stem tissues of subalpine fir, which undergo deep supercooling and remain viable to below −35°C in May, lose their ability to suppress ice formation in June and are then killed at −10°C. However, cells can supercool only to about −40°C, at which temperature ice formation occurs spontaneously. This phenomenon accounts for the low-temperature limits of survival of many alpine and subarctic species that undergo deep supercooling and also explains the altitude of the timberline in mountain ranges, which is at or near the −40°C minimum isotherm.

Woody species in northern Canada, Alaska, and northern Europe and Asia are subject to average annual minima much below −40°C, and their survival does not involve deep supercooling. Instead, ice formation starts at −3 to −5°C in the cell walls, where the crystals continue to grow. Water is gradually withdrawn from the protoplast, which remains unfrozen. Resistance to freezing temperatures depends on the capacity of the extracellular spaces to accommodate growing ice crystals and on the ability of the protoplast to withstand dehydration. This may explain why some woody species that are resistant to freezing are also resistant to water deficit during the growing season. Species of willow (*Salix*), white birch (*Betula papyrifera*), quaking aspen (*Populus tremuloides*), *Prunus pensylvanica*, *Prunus virginiana*, and lodgepole pine (*Pinus contorta*), for example, all tolerate very low temperatures by limiting formation of ice crystals to the cell walls. However, acquisition of resistance depends on slow cooling so that extracellular ice formation and protoplast dehydration are gradual; sudden exposure to very cold temperatures before full acclimation causes intracellular freezing and cell death.

Some Bacteria Living on Leaf Surfaces Increase Frost Damage

When leaves are cooled to temperatures in the range −3 to −5°C, formation of ice crystals on the surface (frost) is accelerated by certain bacteria that naturally inhabit the leaf surface, as *Pseudomonas syringae* and *Erwinia herbicola*. These bacteria act as ice nucleation points at these relatively warm temperatures. When artificially inoculated with cultures of these bacteria, leaves of frost-sensitive species freeze at higher temperatures than leaves that are bacteria-free (Lindow et al., 1982). The surface ice quickly spreads into the cell walls, and cells that lack the ability to halt the spread of extracellular ice crystals into the protoplast are killed. Bacterial strains can be genetically modified so that they are no longer active in ice nucleation, and such strains have been used commercially in foliar sprays to minimize the number of potential ice nucleation points on valuable frost-sensitive crops like strawberry.

Heat Stress and Heat Shock

Turning to the other temperature extreme, to what extent are plants able to resist excessively high temperatures? Few higher plant species survive a steady temperature above 45°C. Cells or tissues of higher plants that are dehydrated and nongrowing can survive much higher temperatures than hydrated, vegetative, growing cells (Table 14.3). Actively growing tissues rarely survive temperatures above 45°C, but dry seeds can endure 120°C and pollen grains 70°C. In general, above 50°C only single-celled organisms will complete their life cycles, and above 60°C only prokaryotes.

Some succulent higher (CAM) plants, such as *Opuntia* and *Sempervivum*, are adapted to hot conditions and can tolerate tissue temperatures of 60–65°C during intense solar radiation in summer. CAM plants keep their stomata closed during the day and so cannot cool by transpiration; instead, they depend on emission of long-wave (infrared) radiation and loss of heat by conduction or convection to the surrounding air (Chapter 10). For typical mesophytic plants that depend on transpirational cooling, a rise in leaf temperature of 4–5°C above ambient air can readily occur in bright sunlight near midday if wind speed is low, particularly if a soil water deficit causes partial stomatal closure. Increases in leaf temperature during the day can be even more pronounced for plants in arid and semiarid regions during drought and with high irradiance from sunshine. Thus, water deficit is invariably associated with some heat stress, and we can see from Table 14.3 that a rise in temperature may shift leaves from temperatures that are nearly optimal into a range that is nearly lethal. Heat stress is also a potential danger in a greenhouse environment, where low air speed and high humidity minimize the rate of leaf cooling. A moderate degree of heat stress slows the growth of the whole plant. More specific symptoms are "bark-burn," in which the cork cambium (phellogen) is killed in thin-barked trees on the side most exposed to solar radiation;

TABLE 14.3. Heat-killing temperatures for plants.

Plant	Heat-killing temperature (°C)	Time of exposure
Nicotiana rustica (wild tobacco)	49–51	10 min
Cucurbita pepo (squash)	49–51	10 min
Zea mays (corn)	49–51	10 min
Brassica napus (rape)	49–51	10 min
Citrus aurantium (sour orange)	50.5	15–30 min
Opuntia (cactus)	>65	—
Sempervivum arachnoideum	57–61	—
Potato leaves	42.5	1 h
Pine and spruce seedlings	54–55	5 min
Medicago seeds (alfalfa)	120	30 min
Grape (ripe fruit)	63	—
Tomato fruit	45	—
Red pine pollen	70	1 h
Various mosses		
Hydrated	42–51	—
Dehydrated	85–110	—

Abridged from Table 11.2 in Levitt (1980).

girdling of bark and phloem at the soil surface when soil temperatures are high; and "burns" with wound cork on fruits (grape, cherry, tomato, pear).

At High Temperatures, Photosynthesis Is Inhibited Before Respiration

Both photosynthesis and respiration are inhibited at high temperatures but, as temperature increases, photosynthetic rates decrease faster than respiratory rates (Fig. 14.11). The temperature at which the amount of CO_2 fixed by photosynthesis equals the amount of CO_2 released by respiration is called the **temperature compensation point.** Above this point, photosynthesis cannot replace the carbon used as a substrate for respiration. As a result, carbohydrate reserves decline, and fruits and vegetables

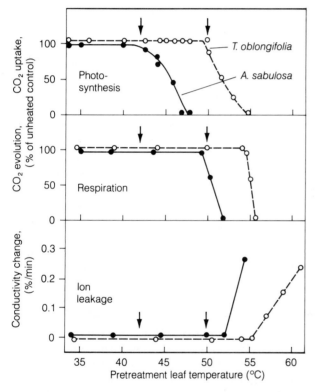

FIGURE 14.11. Response of *Atriplex sabulosa* and *Tidestromia oblongifolia* to heat stress. Photosynthesis and respiration were measured on attached leaves, and ion leakage was measured with leaf slices submersed in water. Control rates were measured at a noninjurious 30°C. Attached leaves were then exposed to the indicated temperatures for 15 min and returned to the initial control conditions before the rates were recorded. Arrows indicate the temperature thresholds for inhibition of photosynthesis. Photosynthesis, respiration, and membrane permeability were all more sensitive to heat damage in *A. sabulosa* than in *T. oblongifolia*. In both species, however, photosynthesis was more sensitive to heat stress than either of the other two processes and was completely inhibited at temperatures that were noninjurious to respiration. (From Björkman et al., 1980.)

lose sweetness. This imbalance between photosynthesis and respiration is one of the main causes of the deleterious effects of high temperatures. In the same plant, the temperature compensation point is usually lower for shade leaves than for sun leaves that are normally exposed to light (and heat). Enhanced respiration relative to photosynthesis at high temperatures is more pronounced and detrimental in C3 plants than in C4 or CAM plants, since both photorespiration and dark respiration continue to rise in C3 plants.

Plants Adapted to Cool Temperatures Acclimate Poorly to High Temperatures

The extent to which plants adapted to one temperature range can acclimate to a contrasting temperature range is illustrated by comparing the responses of two C4 species, *Atriplex sabulosa* (a member of the Chenopodiaceae) and *Tidestromia oblongifolia* (Amaranthaceae). *A. sabulosa* is native to the cool climate of coastal northern California, and *T. oblongifolia* is native to the very hot climate of Death Valley, California, where it grows at a range of temperatures that is lethal for most plant species. When these species were grown in a controlled environment and their growth rates were recorded as a function of temperature, *T. oblongifolia* barely grew at 16°C, while *A. sabulosa* was at 75% of its maximum growth rate. By contrast, the growth rate of *A. sabulosa* began to decline between 25 and 30°C and growth ceased at 45°C, the temperature at which *T. oblongifolia* growth showed a maximum (Björkman et al., 1980). Clearly, neither species could acclimate to the extreme range of the other.

High Temperature Impairs the Thermal Stability of Membranes and of Proteins

The structure and stability of various cellular membranes are important during high-temperature stress, just as they are during chilling and freezing. Excessive fluidity of membrane lipids at high temperatures is correlated with loss of physiological function. In oleander (*Nerium oleander*) plants, acclimation to high temperatures is associated with a greater degree of saturation of fatty acids in membrane lipids, tending to make the membranes less fluid (Raison et al., 1982). At high temperatures the strength of hydrogen bonds and electrostatic interactions between polar groups of proteins within the aqueous phase of the membrane decrease. Thus, integral membrane proteins (which associate with both hydrophilic and lipid regions of the membrane) tend to associate more strongly with the lipid phase. High temperatures thus modify membrane composition and structure and cause leakage of ions (Fig. 14.11c). Membrane disruption also causes inhibition of processes such as photosynthesis and respiration that depend on the activity of membrane-associated enzymes.

Photosynthesis is especially sensitive to high temperature (Chapter 10). In their study of *Atriplex* and *Tidestromia*, Björkman et al. (1980) found that electron transport in photosystem II was more sensitive to high temperature in the cold-adapted *A. sabulosa* than in the heat-adapted *T. oblongifolia*. In both species the decline in electron transport paralleled the decline in photosynthetic CO_2 fixation. Studies of enzymes extracted from these plants showed that ribulose-1,5-bisphosphate carboxylase, NADP-glyceraldehyde dehydrogenase, and phosphoenolpyruvate carboxylase were less stable at high temperatures in *A. sabulosa* than in *T. oblongifolia*. However, the temperatures at which these enzymes began to denature and lose activity were distinctly higher than the temperatures at which photosynthesis began to decline. These results suggest that early stages of heat injury and decline in photosynthesis are more directly related to changes in membrane properties and uncoupling of the energy transfer mechanism in chloroplasts than to a general denaturation of proteins.

Several Adaptations Protect Leaves Against Excessive Heating

In environments with intense solar radiation and high temperatures, plants avoid excessive heating of their leaves by decreasing their absorption of solar radiation. The anatomical and physiological adaptations that accomplish this are similar to the adaptations to water stress that decrease water use by energy dissipation (see page 354). In warm sunny environments in which a transpiring leaf is near its upper limit of temperature tolerance, any further warming arising from decreased evaporation of water or increased energy absorption can damage the leaf. When energy absorption decreases and the leaf cools, the driving force for evaporation of water also decreases. As a result, both drought resistance and heat resistance depend on the same adaptations: reflective leaf hairs and leaf waxes; leaf rolling and vertical leaf orientation; and growth of small, highly dissected leaves to minimize the boundary layer thickness and thus maximize convective and conductive heat loss (see Chapters 4 and 10). Some desert shrubs (e.g., brittlebush, *Encelia farinosa*) have dimorphic leaves to avoid excessive heating: Green, nearly hairless leaves present in the winter are replaced by white, pubescent leaves in the summer.

Abrupt Increases in Temperature Induce Synthesis of Heat Shock Proteins

Recently, researchers have turned their attention to the induction of proteins that may somehow be involved in acclimation—the **heat shock proteins**. Heat shock proteins (HSPs) were discovered in the fruit fly (*Drosophila melanogaster*) and have since been identified in other animals as well as in plants and microorganisms. When cells or seedlings of soybean are suddenly shifted from 25°C to 40°C (just below the lethal temperature), synthesis of the usual set of proteins is suppressed at the level of translation while synthesis of a set of 30–50 other proteins (HSPs) is enhanced. New HSP transcripts (mRNAs) can be detected after 3–5 min of heat shock (Sachs and Ho, 1986). The molecular masses of the HSPs range from 15 to 104 kDa; the lower-mass HSPs are more abundant in higher plants. Some HSPs are very similar in plants and animals. For instance, there is a 75% homology between structural genes coding for a 70-kDa HSP in higher plants and *Drosophila*. There are also many similarities in the regulatory sequences of these genes in *Drosophila* and soybean. Although the experimentally induced temperature shifts (25 to 40°C) rarely occur in nature, HSPs are known to be induced by more gradual rises in temperature and have been identified in plants under field conditions.

Cells or plants that have been induced to synthesize HSPs show improved thermal tolerance and can tolerate exposure to temperatures that were previously lethal. However, the precise role of the HSPs remains unclear. They appear to act as protective agents, and some are localized in the nucleus or the chloroplast. One HSP that is localized in the chloroplast may protect against photoinhibition during heat shock in the light. HSPs may include enzymes associated with mechanisms of thermal tolerance discussed above, such as membrane lipid saturation or protein stability. Some of the HSPs are not unique to high-temperature stress; they are also induced by widely different environmental stresses or conditions, including water deficit, ABA treatment, wounding, and salinity (Bonham-Smith et al., 1988). This discovery raises the possibility that plants previously exposed to one stress will gain cross-protection against another stress. Some of the HSPs might also play a role in normal metabolism. For example, a class of proteins called **chaperonins** has been found to be essential for the post-translational folding and assembly of other proteins. Some HSPs function as chaperonins, and one of these chaperonins in *Escherichia coli* is homologous to the Rubisco-binding protein in the chloroplast (Goloubinoff et al., 1989). Chaperonins are present in the cell at physiological temperatures, but they can become more abundant after a temperature shock.

Salinity

Under natural conditions, terrestrial higher plants encounter high concentrations of salts close to the seashore and in estuaries where seawater and fresh water mix or replace each other with the tides. Far inland, natural salt seepage from geologic marine deposits can wash into adjoining areas, rendering them virtually unusable for agriculture. However, a much more extensive problem in agriculture is the accumulation of salts from irrigation water. Evaporation and transpiration remove pure water (as vapor) from the soil, and this water loss concentrates solutes in the soil. When the quality of irrigation water is poor (that is, when it contains a high concentration of solutes) and when there is no opportunity to flush out accumulated salts to a drainage system with an occasional excess irrigation, salts can quickly reach levels that are injurious to salt-sensitive species. It is estimated that about one-third of the irrigated land on the earth is salt-affected.

Salt Accumulation in Soils Impairs Plant Function and Soil Structure

In discussing the effects of salts in the soil, we distinguish between high concentrations of Na^+ (**sodicity**) and high concentrations of total salts (**salinity**). The two concepts are often related, but in some areas Ca^{2+}, Mg^{2+}, Cl^-, and SO_4^{2-}, as well as NaCl, can contribute substantially to salinity. The high Na^+ concentration of a sodic soil can not only injure plants directly but also degrade the soil structure, decreasing porosity and water permeability. A sodic clay soil known as **caliche** is so hard and impermeable that dynamite is sometimes required to dig through it!

In the field, the salinity of soil water or irrigation water is measured in terms of its electrical conductivity or in terms of osmotic potential. Pure water is a very poor conductor of electrical current, and the conductivity of a water sample is due to the ions dissolved in it. The higher the salt concentration of water, the greater is its conductivity and the more negative its osmotic potential (higher osmotic pressure, π in MPa). The properties of seawater, which is too salty for all but the most tolerant land plants, and of good-quality irrigation water are compared in Table 14.4.

The quality of irrigation water is often poor in semiarid and arid regions. In the United States, the salt

content of the headwaters of the Colorado River is only 50 mg L^{-1} but some 2000 km downstream, in southern California, the salt content of the same river reaches about 900 mg L^{-1}—enough to preclude growing some salt-sensitive crops such as maize. Water from some wells used for irrigation in Texas may contain as much as 2000–3000 mg L^{-1}. An annual application of irrigation water totaling 1 m from such wells would add 8–12 tons of salts per acre to the soil! These levels of salt are damaging to all but the most resistant crops.

Salinity Depresses Growth and Photosynthesis in Sensitive Species

Plants can be divided into two broad groups on the basis of their response to high concentrations of salts. *Halophytes* are native to saline soils and complete their life cycles in that environment. *Glycophytes* (literally, sweet plants), or nonhalophytes, are not able to resist salts to the same degree as halophytes. Usually, there is a threshold concentration of salt above which glycophytes begin to show signs of growth inhibition, leaf discoloration, and loss of dry weight. Among crops, maize, onion, citrus, pecan, lettuce, and bean are highly sensitive to salt; cotton and barley are moderately tolerant; and sugar beet and date palms are highly tolerant (Maas and Hoffman, 1977; Greenway and Munns, 1980). Some species that are highly tolerant of salt, such as *Suaeda maritima* (a salt marsh plant) and *Atriplex nummularia* (a salt bush), show growth stimulation (in terms of fresh weight gain) at Cl^- concentrations many times greater than the lethal level for sensitive species (Fig. 14.12).

Salt Injury Involves Both Osmotic Effects and Specific Ion Effects

Dissolved solutes in the rooting zone generate a negative osmotic potential that lowers the soil water potential. The general water balance of plants is thus affected, because leaves need to develop a more negative water potential to maintain a "downhill" gradient of water potential between the soil and the leaves (see Chapter 4). This effect of dissolved solutes is similar to that of a soil water deficit. Some plants can adjust osmotically when growing in saline soils and in this way prevent loss of turgor, which would slow extension growth of cells while generating a lower (more negative) water potential.

Specific ion effects occur when injurious concentrations of Na^+, Cl^-, or SO_4^{2-} accumulate in cells. Under nonsaline conditions the cytosol of higher plant cells contains 100–200 mM $[K^+]$ and 1 mM $[Na^+]$, an ionic environment in which many enzymes function optimally. An abnormally high ratio of Na^+ to K^+ and high concentrations of total salts inactivate enzymes and inhibit protein synthesis. At a high concentration, Na^+ can displace Ca^{2+} from the plasma membrane of cotton root

TABLE 14.4. Properties of seawater and of good-quality irrigation water.

Property	Seawater	Irrigation water
Concentration of ions (mM)		
Na^+	457	<2.0
K^+	9.7	1.0
Ca^{2+}	10	0.5–2.5
Mg^{2+}	56	0.25–1.0
Cl^-	536	<2.0
SO_4^{2-}	28	0.25–2.5
HCO_3^-	2.3	<1.5
Osmotic potential (MPa)	−2.4	−0.039
Total dissolved salts (mg L^{-1} or ppm)	32,000	500

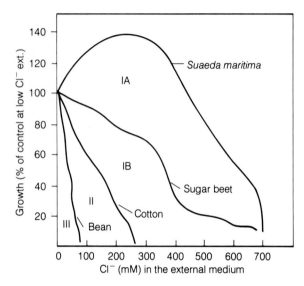

FIGURE 14.12. The growth of different species subjected to salinity relative to that of unsalinized controls. The curves dividing the regions are based on data for different species. Plants were grown for 1–6 months. Representative plant species within the groups are as follows. Group IA, halophytes: *Suaeda maritima*, *Atriplex nummularia*. These species show growth stimulation with Cl⁻ levels below 400 mM. Group IB, halophytes: *Spartina townsendii* and sugar beet (*Beta vulgaris*). These tolerate salt, but growth is retarded. Group II, halophytes and nonhalophytes: This group includes salt-tolerant halophytic grasses without salt glands, such as *Festuca rubra* ssp. *littoralis* and *Puccinellia peisonis*, and nonhalophytes, such as cotton and barley. All are inhibited by high salt concentrations. Within this group, tomato (*Lycopersicon esculentum*) is intermediate, and common bean (*Phaseolus vulgaris*) and soybean (*Glycine max*) are sensitive. Group III, very salt-sensitive nonhalophytes: The species are severely inhibited or killed by low salt concentrations. Included are many fruit trees such as citrus, avocado, and stone fruit. (From Greenway and Munns, 1980.)

hairs, resulting in a change in plasma membrane permeability that can be detected as leakage of K^+ to the bathing solution (Cramer et al., 1985).

Photosynthesis is inhibited when high concentrations of Na^+ and/or Cl^- accumulate in chloroplasts. Since photosynthetic electron transport appears relatively insensitive to salts, either carbon metabolism or photophosphorylation may be affected. Enzymes extracted from salt-tolerant species are just as sensitive to the presence of NaCl as enzymes from salt-sensitive glycophytes. Hence, the resistance of halophytes to salts is not a consequence of a salt-resistant metabolic machinery. Instead, other mechanisms come into play, as discussed in the following section.

Plants Use Different Strategies to Avoid Salt Injury

Plants avoid salt injury by exclusion of ions from the leaves or by compartmentation of ions in vacuoles. Among plants that are sensitive to salt, resistance to moderate levels of salinity in the soil depends on the ability of the roots to prevent potentially harmful ions from reaching the leaves. Sodium ions can enter roots passively (by moving down an electrochemical potential gradient; see Chapter 6), so root cells must use energy for active transport of Na^+ back to the outside solution. By contrast, Cl^- is excluded by the low permeability of root plasma membranes to this ion. Movement of Na^+ into leaves is further minimized by absorption of Na^+ from the transpiration stream (xylem sap) during its movement from roots to shoots. Some salt-resistant plants, such as salt-cedar (*Tamarix* species) and salt bush (*Atriplex* species), do not exclude ions at the root but instead have salt glands in the surface of the leaves. The ions are transported to these glands, where the salt crystallizes and is no longer harmful.

When salts are excluded from leaves, plants use organic substances to lower the solute potential of the cytoplasm and the vacuole and thus lower the leaf water potential. These organic compounds, which do not interfere with cellular metabolism at high concentrations, include glycine betaine, proline, sorbitol, and sucrose. Specific plant families tend to use one or two of these compounds in preference to others (Wyn Jones and Gorham, 1983). The amount of carbon used for the synthesis of these organic solutes can be rather large (about 10% of the plant weight). In natural vegetation this diversion of carbon does not affect survival, but in agricultural crops it can reduce yields.

Many halophytes absorb, rather than exclude, ions and accumulate them in the leaves. However, these ions are sequestered in the vacuoles of leaf cells, where they can contribute to the cell osmotic potential without damaging the salt-sensitive chloroplastic and cytosolic enzymes. In these leaves, water balance is maintained between the cytoplasm and the vacuoles by accumulating organic compounds, such as proline or sucrose, in the cytoplasm. Because the volume of cytoplasm in a mature, vacuolated cell is small compared with the volume of the vacuole, the amount of carbon needed for the synthesis of organic compounds used as osmotica is much less in these plants than in plants that exclude salts.

Salt Stress Induces Synthesis of New Proteins

Exposure to NaCl or to abscisic acid induces synthesis of proteins associated with improved tolerance to NaCl. In tissue culture, cells of *Citrus* species or tobacco (*Nicotiana tabacum*) have been acclimated to tolerate unusually high concentrations of salt. During this acclimation, a number of newly synthesized proteins can be detected by gel electrophoresis. If cells are exposed to low concentrations of ABA before being exposed to high salt concentrations, their ability to acclimate to salt is improved. In addition, ABA stimulates the synthesis of one or more proteins that appear to be the same as those induced during acclimation to NaCl (Singh et al., 1987). Thus, it seems possible that ABA may play some role in acclimation. In intact plants, high salt concentrations generally increase ABA concentrations in leaves, and much of that ABA appears to be transported from roots. This response to salt is similar to the increase in ABA production by roots and transport to shoots described previously in relation to soil water deficit.

Oxygen Deficiency

Roots usually obtain sufficient oxygen for their aerobic respiration (see Chapter 11) directly from the soil. Gas-filled pores in well-drained, well-structured soil readily permit the diffusion of gaseous O_2 to depths of several meters. Consequently, the O_2 concentration deep in the soil is similar to that in humid air. However, soil can become flooded or waterlogged when it is poorly drained or when rain or irrigation is excessive. Water then fills the pores and blocks the diffusion of O_2 in the gaseous phase. Dissolved oxygen diffuses so slowly in water that only a few centimeters of soil near the surface remain oxygenated. When temperatures are low and plants are dormant, oxygen depletion is very slow and relatively harmless. However, when temperatures are higher ($> 20°C$), oxygen consumption by plant roots, soil fauna, and soil microorganisms can totally deplete the oxygen from the bulk of the soil water in as little as 24 h. The growth and survival of many plant species are greatly depressed under such conditions, and crop yields can be severely reduced. Garden pea (*Pisum sativum*) yields can be halved by only 24 h of flooding. However, specialized natural vegetation (plants of marshes and swamps) and crops like rice are well adapted to resist oxygen deficiency in the root environment.

Roots Are Injured in Anaerobic Soil Water

In the absence of O_2, the tricarboxylic acid cycle cannot operate, and ATP can be produced only by fermentation. When the supply of O_2 is insufficient for aerobic respiration, roots begin to ferment pyruvate (formed in glycolysis; see Chapter 11) to lactate. Lactate fermentation, however, is transient because the accumulation of lactate lowers the cellular pH. As the intracellular pH drops, fermentation to ethanol is activated. The *net* yield of ATP in fermentation is only 2 moles of ATP per mole of hexose sugar respired (compared with 36 moles of ATP per mole of hexose in aerobic respiration). Thus, injury to root metabolism by O_2 deficiency originates from a lack of ATP to drive basic metabolic processes.

Nuclear magnetic resonance (NMR) spectroscopy was used to measure the intracellular pH of living maize root tips under nondestructive conditions (Roberts et al., 1984). In healthy cells, the vacuole contents are usually more acidic (pH 5.8) than the cytoplasm (pH 7.4). But under conditions of extreme O_2 deficiency, protons gradually leaked from the vacuole into the cytoplasm, which became increasingly acidic. These changes were associated with the onset of cell death. Apparently, active transport of H^+ into the vacuole by tonoplast ATPases is inhibited by lack of ATP, and without ATPase activity a normal pH gradient cannot be maintained. Abnormal acidity irreversibly disrupts metabolsim in the cytoplasm.

When soil is completely depleted of molecular O_2, anaerobic soil microorganisms (anaerobes) derive their energy from reduction of nitrate to nitrite or to N_2O and N_2, which are lost as gases to the atmosphere, in a process called denitrification. As conditions become more reducing (i.e., lower redox potential), other anaerobes may reduce SO_4^{2-} to H_2S, which is a respiratory poison. Still other anaerobes can reduce Fe^{3+} to Fe^{2+}, which is highly soluble. When anaerobes have an abundant supply of organic substrate, bacterial metabolites such as acetic acid and butyric acid are released into the soil water, and these acids along with reduced sulfur compounds account for the unpleasant odor of waterlogged soil. All of these substances are toxic to plants at high concentrations (Drew and Lynch, 1980).

Interference with root respiration rate and metabolism occurs even before O_2 is completely depleted from the root environment. For a maize root tip at 25°C, the **critical oxygen pressure** (COP) at which the respiration rate is first slowed by O_2 deficiency is about 0.20 atmosphere (20 kPa or 20% O_2 by volume), almost the concentration in ambient air. At this oxygen partial pressure, the rate of diffusion of dissolved O_2 from the solution into the tissue and from cell to cell barely keeps pace with the rate of utilization. When O_2 concentrations are below the COP, the center of the root becomes **anoxic** (completely lacking oxygen) or **hypoxic** (partially deficient in oxygen). The COP is lower when respiration is slow at cooler temperatures, but it also depends on the size of the tissue. For single cells, 0.01 atmosphere O_2 (1% O_2) in the gaseous phase can be adequate because diffusion over short distances ensures an adequate O_2 supply.

The Failure of O_2-Deficient Roots to Function Injures Shoots

Anoxic or hypoxic roots lack sufficient energy to support physiological processes on which the shoots depend (Jackson and Drew, 1984). The failure of roots of wheat or barley to absorb nutrient ions and transport them to the xylem (and thence to the shoot) leads quickly to shortage of ions within developing and expanding tissues. Older leaves senesce prematurely because of reallocation of phloem-mobile elements (N, P, K) to younger leaves. The lower permeability of roots to water often leads to a decrease in leaf water potential and wilting, although this decrease is temporary if stomata close, preventing further water loss by transpiration. Hypoxia accelerates production of the ethylene precursor 1-aminocyclopropane-1-carboxylic acid (ACC) in roots (Chapter 19). In tomato, ACC travels via the xylem sap to the shoot, where, in contact with oxygen, it is converted by the ethylene-forming enzyme (EFE) to ethylene. The upper (adaxial) surfaces of the leaf petioles of tomato and sunflower have ethylene-responsive cells that expand more rapidly when ethylene concentrations are high. This expansion results in **epinasty**, in which the leaves grow downward and appear to droop. Unlike wilting, epinasty does not involve loss of turgor.

In some species (e.g., pea and tomato) flooding induces stomatal closure without detectable changes in leaf water potential. Oxygen shortage in roots, like water deficit or high concentrations of salts, stimulates ABA production, and movement of the ABA to leaves can account for the stomatal response (Zhang and Davies, 1987).

Submerged Organs Can Acquire O_2 Through Specialized Structures

In contrast to flood-sensitive species, wetland vegetation is well adapted to growth for extended periods in water-saturated soil. Even when shoots are partially submerged, they grow vigorously and show no signs of stress. In these plants, the stem and roots develop longitudinally interconnected, gas-filled channels that provide a low-resistance pathway for diffusion of oxygen and other gases. The gases (air) enter through stomata, or through lenticels on woody stems and roots.

In many wetland plants, such as rice, tissues composed of cells separated by prominent, gas-filled spaces are known as **aerenchyma**. These structures develop independently of environmental stimuli. However, in

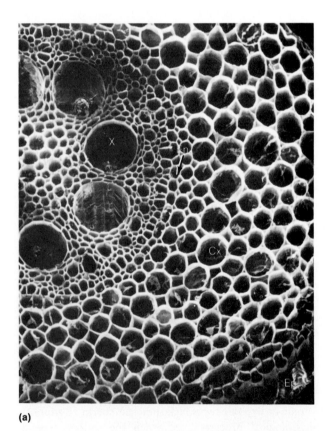

(a)

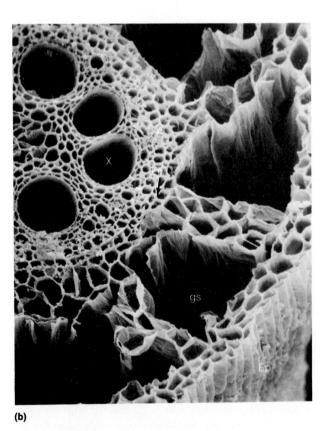

(b)

FIGURE 14.13. Scanning electron micrographs of transverse sections through roots of maize showing changes in structure with oxygen supply (170×). (a) Control root, supplied with air, with intact cortical cells. (b) Oxygen-deficient root growing in a nonaerated nutrient solution. Note the prominent gas-filled spaces (gs) in the cortex formed by degeneration of cells. The stele (all cells interior to the endodermis) and epidermis (Ep) remain intact. Cx, cortex; En, endodermis; X, xylem. (Photographs courtesy of J. L. Basq and M. C. Drew.)

barley, wheat, maize, and other nonwetland plants of the Gramineae, formation of aerenchyma is induced in roots by oxygen deficiency (Fig. 14.13). In the root tip of maize, hypoxia stimulates greater production of ACC and of ethylene, and the latter promotes the breakdown (lysis) of cells in the root cortex. As the root extends into oxygen-deficient soil, the continuous formation of aerenchyma just behind the tip allows oxygen movement within the root to supply the apical zone. In roots of rice and other wetland plants, structural barriers composed of suberized and lignified cells prevent O_2 diffusion outward to the soil. The O_2 thus retained aerates the apical meristem and allows growth to proceed 50 cm or more into anaerobic soil. In contrast, roots of nonwetland species such as maize are leaky to O_2 and fail to conserve it to the same extent. Thus, in the root apex of these plants, internal O_2 becomes insufficient for aerobic respiration, and this severely limits the depth to which such roots can extend into anaerobic soil.

Some Plant Tissues Tolerate Anaerobic Conditions

Most tissues of higher plants cannot survive anaerobically for long. Root tips of maize, for example, remain viable for 20–24 h if they are suddenly deprived of O_2. Under anoxia, some ATP is generated slowly by fermentation, and the energy status of cells gradually declines. By contrast, some plants (or parts of them) can tolerate exposure to strictly anaerobic conditions for an extended period. These include the rhizomes (underground horizontal stems) of *Schoenoplectus lacustris* (giant bulrush), *Scirpus maritimus* (salt-marsh bulrush), and *Typha angustifolia* (narrow-leaved cattail), which can survive for several months and expand their leaves in an anaerobic atmosphere. The embryo and coleoptile of rice and *Echinochloa crus-galli* var *oryzicola* (rice grass) can also survive weeks of anoxia. In nature, the rhizomes overwinter in anaerobic mud at the edges of lakes. In spring, once the leaves have expanded above the mud or water surface, O_2 diffuses down through aerenchyma into the rhizome. Metabolism then switches from an anaerobic (fermentative) to an aerobic mode, and roots begin to grow using the available oxygen. Likewise, during germination of paddy (wetland) rice and of rice grass, the coleoptile breaks the water surface and becomes a diffusion pathway (a "snorkel") for O_2 to the rest of the plant. Even though rice is a wetland species, its roots are as intolerant of anoxia as those of maize.

Acclimation to O_2 Deficit Involves Production of Anaerobic Stress Proteins

When maize roots are made oxygen-deficient, protein synthesis virtually ceases except for about 20 polypeptides that continue to be produced (Sachs and Ho, 1986). Three of these anaerobic stress proteins have been identified as enzymes of the glycolytic pathway, and two others (pyruvate decarboxylase and alcohol dehydrogenase) catalyze fermentation of pyruvate to ethanol. Since ethanolic fermentation is the major metabolic pathway by which ATP is synthesized in anoxic cells of higher plants and higher rates of fermentation correlate with improved energy status of cells, these anaerobic stress proteins are likely to play a physiological role during anoxia.

Air Pollution

The concentration of CO_2 in the atmosphere has been increasing steadily (Chapter 10) because of the growth of industry, urbanization, and the clearing of forests. It is projected to reach 600 μL L^{-1} during the twenty-first century (Schneider, 1989). Carbon dioxide is a natural component of the atmosphere, and its increase to these levels should not directly damage vegetation. Indeed, in C3 plants, CO_2 enrichment accelerates growth through its stimulation of photosynthesis (Chapter 10). However, higher concentrations of CO_2 and other "greenhouse gases" in the atmosphere increase the absorption of infrared radiation emitted by the earth, and this is expected to contribute to future global warming. The effects of such warming on climate, ocean levels, flora, and fauna are of growing concern.

The burning of hydrocarbons in motor vehicle engines gives rise to CO_2, CO, SO_2, NO_x (NO and NO_2 in varying proportions), and C_2H_4 (ethylene), as well as a variety of other hydrocarbons. Additional SO_2 originates from domestic and industrial burning of fossil fuels. Industrial plants, such as chemical works and metal smelting plants, release SO_2, H_2S, NO_2, and HF into the atmosphere. Tall chimney stacks may be used to carry gases and particles to a high altitude and thus avoid local pollution, but the pollutants return to the earth, sometimes hundreds of kilometers from the original source. Photochemical smog is the product of chemical reactions driven by sunlight and involving NO_x of urban and industrial origin and volatile organic compounds from either vegetation (**biogenic** hydrocarbons) or human activities (**anthropogenic** hydrocarbons). The ozone (O_3) and peroxyacetylnitrate (PAN) produced in these complex reactions are injurious to plants and other life forms.

The concentrations of polluting gases or their solutions to which plants are exposed are thus highly variable, depending on location, wind direction, rainfall, and sunlight. In urban areas, concentrations of SO_2 and NO_x in air are typically 0.02–0.5 μL L^{-1}, the upper value being well within the range that is inhibitory to plant growth.

Polluting Gases Inhibit Stomatal Movements, Photosynthesis, and Growth

Polluting gases such as SO_2 and NO_x enter leaves through stomata, following the same diffusion pathway as CO_2. NO_x dissolves in cells and gives rise to nitrite ions (which are toxic at high concentrations) and nitrate ions that enter into nitrogen metabolism as if they had been absorbed through the roots. In some cases, exposure to pollutant gases, particularly SO_2, causes stomatal closure, which protects the leaf against further entry of the pollutant but also curtails photosynthesis. In the cells, SO_2 dissolves to give bisulfite and sulfite ions; sulfite is toxic but at low concentrations is metabolized by chloroplasts to sulfate, which is not toxic. At very low concentrations, bisulfite and sulfite are effectively detoxified by plants, and SO_2 pollution may replace sulfur-containing fertilizers.

In urban areas these polluting gases, especially O_3, may be present in such high concentrations that they cannot be detoxified rapidly. Ozone is highly reactive, binding to plasma membranes and altering metabolism. As a result, stomatal apertures are poorly regulated by guard cells, chloroplast thylakoid membranes are damaged, and photosynthesis is inhibited. Ozone reactions produce toxic oxygen species, including peroxide (OOH^-), superoxide $(OO^{\cdot-})$, singlet oxygen $(O^{\cdot})$, and hydroxyl radical $(OH^{\cdot})$. Plant cells contain very high concentrations of the sulfur-containing tripeptide glutathione, which reacts rapidly with these oxygen species and detoxifies them while being oxidized in the process. Often, exposure of plants to toxic oxygen species stimulates the enzymatic machinery used to regenerate reduced glutathione, thereby providing increased protection.

Many deleterious changes in metabolism caused by air pollution take place without external symptoms of injury, which becomes evident only at much higher concentrations. For example, when plants are exposed to air containing NO_x, lesions on leaves appear at an NO_x concentration of 5 $\mu L\ L^{-1}$, but photosynthesis starts to be inhibited at a concentration of only 0.1 $\mu L\ L^{-1}$. These threshold concentrations characterize the effects of single pollutants. However, two or more pollutants acting together can have a synergistic effect, producing damage at lower concentrations than if they were acting separately. In addition, vegetation weakened by air pollution can become more susceptible to invasion by pathogens and pests.

Polluting Gases, Dissolved in Rainwater, Fall As "Acid Rain"

Unpolluted rain is slightly acidic, with a pH close to 5.6, because the CO_2 dissolved in it produces the weak acid H_2CO_3. Dissolution of NO_x and SO_2 in water droplets in the atmosphere causes the pH of rain to decrease to 3–4,

and in southern California polluted droplets in fog can be as acidic as pH 1.7. Leaching of leaves by a dilute acidic solution can remove mineral nutrients, depending on the age of leaf and the integrity of the cuticle and surface waxes. The total annual contributions to the soil of acid from acid rain (**wet deposition**) and from particulate matter falling on the soil plus direct absorption from the atmosphere (**dry deposition**) may reach 0.4–1.2 kg H^+ acre^{-1} in parts of Europe and the northeastern United States (Schwartz, 1989). In soils that lack free calcium carbonate, and are therefore not strongly buffered, such additions of acid can be harmful to plants. Furthermore, the added acid can result in the release of aluminum ions from soil minerals, causing aluminum toxicity. Air pollution is considered to be a major factor in the decline of forests in polluted areas of Europe and North America.

Summary

Under both natural and agricultural conditions, plants are exposed to unfavorable environments that lead to some degree of stress. Soil water deficits, suboptimal and supraoptimal temperatures, salinity, and poor aeration of soils may each cause some growth restrictions during the growing season, so that the yield of plants at the end of the season expresses only a small fraction of their genetic potential. The same physicochemical factors can become extreme in some habitats, such as deserts or marshes, and only specially adapted vegetation can complete its life cycle in the unusually hostile conditions. In less extreme (mesic) environments, individual plants can become acclimated to changes in water potential, temperature, salinity, and oxygen deficiency so that their fitness for those environments improves. Some species are better able to adapt than others, and various anatomical, structural, and biochemical mechanisms account for acclimation. Recent studies suggest that changes in gene expression at the level of transcriptional or translational control of protein synthesis are involved. Some changes in gene expression during acclimation may be signaled by changes in plant hormone concentrations; thus, abscisic acid has been implicated in acclimation to several stresses, although other plant hormones will undoubtedly be shown to play a role. Some changes in the synthesis of specific mRNAs and the proteins translated from them are induced by a range of different factors—heat shock, salinity, water deficit, and, significantly, exposure to ABA in the absence of stress. Other changes in transcription and translation are induced only by the specific stress to which the plant is exposed. Knowledge of the identity and function of these stress-responsive genes will help in the engineering of plants, through recombinant DNA techniques as well as conventional plant breeding, to improve their ability to contend with environmental stress.

GENERAL READING

Björkman, O., Badger, M. R., and Armond, P. A. (1980) Response and adaptation of photosynthesis to high temperatures. In: *Adaptation of Plants to Water and High Temperature Stress*, Turner, N. C., and Kramer, P. J., eds., pp. 233–249. Wiley, New York.

Boyer, J. S. (1982) Plant productivity and environment. *Science* 218:443–448.

Cherry, J. H., ed. (1989) *Environmental Stress in Plants: Biochemical and Physiological Mechanisms Associated with Environmental Stress Tolerance in Plants*, NATO ASI Series G, Vol. 19. Springer-Verlag, Berlin.

Drew, M. C., and Lynch, J. M. (1980) Soil anaerobiosis, microorganisms and root function. *Annu. Rev. Phytopathol.* 18:37–66.

Fitter, A. H., and Hay, R. K. M. (1987) *Environmental Physiology of Plants*, 2d ed. Academic Press, London.

Greenway, H., and Munns, R. (1980) Mechanisms of salt tolerance in nonhalophytes. *Annu. Rev. Plant Physiol.* 31:149–190.

Jackson, M. B., and Drew, M. C. (1984) Effects of flooding on growth and metabolism of herbaceous plants. In: *Flooding and Plant Growth*, Kozlowski, T. T., ed., pp. 47–128. Academic Press, New York.

Levitt, J. (1980) *Responses of Plants to Environmental Stresses*, Vol. 1, 2d ed. Academic Press, New York.

Sachs, M. M., and Ho, D. T. H. (1986) Alternation of gene expression during environmental stress in plants. *Annu. Rev. Plant Physiol.* 37:363–376.

Schneider, S. H. (1989) The greenhouse effect: Science and policy. *Science* 243:771–781.

Schulze, E.-D. (1986) Carbon dioxide and water vapor exchange in response to drought in the atmosphere and in the soil. *Annu. Rev. Plant Physiol.* 37:247–274.

Schwartz, S. E. (1989) Acid deposition: Unraveling a regional phenomenon. *Science* 243:753–763.

Steward, F. C., ed. (1986) *Plant Physiology: A Treatise*, Vol. IX: *Water and Solutes in Plants*. Academic Press, New York.

Turner, N. C., and Jones, M. M. (1980) Turgor maintenance by osmotic adjustment: A review and evaluation, In: *Adaptation of Plants to Water and High Temperature Stress*, Turner, N. C., and Kramer, P. J., eds., pp. 87–103. Wiley, New York.

Turner, N. C., and Kramer, P. J., eds. (1980) *Adaptation of Plants to Water and High Temperature Stress*. Wiley, New York.

Wyn Jones, R. G., and Gorham, J. (1983) Osmoregulation. In: *Encyclopedia of Plant Physiology*, New Series, Vol. 12C, Lange, O. L., Nobel, P. S., Osmond, C. B., and Ziegler, H., eds., pp. 35–58. Springer-Verlag, Berlin.

CHAPTER REFERENCES

Becwar, M. R., Rajashekar, C., Bristow, K. J. H., and Burke, M. J. (1981) Deep undercooling of tissue water and winter hardiness limitations in timberline flora. *Plant Physiol.* 68:111–114.

Blackman, P. G., and Davies, W. J. (1985) Root to shoot communication in maize plants of the effects of soil drying. *J. Exp. Bot.* 36:39–48.

Blizzard, W. E., and Boyer, J. S. (1980) Comparative resistance of the soil and the plant to water transport. *Plant Physiol.* 66:809–814.

Bohnert, H. J., Ostrem, J. A., and Schmitt, J. M. (1989) Changes in gene expression elicited by salt stress in *Mesembryanthemum crystallinum*. In: *Environmental Stress in Plants*, Cherry, J. H., ed., pp. 159–171. Springer-Verlag, Berlin.

Bonham-Smith, P. C., Kapoor, M., and Bewley, J. D. (1988) A comparison of the stress responses of *Zea mays* seedlings as shown by quantitative changes in protein synthesis. *Can. J. Bot.* 66:1883–1890.

Boyer, J. S. (1970) Leaf enlargement and metabolic rates in corn, soybean, and sunflower at various leaf water potentials. *Plant Physiol.* 46:233–235.

Bozarth, C. S., Mullet, J. E., and Boyer, J. S. (1987) Cell wall proteins at low water potentials. *Plant Physiol.* 85:261–267.

Bray, E. A. (1988) Drought- and ABA-induced changes in polypeptide and mRNA accumulation in tomato leaves. *Plant Physiol.* 88:1210–1214.

Brown, M. S., Pereira, E. S. B., and Finkle, B. J. (1974) Freezing of non-woody plant tissues. 2. Cell damage and the fine structure of freezing curves. *Plant Physiol.* 53:709–711.

Burke, M. J., and Stushnoff, C. (1979) Frost hardiness: A discussion of possible molecular causes of injury with particular reference to deep supercooling of water. In: *Stress Physiology in Crop Plants*, Mussell, H., and Staples, R. C., eds., pp. 197–225. Wiley, New York.

Cornish, K., and Zeevaart, J. A. D. (1985) Movement of abscisic acid into the apoplast in response to water stress in *Xanthium strumarium* L. *Plant Physiol.* 78:623–626.

Cramer, G. R., Laüchli, A., and Polito, V. S. (1985) Displacement of Ca^{2+} by Na^+ from the plasmalemma of root cells. A primary response to salt stress? *Plant Physiol.* 79:207–211.

Goloubinoff, P., Gatenby, A. A., and Lorimer, G. H. (1989) Gro E heat-shock proteins promote assembly of foreign prokaryotic ribulose bisphosphate carboxylase oligomers in *Escherichia coli*. *Nature* 337:44–47.

Hanscom, Z., III, and Ting, I. P. (1978) Responses of succulents to plant water stress. *Plant Physiol.* 61:327–330.

Hartung, W., Radin, J. W., and Hendrix, D. L. (1988) Abscisic acid movement into the apoplastic solution of water-stressed cotton leaves. Role of apoplastic pH. *Plant Physiol.* 86:908–913.

Lindow, S. E., Arny, D. C., and Upper, C. D. (1982) Bacterial ice nucleation: A factor in frost injury to plants. *Plant Physiol.* 70:1084–1089.

Lyons, J. M., Wheaton, T. A., and Pratt, H. K. (1964) Relationship between the physical nature of mitochondrial membranes and chilling sensitivity in plants. *Plant Physiol.* 39:262–268.

Maas, E. V., and Hoffman, G. J. (1977) Crop salt tolerance—current assessment. *J. Irrig. Drainage Div. Am. Soc. Civ. Eng.* 103:115–134.

Matthews, M. A., Van Volkenburgh, E., and Boyer, J. S. (1984) Acclimation of leaf growth to low water potentials in sunflower. *Plant Cell Environ.* 7:199–206.

McCree, K. J., and Richardson, S. G. (1987) Stomatal closure vs. osmotic adjustment: A comparison of stress responses. *Crop Sci.* 27:539–543.

Mohapatra, S. S., Poole, R. J., and Dhindsa, R. S. (1988) Abscisic acid–regulated gene expression in relation to freezing tolerance in alfalfa. *Plant Physiol.* 87:468–473.

Patterson, B. D., Paull, R., and Smillie, R. M. (1978) Chilling resistance in *Lycopersicon hirsutum* Humb. & Bonpl., a wild tomato with a wide altitudinal distribution. *Aust. J. Plant Physiol.* 5:609–617.

Radin, J. W., and Hendrix, D. L. (1988) The apoplastic pool of abscisic acid in cotton leaves in relation to stomatal closure. *Planta* 174:180–186.

Raison, J. K., Pike, C. S., and Berry, J. A. (1982) Growth temperature–induced alterations in the thermotropic properties of *Nerium oleander* membrane lipids. *Plant Physiol.* 70:215–218.

Rao, I. M., Sharp, R. E., and Boyer, J. S. (1987) Leaf magnesium alters photosynthetic response to low water potentials in sunflower. *Plant Physiol.* 84:1214–1219.

Roberts, J. K. M., Callis, J., Jardetzky, O., Walbot, V., and Freeling, M. (1984) Cytoplasmic acidosis as a determinant of flooding intolerance in plants. *Proc. Natl. Acad. Sci. USA* 81:6029–6033.

Singh, N. K., LaRosa, P. C., Handa, A. K., Hasegawa, P. M., and Bressan, R. A. (1987) Hormonal regulation of protein synthesis associated with salt tolerance in plant cells. *Proc. Natl. Acad. Sci. USA* 84:739–743.

Sung, F. J. M., and Krieg, D. R. (1979) Relative sensitivity of photosynthetic assimilation and translocation of ^{14}carbon to water stress. *Plant Physiol.* 64:852–856.

U.S. Department of Agriculture. (1989) *Agricultural Statistics.* U.S. Government Printing Office, Washington, D.C.

Weiser, C. J. (1970) Cold resistance and injury in woody plants. *Science* 169:1269–1278.

Williams, J. P., Khan, M. U., Mitchell, K., and Johnson, G. (1988) The effect of temperature on the level and biosynthesis of unsaturated fatty acids in diacylglycerols of *Brassica napus* leaves. *Plant Physiol.* 87:904–910.

Zhang, J., and Davies, W. J. (1987) ABA in roots and leaves of flooded pea plants. *J. Exp. Bot.* 38:649–659.

Zhang, J., Schurr, U., and Davies, W. J. (1987) Control of stomatal behaviour by abscisic acid which apparently originates in the roots. *J. Exp. Bot.* 38:1174–1181.

UNIT
III

GROWTH AND DEVELOPMENT

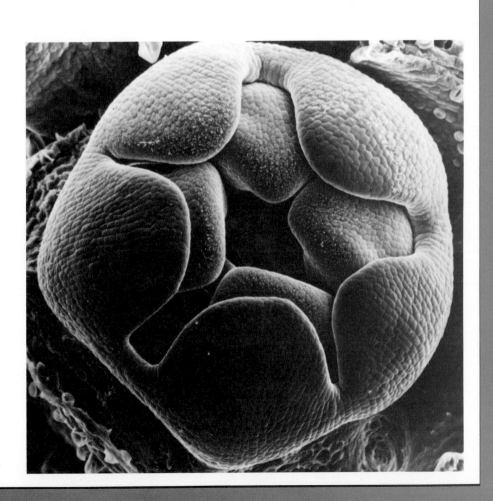

GROWTH AND
DEVELOPMENT

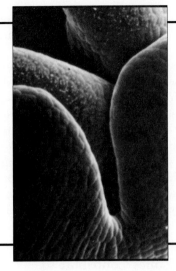

CHAPTER **15**

The Cellular Basis of Growth and Morphogenesis

M UCH OF THE MATERIAL in other chapters of this text concerns the physiological, biochemical, and hormonal processes that occur as plants grow and develop. In the present chapter we will discuss the cellular mechanisms of plant growth and morphogenesis—that is, how plants increase in size and how they generate the specific shapes and patterns of leaves, flowers, and roots. Classical descriptive studies of plant growth established the basic patterns of tissue development and the sites at which growth is concentrated. These early studies provided the foundation for modern cytological investigations using immunofluorescence microscopy, electron microscopy, and powerful new genetic and molecular techniques.

Anatomical and Ultrastructural Aspects of Growth

To place the information derived from the new approaches into perspective, it is necessary to have an appreciation of plant anatomy and patterns of growth. For this reason we will briefly review the basic structure of plants, examining the types of growth that can occur; the sites of cell division, cell expansion, and cell differentiation; and the types of polarity that are generated.

Growth is accompanied by morphogenesis and differentiation. **Morphogenesis** is the development of form or shape of cells and organs. Its course depends to a large extent on two fundamental processes: regulation of the direction of cell expansion and control of the plane of cell division. Integration of the mode of expansion of individual cells within an organ largely determines the shape of the organ itself. Precise control of the plane of cell division allows specific arrangements of cells, such as longitudinal files of vascular elements, to be formed. **Differentiation** is the process by which cells undergo biochemical and structural changes to perform specialized functions. Fully differentiated, mature cells have stopped dividing and have usually ceased expanding.

Plant Growth Is Defined as a Permanent Increase in Size

Growth is manifested as a permanent increase in size. However, size is not the only criterion used to measure growth. For example, the growth of a sample of cells in suspension culture could be assessed by measuring its **fresh weight**—that is, the

373

weight of the living tissue—at selected time intervals. In other cases, fresh weight may fluctuate due to changes in the water status of the plant and so may be a poor indicator of actual growth. In these situations, measurements of **dry weight** are often more appropriate. **Cell number** is a common and convenient parameter by which to measure the growth of unicellular organisms such as the green alga *Chlamydomonas* (Fig. 15.1a), but in multicellular plants, cell division can occur in the absence of growth. For example, during the early stages of embryogeny the zygote divides into progressively smaller cells with no net increase in the size of the embryo. Only when these cells expand does true growth occur. In general, a reliable way to assess growth is to measure one or more size parameters such as length (Fig. 15.1b), height, or width and calculate the area (Fig. 15.1c) or the volume where appropriate.

When growth is measured continuously over time, an S-shaped curve like those in Figure 15.1 is often obtained. Growth curves of this form show that a period of slow growth (the **lag phase**) is followed by a period of rapid growth (the **logarithmic** and **linear phases**), which in turn is succeeded by another period in which growth is slow or absent (the **stationary phase**). Growth curves of this shape may apply to single cells (Fig. 15.1a and b), to plant organs (Fig. 15.1c), or to whole plants. A multitude of factors affect the rate of plant growth and lead to this type of growth kinetics.

Specialized Growth Zones Contribute to Primary and Secondary Growth

Plant growth does not occur uniformly, as in animals, but is instead restricted to certain specialized zones. Two types of growth, called primary and secondary growth, are recognized in plants. **Primary growth** occurs at the tips of shoots and roots and in lateral appendages like leaves and buds. Because the *potential* for primary growth persists at the apices of shoots and roots, plants have the capacity for **indeterminate growth**; that is, they may grow more or less indefinitely. In some plants, such as corn and sunflower, flowering occurs after the production of a fixed number of leaves, and vegetative growth ceases. Such plants exhibit **determinate growth**, although they can be induced to grow in an indeterminate fashion by excising the growing tip from the plant and placing it in tissue culture.

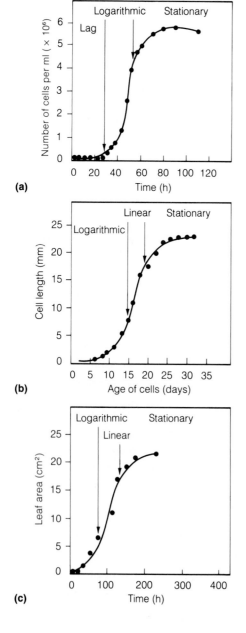

(a)

(b)

(c)

FIGURE 15.1. (left) Measuring plant growth. (a) Growth of the unicellular green alga *Chlamydomonas*. Cells are placed in fresh growth medium under optimum conditions, and growth is assessed by counting the number of cells. An initial lag period, during which cells may synthesize enzymes to metabolize new substrates, is followed by a period of logarithmic or exponential growth during which the rate of increase in cell number increases. Finally, as nutrients and light become limiting or the level of toxic substances builds up, the growth rate decreases and the stationary phase is reached. The number of cells may then decrease as cells die. Growth of unicellular organisms under optimum conditions often lacks the linear phase typically observed for multicellular plants (see parts b and c). (b) Growth of the filamentous green alga *Nitella* assessed by measuring cell length. Growth rate increases rapidly (logarithmic phase) and then remains constant (linear phase) for several days. As the cell approaches maturity, growth slows down and stops (stationary phase). (c) Growth of leaves of radish (*Raphanus sativus*), monitored by measuring leaf area. Again, an S-shaped curve is obtained. Growth is initially slow while cell divisions generate cells that will form the leaf. These cells then expand rapidly (logarithmic and linear phases) until the leaf approaches its mature size and the growth rate decreases (stationary phase). (*Chlamydomonas* data courtesy of Dr. J. D. I. Harper. *Nitella* data courtesy of Dr. G. O. Wasteneys.)

In most plants (an exception being arborescent monocots), primary growth is essentially equivalent to growth in length. The stems and roots of many plants also increase in diameter, and this is termed **secondary growth**. It occurs in regions that have stopped elongating.

Both primary and secondary growth are associated with zones in which cells are rapidly dividing. These permanently embryonic tissues are called **meristems**, and the cells within them characteristically have thin walls, prominent nuclei, and, in most cases, small vacuoles. There are two types of meristems. **Apical meristems** occur at the tips of roots (Fig. 15.2) and shoots (Fig. 15.3) and

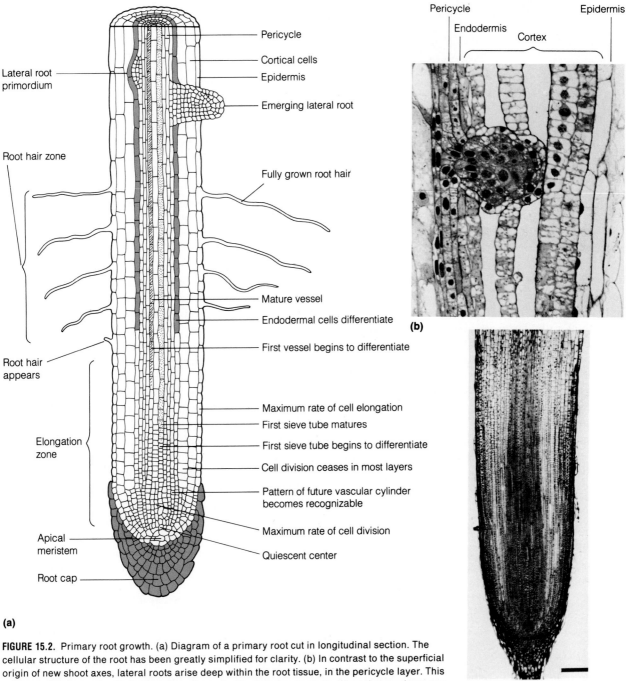

(a)

(b)

(c)

FIGURE 15.2. Primary root growth. (a) Diagram of a primary root cut in longitudinal section. The cellular structure of the root has been greatly simplified for clarity. (b) In contrast to the superficial origin of new shoot axes, lateral roots arise deep within the root tissue, in the pericycle layer. This micrograph shows a longitudinal section of a root of the fern *Ceratopteris*. When the specimen was fixed, the developing lateral root was in the process of pushing its way out through the endodermis, cortex, and epidermis of the root. Bar = 50 μm. (*Ceratopteris* micrograph courtesy of A. Hardham.) (c) Cells in the apical meristem, as seen here in a longitudinal section of a corn (*Zea mays*) root, expand and divide rapidly, generating many files of cells. At increasing distances from the tip, as the rate of cell division falls, the cells become larger and begin to differentiate into specialized cell types. Cells in the meristem and future vascular cylinder are less vacuolate than cells in the cortex and stain more intensely. Section stained with Toluidine Blue O. Bar = 200 μm. (Courtesy of A. Hardham.)

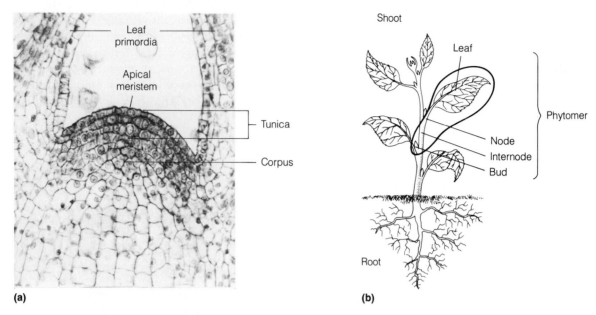

FIGURE 15.3. The shoot apex. (a) Micrograph of a longitudinal section of a shoot apex of *Coleus blumei*. The apical meristem consists of three layers of tunica overlying the corpus. (b) The organization of the shoot into reiterated units called phytomers. Each phytomer consists of a leaf, node, internode, and bud. (Photo © Jack Bostrack.)

give rise to primary growth or growth in length. **Lateral meristems** occur as cylinders of meristematic cells in woody stems and roots and give rise to secondary growth or growth in girth. There are two types of lateral meristems, the vascular cambium and the cork cambium. The **vascular cambium** consists of fusiform initials, which are elongated, highly vacuolate cells that divide longitudinally to form the secondary xylem and phloem, and **ray initials**, small cells that produce the radially oriented rays. The **cork cambium** develops within the cortex and secondary phloem. It gives rise to the **cork cells** that make up the secondary protective layer called the **periderm**. The periderm forms the protective outer surface of the secondary plant body, replacing the epidermis.

Primary Roots Can Be Divided into Four Zones

The root is usually the first organ to appear during seed germination. Classically, the root has been divided into four zones: the root cap, the apical meristem, the zone of elongation, and the root hair zone or zone of maturation (Fig. 15.2). However, except for the root cap, the boundaries of these zones overlap considerably. The cells of the apical meristem are expanding as well as dividing and therefore lie within the elongation zone. Many cells within the elongation zone are also undergoing differentiation, and certain cells, such as root hairs, elongate in the zone of maturation.

Root cap cells are formed by a specialized **root cap meristem**. Continued production of root cap cells by the root cap meristem means that the older cap cells are

progressively displaced toward the tip. As they mature, root cap cells differentiate, and when fully differentiated they are able to perceive gravitational stimuli and secrete mucopolysaccharides (slime) that help the root penetrate the soil. The root cap also protects the apical meristem from mechanical injury.

Divisions of the cells in the apical meristem give rise to progenitors for all the cells of the root. No lateral appendages are formed close to the apex of roots, and in longitudinal sections individual files of cells can sometimes be traced down the length of the root to their apparent point of origin within the apical meristem. Cells in the apical meristem are small and roughly cuboidal. Cells formed at the meristem expand and, for a period, continue to divide. In some roots, a group of meristematic cells called the **quiescent center** expands and divides more slowly than surrounding cells and is thought to serve as a reservoir of cells that are less prone to replication-associated damage. However, the function and general occurrence of the quiescent center are still far from clear.

The **elongation zone**, as its name implies, is the site of rapid root growth. Cells may continue to divide while they extend. After division and elongation have ceased, cells undergo **maturation**, producing files of cells of different types (Fig. 15.2). Differentiation may actually begin much earlier, but the cells do not reach the mature state until they reach this zone. Later in the chapter, we will examine the differentiation and maturation of one of these cell types, the tracheary elements. The central cylinder of vascular tissue that is formed is bordered by the endodermis, which in turn is surrounded by layers of cortical cells and the epidermis. Root hairs develop in this region, a clear indication that extension has ceased. If it had not, the root hairs would be quickly destroyed as they were pulled through the soil.

Lateral or branch roots arise from cell divisions in the pericycle layer at the periphery of the vascular cylinder (Fig. 15.2a and c). They grow out at right angles to the main vascular bundle, compressing the cortical cells and eventually breaking through the epidermis.

Primary Shoot Growth Gives Rise to Phytomers: Nodes and Internodes

Unlike the capped apical meristem of the root, the apical meristem of the shoot is strictly terminal (Fig. 15.3a). It is usually protected by developing leaf primordia (embryonic leaves) that surround it. The shoot apical meristem is classically considered to consist of two zones: the **tunica**, the outer one to several cell layers, and the **corpus**, the inner region of cells. Cells of the tunica divide predominantly in the anticlinal plane, that is, perpendicular to the surface. As a result, the tunica gives rise to the epidermal and subepidermal cell layers of the stem and leaves. In contrast, the corpus divides both anticlinally and periclinally (parallel to the surface), resulting in volume expansion. Divisions by a small number of initial cells in the tunica and corpus layers give rise to progenitor cells for the stem, leaf primordia, lateral buds, and floral appendages. The cell layers of the tunica and corpus are sometimes called histogenic (tissue-producing) layers. Branch shoots arise externally from lateral buds located in the axils of leaves, in contrast to the internal origin of branch roots.

Leaf primordia are produced on the flanks of the apical dome (Fig. 15.3a). Their arrangement around the apex (**phyllotaxis**) reflects the future pattern of leaves around the stem. The sites at which leaf primordia join the stem are called **nodes**, and the regions of the stem between such leaf junctions are called **internodes**. At any given time, several of the terminal internodes, up to 10–15 cm in length, may be undergoing primary growth. In grasses, cell division and extension growth occur at the base of each internode, a region known as the **intercalary meristem**.

Cells in young leaf primordia continue to divide and expand, forming the sheath, petiole, midrib, and blade of the leaf. It was once thought that the leaf blade was broadened due to the activity of a meristem along the leaf margins. However, more recent studies have shown that cells throughout the leaf can contribute to the increase in area of the blade (Poethig, 1987).

The node with its attached leaf, the subtending internode, and the axillary bud below form a reiterated unit in shoot development called a **phytomer** (Fig. 15.3b). Attempts have been made to determine whether phytomers arise from distinct assemblages of cells in the meristem, or whether they represent morphological units

without any strict relationship to meristem cell fate. Studies in which mutations were generated in the shoot apical meristem to allow the fate of the meristem cells to be followed suggest that, although there are boundaries delimiting the cells giving rise to phytomers, the boundaries are not strict. The developmental fate of a given meristem cell is not irrevocably associated with a particular phytomer (Sussex, 1989). The implication is that cell fate is not as strictly determined in plants as it is in animals.

Plant Cells Expand by Two Mechanisms: Tip Growth and Diffuse Growth

Expansion of root hairs, rhizoids, pollen tubes, and fungal hyphae is confined to an apical dome at one end of the cell and is referred to as **tip growth** (Fig. 15.4a). The expanding dome leaves behind it a long cylindrical cell that may or may not divide subsequently. In contrast, growth of cells within multicellular organs typically involves the uniform expansion of a large part of the surface area of the primary cell wall and is called **diffuse growth** (Fig. 15.4b).

Because the rigidity of cell walls limits the expansion of the protoplast, the size and shape of individual cells are determined by the mechanical properties of the walls that surround them. In tip-growing cells, cell wall synthesis and cell expansion are essentially synonymous; growth occurs by the deposition of new wall material at the tip. However, in diffuse-growing cells, expansion is thought to occur by a combination of cell wall loosening and cell wall synthesis. The processes of cell wall loosening and

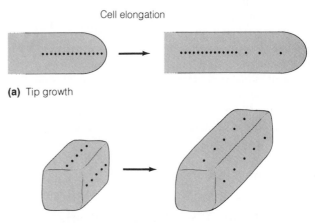

Cell elongation

(a) Tip growth

(b) Diffuse growth

FIGURE 15.4. Surface expansion during tip growth and diffuse growth. (a) Expansion of a tip-growing cell is confined to an apical dome at one end of the cell. If marks are painted on the cell surface and the cell is allowed to continue to grow, only the marks that were initially within the apical dome become separated. (b) Most cells in multicellular plants grow by diffuse growth. If marks are painted on the surface of a diffuse-growing cell, the distance between all the marks increases as the cell grows.

cell wall synthesis are under cellular control (Chapter 1). As we shall see in this and subsequent chapters, diffuse growth is frequently also under hormonal control.

Polarity Is a Property of Organs and Cells

Plants are both **axial** and **polar** structures. An axial structure is symmetrically arranged about a line, or axis. In a polar structure, the opposite ends of the axis are different. Plants have many axes. The main axis runs between the shoot and the root; subsidiary axes include branches and lateral roots. These structures are also polar and display different forms of polarity at different structural levels. The tip of a root is morphologically different from the base and thus shows polarity at the organ level. Within the root, polarity along files of cells can be seen in the form of a gradient of differentiation from apex to base.

Polarity can also exist at the level of an individual cell. Tip-growing cells are obviously polar, but so are many cells that are or have been growing diffusely. A mature root cap cell, for example, secretes slime across its outer wall only. Thus, the side of the cell at the root-soil interface is functionally different from the sides that are adjacent to other root cap cells.

Axiality and polarity are key attributes of plant development, and it is useful to know how and when they are established. Ultimately, the plant genome determines the pattern and dimensions of the various plant axes. However, unlike the situation in animals, growth form in plants is not inflexibly determined; it can also be influenced by the plant's local environment. In contrast to the relative plasticity of growth form, polarity appears at least in some circumstances to be a stable phenomenon. If a cutting is taken from a willow stem, it will produce roots from the base of the segment and shoots from the top, and this manifestation of polarity is maintained even if the segment is inverted before regeneration occurs (Burgess, 1985). The hormonal basis of this phenomenon is described in Chapter 16. At the cellular level, establishing and maintaining polarity is an integral part of the control of cell expansion and cell division. An understanding of this process is thus a vital part of our understanding of plant growth and morphogenesis.

Polarity in Tip-Growing Cells

Plants, like animals, perceive and act upon information coming from their environment. In plants, the responses usually involve directed growth, for which the cells must be able to establish an inherent polarity that can be used to orient subsequent expansion. For the analysis of the induction and establishment of polarity, apolar cells such as spores and pollen grains, which, on germination, grow by tip growth, have been especially valuable. A wide variety of external gradients, including temperature, light intensity, pH, ions, and electrical potential gradients, can orient the axis of growth of these cells. Thus, polarity can be experimentally manipulated without the complexities associated with multicellular tissues.

A Transcellular Current Is the Earliest Manifestation of Polarity

In the tip-growing cells that have been investigated, the earliest manifestation of polarity is the appearance of an ionic current that flows through the cell, looping back through the external medium (see Fig. 15.6) (Weisenseel and Kicherer, 1981). This **transcellular current**, which is largely due to calcium ions, is of the order of 100 pA (1 picoampere = 10^{-12} ampere) and occurs in advance of any morphological sign of polarity. The relevance of this phenomenon to the establishment of cell polarity is indicated by the fact that the site at which positive charge flows into the cell is the site of subsequent cell outgrowth. During tip growth, current continues to flow in at the apex of the filament. Transcellular ionic currents have been observed in a number of different tip-growing plant cells, including root hairs and germinating pollen grains, but have not yet been detected in plant cells that expand by diffuse growth.

The origin and role of these ionic currents in establishing and maintaining cell polarity are unclear. They could arise through the opening, closing, or redistribution of ion channels or pumps in the plasma membrane (Chapter 6). The ionic current generates an electrical potential difference across the membrane that could alter other membrane properties, such as the distribution of membrane proteins. The current also alters the ionic composition of the cytosol, the magnitude of the effect depending on the ions forming the current, which vary in different cell types. Because the Ca^{2+} concentration in the cytosol is usually very low (between 10^{-6} and 10^{-8} M), even a small increase in intracellular Ca^{2+} concentration could have a marked effect. In addition, the transcellular current could generate an electric field within the cell that could induce the movement of charged molecules or organelles.

Actin Microfilaments and Cell Wall Deposition Stabilize the Polarity of the Axis

In many tip-growing cells, such as the germinating spore shown in Figure 15.5, the first morphological feature associated with the establishment of cell polarity is the stratification of cell organelles, including the accumulation of small vesicles at the site of the future outgrowth (Quatrano, 1978). In zygotes of brown algae, such as *Fucus*, asymmetric organelle distributions have been observed before the polarity of the cell becomes fixed or stabilized. Exposure of the zygotes to unilateral light

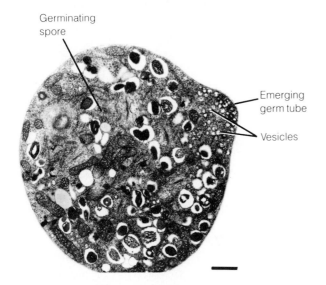

Germinating spore

Emerging germ tube

Vesicles

FIGURE 15.5. One of the first morphological signs of polarity in a germinating spore is the accumulation of small vesicles at the site of the future protuberance. This clustering of vesicles is seen near the outgrowth of a germinating spore of the fungus *Phytophthora cinnamomi*. Bar = 1 μm. (Courtesy of A. Hardham.)

induces inflow of positive current on the shaded side of the cell (Fig. 15.6). For a number of hours after its induction, stratification can be changed by subsequent treatment with light from a different direction. During this phase, cytoplasmic vesicles accumulate at the prospective site of rhizoid formation on the shaded side of the cell. Subsequently, the vesicles fuse with the plasma membrane and secrete their contents, leading to rhizoid emergence.

Inhibition of protein synthesis by treatment with the drug cycloheximide delays rhizoid formation but does not

stop the development of polarity. This suggests that although rhizoid formation requires the synthesis of new proteins, the establishment of polarity does not, so it must involve a redistribution of existing components. However, the stabilization of polarity can be prevented by treatment with another drug, cytochalasin B, which destroys microfilaments, one component of the cell's cytoskeleton (Chapter 1). The fixation of light-induced polarity in *Fucus* zygotes also requires the presence of the cell wall. Removal of the wall with cell wall–digesting enzymes does not prevent the induction of polarity by unilateral light but does prevent stabilization of the axis. The orientation of the embryonic axis remains labile until a new cell wall is synthesized (Kropf et al., 1988).

In animal cells there is evidence that the major signal for exocytosis in many cell types is an increase in the concentration of cytosolic calcium. Usually this involves the sudden release of accumulated secretory vesicles in response to an external stimulus, as in the case of neurotransmitter release. In tip-growing plant and fungal cells the situation is quite different; exocytosis occurs more or less continuously while the tip extends. Nevertheless, there is some evidence that calcium gradients may play a role in regulating vesicle fusion at the tip (Steer, 1988). For example, gradients in membrane-bound calcium have been demonstrated in a variety of tip-growing cells using the dye chlorotetracycline. However, this may not reflect the concentrations of cytosolic (free) calcium.

The fluorescent calcium-indicator dye fura-2 has been used to detect intracellular cytosolic calcium gradients in both plant and animal cells. Since the excitation spectrum of fura-2 shifts upon binding calcium, it is

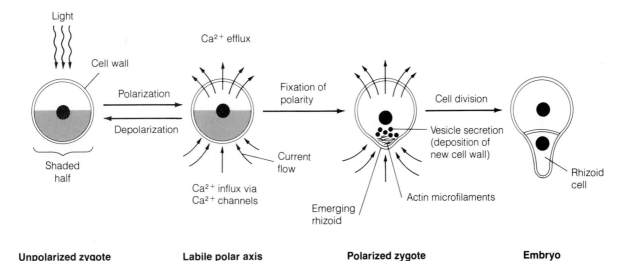

FIGURE 15.6. A summary of the events involved in the establishment of polarity in zygotes of the brown alga *Fucus*. The zygote is polarized by an asymmetric stimulus from the environment, such as unilateral light. A positive current enters the polarized but still spherical zygote at the site at which the rhizoid will emerge. The current, largely composed of calcium ions, flows out on the opposite side. The polarity of the cell becomes fixed when actin microfilaments assemble at the site of rhizoid emergence. Finally, cell division occurs and the rhizoid cell grows at the tip.

possible to determine the intracellular concentrations of both forms of the dye (plus or minus bound calcium) by exciting with the appropriate two wavelengths. The ratio of the two emissions provides a measure of the calcium concentration that is independent of the dye concentration. Using this technique to measure the cytosolic Ca^{2+} gradient in growing *Fucus* rhizoids, a gradient from 450 nM at the tip to 100 nM at the base was observed (Brownlee and Pulsford, 1988). Since secretion in animal cells generally requires 1 to 10 μM Ca^{2+}, the observed concentration of 450 nM at the tip would seem to be too low. These measurements suggest that the model of calcium-triggered exocytosis in animal cells may not be applicable to plant cells. Thus, although there is general agreement that calcium ion influx is an important component of the transcellular currents observed in tip-growing cells, and that calcium influx is required for axis formation, the role of calcium gradients in stimulating exocytosis is still controversial.

From the above observations it is possible to envisage a chain of events that might occur during the establishment of cell polarity (Fig. 15.6). The ionic currents that arise early in the process may initiate the formation of an intracellular Ca^{2+} gradient. The increase in intracellular calcium at the tip may help to organize and stabilize the microfilament components of the cytoskeleton, which, in turn, direct vesicle traffic to the growing tip. The possibility also exists that a calcium gradient, perhaps immediately adjacent to the plasma membrane, may promote vesicle fusion, although this remains to be demonstrated. Finally, cell wall deposition in some way fixes the polarity of the cell, possibly by interacting with transmembrane proteins.

The Position of the Nucleus Can Also Influence Polarity

Cell polarity is influenced not only by external cues but also by internal factors. The large unicellular alga *Acetabularia* has been used extensively to study the role of the nucleus in cell polarity and morphogenesis (Schweiger and Berger, 1981). During normal vegetative growth, the spherical zygote forms a rhizoid, a stalk, and a species-specific cap (Fig. 15.7a). If a segment of stalk is excised and a nucleus is implanted in the segment, a rhizoid will form adjacent to the nucleus even if the site was formerly at the apical end of the segment (Fig. 15.7b). That is, the change in position of the nucleus changes the polarity of the segment.

If the cap is excised, the cell can regenerate a new cap even in the absence of a nucleus (Fig. 15.7c). However, if the nucleus is replaced by one from another species of *Acetabularia*, the cap that is regenerated has a morphology characteristic of the species of the implanted nucleus (Fig. 15.7d). It is thought that the morphogenetic

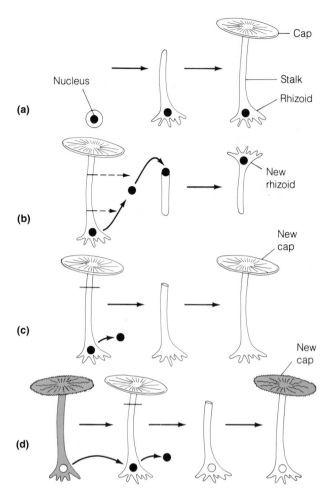

FIGURE 15.7. Diagrammatic representations of polarity and morphogenesis in the alga *Acetabularia*. (a) During vegetative growth, the zygote develops into a cell with a rhizoid and a stalk and then into a mature organism possessing a species-specific cap at the apical end of the cell. (b) An excised segment of the stalk regenerates a rhizoid adjacent to an implanted nucleus. (c) The cell can regenerate a new cap even if the nucleus is removed. (d) If the nucleus is replaced with one from another species, the cap that regenerates has the morphological characteristics of the species of the implanted nucleus.

signal involved in cap formation is stable mRNA. Thus, in *Acetabularia* cell polarity can be maintained and predetermined morphogenetic events can occur in the absence of a nucleus. However, when present, the nucleus has an overriding influence on both polarity and morphogenesis.

Nuclear control of the site of growth in a tip-growing cell has also been shown in the filamentous protonemata of the moss *Funaria* (Sievers and Schnepf, 1981). During branch formation a protuberance forms opposite the nucleus in a subapical cell of the filament (Fig. 15.8a). The position of the nucleus can be altered experimentally by gentle centrifugation. If the cell is centrifuged while in interphase, during recovery the nucleus moves back to its normal position and a branch forms in the usual location (Fig. 15.8b). Recovery of the nucleus can be inhibited by

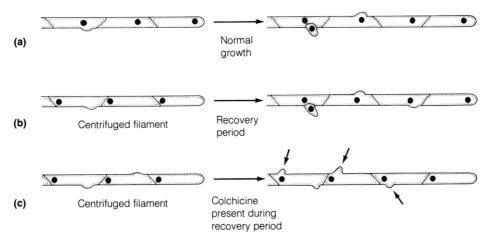

FIGURE 15.8. Diagrammatic representation of side-branch formation in protonemata of the moss *Funaria hygrometrica*. (a) During normal growth, a protuberance is initiated at a site adjacent to the nucleus, the nucleus divides, and one of the daughter nuclei enters the branch, which is then sealed off by a cell wall. (b) Nuclei can be displaced from their normal positions by gentle centrifugation. If allowed to recover, the nucleus will move back to its original position, and branch formation occurs at the normal site. (c) If the nucleus is displaced by centrifugation and movement during recovery is inhibited by the antimicrotubule drug colchicine, the side branch is initiated near the nucleus in an abnormal position at the basal wall.

colchicine, a drug that disrupts microtubules, another component of the cell's cytoskeleton. If movement is inhibited, the side branch emerges near the nucleus but in an abnormal position near the base of the cell (Fig. 15.8c).

These results show that in *Funaria*, as in *Acetabularia*, the nucleus influences the polarity of growth. They also suggest that there are components that are not affected by gentle centrifugation and that retain a polar organization. Since the only part of the cell that is not visibly disrupted is a thin layer of peripheral cytoplasm, it seems likely that the resistant components reside there, in either the cytoplasm or the plasma membrane. Cytoskeletal elements such as microtubules or microfilaments often occur in the peripheral cytoplasm of plant cells. Although there is no direct evidence, it is likely that one or both of these structures are capable of maintaining cell polarity in *Funaria* and other tip-growing cells. There is extensive evidence that microtubules are involved in the control of the polarity of diffuse-growing plant cells.

Polarity in Diffuse-Growing Cells

Most cells of the plant body expand over their entire surface, a process called diffuse growth (see Fig. 15.4). Cells in the stem and root tend to elongate, leading to growth in length. Elongating cells have the potential to alter their direction of expansion under specific environmental or hormonal conditions. In tissues such as the fleshy parts of fruits or storage organs, cells may undergo lateral expansion as a normal part of development. Both types of cell expansion are considered to be primary growth as long as the cells are derived from an apical rather than a lateral meristem. In this section we discuss the role of microtubules and microfibrils in regulating expansion in such cells.

The Direction of Cell Expansion in Diffuse-Growing Cells Is Determined by the Orientation of Cellulose Microfibrils

As we have seen in Chapter 3, the cell wall resists water uptake by the cell, causing a buildup of positive turgor pressure. This internal hydrostatic pressure is directed outward, equally in all directions. If the primary cell wall tended to stretch equally in all directions because of an **isotropic** (random) arrangement of its structural components, the cell would expand radially to generate a sphere (Fig. 15.9a). However, most plant cell walls consist of an **anisotropic** (nonrandom) arrangement of cellulose microfibrils (Chapter 1). In the side walls of cylindrically shaped cells, such as those in stems and roots or the giant internode cells of the filamentous green alga *Nitella*, cellulose microfibrils are deposited circumferentially (transversely) with respect to the long axis of the cell. Cellulose microfibrils have great tensile strength, and wall expansion occurs at right angles to the predominant direction of the microfibrils (Fig. 15.9b). Circumferentially reinforced cells elongate, with a minimal increase in girth. Each successive wall layer is stretched and thinned during cell expansion, so the microfibrils become passively reoriented in the longitudinal direction, that

FIGURE 15.9. The orientation of newly deposited cellulose microfibrils determines the direction of cell expansion.
(a) If cellulose is deposited randomly in the walls of a cell, the cell expands equally in all directions forming a sphere.
(b) If cellulose is deposited transversely (circumferentially) around the cell, lateral expansion is inhibited and the cell expands in the longitudinal direction.
(c) If cellulose is deposited longitudinally, the cell expands in a transverse direction.

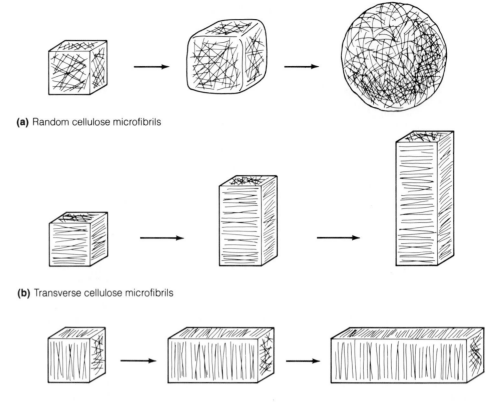

(a) Random cellulose microfibrils

(b) Transverse cellulose microfibrils

(c) Longitudinal cellulose microfibrils

is, in the direction of growth. Successive layers of microfibrils thus show a gradation in their degree of reorientation across the thickness of the wall, with those in the outer layers longitudinally oriented as a result of wall stretching. Because of thinning, these outer layers have much less influence on the direction of cell expansion than do the newly deposited inner layers. There is evidence that it is the inner 25% of the wall that is responsible for determining the direction of cell growth.

The angle at which cellulose microfibrils are deposited is an important part of plant growth and morphogenesis and must be under precise cellular control. It seems unlikely that the cellulose synthetase enzyme complexes in the plasma membrane (see Chapter 1) are solely capable of producing the observed patterns of microfibril deposition. Evidence suggests that alignment is achieved through the participation of microtubules located near the inner surface of the plasma membrane. Their involvement in the control of cellulose deposition is examined in detail in the next section.

Microtubule Orientation Mirrors the Orientation of Newly Deposited Cellulose Microfibrils

Microtubules were discovered in the peripheral cytoplasm, or cortex, of plant cells in 1963 by Myron Ledbetter and Keith Porter at Harvard University

(Ledbetter and Porter, 1963). Since then, the main evidence for the involvement of microtubules in cellulose microfibril deposition has come from observations of their coalignment with newly deposited wall cellulose microfibrils and from studies of wall deposition after treatment with drugs that cause microtubule breakdown. More recently, immunofluorescent and ultrastructural analyses of the arrays have augmented our understanding of microtubule organization in a variety of cell types.

The angle of the microtubules in the peripheral cytoplasm of plant cells almost invariably mirrors that of the newly deposited microfibrils in the adjacent cell wall. Coalignment in the transverse direction is the most commonly observed arrangement (Fig. 15.10). In some cell types, such as collenchyma, the microfibrils in the wall alternate between transverse and longitudinal orientations, and in such cases the microtubules are always observed parallel to the microfibrils of the most recently deposited wall layer.

When plant cells are treated with drugs that cause microtubule disassembly, such as colchicine, the cells lose control over the orientation of cellulose deposition (Fig. 15.11). In higher plant cells, loss of microtubules leads to the deposition of more or less randomly oriented cellulose microfibrils. Oriented cell extension is thus abolished, and the cells expand equally in all directions.

How cortical microtubules influence the orientation of cellulose deposition is still not known, although a

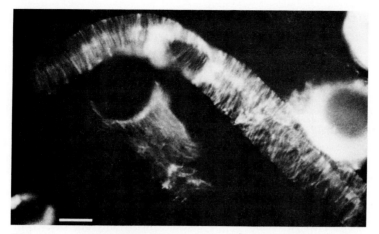

(a)

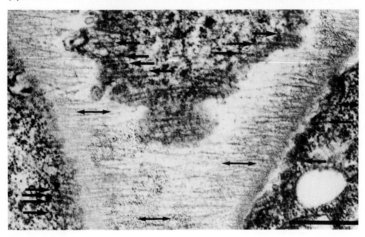

(b)

FIGURE 15.10. The orientation of microtubules in the cell cortex. (a) Staining of isolated cells with fluorescently labeled antibodies to tubulin reveals the arrangement of microtubules in the cells. In this long cylindrical cell from an onion (*Allium cepa*) root, the microtubules run transversely (circumferentially) around the cell. Bar = 10 μm. (b) The alignment of cellulose microfibrils in the cell wall can sometimes be seen in grazing sections in the electron microscope, as in this micrograph of a developing sieve-element cell in a root of *Azolla*. The longitudinal axis of the root and of the sieve element runs vertically. Both the wall microfibrils (double-headed arrows) and the cortical microtubules (arrows) are aligned transversely. Bar = 0.5 μm. (Photos courtesy of A. Hardham.)

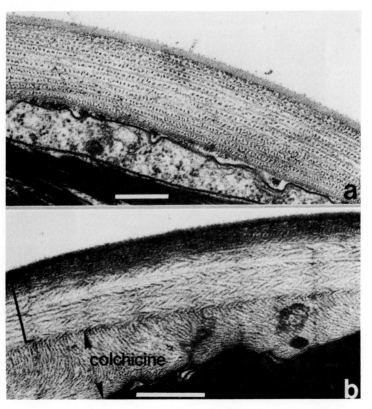

FIGURE 15.11. The cell wall of the unicellular green alga *Oocystis apiculata*. (a) The cells normally deposit a polylamellate (many-layered) secondary wall in which the cellulose microfibril are parallel within each layer and mutually perpendicular in adjacent layers. The microfibrils in this species are especially large and appear as dots and lines. During wall formation, microtubules (not evident in this micrograph) lie parallel to the microfibrils. Bar = 0.25 μm. (b) When the cells are treated with colchicine, a drug that destroys microtubules, the cells no longer produce wall microfibrils with alternating orientations but instead deposit microfibrils with the same orientation. The wall deposited before colchicine treatment is enclosed in the bracket; the herringbone pattern is due to oblique sectioning of the alternating layers. The wall deposited during colchicine treatment, which lies between the arrowheads, has lost the herringbone pattern. Bar = 0.5 μm. (Photos courtesy of A. Hardham.)

number of possible mechanisms have been suggested (Heath and Seagull, 1982). All proposals recognize that cellulose synthetase complexes must move in the plane of the plasma membrane, trailing cellulose microfibrils behind them, and in all cases the microtubules are considered to be linked to components in the plasma membrane. In some models, microtubules act as guidance elements, creating oriented channels or barriers within the membrane along which the synthetase complexes travel, propelled by forces generated by the polymerization and crystallization of the cellulose microfibrils. In other models, microtubules participate in motility generation as well as guidance. Very little is known about the molecular details of the microtubules–plasma membrane association or of the properties of the plasma membrane of plant cells. To date, the main approach in this area of research has been to study the microtubule arrays.

Microtubules are dynamic structures, and cortical microtubules are not present throughout the entire cell cycle. As the cell enters mitosis, the microtubule array is disassembled, to be reinstated again at the end of cell division. Initiation of microtubule polymerization within a cell usually occurs at specific nucleating sites called **microtubule organizing centers** (MTOC). The best-documented MTOCs in eukaryotic cells are the centrosomes and centrioles of animal cells. In plant cells, nucleation of cortical microtubules may occur at the nuclear envelope or in the cell cortex (Gunning and Hardham, 1982). Molecular details of nucleating sites in

plants are largely unknown, although the highest densities of microtubules in cortical arrays have been observed adjacent to newly formed walls, at sites of wall thickening, and at sites where the angle of microfibril deposition is being altered.

Control of the Plane of Cell Division

In the preceding sections we have examined the processes involved in regulating the direction of cell expansion. However, this is only one of two major types of morphogenetic regulation that occur during development. The other is the control of the plane of cell division.

In almost all plant cells, nuclear division is followed by the formation of a new cell wall that fuses with the parent cell walls and separates the two daughter cells. The walls surrounding each cell in the plant are fixed in position with respect to neighboring cells. Since plant cells cannot move and adjust their spatial relationships, as can most animal cells, each new cell wall must have the correct orientation and fuse with parent walls at the correct sites. Thus, there is little doubt that new walls must be inserted with great precision. Although the control of the plane of cell division is poorly understood, one thing is clear: mitosis and cell division could not occur without the participation of cytoskeletal microtubules.

FIGURE 15.12. Immunofluorescent staining of onion (*Allium cepa*) root tip cells with an antibody to tubulin shows the sequence of arrays of microtubules that are formed during cell division. (a) Early prophase. The preprophase band of microtubules appears as two bright spots on the sides of the nucleus (the unstained central region), with a faint image of the band as it passes across the nucleus. (b) Late prophase. Spindle microtubules emanate from the spindle poles and form the spindle fibers. (c) Metaphase. The spindle fibers hold the chromosomes (dark oblique line) in the metaphase plate. In this cell the alignment of the plate is oblique, which is relatively unusual. (d) Early anaphase. The chromosomes are moved poleward within the array of spindle microtubules. (e) Telophase. A dense array of microtubules forms the phragmoplast in the center of the cell at an early stage of cell plate formation. (f) Late stage of cytokinesis. Phragmoplast microtubules form an annular ring around the perimeter of the cell plate and appear only at the edges of the cell. Bar = 10 μm. (Courtesy of A. Hardham.)

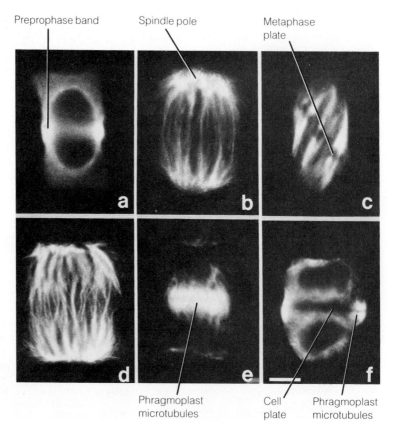

Preprophase band Spindle pole Metaphase plate

Phragmoplast microtubules Cell plate Phragmoplast microtubules

During Cell Division, Microtubules Form the Preprophase Band, the Mitotic Spindle, and the Phragmoplast

In most plant cells entering mitosis, the interphase array in the cell cortex is replaced by a band of microtubules that encircles the nucleus, the **preprophase band** (Fig. 15.12). Like the dispersed interphase array, the preprophase band microtubules lie in the cell cortex near the plasma membrane. The band is generally 2–3 μm wide and may contain from 10 to over 100 microtubules spread in a monolayer or densely packed two or three deep. As mitosis progresses, microtubules begin to appear at the surface of the nucleus. The nuclear envelope breaks down, and the microtubules invade the nuclear area and form the **mitotic spindle**. The spindle is a barrel-shaped structure consisting of two interdigitating arrays of microtubules focused on opposite sides of the nucleus. As the spindle forms, the preprophase band disappears (Fig. 15.12b–d). The chromosomes condense and become aligned along the metaphase plate at the center of the spindle. During anaphase, many of the spindle microtubules shorten and disappear and the sister chromatids separate and move to the spindle poles.

At telophase the nuclear envelope reforms around the daughter nuclei, between which an array of microtubules called the **phragmoplast** appears (Fig. 15.12e). Within the phragmoplast the new cell wall begins to be assembled. At this stage it is called a **cell plate**. The microtubules of the phragmoplast are aligned at right angles to the plane of the cell plate, with two populations of microtubules overlapping in the midzone where the plate is constructed (Fig. 15.13). Small vesicles accumulate along this plane and fuse to form the young cell plate, which grows outward toward the walls of the parent cells. Cell division is completed when the cell plate fuses with the parental walls. The interphase array is then reestablished in the cortex of the two daughter cells.

The Plane of Cell Division Is Determined Before Mitosis

Because the insertion of new cell plates in precisely the right place is one of the key events in plant morphogenesis, evidence for the mechanism of cellular control over the plane of cell division has long been sought (Gunning, 1982). In some plant cells and tissues it has been possible to show that the plane of cell division is influenced by light, gravity, physical pressure, and the plant hormone auxin.

It is clear that in plants the division site is determined *before* mitosis. Some of the early evidence for this came from observations of premitotic nuclear migration into the plane of the future cell plate. Another early sign was seen in highly vacuolate cells in which a raft of cytoplasm called a **phragmosome** forms in the future division plane

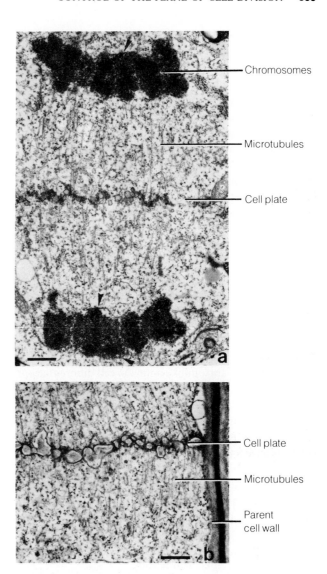

Chromosomes

Microtubules

Cell plate

Cell plate

Microtubules

Parent cell wall

FIGURE 15.13. Phragmoplast microtubules and cell plate formation in cells of the stem of thorn apple (*Datura stramonium*). (a) An early stage of cell plate formation. Small vesicles and electron-dense material have collected in the midzone of the phragmoplast between the two chromatin masses. The nuclear envelope is beginning to reform (arrowheads) around the decondensing chromosomes. The cell plate consolidates as the small vesicles fuse together. (b) At this stage of cell plate formation, the plate has reached the walls of the parent cell. The phragmoplast microtubules are confined to an annular ring around the edge of the plate. They still form two arrays that overlap in the plane of the plate. The microtubules may guide the small vesicles, which contain wall material, to the cell plate. Bar = 1 μm. (Micrographs courtesy of Dr. D. J. Flanders.)

prior to mitosis. However, the factors responsible for these premitotic events were, and still are, unknown.

The preprophase band of microtubules was first observed in 1966 by Jeremy Pickett-Heaps and Donald Northcote at Cambridge University (Pickett-Heaps and Northcote, 1966). It was hoped that studies of the preprophase band might shed some light on how the plane of cell division is controlled, but this has not yet come to pass. The function of the band itself is still an enigma. It disappears long before the cell plate develops, indicating that the microtubules in the band do not play a direct role in guiding the advancing perimeter of the cell plate. It has been suggested that the preprophase microtubules may be precursors for the spindle microtubules, may guide the deposition of extra wall material to mark the division site, or may position or orient the nucleus prior to mitosis. Whatever the function of the band may be, its formation clearly reflects the fact that the site in the cell cortex underlying it has become specialized in some way.

Differentiation of Selected Cell Types

Once a meristematic cell is outside the apical meristem, it will begin to differentiate. **Differentiation** is the process by which a daughter cell acquires metabolic, structural, and functional properties distinct from those of its progenitor cell. In plants, unlike animals, cell differentiation is frequently reversible, particularly when plant tissue is excised and maintained in culture. Such dedifferentiated cells can reinitiate cell division and, given the appropriate nutrients and hormones, even regenerate whole plants. Thus, differentiated cells, with some exceptions, retain all the genetic information (encoded in DNA) required for the development of a complete plant—a property termed **totipotency**. This is because most cells differentiate by regulating gene expression, not by altering their genome. Exceptions to this rule include cells that lose their nuclei, such as phloem sieve-tube members, and cells that are dead at maturity, such as tracheids.

We will examine the process of differentiation in two types of cells: tracheary elements and the guard cells of stomatal complexes. The development of these cells from the meristematic to the fully differentiated state illustrates the fine controls plants exercise over the architecture of their cell walls and over the plane in which cells divide.

In Tracheary Elements, Microtubules Determine the Pattern of Secondary Wall Thickenings

Water and solutes move through the plant via files of tracheary elements. In the final state of differentiation of tracheary elements, only the cell wall remains, the cell (protoplast) having been completely degraded (Chapter 1). This is an example of programmed cell death. During differentiation, deposition of secondary wall material at specific sites on the longitudinal primary wall leads to the formation of wall thickenings with specific patterns. The thickenings have a higher content of cellulose than the primary walls, and they are stiffened

and strengthened by lignin (see Chapter 1). In rapidly growing regions, the secondary wall material is deposited as discrete annular rings separated by bands of primary wall (Fig. 15.14). As the cell grows, the primary wall extends and the rings are pulled apart.

Differentiation of tracheary elements involves specific modification of the cell walls and subsequent degradation of the protoplast. The pattern and architecture of the wall modifications require the participation of microtubules (Hepler, 1981). Before any alteration in the pattern of wall deposition is evident, cortical microtubules change from being more or less evenly distributed along the longitudinal walls of the cell and become clustered into bands (Fig. 15.15a). Wall thickening then occurs beneath the clusters (Fig. 15.15b), the transverse orientation of the microtubules being reflected by the transverse alignment of the cellulose microfibrils. As the thickenings develop, the number of microtubules adjacent to the longitudinal walls increases, reaching a maximum when the thickenings are about half formed.

These observations suggest that cortical microtubules determine both the sites of secondary wall de-

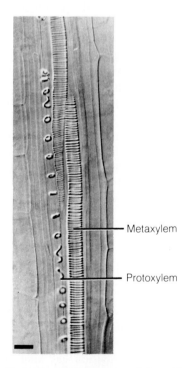

FIGURE 15.14. The pattern of secondary wall deposition during tracheary element development varies according to the rate of cell elongation. In this root of water hyacinth (*Salvinia auriculate*), the first file of tracheary elements to differentiate—the protoxylem—is seen on the left side of the vascular bundle. During differentiation, annular rings or short spiral thickenings of secondary wall material were deposited in the protoxylem elements. Subsequent elongation pulled adjacent rings apart and stretched the short spiral thickenings. The two elements on the right side, the metaxylem, differentiated after the protoxylem, and the closely apposed rings or tight spirals have not been extended by cell elongation. Bar = μm. (Courtesy of A. Hardham.)

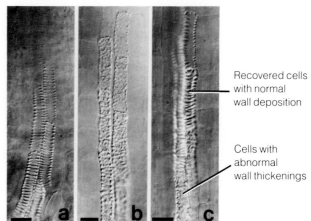

FIGURE 15.16. Colchicine treatments that destroy microtubules inhibit the normal formation of secondary wall thickenings in tracheary elements. (a) During normal growth of roots of *Azolla*, the wall thickenings are spaced evenly along the side walls. (b) In the presence of colchicine, irregular patterns of secondary wall material are deposited. (c) When transferred to fresh medium lacking colchicine, the roots are able to recover, and new tracheary elements are formed with normal annular thickenings. Bar = 10 μm. [(a) From Hardham and Gunning, 1979. (b and c) From Hardham and Gunning, 1980.]

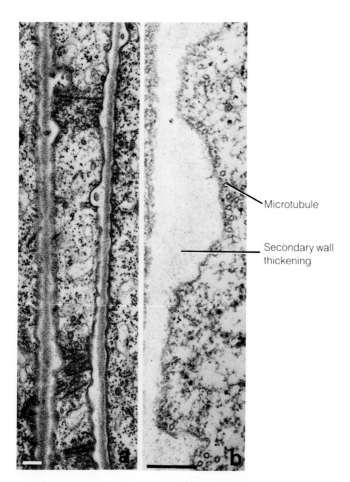

FIGURE 15.15. Development of secondary wall thickenings in tracheary elements in roots of the water fern *Azolla*. (a) Microtubules in the cell cortex indicate the site of wall thickening by grouping into bands before secondary wall formation begins. In this grazing section, the bands of microtubules are cut near the edge of the cell. Many small vesicles lie among the microtubules. (b) Annular thickenings develop beneath the bands of microtubules and are hemispherical in profile. Bar = 0.2 μm. (From Hardham and Gunning, 1979.)

position and the orientation of cellulose microfibrils in the thickenings. Further evidence that microtubules play these roles comes from experiments with drugs that cause microtubule depolymerization (Fig. 15.16). In the absence of microtubules, the normal pattern of wall deposition is lost and the cellulose microfibrils are no longer deposited with a transverse orientation. If the drug is removed and the microtubules are allowed to reassemble, normal thickening can resume (Fig. 15.16c) (Hardham and Gunning, 1979, 1980).

The Differentiation of Stomatal Complexes Depends On the Precise Placement of Cross-Walls During Cytokinesis

As we saw in Chapter 4, the surfaces of leaves contain pores through which gases are exchanged between the atmosphere and spaces within the leaf. The pores form

between the thickened walls of specialized epidermal cells called guard cells. The aperture of each pore can be adjusted by changing the turgor pressure of the guard cells and of surrounding subsidiary cells, if any are present. Guard cells are commonly elliptical or kidney-shaped, although those of grasses have a dumbbell-shaped outline (Chapter 4, Fig. 4.12a).

As pairs of young guard cells differentiate, the central portion of the anticlinal wall between them thickens (Palevitz, 1982). The wall thickening fans out onto the walls parallel to the leaf surface, that is, the inner and outer periclinal walls. Within the thickened walls the cellulose microfibrils are precisely aligned (Chapter 4, Fig. 4.14b), running anticlinally along the anticlinal wall between the two guard cells and radiating out from the pore on the periclinal walls. As in developing tracheary elements, the site of wall thickening is indicated by an accumulation of cortical microtubules. Throughout development, the orientation of the microtubules mirrors that of the cellulose microfibrils in the adjacent wall.

The sequence of divisions involved in converting a small group of epidermal cells to cells of the stomatal complex provides a good example of the controls that are exerted during the placement of new walls at cytokinesis. First, an epidermal cell in the middle row divides asymmetrically in a plane transverse to the long axis of the leaf, generating a large cell and a small cell (Fig. 15.17a and b). The small cell becomes the guard mother cell, the progenitor of the two guard cells. Second, the epidermal cells surrounding the **guard mother cell** divide asymmetrically to form lens-shaped **subsidiary cells** next to the guard mother cell (Fig. 15.17c–e). Third, the guard

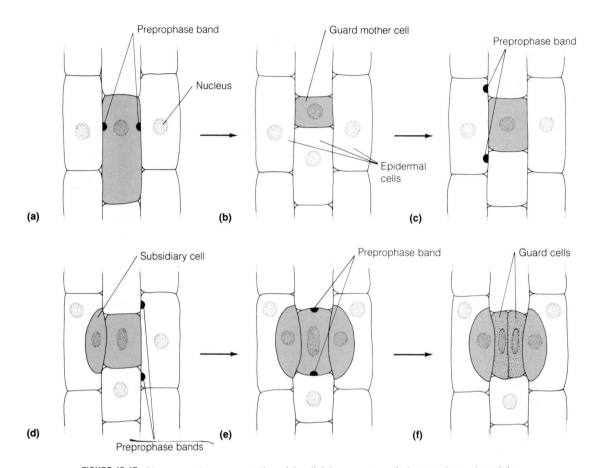

FIGURE 15.17. Diagrammatic representation of the division sequence that generates a stomatal complex containing two subsidiary cells. (a) A preprophase band (filled semicircles) forms in an epidermal cell in the middle of the three cell files. The nucleus moves so that it is encircled by the preprophase band. (b) The epidermal cell divides asymmetrically and the cell plate fuses with the parent walls at the sites previously occupied by the preprophase band. The division generates a large cell and a small cell. The latter is the guard mother cell. (c and d) The nuclei in the lateral epidermal cells migrate to a location near the guard mother cell. Preprophase bands of microtubules accurately predict the site of fusion of the cell plates during these highly asymmetric divisions that form the subsidiary cells. (e) A preprophase band oriented parallel to the leaf axis develops in the guard mother cell. (f) A symmetrical longitudinal division in the guard mother cell yields two guard cells.

mother cell divides symmetrically, producing the two guard cells (Fig.15.17e and f).

One symmetrical and three asymmetrical divisions are required to form the four-celled stomatal complex. Preprophase bands of microtubules form prior to all divisions, accurately predicting the site of fusion of the cell plate with the walls of the parent cell (Fig. 15.7a−e).

The apparent ability of the cell plate to recognize the preprophase band zone has been demonstrated by B. Galatis and co-workers at the University of Athens, Greece. In their experiments, the nucleus in the epidermal cell that will give rise to a subsidiary cell is displaced from its normal position by gentle centrifugation (Fig. 15.18) (Galatis et al., 1984). The preprophase band forms at the correct position near the guard mother cell (Fig. 15.18b) but, because the nucleus and the ensuing mitotic figure are displaced, the cell plate begins to form at some distance from the guard mother cell and usually fuses at

an abnormal site (Fig. 15.18c). However, if by chance the edges of the plate come within 1−2 μm of the preprophase band zone, the plate turns abruptly and fuses with the parent wall at the preprophase band zone (Fig. 15.18d). All these observations suggest that the preprophase band site possesses some means of guiding cell plate growth.

It is clear from studies of developing stomatal complexes that the guard mother cell influences the division plane of the adjacent epidermal cells. Some form of cell-cell communication must occur between guard mother cells and adjacent epidermal cells. In fact, communication between cells is of paramount importance for the correct development and life of all multicellular plants. In the next section, we will examine some aspects of the exchange of information at cell and tissue levels, before addressing the global and more complex issue of morphogenesis of plant organs.

FIGURE 15.18. Experimental studies of cell division during stomatal development. (a) Developing complex in which one subsidiary cell has formed. (b) The nuclei in the epidermal cells have been displaced by gentle centrifugation, but the preprophase band in the cell at the right forms in the usual location next to the guard mother cell. (c and d) During development under continued centrifugation, the epidermal cell nucleus divides, and a cell plate forms between the two daughter nuclei. The cell plate is highly irregular and meanders through the cell, usually fusing with the parent cell walls at sites well removed from the guard mother cell (c). However, if the expanding plate comes close to the zone previously occupied by the preprophase band, it abruptly changes direction and grows directly toward the part of the parent wall that lay under the preprophase band (d).

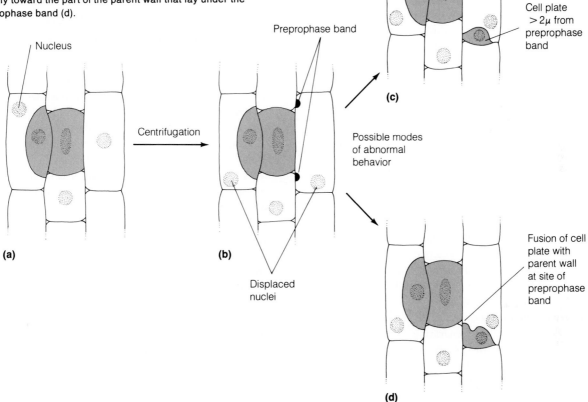

Morphogenesis in Roots and Shoots

We have now examined in detail the structural basis of cell polarity, cell expansion, and cell division and have looked at the contributions these processes make during the development of two selected cell types. Throughout this discussion, the role of microtubules has been a recurring theme. In general, we have focused at the cellular level, for it is there that morphogenesis has its origins. However, cells rarely grow in isolation. On the contrary, cells in multicellular plants are usually in close contact with others around them, and the behavior of each cell is carefully coordinated with that of its neighbors throughout the life of the plant. In this, the final major topic in this chapter, we will examine how cellular controls of cell expansion and the plane of cell division are integrated and coordinated during the growth and development of plant organs such as roots and shoots.

Coordination of cellular activity requires cell-cell communication. This communication can occur either symplastically via plasmodesmata or apoplastically through the cell wall. We will see that, although the development of plant organs usually involves precise and reproducible patterns of division and expansion, developmental strategies must be flexible enough to respond to changing conditions. Changes may be communicated to individual cells by plant hormones or by other means, and, once again, microtubules are seen to play a pivotal role in orchestrating cellular responses.

Cell-Cell Communication Occurs Symplastically via Plasmodesmata or Apoplastically Through the Cell Wall

Almost all living cells in a plant are connected symplastically to their neighbors by cytosolic continuity through plasmodesmata that pass through pores in the cell walls (see Chapter 1). Molecules can be transferred between cells either along the plasma membrane that lines the pore, along the endoplasmic reticulum that forms a central tubule (desmotubule) through the plasmodesma, or through the cytoplasmic sleeve between the plasma membrane and the desmotubule. By following the passage of fluorescent dye molecules through plasmodesmata connecting leaf epidermal cells, the limiting molecular mass for transport was determined to be about 700–1000 Da—equivalent to a molecular size of about 1.6 nm. Since the width of the cytoplasmic sleeve is approximately 5–6 nm, it might be expected to permit the passage of molecules much larger than 1.6 nm. However, the cytoplasmic sleeve is occluded by protein subunits attached to the surface of the desmotubule (Robards and Lucas, 1990). The presence of these particles may act to restrict the size of molecules that can pass through the pore, although this has not yet been demonstrated.

Experiments with plant viruses indicate that the effective diameter of the annulus could be under cellular control. Viruses may be able to increase the effective diameter of plasmodesmata, allowing them to move more easily from cell to cell in the plant. A 30-kDa **movement protein** (MP) produced by tobacco mosaic virus increases the effective plasmodesma diameter of tobacco to 2.4 nm. This was demonstrated by generating transgenic plants (i.e., plants containing foreign genes) that were able to synthesize the viral protein. By monitoring the movement of microinjected fluorescent dyes from cell to cell, it was shown that dye molecules as large as 9400 Da could cross through plasmodesmata in the transgenic plants, compared with 700–800 Da in the control plants (Wolf et al., 1989). It is possible that the viral MP can mimic a plant protein that plays a similar role during development.

The number of plasmodesmata per unit wall area is specifically controlled during development. In axial structures such as shoots and roots the frequency of plasmodesmata is often higher in transverse walls than in longitudinal walls, suggesting that communication between cells along a cell file is greater than that between cells in adjacent files. The number of plasmodesmata in a wall is largely determined by the number inserted into the developing cell plate at cytokinesis, but it can be increased by the de novo formation of plasmodesmata or decreased by their closure.

There is also good evidence that some developmentally important molecules move in the apoplast, which consists mostly of the cell wall space. The apparent pore size of cell walls is 3.5–5.0 nm, which means that only molecules with masses less than approximately 15 kDa diffuse freely across and within the wall. If the wall is impregnated with adcrusting substances such as lignin or suberin (see Chapter 13), the pore size can be so reduced that essentially all movement is forced to occur through the symplast.

Transport of the plant hormone auxin involves movement in both the symplast and the apoplast. Auxin transport is highly polar, occurring preferentially from apex to base, and does not involve plasmodesmata (see Chapter 16).

Roots of the Water Fern *Azolla* Have Been Used for Histological Analyses of Cell Lineages

The longitudinal section of a corn root in Figure 15.2b shows that files of cells can be traced back to the meristematic region at the tip of the root, but it is usually not possible to identify the cells at the apex that give rise to a particular file of cells. Even when young, most roots are made up of many hundreds of thousands of cells and have a complex structure. The regular cell files give the impression that development must follow a very ordered sequence of divisions, and yet surgical experiments have shown that roots of completely normal structure can be formed even when much of the meristem is destroyed or removed. Once again, we have evidence that cell development is not locked into a predetermined pathway but continually responds to the environment.

Despite the complexity of roots, the ultimate origin of all files of cells can be traced back to a group of **initial cells**, comparable to the "stem cells" of animals. When an initial cell divides, it renews itself and also produces a derivative that can embark on its own particular developmental pathway (Fig. 15.19). The derivatives so formed complete a number of divisions that give rise to the cell files in the root. The divisions that form the cell files are called **formative** divisions and are usually longitudinally oriented. The derivatives of the initial cells then divide transversely, resulting in proliferation of the cells within the cell files. These are called **proliferative divisions**.

The structure of roots of many lower plants is less complex and more regular than that of higher plant roots. A detailed study of the morphogenesis of roots of the water fern *Azolla* was possible because of the small size and relatively small number of constituent cells (about 9000 in the mature root) (Fig. 15.20) (Gunning, 1981). The roots have a single initial cell, which is usually called an **apical cell** in ferns (Fig. 15.20a). Divisions parallel to the outer curved face of the apical cell give rise to the progenitors of the two-layered root cap. Thereafter, the apical cell divides parallel to the three internal triangular-shaped faces, cycling in either a clockwise or an anticlockwise direction depending on the side of the shoot axis from which the root originates. These divisions form

Clonal Analysis of Shoot Development

HISTOLOGICAL STUDIES PROVIDE static images of the organization of apical meristems and their derivative cells at a moment in time. Despite the lack of dynamic information, this approach has been useful in the case of roots, which produce linear files of derivative cells unperturbed by the presence of lateral appendages. The situation in shoots is much more complex. Because the cell division pattern is more varied, it is not always possible to trace the lineage of a cell in a leaf or axillary bud back to its origin in the meristem. How, then, can we determine the lineage, or *fate map*, of a shoot cell? To do so, we need a method for marking a single cell in the meristem so that the mark will be visible in all of its derivatives or clones. Because of the latter requirement, genetic markers that cause visible alterations in pigmentation have been favored. This approach to development, called **clonal analysis**, has been widely applied to animal systems and is now providing insights into plant development as well.

Two types of genes have been used as genetic markers: genes involved in anthocyanin production and those involved in chlorophyll synthesis. Plants that are heterozygous at one or more of these loci are treated with x-rays. If the dominant allele of a meristem cell is damaged by the x-rays, the cell loses its ability to synthesize the pigment, as do all of its prog-

eny. The result is an area of unpigmented cells that can be readily detected. A basic assumption in clonal analysis is that the earlier in development the mutation occurs, the more cells exhibit the mutant phenotype. If the mutation occurred at the zygote state, for example, all cells of the plant would be affected. If a meristem initial cell is mutated, a longitudinal **sector** is produced representing all the derivatives of that initial cell. At progressively later stages in development, fewer and fewer progeny are produced, resulting in smaller and smaller **patches** of mutant cells.

Clonal analysis has been used to determine the number of initial cells in the shoot apical meristem. Based on the observation that mutant sectors usually occupy about one-third of the circumference of the stem, it was concluded that three initial cells are present in each of the histogenic layers of the tunica and corpus of most dicots and some monocots (Sussex, 1989).

Another important conclusion based on clonal analysis is that cell fate in plants is far more variable than that in animals. A mutation in one histogenic layer may occasionally show up among the progeny of a different layer. For example, cells of the tunica usually divide anticlinally to produce the epidermis but may occasionally divide periclinally, giving rise to subepidermal cells. That such events occur has been confirmed by clonal analysis. Thus, the developmental fate of meristem cells can be assigned only in a general way (Sussex, 1989).

wedge-shaped cells called **merophytes**. Roots of *Azolla* are unusual in that they do not grow throughout their lifetime. Instead, the apical cell divides only about 50–55 times, generating three tiers of 17 or 18 merophytes.

It has been possible to trace the sequence of formative divisions within each merophyte in *Azolla* roots (Fig. 15.20). At first the merophytes behave similarly and have the same order of divisions, culminating in the generation of 18 derivatives. Almost all the initial cell derivatives complete a number of rounds of transverse, proliferative divisions, the number being characteristic for the cell type.

Longitudinal, formative divisions within the merophytes give rise to progenitor cells for the files of cells in the root. In longitudinal section, a formative division is marked by the site at which a longitudinal wall first begins near the apex. For example, the division that generates the endodermis and inner cortex cells occurs in the fourth

merophyte on both sides of the root, as do the divisions that delineate the tracheary element, pericycle, outer cortex, and root hair–epidermis cell files. Once formed, the progenitor cells usually complete a number of transverse, proliferative divisions before the progeny differentiate. The sites of these divisions can be seen by counting the number of cells within a merophyte along a file. For example, in Figure 15.20e, the first proliferative division for the pericycle on the left-hand side occurs in merophyte 7, the second in merophyte 9, and the third in merophyte 12. On the right-hand side, the endodermis progenitor cell divides proliferatively in merophyte 7 and again in merophyte 10. Only tracheary elements do not proliferate before differentiation begins, so their length is equal to that of the merophyte of which they are part. The transverse section in Figure 15.20d was made in the region of formative divisions; only the upper merophyte has completed the formative division that generates the

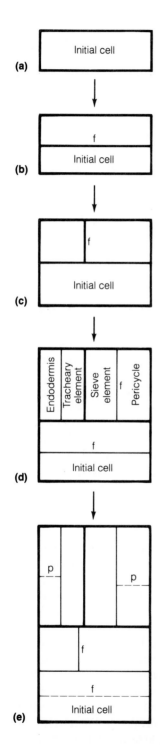

FIGURE 15.19. (above) Diagram illustrating a possible pattern of formative (f) divisions that give rise to progenitor cells for the endodermis, tracheary element, sieve element, and pericycle cell files. In part e, the endodermis and pericycle cells have each completed a transverse, proliferative (p) division.

endodermis and inner cortex initials. The section in Figure 15.20c was taken farther from the root apex, in the zone of proliferation and differentiation, and is redrawn and labeled in Figure 15.20f. Although the root increases in girth near the apex, this expansion is confined to the cortical and root hair–epidermis files, and the diameter of the stele, delimited by the endodermis, remains essentially constant.

The boundaries between merophytes and the age sequence they represent show that wall placement for both formative and proliferative divisions occurs with submicrometer accuracy. The cell plate may be flat or curved. It may or may not meet the parent walls at right angles or avoid previously established wall junctions. The divisions may be symmetrical or asymmetrical, regardless of whether they are formative or proliferative. However, for every division the location at which the cell plate fuses with the walls of the parent cell is precisely predicted by the preprophase band of microtubules, and the site previously occupied by the band is approximately bisected by the cell plate.

From the studies of *Azolla* roots, we have learned a great deal about the lineages of cell divisions that occur during root development, but many basic questions still remain. Foremost among these is the question of what determines the timing and plane of cell division. It seems most likely that regulation of cell division is achieved through an interplay of internal information, derived from the cell's genes, with some type of information from outside the cell. The nature of this extracellular information remains to be determined.

Microtubule and Microfibril Rearrangements on the Outer Epidermal Wall Precede the Formation of a New Axis

The formation of leaf primordia requires coordinated changes in the patterns of cell divisions and the direction of expansion of cells in the region of the new outgrowth. The cell divisions involved were described many years ago. A cluster of epidermal cells divide anticlinally, generating the increased numbers of epidermal cells that will be needed to cover the enlarged surface area. Periclinal

FIGURE 15.20. (right) An especially precise pattern of cell divisions occurs during root development in the water fern *Azolla pinnata*. The single apical cell in (a), shaped like an octant of a sphere, divides at its three triangular faces to form wedge-shaped cells called merophytes. In the young root in (a) there are five merophytes in the left-hand tier and six in the right-hand tier. The merophytes in the root in (b) are drawn and numbered in the diagram (e), where 12 are shown on each side of the root. By recognizing merophyte boundaries, one can see the extent of elongation. Merophyte boundaries seen in transverse section are indicated by the arrows in (c) and (d) and the heavy lines in (f). (a) Bar = 10 μm; (b–d) bar = 20 μm. [(a, c, d) From A. R. Hardham. (b) From Hardham and Gunning, 1979.]

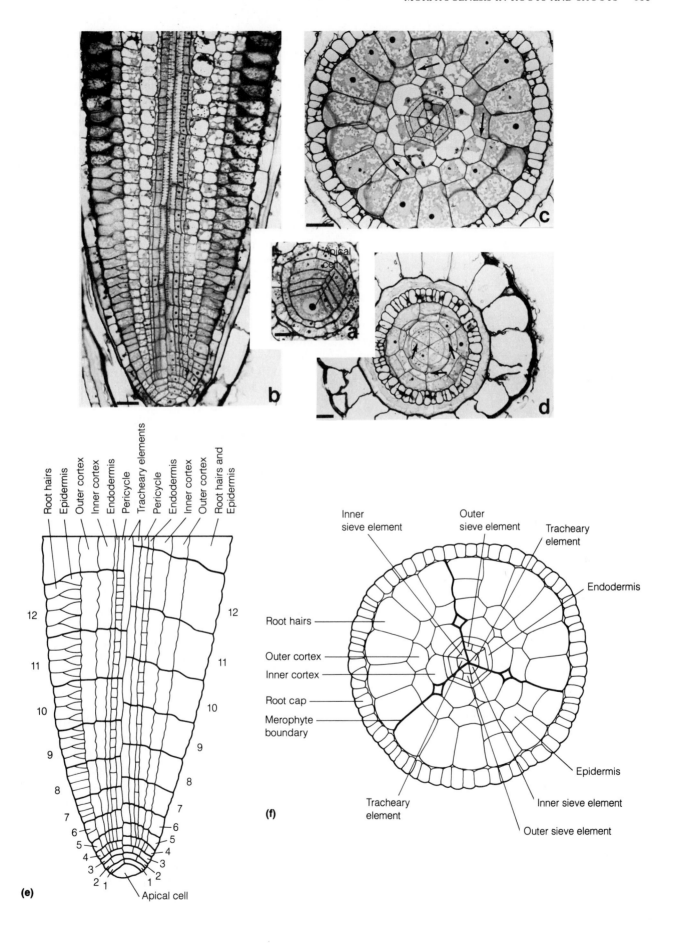

(b)

(c)

Apical cell

(a)

(d)

(e)

Root hairs
Epidermis
Outer cortex
Inner cortex
Endodermis
Pericycle
Tracheary elements
Pericycle
Endodermis
Inner cortex
Outer cortex
Root hairs and Epidermis

12 — 12
11 — 11
10 — 10
— 9
9 — 8
8 — 7
7 — 6
6 — 5
5 — 4
4 — 3
3 — 2
2 1 — 1

Apical cell

(f)

Inner sieve element
Outer sieve element
Tracheary element
Endodermis
Root hairs
Outer cortex
Inner cortex
Root cap
Merophyte boundary
Tracheary element
Epidermis
Inner sieve element
Outer sieve element

divisions in the two or three cell layers under the epidermis form new cells whose growth will contribute to the internal tissues of the new leaf. All these divisions are important, but alone they will not cause the surface to bulge and the new outgrowth to emerge. The direction of cell expansion in the parent tissues must also be altered. Because of the complexity of most shoot apices, we have little information on wall characteristics and microtubule patterns. However, the accessibility of shoot-forming meristematic areas at the base of leaves of some succulent plants, such as *Graptopetalum*, has permitted study of correlated changes in cellulose microfibril deposition and the arrangement of microtubules during the formation of leaf primordia.

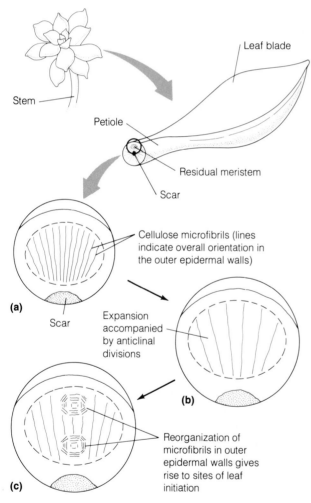

FIGURE 15.21. Two young leaf primordia forming in the meristem at the base of a leaf of the succulent *Graptopetalum paraguayense*. Anticlinal divisions in the epidermis of the residual meristem increase the number of epidermal cells (b). In the underlying tissue, periclinal divisions generate additional cell layers that will form the new leaves. Leaf primordia emerge at the sites where the cellulose microfibrils of the outer epidermal wall have reorganized into circumferential arrangements (c). (From Hardham et al., 1980.)

Following removal of a leaf from the stem of the parent plant, small hemispherical leaf primordia emerge from a flat **residual meristem** surface at the base of the petiole. The cellulose microfibrils in the outer epidermal walls are initially oriented parallel to each other (Fig. 15.21a and b), but around the emerging primordium the angle of cellulose deposition becomes locally reorganized to yield an approximately circumferential arrangement (Fig. 15.21c). A similar reorganization is seen in the microtubule arrays under the outer epidermal wall. Initially the cortical microtubules are oriented across the meristem, but arrays at right angles to this appear in cells that will be located at the sides of the leaf primordia. Similar patterns of microtubule reorganization during the formation of leaf primordia in the shoot apex of ivy have been described (Marc and Hackett, 1989). It seems likely that this type of reorganization is a general occurrence during the formation of leaves and buds.

To understand how the plant produces new leaf primordia, it is important to know the order in which these changes occur. However, it is difficult to analyze the microtubules and microfibrils in cells that will form the primordium before the primordium emerges, because the exact site of emergence is unknown. Nevertheless, alterations in microtubule arrays have occasionally been seen before any bulging of the surface of the meristem was evident. Therefore it seems that reorganization of the microtubules leads to an altered mode of cellulose deposition, which in turn leads to a change in the direction of cell expansion and the emergence of the leaf primordium.

The Cell Cycle May Influence the Ability of a Cell to Differentiate

During normal development, cell divisions in the meristematic region are rapidly followed by differentiation into specialized cells and tissues in the maturation zone. Does this mean that mitosis is a necessary prologue to differentiation? To address this question, we need to study differentiation in nondividing cells. The regeneration of vascular tissue after wounding provides such an experimental system.

Maintenance of intact and functional vascular tissue is crucial to the life of the plant. If vascular strands are severed, many plants are able to reestablish continuity by forming new vascular elements that bypass the damaged area. To accomplish this, differentiated cells, usually cortical parenchyma cells, enter a new phase of differentiation and become specialized to serve as cells of xylem or phloem tissues.

Wounding initiates several rounds of cell division in the cortical cells around the damaged area (Fig. 15.22). These divisions lead to the formation of files of tracheary and sieve elements that connect one end of the severed vascular strand to the other and that often lie parallel to the contours of the wound itself (Fig. 15.22). In this situation, as in normal development, the plane of cell division is carefully controlled. Preprophase bands are formed and cell plates in adjacent cells are precisely aligned so that smooth files of cells are generated.

The question of whether cell division must occur prior to redifferentiation has been addressed principally by using drugs that cause microtubule breakdown and thus inhibit cell division. When this was done, conflicting results were obtained. In some plants differentiation does not occur, but in others differentiation proceeds despite the absence of preceding division. In the latter cases, loss of microtubules causes the new vascular elements to have abnormal shapes; the files of cells are not smooth, and the tracheary elements have irregular patterns of secondary wall thickening. Nevertheless, differentiation produces bridges of xylem and phloem elements to connect the original vascular strands.

Isolated mesophyll cells from *Zinnia* leaves can redifferentiate into tracheary elements without first dividing, and Fukuda and Kommamine (1981) used this elegant system to document the rise in mRNA and protein synthesis that accompanies the initiation of redifferentiation. Tubulin, the component of microtubules, is one of the new proteins synthesized in large amounts.

The apparently conflicting results obtained in studies of different plants can be reconciled by considering the stage of the cell cycle at which the cells are resting at the time of wounding (Dodds, 1981). The G1 stage of the cell cycle can be divided into an early and a late phase, and it has been postulated, in accord with similar studies of animal cells, that cells in early G1 can respond immediately to the signal that triggers redifferentiation. Cells resting in late G1 or in G2 must pass through the cell cycle until they reach early G1, at which point they can commence their new mode of differentiation.

These results indicate that a fully differentiated cell can be reprogrammed to differentiate into another cell type in the absence of DNA duplication and without undergoing cell division. Differentiation does require the production of new mRNA molecules and new protein synthesis.

The ability to control the plane of cell division is especially interesting in these "unprogrammed" divisions following wounding. They are unprogrammed because the cell could not have anticipated the need to divide in the required plane and instructions for their placement could not have been incorporated into the cell's genome. Instead, the cell must be responding to external information. In this case, there is good evidence that auxin is involved (see Chapter 16). The following chapters will examine the effects of light and hormones on plant development.

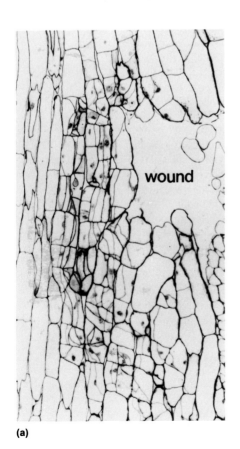

(a)

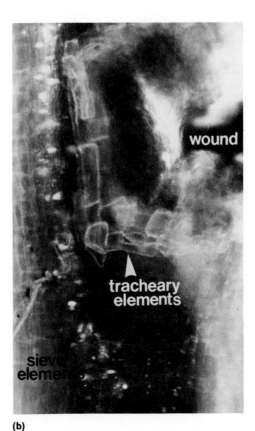

(b)

FIGURE 15.22. Renewed cell division following wounding in pea (*Pisum sativum*) roots. (a) Severing of the vascular tissue induces nearby parenchyma cells to divide and redifferentiate to form new files of vascular cells. The boundaries of the large cortical cells stain strongly and can still be identified. Subdivisions within these cortical cells are denoted by the thinner walls. Bar = 100 μm. (b) Fully differentiated sieve and tracheary elements bridge the wound made through the vascular cylinder. The files of sieve elements are revealed by staining with the fluorochrome aniline blue, which reacts with callose in the sieve tubes. The lignified walls of the tracheary elements have been stained with basic fuchsin. The pattern of differentiation of the tracheary elements often closely follows the contours of the wound. (From Hardham and McCully, 1982.)

Summary

Plants increase in size and develop their characteristic forms through the growth and division of cells in specialized regions called meristems. Cells in or emerging from these regions exert precise spatial and temporal controls over their division, expansion, and differentiation. In all plant organs, the constituent cells have characteristic sizes and shapes. Cell size is governed by the relative rates of cell division and cell expansion, and cell shape by the planes of cell division and the direction of cell expansion. The direction of cell expansion is determined by the orientation of cellulose deposition in the cell walls. In axial structures, such as roots and stems, cellulose microfibrils are usually deposited transversely around the cell, that is, at right angles to the long axis of the organ. This arrangement inhibits lateral expansion and promotes cell elongation. The mode of expansion of the organ is a result of the integration of the expansion characteristics of its constituent cells.

Special sequences of formative divisions at apical meristems create initial cells for all the files of cells in the organ. Once formed, these initial cells may divide proliferatively to increase the numbers of cells within the files. Differentiation into specialized cell types occurs after cells have completed their proliferative divisions. Under special circumstances, cells can redifferentiate into new cell types without dividing.

The establishment of cell polarity is an important part of growth and morphogenesis. Ionic currents and calcium gradients are associated with early stages of development of polarity in tip-growing cells, but the ubiquity and role of these phenomena remain to be determined. They have not been observed in diffuse-growing cells.

The importance of cell communication has been a recurring theme throughout discussions of morphogenesis of shoots and roots. Although much of the information needed for growth and development is encoded in the cell's DNA, it is evident that cells must continually monitor their environment, communicating with their neighbors and responding to information they receive. We are only beginning to identify the molecules that act as important morphogenetic signals. Plant hormones are one group of such molecules, but there are likely to be many others.

Microtubules are versatile cell components. They are involved in establishing polarity by positioning the nucleus and other organelles. During cell division, they form the preprophase band, the spindle, and the phragmoplast. The function of the preprophase band of microtubules is still unknown. It appears before prophase and disappears before the spindle is complete. Its position marks the future plane of cytokinesis, indicating that the plane of cell division is determined before mitosis.

During interphase, arrays of microtubules in the cell cortex line the plasma membrane. Their arrangement corresponds closely to the orientation of cellulose deposition in the adjacent wall. Their density also reflects the rate of wall deposition, with high densities being seen at sites of rapid wall synthesis. When the microtubules are destroyed experimentally, cellulose deposition is disturbed. Nevertheless, we do not understand the exact role of cortical microtubules in cellulose alignment.

Clearly, microtubules and other cytoskeletal components are important in all aspects of plant growth and morphogenesis. As more is learned about the properties of the cytoskeleton, we will gain a better understanding of plant growth and morphogenesis.

GENERAL READING

Alberts, B., Bray, D., Lewis, J., Raff, M., Roberts, K., and Watson, J. D. (1983) *Molecular Biology of the Cell.* Garland Publishing, New York.

Burgess, J. (1985) *An Introduction to Plant Cell Development.* Cambridge University Press, Cambridge, England.

Esau, K. (1965) *Plant Anatomy,* 2d ed. Wiley, New York.

Gunning. B. E. S., and Hardham, A. R. (1982) Microtubules. *Annu. Rev. Plant Physiol.* 33:651–698.

Gunning, B. E. S., and Steer, M. W. (1975) *Ultrastructure and the Biology of Plant Cells.* Edward Arnold, London.

O'Brien, T. P. (1981) The primary xylem. In: *Xylem Cell Development,* Barnett, J. R., ed., pp. 14–46. Castle House Publications, Tunbridge Wells, England.

Quatrano, R. S. (1978) Development of cell polarity. *Annu. Rev. Plant Physiol.* 29:487–510.

Raven, P. H., Evert, R. F., and Curtis, H. (1981) *Biology of Plants,* 3d. ed. Worth, New York.

Ray, P. M., Steeves, T. A., and Fultz, S. A. (1983) *Botany.* Saunders, Philadelphia.

Robards, A. W., and Lucas, W. J. (1990) Plasmodesmata. *Annu. Rev. Plant Physiol. Plant Mol. Biol.* 41:369–419.

Robinson, D. G., and Quader, H. (1982) The microtubule-microfibril syndrome. In: *The Cytoskeleton in Plant Growth and Development,* Lloyd, C. W., ed., pp. 109–126. Academic Press, London.

Steer, M. W. (1988) The role of calcium in exocytosis and endocytosis in plant cells. *Physiol. Plant.* 72:213–220.

Sussex, I. M. (1989) Developmental programming of the shoot meristem. *Cell* 56:225–229.

Weisenseel, M. H., and Kicherer, R. M. (1981) Ionic currents as control mechanism in cytomorphogenesis. In: *Cytomorphogenesis in Plants,* Kiermayer, O., ed., pp. 379–399. Springer-Verlag, Berlin.

CHAPTER REFERENCES

Brownlee, C., and Pulsford, A. L. (1988) Visualization of the cytoplasmic Ca^{2+} gradient in *Fucus serratus* rhizoids: Correlation with cell ultrastructure and polarity. *J. Cell Sci.* 91:249–256.

Dodds, J. H. (1981) The role of the cell cycle and cell division in xylogenesis. In: *Xylem Cell Development*, Barnett, J. R., ed. pp. 168–191. Castle House Publications, Tunbridge Wells, England.

Fukuda, H., and Komamine, A. (1981) Relationship between tracheary element differentiation and the cell cycle in single cells isolated from the mesophyll of *Zinnia elegans*. *Physiol. Plant.* 52:423–430.

Galatis, B., Apostolakos, P., and Katsaros, C. (1984) Experimental studies on the function of the cortical cytoplasmic zone of the preprophase microtubule band. *Protoplasma* 122:11–26.

Gunning, B. E. S. (1981) Microtubules and cytomorphogenesis in a developing organ: The root primordium of *Azolla pinnata*. In: *Cytomorphogenesis in Plants*, Kiermayer, O., ed., pp. 301–325. Springer-Verlag, Berlin.

Gunning, B. E. S. (1982) The cytokinetic apparatus: Its development and spatial regulation. In: *The Cytoskeleton in Plant Growth and Development*, Lloyd, C. W., ed., pp. 229–292. Academic Press, New York.

Hardham, A. R., and Gunning, B. E. S. (1979) Interpolation of microtubules into cortical arrays during cell elongation and differentiation in roots of *Azolla pinnata*. *J. Cell Sci.* 37:411–442.

Hardham, A. R., and Gunning, B. E. S. (1980) Some effects of colchicine on microtubules and cell division in roots of *Azolla pinnata*. *Protoplasma* 102:31–51.

Hardham, A. R., Green, P. B., and Lang, J. M. (1980) Reorganization of cortical microtubules and cellulose deposition during leaf formation in *Graptopetalum paraguayense*. *Planta* 149:181–195.

Hardham, A. R., and McCully, M. E. (1982) Reprogramming of cells following wounding in pea (*Pisum sativum* L.) roots. I. Cell division and differentiation of new vascular elements. *Protoplasma* 112:143–151.

Heath, I. B., and Seagull, R. W. (1982) Oriented cellulose fibrils and the cytoskeleton: A critical comparison of models. In: *The Cytoskeleton in Plant Growth and Development*, Lloyd, C. W., ed., pp. 163–182. Academic Press, London.

Hepler, P. K. (1981) Morphogenesis of tracheary elements and guard cells. In: *Cytomorphogenesis in Plants*, Kiermayer, O., ed., pp. 327–347. Springer-Verlag, Berlin.

Kropf, D. L., Kloareg, B., and Quatrano, R. S. (1988) Cell wall is required for fixation of the embryonic axis in *Fucus* zygotes. *Science* 239:187–190.

Ledbetter, M. C., and Porter, K. R. (1963) A "microtubule" in plant cell fine structure. *J. Cell Biol.* 19:239–250.

Marc, J., and Hackett, W. P. (1989) A new method for immunofluorescent localization of microtubules in surface cell layers: Application to the shoot apical meristem of *Hedera*. *Protoplasma* 148:70–79.

Palevitz, B. A. (1982) The stomatal complex as a model of cytoskeletal participation in cell differentiation. In: *The Cytoskeleton in Plant Growth and Development*, Lloyd, C. W., ed., pp. 345–376. Academic Press, London.

Pickett-Heaps, J. D., and Northcote, D. H. (1966) Organization of microtubules and endoplasmic reticulum during mitosis and cytokinesis in wheat meristems. *J. Cell Sci.* 1:109–120.

Poethig, R. S. (1987) Clonal analysis of cell lineage patterns in plant development. *Am. J. Bot.* 74:581–594.

Schweiger, H. G., and Berger, S. (1981) Pattern formation in *Acetabularia*. In: *Cytomorphogenesis in Plants*, Kiermayer, O., ed., pp. 119–145. Springer-Verlag, Berlin.

Sievers, A., and Schnepf, E. (1981) Morphogenesis and polarity of tubular cells with tip growth. In: *Cytomorphogenesis in Plants*, Kiermayer, O., ed., pp. 265–299. Springer-Verlag, Berlin.

Wolf, S., Deom, C. M., Beachy, R. N., and Lucas, W. J. (1989) Movement protein of tobacco mosaic virus modifies plasmodesmatal size exclusion limit. *Science* 246:377–379.

CHAPTER **16**

Auxins:
Growth and Tropisms

THE FORM AND FUNCTION of a multicellular plant or animal depend on communication among the cells that constitute the organism. Today we know that in higher plants the regulation and coordination of metabolism, growth, and morphogenesis are often dependent on signals from one part of the plant to another. But a hundred years ago this was a new and revolutionary concept. This idea is largely the result of the pioneering work of the German botanist Julius von Sachs. Sachs proposed that chemical messengers are responsible for the formation and growth of different plant organs. He also suggested that external factors such as gravity could affect the distribution of these substances within a plant. Although Sachs did not know the identity of these chemical messengers, his ideas led to their eventual discovery.

Intercellular communication in higher plants (as in animals) is mediated by the action of chemical messengers called **hormones**. In the cells, hormones interact with specific proteins called **receptors**. The hormone-receptor complex is the active form of the hormone. Animal hormones fall into four general categories: proteins, small peptides, amino acid derivatives, and steroids. They range in size and complexity from the large protein hormones such as somatotropin (growth hormone), which consists of a single chain of 191 amino acids (molecular mass approximately 25,000 Da), to the relatively small amino acid derivatives such as epinephrine (adrenalin; 183 Da). Plant hormones are grouped into five classes: auxins, gibberellins, cytokinins, ethylene, and abscisic acid. All of these plant hormones are relatively small molecules, ranging in size from ethylene (28 Da) to gibberellin (346 Da). Like animal hormones, they are effective in very small quantities, such as 10^{-6} to 10^{-8} M. Most animal hormones are synthesized and secreted in one part of the body and are transferred, typically via the bloodstream, to specific target sites in another part of the body. Although some plant hormones are known to act at points remote from their site of synthesis, it is not always possible to demonstrate this.

The field of plant development can be viewed from several different aspects. The cellular mechanisms of plant development were addressed in Chapter 15. In the present chapter and other chapters in this unit, the regulation of these mechanisms by plant hormones will be discussed and many developmental phenomena involving these hormones will be presented.

The focus of this chapter will be on auxin—the first plant hormone to be discovered and probably the best-known of all the plant hormones. Since auxin plays a major role in the regulation of plant cell **elongation** and in the growth responses of plants to unidirectional stimuli, known as **tropisms**, this chapter will emphasize these two aspects of plant development.

Chemistry, Metabolism, and Transport

During the latter part of the nineteenth century, Charles Darwin and his son Francis studied plant growth phenomena involving tropisms. One of their interests was the bending of plants toward light. This phenomenon, which is caused by asymmetric growth, is called **phototropism**. In some experiments the Darwins used seedlings of canary grass (*Phalaris canariensis*), which, like many grasses, have their youngest leaves sheathed in a protective organ called a **coleoptile** (see Fig. 16.1). Coleoptiles are very light-sensitive. In fact, if they are illuminated on one side for several hours, they will bend (grow) toward the source of light. The Darwins found that the tip of the coleoptile perceived the light, for if they covered the tip with foil, the coleoptile would not bend. But the growth zone of the coleoptile (the region that is responsible for the bending toward the light) occurs several millimeters below the tip. Thus, they concluded that some sort of growth signal is produced in the tip, travels to the growth zone, and causes the shaded side to grow faster than the side facing the light. The results of their experiments were published in 1881 in a book entitled *The Power of Movement in Plants*. Nearly 50 years after this, it was found that the growth signal produced in the tip of the coleoptile is auxin.

A long period of experimentation on the nature of the growth stimulus in coleoptiles followed the insightful conclusions of Sachs and Darwin. This research culminated in the demonstration in 1926 by Frits Went that a growth-promoting chemical, later to be named auxin, was present in the tip of oat (*Avena sativa*) coleoptiles. If the tip of a coleoptile was removed, coleoptile growth ceased. Went's major discovery was that the growth-promoting substance from excised coleoptile tips would diffuse into agar blocks. These blocks could then be used to restore growth in decapitated coleoptiles. If an agar block containing auxin was placed on one side of a coleoptile stump, the coleoptile bent away from the side containing the block (see Fig. 16.1). This curvature occurred because the increase in auxin on one side stimulated cell elongation, and the decrease in auxin on the other side caused a decrease in the growth rate. Went found that he could estimate the amount of auxin in a sample by measuring the resulting coleoptile curvature. This technique is called the **Avena curvature test** because Went used *Avena sativa* coleoptiles. It was the first use of a bioassay to quantify a plant growth hormone. A

bioassay is a determination of the effect of a known or suspected biologically active substance on living material. Went's *Avena* curvature test proved to be a valuable tool in the subsequent chemical identification of auxin.

The Principal Auxin in Higher Plants Is Indole-3-Acetic Acid

In the mid-1930s two groups, one in the Netherlands and the other in the United States, discovered the chemical nature of auxin. Work done primarily by F. Kogl and A. J. Haagen-Smit in Holland and Kenneth Thimann in the United States led to the discovery that auxin is indole-3-acetic acid (IAA). There are several other naturally occurring auxins in higher plants, although IAA is by far the most important (Fig. 16.2). Many different synthetic auxins have also been produced (Fig. 16.3a). The term "auxin" is generally used to describe both natural and synthetic chemical substances that stimulate elongation growth in coleoptiles and many stems. As we will see, auxins also affect other plant physiological processes, such as root initiation and lateral bud dormancy.

How can such a diverse group of chemicals be active auxins? A comparison of compounds that possess auxin activity reveals that at neutral pH they have a strong negative charge that is separated from a weaker positive charge on the ring structure by a distance of about 5.5 angstroms (1 angstrom = 10^{-10} m = 0.1 nm) (Fig. 16.3b). This charge separation may be an essential structural requirement for auxin activity. The negative charge arises from the dissociation of the proton from the carboxyl group at neutral pH. The weak fractional positive charge on the indole ring is located at the nitrogen atom. In 2,4-D, the electron-attracting capacity of the chlorine atoms generates a fractional positive charge on the carbon at position 6 (Fig. 16.3b). It was once thought that a ring structure was required for activity, but the existence of a compound with weak auxin activity that lacks a ring, N,N-dimethyl ethylthiocarbamate, suggests that a ring is not absolutely necessary.

Antiauxins are another class of synthetic auxin analogs. These compounds, such as α-(*p*-chlorophenoxy)isobutyric acid or PCIB (Fig. 16.3), have little or no auxin activity but specifically inhibit the effects of auxin. When applied to plants, antiauxins may compete with IAA for specific receptors, thus inhibiting normal auxin action. One can overcome the inhibition of an antiauxin by adding excess IAA.

Three Techniques Are Used for Detection and Analysis of Auxins

Auxin bioassays were invaluable in the early investigations of the nature of auxin and are still used today to detect the presence of auxin activity in a sample. Some bioassays can also provide a rough quantitative estimate

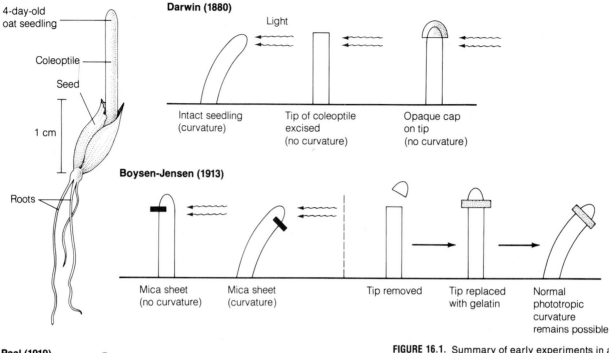

4-day-old
oat seedling

Coleoptile

Seed

1 cm

Roots

Darwin (1880)

Light

Intact seedling
(curvature)

Tip of coleoptile
excised
(no curvature)

Opaque cap
on tip
(no curvature)

Boysen-Jensen (1913)

Mica sheet
(no curvature)

Mica sheet
(curvature)

Tip removed

Tip replaced
with gelatin

Normal
phototropic
curvature
remains possible

Paal (1919)

Tip removed

Tip replaced
on one side of
coleoptile stump

Growth curvature
developed without
a unilateral light
stimulus

FIGURE 16.1. Summary of early experiments in auxin research. From experiments on coleoptile phototropism, Darwin concluded that a growth stimulus is produced in the coleoptile tip and is transmitted to the growth zone. In 1913, Boysen-Jensen discovered that the growth stimulus passes through material such as gelatin but not through water-impermeable substances such as mica. In 1919, Paal provided evidence that the growth-promoting stimulus produced in the tip was chemical in nature. In 1926, Went showed that the active growth-promoting substance can diffuse into an agar block. He also devised a coleoptile-bending bioassay for quantitative auxin analysis.

Went (1926)

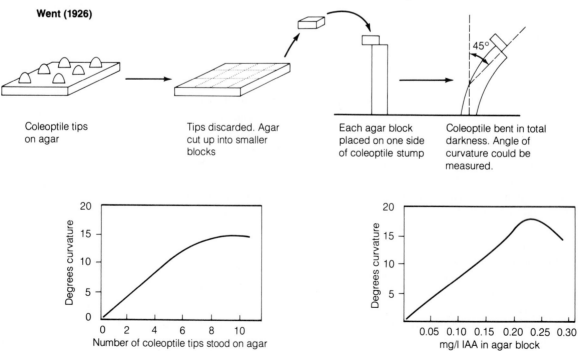

Coleoptile tips
on agar

Tips discarded. Agar
cut up into smaller
blocks

Each agar block
placed on one side
of coleoptile stump

45°

Coleoptile bent in total
darkness. Angle of
curvature could be
measured.

Degrees curvature

Number of coleoptile tips stood on agar

Degrees curvature

mg/l IAA in agar block

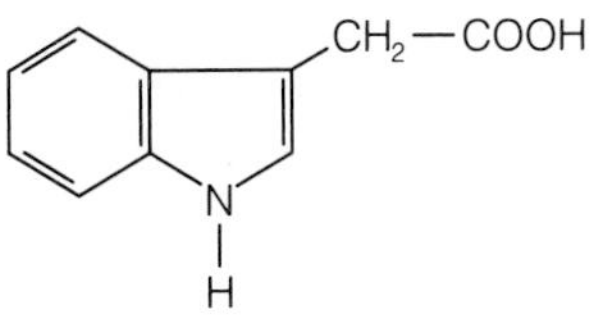
Indole-3-acetic acid
(IAA)

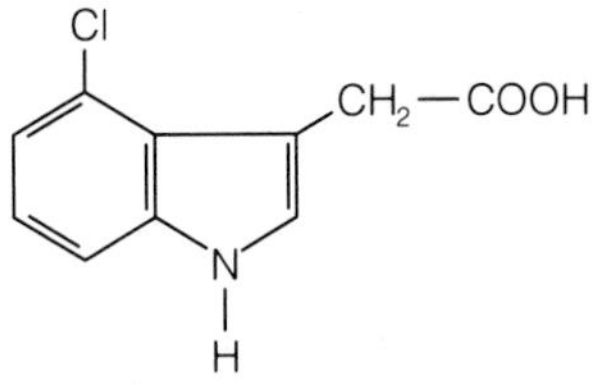
4-Chloroindole-3-acetic acid
(4-Cl-IAA)

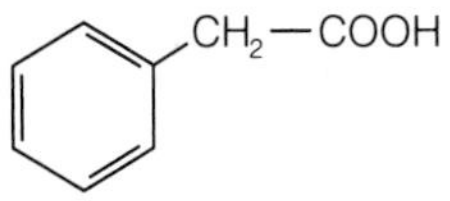
Phenylacetic acid
(PAA)

FIGURE 16.2. Structure of some natural auxins. Although indole-3-acetic acid (IAA) occurs in all plants, there are other substances in plants that have auxin activity. Pea plants, for example, contain 4-chloroindole-3-acetic acid. Other compounds that are not indoles, such as phenylacetic acid, also possess auxin activity. It is not clear what roles, if any, these other natural auxins play in plant development.

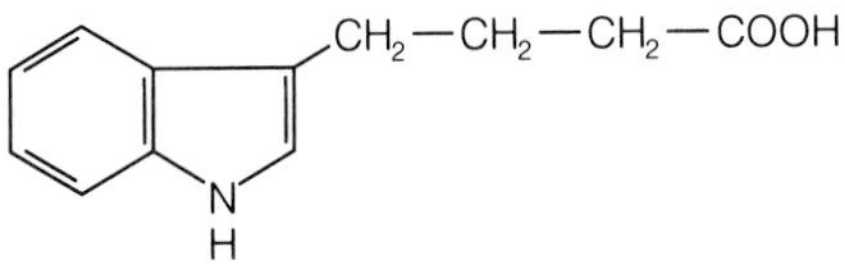
Indole-3-butyric acid
(IBA)

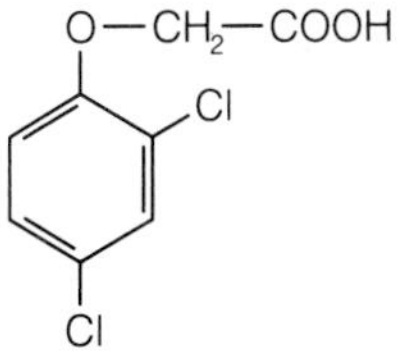
2,4-Dichlorophenoxyacetic acid
(2,4-D)

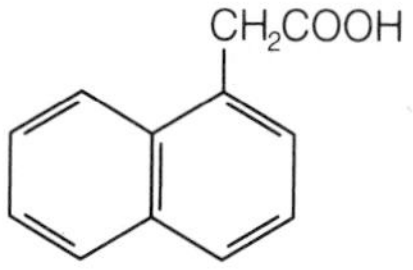
α-Naphthalene acetic acid
(α-NAA)

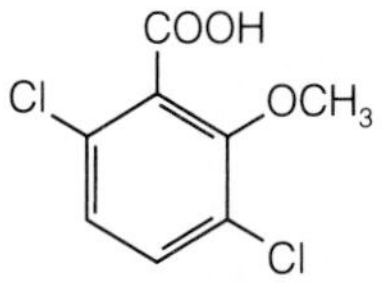
2-Methoxy-3,6-dichlorobenzoic acid
(dicamba)

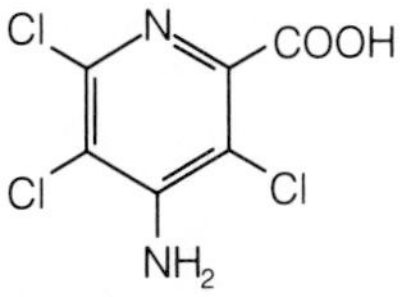
4-Amino-3,5,6-trichloropicolinic acid
(tordon or picloram)

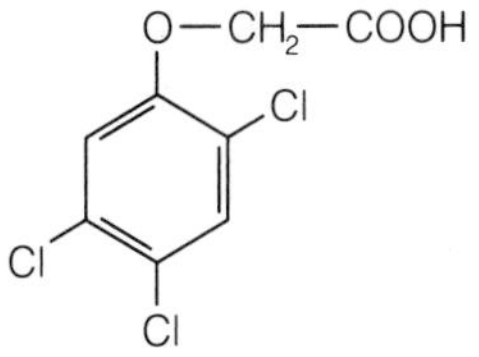
2,4,5-Trichlorophenoxyacetic acid
(2,4,5-T)

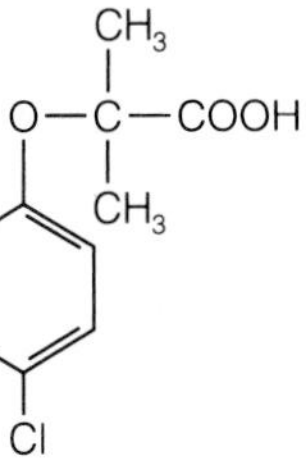
α-(*p*-Chlorophenoxy) isobutyric acid
(PCIB, an antiauxin)

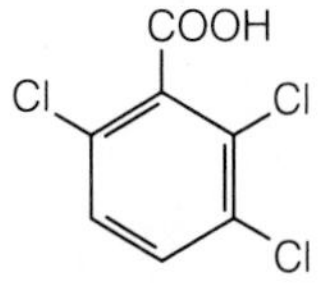
2,3,6-Trichlorobenzoic acid

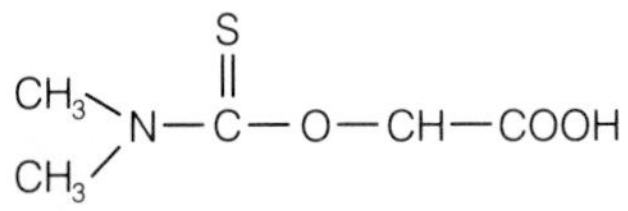
N,N-Dimethylethylthiocarbamate

(a)

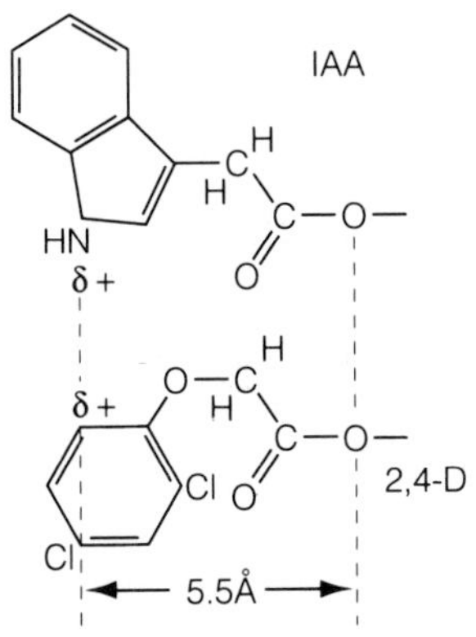

(b)

FIGURE 16.3. (a) Structure of some synthetic auxins. Most of these synthetic auxins are used as herbicides in horticulture and agriculture. The most widely used are probably 2,4,5-T and 2,4-D, which are not subject to breakdown by the plant and are very stable. They are very effective at low concentrations on broad-leaved plants. (b) The dissociated forms of IAA and 2,4-D showing the negative charge on the carboxyl and the fractional positive charge on the ring, separated by a distance of 5.5 Å.

of auxin. The *Avena* coleoptile curvature assay, first used by Frits Went over 60 years ago, is a sensitive measure of auxin activity (it is effective for IAA concentrations of about 0.02–0.2 mg L^{-1}). Another method used for auxin assays and auxin research is the coleoptile straight growth test. This assay is based on the ability of auxin to stimulate the elongation of coleoptile sections floated in a buffered solution. Both of these bioassays can establish the presence of an auxin in a sample, but they cannot be used for precise quantitation or identification of the specific compound. For this, one must turn to other methods.

Over the past 50 years, combinations of physical and chemical methods have been developed for the precise analysis of auxins. Methods used for the purification of auxins include thin-layer chromatography (TLC), high-performance liquid chromatography (HPLC), and gas chromatography (GC). For the precise identification of auxins and other plant hormones, the mass spectrometer,

a powerful analytical instrument, has been used (see Chapter 17). Some of these techniques can detect as little as 10^{-12} g (1 pg) of IAA—well within the amount of auxin found in a single pea stem section or corn kernel. These sophisticated techniques have enabled researchers to analyze accurately auxin precursors, auxin turnover, and auxin distribution within the plant (Brenner, 1981).

Physiological levels (10^{-9} g; 1 ng) of IAA in plant tissues can also be measured using **radioimmunoassay (RIA)**. Because of the sensitivity and selectivity of the antibodies (immunoglobulins) used in this technique, it is as sensitive as some of the best physicochemical methods (Pengelly and Meins, 1977; Pengelly et al., 1981). To perform a specific radioimmunoassay for IAA, one must produce antibodies to IAA (Fig. 16.4). Other types of immunoassays have also been developed in which anti-IAA antibodies are used to identify and quantitate IAA from plant extracts (Weiler et al., 1986).

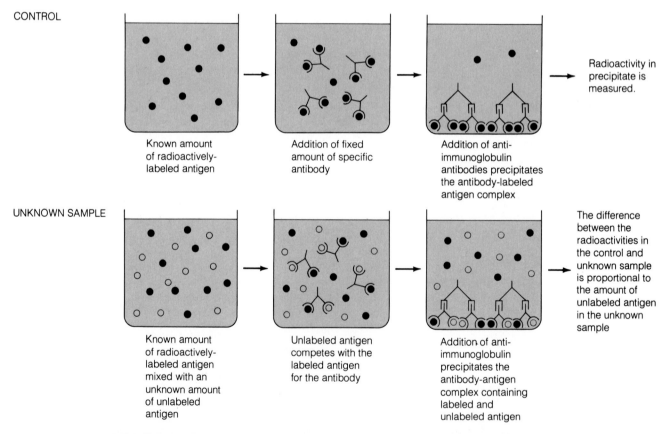

FIGURE 16.4. A radioimmunoassay (RIA) can be used to measure the amount of a substance, in this case, the amount of IAA in a sample of plant tissue. First, specific antibodies must be produced against the substance, which is referred to as the antigen. Since a compound such as IAA is too small by itself to induce a rabbit or a mouse to produce antibodies against it, IAA is chemically bonded to a large protein molecule. When the protein is injected into a mouse or rabbit, the animal will produce antibodies to the antigenic determinants on the foreign substance, including IAA. The anti-IAA antibodies can then be isolated and used in a radioimmunoassay. As shown in the diagram, unlabeled antigen (e.g., IAA) competes with a known amount of radioactively labeled antigen for binding to antibodies. The larger the amount of unlabeled antigen present, the less radioactivity (from the radioactive antigen) is found in the antibody precipitate.

Plants Contain Bound Auxins

Free auxins, such as IAA, are the plant hormones that regulate such processes as cell enlargement. Covalently bonded conjugates of IAA have also been found in all higher plants and are called **bound** or **conjugated auxins** (Cohen and Bandurski, 1982). IAA has been found to be conjugated to both high- and low-molecular-weight compounds. Examples of low-molecular-weight bound auxins include esters of IAA with glucose or *myo*-inositol and amide conjugates such as IAA-*N*-aspartate (Fig. 16.5). High-molecular-weight IAA conjugates include IAA-glucan (7–50 glucose units per IAA) and IAA-glycoproteins found in cereal seeds. The bound

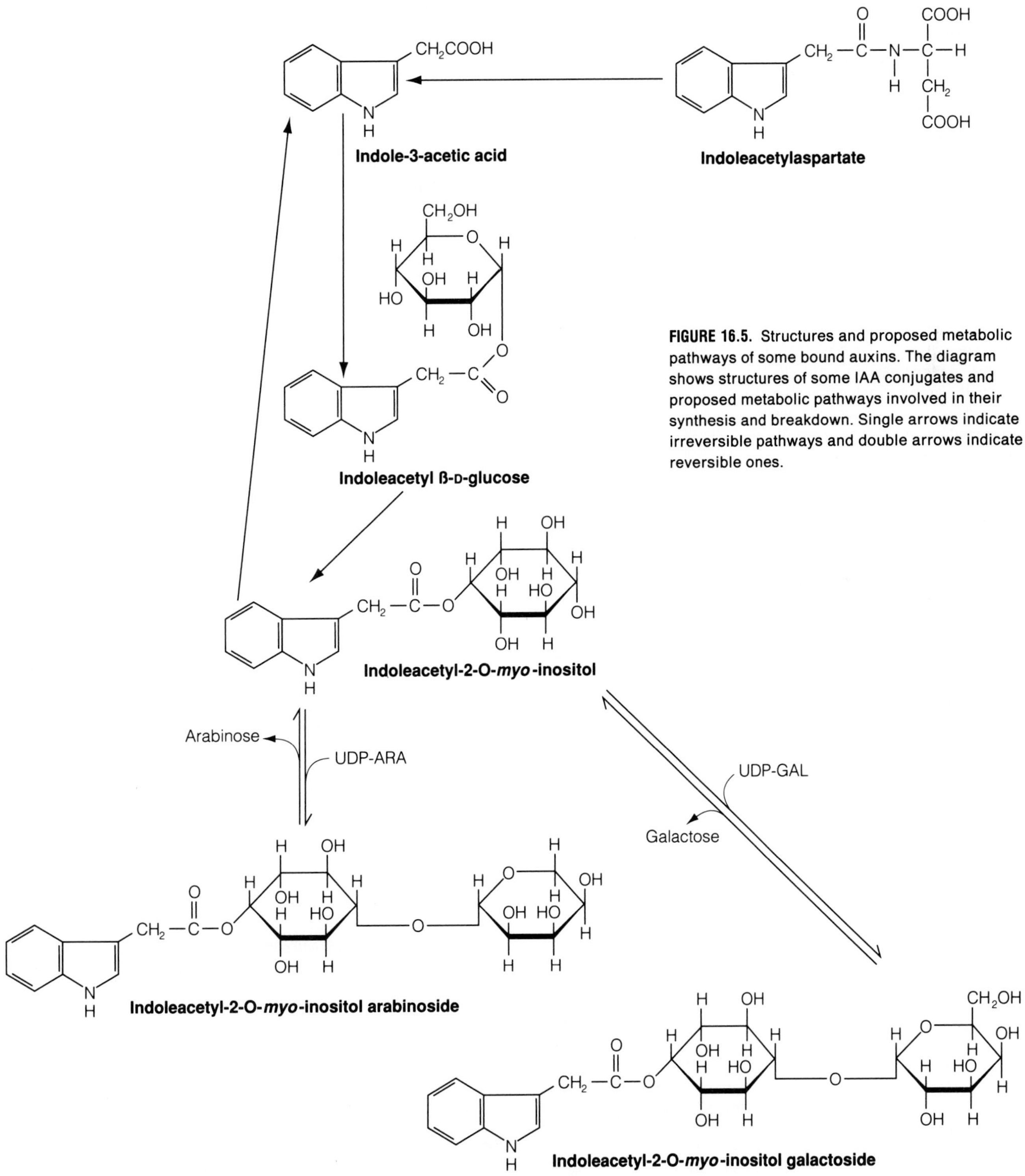

FIGURE 16.5. Structures and proposed metabolic pathways of some bound auxins. The diagram shows structures of some IAA conjugates and proposed metabolic pathways involved in their synthesis and breakdown. Single arrows indicate irreversible pathways and double arrows indicate reversible ones.

auxins themselves probably do not stimulate plant cell growth. Their biological activity has been correlated with their breakdown to free auxin.

The highest concentrations of free auxin in the living plant are in the apical meristems of shoots and in young leaves, since these are the primary sites of auxin synthesis. However, auxins are widely distributed in the plant. Metabolism of bound auxin may also be a major contributing factor in the regulation of levels of free auxin. For example, environmental stimuli such as light and gravity have been shown to influence the rate of auxin conjugation (removal of free auxin) and release (hydrolysis of bound auxin). In addition, the bound auxins may have a number of other functions, including the storage, protection against oxidative degradation, and transport of IAA.

Auxin Levels Are Controlled by the Rates of Synthesis and Degradation

Auxin metabolism includes the synthesis, degradation, and deactivation of IAA. As noted above, these processes help to control the steady-state auxin levels within a plant and thus influence the physiological processes affected by IAA. Studies of IAA metabolism have revealed a complex array of pathways for the synthesis, oxidation, and conjugation of this plant hormone.

The exact mechanism of IAA biosynthesis is still uncertain, although in most plants IAA is synthesized from the amino acid tryptophan. Several pathways from tryptophan to IAA are known (Fig. 16.6). The pathway involving indole-3-pyruvic acid and indole-3-acetaldehyde probably occurs in most higher plants. The

FIGURE 16.6. Metabolic pathways from tryptophan to IAA. There are at least two biosynthetic pathways by which the amino acid tryptophan may be converted to IAA in plants. (The name of the enzyme involved in catalyzing each reaction is shown next to the arrow.) Each pathway leads to indole-3-acetaldehyde, the presumed immediate precursor of IAA. One or both of these pathways has been detected in a variety of plants.

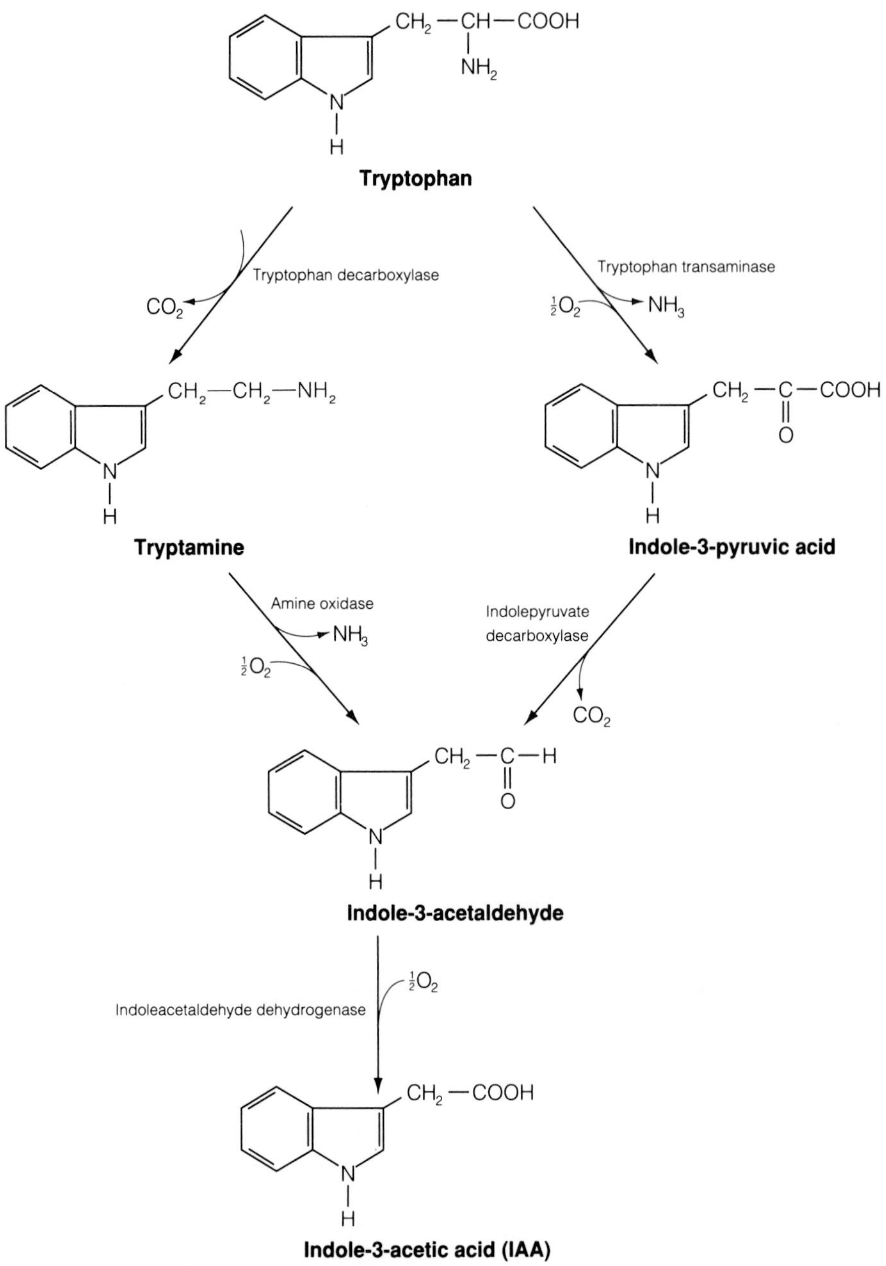

pathway with tryptamine as an intermediate is found in some plant species but not in others. Plants also contain indoles, which may serve as IAA precursors. Knowledge of the exact sequence of reactions leading to the biosynthesis of IAA in plants eventually may come from studies of mutant plants, each with a deficiency in a single step in the pathway (Last and Fink, 1988).

In general, IAA biosynthesis within a plant is associated with sites of rapid cell division, especially in shoots. Apical meristems of shoots, young leaves, and developing fruits are the primary sites of IAA synthesis in higher plants. Although IAA may also be produced in mature leaves and in root tips, the levels of IAA production there are usually lower.

Like IAA biosynthesis, the enzymatic breakdown (oxidation) of IAA may involve more than one pathway. For some time it has been thought that peroxidative enzymes are chiefly responsible for IAA oxidation, primarily because these enzymes are ubiquitous in higher plants. In some cases, however, peroxidase activity has been shown to be *inversely* proportional to the free IAA content of the tissue. More recently another pathway from IAA to oxindole-3-acetic acid has been implicated in the controlled degradation of IAA (Fig. 16.7).

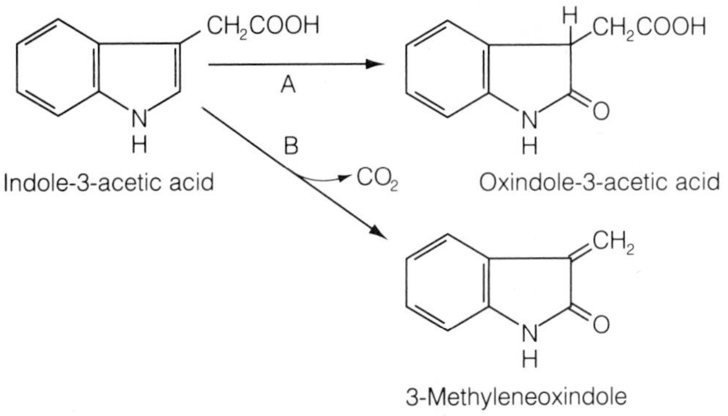

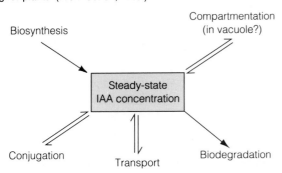

FIGURE 16.7. Biodegradation of IAA. There are at least two pathways in plants for the biodegradation of indole-3-acetic acid. Only the first stable reaction products are shown. Reaction A has been demonstrated in corn (*Zea mays*) seedlings, and reaction B is catalyzed in vitro by peroxidase, which is commonly found in higher plants. (From Cohen, 1983.)

FIGURE 16.8. Factors that influence the steady-state IAA levels in plant cells. Note that biosynthesis can lead only to an increase and biodegradation can lead only to a decrease in free IAA. The other factors can contribute to either a decrease or increase in free IAA.

In vitro, IAA is oxidized when exposed to light, and its photodestruction may be promoted by plant pigments such as riboflavin. The products of such photooxidation have been isolated from plants, but the pathway of IAA photooxidation and its role, if any, in plant growth regulation are not known.

In addition to synthesis, degradation, and conjugation of IAA, its compartmentation (e.g., in the vacuole) and transport may help to regulate free IAA levels in the cytosol. This is shown schematically in Figure 16.8.

Auxin Is Transported Both Passively and Actively

Because IAA transport plays a significant role in the regulation of plant cell growth and differentiation, it is important to understand how IAA is moved in the plant. Studies have revealed that the nature of IAA transport may depend on the developmental stage of the plant as well as on the plant organ or tissue in question. There appear to be at least two basic systems for IAA transport in higher plants: (1) an energy-requiring, unidirectional polar transport system and (2) a passive nonpolar transport system via the phloem.

Over 50 years ago it was discovered that in isolated oat coleoptile sections IAA moves mainly from the apex to the base (basipetal direction). This type of unidirectional transport is often described as **polar transport**. To study IAA transport, researchers have employed the donor-receiver agar block method. An agar block containing auxin (donor block) is placed on one end of a tissue segment and a receiver block is placed on the other end. Auxin movement through the tissue into the receiver block over time can then be determined. Before radioisotope-labeled IAA was available, investigators used auxin bioassays to quantitate the auxin in the receiver block and in this way determined the rate and direction of auxin transport in a variety of tissues.

From these and other studies, a general picture of polar IAA transport has emerged. In coleoptiles and in vegetative shoots, basipetal transport predominates. In roots, IAA transport is mainly toward the root tip (acropetal). Radioactive IAA applied to the shoot can be found in the roots, but it is not known exactly how much of the natural supply of IAA to the roots is provided by the shoot. The polar movement of IAA is an active process that requires energy, since poisons that inhibit ATP synthesis, such as dinitrophenol (DNP), also inhibit polar IAA transport.

The IAA that is synthesized in mature leaves appears to be transported in a nonpolar fashion via the phloem. Auxin, along with the other components of the phloem, can move from these leaves up or down the plant at velocities much higher than those of polar transport (see Chapter 7). Phloem transport is largely passive, not directly requiring energy, and it is not known how important it is for the long-distant movement of IAA. In corn seedlings, nonpolar movement via the phloem may also account for the export and distribution of bound IAA from the germinating seed kernel to the coleoptile tip, where the bound IAA is degraded to release free IAA.

Polar Auxin Transport Involves Basally Located Carrier Proteins

In coleoptiles, polar auxin transport may occur in all the cells. In stems, however, it may take place in **bundle sheath parenchyma** cells adjacent to the vascular bundles. Polar auxin transport requires metabolic energy and moves auxin at a rate of about 1 cm h^{-1}. It is specific for IAA and can be inhibited by the synthetic compounds naphthylphthalamic acid (NPA) and 2,3,5-triiodobenzoic acid (TIBA) (Fig. 16.9). This polar movement of IAA is largely independent of the orientation of the tissue, so it is not greatly affected by gravity. Auxin probably does not travel via the symplast but crosses both the plasma membrane and cell wall in moving from cell to cell.

According to the currently accepted chemiosmotic model of IAA transport (Goldsmith, 1977), plant cells expend metabolic energy to maintain a pH gradient (outside more acidic) and electrical potential (outside positive) across the plasma membrane (Fig. 16.10) to drive polar IAA transport and various other transport processes (described in Chapter 5). Early evidence for this model was provided by P. H. Rubery and P. A. Sheldrake at the University of Cambridge, England (Rubery and Sheldrake, 1973), who showed that IAA uptake by plant cells increases as the extracellular pH is lowered from a neutral to a more acidic value.

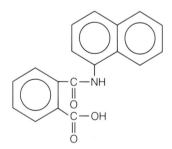

Naphthylphthalamic acid (NPA)

2,3,5-Triiodobenzoic acid (TIBA)

FIGURE 16.9. Structure of auxin-transport inhibitors. 2,3,5-Triiodobenzoic acid (TIBA) is also a weak auxin.

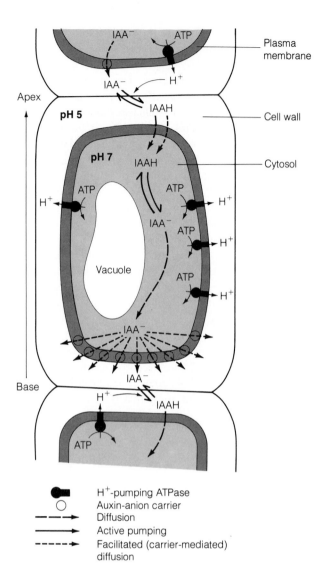

- H^+-pumping ATPase
- ○ Auxin-anion carrier
- ---→ Diffusion
- ——→ Active pumping
- ----→ Facilitated (carrier-mediated) diffusion

FIGURE 16.10. Hypothetical scheme for polar auxin transport. The chemiosmotic model of polar auxin transport is illustrated above, showing one cell in a column of auxin-transporting cells in a stem or coleoptile. (For clarity the volume of the vacuole has been reduced and the thickness of the plasma membrane has been increased.) The proton-pumping ATPases in the plasma membrane use the energy of ATP hydrolysis to maintain a transmembrane pH gradient. This causes the extracellular space to be acidic (pH 5) and leads to an electrical potential (outside positive) across the plasma membrane. Both of these factors play a role in the chemiosmotic model of polar auxin transport. Note that the IAA$^-$ carriers are located at the basal end of the cell. (From Jacobs, 1983.)

Since the pK$_a$ of IAA is 4.75, a significant proportion of IAA in an acidic environment (pH 5 and below) is in the undissociated form IAAH, in which the carboxyl group is protonated. IAAH is relatively lipophilic; that is, it can easily cross lipid bilayer membranes. Assuming that the solution in the apoplast is acidic, IAAH would diffuse from the extracellular space across the plasma membrane into the cytosol. Because of the neutral pH of the cytosol, most of the IAAH would dissociate into the auxin-anion form (IAA$^-$). Auxin tends to accumulate inside the cell

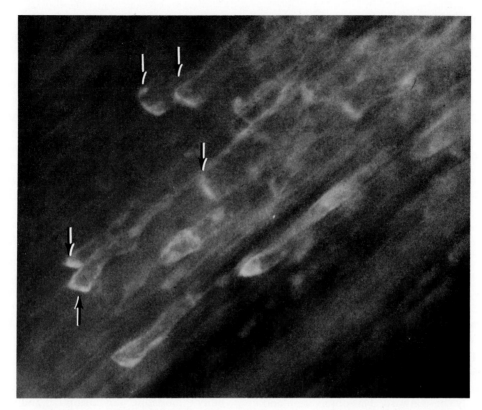

FIGURE 16.11. An immunofluorescence photomicrograph of a median longitudinal fresh section of etiolated pea stem tissue that has been treated with monoclonal antibodies raised against the NPA-binding protein. The NPA-binding protein appears to be most concentrated in the plasma membrane at the basal ends of bundle sheath parenchyma cells (arrows). This is believed to correspond to the location of the auxin-anion efflux carrier, responsible for the polarity of auxin transport. ($\times$ 200) (Photograph courtesy of Mark Jacobs; from Jacobs and Gilbert, 1983.)

because the membrane is less permeable to IAA$^-$ than to IAAH.

Polar auxin transport is probably due to the presence of specific auxin-anion transport proteins in the plasma membrane at the basal end of the cells. Evidence for this was obtained by Jacobs and Gilbert (1983) using monoclonal antibodies against auxin-anion transport proteins. The monoclonal antibodies were initially selected for their ability to compete with radioactively labeled NPA (a polar transport inhibitor) for binding to isolated plasma membrane vesicles. By means of an immunofluorescence technique, the location of the presumed NPA-binding proteins was identified in the plasma membrane at the basal ends of parenchyma cells sheathing the vascular bundles (Fig. 16.11). These transport proteins facilitate the passive auxin efflux that is favored by both the concentration gradient of IAA$^-$ across the plasma membrane and the cell membrane potential (outside positive). As the IAA$^-$ molecules move into the acidic cell wall region, they are protonated to IAAH, which diffuses down a concentration gradient into the next cell in the series. Polar auxin transport in plant tissues probably involves repetition of this process down a column of auxin-transporting cells.

How is polar auxin transport regulated? The synthetic compound NPA (see Fig. 16.9) blocks polar auxin transport, presumably by interfering with the auxin-anion efflux step and causing a net accumulation of IAA$^-$ in the transporting cells. NPA does *not* directly compete with IAA, but it does bind to a protein in the plasma membrane. It is likely that NPA binds to either the auxin-anion transport protein (at a site other than the IAA$^-$ binding site) or another plasma membrane protein that interacts with the IAA$^-$ transporter. Jacobs and Rubery (1988) found that a group of flavonoids, including quercetin, apigenin, and kaempferol, can specifically compete with NPA for binding to its receptor. Because these flavonoids are widely distributed in the plant kingdom, they are excellent candidates for natural regulators of polar auxin transport in plants.

Physiological Effects of Auxin

Auxin has a wide variety of effects on plant growth and morphogenesis. It promotes the elongation growth of stems and coleoptiles, yet it may inhibit root elongation. It promotes cell division in stems but may inhibit it in lateral buds. The development of fruit may also depend on the presence of auxin. The varied effects of auxin often depend on a number of factors, including (1) the developmental stage of the tissue or organ, (2) the concentration of auxin, (3) the type of auxin (natural or synthetic), (4) the involvement of other plant hormones, and (5) the use of intact versus excised tissue for experiments. The fragmentary experimental evidence presented in this section reflects the present state of the research.

Auxins Induce Cell Elongation in Stems and Coleoptiles

We have already seen that auxins promote elongation growth in young stems and coleoptiles. When the source of auxin is removed from these tissues, the growth rate decreases. When auxin is restored, growth usually resumes within 10–20 min. The optimal auxin concentration for elongation growth is typically 10^{-5} to 10^{-6} M. As shown in Figure 16.12, high (supraoptimal) levels of added IAA may actually inhibit growth. This effect is usually attributed to auxin-induced production of the plant hormone ethylene, which suppresses elongation growth (see Chapter 19).

When stem sections from growing regions of a shoot are first split longitudinally and treated with auxin, the two halves typically curve inward. This observation and related phenomena have been interpreted to indicate that auxin exerts its effect primarily on the epidermis of stems. It suggests that the epidermis limits the rate of elongation of stem tissues.

The mechanism of auxin-induced elongation in plants has been difficult to discern. Auxin has multiple effects on plant cells, and it is difficult to determine which

of these effects are directly related to growth promotion and which are not. In addition, auxin may indirectly influence cell enlargement through its effects on a number of biophysical parameters, including cell wall mechanical properties, turgor pressure, osmotic pressure, and water permeability (hydraulic conductivity). A detailed discussion of these parameters is presented in the box on plant cell expansion. In general, for growth to occur the wall must yield to the force of turgor pressure.

The expansion of cells during plant growth is an irreversible process, requiring both water absorption from the environment and permanent stretching of the wall surrounding each cell. It is through water absorption that cells increase in volume (about 95% of the mass of growing cells is composed of water). New surface area to accommodate the additional cell volume is generated by irreversible wall expansion. Continued growth is, of course, also dependent on the synthesis of new wall material.

Auxin Increases the Extensibility of the Cell Wall

IAA causes a fairly rapid increase in cell wall extensibility ("loosening") in coleoptiles and young developing stems (see the discussion of diffuse-growing cells in Chapter 15). Auxin does not directly bind to the cell walls, but acts at the plasma membrane or within the cell. This means that, in response to IAA, plant cells must export some wall-loosening factor (WLF) that promotes wall extensibility. One candidate for the WLF is the hydrogen ion (H^+).

That hydrogen ions can act as the intermediate (the WLF) between auxin and cell wall loosening was first demonstrated independently by Rayle and Cleland (1970) at the University of Washington (Seattle) and Hager, Menzel, and Krauss (1971) at the University of Münster in West Germany. These results led to the **acid growth theory** of auxin-stimulated plant cell elongation (Rayle and Cleland, 1977). According to the theory, auxin causes responsive cells to extrude protons actively into the cell wall region, and the resulting decrease in pH activates wall-loosening enzymes that promote the breakage of key cell wall bonds, increasing wall extensibility. In coleoptiles and a variety of young dicot stems, a pH below 5.5 can partially substitute for auxin in stimulating cell elongation. When added to these tissues, auxin causes proton extrusion within 10–15 min. The fungal phytotoxin fusicoccin (Fig. 16.13) also stimulates rapid H^+ extrusion by plant cells and leads to rapid cell enlargement (Marré, 1979). These and other observations support the theory that acidification of the cell wall is a critical step, not only in auxin stimulation of cell elongation but also in the control of plant cell enlargement in general.

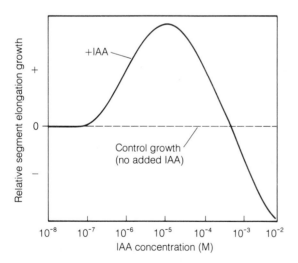

FIGURE 16.12. Dose-response relationship for IAA-induced growth in stem or coleoptile segments. Elongation growth of excised segments of coleoptiles or young stems is plotted versus increasing amounts of added IAA. As supraoptimal levels of IAA are added, elongation growth actually falls below the control level of growth (no added auxin).

The Biophysics of Plant Cell Expansion

WATER UPTAKE FOR cell growth is a passive phenomenon. The driving force leading to water absorption by the cell is a difference in water potential across the plasma membrane. We define this difference as $\Delta\psi = \psi^o - \psi^i$, where ψ^o is the outside water potential and ψ^i is the inside water potential. Water flows passively from a region of high water potential to one of lower water potential. For example, it would flow from a compartment containing pure water, which has a water potential of zero, into a cell with a water potential of, say, -0.5 MPa. Recall that the water potential of the cell is given by $\psi = P - \pi$, where P is the turgor pressure and π is the osmotic pressure of the cell. $\Delta\psi$ is given by $\Delta\pi - P$, assuming that the outside solution is at atmospheric pressure.

The rate of absorption also depends on the surface area of the cell and on the ease with which water is translocated across the membrane. Thus, we have

$$\frac{dV}{dt} = A\,Lp\,(\Delta\psi) \qquad (16.1)$$

where dV/dt is the rate of increase in volume (in units of cm^3 s^{-1}), A is the surface area (in cm^2), and Lp is the specific hydraulic conductivity of the membrane (in cm s^{-1} MPa^{-1}). The growth rate here is the rate of volume increase, dV/dt, and is an **absolute growth rate**. Frequently it is more useful and convenient to work with a **relative growth rate**, which is the absolute growth rate divided by the size (V) of the growing cell or plant:

$$\text{Relative growth rate} = \frac{1}{V}\frac{dV}{dt} = \frac{\bar{V}}{V} \qquad (16.2)$$

We use $\bar{V}/V$ to symbolize the relative growth rate. By dividing both sides of Equation 16.1 by V and substituting $(\Delta\pi - P)$ for $\Delta\psi$, we obtain

$$\frac{\bar{V}}{V} = \frac{A\,Lp}{V}(\Delta\psi) = L(\Delta\pi - P) \qquad (16.3)$$

where L is a new term defined as the **relative cell hydraulic conductance** and is a measure of the relative ease with which water can enter the cell. It has units of s^{-1} MPa^{-1}. For the right-hand side of Equation 16.3 if we assume that the solution outside the cell is pure water at atmospheric pressure, so that

P and π outside are both zero, this assumption allows us to equate $\Delta\psi$ with the absolute value of π inside minus P inside. Equation 16.3 states that *the relative growth rate of a cell depends on the relative cell hydraulic conductance, the osmotic pressure of the cell, and the turgor pressure of the cell.*

In a nongrowing cell, water absorption in response to a $\Delta\psi$ would normally increase cell volume, causing the cell contents to push harder against the cell wall and to increase turgor pressure. This increase in P would quickly increase the cell ψ, $\Delta\psi$ would fall to zero, and net water uptake would cease.

In a growing cell, however, $\Delta\psi$ is prevented from reaching zero because the cell wall is "loosened" and distends or stretches irreversibly, causing a reduction in cell turgor pressure. Such stretching of the cell wall should not be confused with the reversible, elastic stretching that occurs in nongrowing cells, such as guard cells, when they absorb water and swell. In growing cells, the wall undergoes a process termed **stress relaxation**, which reduces the stress in the wall and is responsible for the maintenance of $\Delta\psi$ in growing cells.

Stress relaxation can be understood as follows. In a turgid cell, the cell contents, mostly water, push against the cell wall, causing the wall to stretch elastically (that is, reversibly) and giving rise to a stress (force/area) in the wall. Wall stress is the equal and opposite reaction to the hydrostatic pressure of the cell contents, that is, to turgor pressure. In a growing cell, biochemical loosening of the cell wall enables the wall to yield inelastically to the wall stress. Because water is nearly incompressible, only a minuscule expansion of the wall is needed to reduce cell turgor pressure and, simultaneously, wall stress. Thus, *stress relaxation is a decrease in wall stress with nearly no change in wall dimensions.* As a consequence of stress relaxation, the cell water potential is reduced and water flows into the cell, causing a measurable extension of the cell wall and increasing the surface area and cell volume. Sustained growth of plants entails simultaneous stress relaxation of the wall, which tends to reduce turgor pressure, and water absorption, which tends to increase turgor pressure.

Empirically, it has been found that wall relaxation and expansion depend on turgor pressure. As turgor pressure is reduced, wall relaxation and growth slow down. At some point before P reaches zero, growth ceases entirely. The value of P at which growth ceases is the minimum turgor or threshold turgor necessary for growth. Frequently it is called **the yield threshold** and denoted by the symbol Y.

This dependence of cell expansion on turgor pressure is embodied in the following empirical relation, which to a first-order approximation describes the growth rate as a function of turgor pressure:

$$GR = m(P - Y) \qquad (16.4)$$

where GR is the growth rate of the cell, P is turgor pressure (in units of MPa), Y is the yield threshold (MPa), and m is the coefficient relating the growth rate to turgor in excess of Y ($s^{-1}\,MPa^{-1}$). The coefficient m is usually called the **irreversible extensibility** of the cell wall, or the **wall extensibility** for short. From this equation we see that the growth rate becomes zero when $P = Y$. At higher values of P, wall growth is governed by m and P (in excess of Y), that is, $(P - Y)$.

Under conditions of sustained and steady-state growth, the rate of cell wall expansion equals the rate of water absorption. Setting the two GRs in equations 16.3 and 16.4 equal (using π_{cell} instead of $\Delta\pi$) and solving for P, we obtain

$$P = \frac{L\pi + mY}{L + m} \qquad (16.5)$$

Substituting this relation into either Equation 16.3 or 16.4, we find that the relative growth rate $\bar{V}/V$ is given by

$$\text{Growth rate} = \frac{\bar{V}}{V} = \frac{Lm}{L + m}(\pi - Y) \qquad (16.6)$$

Equation 16.6 is the general equation governing the expansion of growing plant cells. It shows that

growth may be influenced by the parameters controlling water absorption (L and π) and by those controlling irreversible wall expansion (m and Y). When a stimulus such as light or a plant hormone alters the growth rate of a plant, it must do so by affecting one or more of the parameters in Equation 16.6.

Cells can regulate their osmotic pressure by altering the concentration of solutes in the cytosol. Solute pumps in the plasma membrane may be activated, or stores of osmotically inactive polymeric molecules may be synthesized or degraded. The former mechanism apparently provides a more rapid means of adjustment than the latter.

Regulation of water absorption by changing L or π will influence the growth rate, but localized and directed cell expansion can be achieved only by precise control of cell wall properties affecting the values of m and Y. Wall yielding may be spatially restricted within the cell, as in tip-growing cells, or it may occur diffusely over the cell wall area as in the majority of plant cells. The parameter m is a measure of the relative irreversible deformability of the cell wall. Experiments indicate that m is not an intrinsic property of the cell wall; instead, it depends on the rate of some biochemical processes that loosen the structure of the wall and permit wall relaxation. These processes require metabolic energy. The rigidity of the wall may be decreased, for example, by breaking cross-links between various wall components. Experiments suggest that auxin may enhance the growth rate primarily by increasing m. As previously noted in Chapter 14, under water stress conditions m usually decreases and Y increases. Some of the effects of water stress may be caused by an enhanced synthesis of abscisic acid (Chapter 19).

Cell wall acidification is not the only way in which auxin induces plant cell elongation. Acid-stimulated wall loosening alone cannot sustain elongation over a period of many hours, as auxin can. Auxin must also affect other factors important to plant cell growth, such as (1) the uptake or generation of osmotic solutes, (2) the maintenance of cell wall structure, and (3) the hydraulic conductivity of the cell membrane. The precise role of auxin in regulating these factors is not clear. Although auxin increases the uptake of certain osmotic solutes, it does not appear to affect the cell's osmoregulation mechanism. Auxin may help maintain the cell wall loosening capacity by stimulating the secretion of new cell wall material. In

fact, auxin is known to increase the activity of certain enzymes involved in wall polysaccharide synthesis. IAA may also promote the synthesis of cell wall proteins required for growth.

Does Auxin Regulate Growth in Roots and Leaves?

Acidification of the cell wall region also appears to play a role in cell enlargement in roots and leaves. Does this mean that auxin regulates growth in these organs as well as in stems and coleoptiles?

The control of root elongation growth is not well

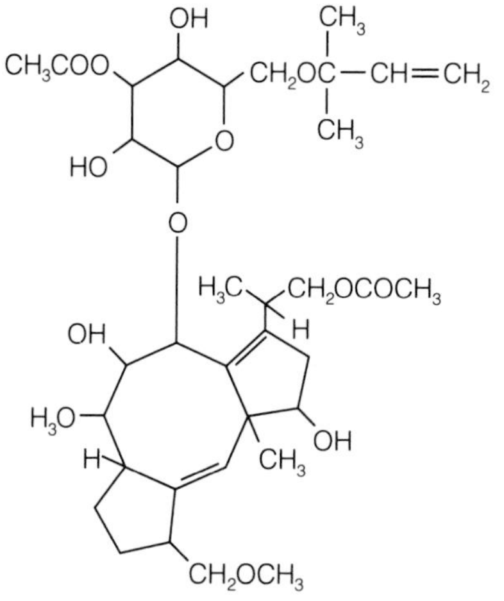

Fusicoccin

FIGURE 16.13. Chemical structure of the fungal phytotoxin fusicoccin. This toxin, associated with the wilt diseases of almond and peach trees, causes massive release of protons from plant cells. The resulting acidification of the cell wall induces wall loosening and rapid cell expansion. Note that the compound is structurally unrelated to auxin and presumably acts via a different mechanism. However, the effect of fusicoccin supports a role for acid excretion in cell growth. (From Marré, 1979.)

understood. Originally it was hypothesized that the auxin responses in roots and shoots are qualitatively the same but that the optimal auxin concentration is much lower in roots. The shoot supplies auxin to the root, although there is evidence that some auxin may be synthesized in the root tip as well. Reports of auxin-induced root growth have been contradictory. A possible source of confusion is the fact that auxin induces the synthesis in roots of the plant hormone ethylene, a potent root-growth inhibitor (Chapter 19). If ethylene synthesis is specifically blocked, low concentrations (10^{-10} to 10^{-9} M) of auxin promote growth of intact roots, but higher concentrations (10^{-6} M) still inhibit growth.

Young expanding leaves are a rich source of free auxin for the growing plant. Although auxin may regulate the growth of leaf veins, the role of auxin and other plant hormones in regulating leaf growth is poorly understood. Apart from auxin, light has been shown to be an important factor stimulating leaf expansion as well as proton extrusion (Van Volkenburgh and Cleland, 1979).

Tropisms Are Growth Responses to Directional Stimuli

In addition to regulating cell elongation in the normal course of plant development, auxins may mediate the effects of light and gravity on plant growth. Plants can orient themselves favorably with respect to the environment—the leaves of a houseplant facing a window; rows of corn plants growing perfectly straight, perpendicular to the earth's surface; tap roots growing toward the center of the earth; the young shoots of a toppled tree bending upward. These are all **tropic responses**, that is, growth in response to directional light (**phototropism**) or gravity (**gravitropism**). When a shoot, coleoptile, or leaf is illuminated from one side, the growing zone of the shoot, the coleoptile, or the petiole of the leaf may bend toward the direction of the light. This bending toward light is called **positive phototropism**. Many roots bend away from the light, displaying **negative phototropism**. When a plant is placed horizontally, the growing zone of the root curves until the root tip is again pointed vertically downward (**positive gravitropism**) and the growing zone of the shoot bends vertically upward (**negative gravitropism**). How the physical effects of light and gravity are perceived and transduced into a differential growth response in higher plants is beginning to be understood. Many explanations have involved auxin.

The discovery of auxin began with studies of tropisms in coleoptiles. Because dark-grown (etiolated) coleoptiles are very sensitive to light and are relatively easy to grow and work with, much of our information on the mechanism of phototropism in plants has come from studies of coleoptiles. Much less is known about phototropism in shoots, leaves, and roots (Dennison, 1979).

Action Spectra for Phototropism Have Peaks in the Visible and the Near-Ultraviolet

Based on the experiments of the Darwins, as well as on more recent studies, the coleoptile tip is regarded as the site of light perception. What wavelengths are the most effective in inducing phototropic bending? In etiolated coleoptiles the action spectrum of positive phototropic curvature indicates that there are two major peaks: a primary one in the visible part of the spectrum (between 420 and 480 nm) and a secondary peak in the near-ultraviolet region (between 340 and 380 nm) (Fig. 16.14). The primary peak exhibits some fine structure, with minor peaks at 425, 445, and 472 nm. A number of responses in plants and fungi, including phototropism in *Phycomyces*, carotenoid biosynthesis in *Neurospora*, stomatal opening, and chloroplast rearrangements, have very similar action spectra and are collectively referred to as **blue-light responses**.

The identity of the pigment (or pigments) involved in phototropism is still unknown. Because the action spectrum shows a major peak in the blue part of the spectrum, the photoreceptor is likely to be a yellow pigment. The absorption spectra of two types of yellow pigments, carotenoids (e.g., β-carotene) and flavins (e.g., riboflavin), are shown in Figure 16.14. Note that neither pigment's absorption spectrum exactly matches the action spectrum of phototropism. β-Carotene (in ethanol) is missing the secondary peak in the near-UV region, while riboflavin (in water) is missing the fine structure in the blue/violet range. Because light absorption is influenced by the

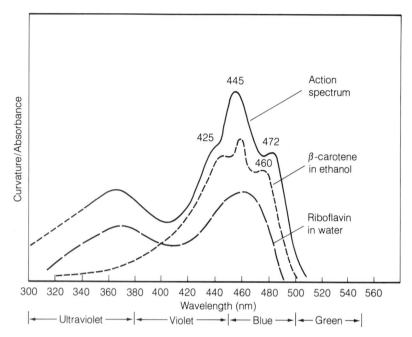

FIGURE 16.14. Action spectrum for phototropism of oat coleoptiles (solid line) and the absorption spectra of β-carotene and riboflavin (dashed lines). The relative growth response of oat coleoptiles bending toward unidirectional light (positive phototropism) is plotted versus the wavelength of the light. This type of diagram is called an *action spectrum* because it shows the wavelengths of light that are most active in eliciting a biological response. Thus, it can provide clues to the identity of the pigment involved in a particular response. (Dashed lines) Absorption spectra of two pigments that may be involved in coleoptile phototropism. These spectra show the relative amounts of light over the wavelength range 300–550 nm absorbed in vitro by both pigments. Comparison of the action spectrum with the absorption spectra of the pigments shows that they may be involved in the phototropic response.

molecular environment, the absorption spectra of both pigments can be made to approximate the action spectrum by altering the solvent. For example, riboflavin in benzene shows fine structure in its primary absorption peak similar to that of the action spectrum. This makes it difficult to use absorption spectra to identify the photoreceptor.

Most workers in the field believe that the blue-light photoreceptor is probably not a carotenoid. For example, carotenoidless albino mutants of the fungus *Phycomyces* exhibit a normal phototropic response (Presti et al., 1977). In plants, albino mutants as well as seedlings treated with inhibitors of carotenoid biosynthesis retain their ability to bend toward unidirectional light. The fact that most, if not all, carotenoids in plant cells occur in plastids also seems inconsistent with a role in phototropism.

On the other hand, there is limited evidence supporting a role for a flavin, most likely a flavoprotein complex. Such a pigment has been identified in the plasma membranes of *Neurospora crassa* and in *Zea mays* coleoptiles (Brain et al., 1977). The pigment was detected by measuring the photoreduction of a *b*-type cytochrome, which results in a change in the absorbance. Importantly, the action spectrum for the light-induced absorbance change (LIAC) is very similar to that of phototropism.

The maximum peak for the LIAC effect (∼465 nm) is a region of the spectrum in which the cytochrome itself shows no appreciable absorption. This suggests that the flavoprotein is in close contact with a *b*-type cytochrome in the plasma membrane, enabling the flavoprotein to reduce it upon absorbing light of the appropriate wavelength. What happens next to transduce the light signal remains obscure.

Finally, recent work in plants suggests that there is probably more than one blue-light photoreceptor, clouding an already complicated picture (Konjevic et al., 1989).

Phototropism May Be Mediated by the Lateral Redistribution of Auxin

An early explanation of phototropism in etiolated coleoptiles was presented in the 1920s and is known today as the Cholodny-Went theory. According to the theory, the phototropic stimulus somehow results in an unequal lateral distribution of auxin at the tip (higher auxin concentration on the shaded side). This lateral redistribution of auxin is carried to the growth zone by polar auxin transport, resulting in differential growth and phototropic curvature.

We know that positive phototropic bending of coleoptiles involves a decrease in elongation growth on the illuminated side and an increase on the shaded side. An auxin gradient can clearly be measured in response to unilateral light (Fig. 16.15). This unequal distribution of auxin appears to be a result of lateral auxin transport at the tip. It apparently is not a result of the photooxidation of auxin on the illuminated side or of increased IAA synthesis on the shaded side. Lateral transport of both endogenous IAA and radioisotope-labeled IAA added to the tissue has been shown to occur in response to unilateral light. The mechanism of the light-activated lateral transport of IAA is presently unknown.

In shoots, leaves, and roots, the mechanism of phototropism is more complicated. For instance, the perception of phototropically active light in these organs may involve two different photoreceptors—a blue-light receptor and phytochrome, which absorbs red light. Although both blue and red light inhibit stem and root elongation, the nature of their involvement in the phototropic response is not yet understood.

Gravitropism in Shoots Involves Auxin Redistribution

In general, the perception of gravity in plants involves the movement of free-falling bodies within the plant cell. These gravity-sensitive bodies are called **statoliths**. In gravity-sensing plant cells (**statocytes**), the statoliths are probably starch grains located within **amyloplasts**, which are modified plastids. The amyloplasts sediment through the cytosol and come to rest on the lowermost part of the cell (Fig. 16.16), enabling the plant to detect the direction of gravity. It has been suggested that the contact of amyloplasts with the endoplasmic reticulum near the plasma membrane somehow redirects auxin transport (see discussion below). At present, the exact role of the statoliths remains unknown (Sievers and Hensel, 1982).

In coleoptiles, gravitropic curvature is a result of an increase in growth rate on the lower side of the horizontal coleoptile and a decrease on the upper side. The unequal lateral distribution of auxin in gravistimulated coleoptiles has been confirmed using radioisotope-labeled IAA, physicochemical methods, and radioimmunoassay (Mertens and Weiler, 1983; Migliaccio and Rayle, 1984). Early experimental evidence indicated that the tip of the coleoptile is the site of graviperception. For example, if coleoptile tips are placed horizontally, greater amounts of auxin diffuse from the lower side into an agar block. This lateral migration of auxin at the tip fits nicely with the Cholodny-Went theory. However, after removal of the apex, coleoptiles remain graviresponsive and lateral auxin redistribution still takes place. Thus, the apical region may be the principal, but not the only, site of graviperception in coleoptiles.

How does gravity cause the apparent redistribution of auxin in coleoptiles? One theory is that the pressure

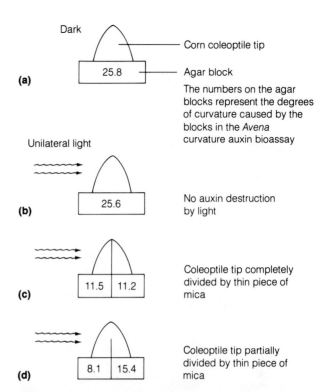

FIGURE 16.15. Evidence for the lateral redistribution of auxin stimulated by unidirectional light in corn coleoptiles. When an agar block is placed under an excised corn coleoptile tip in the dark, auxin diffuses into the block (a). If the coleoptile is exposed to unidirectional light and auxin is allowed to diffuse into the block, about the same amount of auxin is found (b) as in the dark. Apparently auxin is not photooxidized by the light. If the coleoptile and the agar block are completely divided in half, no lateral redistribution of auxin is observed (c), nor is there increased auxin biosynthesis on the shaded side. However, if the tip is only partially divided, lateral redistribution of auxin in response to unidirectional light is observed (d) at the tip.

of the amyloplasts on the plasma membrane affects the electrical potential across the gravistimulated cells such that some auxin is transported laterally. Other theories of gravistimulated auxin redistribution include (1) stimulation of the conversion of bound auxins to free auxins on the lower side of horizontal coleoptiles, (2) liberation of free auxin from intracellular compartments, and (3) increased "leakage" of auxin from the stele into the cortex on the lower side. (The stele contains much higher levels of free auxin than the cortex.)

Some critics of the Cholodny-Went theory have questioned whether the observed unequal lateral auxin distribution is enough to cause the observed growth curvatures (Firn and Digby, 1980). It has been found that gravistimulation leads to an accumulation of calcium on the upper side of the coleoptile. Since calcium can inhibit auxin-induced growth (presumably by rendering the wall less extensible), its accumulation might amplify the effect of a lateral auxin gradient on differential growth.

In general, the evidence tends to favor the explanation that gravity is perceived in cells located in the grow-

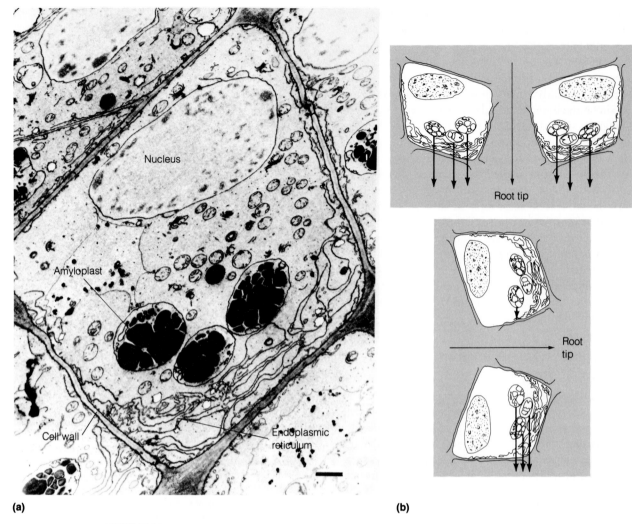

(a)   (b)

FIGURE 16.16. Graviperception in statocytes of *Lepidium* roots. (a) A single statocyte found in the core of the root cap from a primary root of *Lepidium*. Note the starch-laden amyloplasts resting against the endoplasmic reticulum (ER) (Bar = 1μm). (b) The tendency of these amyloplasts to sediment in response to reorientation of the cell and to remain resting against the ER. The dashed arrows point toward the root tip. When the root is oriented vertically (top), the pressure exerted by the amyloplasts on the ER (solid arrows) is equally distributed. In a horizontal orientation (bottom), the pressure on the ER is unequal on either side of the vertical axis of the root. (From Volkmann and Sievers, 1979.)

ing region of young shoots as well as in the apex. Horizontal positioning of the shoot results in rapid net loss of auxin in the peripheral cells on the upper side of the growth zone and simultaneous net gain of auxin in the lower peripheral cells. This redistribution of free auxin is mediated by either the graviresponsive cells or nearby graviperceptive cells. The mechanism of auxin redistribution is unknown, but it may involve lateral auxin transport. Unequal lateral distribution of auxin sets up an asymmetric pH gradient across the shoot, with the lower side more acidic. The asymmetric acid-induced wall loosening that results is partly responsible for the differential growth. The shoot apex serves to reinforce this unequal lateral auxin redistribution.

Statoliths in the Cap Cells Direct the Transport of an Inhibitor to the Lower Side of a Horizontal Root

In the primary root of higher plants, gravity is perceived in the root cap. The graviresponsive amyloplasts are located in the statocytes (Fig. 16.16) in the central cylinder of the root cap. By micromanipulations it is possible to remove the root cap from otherwise intact roots, and this abolishes graviresponsiveness in most plant species tested.

The gravitropic curvature is localized in the elongation zone of the root distal to the root apex (Chapter 15). Therefore, there must be some means of communication between the root cap and the elongation zone. Since roots are positively gravitropic, their gravitropic curvature results from a net increase in growth on the upper side of a horizontal root, the opposite of shoot gravitropism.

As with coleoptiles and shoots, the Cholodny-Went theory has been used for many years to explain root gravitropism (Wilkins, 1979). According to this theory, auxin is synthesized in and released from the root tip and is redistributed laterally toward the lower side when the root is positioned horizontally. This accumulation of auxin on the lower side may result in a supraoptimal auxin concentration and, consequently, inhibit growth. Conversely, the upper side would be exposed to auxin at nearly optimal levels, which would presumably accelerate growth. One or both of these effects would ultimately lead to downward curvature of the root.

The involvement of auxin in the mechanism of root gravitropism has been questioned, however. Most of the auxin in the root has been synthesized in the shoot and transported into the root, where it moves acropetally in the vascular cylinder. Some investigators have suggested that the plant growth inhibitor abscisic acid (ABA, Chapter 19) is the signal produced by the root cap, but more recent evidence does not favor this idea (Moore and Smith, 1984). Others think that the signal may be electrical or electrochemical rather than hormonal. In addition, calcium has been strongly implicated in the mechanism of root gravitropism (Mulkey et al., 1981).

Exogenous (applied) calcium ions (Ca^{2+}), as well as other cations such as aluminum (Al^{3+}), affect IAA transport in primary roots of corn (*Zea mays*) (Hasenstein and Evans, 1988). Placing a block of agar containing calcium ions on the side of the cap of a vertically oriented corn root induces the root to grow toward the side with the agar block. In a variation on the donor-receiver block method for measuring polar auxin transport, the effect of gravity on the transport of radioactive $^{45}Ca^{2+}$ in root caps has been measured. The $^{45}Ca^{2+}$ was preferentially transported to the lower half of the cap of a gravistimulated root. Calcium is known to be required for IAA transport in some plants. In horizontally oriented roots, the gravity-sensing organelles (presumably amyloplasts) may somehow effect the downward transport of both calcium and auxin (Fig. 16.17). Calcium may enhance the loading of auxin into the basipetally moving transport stream. The increased transport of auxin into the lower side of the elongation zone would lead to growth inhibition and, thus, downward curvature.

For over 50 years, the primary involvement of auxin in gravitropism in higher plants—that is, the Cholodny-Went theory—has been upheld. In the mid-1970s this theory was challenged. Since then, there has been a flurry of research activity on plant gravitropism. Some of the results of this research have been contradictory, some have refuted the Cholodny-Went theory, and some have

supported it. It is clear that coleoptiles, shoots, and roots respond to gravistimulation in different ways. A single, all-encompassing explanation of gravitropism may not be possible.

Auxin from the Apical Bud Inhibits the Growth of Lateral Buds

In most higher plants, the growing apical bud inhibits, to varying degrees, the growth of lateral (axillary) buds, a phenomenon called **apical dominance**. Removal of the

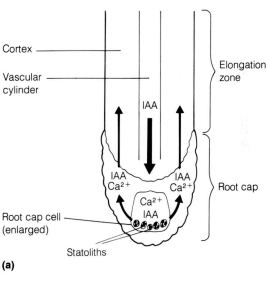

(a)

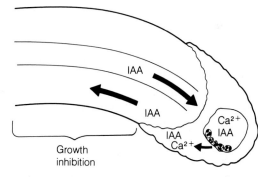

(b)

FIGURE 16.17. Proposed model for the redistribution of calcium and auxin during gravitropism in roots. (a) IAA is synthesized in the shoot and transported to the root via the vascular system. When the root is vertical, the statoliths in the cap cells settle to the basal ends of the cells. Calcium ions and auxin (transported acropetally in the root) are redistributed symmetrically in the lateral direction. The IAA is then transported basipetally within the cortex to the elongation zone, where it promotes elongation. (b) In a horizontal root the statoliths settle to the side of the cap cells, triggering polar transport of calcium and IAA to the lower half of the cap. The IAA is further transported basipetally within the cortex. The arrival of supraoptimal concentrations of auxin inhibits elongation and causes downward curvature. (From Hasenstein and Evans, 1988.)

shoot apex (decapitation) usually results in the growth of one or more of the lateral buds. Fifty years ago it was found that IAA could substitute for the apical bud in maintaining inhibition of lateral buds.

How does auxin from the shoot apex contribute to the inhibition of lateral bud growth? There are several theories. It was originally suggested by K. V. Thimann and F. Skoog that the optimal auxin concentration for bud growth is low, much lower than the auxin concentration normally found in the stem. The level of auxin normally present in the stem was thought to be so high as to inhibit lateral bud growth.

Other theories involve an indirect effect of auxin related to (1) nutrient starvation, (2) the plant hormone cytokinin, or (3) unidentified inhibitors present in the lateral buds. The apical meristem and young developing leaves possess high levels of auxin, and lateral buds do not. The high auxin levels in the apical regions of the stem may attract nutrients and plant hormones involved in the regulation of cell division, such as cytokinins. According to the nutrient starvation theory, the shoot apex acts as a sink for nutrients, diverting them from the lateral buds, possibly because the vascular connections to the lateral buds are incomplete. However, direct application of nutrient solutions to axillary buds fails to reverse apical dominance, and some axillary buds have already developed vascular connections. Cytokinins are known to stimulate lateral bud growth. Removal of the apical bud may increase the levels of cytokinins, nutrients, or both in the lateral buds, thereby stimulating their growth. Finally, the plant hormone abscisic acid has been found in dormant lateral buds in intact plants. When the shoot apex is removed, the ABA levels in the lateral buds decrease. High levels of IAA in the shoot may help to keep ABA levels high in the lateral buds. Removing the apex removes a major source of IAA, which may allow the bud growth-inhibitor levels to fall.

It is clear that although auxin is a basic component of apical dominance, other factors are also involved. Apical dominance is probably due to some combination of the above effects.

Auxin Promotes the Formation of Lateral Roots

Although elongation growth of the primary root is inhibited by auxin concentrations above 10^{-8} M, initiation of lateral (branch) roots and adventitious roots is stimulated by high auxin levels. Lateral or branch roots are commonly found above the elongation and root hair zone and originate from small groups of cells in the pericycle. These cells are stimulated to divide by auxin. The dividing cells gradually form into a root apex, and the lateral root grows through the root cortex and epidermis. Adventitious roots develop from a part of the plant other than the root or in a fashion other than the normal form

of root branching. They can originate in a variety of tissue locations from a cluster of mature cells that renew cell division activity. These dividing cells develop into a root apical meristem in a manner somewhat analogous to lateral root formation.

Because of polar auxin transport, IAA tends to accumulate just above any wound site in shoots and roots. There it promotes the initiation of adventitious root formation and enhances the survival of aerial plant parts after damage at or below ground level. This effect of auxin on root initiation has been very useful horticulturally for the propagation of plants by cuttings.

Auxin Delays the Onset of Leaf Abscission

The shedding of leaves, flowers, and fruits from the living plant is known as **abscission**. Abscission occurs in a region called the **abscission zone**, typically located near the base of the petiole. In most plants, leaf abscission is preceded by the differentiation of a distinct layer of cells, the **abscission layer**, within the abscission zone. During leaf aging, called leaf **senescence**, the walls of the cells in the abscission layer are digested, which causes them to become soft and weak. The leaf eventually breaks off at the abscission layer due to stress on the weakened cell walls.

IAA is known to *delay* the early stages of leaf abscission and to *promote* the later stages. Auxin levels are high in young leaves, progressively decrease in maturing leaves, and are relatively low in senescing leaves. During the early stages of leaf abscission, application of IAA inhibits leaf drop. During the later stages, however, auxin application actually hastens the process, probably by inducing the synthesis of ethylene, which promotes leaf abscission. Young leaves are apparently less sensitive to ethylene than older leaves.

Auxin Regulates Fruit Development

There is much evidence that auxin is involved in the regulation of fruit development. Auxin is produced in pollen and in the endosperm and embryo of developing seeds, and the initial stimulus for fruit growth may result from pollination. Successful pollination initiates ovule growth, which is known as **fruit set**. After fertilization, fruit growth may depend on auxin produced in developing seeds. The endosperm may contribute auxin during the first stage of fruit growth, and the developing embryo may take over as the main auxin source during the latter stages.

In some plant species, seedless fruits may be produced naturally or by treating the unpollinated flowers with auxin. The production of such seedless fruits is called **parthenocarpy**. In stimulating the formation of parthenocarpic fruits, auxin may act primarily to induce fruit set, which in turn may trigger the endogeneous production of auxin by certain fruit tissues to complete the developmental process. It is important to understand that other plant hormones are also involved in fruit growth. For instance, ethylene is known to influence fruit development, and some of the effects of auxin on fruiting may be mediated through the promotion of ethylene synthesis.

Synthetic Auxins Have a Variety of Commercial Uses

Auxins have been used commercially in agriculture and horticulture for over 40 years (Weaver, 1972). The early commercial uses included rooting of cuttings for plant propagation, promotion of flowering in pineapple, prevention of fruit and leaf drop, induction of parthenocarpic fruit, and thinning of fruit. Today, in addition to these applications, auxins are widely used as weed killers. Synthetic auxins are most effective because they are not broken down by the plant as quickly as IAA.

The Mechanism of Auxin Action

In the preceding section we saw how auxin affects certain physiological processes in higher plants. We now turn our attention to auxin action at the cellular level. Once inside the plant cell, how does auxin, in relatively small concentrations, induce certain cells to grow or to divide? More specifically, (1) what are the first detectable effects of auxin on responsive cells? (2) have specific intracellular target sites for auxin been discovered? and (3) does auxin affect gene expression?

Work on the *mechanism* of auxin-induced growth has been done at three levels: the tissue level, the subcellular level, and the gene level. Research using growing tissue segments or intact plants has been concerned with the rapid growth kinetics in response to auxin and other rapid responses to this hormone. Research at the subcellular level has involved the identification and isolation of auxin receptor proteins with the aid of radioisotope-labeled auxins. Finally, investigations of the effects of auxin at the DNA level have involved study of auxin-induced changes in gene expression.

The Lag Time for Auxin-Induced Growth Is 15–20 Minutes

To probe the sequence of events in auxin action, investigators have studied the rapid responses of plant cells to auxin (Evans, 1974). When a stem or coleoptile section is excised and hooked up to a sensitive growth-measuring device, the growth response to auxin can be measured with extremely high resolution. When the segment is perfused with buffer alone, its growth rate slowly decreases to some basal level. Addition of auxin causes a marked increase in the growth rate after only a 15-min lag period (Fig. 16.18). Cell walls isolated from auxin-treated coleoptiles also show enhanced extensibility after a lag time of 15–20 min. The initial phase of auxin-induced growth lasts for about an hour and is followed by a prolonged steady-state phase (Fig. 16.18). Not surprisingly, auxin-induced growth is sensitive to respiratory inhibitors, such as cyanide and azide, indicating that metabolic processes are involved.

The lag time for auxin-induced growth is the same as that for auxin-induced proton extrusion. Observations such as these led researchers in the 1970s to propose that the initial stimulation of growth by auxin is mediated by membrane interactions rather than by specific gene expression. It was assumed (without direct evidence) that 15 min was too short a period for auxin to induce the synthesis of specific proteins. We now know that auxin does, indeed, induce the transcription of a number of specific mRNAs within 15 min. However, these mRNAs may encode proteins involved in the steady-state phase of growth rather than the initial phase, since it is difficult to conceive how the newly transcribed mRNAs could bring about such a large increase in the growth rate within the observed lag time.

Intracellular Second Messengers May Mediate Plant Hormone Action

We assume that auxin, like animal hormones, must bind to a specific receptor protein to elicit a cellular response. Since more is known about animal systems, research on plant hormone receptors has largely been based on animal

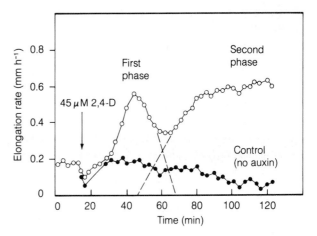

FIGURE 16.18. Two overlapping growth responses to auxin in soybean hypocotyl segments. Hypocotyl segments from young soybean plants were incubated for 1 h in buffered sucrose (to remove endogenous auxin) and then transferred to a growth chamber. After growth was monitored for 15 min, the buffered sucrose solution was replaced with fresh solution with or without 45 μM 2,4-D and growth continued to be monitored. The dashed lines designate the boundaries of the two apparently overlapping but separate growth phases in response to 2,4-D. The first phase is completed after about 1 h of auxin treatment. (From Vanderhoef and Stahl, 1975.)

models. For example, the steroid hormones, which pass freely through membranes because of their hydophobic nature, are thought to bind to cytoplasmic protein receptors. The steroid-receptor complex then migrates to the nucleus, where it exerts its effect on gene expression by binding to specific DNA sequences. In contrast, nonsteroidal extracellular signals, such as peptide growth factors and protein hormones, interact with receptors on the cell surface. The hormone, itself, does not have to enter the cell. Rather, a cascade of enzyme activation steps is initiated, resulting in the production of "second messengers" that directly stimulate the response and amplify the hormone signal. Such a cascade is often referred to as a **signal transduction pathway.**

A common element in a great number of signal transduction pathways is the participation of GTP-binding proteins (**G proteins**) as intermediates between the hormone-receptor complex and the enzyme systems that produce second messengers. As shown in Figure 16.19, binding of a hormone to its receptor on the cell surface results in the activation of a G protein (that is, it exchanges GTP for GDP). The activated G protein then initiates a series of molecular events. Activation of adenylate cyclase leads to cyclic-AMP production; activation of phospholipase C leads to the production of inositol trisphosphate (IP_3) and diacylglycerol (DAG). IP_3 opens calcium channels in cellular organelles, releasing free calcium into the cytosol. The increase in free calcium can affect a host of calcium-dependent reactions. The lipid-soluble DAG may activate protein kinases, which may activate ion exchange proteins in the plasma membrane by phosphorylation.

There is increasing evidence that signal transduction pathways involving G proteins also operate in plant cells (Einspahr and Thompson, 1990). G proteins, phospholipase C, and protein kinases have been identified in plant membranes (Blum et al., 1988; Tate et al., 1989). A single report of auxin-induced IP_3 production (Ettlinger and Lehle, 1988) awaits confirmation. However, there is strong evidence that IP_3 causes the release of calcium from plant mitochondria and vacuoles in a manner similar to its action in animal cells, indicating that it may serve as a second messenger in plant cells.

Several Potential Auxin Receptors Have Been Identified

Earlier efforts to identify auxin receptors relied mainly on in vitro binding studies using radioisotope-labeled auxins (Rubery, 1981). A clear distinction should be made between reported auxin-binding sites and true auxin receptors. Although auxin may specifically "bind" to a component of the cell, one must establish that this component is functionally involved in auxin action, rather than, say, auxin metabolism, before it can be called an auxin receptor. Auxin-binding proteins may be considered potential auxin receptors. To date, many auxin-binding sites have been reported on membranes (endoplasmic reticulum, plasma membrane, and tonoplast) and in the cytosol but have not been unequivocally identified as true auxin receptors. However, two findings suggest that authentic auxin receptors have now been identified. Löbler and Klämbt (1985) have purified a possible auxin receptor by affinity column chromatography. An auxin analog, TIBA, was covalently bound to a matrix and cell proteins were passed through the column. The analog bound to a specific protein, which was eluted from the column. Antibodies to this protein were able to inhibit auxin-induced growth. The protein was localized in the outer epidermis of corn coleoptiles by immunocytochemistry and is thought to be on the outer surface of the plasma membrane. More recently, a cDNA clone encoding the putative auxin receptor was sequenced and shown to be a small protein of about 22 kDa (Tillman et al., 1989). Consistent with its proposed location on the plasma membrane, the protein contained a signal peptide.

Another way to identify potential auxin receptors is to use a photoaffinity auxin analog that is radioactively labeled. For example, Terri Lomax and her colleagues at Oregon State University incubated tomato stem membranes with tritiated azido-IAA. In the presence of ultraviolet light, the azido group covalently binds to proteins. Under these conditions, a polypeptide doublet of 40 and 42 kDa was labeled on the tomato stem membranes. A tomato mutant called *diageotropica*, which exhibits abnormal growth because of its inability to respond to auxin, did not contain the 40 and 42 kDA proteins (Hicks et al., 1989). Perhaps the *diageotropica* mutants grow abnormally because they lack these receptor proteins and thus cannot respond to their endogeneous auxin.

Specific auxin binding to proteins soluble in the cytosol has also been reported. At least one of these soluble proteins may be involved in the process of gene transcription of mRNA, which is used to make new proteins.

FIGURE 16.19. (right) Models of signal transduction in mammalian cells. The hormone binds to its receptor on the extracellular surface of the plasma membrane. The hormone-receptor complex interacts with a G protein. As a result of the interaction, the G protein exchanges a bound GDP for a GTP and becomes activated. The activated G protein can act in various ways. On the left, the activated G protein stimulates the activity of adenylate cyclase, increasing the intracellular concentration of cyclic AMP, which can activate a protein kinase and other enzymes. On the right, the activated G protein activates phospholipase C, which causes the turnover of membrane phospholipids. This causes an increase of diacylglycerol (DAG), which can activate protein kinase C in the membrane, and inositol 1,4,5-triphosphate (IP_3), which can cause the release of free calcium from the ER and vacuoles.

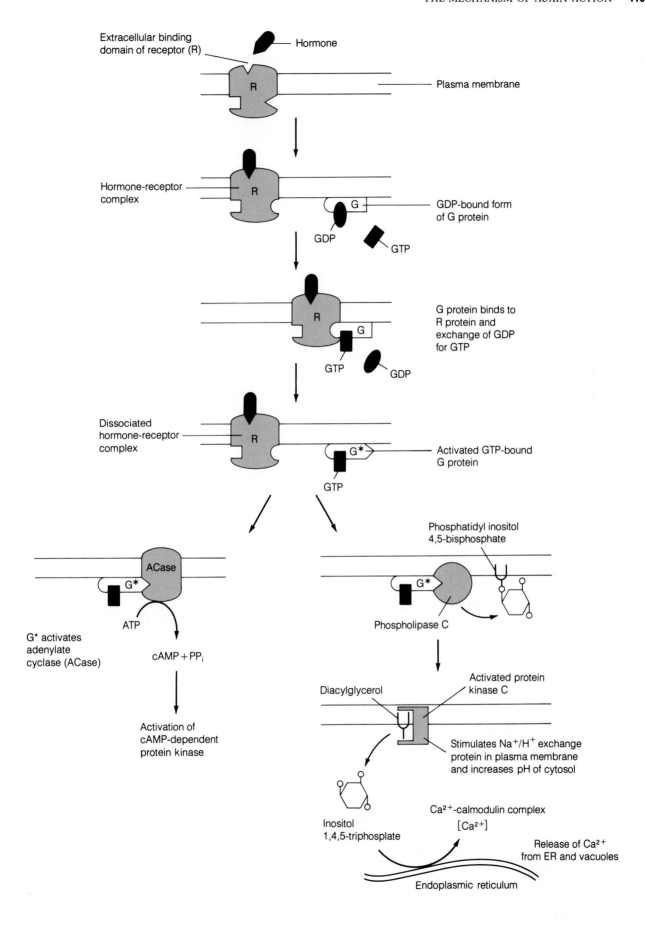

Auxin Action May Involve Calcium Fluxes

Studies indicating that auxin receptors may reside on both plasma membranes and intracellular membranes agree with the idea that auxin action may involve the transport of ions. As we have seen, proton efflux may be a critical step in auxin-induced cell elongation. The action of this hormone may be dependent on calcium ion transport as well.

Proton transport systems may be involved in auxin-induced cell wall acidification (Cleland, 1982). The primary candidates for such systems are proton-pumping ATPases located on the plant plasma membrane. These ATPases use the energy of ATP hydrolysis to generate an electrochemical proton gradient, which in turn drives a variety of transport processes at the plasma membrane (see Chapter 5). Auxins probably do not directly affect these ATPases but may indirectly stimulate the proton pumping. In addition, the phytotoxin fusicoccin probably increases proton extrusion from plant cells by stimulating the proton-pumping ATPases. A fusicoccin-binding site has been identified on the plasma membrane of plant cells and found to differ from the auxin binding site.

The role of calcium in auxin action seems to be very complex and, at this point in time, very hypothetical (Evans, 1985). First, auxin may stimulate the release of calcium from the cytosol into the extracellular space, possibly by activating a plasma membrane calcium pump. Second, auxin may affect calcium ion efflux from the vacuole, across the tonoplast, into the cytosol. The release of calcium from intracellular compartments of animal cells appears to be mediated by IP_3, as illustrated in Figure 16.19, and a similar mechanism may exist in plant cells. Such auxin-induced changes in cytosolic calcium levels could affect the calcium-activated regulator protein calmodulin.

Calmodulin has been found in all living cells examined so far. This protein regulates the activity of a number of enzymes such as protein kinases, which in turn regulate the activity of other enzymes by phosphorylating them. Thus, the activation of calmodulin triggers a cascade of biochemical reactions which could have physiological effects.

Auxin Regulates the Expression of Specific Genes

In the early 1960s, it was suggested that auxin may act by inducing the expression of specific genes. However, failure to detect rapid effects of auxin on de novo mRNA and protein synthesis appeared to preclude the involvement of specific gene transcription in the primary action of this hormone. In the 1970s research on auxin-induced gene activation waned. It was replaced by interest in theories (such as the acid growth theory) concerned primarily with auxin activation of preexisting systems within responsive plant cells. With the technical advances in molecular biology of the late 1970s and early 1980s, it became possible to detect rapid changes in translatable mRNAs, that is, mRNAs that are fully processed and able to direct protein synthesis on ribosomes (Fig. 16.20). Such changes may include an increase or decrease in mRNA synthesis, an increase or decrease in mRNA turnover, and the activation of "masked" mRNAs.

What are some of the effects of auxin on translatable mRNAs? Several mRNA sequences increase in amount (or translation activity) within about 20 min of exposure of dark-grown soybean or pea tissues to IAA or 2,4,-D (Theologis and Ray, 1982; Theologis et al., 1985; Hagen et al. 1984); other mRNAs substantially increase in amount after 1–2 h of auxin treatment. Some mRNAs are repressed after long-term auxin treatments (>2 h).

The level of translatable mRNA does not necessarily correspond to the abundance of mRNA, since RNAs may be incompletely processed or slightly degraded. To measure abundance, specific cDNA probes are used in hybridization studies. In this regard, the use of "plus-and-minus" screening methods for the identification of auxin-regulated mRNAs (Fig. 16.21) has greatly facilitated research. The procedure allows one to clone cDNA sequences that are specifically stimulated by auxin. Once such auxin-specific clones have been isolated, they can be used to prepare radioactive cDNA probes. By hybridizing such radioactive cDNA probes to total mRNA extracted from tissues at various times after auxin treatment, it is possible to determine the kinetics of auxin stimulation of that mRNA. This is typically done on a "Northern blot," that is, on RNA that has been size-separated by electrophoresis and then blotted onto nitrocellulose paper. The amount and length of the auxin-stimulated mRNA can

---

FIGURE 16.20. (right) Procedure for studying the effects of auxin on translatable messenger RNAs. This simplified diagram, based on the methods of Theologis and Ray (1982), shows the procedure used to detect auxin-induced changes in translatable mRNAs. The tissue sections are first depleted of endogenous auxin by incubation in buffered sucrose, then incubated in the presence and absence of IAA and frozen in liquid nitrogen after various incubation times. Nucleic acids are extracted from the frozen segments, and polyadenylated mRNAs (translatable mRNAs) are isolated. These mRNAs are translated into proteins in vitro using a cell-free protein synthesis system prepared from wheat germ. The protein synthesis system consists of ribosomes, enzymes, and tRNA (the "hardware" needed for protein synthesis), to which are added 20 amino acids (protein building blocks) including radioactive methionine. The reactions are driven by energy from ATP. After 1 h the in vitro translation is terminated and the newly synthesized polypeptides are separated by two-dimensional gel electrophoresis (i.e., separated by net electrical charge in the first dimension and by molecular weight in the second dimension). Since newly synthesized polypeptides are radioactive, when the gel is exposed to photographic film the polypeptides show up as black spots on the developed film, or *autoradiograph*. Within 2 h, auxin may cause at least five mRNA sequences to increase in translational activity (numbered 1–6).

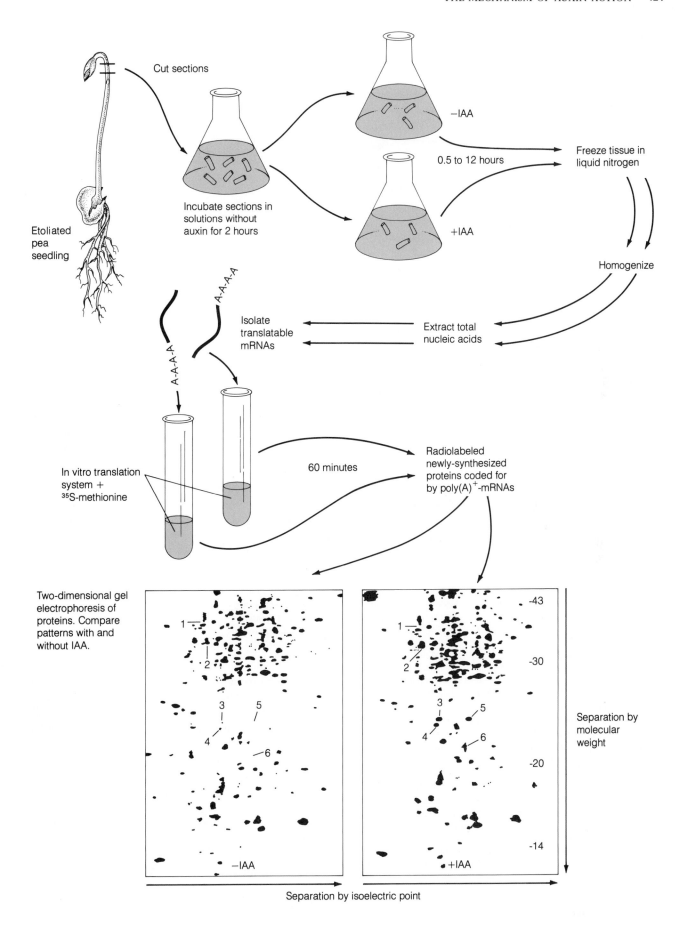

Cut sections

−IAA

0.5 to 12 hours

Freeze tissue in liquid nitrogen

Incubate sections in solutions without auxin for 2 hours

+IAA

Homogenize

Etoliated pea seedling

A-A-A-A

A-A-A-A

Isolate translatable mRNAs

Extract total nucleic acids

In vitro translation system + 35S-methionine

60 minutes

Radiolabeled newly-synthesized proteins coded for by poly(A)+-mRNAs

Two-dimensional gel electrophoresis of proteins. Compare patterns with and without IAA.

−IAA

+IAA

−43

−30

−20

−14

Separation by molecular weight

Separation by isoelectric point

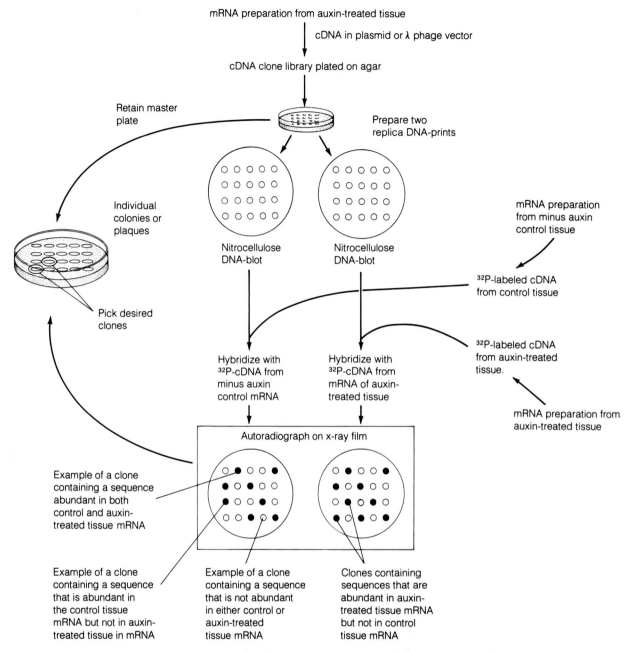

mRNA preparation from auxin-treated tissue

cDNA in plasmid or λ phage vector

cDNA clone library plated on agar

Retain master plate

Prepare two replica DNA-prints

Individual colonies or plaques

mRNA preparation from minus auxin control tissue

Nitrocellulose DNA-blot

Nitrocellulose DNA-blot

³²P-labeled cDNA from control tissue

Pick desired clones

³²P-labeled cDNA from auxin-treated tissue.

Hybridize with ³²P-cDNA from minus auxin control mRNA

Hybridize with ³²P-cDNA from mRNA of auxin-treated tissue

mRNA preparation from auxin-treated tissue

Autoradiograph on x-ray film

Example of a clone containing a sequence abundant in both control and auxin-treated tissue mRNA

Example of a clone containing a sequence that is abundant in the control tissue mRNA but not in auxin-treated tissue in mRNA

Example of a clone containing a sequence that is not abundant in either control or auxin-treated tissue mRNA

Clones containing sequences that are abundant in auxin-treated tissue mRNA but not in control tissue mRNA

FIGURE 16.21. The plus-and-minus method of screening for auxin-stimulated mRNAs. cDNA is prepared from mRNA extracted from auxin-treated tissue and cloned into a plasmid vector or λ phage. The cDNA "library" is plated out on agar, and two nitrocellulose prints are made which contain colonies or plaques in the same positions as on the master plate. The filters are further treated to release and denature the DNA and to bind it to the filter. Radioactively-labeled cDNA synthesized from mRNA of control and auxin-treated tissue is hybridized to each of the replica prints. After washing, the prints are placed under x-ray film. By comparing the spots that develop on the film, it is possible to identify auxin-specific clones. Such clones can then be isolated from the master plate. Typically, the clones selected initially must be rescreened several times to obtain single purified clones.

be estimated from the size and location of the spot that develops on exposure of the blot to x-ray film. This approach has been used to demonstrate that auxin rapidly and specifically alters the abundance of a few RNA species.

In soybean hypocotyls, a group of small homologous RNAs have been shown to increase within 2.5 min of treatment with 2,4,-D. These RNAs have been termed SAURs, for "small auxin up-regulated RNAs." By use of a technique called nuclear run-off (see Chapter 17), the increase in these small RNAs was shown to be caused at least in part by auxin stimulation of transcription (McClure et al., 1989). Interestingly, the SAUR genes are clustered

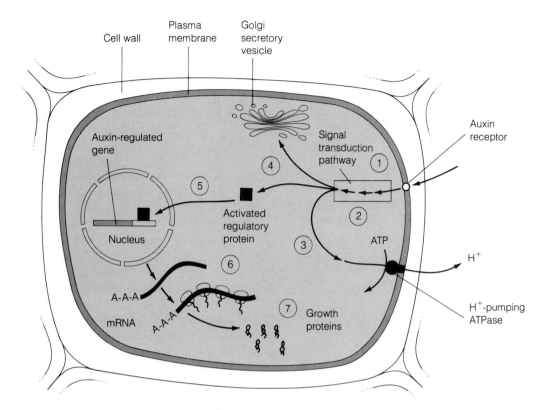

Cell wall   Plasma   Golgi
membrane   secretory
vesicle

FIGURE 16.22. Possible steps in the action of auxin. (1) IAA binds to its receptor (shown here on the plasma membrane); (2) the IAA-receptor complex interacts with other ligands, initiating a chain of biochemical events called the signal transduction pathway; (3) the plasma membrane proton pump becomes activated, acidifying the wall and causing wall loosening; (4) cell wall synthesis and secretion are stimulated; (5) regulatory proteins migrate from the cytosol to the nucleus; (6) regulatory proteins bind to regulatory sites on specific genes, stimulating transcription; (7) translation of auxin-regulated mRNA leads to proteins required for sustained growth.

together in the soybean genome. Do these small RNAs encode proteins involved in growth? In an experiment on gravitropism in soybean hypocotyls, it was shown that SAURs accumulate in the lower half of a hypocotyl that is bending in response to gravity. In vertical seedlings, auxin-regulated RNAs are symmetrically distributed. Within 20 min after the seedling is oriented horizontally, SAURs begin to accumulate in the lower half of the hypocotyl (McClure and Guilfoyle, 1989). Under these conditions gravitropic bending first becomes evident after 45 min. This experiment suggests that SAURs do play a role in regulating growth; it also provides indirect evidence for the asymmetric distribution of auxin during gravitropic bending.

Do the rapidly induced mRNAs encode proteins that are required for auxin-induced growth? Some experiments indicate that "growth-limiting proteins" are indeed required for auxin-induced growth. For example, the protein synthesis inhibitor cycloheximide prevents auxin stimulation of elongation of corn coleoptiles (which have been abraded to allow rapid entry of the inhibitor) if given within 5 min of auxin treatment (Edelmann and Schopfer, 1989). However, it has been difficult to connect

the early transcripts with the "growth-limiting proteins." For example, the rapid (~30 min) auxin stimulation of specific mRNAs in dark-grown pea stem sections was not observed in pea stems grown in dim red light (Dietz et al., 1990). In light-grown peas, the earliest auxin enhancement of mRNA occurs after 1 h, even though growth begins within 15 min of auxin treatment. In addition, most of the auxin-stimulated mRNAs were not specific for the epidermis, the tissue in which auxin produces its effects. These results suggest that auxin-stimulated mRNAs encode proteins involved in steady-state growth during the second or long-term phase of cell enlargement. Perhaps the production of "growth-limiting proteins" is the result of auxin acting at a posttranscriptional step. Further research will be necessary to resolve these questions.

A diagram summarizing the various steps in auxin action is shown in Figure 16.22.

Summary

In plants, development is regulated by five hormones, one of which is auxin. The most common form of auxin in the plant kingdom is indole-3-acetic acid (IAA), although there are several other natural and synthetic auxins. One of the most important roles of auxin in higher plants is the regulation of elongation growth in young stems and coleoptiles. Auxin may also be responsible for mediating the plant's responses to unidirectional light (phototropism) and gravity (gravitropism).

Accurate measurement of the amount of auxin in plant tissues is critical for understanding the role of this hormone in plant physiology. Early methods for qualitative and quantitative analyses of auxin involved coleoptile-based bioassays. At present, plant growth substances can be analyzed by several techniques, including physicochemical methods and immunoassay.

Regulation of growth in plants may depend in part on the amount of free auxin present in plant cells, tissues, and organs. Free auxin levels can be modulated via several pathways, including synthesis and breakdown of conjugated IAA (bound auxins), active transport of IAA, and IAA metabolism.

Biophysical parameters such as water uptake and irreversible cell wall expansion must be considered when discussing current models for auxin-stimulated cell expansion. In young developing stems and coleoptiles, auxin promotes elongation growth primarily by increasing the cell wall extensibility in auxin-sensitive cells. This increased cell wall extensibility or loosening occurs only when the cell wall region is exposed to an acidic pH. Thus, one of the important actions of auxin may be to induce cells to excrete protons into the cell wall region. In addition to this acidification, auxin may promote the synthesis and deposition of polysaccharide and protein material needed to maintain the wall-loosening capacity. Auxin may help to sustain cell elongation by promoting the uptake of solutes into the cell.

Auxin affects a number of tropisms in plants. Tropisms can be divided into three stages: detection of stimulus, signal transduction, and response. According to the Cholodny-Went theory, auxin is the primary signal molecule in both phototropism and gravitropism. In addition to its role in tropisms, auxin promotes lateral root formation and fruit development. It may also inhibit lateral bud development and leaf abscission. Because of this wide variety of effects on plant development, the commercial use of synthetic auxins is widespread in agriculture and horticulture.

Studies of auxin binding have revealed potential subcellular targets for auxin with regard to both hormone transport and physiological action. Membrane-bound auxin receptors may regulate ion transport or other aspects of the initial stages of auxin action. Soluble auxin-binding proteins may be involved in gene transcription.

At the biochemical level, there may be two major phases of auxin action. First, there is rapid stimulation of growth, which may be mediated by proton efflux and possibly by a change in cytosolic calcium. In this case auxin acts on cellular components already in place. The second phase of auxin action may involve the activation or repression of specific genes. This may be important in maintaining the growth response, and it is certainly involved in some of the physiological effects of auxin other than cell elongation.

GENERAL READING

Brenner, M. L. (1981) Modern methods for plant growth substance analysis. *Annu. Rev. Plant Physiol.* 32:511–538.

Cohen, J. D., and Bandurski, R. S. (1982) Chemistry and physiology of the bound auxins. *Annu. Rev. Plant Physiol.* 33:403–430.

Darwin, C. (assisted by F. Darwin) (1881) *The Power of Movement in Plants.* Appleton-Century-Crofts, New York.

Davies, P. J., ed. (1987) *Plant Hormones and Their Role in Plant Growth and Development.* Martinus Nijhoff, Dordrecht, The Netherlands.

Dennison, D. S. (1979) Phototropism. In: *Encyclopedia of Plant Physiology*, New Series, Vol. 7, Haupt, W., and Feinleib, M. E., eds., pp. 506–566. Springer-Verlag, Berlin.

Evans, M. L. (1974) Rapid responses to plant hormones. *Annu. Rev. Plant Physiol.* 25:195–223.

Evans, M. L. (1985) The action of auxin on plant cell elongation. *Crit. Rev. Plant Sci.* 2:213–265.

Firn, R. D., and Digby, J. (1980) The establishment of tropic curvatures in plants. *Annu. Rev. Plant Physiol.* 31:131–148.

Goldsmith, M. H. M. (1977) The polar transport of auxin. *Annu. Rev. Plant Physiol.* 28:439–478.

Marré, E. (1979) Fusicoccin: A tool in plant physiology. *Annu. Rev. Plant Physiol.* 30:273–288.

Skoog, F., ed. (1951) *Plant Growth Substances.* University of Wisconsin Press, Madison.

Theologis, A. (1986) Rapid gene regulation by auxin. *Annu. Rev. Plant Physiol.* 37:407–438.

Wareing, P. F., and Phillips, I. D. J. (1981) *Growth and Differentiation in Plants*, 3d ed. Pergamon, Oxford.

Weaver, R. J. (1972) *Plant Growth Substances in Agriculture.* Freeman, San Francisco.

Weiler, E. W., et al. (1986) Antisera- and monoclonal antibody–based immunoassay of plant hormones. In: *Immunology in Plant Science*, Wang, T. L., ed., pp. 27–58. Cambridge University Press, Cambridge, England.

Wilkins, M. B. (1979) Growth-control mechanisms in gravitropism. In: *Encyclopedia of Plant Physiology*, New Series, Vol. 7, Haupt, W., and Feinleib, M. E., eds., pp. 601–626. Springer-Verlag, Berlin.

CHAPTER REFERENCES

Blum, W., Hinsch, K.-D., Shultz, G., and Weiler, E. W. (1988) Identification of GTP-binding proteins in the plasma membrane of higher plants. *Biochem. Biophys. Res. Commun.* 156:954–959.

Brain, R. D., Freeberg, J. A., Weiss, C. V., and Briggs, W. R. (1977) Blue light–induced absorbance changes in membrane fractions from corn and *Neurospora*. *Plant Physiol.* 59:948–952.

Cleland, R. E. (1982) The mechanism of auxin-induced proton efflux. In: *Plant Growth Substances*, Wareing, P. F., ed., pp. 23–31. Academic Press, London.

Cohen, J. D. (1983) Metabolism of indole-3-acetic acid. *What's New in Plant Physiology* 14:41–44.

Dietz, A., Kutschera, U., and Ray, P. M. (1990) Auxin enhancement of mRNAs in epidermis and internal tissues of the pea stem and its significance for control of elongation. *Plant Physiol.* 93:432–438.

Edelmann, H., and Schopfer, P. (1989) Role of protein and RNA synthesis in the initiation of auxin-mediated growth in coleoptiles of *Zea mays* L. *Planta* 179:475–485.

Einspahr, K. J., and Thompson, G. A. (1990) Transmembrane signaling via phosphatidylinositol 4,5-bisphosphate hydrolysis in plants. *Plant Physiol.* 93:361–366.

Ettlinger, C., and Lehle, L. (1988) Auxin induces rapid changes in phosphatidylinositol metabolites. *Nature* 331:176–178.

Hagen, G., Kleinschmidt, A., and Guilfoyle, T. (1984) Auxin-regulated gene expression in intact soybean hypocotyl and excised hypocotyl sections. *Planta* 162:147–153.

Hager, A., Menzel, H., and Krauss, A. (1971) Versuche und hypothese zur primarwirkung des auxins beim streckungswachstrum. *Planta* 100:47–75.

Hasenstein, K. H., and Evans, M. L. (1988) Effects of cations on hormone transport in primary roots of *Zea mays*. *Plant Physiol.* 86:890–894.

Hicks, G. R., Rayle, D. L., and Lomax, T. L. (1989) The *diageotropica* mutant of tomato lacks high specific activity auxin binding sites. *Science* 245:52–54.

Jacobs, M. (1983) The localization of auxin transport carriers using monoclonal antibodies. *What's New in Plant Physiology* 14:17–20.

Jacobs, M., and Gilbert, S. F. (1983) Basal location of the presumptive auxin transport carrier in pea stem cells. *Science* 220:1297–1300.

Jacobs, M., and Rubery, P. H. (1988) Naturally occurring auxin transport regulators. *Science* 241:346–349.

Konjevic̀, R., Steinitz, B., and Poff, K. (1989) Dependence of the phototropic responses of *Arabidopsis thaliana* on fluence rate and wavelength. *Proc. Natl. Acad. Sci. USA* 86:9876–9880.

Last, R. L., and Fink, G. R. (1988) Tryptophan-requiring mutants of the plant *Arabidopsis thaliana*. *Science* 240:305–310.

Löbler, M., and Klämbt, D. (1985) Auxin-binding protein from coleoptile membranes of corn (*Zea mays* L.). *J. Biol. Chem.* 260:9848–9853.

McClure, B. A., and Guilfoyle, T. (1989) Rapid redistribution of auxin-regulated RNAs during gravitropism. *Science* 243:91–93.

McClure, B. A., Hagen, G., Brown, C. S., Gee, M. A., and Guilfoyle, T. (1989) Transcription, organization, and sequence of an auxin-regulated gene cluster in soybean. *Plant Cell* 1:229–239.

Mertens, R., and Weiler, E. W. (1983) Kinetic studies of the redistribution of endogenous growth regulators in gravi-reacting plant organs. *Planta* 158:339–348.

Migliaccio, F., and Rayle, D. L. (1984) Sequence of key events in shoot gravitropism. *Plant Physiol.* 75:78–81.

Moore, R., and Smith, J. D. (1984) Growth, graviresponsiveness, and abscisic-acid content of *Zea mays* seedlings treated with fluridone. *Planta* 162:342–344.

Mulkey, T. J., Kuzmanoff, K. M., and Evans, M. L. (1981) Correlations between proton-efflux and growth patterns during geotropism and phototropism in maize and sunflower. *Planta* 152:239–241.

Noggle, G. R., and Fritz, G. J. (1983) *Introductory Plant Physiology*. Prentice-Hall, Englewood Cliffs, N. J.

Pengelly, W. L., Bandurski, R. S., and Schulze, A. (1981) Validation of radioimmunoassay for indole-3-acetic acid using gas chromatography–selected ion monitoring–mass spectrometry. *Plant Physiol.* 68:96–98.

Pengelly, W. L., and Meins, F., Jr. (1977) A specific radioimmunoassay for nanogram quantities of the auxin, indole-3-acetic acid. *Planta* 136:173–180.

Presti, D., Hsu, W.-J., and Delbrück, M. (1977) Phototropism in *Phycomyces* mutants lacking β-carotene. *Photochem. Photobiol.* 26:403–405.

Rayle, D. L., and Cleland, R. E. (1970) Enhancement of wall loosening and elongation by acid solutions. *Plant Physiol.* 46:250–253.

Rayle, D. L., and Cleland, R. E. (1977) Control of plant cell enlargement by hydrogen ions. *Curr. Top. Dev. Biol.* 11:187–214.

Rubery, P. H. (1981) Auxin receptors. *Annu. Rev. Plant Physiol.* 32:569–596.

Rubery, P. H., and Sheldrake, P. A. (1973) Effect of pH and surface charge on cell uptake of auxin. *Nature (New Biol.)* 244:285–288.

Sievers, A., and Hensel, W. (1982) The nature of graviperception. In: *Plant Growth Substances*, 1982, Wareing, P. F., ed., pp. 499–506. Academic Press, London.

Tate, B. F., Schaller, G. E., Sussman, M. R., and Crain, R. C. (1989) Characterization of a polyphosphoinositide phospholipase C from the plasma membrane of *Avena sativa*. *Plant Physiol.* 91:1275–1279.

Theologis, A., Huynh, T. V., and Davis, R. W. (1985) Rapid induction of specific mRNAs by auxin in pea epicotyl tissue. *J. Mol. Biol.* 183:53–68.

Theologis, A., and Ray, P. M. (1982) Early auxin-regulated polyadenylated mRNA sequences in pea stem sections. *Proc. Natl. Acad. Sci. USA* 79:418–421.

Tillmann, U., Viola, G., Kayser, B., Siemeister, G., Hesse, T., Palme, K., Löbler, M., and Klämbt, D. (1989) cDNA clones of the auxin-binding protein from corn coleoptiles (*Zea mays* L.): Isolation and characterization by immunological methods, *EMBO J.* 8:2463–2467.

Vanderhoef, L. N., and Stahl, C. A. (1975) Separation of two responses to auxin by means of cytokinin inhibition. *Proc. Natl. Acad. Sci. USA* 72:1822–1825.

Van Volkenburgh, E., and Cleland, R. E. (1979) Separation of cell enlargement and division in bean leaves. *Planta* 146:245–247.

Volkmann, D., and Sievers, A. (1979) Graviperception in multicellular organs. In: *Encyclopedia of Plant Physiology*, New Series, Vol. 7, Haupt, W., and Feinleib, M. E., eds., pp. 573–600. Springer-Verlag, Berlin.

CHAPTER **17**

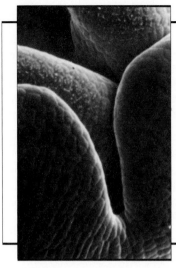

Gibberellins

F OR NEARLY 30 YEARS after the discovery of auxin in 1927 and over 20 years after its structural elucidation as indole-3-acetic acid, Western plant scientists tried to ascribe all developmental phenomena in plants to auxin. However, as we shall see in this and subsequent chapters, plant growth and development are regulated by several different types of hormones acting individually and in concert. Only with the discovery of other hormonal compounds did the nature of this control become evident. In the 1950s the second group of hormones, the gibberellins, were characterized. The gibberellins are a large group of related compounds (over 80 are known) defined not by their biological action but by their chemical structure; in fact, some are *biologically inactive*. Their main effect is that they are responsible for tallness: the stem of a tall plant contains more *biologically active* gibberellins than does the stem of a dwarf plant. They are a group of hormones whose synthesis and level are under clear genetic control. Mendel's tall/dwarf alleles in peas are genes that control gibberellin metabolism.

Discovery

Though gibberellins became known to American and British scientists in the 1950s, they were discovered much earlier by Japanese scientists. Rice farmers in Asia had long known of a disease that made the rice plants grow tall but eliminated seed production. In Japan this disease was known as the "foolish seedling," or *bakanae*, disease. Plant pathologists investigating the disease found that the plants' tallness was induced by a chemical secreted by a fungus that had infected the tall plants. This chemical was isolated from culture filtrates of the fungus and called gibberellin after *Gibberella fujikuroi*, the name of the fungus. In the 1930s Japanese scientists succeeded in obtaining impure crystals of the fungal growth-active compounds, which they termed gibberellin A, but because of communication barriers and World War II the information did not reach the West. It was not until the mid-1950s that two groups, headed by Brian Cross at the Imperial Chemical Industries (ICI) research station at Welyn in Britain and Frank Stodola at the U.S. Department of Agriculture (USDA) in Peoria, Illinois, succeeded in elucidating the structure of material that they purified from fungal culture filtrates, which they named gibberellic acid (Fig. 17.1b). At about the same time, Nobutaka Takahashi and Saburo Tamura at Tokyo University isolated three gibberellins from the original gibberellin A and named them gibberellin A_1,

FIGURE 17.1. Structures of some important gibberellins, gibberellin derivatives, and gibberellin biosynthesis inhibitors.

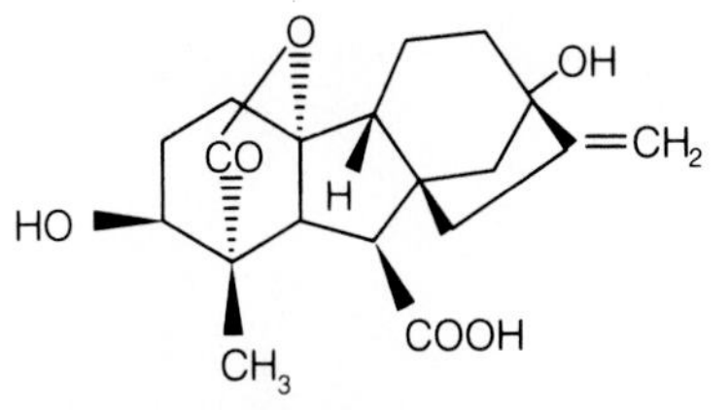

(a) Gibberellin A₁ (GA₁)

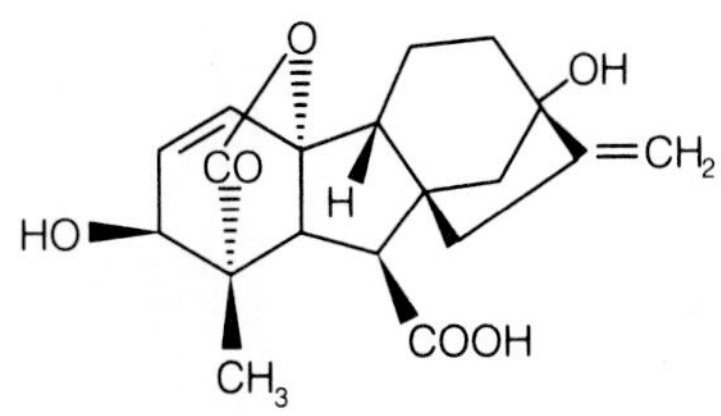

(b) Gibberellic acid (GA₃)

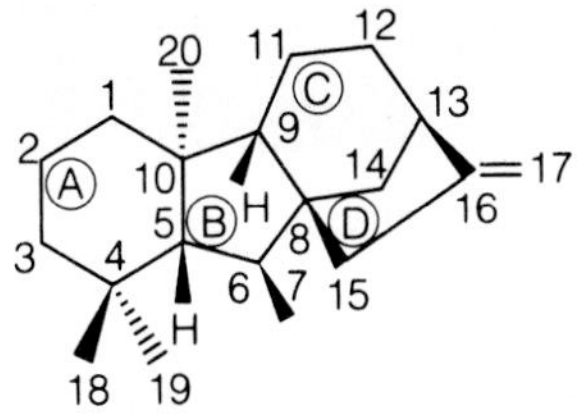

(c) *ent*-Gibberellane skeleton

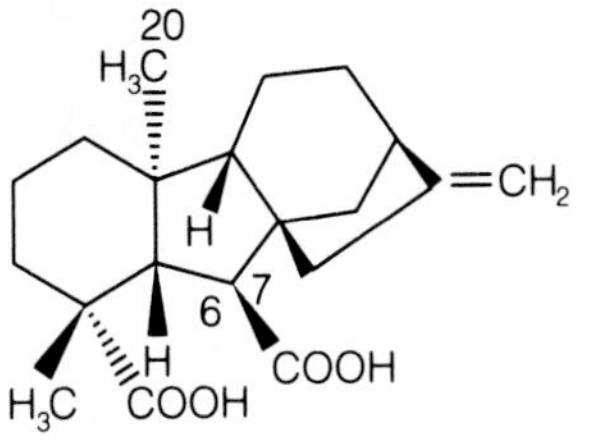

(d) GA₁₂ (a C₂₀ – GA)

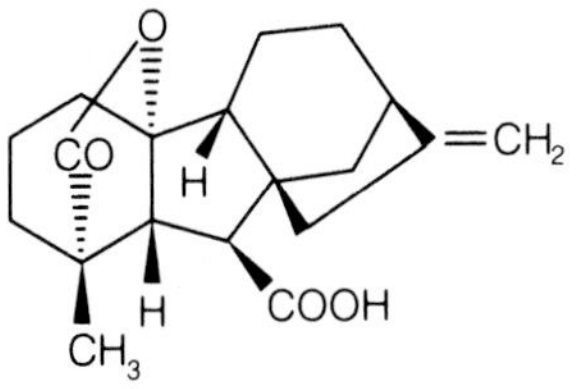

(e) GA₉ (a C₁₉ – GA)

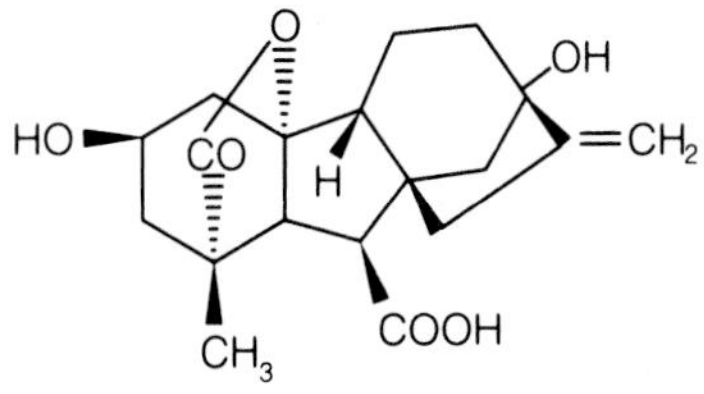

(f) GA₂₉

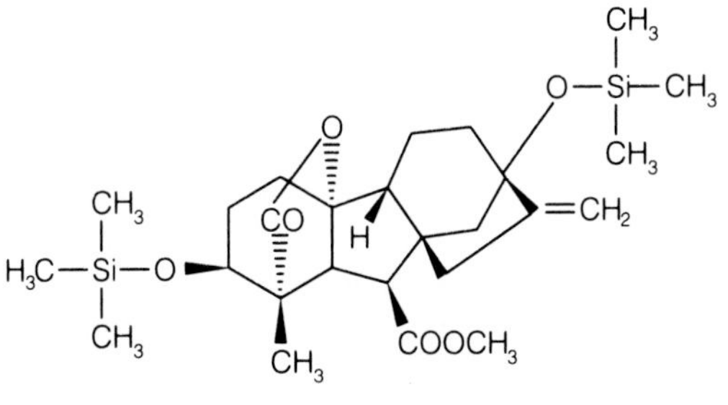

(g) GA₁ methyl ester-trimethylsilyl ether

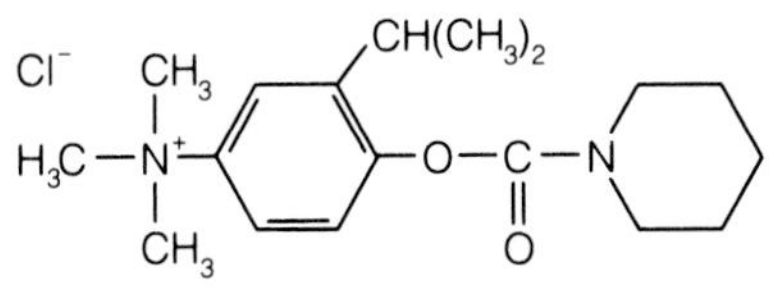

(h) AMO-1618

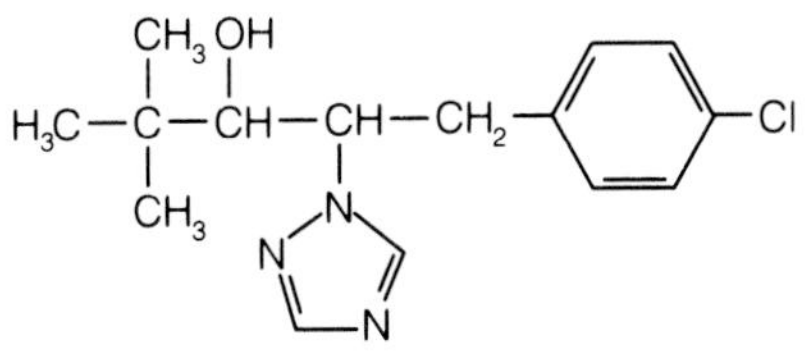

(i) Paclobutrazol

gibberellin A₂, and gibberellin A₃. Gibberellin A₃ and gibberellic acid proved to be identical. It became evident that a whole family of gibberellins exists and that in each fungal culture different gibberellins predominate, though gibberellic acid is always a principal component.

With the availability of gibberellic acid, physiologists rapidly began testing it on a wide variety of plants. Spectacular responses were obtained in the elongation growth of dwarf and rosette plants, particularly in genetically dwarf peas (*Pisum sativum*) and dwarf maize (*Zea mays*) (Fig. 17.2) and in biennial rosette plants (Fig. 17.3). At the same time, plants that were genetically

very tall showed no further response to applied gibberellins. More recently, experiments with dwarf peas and dwarf corn have confirmed that the natural elongation growth of plants is in fact regulated by gibberellins, as we shall describe later.

Since applications of gibberellins could increase the height of dwarf plants, it was natural to ask whether plants contained their own gibberellins. Shortly after the discovery of the growth effects of gibberellic acid, Margaret Radley at ICI and Bernard Phinney and coworkers at the University of California–Los Angeles showed that gibberellin-like substances could be isolated

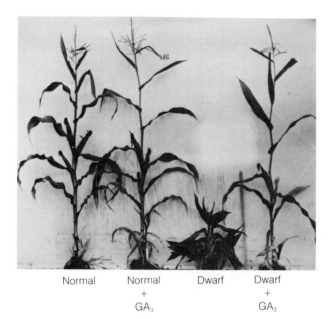

Normal   Normal   Dwarf   Dwarf
               +   +
             GA_3 GA_3

FIGURE 17.2. The effect of gibberellic acid on normal and dwarf corn. Left to right: normal control, normal plus gibberellin, dwarf control, dwarf plus gibberellin. (From Phinney and West. 1960.)

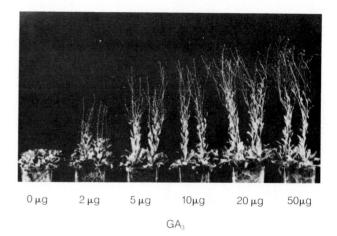

0 µg   2 µg   5 µg   10µg   20 µg   50µg

GA_3

FIGURE 17.3. Gibberellic acid applications cause bolting and sometimes flowering in plants that require long days to flower. Here *Samolus parviflorus* has been induced to flower under noninductive short days by applications of gibberellin. From left to right, plants were given 0, 2, 5, 10, 20, or 50 µg per plant each (From Lang, 1957.)

from several species of plants. "Gibberellin-like substance" refers to a semipurified extract of a plant part whose biological activity is similar to the effect of applied gibberellic acid. Such a response indicates, but does not prove, that the tested substance is or is similar to gibberellic acid. In 1958 the first conclusive identification of a gibberellin (gibberellin A_1; Fig. 17.1a) from higher plants was made by Jake MacMillan, also working at ICI. As the concentration of gibberellins in immature seeds far exceeds that in vegetative tissue, such seeds were the tissue of choice. However, since the concentration of gibberellins in plants is very low (a few parts per billion in vegetative tissue and up to 1 part per million in seeds), MacMillan and co-workers had to use kilograms of runner bean (*Phaseolus coccineus*) seeds. A good proportion of the research station staff had to shell truckloads of beans in order to provide enough material for the chemical purification and identification methods available at the time. With today's sophisticated methodology we can chemically identify gibberellins isolated from less than 1 g of seed tissue.

As more and more gibberellins from fungal sources were characterized, they were numbered as gibberellin A_x (or GA_x) in the order of their discovery. This scheme was universally adopted for all gibberellins in 1968. It must be emphasized, however, that the number of a gibberellin is simply a cataloging convenience, designed to prevent chaos in the naming of the gibberellins. It does not imply any close chemical similarity or metabolic relationship between gibberellins with adjacent numbers.

A wonderful personal account of the history of the discovery of gibberellins has been written by Phinney (1983).

Biosynthesis

The highest gibberellin levels occur in seeds, and they are almost certainly synthesized in situ. However, seed tissues are not uniform, and their gibberellin contents may differ. This is particularly true for the seed coat and embryo, which represent maternal tissues and tissues of the next generation, respectively. There is also evidence that the gibberellin may be transferred between tissues at different points of the metabolic pathway. For example, GA_{29} (an inactive, penultimate end product of metabolism) is formed in pea cotyledons, but the next step of metabolism (to GA_{29}-catabolite) occurs only after transfer to the seed coat. We have no real understanding of the function of the high levels of gibberellins in developing seeds. In very young embryos they appear to play a role in the growth of the suspensor (cells attached to the very young embryo that probably facilitate its uptake of nutrients); in older seeds they may be exported to the fruit, where they stimulate fruit development. There is no evidence that the fully synthesized gibberellins present in seeds are used for the growth of the young seedlings, as the gibberellin level decreases to zero in mature seeds (Fig. 17.4). Mature seeds do, however, contain GA_{12}-aldehyde, the immediate gibberellin precursor, and this is converted into growth-active gibberellins during the early stages of germination (Graebe, 1986). Gibberellin synthesis inhibitors, which cause dwarfing by blocking the synthesis of GA_{12}-aldehyde, are ineffective if applied during germination, but their application during seed development results in dwarfed seedlings when the seeds subsequently germinate. However, in wild oat (*Avena fatua*) and morning glory (*Pharbitis nil*) these compounds do not inhibit germination, indicating that gibberellin is not essential for germination in these species.

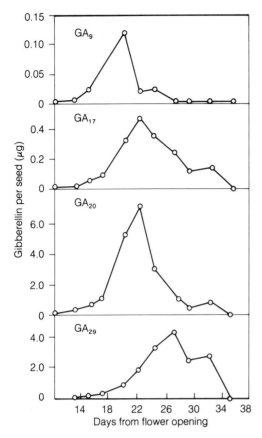

FIGURE 17.4. The levels of GA_9, GA_{17}, GA_{20}, and GA_{29} in seeds of pea throughout growth and maturation. As seeds near maturation, the amounts of all gibberellins decrease. (Note differing scales on the ordinates.) (From Frydman et al., 1974.)

Lower levels of gibberellins occur in vegetative tissue of plants, primarily in the young leaves, buds, and upper stem. These tissues also appear to be the sites of gibberellin synthesis (Coolbaugh, 1985). The initial steps of gibberellin production may occur in one tissue and metabolism to active gibberellins in another. There is some indication that gibberellins are synthesized in roots, but this has not been studied very thoroughly.

There is no evidence that gibberellin biosynthesis is confined to any particular organelle. Studies of enzyme extracts, particularly from wild cucumber (*Marah macrocarpus*), indicate that the enzymes of gibberellin biosynthesis are both soluble and membrane-bound. Enzymes that catalyze the conversion of the water-soluble phosphorylated intermediates are soluble, whereas those that oxidize extremely hydrophobic intermediates (kaurenoids) are embedded in the lipid bilayer of membranes.

Gibberellins Are Made Up of Isoprene Units

A terpenoid is a compound made up of five-carbon

$$CH$$
$$|$$

(isoprene) building blocks ($—CH_2—C=CH—CH_2—$)

joined head to tail. Gibberellins are terpenoids consisting of 20 carbons derived from four isoprenoid units. **Mevalonic acid** is considered to be the starting compound for terpenoid biosynthesis and is itself synthesized from acetyl CoA (you will recall that acetyl CoA is an intermediate in the respiratory pathway and an important precursor for many compounds). Mevalonic acid (a six-carbon compound) is phosphorylated by ATP, then decarboxylated to form isopentenyl pyrophosphate, the first isoprenoid compound on the pathway. These isopropene units are then added successively to produce geranyl pyrophosphate (C_{10}) (the precursor of geraniol, or geranium oil), farnesyl pyrophosphate (C_{15}), and geranylgeranyl pyrophosphate (C_{20}) (Fig. 17.5). Up to this stage, the synthetic pathway is shared by all terpenoids (including essential oils, carotenoids, and steroids), and the compounds are still linear chains of C_5 units, albeit in a folded form. The geranylgeranyl pyrophosphate is then cyclized to form the first product specific to gibberellin biosynthesis, namely *ent*-kaurene (the term *ent* refers to the enantiomeric form of (+)-kaurene).

The methyl group on carbon 19 of *ent*-kaurene is oxidized to carboxylic acid (Fig. 17.6), followed by the contraction of the B ring from a six- to a five-carbon ring to give GA_{12}-aldehyde, which is the first gibberellin formed in all plants and thus the precursor of all the other gibberellins (Sponsel, 1987).

A whole range of dwarfing or growth-retarding chemicals have been shown to act by inhibiting gibberellin biosynthesis. AMO-1618 (Fig. 17.1h), for example, inhibits *ent*-kaurene synthesis, while the more recently developed, highly specific growth retardant paclobutrazol (Fig. 17.1i) inhibits the further metabolism of *ent*-kaurene.

Oxidations and Hydroxylations of GA_{12} Yield All the Other Gibberellins

All gibberellins are based on the *ent*-gibberellane skeleton (Fig. 17.1c). Some gibberellins have the full compliment of carbons (i.e., C_{20}-GAs) (Fig. 17.1d), while some have only 19 (C_{19}-GAs), having lost one carbon to metabolism. In almost all C_{19}-GAs the carboxylic acid at carbon 19 bonds to carbon 10 to give a lactone bridge (Fig. 17.1e). Beyond this, there are variations in the basic structure, particularly in the oxidation state of carbon 20 (in C_{20}-GAs) and the number and position of hydroxyl groups on the molecule. The location of the hydroxyl groups and their stereochemistry (designated by ‖‖‖‖ or ► for α or β bonds—behind or in front of the formula as viewed on a page, respectively) have a strong bearing on their biological activity. For example, hydroxylation in the β configuration at carbon 2 always eliminates biological activity (Fig. 17.1f).

With this broad range of structural possibilities (presently there are 84 known GAs), gibberellins and

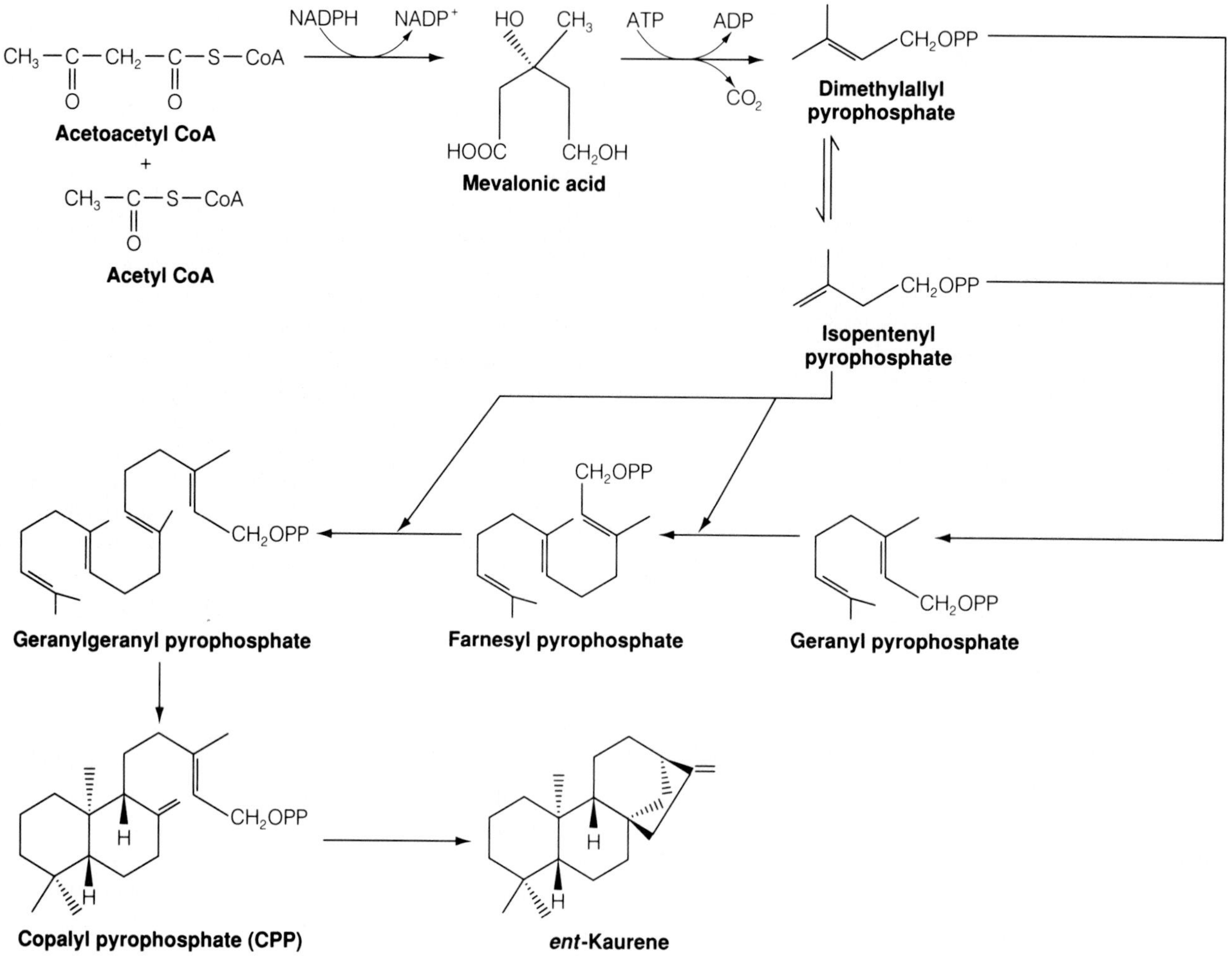

FIGURE 17.5. The gibberellin biosynthesis pathway from mevalonic acid to *ent*-kaurene.

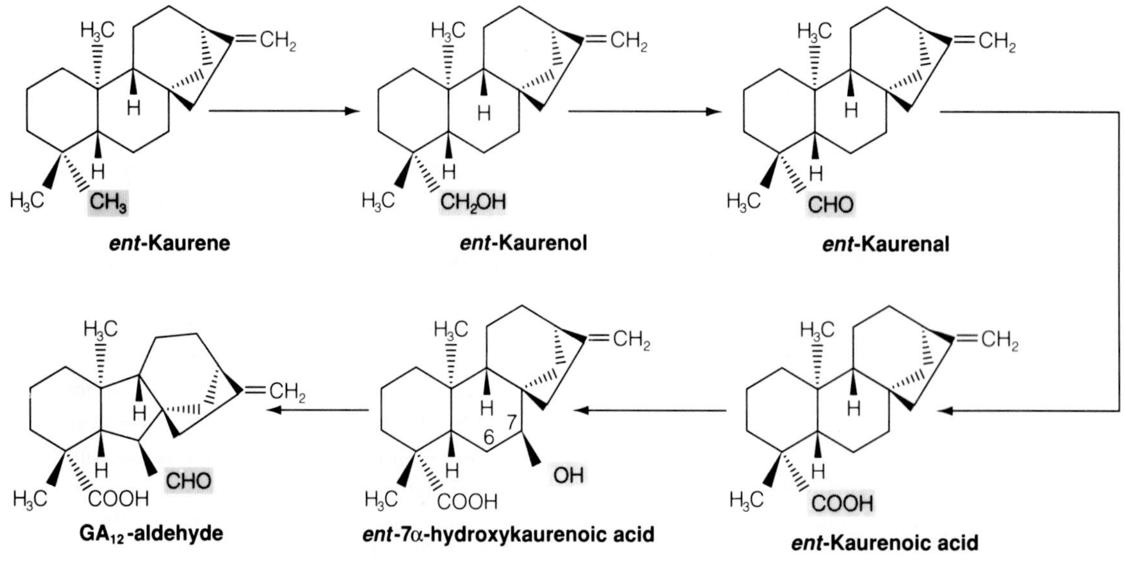

FIGURE 17.6. The gibberellin biosynthesis pathway from *ent*-kaurene to gibberellin A_{12}-aldehyde.

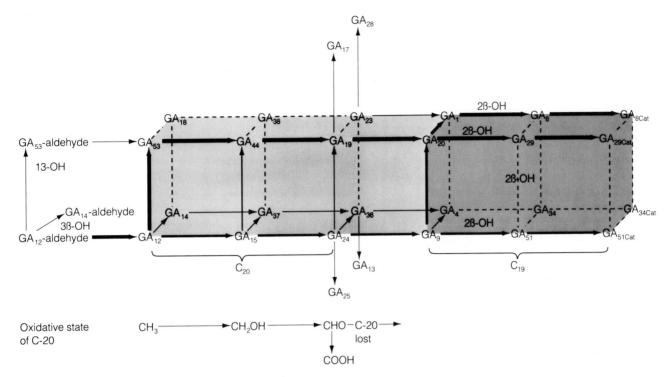

FIGURE 17.7. A grid of gibberellin metabolism common in higher plants, starting from GA_{12}-aldehyde (lower left). Movement to the right follows increasing oxidation of C-20 on the GA molecule, as indicated below the grid. This is followed by loss of the C-20 to give 19-carbon GAs (indicated by a darker shade), then hydroxylation in the 2β position, and finally metabolism to specific GA "catabolites." Metabolism takes place with no other hydroxyls on the molecules (lower line) or with hydroxyls at the 13 or 3β positions, or both, as represented by upward or diagonal lines. The major pathway is indicated by thick arrows, the less common pathway by medium arrows, and the minor pathways by thin arrows. Some interconversions (represented by dashed lines) have not actually been demonstrated. The most common sequence is for 13-hydroxylation to occur early as shown in Figure 17.8. Cat = catabolite. (From Sponsel, 1987.)

their interconversions have tended to be viewed as a difficult group by nonspecialists. As various interconversion systems have been studied, however, a more general pattern has emerged. Two basic chemical changes occur in most plants: a successive oxidation at carbon 20 ($CH_2 \rightarrow CH_2OH \rightarrow CHO$), followed by loss of carbon 20 as CO_2 in combination with hydroxylation at carbon 13 or carbon 3, or both. With this in mind, we can build up a metabolic grid comprising most of the gibberellins present in higher plants (Fig. 17.7). This grid does not tell us at which points the hydroxylations may actually occur, but it does show relationships. In peas, maize, and spinach (*Spinacia oleracea*), for example, the early 13-hydroxylation pathway predominates, with GA_{12} being 13-hydroxylated to GA_{53} prior to the oxidation of carbon 20 (Fig. 17.8) (Sponsel, 1987; Graebe, 1987). A subsequent hydroxylation at carbon 3 to form GA_1 occurs only after the elimination of carbon 20. In all these plants oxidation of carbon 20 with no hydroxylation, to form GA_9, also occurs but to a lesser extent. Exogenously supplied GA_9 can be converted by both shoots and seeds of pea to GA_{20}, but it is uncertain whether endogenous GA_9 is converted, as it might be in

a different cellular location from that reached by exogenously supplied hormone. Further metabolism of C_{19}-GAs consists of a hydroxylation in the 2β position that eliminates all biological activity; therefore, this is a metabolic inactivation. The 2β-hydroxylated compounds are then metabolized to form specific "catabolite" compounds. Further metabolism has not been characterized.

Most Gibberellins Are Precursors of Bioactive Gibberellins

When gibberellins are applied to a bioassay system that gives a biological response such as growth, many produce a response of varying intensity while others are inactive. Are the gibberellins that produce a response active because they are converted to other gibberellins, which are the active ones, or are they active in their own right? This is a difficult question, and it can be answered conclusively only if absolute chemical or genetic blocks can be found in the gibberellin metabolic pathway. We do know that most gibberellins are metabolized to other, biologically active gibberellins in bioassay systems. On the 13-hydroxy (13-OH) pathway, GA_{53}, GA_{44}, GA_{19},

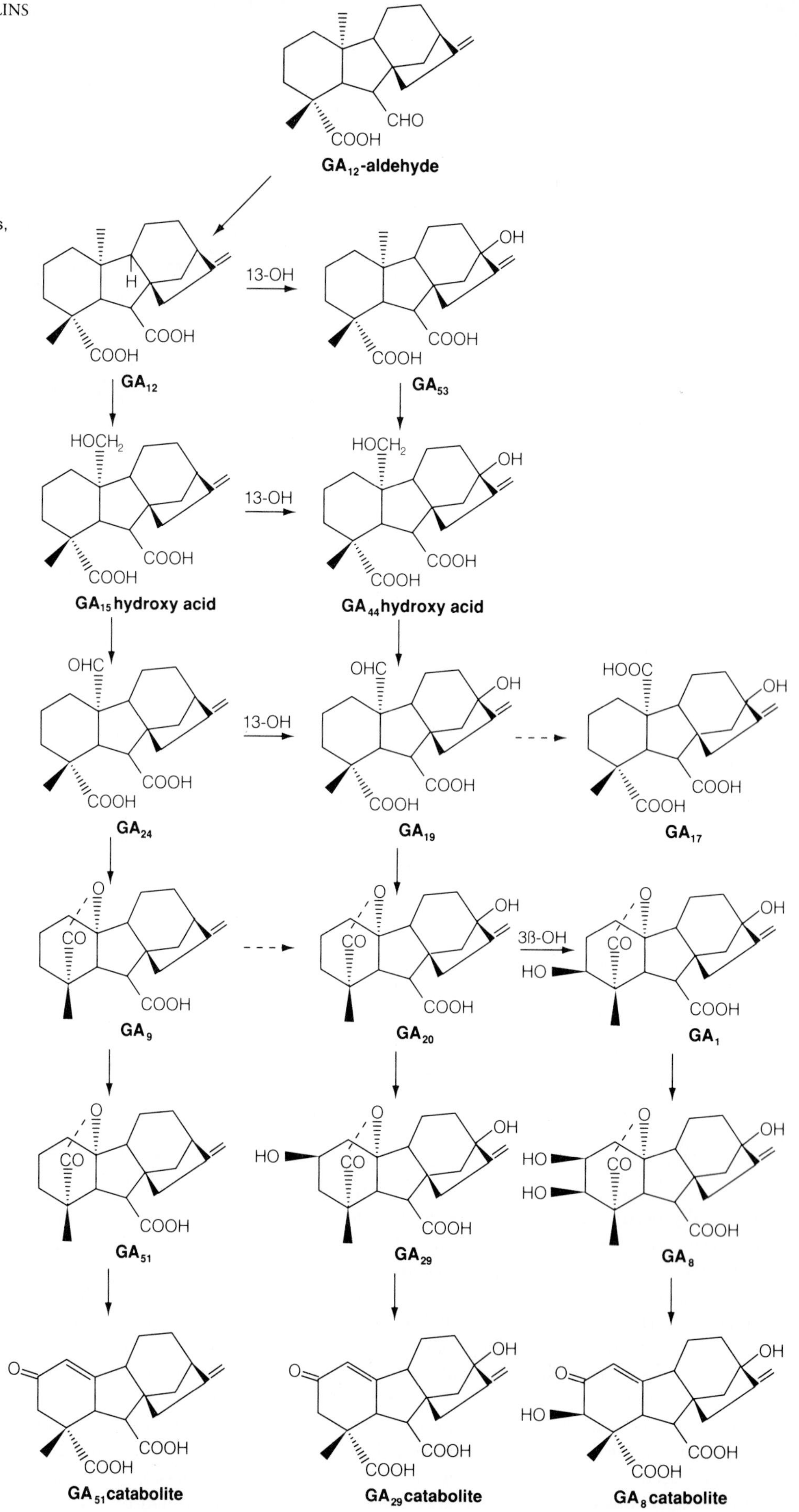

FIGURE 17.8. The pathway of gibberellin metabolism in peas. The same pathway occurs in numerous other species. Conversions shown with black arrows have been demonstrated. Conversions with broken arrows occur following application of GAs, but it is not known whether they occur naturally.

GA_{20}, and GA_1 are all biologically active, but only GA_1 is not converted to other biologically active gibberellins. In fact, it is now thought that only GA_1 is truly active per se in normal stem elongation (Phinney, 1985). The evidence for this is a combination of several factors. First, tall plants contain GA_1, whereas dwarf plants (such as dwarf peas) contain no measurable level of GA_1. However, dwarf plants do contain GA_{20} and show some stem growth, and they are considerably taller than dwarfs completely devoid of all gibberellins (**nana** mutants; see page 440). Presumably, then, GA_1 is the main active gibberellin, but GA_{20} has some activity. However, studies with isotopically labeled gibberellin have indicated that even in dwarf plants there is some production of GA_1, but the GA_1 is rapidly metabolized to GA_8 and GA_8-catabolite so its level is too low to be detected. In fact, the degree of stem elongation was found to correspond to the amount of GA_{20} metabolized on the GA_1 pathway. The step between GA_{20} and GA_1 is genetically controlled (see later), but even in dwarf plants in which the gene is nominally switched off the process still goes on to a small extent (the process is said to be "leaky"). From this it has been concluded that, at least with regard to stem elongation, the only active gibberellin is GA_1 and that all other gibberellins are precursors. GA_3, which is uncommon in higher plants, would probably also be active, as it differs from GA_1 only in having one double bond. Whether some of the other biologically active gibberellins are active in their own right in other plant responses is unknown, as the crucial studies with labeled gibberellins have not been done. In any case, the number of truly active gibberellins is likely to be quite small.

Inactive gibberellins are normally metabolic products of active gibberellins. Inactivating steps are 2β-hydroxylation (e.g., GA_{20} to GA_{29}, or GA_1 to GA_8) and oxidation of carbon 20 to a carboxyl group. There is no evidence that these inactivating steps are reversible. They appear to represent a mechanism for getting rid of a highly biologically active compound.

Gibberellins May Also Be Conjugated to Sugars

Numerous **gibberellin glycosides** are known and are particularly prevalent in some seeds. The conjugating sugar is usually glucose, and it may be attached to the gibberellin via a carboxyl group, giving a gibberellin glycoside, or a hydroxyl group, giving a gibberellin glycosyl ether. When gibberellins are applied to a plant, a certain proportion usually become glycosylated. Glycosylation may therefore represent another form of inactivation. In some cases, fed glucosides are metabolized back to the free GAs, so glucosides may also be a storage form of gibberellins (Schneider and Schmidt, 1990). The extent to which they fulfill this function is presently unknown.

Detection and Assay

As with all plant hormones, the original means of detecting and assaying gibberellins was bioassay, because of its sensitivity and specificity. Measuring gibberellins physically or chemically was extremely difficult as they do not absorb ultraviolet light, do not fluoresce, and do not have any distinguishing chemical characteristics that would form the basis of a specific chemical assay. Thus, except for the highly complex characterization of crystallized gibberellins obtained from vast quantities of plant material, virtually all identifications in plants in the past consisted of varying degrees of purification followed by some form of chromatography, often paper or thin-layer chromatography (TLC). More recently, **high-performance liquid chromatography** (HPLC) followed by bioassay of the chromatographic fractions has been used. Some form of chromatography was essential to separate out the various gibberellins, but even then the identification was often equivocal and numerous misidentifications can be found in the literature. Because of the uncertainty in identification, some journals insisted on use of the term "gibberellin-like." With the increasing availability of **gas chromatography** combined with **mass spectrometry** (GC-MS), this has now become the method of choice. Mass spectrometry allows the simultaneous identification and quantitation of the gibberellins present in a sample purified from plant material. The gas chromatography preceding mass spectrometry acts as a final stage of near-absolute purification.

Immunoassay has limited use for gibberellins because of the vast array of molecular structures and the cross-reactivity inherent in immunoassays. However, immunological methods promise to be useful in the purification of gibberellins from plant extracts and in the study of gibberellin receptors (Pence and Caruso, 1987).

Bioassays Detect and Measure Gibberellin-like Compounds

There are many bioassays for gibberellins (Reeve and Crozier, 1975), but three are probably the most widely used. All rely on a plant response that has been shown to be, or is presumed to be, naturally regulated by endogenous gibberellins. Bioassays are sensitive to the presence of inhibitory compounds, so some purification of the plant extract is essential. A chromatographic step is also needed to separate out the different groups of gibberellins. The sensitivity of the bioassays to the different gibberellins varies, though all are highly responsive to GA_1 and GA_3. Bioassays cannot provide conclusive information on identity.

Lettuce Hypocotyl Elongation Bioassay. When lettuce (*Lactuca sativa*) seedlings are germinated on filter paper in bright light their hypocotyls are very short (about

The Mass Spectrometer

A MASS SPECTROMETER BOMBARDS molecules with electrons, which shatter the molecules in a characteristic way and at the same time put an electrical charge on each fragment or intact molecule. Then it accelerates the fragments through a magnetic, electrical, or radio-frequency field to a detector, varying the field to allow only fragments of a certain mass to pass to the detector at any one time.

A basic plan of a mass spectrometer is shown in Figure 17.A. The molecules exit from the end of the column of a gas chromatograph (at the left in the figure) and pass through holes in a positively charged plate (the repeller) into an ionizing chamber, where they are bombarded by a focused beam of high-energy electrons produced by an electrically heated fine wire filament. These electrons cause some of the molecules to fragment, with each fragment losing an electron in the process and acquiring a positive charge. The positively charged repeller causes the fragments to be shot away at great speed through a series of electromagnetic lenses, which serve to focus the beam of ionized fragments. The beam then passes between four rods (quadrupole magnets), which generate an oscillating electrical and radio-frequency field. The smaller the mass of the fragment, the more it moves as the field oscillates. The field is electronically adjusted so that only fragments of a certain mass-to-charge ratio hit the center at the end of the quadropole rods at any one time. As the fragments have the same charge, this means that they are separated according to mass. By varying the field strength in a stepwise fashion, ions are sequentially selected according to mass to strike the detector (called an electron multiplier tube). At the detector they produce an electrical signal that is magnified for analysis by a computer. The scan of all the mass fragments present, from low to high, is repeated about every second.

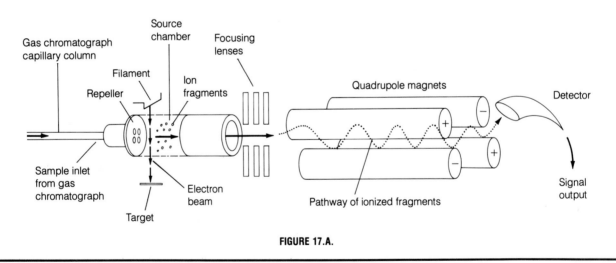

FIGURE 17.A.

5 mm), but with the incorporation of gibberellin in the water on the filter paper the hypocotyls are stimulated to elongate up to 30 mm, as compared to 5 mm in the absence of gibberellin. The assay is sensitive to about 1 to 100 ng in 0.2 ml of water (for 10 seeds on a piece of filter paper 1.5 cm in diameter). It has the disadvantage that the seedlings are very sensitive to inhibitors in incompletely purified extracts and can be easily killed by such compounds.

Dwarf Rice/Microdrop Bioassay. Rice seedlings of a dwarf rice (*Oryza sativa*, cultivar Tanginbozu) are germinated in small cups of agar in the light, and then 0.5-μl drops of solution to be tested are placed on the tip of the second emerging leaf of each seedling (Fig. 17.9). After two more days of growth, the elongation of the second leaf sheath is measured. Dramatic stimulation of elongation of the leaf sheath (from 10 mm with no GA_3 up to 65 mm) can be obtained with 0.1 to 100 ng of gibberellic acid per plant, and the assay is less sensitive to inhibitors than the lettuce assay. The dwarf rice cultivar Waito-C can be used to detect specifically gibberellins hydroxylated at carbon 3 (e.g., GA_1, GA_3) as this variety has a block at the natural 3-hydroxylation step.

0   0.1 ng   1 ng   10 ng   100 ng

Amount of GA₃ applied in 0.5 µl to each plant

FIGURE 17.9. Gibberellin causes elongation of the left sheath of rice seedlings, and this is used as a bioassay. (Photo courtesy of Peter Davies.)

α-Amylase Production in Germinating Cereal Grain. As will be discussed later, the presence of gibberellin in cereal grains stimulates the production of the starch-degrading enzyme α-amylase, with resulting production of sugars from starch. Hydrated cereal grains (usually barley, *Hordeum vulgare*) are cut in half laterally, and the half without the embryo is incubated in a gibberellin-containing solution. The resulting α-amylase production causes solubilization of the starch of the endosperm. The reducing sugars that are released can be measured colorimetrically or by refractive index, or exogenous soluble starch can be added and its disappearance measured colorimetrically with iodine. Sensitivities of 0.1 to 100 ng ml^{-1} can be obtained.

Gas Chromatography–Mass Spectrometry Provides Definitive Analysis of Gibberellins

Gas chromatography–mass spectrometry (GC-MS) is by far the method of choice for gibberellin analysis. Data produced by a mass spectrometer are as specific as a fingerprint, and the intensity of the signal can be used to calculate the amount of the substance present (Fig. 17.10). When a gas chromatograph, which can separate up to dozens of compounds in a mixture, is placed in front of the mass spectrometer, the purified compounds are fed one at a time to the mass spectrometer for analysis (Fig. 17.11) (Horgan, 1987).

Purification. Some purification and concentration of the gibberellin are needed prior to GC-MS analysis. Because

gibberellins have no physical characteristics (e.g., ultraviolet absorption) that can be used to monitor the purification, the method of choice is to add a known amount of radiolabeled gibberellin to the crude plant extract and follow the radioactivity during purification (Maki et al., 1986). As the endogenous and radiolabeled molecules behave similarly, they will be collected together. In the most accurate methods, gibberellins that contain both radiolabeled atoms and atoms of stable isotopes are added. The presence of stable isotopes enables the. internal standard to be separated and detected in the mass spectrometer and, because the amount is known, it can also be used to calculate the amount of endogenous gibberellin present.

Following solvent extraction of the plant tissue and some preliminary purification steps, the plant extract is injected into an HPLC column (Horgan, 1987). An HPLC machine is essentially a highly accurate column with special pumps designed to pump a liquid of known and changing composition through the column at pressures of the order of 7 MPa (1000 psi). The sample containing gibberellins is passed onto the column in water, and the gibberellins bind to the column material. The polarity of the solvent in the column is then slowly changed. As the solvent polarity decreases, the gibberellins partition between the column packing and the solvent, slowly moving down the column and separating from one another in accordance with their polarity. Polar gibberellins (those with three hydroxyls) elute from the column first, with the others following in order of decreasing polarity (see Fig. 17.19).

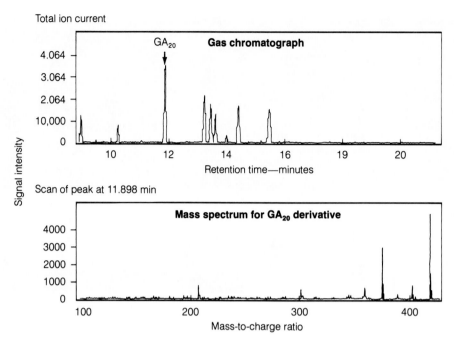

FIGURE 17.10. The plot from a mass spectrometer attached to a gas chromatograph provides (above) a total ion current trace in which each peak represents a compound and (below) a mass spectrum of each peak, in this case the peak at 11.89 min, which represents GA_{20} methyl ester–trimethylsilyl ether (Fig. 17.1g).

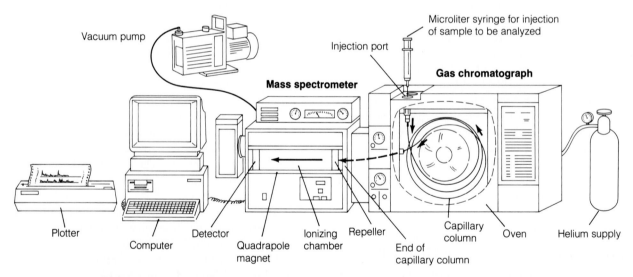

FIGURE 17.11. A gas chromatograph–mass spectrometer consists of a heated column, on which compounds are separated, connected to a mass spectrometer, in which the molecules from the column are split into characteristic fragments that are detected and analyzed by computer.

The semipurified gibberellins must be chemically modified before GC-MS to make them volatile. This is usually accomplished through simple chemical reactions that put a methyl group on all carboxyl groups and then a trimethylsilyl group on all hydroxyl groups (Fig. 17.1g). One microliter of solution containing the gibberellins is then injected into a helium gas stream flowing through a silica column in a gas chromatograph. The gibberellins and other compounds partition between the helium and the column surface, largely in order of increasing boiling point, with more passing into the gas as the temperature of the oven surrounding the column is steadily raised. The compounds separate from each other as they migrate through the column, so that each emerges from the column purified from all the others. A typical run takes about 20 min, and as each compound emerges over a period of only about 5 s, dozens of compounds can be analyzed in a mixture. From the gas chromatograph, the sample passes directly into the mass spectrometer for analysis and a mass spectrum, which shows the ions present and their intensities, is displayed on a computer. The mass spectrum of the sample can be compared with the mass spectrum produced by a given amount of a known gibberellin. If a particular gibberellin is sought and its spectrum is known, the mass spectrometer can be set to look for only certain specific ions; in this case there is an increase in sensitivity over that of a mass spectrum showing many ions.

GC-MS can be used to identify the gibberellins present in a sample even if no standard is available, as the spectra of all the gibberellins have been published. This is useful because only a few gibberellins can be easily purchased or obtained from other researchers; others are so rare that only specialized laboratories have even a few micrograms. If the electrical signals produced by all the ions in the scans of the peak of a gibberellin are summed (to give the total ion current) (Fig. 17.10), then that can be used to determine the amount of the gibberellin present.

The Physiological Effects of Gibberellins

With the availability of gibberellic acid in the late 1950s, plant scientists applied it to a wide range of plant systems and sometimes obtained amazing results, producing giants of the plant kingdom (Fig. 17.12). For plant biologists, however, the question is whether the endogenous compound produces similar effects in the plant and whether it is an important natural regulator of plant growth and development. Proving a natural role is a difficult task. The first step is usually to apply a gibberellin biosynthesis inhibitor and show that the purported endogenous gibberellin effect is eliminated, or at least reduced, and that the inhibitor can be overcome with applied gibberellin. Then the gibberellin must be shown to be present in a physiological concentration in the plant. But what is a physiological concentration? Commonly whole tissues (e.g., shoots) are extracted, yet the gibberellin may be present only in certain parts, cells, or even subcellular compartments within that tissue. At present we have no way to localize the gibberellin in any compartment too small to be dissected under a microscope, and it is often difficult to accumulate enough of such tissue to provide sufficient gibberellin for analysis. The most definitive proof has come with the use of **single gene mutants** that can be shown to affect both a specific developmental event and a step in gibberellin biosynthesis. Such a proof has been elegantly produced for stem elongation, as described below. But for the vast majority of events thought to be regulated by gibberellins, the evidence has been simply correlations providing varying degrees of certainty and sometimes producing large amounts of dispute among plant scientists.

FIGURE 17.12. Cabbage, a long-day plant, remains as a rosette in short days but can be induced to bolt and flower by applications of gibberellin. In the case illustrated, giant flowering stalks were produced. (Photo courtesy of S. Wittwer, Michigan State University.)

Gibberellins are extremely active molecules. In stem elongation, responses to GA_3 can be seen down to levels of 10^{-10} g (0.1 ng) for lettuce or rice seedlings. Sensitivities to even smaller amounts have been recorded. The methyl ester of GA_{73} has been found in the fern *Lygodium japonicum*, where it induces antheridium formation in dark-grown protonemata at a concentration as low as 10^{-14} M, which is about 3×10^{-15} g ml^{-1} (3 femtograms) or one-thirtieth of a part per trillion!

Applied Gibberellins Produce a Range of Effects in Plants

Virtually all the effects described in this section are produced by gibberellic acid (GA_3). As GA_3 is close in structure and bioactivity to GA_1, which is more common in higher plants, endogenous gibberellins may have these effects even if GA_3 has not been shown to be present in the plant. Sometimes applied GA_3 has little activity in a system, yet nonpolar gibberellins, such as the commercially available $GA_4 + GA_7$ mixture, may produce an effect. This suggests that while GA_1 is the only gibberellin active in stem elongation, other gibberellins may produce other effects without conversion to GA_1. However, one must always keep in mind that an applied gibberellin has to get to its site of action to function, and since a nonpolar gibberellin will diffuse more easily across membranes, the polarity of an active gibberellin may be more indicative of mobility within the plant than of activity.

Stem Elongation of Dwarf Plants. GA_3 causes such extreme stem elongation in dwarf plants that they resemble the tallest varieties of the same species (see Fig. 17.2). Accompanying this effect are a decrease in stem thickness, a decrease in leaf size, and a light green color of the leaves.

Bolting in Long-Day Plants. Some plants assume a rosette form in short days and bolt (grow a long flower stalk) and flower only in long days (see Chapter 21). Gibberellin application results in bolting of plants kept in short days, and natural bolting is regulated by gibberellin (see Figs. 17.3 and 17.12). In addition, many long-day plants also have a cold requirement that is overcome by the applied gibberellin. It is unlikely that the cold period itself leads to production of endogenous gibberellins, though it may in some way facilitate their production in subsequent long days. In many instances applied gibberellins cause stem elongation but not flowering. Thus, it is doubtful that the currently available gibberellins constitute the elusive flowering hormone florigen (Metzger, 1987).

Modification of Juvenility. Many woody perennials do not flower until they reach a certain stage of maturity; up to that stage they are said to be juvenile. The juvenile and mature stages often have different leaf forms. Applied gibberellins can regulate this juvenility in both directions, depending on the species. In English ivy (*Hedera helix*) GA_3 can cause a reversion from a mature to a juvenile

| Juvenile | GA_3 | GA_3 ABA | Mature |

FIGURE 17.13. Exogenous gibberellin causes rejuvenation of mature *Hedera helix* (English ivy) plants. Note the distinct differences in leaf shape and internode length between juvenile and mature forms. Left to right: juvenile control; GA-induced reversion to the juvenile form of a mature shoot treated with 5 nmol GA_3; abscisic acid (ABA, 5 μmol) prevention of rejuvenation by 5 nmol GA_3; mature control shoot. (From Rogler and Hackett, 1975a, 1975b).

state (Fig. 17.13), while many juvenile conifers can be induced to enter the reproductive phase by applications of nonpolar gibberellins such as $GA_4 + GA_7$ (Metzger, 1987). The latter is one instance in which GA_3 is not effective.

Induction of Maleness in Flowers. In plants that have flowers of a single sex, with both sexes on a single plant (e.g., cucumber) or on separate plants (e.g., spinach), gibberellin applications result in an increasing tendency to produce male flowers (Metzger, 1987).

Fruit Setting and Growth. Exogenous applications of gibberellins can cause fruit set (the start of fruit growth following pollination) and growth of some fruit, where auxin may be without effect. Gibberellin stimulation of fruit set has been observed in apple (*Malus sylvestris*). The most notable example of growth stimulation is in grapes (*Vitis vinifera*), where it finds a commercial use.

Induction of Seed Germination. Some seeds, particularly of wild plants, require light or cold in order to induce germination. In such seeds the dormancy can often be overcome without light or cold by applying gibberellin. As changes in gibberellin levels are often, but not always, seen in response to chilling of seeds, gibberellins may represent a natural regulator of one or more of the processes involved in germination.

Enzyme Production During Germination. Gibberellin application stimulates the production of numerous hydrolases, notably α-amylase, by the aleurone layers of germinating cereal grains. As this is the principal system in which gibberellin action has been analyzed, it will be discussed much more fully later.

Endogenous GA₁ Regulates Stem Elongation

When applications of gibberellin to dwarf plants made them grow tall, it was presumed that gibberellins must be the natural regulators of plant stem growth, but proof seemed to be lacking. Attempts to demonstrate that tall plants had more, or more active, gibberellins than dwarf plants were unsuccessful. In the early 1980s, however, James Reid and co-workers at the University of Tasmania in Australia, in collaboration with Jake Macmillan's group at the University of Bristol in England, finally demonstrated that tall stems do indeed contain more bioactive gibberellin than dwarf stems and that the level of this gibberellin mediates the genetic control of tallness (Reid, 1987). This breakthrough can be attributed to two factors. First, tall and dwarf plants of known genetic makeup were compared. Genes regulating tallness were identified, and the plants used were genetically identical except for the gene that is primarily responsible for

tallness. Second, gibberellin chemistry and instrumental analysis had advanced to the level at which unequivocal identifications could be made and custom-tailored gibberellin molecules, labeled with both stable and radioactive isotopes, could be used in sophisticated studies of metabolism.

The reasons for the earlier failure to show any relationship between stem growth and gibberellin content are numerous. The fact that the dwarf plants contained mainly inactive gibberellins was not noticed because bioassay plants have the ability to convert some inactive gibberellins to active gibberellins. In addition, whole stems were examined in the early investigations, whereas we know now that the highly biologically active gibberellin in tall plants, which turned out to be GA_1, is found only in the expanding internodes and is not present in older stem tissue.

The Discovery of GA₁ in Tall Plants. Tall plants containing the Le^1 allele were compared with isogenic dwarf plants (plants with the same genetic makeup except for the genes mentioned) containing the *le* allele. These are the two alleles of the gene regulating tallness in peas, which was first investigated by Gregor Mendel in the pioneering study in genetics in 1866. Gibberellin extracts of expanding internodes of these plants were separated into the different gibberellins by column chromatography, and the resulting fractions were assayed. The tall plants contained much more activity than the dwarf plants for one particular gibberellin that ran at the same position on the chromatographic column as GA_1 (Potts et al., 1982). This was the first indication that tall plants contained more GA_1 than dwarf plants. GC-MS was then used to prove that the particular gibberellin was indeed GA_1 and that the levels were higher in tall plants (Ingram et al., 1983).

The precursor of GA_1 in higher plants is GA_{20} (GA_1 is 3β-OH GA_{20}). If GA_{20} is applied to dwarf (*le*) pea plants, they fail to respond, but they do respond to applied GA_1, so that the *Le/le* gene difference is masked. The implication was that the *Le* gene conferred on the plants the ability to convert GA_{20} to GA_1. This was proved by using GA_{20} labeled with both ^3H (for following during purification) and ^{13}C (for identification of the products by GC-MS) (Ingram et al., 1984). The $[^{13}C, ^3H]GA_{20}$ was metabolized to GA_1, GA_8 (2β-OH GA_1), and GA_{29} (2β-OH GA_{20}) in *Le* plants but only to GA_{29} and GA_{29}-catabolite in *le* plants. Thus, it was conclusively demonstrated that the *Le* gene, which regulates tallness (or stem length) in peas, does so by causing the synthesis of an enzyme that 3β-hydroxylates GA_{20} to produce GA_1. In the absence of this enzyme, no GA_1 is produced and the plants are dwarf.

---

1. All alleles mentioned in this section are homozygous.

Correlations Between GA₁ and Tallness. Dwarf plants, although shorter than tall plants, do have some stem length. As previously noted, there are also nana mutants of pea in which the stem length is extremely short (1 mm, compared with about 3–5 cm in mature dwarf plants and 15 cm in mature tall plants) (Fig. 17.14). These extremely dwarf plants have the allele *na* (normal = *Na*), which blocks gibberellin biosynthesis between *ent*-kaurene and

GA_{12}-aldehyde (Ingram and Reid, 1987), and are virtually gibberellin deficient. At first glance, this seems to indicate that GA_1 precursors, like GA_{20}, present in dwarf plants must also regulate stem growth. Further investigations, however, showed that this is not the case because the gene *le* turned out to be "leaky"—that is, it did not block GA_1 formation completely. The amount of GA_1 measurable in dwarf plants was negligible, but studies following the application of isotopically labeled GA_{20} to the plants showed that some had been converted into GA_8, so it must have passed through GA_1. When the metabolism through the 3β-hydroxylation pathway was compared in peas differing in tallness, it turned out that tallness corresponded closely to the extent of 3β-hydroxylation (Fig. 17.15). From this it has been concluded that only GA_1 regulates tallness in peas (Ingram et al., 1986). Similar conclusions have been reached for maize, also by using genotypes that have blocks in the gibberellin biosynthesis pathway (Phinney, 1985), and it appears that the control of stem elongation by GA_1 may be universal.

Other elegant experiments also point toward the central role of GA_1 in the regulation of tallness. Plants of the genotype *na Le* are nana as they have virtually no gibberellins (Potts and Reid, 1983). However, if they are fed GA_{20} they become tall, whereas *na le* plants respond only weakly (Ingram et al., 1986). In addition, it had always been said that tallness in peas is nongraftable; that is, a dwarf plant containing the *le* allele grafted onto a tall plant remains dwarf. The reason for the failure to graft tallness in peas in this case is that *GA₁ is synthesized from GA₂₀ only in the stem apex*. In contrast, the production of gibberellins under the influence of *Na* occurs in mature leaf tissue. Thus, if a shoot of the genotype *na Le* (which is nana) is grafted onto a dwarf plant of

FIGURE 17.14. Phenotypes and genotypes of peas that differ in the gibberellin content of their vegetative tissue. Left to right: Nana (*na*), an ultradwarf containing no detectable GAs; Dwarf (*Na le*) containing GA₂₀; Tall (*Na Le*) containing GA₁; Slender (*la cryˢ*), an ultratall also containing no detectable GAs but with a "switched-on" GA receptor or some subsequent growth-regulating process. (All alleles are homozygous.) (Redrawn from Davies, 1987.)

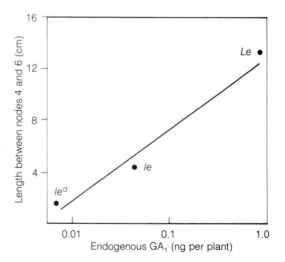

FIGURE 17.15. Stem elongation corresponds closely to the level of GA₁. Here the GA₁ content in peas with three different alleles at the *Le* locus is plotted against the internode elongation in plants with those alleles. There is a close correlation between the GA₁ level and internode elongation (From Ross et al., 1989.)

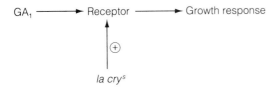

FIGURE 17.16. Model for the mechanism of the Slender mutation in peas. In normal peas, GA_1 binds to its receptor, leading to a growth response. In the Slender mutants the *la cry^s* alleles are thought to activate the receptor directly, stimulating growth without the participation of GA_1.

genotype *Na le*, the resulting plant is tall (Reid et al., 1983). In this case the stock (rooted portion) synthesizes GA_{20}, which is transported to the scion (grafted shoot), where it is converted into GA_1.

Some Plants Grow Unusually Tall in the Absence of Gibberellins

Not all tallness is regulated by GA_1. The unusually tall pea mutant called Slender (Fig. 17.14) can exhibit the tall phenotype in the absence of GA_1 or any gibberellins at all. Slender is determined by the alleles *la cry^s* (normal is *La Cry*), which in some as yet unknown way switch on the tallness response. The plants look the same whether they possess *Le* or *le*, *Na* or *na*, and whether or not GA_1 or any other gibberellin is present (Potts et al., 1985). It has been proposed that *la cry^s* regulates a hypothetical gibberellin receptor or a process subsequent to gibberellin binding to a receptor, so that the plant acts as if it had been switched on by GA_1 (Fig. 17.16). The occurrence of growth without GA_1 does not alter the fundamental fact that tallness in peas is regulated by GA_1; it simply shows that there are many steps at which any physiological process can be regulated. Slender-type plants are rather rare. They have also been found in barley but not in maize.

Photoperiod May Regulate Gibberellin Metabolism

As described earlier, long-day plants bolt in long days, and the bolting, if not the flowering, can be duplicated by applications of gibberellins. Gibberellins are further implicated in the mediation of photoperiodism by the finding that gibberellin synthesis inhibitors can negate the long day–induced bolting process. The question of whether photoperiod acts by regulating the level of gibberellins has been investigated in spinach and peas.

To study the effects of photoperiodism on gibberellin levels, gibberellins in the plant must be characterized and their levels determined. In spinach (*Spinacea oleracea*), Metzger and Zeevaart (1980) found that the gibberellins of the 13-OH pathway were present ($GA_{53} \rightarrow GA_{44} \rightarrow GA_{19} \rightarrow GA_{20}$). In short days, when the plants maintained a rosette form (Fig. 17.17), the level of GA_{20} was relatively low and that of GA_{19} was relatively high. On transfer to long days, the level of GA_{19} decreased and that of GA_{20} increased, while the other GAs remained at the same level (Fig. 17.18). This indicates that long days control the step from GA_{19} to GA_{20}. In short days GA_{19} builds up because its conversion to GA_{20} is restricted. With the advent of long photoperiods, the pool of GA_{19} is converted to GA_{20}, and this is associated with bolting. The existence of photocontrol of the conversion of GA_{19} to GA_{20} was subsequently verified by feeding deuterated GA_{53} ($[^2H]GA_{53}$) to the shoots and following the appearance of the deuterium in the subsequent gibberellins by GC-MS (Gianfagna et al., 1983). Once again, $[^2H]GA_{19}$ built up in short days but was metabolized to $[^2H]GA_{20}$ in long days. Whether GA_{20} is the active molecule in spinach or whether there is a subsequent conversion to GA_1 that causes the stem elongation has not been investigated, though the endogenous level of GA_1 appears to be low.

In peas the dominant alleles *Sn* and *Hr* prevent the senescence of the plants when they are grown in short

Short day   Short day   Long day
                   + GA_3

FIGURE 17.17. Spinach plants bolt only in long days, remaining in a rosette form in short days. GA applications will cause plants in short days to assume the same morphology as those exposed to long days. As shown in Figure 17.18, long days cause changes in the gibberellin contents of the plant. (Photo courtesy of J. A. D. Zeevaart.)

FIGURE 17.18. With increasing numbers of long days, the stem length of spinach plants increases. This is preceded by changes in the levels of gibberellins as measured by GC-MS. (The level of each gibberellin is expressed as a percentage of the maximum level present.) The level of GA_{19} (the precursor of GA_{20}) falls while that of GA_{20} rises. The level of GA_{29}, a product of GA_{20}, also rises but later than the increase in GA_{20}. GA_{44}, a precursor of GA_{19}, and GA_{17}, a branch product of GA_{19}, are unaffected. These results show that long days cause the conversion of GA_{19} to GA_{20}. The GA levels were measured by the relative intensity (as a percentage of the maximum) of the signal for one characteristic mass fragment of each GA. (From Metzger and Stewart, 1980.)

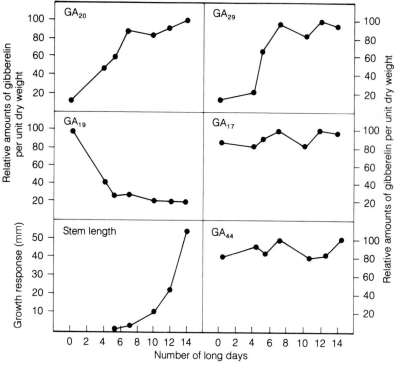

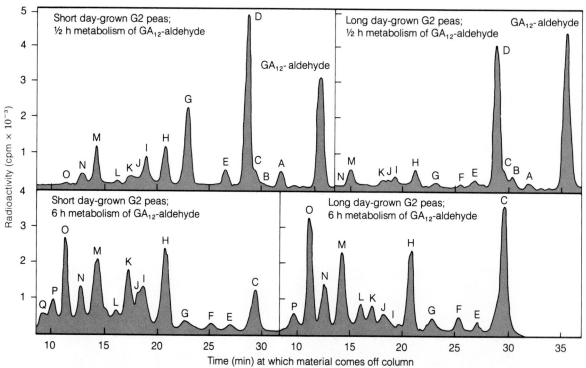

FIGURE 17.19. HPLC traces of radioactive gibberellins (appearing as peaks) produced from the metabolism of GA_{12}-aldehyde by shoots of peas of the genetic line G2 grown in short days or long days. The GA_{12}-aldehyde comes off the column last (at 35 min). Metabolism produces successively more polar products that come off the column earlier. Metabolism was allowed to proceed for $\frac{1}{2}$ h or 6 h. Note that peak G (GA_{53}) is prominent in short days at $\frac{1}{2}$ h but is almost lacking in long days. After 6 h, peaks I, J, and K (GA_{44}, GA_{19}, and GA_{20}, respectively) are also larger in short days than in long days. These results show that photoperiod controls one of the two steps between GA_{12}-aldehyde and GA_{53}, with GA_{53} and the subsequent 13-OH GAs (44, 19, 20) appearing in short days as metabolism progresses. Since G2 peas senesce in long days but continue growing in short days, high levels of 13-OH GAs correlate with, and may be responsible for, continued growth and the prevention of senescence. (From Davies et al., 1986.)

days. One of the actions of these genes has been shown to be on gibberellin metabolism. If ^{14}C-labeled GA_{12}-aldehyde is supplied to pea shoots, it is metabolized to a whole range of gibberellins including those of the 13-OH pathway. In short photoperiods the metabolism through this pathway is considerably more than in long days. The point of photoperiodic control is one of the two steps between GA_{12}-aldehyde and GA_{53} (GA_{12}-aldehyde → GA_{12} → GA_{53}), as the level of GA_{53} and all subsequent gibberellins is influenced by photoperiod (Fig. 17.19) (Davies et al., 1986). The metabolism of GA_{12} is so rapid that little builds up, making it difficult to determine whether the regulated step is before or after GA_{12}. When genotypes possessing one or both of the recessive alleles (*sn Hr, Sn hr,* or *sn hr*) were tested, they all showed lower metabolism of GA_{12}-aldehyde to 13-OH GAs similar to that displayed by *Sn Hr* plants in long days. Thus, the presence of both *Sn* and *Hr* is needed for enhanced production of 13-OH gibberellins in short days. The end result is that in short days GA_{19} and GA_{20} build up. Whether the higher levels of these gibberellins are responsible for the prevention of senescence in short days has yet to be conclusively demonstrated.

Given the widespread implications of photoperiodic or temperature control of development by gibberellins, it is likely that more examples of a direct environmental regulation of one or more steps of gibberellin metabolism will be forthcoming.

The Mechanism of Gibberellin Action

The effect of gibberellins applied to intact dwarf plants is so dramatic it would seem to be a simple task to determine how they act. Unfortunately, this is not the case, because, as we have seen in the case of auxin, so much about plant cell growth is not understood. We do know some characteristics of gibberellin-induced stem elongation. Gibberellin increases both cell division and cell elongation, because increases in cell number and cell length have been noted in response to gibberellin applications. There is a particularly striking increase in mitosis in the subapical meristem of rosette long-day plants following treatment with gibberellin (Fig. 17.20). In addition, internodes of tall peas have more cells of greater length than dwarf peas. Thus, the effects of gibberellin on stem elongation are multiple.

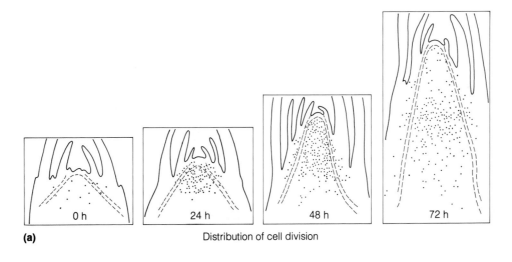

(a) Distribution of cell division

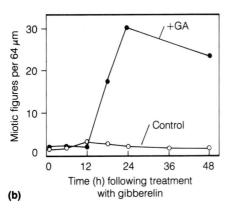

(b)

FIGURE 17.20. Gibberellin applications to rosette plants induce bolting in part by increasing cell division. (a) Longitudinal sections through the axis of *Samolus parviflorus* show an increase in cell division following application of GA. (Each dot represents one mitotic figure in a 64-μm-thick slice. (b) The number of such mitotic figures with and without GA in stem apices of *Hyoscyamus niger*. (From Sachs, 1965.)

Gibberellins Increase Cell Wall Extensibility in Gibberellin-Responsive Stems

How does GA_1 regulate cell elongation? How similar is gibberellin-induced growth to auxin-induced growth? The answers to these questions are still forthcoming, although the possibilities have been narrowed somewhat by work over the past decade. As discussed in Chapter 16, the elongation rate can be influenced by both cell wall extensibility and the osmotically driven rate of water uptake. Which of these parameters is affected by gibberellin? In lettuce hypocotyls, GA_3 had no effect on the osmolality of the expressed cell sap, but did increase cell wall extensibility (Métraux, 1987). An analysis of pea genotypes differing in gibberellin content or sensitivity showed that gibberellin decreases the wall yield threshold, which is the minimum force that will cause wall extension (Behringer et al., 1990). Thus, GA and auxin both seem to exert their effects by modifying cell wall properties.

In the case of auxin, cell wall loosening appears to be mediated in part by cell wall acidification (see Chapter 16). However, this does not appear to be the mechanism of gibberellin action. The initial response to applied gibberellin is less pronounced than the response to auxin, and, despite much searching, no evidence for wall acidification has been found. Although the response to gibberellin application in some plants, such as lettuce, has a lag similar to that of auxin in pea stem segments (10–15 min), the growth rate increases slowly rather than showing the abrupt change typical of auxin (see Chapter 16). Also, the lag before growth stimulated by

gibberellin is often somewhat longer. Oat stems show a most pronounced response to applied GA_3, with a lag time of about 45 min (Fig. 17.21). These different responses point to a growth-promoting mechanism different from that of auxin, at least for short-term growth. Various suggestions have been made and all have some experimental support (Métraux, 1987), but as yet none provide a clear-cut answer.

Gibberellin may induce growth by altering the distribution of calcium in the tissue. Calcium ions inhibit the growth of lettuce hypocotyls, and this inhibition can be reversed by GA_3. Calcium is well known to reduce wall extensibility in dicots (but not in monocots), so gibberellin may in some way lower the calcium concentration of the wall, possibly by stimulating calcium uptake into the cell. Gibberellins may also stimulate cell wall synthesis, but measurements to date indicate that wall synthesis follows rather than precedes the stimulation of cell elongation. One suggestion is that gibberellin prevents reactions that would otherwise cause stiffening of the cell wall. It has been proposed that cell wall stiffening is due to cross-linking of phenolic cell wall components related to lignin (see Chapter 13), such as ferulic acid residues, under the influence of the enzyme peroxidase. Gibberellin appears to inhibit cell wall peroxidase activity, but whether this is the mechanism by which gibberellins enhance growth is still uncertain. Clearly, much more will have to be known about the basic mechanisms of cell wall extension before the regulatory role of hormones can be understood.

Gibberellins from the Embryo Stimulate α-Amylase Production in Cereal Seeds

Cereal "seeds" (technically fruits or caryopses) can be divided into two parts: the diploid embryo and the triploid endosperm (see Chapter 1). The **embryo** consists of the embryo proper and a specialized absorptive organ, the **scutellum**. The endosperm is composed of two tissues, the centrally located starchy endosperm and the aleurone layer (Fig. 17.22a). The **starchy endosperm**, typically nonliving at maturity, consists of thin-walled cells filled with starch grains. The **aleurone layer** surrounds the starchy endosperm and is cytologically and biochemically quite distinct. Aleurone cells are enclosed in thick cell walls and contain large numbers of protein-storing organelles called aleurone grains, or protein bodies, as well as lipid-storing spherosomes (Fig. 17.22b).

During germination, the stored food reserves of the starchy endosperm, chiefly starch, are broken down by hydrolytic enzymes, and the solubilized sugars, amino acids, and other products are transported to the growing embryo. The two enzymes responsible for starch degradation are α- and β-amylase. α-Amylase hydrolyzes

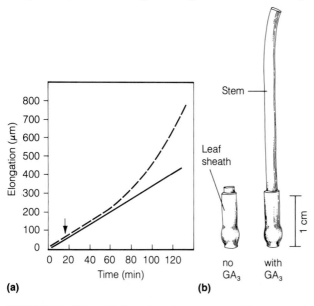

FIGURE 17.21. Gibberellin induction of stem growth. (a) In the elongation of oat (*Avena sativa*) internodes, there is a lag of about 45 min from the time of application of 30 μM GA_3 (at arrow). (b) The growth of the oat stems after 60 h without (left) or with (right) 30 μM GA_3. (After Adams et al., 1973.)

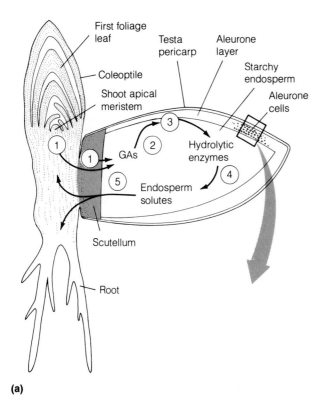

(a)

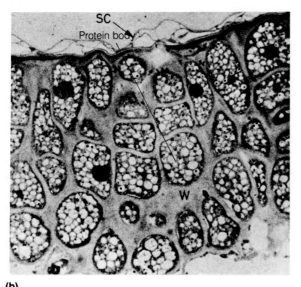

(b)

FIGURE 17.22. The structure of barley seeds and the functions of various tissues during germination. (a) Diagram of the major tissues. The seed is enclosed in a fused seed coat–fruit wall (testa-pericarp). (1) Gibberellins are synthesized by the coleoptile and scutellum of the embryo and released into the starchy endosperm; (2) gibberellins diffuse to the aleurone layer; (3) the aleurone layer is induced to synthesize and secrete α-amylase and other hydrolases into the starchy endosperm; (4) starch and other macromolecules are broken down to small substrate molecules; (5) the endosperm solutes are absorbed by the scutellum and transported to the growing embryo (indicated by arrows). (b) Light micrograph of aleurone cells from the area enclosed in box in (a). Note the presence of thick cell walls (W) and the numerous protein bodies (aleurone grains) within each cell. (From Jones and MacMillan, 1984.)

starch chains internally to produce oligosaccharides or limit dextrins consisting of α-1,4-linked glucose residues. β-Amylase degrades starch from the ends to produce maltose, a disaccharide. Maltose can be further degraded to two glucose residues by the enzyme maltase. This process is of considerable importance to the brewing industry, which makes use of the sprouted barley seed as a source of malt for making beer. Malt provides the glucose for the growth of yeast and the production of alcohol by fermentation.

Amylase is secreted into the starchy endosperm of cereal seeds by two tissues: the scutellum and the aleurone layer (Fig. 17.22a). Although the primary function of the scutellum is to absorb the sugars produced during starch degradation, it also can release α-amylase and thereby play a role in starch degradation. The aleurone layer of graminaceous monocots (e.g., barley, wheat, rice, rye, and oats), in contrast, is a highly specialized, terminally differentiated secretory tissue whose sole function appears to be the synthesis and release of hydrolytic enzymes. Upon completion of this function, the cells of the aleurone layer are depleted of their contents and die.

Experiments carried out in the 1960s confirmed Haberlandt's observation in 1890 that the secretion of starch-degrading enzymes by barley aleurone layers depends on the presence of the embryo. When the embryo was removed to produce a deembryonated half-seed, no starch degradation occurred even after prolonged hydration of the half-seed on agar. However, when the half-seed was incubated in close proximity to the excised embryo, starch digestion occurred. Thus, the embryo produced a diffusible substance that triggered α-amylase production by the aleurone layer.

It was soon discovered that gibberellic acid could substitute for the embryo in stimulating starch degradation. When half-seeds were incubated in buffered solutions in the presence of gibberellic acid, there was a large stimulation of α-amylase release into the medium (relative to the control half-seeds incubated in the absence of gibberellic acid) after an 8-h lag period. The significance of the GA$_3$ effect became clear when it was shown that the embryo synthesizes and releases gibberellins into the endosperm during germination. Thus, the cereal embryo efficiently regulates the mobilization of its own food reserves through the secretion of gibberellins, which stimulate the digestive function of the aleurone layer (Fig. 17.22a).

Since the 1960s, investigators have utilized isolated aleurone layers or even aleurone cell protoplasts rather than half-seeds. The response of isolated aleurone layers to GA is illustrated in Figure 17.23a. The isolated aleurone layer, consisting of a homogeneous population of target cells, provides a unique opportunity to study the molecular aspects of hormone action in the absence of nonresponding cell types.

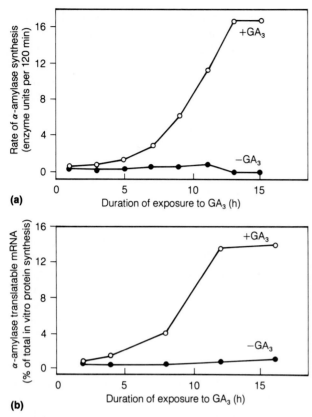

(a)

(b)

FIGURE 17.23. Gibberellin effects on enzyme synthesis and mRNA. (a) α-Amylase synthesis by isolated barley aleurone layers is quite evident after 6–8 h of treatment with GA$_3$ (10^{-6} M). (b) mRNA extracted from aleurone cells and translated in vitro showed an increasing presence of α-amylase mRNA, with the appearance of the mRNA preceding the release of the α-amylase from the aleurone cells by 1–2 h. The α-amylase mRNA in this case was measured by the in vitro production of α-amylase as a percentage of the protein produced by the translation of the bulk mRNA. (From Higgins et al., 1976.)

Gibberellic Acid Enhances the Transcription of α-Amylase mRNA

Prior to the development of molecular biological approaches, there were already indications that gibberellic acid might enhance α-amylase production at the level of transcription. First, it was demonstrated by both radioactive and heavy-isotope labeling studies (using ^{14}C-labeled amino acids and $H_2{}^{18}O$) that the stimulation of α-amylase activity by gibberellic acid is due to de novo synthesis (i.e., synthesis from amino acids) of the enzyme rather than to activation of preexisting enzyme. Second, GA$_3$-stimulated α-amylase production is blocked by inhibitors of transcription and translation.

There is now definitive evidence that GA acts primarily by inducing the expression of the gene for α-amylase. (The situation is actually more complicated than this because α-amylase belongs to a multigene family, but for the sake of clarity we will consider these genes as a single group.) First, it was shown that GA$_3$ enhances

the level of translatable mRNA for α-amylase in aleurone layers. Total mRNA was extracted from GA$_3$-treated and control tissue and translated in vitro in the presence of the radioactively labeled amino acid [^{35}S]methionine. Upon precipitation of the reaction products with α-amylase antibody, the radioactivity in α-amylase was quantified. Based on the amount of radioactivity in the immunoprecipitated protein, more α-amylase was synthesized from mRNA extracted from GA-treated tissue than from the mRNA of the controls, and the increase in the presumptive α-amylase mRNA preceded the appearance of α-amylase in the medium (Higgins et al., 1976) (Fig. 17.23b).

The method just described measures only translatable α-amylase mRNA. To obtain a direct measurement of the total amount of amylase mRNA, α-amylase cDNA clones were isolated and used to make ^{32}P-labeled cDNA probes. By hybridizing the radioactively labeled cDNA probes to "Northern" or RNA blots, it was shown that the level of α-amylase mRNA is strongly enhanced by gibberellic acid (Chandler et al., 1984).

There are two ways in which the level of α-amylase mRNA could be increased: by stimulating transcription or by decreasing mRNA turnover. To discriminate between these alternatives, a **nuclear run-off** experiment was performed (Jacobsen and Beach, 1985) (see Chapter 20, Fig. 20.17). Isolated nuclei, although incapable of initiating transcription, can complete the transcripts already being synthesized at the time of their isolation if provided with the appropriate conditions and substrates. Transcriptionally active nuclei were isolated from aleurone cell protoplasts that had been incubated in the presence or absence of gibberellic acid. The isolated nuclei were then allowed to "run off" their transcripts in the presence of ^{32}P-labeled uridine triphosphate (UTP). This resulted in the specific labeling of the mRNA sequences being transcribed at the time the nuclei were isolated. The incorporation of label into α-amylase mRNA was quantified by hybridizing the total labeled mRNA to cloned α-amylase cDNA on nitrocellulose paper and counting the bound radioactivity. The results showed that only nuclei isolated from GA-treated cells synthesized significant levels of α-amylase mRNA, demonstrating that gibberellin promotes the transcription of α-amylase mRNA (Jacobsen and Beach, 1985).

DNA-Binding Proteins Regulate the Transcription of α-Amylase

The promotion of transcription by gibberellin is thought to be mediated by DNA-binding proteins, similar to the regulatory proteins of prokaryotic operons (Chapter 2). To demonstrate such DNA-binding proteins in rice, a technique called **gel retardation assay** has recently been applied. Using α-amylase cDNA clones from barley, the

DNA fragments from rice
α-amylase gene, plus:

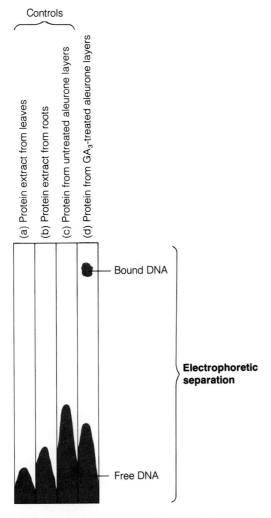

FIGURE 17.24. Diagram of a mobility shift DNA-binding assay for the detection of gibberellin-induced DNA-binding protein. Labeled DNA fragments from the 5'-upstream region of the rice gene for α-amylase were incubated in the presence of protein extracts from GA-treated and control tissues. The DNA-protein mixture was then placed at the top of a polyacrylamide gel and electrophoresed downward. The free DNA is seen as smears at the bottom of each lane, while the DNA-protein complex forms a spot near the top of lane D because its mobility has been shifted (retarded). (a) Control protein from leaves; (b) control protein from roots; (c) control protein from untreated aleurone layers; (d) protein from GA₃-treated aleurone layers. (From Ou-Lee et al., 1988.)

experiment, sometimes called a **mobility shift assay**, is based on the fact that if a protein binds to the DNA, the DNA will be increased in size and will migrate more slowly on an electrophoretic gel. Such a retardation of α-amylase upstream DNA was found only in the presence of proteins isolated from GA_3-treated aleurone cells (Fig. 17.24) (Ou-Lee et al., 1988). To locate the exact DNA sequences involved in the protein binding, the upstream sequence was incubated with an exonuclease (an enzyme that degrades the DNA from the ends only) in the presence of the protein extract from GA_3-treated aleurones. The GA-induced protein was found to protect against digestion by the exonuclease (Fig. 17.25). Analysis of the undigested DNA yielded a base sequence to which the GA-induced protein bound and which had a high degree of similarity in both rice and barley (Fig.17.26). The protected DNA sequence was about

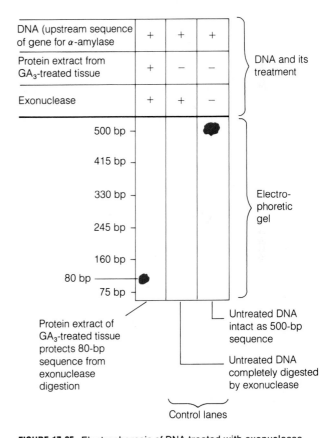

FIGURE 17.25. Electrophoresis of DNA treated with exonuclease and a protein extract from GA₃-treated tissue. Upstream α-amylase DNA fragments were labeled at the 5' end with ³²P and then incubated in the presence of an exonuclease that degrades DNA from the 3' end toward the 5' end. The original DNA had about 500 base pairs (bp), and in the absence of the exonuclease (right lane) the undegraded DNA can be seen at the top of the electrophoretic gel. If exonuclease is added, the DNA is completely degraded (center lane) and migrates to the bottom. When the protein extract from GA₃-treated aleurone tissue is added (left lane), the exonuclease digestion is unable to attack an 80-bp fragment of DNA because a protein has bound to it. (From Ou-Lee et al., 1988.)

gene for α-amylase was isolated from the genomic library of rice. The α-amylase DNA clones from rice included the adjacent 5'-upstream sequence (the DNA upstream of the transcription start site) that presumably acts as the regulatory sequence controlling transcription of the gene. This DNA was end-labeled with [³²P]ATP so that it could be detected. It was then applied to an electrophoretic gel in the presence of proteins extracted from aleurone tissues that had or had not been treated with GA_3. This

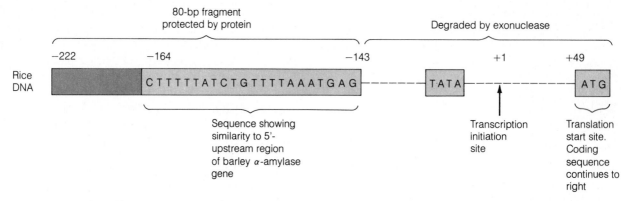

FIGURE 17.26. Analysis of the 80-bp fragment of DNA protected by a protein from GA$_3$-treated aleurone tissue in Figure 17.25 shows that it is upstream of the α-amylase structural gene (the base sequence that actually produces the protein). The boxed ATG represents the translation start site for the protein, which continues to the right. The exonuclease cleaves from the right in the diagram up to the shaded box, where the protein protection starts. This region of approximately 80 bp in rice has considerable homology to a similar region in barley, except that barley has an additional 44 bp inserted. The region protected from exonuclease by the bound protein probably represents the part of the DNA where the turn-on signal is received in the presence of gibberellin. Within the sequence degraded by exonuclease, the vertical arrow indicates the transcription initiation region in barley, and the boxed sequences to the left of the arrow represent the TATA box, where RNA polymerase presumably binds.

150 base pairs upstream of the putative translation initiation site of the rice α-amylase gene. From this we can hypothesize that gibberellin increases the level of a protein that switches on the production of α-amylase mRNA by binding to an upstream regulatory sequence of the α-amylase gene. However, direct proof of the function of the DNA-binding protein as a transcriptional regulator is still needed.

The question of how gibberellin causes the production of the DNA-binding protein now arises. Gibberellin could induce the transcription of its mRNA, but then something (another protein?) would have to regulate the production of this mRNA. A simpler explanation would be that gibberellin causes the modification, release, or possibly synthesis of the regulatory protein so that it is present in an active form only in aleurone cells that have received the hormonal signal.

The Gibberellin Receptor May Be Located on the Plasma Membrane of Aleurone Cells

For a hormone to be detected, it must presumably be bound by some receptor protein (Libbenga and Mennes, 1987). Little is known about gibberellin receptors, although advances are beginning to be made. One approach has been to make an antibody that recognizes GA$_3$ (this is actually done with GA$_3$ bound to a protein such as bovine serum albumin; see Pence and Caruso, 1987) and then make an antibody to the antibody, or **anti-idiotype**. (The binding domain of an antibody is termed its idio-type, and the antibody to the idiotype is called an anti-idiotype.) The idea is that the anti-idiotype will mimic the GA$_3$ but bind more tightly, so that the binding can be observed and measured. When this approach was used, the GA$_3$ antibody was shown to bind to the surface of wild oat aleurone protoplasts (Hooley et al., 1990). Thus, the gibberellin receptor may be in the cell membrane, and its binding of the gibberellin molecule may send a signal to the nucleus. The final signal must be the DNA-binding protein, which is now being characterized. The intermediate steps between the gibberellin receptor and the DNA-binding protein are unknown. The signal released by the membrane may be the receptor itself, but it is more likely that there is a complex sequence of events (signal transduction) that ultimately gives rise to the DNA-binding protein.

Commercial Applications of Gibberellins

The major uses of gibberellins are in the management of fruit crops, the malting of barley, and the extension of sugarcane with a resulting increase in sugar yield (Gianfagna, 1987). In some crops a reduction in height is desirable, and this can be accomplished by use of gibberellin synthesis inhibitors.

Fruit Production. A major use of gibberellins is to increase the size of seedless grapes. Bunch size and shape are also improved because of the stimulation of fruit stalk length. A mixture of benzyladenine (a cytokinin) and GA_{4+7} can cause an elongation of apple fruit and is used to improve the shape of Delicious-type apples under certain conditions. Although this does not affect yield or taste, it is considered commercially desirable. In citrus fruits gibberellins delay senescence, so the fruits can be left on the tree longer to extend the market period.

Malting of Barley. The increase in the production of malt because of an increase in α-amylase has been described above. This effect is not only of scientific interest; it is used commercially to increase the yield of malt prior to fermentation.

Increasing Sugarcane Yields. Gibberellin treatment can increase the yield of raw cane by up to 20 tons per acre and the sugar yield by 2 tons per acre. This increase is a result of the stimulation of internode elongation during the winter season.

Uses in Plant Breeding. The juvenility period in conifers can be detrimental to a breeding program by preventing the reproduction of desirable trees for many years. Application of GA_{4+7} can reduce the time to seed production considerably by causing the formation of cones on very young trees. In addition, the promotion of maleness in cucurbits and the stimulation of bolting in biennial vegetables such as beet (*Beta vulgaris*) and cabbage (*Brassica oleracea*) are valuable attributes of gibberellins that are occasionally used commercially in seed production.

Gibberellin Synthesis Inhibitors. Bigger is not always better, and gibberellin synthesis inhibitors are used commercially to prevent elongation growth. In floral crops, short, stocky plants such as lilies, chrysanthemums, and poinsettias are desirable, and restrictions on elongation growth can be achieved by applications of gibberellin synthesis inhibitors such as ancymidol (known commercially as A-Rest) or paclobutrazole (known as Bonzi). The same is true of cereal crops grown in cool, damp climates, such as in Europe. Restriction of extension growth in roadside shrub plantings can also be achieved with such agents.

Summary

Gibberellins are a family of compounds, now numbering over 80, defined by their structure. Not all are found in higher plants. Some gibberellins are biologically active, while others are not. Gibberellins were originally discovered as products of the fungus *Gibberella* and later found to be present in angiosperms. The gibberellins are terpenoid compounds, made up of isoprene units derived from mevalonic acid. The first compound in the isoprenoid pathway committed to gibberellin biosynthesis is *ent*-kaurene, which is converted to GA_{12}-aldehyde, the precursor of all the other gibberellins. GA_{12}-aldehyde, which has 20 carbon atoms, is converted to the other gibberellins by sequential oxidation of carbon 20, followed by the loss of this carbon to give 19-carbon GAs coupled with hydroxylation at one or more positions on the molecule, notably at C-13 and/or C-3. A subsequent hydroxylation at C-2 eliminates biological activity. Gibberellins may also be glycosylated to give either an inactivated form or a storage form. Gibberellins are identified and quantified by gas chromatography–mass spectroscopy following separation by high-performance liquid chromatography. Bioassay may be used to give an initial idea of the gibberellins present in a sample.

The most pronounced effect of applied gibberellins is stem elongation in dwarf plants. Gibberellins also cause bolting in long-day plants; cause changes in juvenility and flower sexuality; and promote fruit set, fruit growth, and seed germination. Endogenous gibberellins may have all these functions, but only certain roles have been proved. Most notably, it has been established that the level of GA_1 is responsible for tallness. Although we have little idea of how gibberellin causes cell wall loosening, we do know that gibberellin synthesis and metabolism are under strict genetic control. Dwarf plants are blocked in the production of GA_1. Not all tallness, however, corresponds to the GA_1 level, as extremely tall Slender plants may lack gibberellins. In this case a mutation in or subsequent to the gibberellin receptor is thought to mimic the presence of saturating GA levels.

Photoperiod regulates gibberellin metabolism to increase the level of biologically active GAs. In spinach, long days promote conversion of GA_{19} to GA_{20}, while in peas of a certain genotype short days increase the conversion of GA_{12}-aldehyde to GA_{53}.

Gibberellin induction of the enzyme α-amylase in aleurone cells of germinating cereal grains is now well elucidated. The gibberellin from the embryo induces the transcription of the genes for α-amylase mRNA. It appears to do so by causing an unidentified protein factor to bind to the upstream regulatory sequences of the α-amylase gene, thereby switching the gene on. How gibberellin increases the amount or activity of the binding protein is not yet known. Preliminary evidence suggests that the GA receptor may be on the plasma membrane, suggesting the intervention of a signal transduction pathway.

Gibberellins have several commercial applications, mainly in enhancing seedless grape production and the malting of barley. Gibberellin synthesis inhibitors are used as dwarfing agents.

GENERAL READING

Gianfagna, T. J. (1987) Natural and synthetic growth regulators and their use in horticultural and agronomic crops. In: *Plant Hormones and Their Role in Plant Growth and Development*, Davies, P. J., ed., pp. 614–635. Kluwer, Boston.

Graebe, J. E. (1987) Gibberellin biosynthesis and control. *Annu. Rev. Plant Physiol.* 38:419–465.

Horgan, R. (1987) Instrumental methods of plant hormone analysis. In: *Plant Hormones and Their Role in Plant Growth and Development*, Davies, P. J., ed., pp. 222–239. Kluwer, Boston.

Jacobsen, J. V., and Chandler, P. M. (1987) Gibberellin and abscisic acid in germinating cereals. In: *Plant Hormones and Their Role in Plant Growth and Development*, Davies, P. J., ed., pp. 164–193. Kluwer, Boston.

Jones, R. L., and Macmillan, J. (1984) Gibberellins. In: *Advanced Plant Physiology*, Wilkins, M. B., ed., pp. 21–52. Pitman, London.

Libbenga, K. R., and Mennes, A. M. (1987) Hormone binding and its role in hormone action. In: *Plant Hormones and Their Role in Plant Growth and Development*, Davies, P. J., ed., pp. 194–221. Kluwer, Boston.

Métraux, J.-P. (1987) Gibberellins and plant cell elongation. In: *Plant Hormones and Their Role in Plant Growth and Development*, Davies, P. J., ed., pp. 296–317. Kluwer, Boston.

Metzger, J. D. (1987) Hormones and reproductive development. In: *Plant Hormones and Their Role in Plant Growth and Development*, Davies, P. J., ed., pp. 431–462. Kluwer, Boston.

Pence, V. C., and Caruso, J. L. (1987) Immunoassay methods of plant hormone analysis. In: *Plant Hormone and Their Role in Plant Growth and Development*, Davies, P. J., ed., pp. 240–256. Kluwer, Boston.

Phinney, B. O. (1983) The history of gibberellins. In: *The Biochemistry and Physiology of Gibberellins*, Vol. 1, Crozier, A., ed., pp. 19–52. Praeger, New York.

Reeve, D. R., and Crozier, A. (1975) Gibberellin bioassays. In: *Gibberellins and Plant Growth*, Krishnamoorthy, H. N., ed., pp. 35–64. Wiley Eastern, New Delhi.

Reid, J. B. (1987) The genetic control of growth via hormones. In: *Plant Hormones and Their Role in Plant Growth and Development*, Davies, P. J., ed., pp. 318–340. Kluwer, Boston.

Sachs, R. M. (1965) Stem elongation. *Annu. Rev. Plant Physiol.* 16:73–96.

Schneider, G., and Schmidt, J. (1990) Conjugation of gibberellins in *Zea mays* L. In: *Plant Growth Substances, 1988*, Pharis, R. P., and Rood, S. B., eds., pp. 300–306. Springer-Verlag, Heidelberg.

Sponsel, V. M. (1987) Gibberellin biosynthesis and metabolism. In: *Plant Hormones and Their Role in Plant Growth and Development*, Davies, P. J., ed., pp. 43–75. Kluwer, Boston.

CHAPTER REFERENCES

Adams, P. A., Kaufman, P. B., and Ikuma, H. (1973) Effects of GA and sucrose on the growth of oat (*Avena*) stem segments. *Plant Physiol.* 51:1102–1108.

Behringer, F. J., Cosgrove, D. J., Reid, J. B., and Davies, P. J. (1990) Physical basis for altered stem elongation rates in internode length mutants of *Pisum*. *Plant Physiol.* 94:166–173.

Chandler, P. M., Zwar, J. A., Jacobsen, J. V. Higgins, T. J. V., and Inglis, A. S. (1984) The effects of gibberellic acid and abscisic acid on α-amylase mRNA levels in barley aleurone layers. Studies using an α-amylase cDNA clone. *Plant Mol. Biol.* 3:407–418.

Coolbaugh, R. C. (1985) Sites of gibberellin biosynthesis in pea seedlings. *Plant Physiol.* 78:655–657.

Davies, P. J., ed. (1987) *Plant Hormones and Their Role in Plant Growth and Development*. Kluwer, Boston.

Davies, P. J., Birnberg, P. R., Maki, S. L., and Brenner, M. L. (1986) Photoperiod modification of [^{14}C]gibberellin A$_{12}$ aldehyde metabolism in shoots of pea, line G2. *Plant Physiol.* 81:991–996.

Frydman, V. M., Gaskin, P., and MacMillan, J. (1974) Qualitative and quantitative analysis of gibberellins through seed maturation in *Pisum sativum* cv. Progress No. 9. *Planta* 118:123–132.

Gianfagna, T., Zeevaart, J. A. D., and Lusk, W. J. (1983) Effect of photoperiod on the metabolism of deuterium-labeled gibberellin A$_{53}$ in spinach. *Plant Physiol.* 72:86–89.

Graebe, J. E. (1986) Gibberellin biosynthesis from gibberellin A$_{12}$-aldehyde. In: *Plant Growth Substances 1985*, Bopp, M., ed., pp. 74–82. Springer-Verlag, New York.

Higgins, T. J. V., Zwar, J. A., and Jacobsen, J. V. (1976) Gibberellic acid enhances the level of translatable mRNA for α-amylase in barley aleurone layers. *Nature* 260:166–169.

Hooley, R., Beale, M. H., Smith, S. J., and MacMillan, J. (1990) Novel affinity probes for gibberellin receptors in aleurone protoplasts of *Avena fatua*. In: *Plant Growth Substances, 1988*, Pharis, R. P., and Rood, S. B., eds., pp. 145–153. Springer-Verlag, Berlin.

Ingram, T. J., and Reid, J. B. (1987) Internode length in *Pisum*. Gene *na* may block gibberellin synthesis between *ent*-7α-hydroxykaurenoic acid and gibberellin A$_{12}$-aldehyde. *Plant Physiol.* 83:1048–1053.

Ingram, T. J., Reid, J. B., and MacMillan, J. (1986) The quantitative relationship between gibberellin A$_1$ and internode growth in *Pisum sativum* L. *Planta* 168:414–420.

Ingram, T. J., Reid, J. B., Murfet, I. C., Gaskin, P., Willis, C. L., and MacMillan, J. (1984) Internode length in *Pisum*. The *Le* gene controls the 3β-hydroxylation of gibberellin A$_{20}$ to gibberellin A$_1$. *Planta* 160:455–463.

Ingram, T. J., Reid, J. B., Potts W. C., and Murfet, I. C. (1983) Internode length in *Pisum*. IV. The effect of the *Le* gene on gibberellin metabolism. *Physiol. Plant.* 59:607–616.

Jacobsen, J. V., and Beach, L. R. (1985) Control of transcription of α-amylase and ribosomal RNA genes in barley aleurone protoplasts by gibberellin and abscisic acid. *Nature* 316:275–277.

Lang, A. (1957) The effect of gibberellin upon flower formation. *Proc. Natl. Acad. Sci. USA* 43:709–717.

Maki, S. L., Brenner, M. L., Birnberg, P. R., Davies, P. J., and Krick, T. P. (1986) Identification of pea gibberellins by studying [^{14}C]GA$_{12}$-aldehyde metabolism. *Plant Physiol.* 81:984–990.

Metzger, J. D., and Zeevaart, J. A. D. (1980) Effect of photoperiod on the levels of endogenous gibberellins in spinach as measured by combined gas chromatography–selected ion current monitoring. *Plant Physiol.* 66:844–846.

Ou-Lee, T.-M., Turgeon, R., and Wu, R. (1988) Interaction of a gibberellin-induced factor with the upstream region of an α-amylase gene in rice aleurone tissue. *Proc. Natl. Acad. Sci. USA* 85:6366–6369.

Phinney, B. O. (1985) Gibberellin A$_1$ dwarfism and shoot elongation in higher plants. *Biol. Plant.* 27:172–179.

Phinney, B. O., and West, C. A. (1960) Gibberellins in the growth of flowering plants. In: *Developing Cell Systems and Their Control*, Rudnick, D., ed. Ronald Press, New York.

Potts, W. C., and Reid, J. B. (1983) Internode length in *Pisum*. III. The effect and interaction of *Na/na* and *Le/le* gene differences on endogenous gibberellin-like substances. *Physiol. Plant.* 57:448–454.

Potts, W. C., Reid, J. B., and Murfet, I. C. (1982) Internode length in *Pisum*. I. The effect of the *Le/le* gene difference on endogenous GA-like substances. *Physiol. Plant.* 55:323–328.

Potts, W. C., Reid, J. B., and Murfet, I. C. (1985) Internode length in *Pisum*. Gibberellins and the slender phenotype. *Physiol. Plant.* 63:357–364.

Reid, J. B., Murfet, I. C., and Potts, W. C. (1983) Internode length in *Pisum*. II. Additional information on the relationship and action of loci *Le, La, Cry, Na* and *Lm. J. Exp. Bot.* 34:349–364.

Rogler, C. E., and Hackett, W. P. (1975a) Phase change in *Hedera helix*: Induction of the mature to juvenile phase change by GA$_3$. *Physiol. Plant.* 34:141–147.

Rogler, C. E., and Hackett, W. P. (1975b) Phase change in *Hedera helix*: Stabilization of the mature form with abscisic acid and growth retardants. *Physiol. Plant.* 34:148–152.

Ross, J. J., Reid, J. B., Gaskin, P., and MacMillan, J. (1989) Internode length in *Pisum*. Estimation of GA$_1$ levels in genotypes *Le, le,* and *le^d. Physiol. Plant.* 76:173–176.

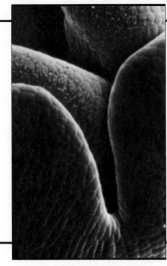

CHAPTER 18

Cytokinins

T HE CYTOKININS WERE DISCOVERED as a result of efforts to find factors that would stimulate plant cells to divide. Subsequently, they were shown to affect many other physiological and developmental processes. These effects include the delay of senescence in detached organs, the mobilization of nutrients, chloroplast maturation, the expansion of cotyledons, and the control of morphogenesis. Cell division is not necessarily a component of these other responses. In the case of chloroplast maturation, delay of senescence, and mobilization of nutrients, cell division does not even occur during the cytokinin-induced response. In other words, cytokinins have a number of different, apparently unrelated regulatory roles in higher plants. Nevertheless, the control of cell division is the process by which the cytokinins are defined, and it is of considerable significance for plant growth and development. For these reasons, we will present what we know about the cytokinins after first learning about how cell division is regulated in plants and how the ability to culture cells and tissues outside the whole plant has contributed to our understanding of the factors regulating plant cell division.

Cell Division in Plant Development

Mature plant cells generally do not divide. They are formed as a result of cell divisions in a primary or secondary meristem, and they enlarge and differentiate after they are formed. However, once they assume their function, whether it be transport, photosynthesis, or support, they normally do not divide again during the life of the plant. In this respect they appear to be similar to animal cells, which are said to exhibit terminal differentiation. However, many examples could be cited to show that this similarity to the behavior of animal cells is only superficial. Virtually every type of plant cell that retains its nucleus at maturity has been shown to be capable of dividing.

Differentiated Plant Cells Can Resume Division

Under some circumstances mature, differentiated plant cells may resume cell division. This can occur in the normal course of development of some types of cells or in response to external stimuli, demonstrating that plant cells are not terminally differentiated, except, of course, when they are dead at maturity. In many species, some mature cells of the cortex and/or phloem resume division to form secondary meristems, such as the vascular cambium or the cork cambium. The abscission zone at the base of a leaf petiole is a region where mature parenchyma cells begin to divide again after a period of mitotic inactivity.

Wounding plant tissues induces cells near the lesion to divide. Even highly specialized cells, such as phloem fibers and guard cells, may be stimulated by wounding to divide at least once. Wound-induced mitotic activity typically is self-limiting; after a few divisions the derivative cells stop dividing and do not divide again. However, when the soil-dwelling bacterium *Agrobacterium tumefaciens* invades a wound it can cause the plant neoplastic (tumor-forming) disease known as **crown gall**, which provides dramatic natural evidence of the mitotic potential of mature plant cells. Crown gall affects most dicots. Without *Agrobacterium* infection, the wound-induced cell division would subside after a few days and some of the new cells would differentiate as a protective layer of cork cells or vascular tissue. However, *Agrobacterium* changes the character of the cells that divide in response to the wound so that they become cancer-like. They do not stop dividing, but continue to divide throughout the life of the plant to produce the unorganized mass of tumor tissue that is called a gall (Fig. 18.1). We will have more to say about this important disease later in this chapter.

Diffusible Factors May Control Cell Divison

The above considerations suggest that mature plant cells that have stopped dividing do so because they no longer receive some signal, possibly a hormone, that is necessary for the initiation of cell division. The idea that cell division may be initiated by a diffusible factor is derived from the work of Gottleib Haberlandt, who, before World War I, demonstrated that vascular tissue contained a soluble substance or substances that would stimulate the division of wounded potato tuber tissue. The effort to determine the nature of this factor (or factors) led to the discovery of the cytokinins in the 1950s.

Plant Tissues and Organs Can Be Cultured

Biologists long have been intrigued by the possibility that organs, tissues, and cells could be grown in culture on a simple nutrient medium, in the way that microorganisms are cultured in test tubes or on Petri dishes. In the 1930s Philip R. White demonstrated that tomato roots could be grown indefinitely in a simple nutrient medium containing only sucrose as an energy source, mineral salts essential for plant nutrition, and a few vitamins. These isolated roots continued to grow in culture for many years, with periodic subculturing. The cells of the apical and lateral meristems divided repeatedly and the derivative cells differentiated, just as they would have done in the intact plant to produce the cells and tissues of the roots, but they did so without stems or leaves attached.

FIGURE 18.1. Tumors form on stems that have been infected with the crown gall bacterium, *Agrobacterium tumefaciens*. Several weeks before this photo was taken, the stem of a Malabar nightshade plant was wounded with a needle contaminated with a virulent strain of the crown gall bacterium. Some of the stem cells near the wound were transformed by DNA from the *Agrobacterium* Ti plasmid so that they continued to proliferate after the wound had healed. The bacterium transferred a portion of the Ti plasmid known as the T-DNA into wounded host cells. The T-DNA subsequently became integrated into the plant's nuclear DNA. The T-DNA contains genes that encode enzymes involved in the synthesis of both auxin and cytokinin. Expression of these genes in a transformed plant cell leads to hormone synthesis, and these hormones in turn stimulate the cells to divide repeatedly to form tumors, which are a type of tissue known as callus. (USDA photo.)

In contrast to the behavior of roots, isolated stem tissues exhibit very little growth in culture on simple media. Some growth may occur if auxin is added to the medium, but usually it is not sustained. Frequently this auxin-induced growth is due to cell enlargement only. With most plants, no growth will occur on the simple medium lacking hormones even if the cultured stem tissue contains apical or lateral meristems, at least until adventitious roots are formed. Once the stem tissue has rooted, growth resumes, but now as an integrated, whole plant. These observations tell us that there is a difference in the regulation of cell division in root and shoot meristems. They also suggest that some root-derived factor(s) may regulate growth in the shoot.

Crown gall tissue is an exception to these generalizations. After a gall has formed on a plant, the bacterium that incited gall formation can be killed by heating the plant to 42°C. The plant will survive and its gall tissue will continue to grow as a bacteria-free tumor (Braun, 1958). Tissues removed from these bacteria-free tumors grow on simple, chemically defined culture medium that would not support the proliferation of normal stem tissue of the same species. However, these stem-derived tissues are not organized. Instead they are a mass of cells, growing without recognizable organs as a tissue called **callus**. Callus tissue sometimes develops on plants in response to wounding, or in graft unions where stems of two different plants are joined. Crown gall tumors are a type of callus, whether they are growing attached to the plant or in culture. The finding that crown gall callus tissue could be cultured demonstrated that mature stem cells were capable of proliferating in culture and that contact with the bacteria may have caused the cells to produce cell division–stimulating factors.

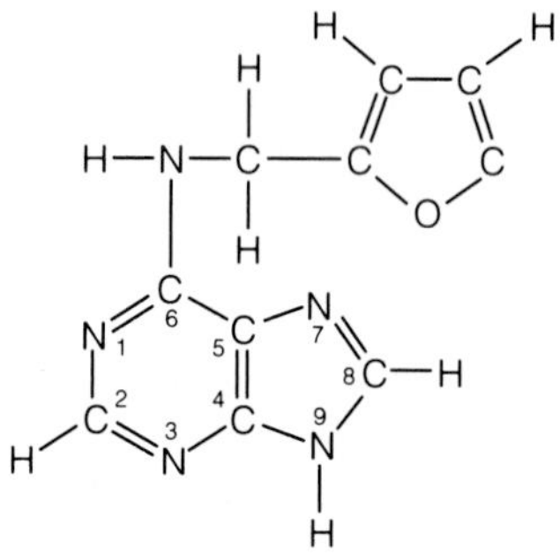

The Discovery and Identification of Cytokinins

A great many substances were tested in an effort to initiate and sustain the proliferation of normal stem tissues in culture. Materials ranging from yeast extract to tomato juice were found to have a positive effect, at least with some tissues. However, the most dramatic stimulations occurred when the liquid endosperm of coconut, also known as coconut milk or water, was added to the culture medium. White's nutrient medium supplemented with an auxin and 10–20% coconut milk will support the division of mature, differentiated stem cells from a wide variety of species and their continued proliferation to form callus tissue. This indicated that coconut milk contains a substance or substances that stimulates mature cells to enter and remain in the cell division cycle. Many attempts were made to determine the chemical nature of the substance in coconut milk responsible for initiating cell proliferation. Ultimately, coconut milk was shown to contain the cytokinin zeatin, but this did not occur until several years after the discovery of the cytokinins (Letham, 1974).

The Discovery of Kinetin, a Synthetic Cytokinin

Folke Skoog at the University of Wisconsin tested many substances for their ability to initiate and sustain the proliferation of cultured tobacco pith tissue. He had observed a slight promotive effect of the nucleic acid base adenine, so he tested the possibility that nucleic acids would stimulate division in this tissue. Surprisingly, autoclaved herring sperm DNA had a powerful cell division–promoting effect. A single chemical substance was isolated from the heat-treated DNA that would, in the presence of an auxin, stimulate tobacco pith paren-

Kinetin

FIGURE 18.2. The chemical structure of kinetin. Kinetin was first isolated from autoclaved herring sperm DNA, which had been shown to stimulate the proliferation of tobacco pith tissue when it was cultured on a medium containing an auxin. Although it is not found naturally in any plant, kinetin is similar to native cytokinins in that it is an aminopurine derivative with a substitution on the nitrogen at position 6 (N^6). It differs from native cytokinins in the kind of group found on the N^6 nitrogen.

chyma tissue to proliferate in culture. This substance was identified as 6-furfurylaminopurine and named **kinetin** (Fig. 18.2) (Miller et al., 1955). Kinetin stimulated cell division in tobacco pith when it was cultured on a medium containing an auxin. No cytokinin-induced growth occurred in the absence of an auxin.

Kinetin is not a naturally occurring plant growth regulator, and it does not occur as a base in the DNA of any species. It is a by-product of the heat-induced degradation of the DNA in which the deoxyribose sugar of adenosine is converted to a furfuryl ring and shifted from the 9 to the 6 position on the adenine ring. The discovery of kinetin was important because it demonstrated that cell division could be induced by a simple chemical substance. More important, when a synthetic molecule initiates a biological response, frequently it does so because the synthetic molecule has many of the same properties as naturally occurring molecules that regulate that response in the organism. Thus, the discovery of kinetin suggested that naturally occurring molecules with a structure similar to kinetin may regulate cell division activity within the plant. Of course, this proved to be correct.

Zeatin Is the Naturally Occurring Cytokinin of Most Plants

Several years after kinetin's discovery, Carlos Miller in the United States and D. S. Letham in Australia independently demonstrated that extracts of the immature endosperm of corn (*Zea mays*) contained a substance that had the same biological effect as kinetin. It stimulated mature plant cells to divide when added to a culture medium along with an auxin. Letham isolated the molecule responsible for this activity and identified it as 6-(4-hydroxy-3-methylbut-*trans*-2-enylamino)purine, which he called **zeatin** (Fig. 18.3) (Letham, 1973). Clearly, its

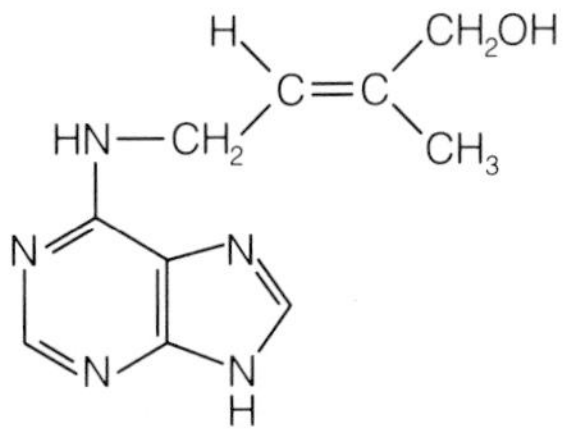

trans-Zeatin

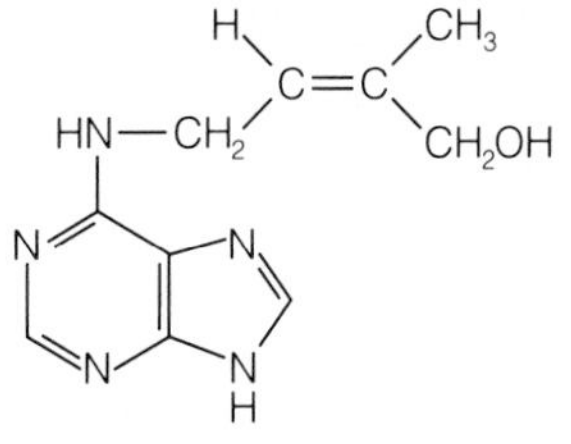

cis-Zeatin

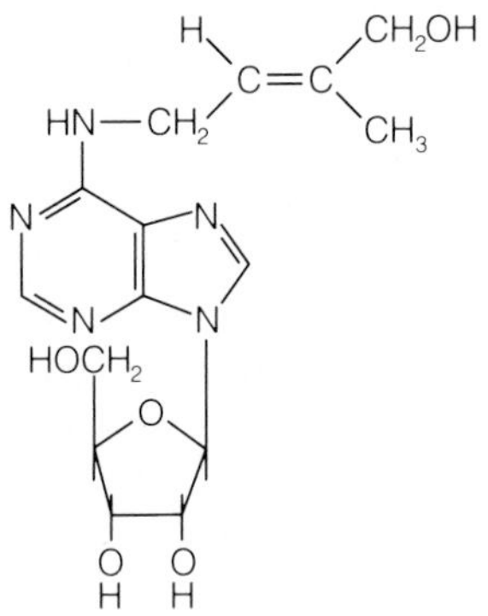

Ribosylzeatin
(zeatin riboside)

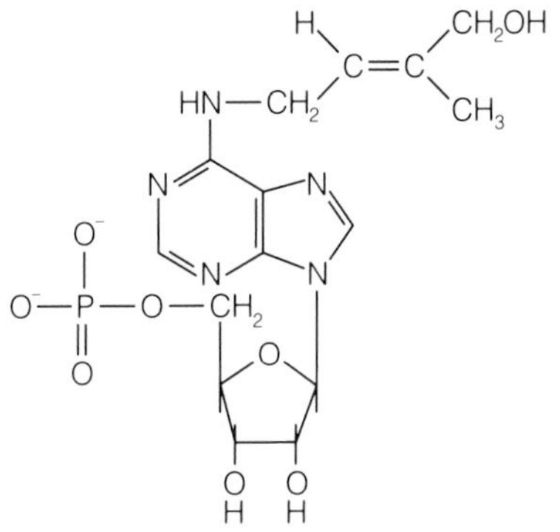

Ribosylzeatin-5'-monophosphate
(zeatin ribotide)

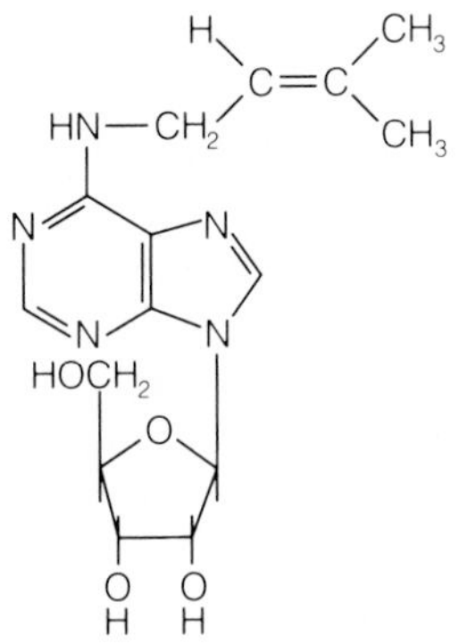

N^6-(Δ^2-Isopentenyl)-adenosine
(i^6Ado)

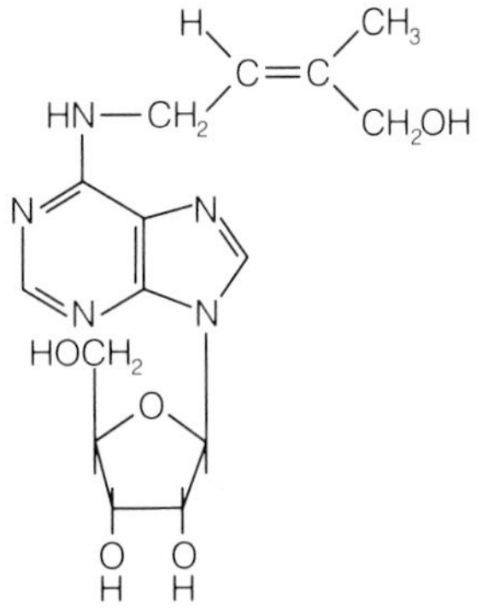

cis-Ribosylzeatin

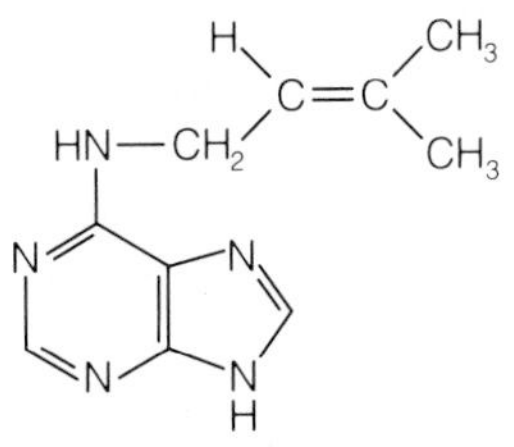

N^6-(Δ^2-Isopentenyl)adenine
(i^6Ade)

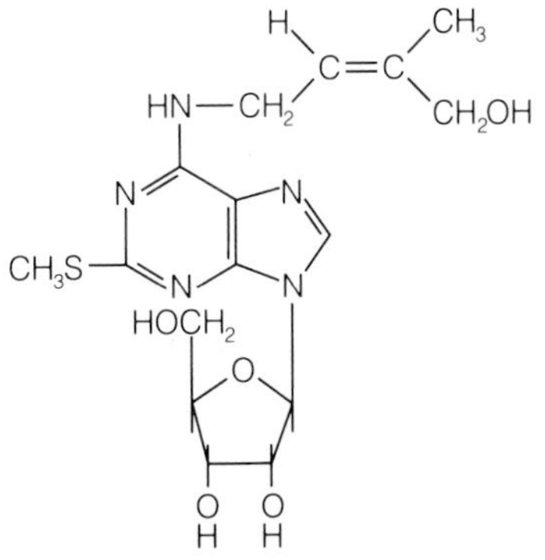

2-Methylthro-*cis*-ribosylzeatin
(bacterial)

FIGURE 18.3. The structures of some naturally occurring cytokinins. Most plants have *trans*-zeatin as the principal free cytokinin, but dihydrozeatin and isopentenyladenine are also native plant cytokinins. Free cytokinins also include the riboside and ribotides of zeatin, dihydrozeatin, and isopentenyladenine, although these may be active as cytokinins by conversion to the respective bases. Free cytokins from bacteria usually are *cis*- or *trans*-zeatin or isopentenyladenine, as well as their ribosides and ribotides.

structure is similar to that of kinetin. The two molecules have different side chains, but both are adenine or aminopurine derivatives with a side chain attached to the N^6 nitrogen. Because the side chain of zeatin has a double bond, it can exist in either the cis or the trans configuration. The naturally occurring free zeatin of higher plants has the trans configuration, although both the cis and trans forms of zeatin are active as cytokinins. Since its discovery in immature maize endosperm, zeatin has been found in many plants and in some bacteria. Zeatin is the most prevalent cytokinin in higher plants, but other substituted aminopurines active as cytokinins have been isolated from many plant and bacterial species. These differ from zeatin in the nature of the side chain attached to the N^6 nitrogen (Fig. 18.3) or in the attachment of a side chain to carbon 2. In addition, each of these can be present in the plant as a riboside (a ribose sugar is attached to the N^9 nitrogen of the purine ring), a ribotide (the ribose sugar moiety is esterified with phosphoric acid), or a glucoside (a glucose sugar molecule is attached to the N^7 or N^9 nitrogen of the purine ring) (Fig. 18.3) (Letham and Palni, 1983).

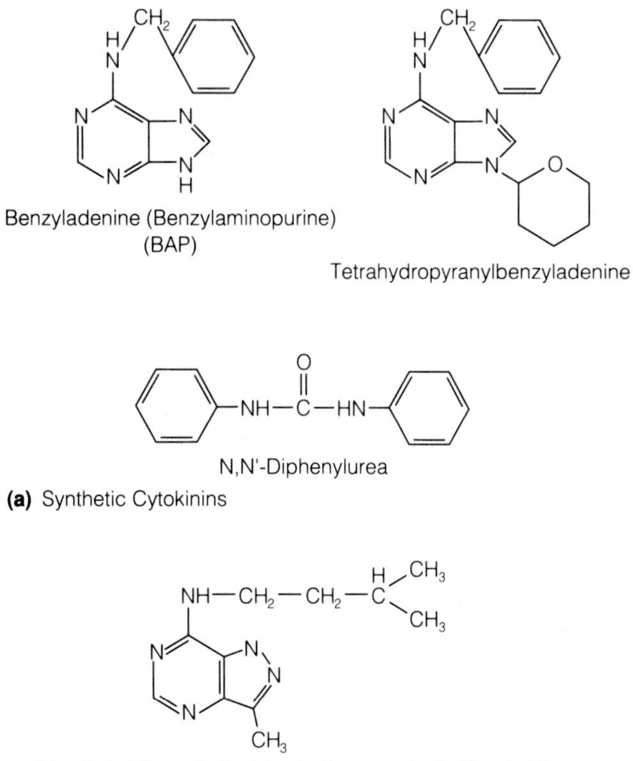

Some Synthetic Compounds Can Mimic or Antagonize Cytokinin Action

A large number of chemical compounds have been synthesized and tested for cytokinin activity. Analysis of these compounds provides insight into the structural requirements for activity. Nearly all compounds active as cytokinins are 6-substituted aminopurines. Certainly all the naturally occurring cytokinins are aminopurine derivatives. **Benzylaminopurine** (BAP), which has found uses in agriculture, is an example of a synthetic 6-substituted aminopurine cytokinin (see Fig. 18.6). Of course, kinetin itself is a synthetic 6-substituted aminopurine cytokinin. The only exceptions to this generalization are certain diphenylurea derivatives that show weak cytokinin activity (Fig. 18.4). These diphenylurea derivatives are synthetic cytokinins but they are not 6-substituted aminopurines, and it is not clear why they are active. One suggestion is that they may stimulate the biosynthesis of the native cytokinins in plant tissues that respond to these compounds (Mok et al., 1979).

In the course of determining the structural requirements for cytokinin activity, some molecules were found to act as cytokinin antagonists. These molecules are able to block the action of cytokinins, and their effects may be overcome by adding more of the cytokinin. One of the most potent cytokinin antagonists is 3-methyl-7-(3-methylbutylamino)pyrazolo[4,3-*d*]pyrimidine (Fig. 18.4).

A Variety of Methods Are Used to Assay for Cytokinins

The naturally occurring molecules with cytokinin activity may be detected and identified by a combination of methods, including chemical methods such as high-performance liquid chromatography and mass spectroscopy (Chapter 17), immunological methods, and bioassays. As discussed in Chapter 17, introduction of combined gas chromatography and mass spectroscopy, along with the use of isotopically labeled cytokinins as internal standards, has resulted in a major advance in measurement of cytokinin levels in plants (MacDonald and Morris, 1985).

Historically, bioassays were the only methods for the identification of cytokinins. Cell proliferation bioassays continue to be important because a cytokinin is defined operationally as a compound "which, in the presence of optimal auxin, induces cell division in tobacco pith or similar tissue cultures" (Letham and Palni, 1983). In addition to tobacco pith tissue, carrot tap root phloem requires a cytokinin for its proliferation in culture. Also,

Benzyladenine (Benzylaminopurine) (BAP)

Tetrahydropyranylbenzyladenine

N,N'-Diphenylurea

(a) Synthetic Cytokinins

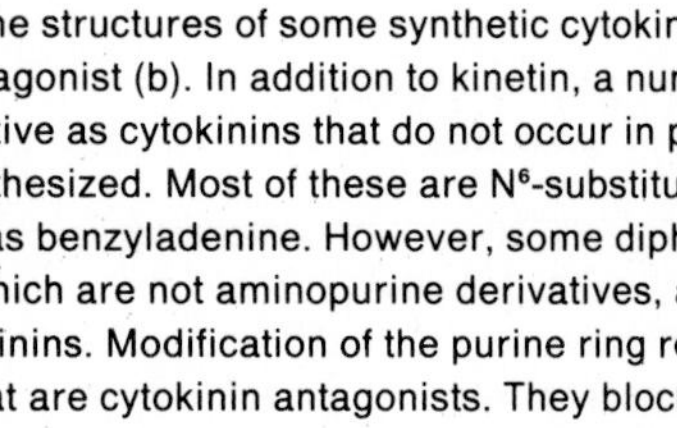

3-Methyl-7-(3-methylbutylamino)pyrazolo[4,3-*d*]pyrimidine

(b) Cytokinin Antagonist

FIGURE 18.4. The structures of some synthetic cytokinins (a) and a cytokinin antagonist (b). In addition to kinetin, a number of compounds active as cytokinins that do not occur in plants have been synthesized. Most of these are N^6-substituted aminopurines, such as benzyladenine. However, some diphenylurea compounds, which are not aminopurine derivatives, are weak synthetic cytokinins. Modification of the purine ring results in compounds that are cytokinin antagonists. They block the action of cytokinin, possibly by competing with cytokinin for binding to its receptor.

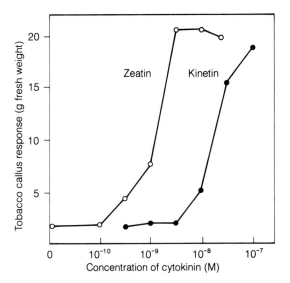

FIGURE 18.5. Cultured tobacco tissues are dependent on cytokinin for their growth in culture. Cultured tobacco callus was transferred to fresh medium that contained an auxin and either zeatin or kinetin at the indicated concentrations. The tissues were weighed after growing for 1 month on the various media. Clearly zeatin is more effective than kinetin in supporting tobacco tissue growth. The maximum response was obtained with 5×10^{-8} M zeatin, and higher zeatin concentrations either did not stimulate additional growth or were slightly inhibitory. The kinetin concentration had to be at least 10-fold higher to achieve the same growth stimulation. (From Leonard et al., 1968.)

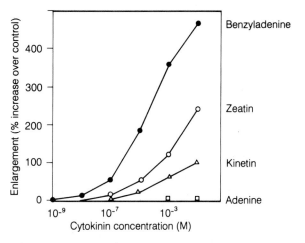

FIGURE 18.6. Cytokinins stimulate the enlargement of cocklebur (*Xanthium*) cotyledons. Pieces of the cotyledons were cut from the seeds and placed on filter paper moistened with water or a solution containing one of the cytokinins indicated over a wide concentration range. The cotyledon pieces were weighed initially and again after 4 days in the dark, and their percentage enlargement was calculated. The four compounds tested varied considerably in effectiveness. Although zeatin was more effective than kinetin, the synthetic cytokinin benzyladenine was the most potent cytokinin in evoking this response. (From Esashi and Leopold, 1969.)

some continuously cultured callus tissues of tobacco and soybean will not grow in culture without cytokinin. These cytokinin-requiring tissues all exhibit a linear increase in growth with increasing cytokinin concentration over a very broad range (Fig. 18.5).

The disadvantage of cell proliferation assays is that it takes nearly a month to see the results. Therefore, a number of alternative bioassays, some of which can be performed in a few days, were developed for the detection of cytokinins in plant extracts. Other cytokinin-stimulated physiological and developmental responses have been adapted as cytokinin bioassays.

For example, in some organs cytokinins induce cells to enlarge without dividing. Bioassays based on this effect were devised using the cytokinin-dependent expansion of radish (*Raphanus*) or cocklebur (*Xanthium*) cotyledons and radish leaf disks (Fig. 18.6). Cytokinins also delay the senescence of detached leaves, and a bioassay based on the retention of chlorophyll in detached oat (*Avena*) leaves has been used to estimate cytokinin levels in plant extracts. The advantage of these alternative assays is their speed. However, they must be used with care, since substances that are not cytokinins also may evoke the response or interfere with it.

Immunological methods may also be used for cyto-kinin measurement. Antibodies against cytokinins can be produced by injecting rabbits or mice with cytokinin

ribosides conjugated to a protein. Some of the antibodies that the animal produces to this complex will be directed against the cytokinin. These antibodies can be used to quantitate the amount of a cytokinin in a sample by means of a radioimmunoassay (Hansen et al., 1984). Plant extracts are first fractionated, usually by high-performance liquid chromatography (HPLC), and the cytokinins in the fractions are detected and their concentrations determined by means of the radioimmunoassay. The cytokinin antibodies can also be used to isolate the hormone from extracts by immunoaffinity chromatography. These immunological methods hold great promise for the identification and quantification of naturally occurring cytokinins because the antibodies are highly specific and more sensitive than most bioassays (Brandon and Corse, 1987). Furthermore, these immunological methods are very fast.

Naturally Occurring Cytokinins Are Both Free and Bound

The naturally occurring cytokinins are found as free molecules (not covalently attached to any macromolecule) in plants and certain bacteria. Cytokinins also exist as modified bases in certain transfer RNA molecules of all organisms.

Free cytokinins have been found in a wide spectrum of angiosperms and probably are universal in this group of plants. They have also been found in algae, diatoms, mosses, ferns, and conifers. Their regulatory role has been demonstrated only in angiosperms, conifers, and mosses, but they may function to regulate the growth, development, and metabolism of all plants. Zeatin is usually the most abundant naturally occurring free cytokinin. In addition to zeatin, **dihydrozeatin** and **isopentenyladenine** (i[6]Ade) are commonly found in higher plants. Numerous derivatives of these three cytokinins have been identified in plant extracts (Fig. 18.3).

Transfer RNA (tRNA) contains not only the four nucleotides used to construct all other forms of RNA but also some unusual nucleotides in which the base has been modified. Some of these "hypermodified" bases act as cytokinins when the tRNA is hydrolyzed and tested in one of the cytokinin bioassays. Some tRNAs of higher plants contain zeatin as a hypermodified base, although it is usually the cis isomer instead of the trans form found as free cytokinin. Higher plant tRNAs may also contain a derivative of zeatin in which a methylthio group is attached to the purine ring (2-methylthio-*cis*-ribosylzeatin) or may contain N^6-(Δ^2-isopentenyl) adenosine (i[6]Ado) (Fig. 18.3). In addition to tRNAs that participate in cytoplasmic protein synthesis, plants have a different set of tRNAs for protein synthesis within chloroplasts. These two classes of tRNA contain different cytokinins.

Cytokinins are not confined to plant tRNAs. They are part of certain tRNAs from all organisms, from bacteria to humans. In bacteria, methylthio-i[6]Ado is the main hypermodified base with cytokinin activity, while i[6]Ado is the principal cytokinin of animal tRNAs.

Some Bacteria Also Secrete Free Cytokinins

Some bacteria and fungi are intimately associated with higher plants. In many cases these microorganisms produce and secrete substantial amounts of cytokinins and/or cause the plant cells to synthesize plant hormones, including cytokinins. The cytokinins produced by microorganisms include zeatin, i[6]Ade, *cis*-zeatin, and their ribosides (Fig. 18.3). Infection of plant tissues with these microorganisms can induce the tissues to divide and, in some cases, to form special structures such as mycorrhizae, in which the microorganism can reside in a mutualistic relationship with the plant. As we saw earlier, some pathogenic bacteria may stimulate plant cells to divide. In addition to the crown gall bacterium *Agrobacterium tumefaciens*, these include *Corynebacterium fascians*, which is a major cause of the growth abnormality known as witches'-broom. The shoots of plants infected by *C. fascians* begin to resemble an old-fashioned straw broom after the infection has proceeded for several months because the lateral buds, which normally would remain dormant, are stimulated to grow by the bacterial cytokinin.

Biosynthesis, Metabolism, and Transport of Cytokinins

The side chains of naturally occurring cytokinins are chemically related to rubber, carotenoid pigments, the plant hormones gibberellin and abscisic acid, and some of the plant defense compounds called phytoalexins. All of these are constructed, at least in part, from units of isoprene (see Chapter 13). Isoprene is similar in structure to the side chain of zeatin, 2-i[6]Ade, and other cytokinins.

$$CH_3-\underset{\underset{\text{Isoprene}}{|}}{\overset{\overset{CH_3}{|}}{C}}=CH-CH_3$$

These cytokinin side chains are synthesized from an isoprene derivative. In the case of rubber or the carotenoids, large molecules are constructed by the polymerization of many isoprene units, whereas cytokinin contains just one of these units. The precursor for the formation of these isoprene structures is mevalonic acid. Mevalonic acid is converted to Δ^2-isopentenyl pyrophosphate (Δ^2-IPP) by a series of enzymatic reactions known as the mevalonic acid pathway (Chapter 13).

Cytokinin Synthase Catalyzes the First Step in Cytokinin Biosynthesis

An enzyme that synthesizes cytokinins has been found in plants. It is called Δ^2-isopentenyl-pyrophosphate:AMP Δ^2-isopentenyltransferase, or **cytokinin synthase**. It is a type of prenyltransferase, similar to the prenyltransferase enzymes active in the synthesis of other isoprenoid compounds. The prenyltransferase active in cytokinin synthesis catalyzes the transfer the isopentenyl group of Δ^2-IPP to adenosine monophosphate (Chen, 1982). The product of this reaction, the ribotide of i[6]Ade, is not a major higher plant cytokinin. However, it is active as a cytokinin in bioassays and is readily converted to zeatin and other cytokinins (Fig. 18.7).

Modification of the Polymerized Base Produces the tRNA Cytokinins

The synthesis of the tRNA cytokinins takes place by an entirely different route. Free cytokinins are not used to any significant extent in the synthesis of the cytokinin-active bases in tRNAs. Instead, tRNAs are made from the four conventional nucleotides during the transcription of the genes encoding their sequences. The first product is a tRNA precursor that lacks the hypermodified bases and is somewhat larger than the final product. This precursor is processed, much as messenger RNA is processed, to yield the functional tRNA molecule. In this processing, specific adenine residues of some of the tRNAs are modified to

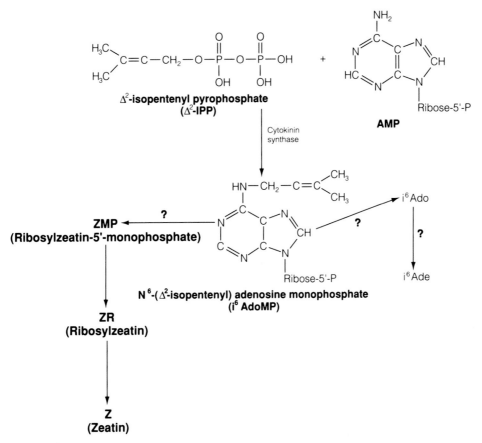

FIGURE 18.7. Scheme for the biosynthesis of some cytokinins. The key enzyme in cytokinin biosynthesis, cytokinin synthase, catalyzes the transfer of an isopentenyl group from isopentenyl pyrophosphate to the N^6 nitrogen of adenosine monophosphate (AMP). The product, i^6AdoMP, can be readily converted to the cytokinin zeatin or i^6Ade in many plant tissues, although the enzymes catalyzing these steps have not been characterized (Chen, 1982). The steps for enzymes that have not yet been identified are indicated by question marks.

give the cytokinin. As with the free cytokinins, Δ^2-isopentenyl groups are transferred to the adenine molecules from Δ^2-isopentenyl pyrophosphate by a prenyltransferase. This enzyme is unlikely to be the same transferase that was involved in the synthesis of the free cytokinins. The transferase that synthesizes the tRNA cytokinins must be able to recognize a specific base sequence in the tRNA and transfer the isopentenyl group to the adenosine nearest the 3′ end of the anticodon (Fig. 18.8). It is unlikely that it would also utilize free AMP as a substrate, but the enzyme has not been isolated and characterized.

Cytokinins Synthesized in the Root Are Transported to the Shoot in the Transpiration Stream

Root apical meristems are major sites of synthesis of the free cytokinins in whole plants. The cytokinins synthesized in roots appear to move through the xylem into the shoot. This movement has not been proved, but there is circumstantial evidence to support it. When the shoot is cut from a rooted plant near the soil line, the xylem sap may continue to flow from the cut stump for some time. This xylem exudate contains cytokinins. If the soil covering the roots is kept moist, the flow of xylem exudate can continue for several days. Since its cytokinin content does not diminish, the cytokinins found in the exudate may be synthesized by the roots. Also, environmental factors that interfere with root function, such as water stress, reduce the cytokinin content of the xylem exudate (Torrey, 1976). Using antibodies raised against zeatin riboside and isopentenyladenosine, it has been shown by immunocytochemical staining methods that cytokinins are concentrated in the apical meristems of both roots and shoots of tomato (Sossountzov et al., 1988). The staining pattern also showed a basipetally decreasing gradient of cytokinins along both the root and stem. This observation suggests that tomato shoots may produce much of their own cytokinins. Thus, the importance of root-derived cytokinins for shoot development remains to be elucidated.

Based on their cytokinin content and behavior in culture, embryos may be sites of cytokinin biosynthesis.

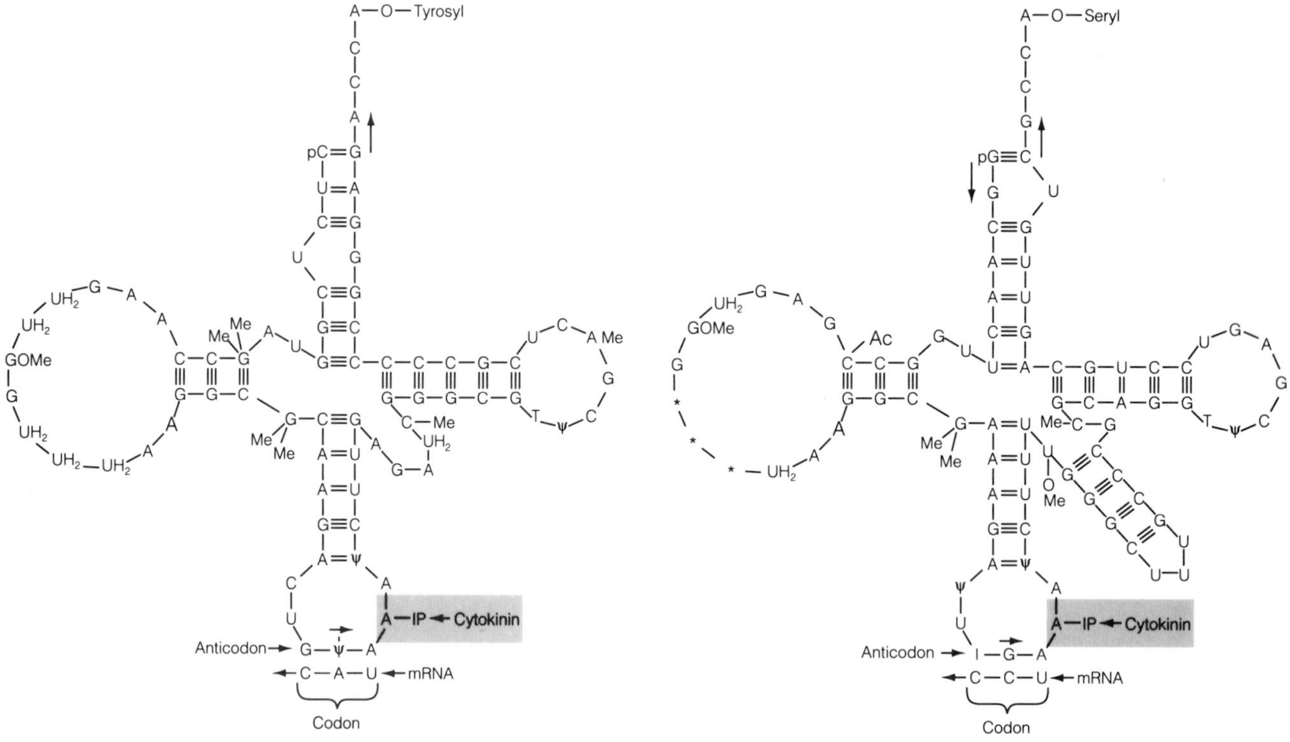

FIGURE 18.8. The structure of yeast tRNAs for tyrosine and serine, both of which contain a cytokinin. The cytokinin in either case is i⁶Ado, indicated as A-IP in the figure. Note that the cytokinin is the base next to the 3′ end of the anticodon, a position in which the side chain could be expected to play a role in codon-anticodon recognition (Hall, 1970).

Unlike the shoot apical meristem, embryos removed from the plant will continue to grow and develop normally when cultured in medium lacking hormones. It is not clear exactly where in embryonic development cytokinin autonomy begins, although it is probable that very young embryos lack the ability to synthesize cytokinins since this would explain the high cytokinin levels present in immature endosperm tissue.

Cultured Cells Can Acquire the Ability to Synthesize Cytokinins

When cultured normal callus tissues of many species have been subcultured repeatedly over a long period, they may become hormone autonomous; that is, they will grow on culture medium lacking an auxin or a cytokinin (Meins et al., 1980). This phenomenon is known as **habituation**. Most likely, a plant tissue becomes habituated because it acquires the capacity to synthesize its own hormones. Habituated callus tissues have been shown to contain cytokinins.

Frederick Meins and his colleagues in Basel, Switzerland, have shown that different tissues of tobacco plants exhibit different capacities to synthesize cytokinins in tissue culture. Tobacco pith tissue is cytokinin-dependent but capable of being habituated. In contrast, tissues cultured from the cortex of the stem are cytokinin-autonomous; they can be cultured indefinitely on medium containing auxin as the only hormone supplement. Finally, tobacco leaf tissue has an absolute requirement for cytokinins and cannot be habituated (Meins, 1989). These observations indicate that the ability to synthesize cytokinins can be switched on or off during development and that the on and off states can be stably maintained during cell divisions. Such developmental control of cytokinin synthesis may have important effects on morphogenesis.

Cytokinin Synthesis by Crown Gall Tissues Results from Genetic Transformation of Plant Cells

Bacteria-free tissues from crown gall tumors proliferate in culture in the absence of any hormone additions to the culture medium. Substantial amounts of free cytokinins, in addition to auxins, are found in these crown gall tissues. Furthermore, when radioactively labeled adenine is fed to periwinkle (*Vinca rosea*) crown gall tissues, it is incorporated into both zeatin and zeatin riboside, demonstrating that active cytokinin biosynthetic machinery is present in the gall tissue. Control stem tissue, which has not been associated with the crown gall organism, does not incorporate labeled adenine into cytokinins.

During *Agrobacterium tumefaciens* infection, plant cells incorporate some DNA from the bacteria into their chromosomes. The virulent strains of *Agrobacterium* contain a large plasmid known as the **Ti plasmid.** Plasmids are circular pieces of extrachromosomal DNA that are not essential for the life of the bacterium. However, plasmids frequently contain genes that enhance the ability of the bacterium to survive in special environments.

A small portion of the Ti plasmid, known as the **T-DNA**, is incorporated into the host plant cell's nuclear DNA (Fig. 18.9) (Chilton, 1983). The T-DNA contains genes necessary for the biosynthesis of *trans*-zeatin and auxin, as well as a member of a class of unusual nitrogen-containing compounds called **opines.** Opines are not synthesized by plants except after crown gall transformation. The T-DNA gene involved in cyto-kinin biosynthesis encodes a prenyltransferase that uses AMP as its substrate, similar to the cytokinin synthase that was identified in normal tobacco tissue. The T-DNA also contains two genes encoding enzymes that convert tryptophan to the auxin indoleacetic acid (IAA). The pathway of tryptophan conversion differs from that in untransformed cells and involves indoleacetamide as an intermediate. None of these T-DNA genes are expressed in the bacterium, but when they are inserted into the plant chromosomes the cells begin to produce zeatin, auxin, and an opine. The bacterium can utilize the opine as a nitrogen source, but higher plant cells cannot. Thus, by transforming the plant cells, the bacterium provides itself with an expanding environment (the gall tissue) in which the host cells are directed to produce a substance (the opine) that only the bacterium can utilize for its nutrition (Bomhoff et al., 1976).

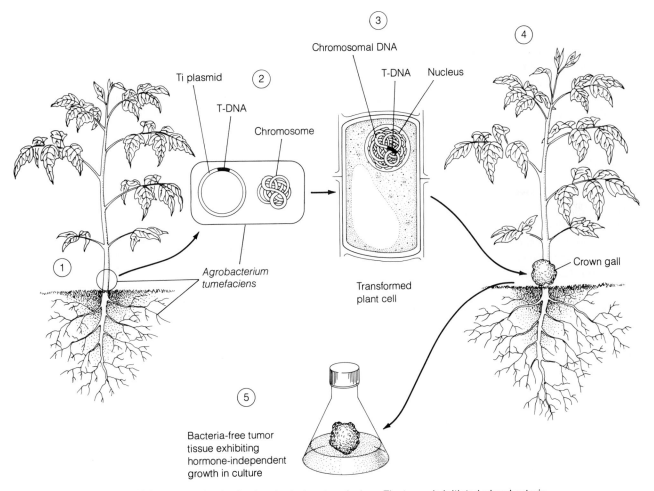

FIGURE 18.9. Tumor induction by *Agrobacterium tumefaciens*. The tumor is initiated when bacteria enter a lesion, which is usually near the crown of the plant (the junction of root and stem), and attach themselves to cells (1). A virulent bacterium carries, in addition to its chromosomal DNA, a Ti plasmid (2). The plasmid's T-DNA is introduced into a cell and becomes integrated into the cell's chromosomal DNA (3). Transformed cells proliferate to form a crown gall tumor (4). Tumor tissue can be "cured" of bacteria by incubation at 42°C. The bacteria-free tumor can be cultured indefinitely in the absence of hormones (5). (From Chilton, 1983.)

The Ti Plasmid and Plant Genetic Engineering

T HE TI PLASMID has been developed as a vehicle for introducing foreign genes into plants (Fig. 18.A). For this purpose it is necessary to disarm the plasmid so that it does not cause tumors. This has been accomplished by deleting the genes in the T-DNA that encode the enzymes controlling auxin and cytokinin synthesis. (These genes are referred to as phyto-oncogenes because they induce neoplastic, or tumor-producing, growth.) In addition, it is necessary to introduce a gene into the T-DNA that will enable the investigator to select the transformed cells. Genes for antibiotic resistance are normally used for this purpose (Matzke and Chilton, 1981; An et al., 1985). A cloned gene can then be inserted into the T-DNA of the engineered Ti plasmid, and the plasmid can be used to infect either cultured cells, leaf disks, or root slices. The infected cells are placed on a culture medium containing auxin and cytokinin (to induce growth) and the antibiotic. Only the transformed cells can grow in the presence of the antibiotic, because they have received the T-DNA containing not only the foreign gene but also the gene for antibiotic resistance. To obtain a plant containing the foreign gene, it is necessary to regenerate plants from the cultured, transformed cells. Fortunately, methods for accomplishing this have been developed for many plants, although not yet for all important crop species. The regeneration involves adjusting the cytokinin/auxin ratios to bring about both shoot and root formation. At present, both the transformation of the cereal grains with *Agrobacterium* and their regeneration from cultured cells are very difficult. Nevertheless, there have been some remarkable successes with this approach. Investigators have introduced a number of foreign genes into plants such as tobacco, including soybean storage protein genes and genes for herbicide resistance (see Chapter 2). In the future, genes for such desirable characteristics as disease resistance and salt tolerance may be transferred to crop plants by this technique. This approach also is very useful for the study of DNA sequences that are involved in the control of gene expression.

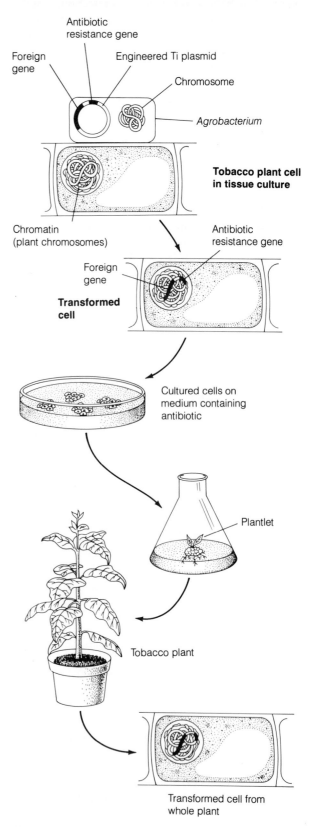

FIGURE 18.A. (From Chilton, 1983.)

There are important differences between the control of cytokinin biosynthesis in crown gall tissues and that in normal tissues, since the T-DNA genes for plant hormone synthesis are expressed even in cells in which the native plant genes for cytokinin biosynthesis are completely shut down. The regulation of the expression of the T-DNA gene encoding crown gall cytokinin synthase may be very different from that of the native higher plant gene for the analogous enzyme.

Cytokinin Nucleotides Are the Predominant Form in the Xylem

Cytokinins are passively transported from roots into the shoot through the xylem. They move through the plant with the transpiration stream, along with water and minerals taken up by the roots. Since the cytokinins present in the xylem exudate are predominantly nucleotides, it has been suggested that the nucleotides may be the form in which cytokinins are transported. Once they reach the leaves, the nucleotides may be converted to the free base or to glucosides. Cytokinin glucosides accumulate to high levels in leaves, and substantial amounts may be present even in senescing leaves. Although the glucosides are active as cytokinins in bioassays, they appear to lack hormonal activity after they are formed in leaf cells, possibly because they are compartmentalized in such a way that they are unavailable. Compartmentalization may explain the conflicting observations that cytokinins are transported readily by the xylem but that radioactive cytokinins given to leaves in intact plants do not appear to move from the site of application.

The different chemical forms of the cytokinins are rapidly metabolized by most plant tissues. Cytokinin bases, when given to many plant tissues, are converted to the respective nucleotides: zeatin to zeatin ribonucleotide, i^6Ade to i^6Ade ribonucleotide, and so forth. They also may be converted to their glucosides.

Many plant tissues have been found to contain the enzyme **cytokinin oxidase**, which converts zeatin, zeatin riboside, and i^6Ade to adenine or its derivatives. This enzyme may be responsible for the inactivation of the hormone, possibly as a mechanism for preventing its accumulation to toxic levels.

The Free Base Is Probably the Active Form of the Hormone

The active form of cytokinin is not known, although circumstantial evidence suggests that it is the free base. For example, excised radish cotyledons grow when they are cultured in a solution containing the synthetic cytokinin base benzylaminopurine. The cultured cotyledons readily take up the hormone and convert it to various BAP glucosides, BAP ribonucleoside, and BAP ribonucleotide. When the cotyledons are transferred back to a medium lacking a cytokinin, their growth rate declines, as do the concentrations of BAP, BAP ribonucleoside, and BAP ribonucleotide in the tissues. However, the level of the BAP glucosides remains constant. This suggests that the glucosides cannot be the active form of the hormone. In cultured tobacco cells, growth does not occur unless cytokinin ribosides supplied in the culture medium are converted to the free base. Thus, the free base is the most likely candidate for the active form of the hormone.

The Biological Role of Cytokinins

Plant hormones do not seem to work alone. Even in cases in which a response can be evoked by application of a single hormone, the tissue may contain additional endogenous hormones that contribute to the response. In some cases we know that a response is evoked by two or more hormones. However, until we have a way to block the action of the endogenous hormones, we cannot be certain that the actual regulation of the phenomenon being studied is as simple as it appears to be. Nevertheless, the cytokinins can evoke a variety of physiological, metabolic, biochemical, and developmental processes when they are applied to higher plants, and it is probable that they play an important role in the regulation of these events in the intact plant. Some of these responses were mentioned in connection with the bioassays used to detect cytokinins. Here we will outline some of the major biological roles of the cytokinins, but the reader should be aware that there are more.

The Cell Cycle in Plants Has Two Control Points

As discussed in Chapter 15, the cell cycle in eukaryotes is a complicated process that involves a number of distinct biochemical and cytological events, all of which are integrated in some manner. In a dividing cell population, such as an actively growing meristem, the average cell size remains constant. This means that, after it is formed, each cell approximately doubles its cytoplasmic mass, including all of its organelles, before it divides again. Most of the growth in cytoplasmic mass occurs in the G1 phase of the cell cycle, whereas the replication of the DNA and other components of the nucleus as well occurs in the S phase.

In the mammalian cell cycle there is a key regulatory point somewhere early in G1. At this point the cell becomes committed to the initiation of DNA synthesis. Once a cell has initiated DNA synthesis, it is irreversibly committed to completing the cell cycle through mitosis and cytokinesis. After the cell has completed mitosis, it may initiate another complete cycle (G1 through mitosis) or may leave the cell cycle and differentiate. This choice is made at the critical G1 point, before the cell begins to replicate its DNA. DNA replication and mitosis are deterministically linked in mammalian cells.

The regulation of the plant cell cycle appears to be more complex. Plant cells can leave the proliferative cycle either before or after they have replicated their DNA. This suggests that there are two control points in the plant cell cycle, one regulating the initiation of DNA replication and the other governing the initiation of mitosis. As a consequence, whereas most animal cells are diploid (having two sets of chromosomes), plant cells frequently become polyploid (having more than two sets of chromosomes) because they may go through one or more cycles of nuclear DNA replication without mitosis (Van't Hof, 1985).

Hormones May Regulate the Plant Cell Cycle

Earlier in this chapter we saw that cytokinin triggers cell proliferation in tissues that contain or are supplied with an optimal level of auxin. There is evidence that both hormones participate in the regulation of the cell cycle, but that auxin may regulate events leading to DNA replication while cytokinin regulates events leading to mitosis. Cultured tobacco tissue, when treated with auxin alone, may initiate DNA synthesis, but the cells do not divide unless they also are given a cytokinin. Similarly, in tissues treated with cytokinin alone, a few cells that had previously replicated their DNA will be stimulated to divide (Jouanneau and Tandeau de Marsac, 1973). Both hormones are necessary for the cells to remain in the cell cycle. These observations do not mean that auxin directly triggers DNA replication or that cytokinin directly initiates mitosis in G2-phase cells. We do not know enough about either the plant cell cycle or the mechanism

of hormone action to do more than note that these hormones permit the respective processes to occur. This caveat holds for all the other responses that are evoked by cytokinin as well. The state of our knowledge is such that we rarely can do more than make correlations between the application of the hormone and the appearance of a given response.

The Auxin/Cytokinin Ratio Regulates Morphogenesis in Cultured Tissues

Shortly after the discovery of kinetin, it was observed that cultured tobacco pith segments or tobacco callus tissues produce roots or shoots, depending on the ratio of auxin to cytokinin in the culture medium. High levels of auxin relative to kinetin stimulated the formation of roots, whereas high levels of the cytokinin relative to auxin led to the formation of shoots. At intermediate levels the tissue grew as an undifferentiated callus (Fig. 18.10) (Skoog and Miller, 1965).

More recently, investigators have used molecular genetic methods to investigate the significance of the auxin/cytokinin ratio in regulating morphogenesis in crown gall tissues. Mutations have been made in the T-DNA of the *Agrobacterium* Ti plasmid to observe the effects of different T-DNA genes on the growth and development of tumors. In some cases, bacteria containing Ti plasmids with mutated T-DNA genes induced the formation of a mass of proliferating shoots or roots instead of the usual tumors. These partially differentiated tumors are said to be **teratomas**. When the mutations occurred in one region of the T-DNA, the teratomas consisted of an abnormal proliferation of roots. Mutations in a different region of the T-DNA resulted in an abnormal proliferation of shoots. The first class of mutants were said to be "rooty," while the second were called "shooty." Subsequently it was proved that shooty

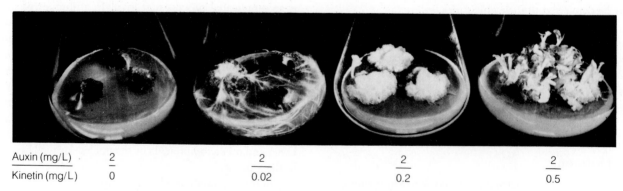

| Auxin (mg/L) | $\dfrac{2}{0}$ | $\dfrac{2}{0.02}$ | $\dfrac{2}{0.2}$ | $\dfrac{2}{0.5}$ |
| Kinetin (mg/L) | | | | |

FIGURE 18.10. Regulation of growth and organ formation in cultured tobacco callus by auxin and cytokinin. The tissues were cultured for approximately 1 month on media that contained an auxin (indoleacetic acid) at a concentration of 11.4 μM and a cytokinin (kinetin) at some concentration between 0 and 50 μM. At high ratios of auxin to cytokinin (A/C = 114) very little growth as callus tissue occurred, but roots were produced. Shoots were formed at intermediate ratios (A/C between 2 and 5), while the tissue grew only as unorganized callus at low ratios (A/C below 2). (From Skoog and Miller, 1965.)

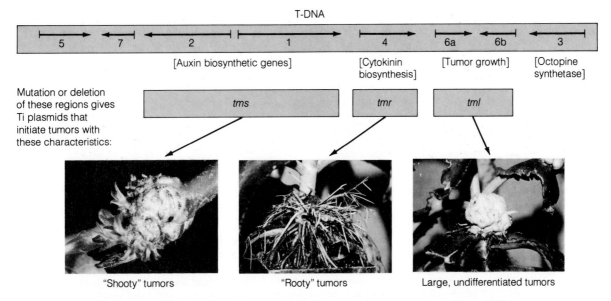

FIGURE 18.11. Map of the T-DNA from an *Agrobacterium* Ti plasmid, showing the effects of T-DNA mutations on crown gall tumor morphology. The shaded bar represents the T-DNA. The numbered arrows represent the locations of genes in the T-DNA that are expressed in plant cells transformed with the T-DNA. Genes 1 and 2 encode two different enzymes involved in auxin biosynthesis, while gene 4 encodes a cytokinin biosynthetic enzyme. Mutations or deletions that inactivate genes 1 and 2 (the *tms* loci) result in a T-DNA that will produce "shooty" tumors. Mutations that inactivate gene 4 (the *tmr* locus) result in a T-DNA that causes "rooty" tumors to form on infected plants. Mutations in gene 6 (the *tml* locus) cause the formation of large, undifferentiated tumors. Gene 3 encodes octopine synthase, the enzyme responsible for the formation of octopine by the transformed cells. Octopine is an opine, one of the unusual amino acids that are produced only by crown gall tissues as a result of their transformation by *Agrobacterium*-derived T-DNA. The bacteria are able to utilize these opines as an energy source. (From Morris, 1986.)

mutations had inactivated genes necessary for the synthesis of the auxin IAA. Similarly, the rooty mutants lost the function of the T-DNA gene necessary for the synthesis of zeatin (Fig. 18.11). Tissues transformed by rooty Ti plasmids contained high levels of auxin relative to cytokinin, whereas those transformed by shooty mutant plasmids had high levels of cytokinin and low levels of auxin (Akiyoshi et al., 1983). These findings further demonstrate the importance of the auxin/cytokinin ratio in regulating morphogenesis.

Cytokinins Delay Senescence and Stimulate Nutrient Mobilization

Leaves detached from the plant slowly lose chlorophyll, RNA, lipids, and protein, even if they are kept moist and provided with minerals. This programmed aging process, leading to death, is termed **senescence**. Leaf senescence is more rapid in the dark than in the light. Treating isolated leaves of many species with cytokinins will delay their senescence. Although cytokinins do not prevent senescence completely, their effects can be quite dramatic, particularly when the cytokinin is sprayed directly on the intact plant. If only one leaf is treated, it remains green after other leaves of similar developmental age have yellowed and dropped off the plant (Leopold and Kawase, 1964). Even a small spot on a leaf will remain

green if treated with a cytokinin, while the surrounding tissues on the same leaf begin to senesce.

In an effort to account for this phenomenon, investigators have studied the effect of cytokinin treatment on the movement of nutrients into treated leaves. Radioisotope-labeled nutrients (sugars, amino acids, etc.) are fed to the plants after one leaf or part of a leaf is treated with a cytokinin. By subjecting the leaves to autoradiography, the pattern of movement and the sites at which the nutrients accumulate can be determined. Experiments of this nature have demonstrated that nutrients are preferentially transported to and accumulate in the cytokinin-treated tissues. It has been postulated that the hormone causes nutrient mobilization by creating a new source-sink relationship. The metabolism of the treated area may be stimulated by the hormone so that nutrients move toward it. However, it is not necessary for the nutrient itself to be metabolized in the sink cells, since even nonmetabolizable substrate analogs are mobilized by cytokinins (Fig. 18.12).

Cytokinins Promote the Maturation of Chloroplasts

Although seeds can germinate in the dark, the morphology of dark-grown seedlings is very different from that of light-grown seedlings. Dark-grown seedlings are

Seedling A   Seedling B   Seedling C

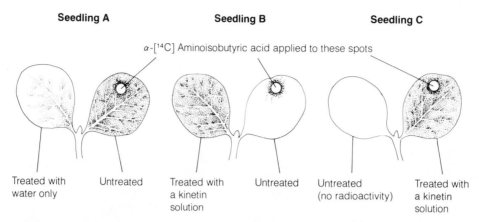

α-[¹⁴C] Aminoisobutyric acid applied to these spots

Treated with   Untreated   Treated with   Untreated   Untreated   Treated with
water only   a kinetin   (no radioactivity) a kinetin
                                          solution   solution

FIGURE 18.12. Illustration of the effect of cytokinin on the movement of an amino acid in cucumber seedlings. A radioactively labeled amino acid that cannot be metabolized, such as aminoisobutyric acid, was applied as a discrete spot on the right cotyledon of each of these seedlings. In seedling A, the left cotyledon was sprayed with water as a control. The left cotyledon of seedling B and the right cotyledon of seedling C were each sprayed with a solution containing 50 μM kinetin. Later, the location of the radioactivity was detected by autoradiography. The dark stippling represents the distribution of the radioactive amino acid as revealed by autoradiography. The results show that kinetin applied to the opposite cotyledon (seedling B) draws α-[¹⁴C]aminoisobutyric acid away from the site of application. The opposite cotyledon has become a nutrient sink. When the labeled cotyledon is treated with kinetin (seedling C), radioactivity is retained in the treated cotyledon. (Drawn from data obtained by K. Mothes.)

said to be **etiolated**. The internodes of etiolated seedlings are more elongated, leaves do not expand, and chloroplasts do not mature. Instead, the proplastids of dark-grown seedlings develop into **etioplasts**, in which the inner membranes form a compact, highly regular lattice known as the prolamellar body (Chapter 1). Etioplasts contain some carotenoids, which is why the etiolated seedlings appear yellow, but they do not synthesize chlorophyll or most of the enzymes and structural proteins required for the formation of the chloroplast thylakoid system and photosynthetic machinery. When seedlings are germinated in the light, chloroplasts mature directly from the proplastids present in the embryo, but etioplasts also can mature into chloroplasts when etiolated seedlings are illuminated. If the etiolated leaves are treated with cytokinin before they are illuminated, they form chloroplasts with more extensive grana. Also, chlorophyll and photosynthetic enzymes are synthesized at a greater rate after illumination (Lew and Tsuji, 1982). Cytokinin cannot promote these changes in the dark, but the stimulation of light-initiated chloroplast maturation by cytokinin is quite pronounced. These results suggest that cytokinins promote the synthesis of photosynthetic proteins.

Cytokinins Can Stimulate Cell Enlargement

Cytokinins can promote cell enlargement in certain tissues and organs. This effect is most clearly seen in dicots with leafy cotyledons, such as mustard, cucumber, and sunflower. The cotyledons of these species expand

as a result of cell enlargement during seedling growth. Cytokinin treatment promotes additional cell expansion with no increase in the dry weight of the treated cotyledons (Fig. 18.13). Cleon Ross and his colleagues at Colorado State University have shown that cytokinin stimulated the growth of the cotyledons by increasing the plasticity of the cell walls without changing their elasticity (Rayle et al., 1982). At noted in the discussion of auxins, the plasticity of the wall is a measure of its ability to be stretched permanently by turgor pressure (Chapter 16). Leafy cotyledons expand to a much greater extent when the seedlings are grown in the light instead of the dark. However, cytokinins promote the growth of cotyledons in both the light and the dark. Superficially, at least, this cytokinin-promoted growth appears to be similar to the auxin-induced promotion of stem cell elongation. However, cytokinin did not increase plasticity by acidification of the cell wall, as auxin appears to do (Ross and Rayle, 1982). In addition, neither auxin nor gibberellin promotes cotyledonary cell expansion.

These observations illustrate an important point about plant hormones: their effects are not universal. Instead, a given hormone elicits a specific response from a specific tissue in a particular species or group of species when the hormone is applied under carefully controlled conditions. In a different species the same response may be elicited only under different conditions, or it may be elicited by a different hormone. For example, we usually think of auxins and gibberellins as hormones that regulate stem elongation. However, cytokinins promote the elongation of young wheat coleoptiles and water-

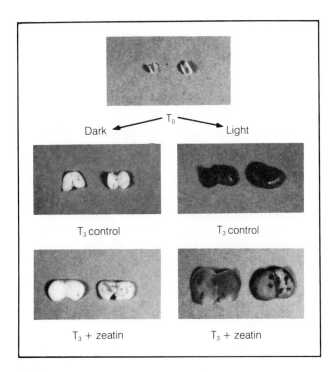

FIGURE 18.13. The effect of cytokinin on the expansion of radish cotyledons. The upper row shows germinating radish seedlings before the experiment began (T_0). The two cotyledons were removed from the seedlings and incubated for 3 days in a phosphate buffer that either lacked a cytokinin (−zeatin) or contained zeatin at 2.5 μM (+zeatin). The cotyledons treated with zeatin have expanded to a much greater extent than the controls. (From Huff and Ross, 1975.)

melon hypocotyls, in both cases by increasing cell elongation with only a minor effect on cell number. In both cases this effect was observed only when the cytokinin was applied under specific circumstances: only at an early stage of development in the wheat coleoptiles, and only when applied to the roots or to the shoot tip of intact watermelon seedlings. Most other studies of the effects of applied cytokinins on stem elongation have used isolated stem segments, in which cytokinins have a strong inhibitory effect on auxin-promoted cell elongation, as in the case of soybean hypocotyls.

The Mechanism of Cytokinin Action

As noted in earlier chapters, it is generally assumed that plant hormones interact with specific receptors that reside either on the cell surface or within the cytoplasm. In animal systems in which they have been characterized, hormone receptors are proteins to which the hormone binds with high affinity. Binding of a steroid hormone to its cytoplasmic receptor produces an active hormone-receptor complex that moves into the nucleus, where it binds preferentially to particular DNA sequences and leads to the expression of specific genes. As we discussed

in Chapter 16, the receptors for nonsteroid animal hormones, such as insulin, are found on the cell surface, usually as integral proteins in the plasma membrane. In this case, the binding of the hormone to its receptor may stimulate the formation in the cytoplasm of a second messenger that alters some aspect of cellular metabolism. The hormone-receptor complex need not enter the cell to trigger the formation of the second messenger and a biological response. These two models for animal hormone action are useful in designing experiments to examine plant hormone action, but so far there has been little direct evidence that the cytokinins act by a mechanism similar to either of these models.

Plant Cells Contain Cytokinin-Binding Proteins

Specific cytokinin binding sites have been identified in some plants, and cereal grains contain factors that bind cytokinins. In the case of a protein factor called CBF-1 (CBF = cytokinin-binding factor), this binding has the expected characteristics of hormone-receptor binding. The binding sites can be saturated by increasing concentrations of the hormone, cytokinins bind to CBF-1 with moderately high affinity, and the binding is specific for the cytokinins. The CBF-1 was obtained from salt washes of ribosomes, suggesting that it may be a ribosomal protein (Erion and Fox, 1981). Although no function was proposed for the CBF-1–cytokinin complex, the location of the binding factor suggests that the complex may regulate protein synthesis.

Cytokinins Regulate Protein Synthesis

There is good evidence that cytokinins play a role in regulating protein synthesis. Polyribosomes are the protein synthetic machinery. They consist of messenger RNA molecules to which numerous ribosomes are attached. Each ribosome is in a different stage of translating the mRNA into a polypeptide. When cultured soybean cells are treated with cytokinin, within 15 min there is an increase in the polyribosome content of the cells. This increase comes about as a result of the movement of free monoribosomes (monosomes) into polyribosomes (polysomes) (Fig. 18.14) (Tepfer and Fosket, 1978).

Cytokinin can not only increase the rate of protein synthesis but also change the spectrum of proteins made by plant tissues. This has been demonstrated by labeling the proteins synthesized by cultured soybean cells with [^{35}S]methionine and separating them by polyacrylamide gel electrophoresis. Autoradiograms of the gel show that cytokinin treatment enhances the synthesis of certain proteins while inhibiting the synthesis of others (Fosket and Tepfer, 1978). Tobacco tissues treated with cytokinin also show altered patterns of protein synthesis, suggesting that the hormone is necessary for the synthesis

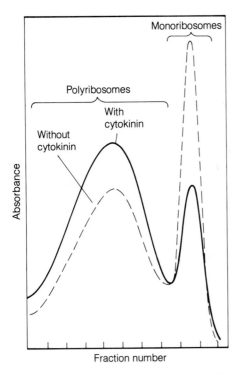

FIGURE 18.14. Cytokinin stimulation of polyribosome formation in cultured soybean tissues. Cultured soybean cells were grown with or without zeatin for 3 h. The cells were then homogenized and their monoribosomes and polyribosomes were separated by centrifugation through a sucrose gradient. The relative distribution of monoribosomes and polyribosomes is shown by the solid line for the cytokinin-treated cells and the broken line for the control. The cytokinin-treated cells have fewer monoribosomes and more polyribosomes than the control. The effect can be detected within 15 min. Since polyribosomes represent ribosomes actively synthesizing proteins, the results suggest that cytokinins stimulate total protein synthesis in these cells. (From Tepfer and Fosket, 1978.)

of some proteins and inhibitory to the synthesis of others (Eichholz et al., 1983). The specific nature of the proteins whose synthesis is enhanced or suppressed by cytokinin is not known in either of these tissues.

Cytokinins Affect a Posttranscriptional Step in *Lemna*

Elaine Tobin and her colleagues at the University of California–Los Angeles have studied the role of cytokinin in chloroplast maturation in the small aquatic duckweed *Lemna gibba*. *Lemna* can be grown heterotrophically in the dark for long periods of time by adding sucrose to the culture medium. When green *Lemna* plants are placed in the dark, the nuclear-encoded mRNAs for two important chloroplast proteins—the small subunit of ribulose-1,5-bisphosphate carboxylase (SSU) and the major chlorophyll *a/b*–binding polypeptide of light harvesting complex II (LHCP)—decrease in abundance over a week. As a result, the synthesis of SSU and LHCP drastically decreases. However, the decrease in abun-

dance of these mRNAs can be inhibited either by a pulse of dim red light or by the addition of cytokinin to the culture medium 24 h prior to harvest. The abundance of LHCP mRNA, for example, is about 5-fold greater in the cytokinin-treated plants than in the controls. The combination of cytokinin and red light caused a 9.5-fold increase relative to the controls.

How does cytokinin prevent the decrease in these two important mRNAs in the dark? To determine whether cytokinin treatment influences the LHCP mRNA levels by increasing the rate of transcription or by decreasing the rate of mRNA turnover, Flores and Tobin (1988) analyzed the transcripts produced by nuclei isolated from cytokinin-treated and control *Lemna* in nuclear run-off experiments (described in Chapter 20). Briefly, isolated nuclei are allowed to extend their transcripts (initiated in vivo) in the presence of ^{32}P-labeled nucleotides. The labeled RNA is then hybridized to filters containing cloned LHCP sequences. The amount of radioactivity bound corresponds to the amount of transcript synthesized. As shown in Figure 18.15, cytokinin caused only a 1.5-fold enhancement of transcription while a 1-min exposure to red light stimulated transcription 4-fold. When

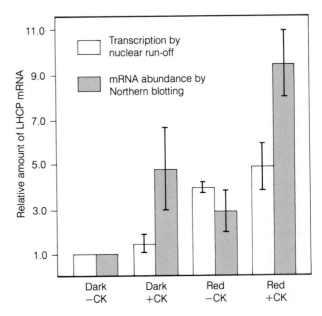

FIGURE 18.15. Comparison of the effects of a cytokinin (benzyladenine) and dim red light on the transcription versus total abundance of LHCP mRNA in *Lemna gibba*. Transcription was determined by nuclear run-off experiments with isolated nuclei. Total abundance was measured by hybridizing ^{32}P-labeled cDNA probes to total RNA on Northern blots. In the dark, cytokinin (CK) caused a 5-fold increase in the total mRNA but only a 1.5-fold increase in transcription. In contrast, red light alone caused a 4-fold enhancement of transcription. The results indicate that red light alters transcription, whereas cytokinin affects a posttranscriptional step. The error bars represent the standard error. LHCP = chlorophyll *a/b*–binding polypeptide of light harvesting complex II. (Data from Flores and Tobin, 1988.)

cytokinin was given along with red light, a 4.9-fold increase in transcription was observed. These results suggest that transcription of the gene is mainly under the control of the phytochrome system (Chapter 20). Cytokinin appears to increase the abundance of LHCP mRNA at a posttranscriptional level, possibly by making the mRNA more stable.

The Cytokinins in tRNA May Regulate Protein Synthesis

The relationship between the free cytokinins and the cytokinin bases in tRNA is unknown. There is no evidence that the free cytokinins act through a mechanism that involves the tRNAs. Indeed, the strongest argument against such a mode of action is that the cytokinins of tRNA are synthesized by the addition of a side chain to the polymerized base, rather than by the direct incorporation of free cytokinin. Nevertheless, the tRNA cytokinins could play an important role in protein synthesis. In the tRNAs that contain them, cytokinins occupy a critical position adjacent to the anticodon, where they may influence the binding of the tRNA to mRNA (Fig. 18.8).

There are different tRNA molecules for each amino acid, the number of tRNAs varying from two to six, depending on the amino acid. Each tRNA has a different anticodon, consisting of a three-base sequence comple-mentary to the three bases of the codon that specifies it. In the bacterium *Escherichia coli*, only the tRNAs whose anticodons recognize codons beginning with the letter U have been found to contain cytokinins (Fig. 18.16). The tRNAs that contain a cytokinin may require it for proper codon-anticodon recognition. Different proteins have quite different codon usages. One protein may use the codon CUU to specify leucine while another uses the codon UUA to specify the same amino acid. If the cytokinin-containing leucyl tRNA responding to the UUA codon was not present or was inactive because the adenosine base next to the codon had not been modified to make it a cytokinin, the second protein would be made and the first would not, even though the mRNAs for these two proteins had the same abundance. Thus, the tRNA processing step could play an important regulatory role in determining the nature of the proteins that are made by a cell.

Finally, although most of the free cytokinins in plants are thought to be synthesized as the free molecules, it is possible that turnover of tRNA can contribute to the total cytokinin activity of cells.

Cytokinins Regulate Calcium Concentration in the Cytosol

One of the many biological effects of the cytokinins is stimulation of bud formation from the protonemata of the moss *Funaria hygrometrica*. The protonemata grow

1st letter of codon	2nd Letter of codon				3rd letter of codon
	U	C	A	G	
U	PHE	SER	TYR	CYS	U
	PHE	SER	TYR	CYS	C
	LEU	SER	C.T.	C.T.	A
	LEU	SER	C.T.	TRY	G
C	LEU	PRO	HIS	ARG	U
	LEU	PRO	HIS	ARG	C
	LEU	PRO	GLN	ARG	A
	LEU	PRO	GLN	ARG	G
A	ILEU	THR	ASN	SER	U
	ILEU	THR	ASN	SER	C
	ILEU	THR	LYS	ARG	A
	MET	THR	LYS	ARG	G
G	VAL	ALA	ASP	GLY	U
	VAL	ALA	ASP	GLY	C
	VAL	ALA	GLU	GLY	A
	VAL	ALA	GLU	GLY	G

FIGURE 18.16. The genetic code. The shaded boxes have corresponding tRNAs that contain a hypermodified base with cytokinin activity at the 3′ end of their anticodons in the bacterium *Escherichia coli*. C.T. = codon terminator. (Skoog and Armstrong, 1970).

at their tips to form a one-cell-thick filament of cells. Bud formation begins with an asymmetric division of a target cell several cells back from the tip (see Chapter 15). This step is initiated by cytokinin-binding to an unidentified receptor within the target cell. The new bud begins as protuberance at the basal end of the cell, and its further development requires the continuous presence of cytokinin. Secretory vesicles accumulate at the site of the new bud.

Cytokinin is ineffective in inducing moss bud formation in calcium-free medium, however, and the calcium ionophore A23187 can substitute for cytokinin to initiate bud development (Saunders and Hepler, 1982). Ionophores are molecules that insert into membranes, opening pores through which specific ions may freely pass, either alone or in exchange for another ion. Most cells maintain a cytosolic calcium concentration between 0.1 and 1.0 μM by actively pumping calcium outside the cell or into the vacuole. The calcium concentrations of the vacuole and the cell wall are on the order of 1000 times greater than that of the cytosol. The calcium ionophore A23187 increases the cytosolic calcium concentration by exchanging external Ca^{2+} for internal hydrogen ions.

These experiments with the moss *Funaria* indicate that cytokinin may act, at least in part, by increasing the cytosolic concentration of calcium ions by promoting calcium uptake from the medium (since external Ca^{2+} is required). How, then, does a change in the concentration of cytosolic calcium bring about bud initiation? As we discussed in Chapter 15, calcium can affect the cytoskeleton, which can regulate exocytosis. In addition, calcium ions have been shown to act through the regulatory protein calmodulin (Chapter 16). Each calmodulin has four high-affinity calcium binding sites. Calmodulin alone is inactive as an regulator, but the calmodulin-calcium complex can bind to and activate a number of enzymes including protein kinases—enzymes that add phosphorus to the serine or tyrosine hydroxyl groups of proteins. The phosphorylation of enzymes can change their activity. Thus, the calcium-calmodulin complex would be in a position to act as a master switch regulating alternative metabolic pathways within the cell.

The finding that calcium ions play a role in cytokinin-induced responses does not provide us with the mechanism of action of the hormone. For example, buds initiated by treatment with calcium and A23187 fail to develop normally in the absence of cytokinin. The results do demonstrate that calcium ions may act as a second messenger, transforming the hormonal signal into a biochemical switch regulating the initial stages of bud formation. Further development of the bud almost certainly involves changes in protein synthesis, possibly via cytokinin-induced alterations of transcription, mRNA stability, or translation.

Summary

Mature plant cells generally do not divide in the intact plant, but they can be stimulated to divide by wounding, infection with certain bacteria, and plant hormones. Cytokinins are N^6-substituted aminopurines that will initiate cell proliferation in mature tobacco pith tissue when it is cultured on a medium that also contains an auxin. The principal cytokinin of higher plants, zeatin, is 16-(4-hydroxy-3-methylbut-*trans*-2-enylamino)purine. Zeatin is also present in plants as a riboside or ribotide and as glucosides. These forms are also active as cytokinins in bioassays, possibly through their conversion to the free zeatin base by plant tissue.

Cytokinins may be detected in plant extracts by a number of bioassays, but in recent years sophisticated chemical methods for cytokinin identification have been developed, including immunochemical methods in which antibodies to cytokinins are used to detect and quantitate cytokinins in plant extracts.

A prenyltransferase enzyme, cytokinin synthase, catalyzes the transfer of the isopentenyl group from Δ^2-isopentenyl pyrophosphate to the N^6 nitrogen of adenosine monophosphate. The product of this reaction is readily converted to zeatin and other cytokinins. Cytokinin synthesis occurs in roots, in developing embryos, and in crown gall tissues. Crown galls originate from plant tissues that have been infected with a virulent strain of the Ti plasmid–bearing bacterium *Agrobacterium tumefaciens*. The bacterium injects a specific region of the plasmid known as the T-DNA into wounded plant cells, and the T-DNA is incorporated into the host nuclear genome. The T-DNA contains a gene for cytokinin biosynthesis, as well as genes for auxin biosynthesis. These phyto-oncogenes are expressed in the plant cells, leading to hormone synthesis and unregulated proliferation of the cells to form the gall.

Cytokinins are concentrated in the young, rapidly dividing cells of the shoot and root apical meristems. They do not appear to be actively transported through living plant tissues. Instead, they are transported passively into the shoot from the root through the xylem, along with water and minerals. Cytokinins participate in the regulation of many plant processes, including cell division, morphogenesis of shoots and roots, chloroplast maturation, cell enlargement, and senescence. These responses are not universal, however. Instead, cytokinin will affect a given response in a particular tissue at a particular stage of development.

The mechanism of action of cytokinin is unknown. Cytokinin has a profound effect on the rate of protein synthesis and on the kinds of proteins made by plant cells. In particular, cytokinins stimulate the synthesis of specific chloroplast proteins that are encoded by nuclear genes and synthesized by cytoplasmic ribosomes. They appear to do this by stabilizing specific mRNAs and by slowing their degradation. In some systems, the initial response to cytokinins may be mediated by an increase in the cytosolic calcium concentration. Thus, like other hormones, cytokinin probably acts through a receptor and a signal transduction pathway.

GENERAL READING

Chilton, M.-D. 1983. A vector for introducing new genes into plants. *Sci. Am.* 248:50–59.

Letham, D. S., and Palni, L. M. S. (1983) The biosynthesis and metabolsim of cytokinins. *Annu. Rev. Plant Physiol.* 34:163–197.

Morris, R. O. (1986) Genes specifying auxin and cytokinin biosynthesis in phytopathogens. *Annu. Rev. Plant Physiol.* 37:509–538.

Nester, E. W., Gordon, M. P., Amasino, R. M., and Yanofsky, M. F. (1984) Crown gall: A molecular and physiological analysis. *Annu. Rev. Plant Physiol.* 35:387–413.

Skoog, F., and Armstrong, D. J. (1970) Cytokinins. *Annu. Rev. Plant Physiol.* 21:359–384.

Skoog, F., and Miller, C. O. (1965) Chemical regulation of growth and organ formation in plant tissues cultured *in vitro*. In: *Molecular and Cellular Aspects of Development*, Bell, E., ed., pp. 481–494. Harper & Row, New York.

Torrey, J. G. (1976) Root hormones and plant growth. *Annu. Rev. Plant Physiol.* 27:435–459.

Van't Hof, J. (1985) Control points within the cell cycle. In: *The Cell Division Cycle in Plants*, Bryant, J. A., and Francis, D., eds, pp. 1–13. Cambridge University Press, London.

CHAPTER REFERENCES

Akiyoshi, D. E., Morris, R. O., Hinz, R., Mischke, B. S., Kosuge, T., Garfinkel, D. J., Gordon, M. P., and Nester, E. W. (1983) Cytokinin/auxin balance in crown gall tumors is regulated by specific loci in the T-DNA. *Proc. Natl. Acad. Sci. USA* 80:407–411.

An, G., Watson, B. D., Stachel, S., Gordon, M. P., and Nester, E. W. (1985). New cloning vehicles for transformation of higher plants. *EMBO J.* 4:277–284.

Bomhoff, G., Klapwijk, P. M., Kester, H. C. M., and Schilperoort, R. A. (1976) Octopine and nopaline synthesis and breakdown genetically controlled by a plasmid of *Agrobacterium tumefaciens*. *Mol. Gen. Genet.* 145:177–181.

Brandon, D. L., and Corse, J. W. (1987) Monoclonal antibody–based approach to distinguish classes of cytokinins. In: *Molecular Biology of Plant Growth Control*, Fox, J. E., and Jacobs, M., eds., pp. 209–217. Alan R. Liss, New York.

Braun, A. C. (1958) A physiological basis for autonomous growth of the crown-gall tumor cell. *Proc. Natl. Acad. Sci. USA* 44:344–349.

Caplin, S. M., and Steward, F. C. (1948) Effect of coconut milk on the growth of the explants from carrot root. *Science* 108:655–657.

Chen, C.-M. (1982) Cytokinin biosynthesis in cell free systems. In: *Plant Growth Substances 1982*, Wareing, P. F., ed., pp. 155–163. Academic Press, New York.

Eichholz, R., Harper, J., Felix, G., and Meins, F., Jr. (1983) Evidence for an abundant 33,000-dalton polypeptide regulated by cytokinin in cultureds tobacco tissues. *Planta* 158:410–415.

Erion, J. L., and Fox, J. E. (1981) Purification and properties of a protein which binds cytokinin-active 6-substituted purines. *Plant Physiol.* 67:156–162.

Esashi, T., and Leopold, A. C. (1969) Cotyledon expansion as a bioassay for cytokinins. *Plant Physiol.* 44:618–620.

Flores, S., and Tobin, E. M. (1988). Cytokinin modulation of LHCP mRNA levels: The involvement of post-transcriptional regulation. *Plant Mol. Biol.* 11:409–415.

Fosket, D. E., and Tepfer, D. A. (1978) Hormonal regulation of growth in cultured plant cells. *In Vitro* 14:63–75.

Hall, R. H. (1967). Cytokinins in the transfer-RNA: Their significance to the structure of t-RNA. In: *Biochemistry and Physiology of Plant Growth Substances*, Wightman, F., and Setterfield, G., eds., pp. 47–56. Runge Press, Ottawa.

Hansen, C. E., Wenzler, H., and Meins, F., Jr. (1984) Concentration gradients of *trans*-zeatin riboside and *trans*-zeatin in the maize stem. *Plant Physiol.* 75:959–963.

Huff, A. K., and Ross, C. W. (1975) Promotion of radish cotyledon enlargement and reducing sugar content by zeatin and red light. *Plant Physiol.* 56:429–433.

Jouanneau, J. P., and Tandeau de Marsac, N. (1973) Stepwise effects of cytokinin activity and DNA synthesis upon mitotic cycle events in partially synchronized tobacco cells. *Exp. Cell Res.* 77:167–174.

Leonard, N. J., Hecht, S. M., Skoog, F., and Schmitz, R. Y. (1968) Cytokinins: Synthesis of 6-(3-methyl-3-butenylamino)-9-β-D-ribofuranosylpurine (3iPA), and the effect of side-chain unsaturation on the biological activity of isopentenylaminopurines and their ribosides. *Proc. Natl. Acad. Sci. USA* 59:15–21.

Leopold, A. C., and Kawase, M. (1964) Benzyladenine effects on bean leaf growth and senescence. *Am. J. Bot.* 51:294–298.

Letham, D. S. (1973) Cytokinins from *Zea mays*. *Phytochemistry* 12:2445–2455.

Letham, D. S. (1974) Regulators of cell division in plant tissues. XX. The cytokinins of coconut milk. *Physiol. Plant.* 32:66–70.

Lew, R., and Tsuji, H. (1982) Effect of benzyladenine treatment duration on delta-aminolevulinic acid accumulation in the dark, chlorophyll lag phase abolition, and long-term chlorophyll production in excised cotyledons of dark-grown cucumber seedlings. *Plant Physiol.* 69:663–667.

MacDonald, E. M. S., and Morris, R. O. (1985) Isolation of cytokinin by immunoaffinity chromatography and analysis by high-performance liquid chromatography radioimmunoassay. *Methods Enzymol.* 110:347–358.

Matzke, A. J. M., and Chilton, M.-D. (1981) Site-specific insertion of genes into T-DNA of the *Agrobacterium* tumor-inducing plasmid: An approach to genetic engineering of higher plants. *J. Mol. Appl. Genet.* 1:39–49.

Meins, F., Jr. (1989) A biochemical switch model for cell-heritable variation in cytokinin requirement. In: *Molecular Basis of Plant Development*, Goldberg, R., ed., pp. 13–24. Alan R. Liss, New York.

Meins, F., Jr., Lutz, J., and Binns, A. N. (1980) Variation in the competence of tobacco pith cells for cytokinin-habituation in culture. *Differentiation* 16:71–75.

Miller, C. O., Skoog, F., von Saltza, M. H., and Strong, F. M. (1955) Kinetin, a cell division factor from deoxyribonucleic acid. *J. Am. Chem. Soc.* 77:1392.

Mok, M. C., Kin, S.-G., Armstrong, D. J., and Mok, D. W. (1979) Induction of cytokinin autonomy by *N,N'*-diphenylurea in tissue cultures of *Phaseolus lunatus* L. *Proc. Natl. Acad. Sci. USA* 76:3880–3884.

Morel, G. (1970) Le problème de la transformation tumorale chez les végètaux. *Physiol. Veg.* 8:189–204.

Murashige, T., and Skoog, F. (1962) A revised medium for rapid growth and bioassays with tobacco tissue cultures. *Physiol. Plant.* 15:473–497.

Rayle, D. L., Ross, C. W., Robinson, N. (1982) Estimation of osmotic parameters accompanying zeatin-induced growth of detached cucumber cotyledons. *Plant Physiol.* 70:1634–1636.

Ross, C. W., and Rayle, D. L. (1982) Evaluation of H^+ secretion relative to zeatin-induced growth of detached cucumber cotyledons. *Plant Physiol.* 70:1470–1474.

Saunders, M. J., and Hepler, P. K. (1982) Calcium ionophore A23187 stimulates cytokinin-like mitosis in *Funaria*. *Science* 217:943–945.

Sossountzov, L., Maldiney, R., Sotta, B., Sabbagh, I., Habricot, Y., Bonnet, M., and Miginiac, E. (1988) Immunocytochemical localization of cytokinins in Craigella tomato and a sideshootless mutant. *Planta* 175:291–304.

Tepfer, D. A., and Fosket, D. E. (1978) Hormone-mediated translational control of protein synthesis in cultured cells of *Glycine max. Dev. Biol.* 62:486–497.

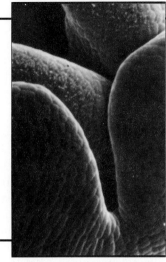

CHAPTER 19

Ethylene and Abscisic Acid

ISCUSSIONS FROM PREVIOUS CHAPTERS have shown that plant hormones can be classified into two broad groups: growth stimulators, including auxins, gibberellins, and cytokinins, which are involved in processes such as cell division, cell elongation, organ initiation, and differentiation; and growth inhibitors or antagonists to the known growth stimulators, including abscisic acid (ABA) and ethylene.

ABA and ethylene control processes that are characteristic of the last stages of plant development, such as senescence, abscission, flower fading, and fruit ripening. In addition, both hormones control growth rates under unfavorable environmental conditions (Chapter 14) by inhibiting cellular processes such as growth, protein synthesis, and ion transport. Growth is markedly altered when tissues are exposed to exogenous ABA or ethylene. Moreover, many studies have shown that the endogenous levels of these plant hormones increase during senescence, abscission, and fruit ripening as well as during environmental stress. ABA, which is known as the plant stress hormone, regulates growth, stomatal closure, protein synthesis, and other biochemical processes, mainly under drought conditions (Chapter 14). Endogenous levels of ethylene, on the other hand, increase not only under water stress but also in response to mechanical tissue wounding, air pollution, pathogens, insect damage, flooding, and anaerobic conditions.

Ethylene

During the nineteenth century, when coal gas was used for street illumination, it was observed that trees in the vicinity of street lamps defoliated more extensively than other trees. Eventually it became apparent that coal gas and air pollutants brought about plant tissue damage, and ethylene was identified as the active component of coal gas by a Russian student in 1901. As a graduate student at the Botanical Institute of St. Petersburg, Dimitry N. Neljubow observed that dark-grown pea seedlings in the laboratory showed reduced stem elongation, increased lateral growth (swelling), and abnormal horizontal growth (negative gravitropism)—conditions that were later termed the *triple response*. When the "laboratory air" was removed and the plants were allowed to grow in fresh air, they regained their normal rate of growth. Ethylene, which was present in the laboratory air, was identified as the molecule causing the response.

The first indication that ethylene is a natural product of plant tissues was reported by H. H. Cousins in 1910. Cousins noticed that when oranges were packed and shipped in the same container with bananas, the bananas ripened prematurely. In 1934, R. Gane identified ethylene chemically as a natural product of plant metabolism, and because of its effects on the plant it was classified as a hormone.

For 25 years ethylene was not recognized as an important plant hormone, mainly because many physiologists believed that the effects of ethylene could be mediated by auxin, the first plant hormone to be discovered (Chapter 16). It was thus thought that auxin was the main plant hormone and that ethylene had only an insignificant and indirect physiological role. Work on ethylene was also hampered by the lack of chemical techniques for its quantitation. However, after gas chromatography was introduced in ethylene research in 1959, ethylene was rediscovered and its physiological significance as a plant growth regulator was recognized (Burg and Thimann, 1959).

The Properties of Ethylene Are Deceptively Simple

Ethylene is the simplest known olefin (its molecular weight is 28) and is lighter than air under physiological conditions.

Ethylene

It is flammable and readily undergoes oxidation. Ethylene can be oxidized to ethylene oxide,

Ethylene oxide   Ethylene glycol

and ethylene oxide can be hydrolyzed to ethylene glycol. In most plant tissue, ethylene can also be completely oxidized to CO_2 (Beyer, 1979).

Ethylene   Ethylene oxide   Oxalic acid   Carbon dioxide

Ethylene is released easily from the tissue and diffuses in the gas phase through the intercellular spaces and outside the tissue. At an ethylene concentration of $1\ \mu l\ L^{-1}$ in the gas phase at 25°C, the concentration of ethylene in water is 4.4×10^{-9} M. Since the ethylene gas is easily lost from the tissue and may affect other tis-

sues or organs, ethylene-trapping systems are used during the storage of fruits, vegetables, and flowers. $KMnO_4$ is an effective absorbent of ethylene and can reduce its concentration in apple storage areas from 250 to $10\ \mu l\ L^{-1}$, markedly extending the storage life of the fruit.

Bacteria, Fungi, and Plant Organs Produce Ethylene

Even away from cities and industrial air pollutants, the environment is seldom free of ethylene contamination from biological sources. The highest levels of ethylene production occur in senescing tissues or ripening fruits (> 1.0 nl per gram fresh weight per hour), but all organs of higher plants can synthesize ethylene. Ethylene is biologically active at very low concentrations—less than 1 ppm ($1\ \mu l\ L^{-1}$). The internal ethylene concentration in a ripe apple has been reported to be as high as $2500\ \mu l\ L^{-1}$. Young developing leaves produce more ethylene than do fully expanded leaves. In bean (*Phaseolus vulgaris*), young leaves produce 0.4 nl $g^{-1}\ h^{-1}$, compared with 0.04 nl $g^{-1}\ h^{-1}$ for older leaves. With few exceptions, nonsenescent tissues that are wounded or mechanically perturbed will temporarily increase their ethylene production by severalfold within 25–30 min. Ethylene levels later return to normal. Gymnosperms and lower plants, including ferns, mosses, liverworts, and certain cyanobacteria, have all shown ability to produce ethylene when tested. Ethylene production by fungi and bacteria contributes appreciably to the ethylene content of soil. In culture, the capacity of microorganisms to synthesize ethylene appears to depend on the nature of the medium on which they grow. Certain strains of the common enteric bacterium *Escherichia coli* and of yeast produce large amounts of ethylene from methionine. There is no compelling evidence that healthy mammalian tissues produce ethylene, nor does it appear to be a metabolic product of invertebrates.

Biosynthesis and Catabolism Determine the Physiological Activity of Ethylene

In higher plants, the amino acid **methionine** is the precursor of ethylene (Fig. 19.1). Methionine is converted to ethylene in a series of reactions:

Methionine → S-adenosylmethionine (SAM) →

1-aminocyclopropane-1-carboxylic acid (ACC) → C_2H_4

The detailed biosynthetic pathway of ethylene became known only in 1979. Previous efforts had been unsuccessful because of the lack of a tissue homogenate or a cell-free system capable of producing ethylene. An additional difficulty was the fact that ethylene can be synthesized from a variety of compounds known to be present in plant

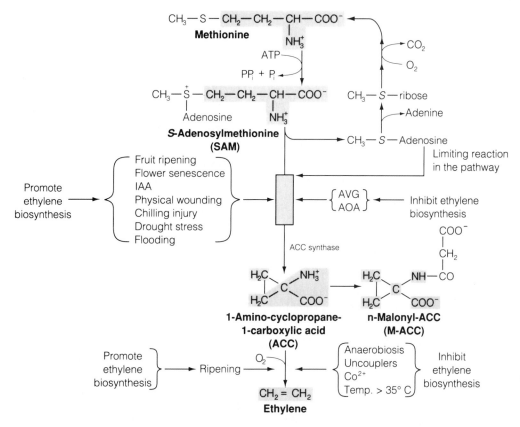

FIGURE 19.1. Ethylene biosynthesis and its regulation. The amino acid methionine is the precursor of ethylene. The rate-limiting step of the pathway is the conversion of *S*-adenosylmethionine (SAM) to the immediate precursor of ethylene, 1-aminocyclopropane-1-carboxylic acid (ACC), catalyzed by ACC synthase. The CH_3—S group of the methionine is recycled and thus conserved for continued synthesis of methionine. Besides its conversion to ethylene, ACC can also be conjugated to form *N*-malonyl ACC (M-ACC). Other abbreviations: AVG, aminoethoxyvinylglycine; AOA, aminooxyacetic acid. (From Beyer et al., 1984.)

tissues, so there were many candidates for the natural precursors of ethylene.

The breakthrough in the elucidation of ethylene biosynthesis occurred when it was shown that methionine could be a substrate for ethylene production in a cell-free tissue system (Lieberman and Mapson, 1964). In vivo experiments by M. Lieberman and his co-workers at the Agricultural Research Service in Beltsville, Maryland, showed that various plant tissues can convert L-[^{14}C]methionine to [^{14}C]ethylene and that the ethylene is derived from carbons 3 and 4 of methionine. Other experiments showed that the S group of methionine is recycled in the tissue. Without this recycling, the amount of reduced sulfur present would limit the available methionine and the synthesis of ethylene (Burg and Clagett, 1967). Subsequent work showed that *S*-adenosylmethionine (SAM), which is synthesized from methionine and ATP, is an intermediate in the ethylene biosynthetic pathway.

Fourteen years after it was discovered that methionine is a precursor of ethylene in higher plants, the final step in the pathway became clear. The immediate precursor of ethylene was found to be **1-aminocyclopropane-1-carboxylic acid (ACC)** (Adams and Yang, 1979). The role of ACC became evident in experiments in which plants were treated with ^{14}C-labeled methionine. Under anaerobic conditions, ethylene was not produced from the [^{14}C]methionine and labeled ACC accumulated in the tissue, but on exposure to oxygen a surge of ethylene production took place. The labeled ACC was rapidly converted to ethylene by various plant tissues, suggesting that ACC is the immediate precursor of ethylene in higher plants (Fig. 19.1).

When ACC is supplied exogenously to plant tissue that normally produces very little ethylene, a substantial increase in ethylene production occurs. This observation indicates that the synthesis of ACC is usually the metabolic step limiting ethylene production in plant tissues.

ACC synthase, the enzyme that catalyzes the conversion of SAM to ACC, (Fig. 19.1) has been characterized in many types of tissues of various plants (Kende and Boller, 1981). Its activity is regulated by several environmental and internal factors. In most cases, the ACC synthase reaction has been found to be the enzymatic step limiting ethylene production.

Because ACC synthase is present in such low amounts in plant tissues (0.0001% of the total protein of ripe tomato), it has been difficult to purify the enzyme for biochemical analysis. This obstacle was recently overcome using advanced techniques of molecular biology. Partially purified ACC synthase from zucchini was used to produce antibodies against the enzyme, and the antibodies were used to isolate the gene coding for the enzyme (Sato and Theologis, 1989). The isolated gene was later expressed in the bacterium *E. coli*, making it possible to isolate purified ACC synthase in large amounts.

The last step in ethylene biosynthesis—the conversion of ACC to ethylene—has several characteristics typical of enzyme-catalyzed reactions, but the enzyme involved, called **ethylene forming enzyme** (EFE), has yet to be isolated. Despite this limitation, the conversion of ACC to ethylene has been intensively investigated in recent years (Yang, 1987).

Methionine is found at quite low, nearly constant concentrations in plant tissues, including those that produce large amounts of ethylene, such as ripening fruits. Since methionine is the sole precursor of ethylene in higher plants, tissues with high rates of ethylene production require a continuous supply of methionine. This supply is ensured by methionine recycling, as shown in Figure 19.1.

Not all the ACC found in the tissue is converted to ethylene. ACC can also be converted to a nonvolatile compound, *N*-malonyl ACC (Fig. 19.1) (Amhrein et al., 1982; Hoffman et al., 1982), which does not break down and seems to accumulate in the tissue. Since *N*-malonyl ACC cannot be converted to ethylene, its formation is thought to play an important role in the control of ethylene biosynthesis, mainly by preventing overproduction of ethylene.

The catabolism of ethylene has been studied by supplying $^{14}C_2H_4$ to plant tissues and tracing the radioactive compounds. Carbon dioxide, ethylene oxide, ethylene glycol, and the glucose conjugate of ethylene glycol have been identified as metabolic breakdown products. Possible roles for ethylene catabolism include a reduction of ethylene activity in tissue, oxidation of ethylene at the receptor or binding site as a requirement for ethylene function, and an increase in tissue sensitivity to ethylene caused by products of ethylene oxidation (Beyer et al., 1984).

Environmental Stresses and Auxins Promote Ethylene Biosynthesis

Ethylene biosynthesis is stimulated by several factors including developmental state, environmental conditions, other plant hormones, and physical and chemical injury.

Fruit ripening is one of the best-studied developmental processes regulated by ethylene. Levels of ethylene and ACC in the tissue, of ethylene biosynthesis, and of EFE activity all increase as the fruits mature (Fig. 19.2a and b). Application of ACC to unripe fruits only slightly enhances ethylene production, indicating that an increase in the activity of EFE is a critical step in ripening (Yang, 1987).

Stress-Induced Ethylene Production. Ethylene biosynthesis is increased by stress conditions such as drought, flooding, chilling, or mechanical wounding (Chapter 14). In all these cases ethylene is produced by the usual biosynthetic pathway, the most important reaction being the conversion of SAM to ACC. This "stress ethylene" is involved in the onset of stress responses such as abscission, senescence, wound healing, and increased disease resistance (Chapter 14).

Auxin-Induced Ethylene Production. In some instances, auxins and ethylene can cause similar plant responses, such as induction of flowering in pineapple and inhibition of stem elongation (Abeles, 1973). These responses might be due to the ability of auxins to promote ethylene synthesis by enhancing the conversion of SAM to ACC. These observations suggest that some responses previously attributed to auxin are in fact mediated by the ethylene produced in response to auxin. Inhibitors of protein synthesis block both ACC synthesis and IAA-induced ethylene synthesis, indicating that the synthesis of ACC synthase caused by auxins brings about the marked increase in ethylene production (Imaseki et al., 1982; Yang et al., 1982). Recently, the levels of mRNA encoding ACC synthase of zucchini (*Cucurbita pepo*) fruits were measured using recombinant DNA techniques (Sato and Theologis, 1989). The amount of ACC synthase mRNA in the fruits increased in response to both wounding and IAA treatment. Whether this increase in ACC synthase mRNA is caused by an enhanced rate of transcription or a decrease in turnover rates remains to be determined.

Both Ethylene Production and Ethylene Action Can Be Blocked by Specific Inhibitors

Inhibitors of hormone synthesis or action are valuable for the study of the biosynthetic pathways and physiological roles of hormones. Inhibitors are particularly helpful when it is difficult to distinguish between different hormones that have identical effects in plant tissue or when a hormone affects the synthesis or the action of another hormone. Ethylene mimics high concentrations of auxins by causing stem growth inhibition and epinasty (a downward curvature of leaves). Use of specific inhib-

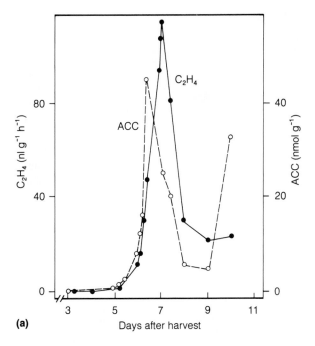

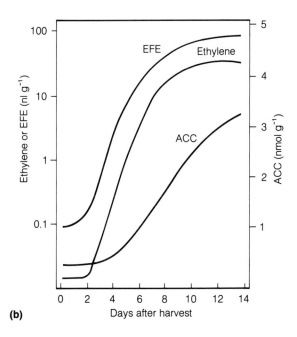

FIGURE 19.2. Changes in ethylene and ACC contents and ethylene forming enzyme (EFE) activity during ripening. (a) Simultaneous changes in ethylene and ACC contents in avocado. The avocado fruits do not ripen as long as they are attached to the tree. Changes in ACC and ethylene amounts were monitored after harvesting. (From Hoffman and Yang, 1980.) (b) Changes in the ACC, EFE, and ethylene concentrations of Golden Delicious apple fruits. The data are plotted as a function of days after harvest. Increases in ethylene and ACC concentrations and EFE activity are closely correlated with ripening. (From Yang, 1987.)

itors of ethylene biosynthesis and action made it possible to discriminate between the actions of auxin and ethylene. These studies showed that ethylene is the primary effector and that auxin is indirectly involved by causing a substantial increase in ethylene production. **Aminoethoxyvinylglycine (AVG)** and **aminooxyacetic acid (AOA)** block the conversion of SAM to ACC (see Fig. 19.1). Since AVG and AOA are known to inhibit enzymes that use the cofactor pyridoxal phosphate, this suggests that the ACC synthase is a pyridoxal phosphate enzyme. Cobalt is also an inhibitor of the ethylene biosynthetic pathway, blocking the conversion of ACC to ethylene, the last step in ethylene biosynthesis.

Most of the effects of ethylene can be antagonized by specific ethylene inhibitors. Silver ions applied as $AgNO_3$ or as silver thiosulfate are potent inhibitors of ethylene action. Silver is very specific; the inhibition it causes cannot be induced by any other metal ion. Though less efficient than Ag^+, carbon dioxide at high concentrations (in the range of 5 to 10%) inhibits many effects of ethylene, such as induction of fruit ripening. This effect of CO_2 has often been exploited by storing fruits under elevated CO_2 concentrations to delay ripening. The high concentrations of CO_2 required for inhibition make it unlikely that CO_2 acts as an ethylene antagonist under natural conditions.

Ethylene Can Be Measured in Bioassays or by Gas Chromatography

The triple response of etiolated pea seedlings reported by Neljubow in 1901 is still a reliable bioassay for ethylene because of its specificity, sensitivity to low concentrations, and rapidity. When plant tissues are exposed to various concentrations of ethylene (0.1 $\mu l\ L^{-1}$ and higher) in a sealed environment, inhibition of stem elongation, increased lateral growth (swelling), and horizontal growth of the epicotyl are observed (Fig. 19.3a). The magnitude of the response is proportional to the ethylene concentration present in the sample. Actual concentrations are determined by comparing the response of the sample with that of tissue exposed to known amounts of ethylene. Other bioassays include epinasty of tomato leaves (Fig. 19.3b) and leaf abscission, but these assays are less sensitive than the triple response. All bioassays are diagnostic for the presence of ethylene, but none of them permit quantification over a large range of ethylene concentrations. Such quantification can be accomplished by gas chromatography.

Gas chromatography is the most sensitive and accurate method for ethylene detection. As little as 5 parts per billion (ppb) of ethylene can be detected, and the analysis time is only 1–4 min. Usually, the ethylene produced by a plant tissue is allowed to accumulate in a sealed vial and a sample is withdrawn with a syringe. The

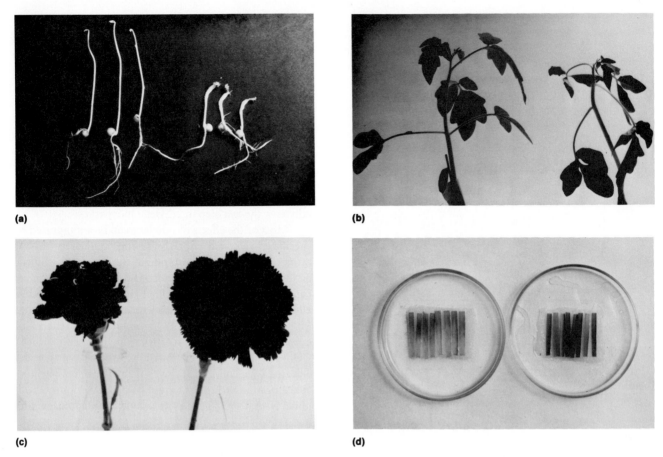

FIGURE 19.3. Some physiological effects of ethylene on plant tissues in various developmental stages. (a) Triple response of etiolated pea seedlings. Six-day-old pea seedlings were treated with 10 μg L^{-1} ethylene (right) or left untreated (left). The treated seedlings show the typical symptoms of the triple response: inhibition of stem elongation, enhancement of radial expansion, and horizontal growth. (b) Epinasty, or downward bending of the tomato leaves (right), is caused by ethylene treatment. Epinasty occurs when the cells on the upper side of the petiole grow faster than those on the bottom. (c) Flower senescence. Treatment of cut carnation flowers with ethylene (10 μg L^{-1}) accelerated their senescence as evident from the premature wilting and shriveling of the petals (left). (d) Leaf senescence. Oat leaf segments incubated with the ethylene precursor ACC showed senescence symptoms (yellowing), while leaf segments incubated without ACC remained green during the 5 days of incubation (right). (Photos courtesy of S. Gepstein.)

sample is injected into a gas chromatograph column in which the different gases are separated and detected by a flame ionization detector. Quantification of ethylene by this method is very accurate.

Ethylene Has Numerous Effects on Different Plant Species and Organs

Fruit Ripening. Ethylene has been recognized for many years as the hormone that accelerates fruit ripening. Addition of ethylene to fruits hastens ripening, and a dramatic increase in ethylene production is closely associated with **initiation** of ripening. Inhibitors of ethylene biosynthesis (such as AVG) or ethylene action (such as CO_2 or Ag^+) delay or even prevent ripening. All these observations strongly indicate that ethylene is the main agent controlling ripening. In many fruits, ripening is characterized by a climacteric rise in respiration and eth-

ylene production (Fig. 19.4). Apples, bananas, avocado, and tomatoes are examples of the climacteric class of fruits. In contrast, fruits such as citrus and grapes do not exhibit the respiration and ethylene production rise and are called nonclimacteric. When unripe climacteric fruits are treated with ethylene, the onset of the climacteric rise is hastened (Yang, 1987). When nonclimacteric fruits are treated in the same way, the magnitude of the respiratory rise increases as a function of the ethylene concentration, but the treatment does not trigger production of endogenous ethylene and does not accelerate ripening. The role of ethylene in ripening of climacteric fruits has resulted in many practical applications aimed at either uniform ripening or delay of ripening.

Abscission. The shedding of leaves, fruits, flowers, and other plant organs is termed **abscission**. The abscission process takes place in specific layers of cells, called ab-

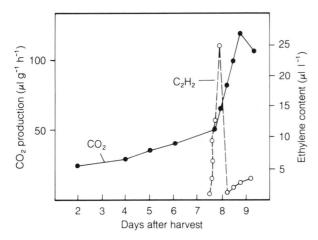

FIGURE 19.4. Relationship between ethylene production and respiration rate in banana. Ripening is characterized by a climacteric rise in respiration. During banana ripening there is a massive increase in CO_2 release, followed by a decrease. A climacteric rise in ethylene production precedes the CO_2 increase. These relationships suggest that ethylene is the hormone that triggers the ripening process. (From Burg and Burg, 1965.)

describes the process in three distinct sequential phases (Fig. 19.5) (Morgan, 1984):

1. **Leaf maintenance phase.** Prior to the perception of any signal (internal or external) that initiates the abscission process, the leaf remains healthy and fully functional in the plant.

2. **Shedding induction phase.** An abscission signal is perceived and transduced into a message, such as changes in the rate of synthesis of hormones in the leaf.

3. **Shedding phase.** The actual abscission events are expressed in biochemical, anatomical, and physiological processes that result in shedding.

scission layers, which become morphologically and biochemically differentiated during organ development. Weakening of the cell walls at the abscission layer depends on cell wall–degrading enzymes such as cellulase and polygalacturonase. Ethylene appears to be the primary regulator of the abscission process, with auxin acting as a suppressor of the ethylene effect.

A model of the hormonal control of leaf abscission

During the early phase of leaf maintenance, auxin prevents abscission by repressing the synthesis of the hydrolytic enzymes involved in abscission. It has long been known that removal of the leaf blade (the site of auxin production) promotes petiole abscission. Application of exogenous auxin to petioles from which the leaf blade has been removed delays the abscission process. In the shedding induction phase, the auxin level decreases and the ethylene level rises. Ethylene appears to decrease the activity of auxin by both reducing its synthesis and transport and increasing its destruction. The reduction in the concentration of free auxin increases the responsivity of specific **target cells** to ethylene. The target cells, located in the abscission zone, accumulate and discharge cytoplasmic vesicles into the cell wall, in conjunction with

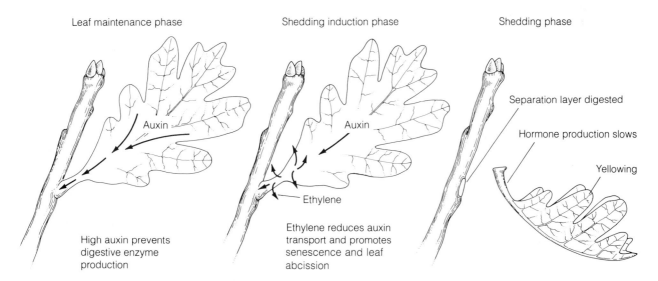

FIGURE 19.5. Schematic view of the hormonal balance during leaf abscission. According to this model, auxin prevents leaf shedding during the leaf maintenance phase. In the shedding induction phase, the level of auxin decreases, and the level of ethylene increases. These changes in the hormonal balance increase the sensitivity of the target cells to ethylene, which triggers the events involved in the shedding phase. During the shedding phase, specific enzymes that hydrolyze the cell wall polysaccharides are synthesized. Wall hydrolysis results in cell separation and leaf abscission. (From Morgan, 1984.)

cellulase synthesis. The shedding phase is characterized by induction of genes encoding specific hydrolytic enzymes of cell wall polysaccharides and proteins. The action of these enzymes leads to cell wall loosening, cell separation, and abscission.

Epinasty. The downward curvature of leaves that occurs when the upper (adaxial) side of the petiole grows faster than the lower (abaxial) side is termed **epinasty** (Fig. 19.3b). Ethylene and high concentrations of auxin induce epinasty, and it has now been established that auxin acts indirectly by inducing ethylene production. Flooding (waterlogging) or anaerobic conditions around tomato roots trigger enhanced synthesis of ethylene in the shoot, leading to the epinastic response (Chapter 14). Since these environmental stimuli are sensed by the roots and the response is displayed by the shoot, a "signal" from the roots must be transported to the shoots. This signal is ACC, the immediate precursor of ethylene (Bradford and Yang, 1980). ACC levels were found to be significantly higher in the xylem sap following flooding of tomato roots for 1–2 days (Fig. 19.6). Because water

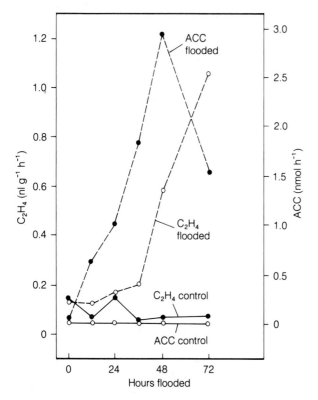

FIGURE 19.6. Changes in the amounts of ACC in the xylem sap and ethylene production in the petiole following flooding of tomato plants. ACC is synthesized in roots but it is converted to ethylene very slowly under anaerobic conditions of flooding. ACC is transported via the xylem to the shoot where it is converted to ethylene. The gaseous ethylene cannot be transported, so it usually affects the tissue near the site of its production. The ethylene precursors, ACC, is the transportable molecule and can produce ethylene far from the site of ACC synthesis. (From Bradford and Yang, 1980).

fills the air spaces in waterlogged soil and O_2 diffuses slowly through water, the concentration of oxygen around flooded roots decreases dramatically. The elevated production of ethylene appears to be caused by the accumulation of ACC in the roots under anaerobic conditions. The conversion of ACC to ethylene requires oxygen (Fig. 19.1), and the ACC accumulated in the anaerobic roots is transported to shoots, where it is readily converted to ethylene.

Seedling Growth. The triple response to ethylene of etiolated pea seedlings has already been described (Fig. 19.3a). At concentrations above 0.1 μl L^{-1}, ethylene changes the growth pattern of seedlings by reducing the rate of elongation and increasing lateral expansion, leading to swelling of the region below the hook. These responses to ethylene are common to growing shoots of most dicots and to coleoptiles and mesocotyls of seedlings of graminaceous plants such as oat or wheat.

The ethylene-induced inhibition of elongation and promotion of lateral expansion of cells appears to be due to an alteration of the mechanical properties of the cell wall, involving changes in microtubule and cellulose microfibril arrangement from a transverse to a longitudinal orientation (Eisinger, 1983).

The typical horizontal growth (Fig. 19.3a) that occurs after exposure to ethylene may have an important role during germination. When physical barriers in the ground prevent seedling emergence, ethylene production induces horizontal growth, allowing the seedlings to find soil conditions that permit their emergence to the surface.

Hook Opening. Etiolated seedlings are usually characterized by the hook shape of the terminal portion of the shoot apex. This shape facilitates displacement of the seedling through the ground and protects the tender apical meristem. Like epinasty, hook closure and opening ensue from ethylene-induced asymmetric growth. The shape of the hook is a result of faster growth of the outer side compared with the inner one. When the hook is exposed to light, it opens because the rate of growth of its inner side increases. Red light induces opening and far-red light prevents opening, indicating that phytochrome (Chapter 20) is the photoreceptor involved in this process. A close interaction between phytochrome and ethylene controls hook opening. As long as ethylene is produced by the hook tissue in the dark, growth of cells on the inner side is inhibited. Red light inhibits ethylene formation, promoting growth on the inner side and hook opening.

Seed and Bud Dormancy. When applied to seeds of cereals, ethylene breaks dormancy and initiates germination. In peanuts, ethylene production and seed germination are closely correlated. In addition to its effect on dormancy, ethylene increases the rate of seed germination of several species. Bud dormancy may also be broken by ethylene, and ethylene treatment is used to promote sprouting in potato tubers and other bulbs.

Growth Promotion. In several monocot species such as rice, ethylene can act as a promoter of stem elongation, in contrast to its common inhibitory effect in most seedlings. Exposure of rice seedlings to ethylene or to flooding conditions that produce an anaerobic environment causes a dramatic increase in internodal growth. In the absence of O_2, ethylene synthesis is diminished, but the available ethylene diffuses very slowly in the waterlogged soil. As a result, the submerged rice plants are exposed to high ethylene concentrations that increase stem elongation.

Induction of Roots. Ethylene is capable of inducing root formation in leaves, stems, flower stems, and even other roots. This response requires unusually high ethylene concentrations ($10 \mu l L^{-1}$).

Flowering. Whereas ethylene inhibits flowering in many species, it induces flowering in pineapple and is used commercially in this species for synchronization of fruit set. Flowering of other species of the Bromeliaceae, such as mango, is also initiated by ethylene. On plants that have separate male and female flowers, ethylene may change the sex of developing flowers; promotion of female flower formation in cucumber is one example of this effect.

Flower and Leaf Senescence. Onset of leaf and flower senescence is hastened by ethylene and significantly delayed by inhibition of ethylene synthesis (by AVG or Co^{2+}) or action (by Ag^+ or CO_2). Furthermore, enhanced ethylene production is associated with chlorophyll loss and color fading, which are characteristic features of leaf and flower senescence (Fig. 19.3).

The Mechanism of Ethylene Action Is Not Well Understood

In animal systems the first step of a hormone response is binding of the hormone to a specific receptor in the target tissue. This binding triggers a series of reactions that result in a physiological response. Many studies of ethylene action have led to the hypothesis that ethylene interacts with a receptor containing Zn or Cu (Beyer, 1976; Burg and Burg, 1965). Ethylene binding was studied in an isolated cell-free system from cotyledons of *Phaseolus vulgaris*. Following treatment with ^{14}C-labeled ethylene, the bound radioactive ethylene was found to be associated with Golgi bodies or with the endoplasmic reticulum (Bengochea et al., 1980). The binding is heat-labile and is inhibited by proteolytic enzymes, suggesting that the hormone receptor is an integral membrane protein.

Chemical evidence indicates that the receptor to which ethylene binds is a copper-containing protein that might be involved in ethylene oxidation. The mode of action of ethylene may be connected to its conversion into oxidation products, ethylene oxide and ethylene glycol. However, direct evidence for this hypothesis remains unavailable.

Fruit ripening is the most complex process regulated by ethylene. Ripening comprises a series of metabolic transitions that bring about changes in fruit texture, color, and flavor and other metabolic changes that lead to fruit senescence. Underlying these morphological and biochemical changes are changes in gene expression that trigger the ripening process (Kende and Boller, 1981). The softening of cell walls associated with ripening correlates with increasing activity of cellulase and polygalacturonase, which catalyze the hydrolysis of cellulose and pectin, the main cell wall constituents. During fruit ripening in avocado and tomato, ethylene causes the accumulation of both cellulase and polygalacturonase mRNAs (Christoffersen et al., 1984). These observations indicate that ethylene regulates the transcription of genes encoding the cell wall–digesting enzymes (Fig. 19.7) (Christoffersen, 1987). The regulation of gene expression by ethylene is a subject of active investigation (Lincoln et al., 1987; Margossian et al., 1988).

Ethylene Has Important Commercial Uses

Since ethylene regulates so many physiological processes in plant development, it is one of the most widely used plant hormones in agriculture. Auxins and ACC can trigger the natural biosynthesis of ethylene and in several cases are used in agricultural practice. Because of its high diffusion rate, ethylene proper is very difficult to apply in the field as a gas, but this limitation can be overcome by using an ethylene-releasing compound. The most widely used compound is ethephon or **2-chloroethylphosphonic acid**, which was discovered in the 1960s and is known by its trade name, Ethrel (Fig. 19.8). Ethephon is sprayed in aqueous solution and is readily absorbed and transported within the plant. It releases ethylene slowly by a chemical reaction, allowing the hormone to exert its effects. Ethephon hastens fruit ripening of apples and tomatoes and degreening of citrus, synchronizes flowering and fruit set in pineapples, and accelerates abscission of flowers and fruits. Its application makes it possible to achieve fruit thinning or fruit drop in cotton, cherries, and walnuts. It is also used to promote female sex expression in cucumber, to prevent self-pollination and increase yield, and to inhibit terminal growth of some plants in order to promote lateral growth and compact flowering.

Storage facilities developed to inhibit ethylene production and promote preservation of fruits have a controlled atmosphere of low O_2 concentration and low temperature that inhibits ethylene biosynthesis. A re-

FIGURE 19.7. Hypothetical model for the ethylene-induced synthesis of cellulase during avocado fruit ripening. Hydrolysis of the cell wall, a basic step in fruit ripening, depends on increased activity of cellulase. The higher cellulase activity is due to ethylene-dependent accumulation of the mRNA coding for the cellulase. This model describes the sequence of events that includes binding of ethylene, signal transduction, transcription of the cellulase gene and production of mRNA, and synthesis of the cellulase protein, which is then secreted into the cell wall. (From Christoffersen, 1987.)

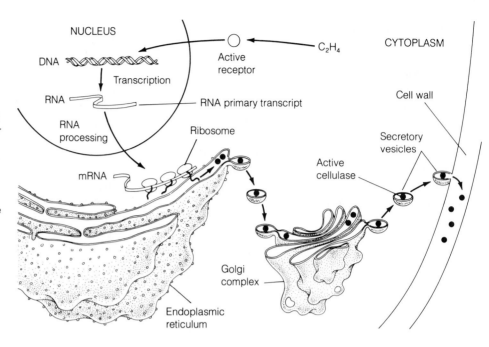

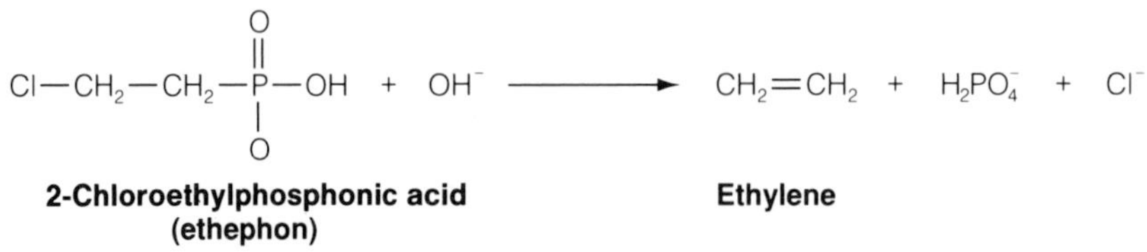

FIGURE 19.8. The ethylene-releasing compound ethephon. When taken up by plant tissue, ethephon is converted to ethylene. Under the trade name of Ethrel, ethephon has many commercial applications, mainly in synchronizing fruit ripening or flowering.

$$Cl-CH_2-CH_2-\overset{\overset{O}{\|}}{\underset{\underset{O}{|}}{P}}-OH \ + \ OH^- \longrightarrow CH_2{=}CH_2 \ + \ H_2PO_4^- \ + \ Cl^-$$

2-Chloroethylphosphonic acid (ethephon)   **Ethylene**

latively high concentration of CO_2 (3–5%) prevents ethylene's action as a ripening promoter. Low pressure is used to remove ethylene and oxygen from the storage chambers, reducing the rate of ripening and preventing overripening. Specific inhibitors of ethylene biosynthesis and action are also useful in postharvest preservation. Silver (Ag^+) is used extensively to increase the longevity of cut carnations and several other flowers. The potent inhibitor AVG retards fruit ripening and flower fading, but its commercial use has not yet been approved by regulatory agencies.

Abscisic Acid

For many years, plant physiologists suspected that the phenomenon of seed or bud dormancy is caused by inhibitory compounds and designed experiments aimed at their isolation. Abscisic acid was found to be one of the inhibitory compounds.

Early experiments involved the use of chromatography for the separation of plant extracts, together with bioassays based on oat coleoptile growth. These efforts revealed an inhibitory material different from auxin (Bennet-Clark and Kefford, 1953). Ten years later, a

substance that promotes abscission of cotton fruits was purified and crystallized and was called abscisin II (Ohkuma et al., 1963). At about the same time, a substance that promotes bud dormancy was purified from sycamore leaves and was called dormin. When dormin was chemically identified, it was found to be identical to abscisin II, and the compound was renamed **abscisic acid** (ABA) because of its involvement in the abscission process. For several years, the induction of abscission was thought to be one of the primary roles of ABA. However, we now know that ethylene is the major factor causing organ abscission.

ABA Is Widely Distributed in Nature

Abscisic acid has been found to be a ubiquitous plant hormone in vascular plants. It has been detected in mosses but appears to be absent in liverworts. Several genera of fungi make ABA as a secondary metabolite (Zeevaart and Creelman, 1988). In algae and liverworts, a compound similar to ABA and named **lunularic acid** (Fig. 19.9) appears to play a physiological role similar to that of ABA in higher plants. Within the plant, ABA has been detected in every major organ or living tissue from the root cap to the apical bud (Milborrow, 1984). ABA is synthesized in

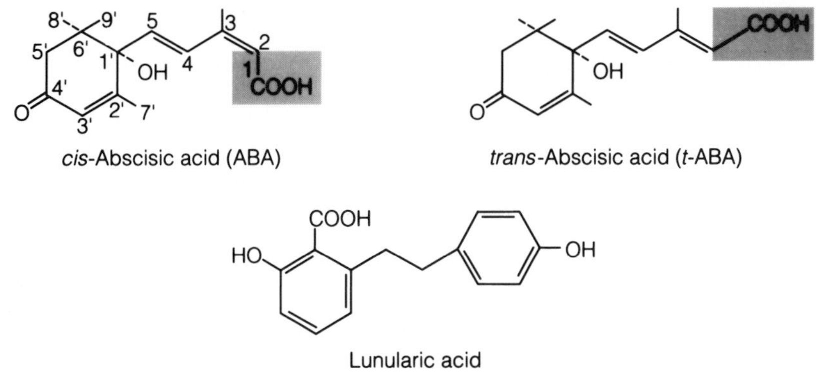

cis-Abscisic acid (ABA)

trans-Abscisic acid (t-ABA)

Lunularic acid

FIGURE 19.9. The chemical structures of *cis-* and *trans-*abscisic acid and lunularic acid. The numbers in the formula for ABA identify the carbon atoms.

almost all cells containing chloroplasts or amyloplasts. Surprisingly, ABA was recently found in mammalian brain, but it is unclear whether it originates from ingested food or is a regular brain metabolite.

The Chemical Structure of ABA Determines Its Physiological Activity

The structure of ABA (Fig. 19.9) resembles the terminal portion of some carotenoid molecules (Figs. 8.5 and Fig. 19.10). The 15 carbon atoms of ABA configure an aliphatic ring with one double bond, two methyl groups, and an unsaturated chain with a terminal carboxyl group (Fig. 19.9). The position of the protons at C-2 and C-4 and the ensuing orientation of the carboxyl group at C-2 determine the cis and trans isomers of ABA (Fig. 19.9). Nearly all the naturally occurring ABA is in the cis form, and, by convention, the name abscisic acid refers to that isomer. In addition, ABA also has an asymmetric carbon atom at position 1′ in the ring, resulting in the (+) and

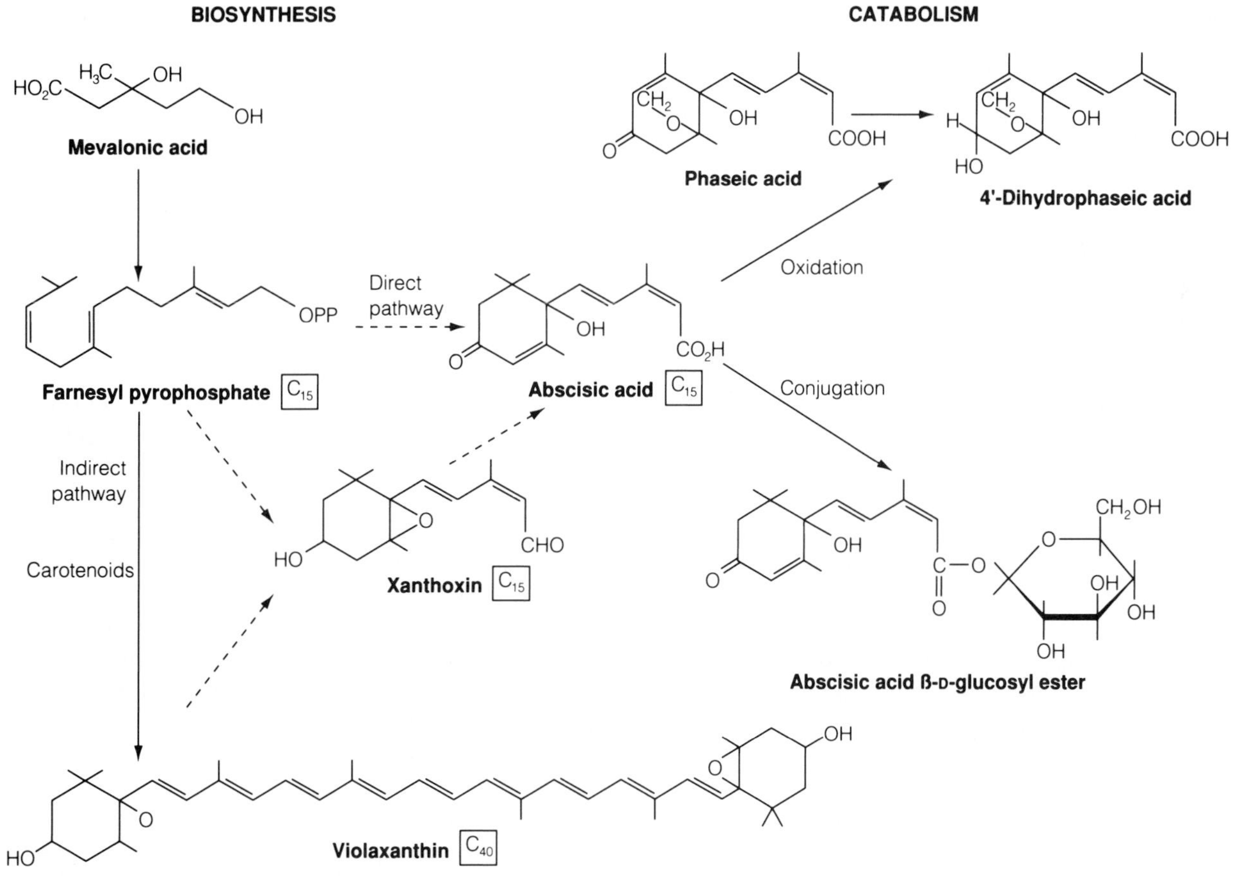

FIGURE 19.10. ABA biosynthesis and metabolism. There are two alternative biosynthetic pathways for ABA. In the direct pathway, ABA is derived from a 15-carbon precursor, farnesyl pyrophosphate. In the indirect pathway, ABA is derived from a 40-carbon compound, violaxanthin. The pathways for ABA catabolism include conjugation to form the ABA β-D-glucosyl ester or oxidation to form phaseic acid and dihydrophaseic acid.

$(-)$ (or S and R, respectively) enantiomers. The $(+)$ enantiomer is the natural form; commercially available synthetic ABA is a mixture of approximately equal amounts of the $(+)$ and $(-)$ forms. The $(+)$ enantiomer is the only one active in fast responses to ABA, such as stomatal closure (Chapter 14). In long-term responses, such as changes in protein synthesis, both enantiomers are active. In contrast to the cis-trans isomers, the $(+)$-$(-)$ ones cannot be interconverted in the plant tissue.

Studies of the structural requirements for biological activity of ABA have shown that almost any change in the molecule results in loss of activity. Some of the features shown to be essential for biological activity include the carboxyl group, the tertiary hydroxyl group, and the 2-cis and ring double bonds. Catabolic products of ABA present in the tissue and devoid of any of these groups are biologically inactive.

ABA Is Assayed by Biological, Physicochemical, and Immunological Methods

Coleoptile growth, the classical bioassay devised for auxins (Chapter 16), is also used for ABA detection in plant extracts by measuring the coleoptile growth inhibition. This bioassay has adequate sensitivity (minimum detectable level, 10^{-7} M) and shows a linear response in the range 10^{-7}–10^{-5} M, but it has some disadvantages. Plant extracts have other promoters and inhibitors, which decrease the specificity of the bioassay and make preliminary purification steps necessary. Other available bioassays for ABA include inhibition of germination, inhibition of gibberellic acid–induced amylase synthesis in aleurone layers, induction of leaf abscission, and stomatal closure. Stomatal closure is highly specific for ABA because it is affected little by other plant growth regulators. Additional advantages of this bioassay include a fast response of guard cells to ABA, high sensitivity (minimum detectable level, 10^{-9} M), and a linear response over a wide range of concentration. Bioassays are still in use because of their simplicity.

Physical methods are much more reliable than bioassays because of their specificity and suitability for quantitative analysis. The most widely used techniques are those based on gas chromatography.

Gas chromatography allows detection of as little as 10^{-13} g ABA, but it requires several preliminary purification steps, including thin-layer chromatograph.

Another way to purify ABA in plant extracts is by **immunoassay**. This method relies on the specific recognition of ABA by antibodies obtained from rabbits or mice injected with the growth regulator. Immunoassays can detect 10^{-13} g of ABA in crude or partially purified extracts. These specific and highly sensitive assays are currently available for ABA, auxins, gibberellins, and cytokinins (Weiler, 1984).

The Amount of ABA in a Tissue Depends On Its Biosynthesis, Metabolism, and Transport

As with other plant hormones, the regulatory effects of ABA depend on its concentration in the tissue. Endogenous concentrations of ABA vary with growth and environmental conditions, so there is a need to understand the regulation of the biosynthesis and catabolism of the growth regulator.

The complete biosynthetic pathway of ABA remains to be characterized. Two alternative metabolic routes have been intensively investigated (Zeevaart and Creelman, 1988). One pathway involves mevalonic acid, which is a precursor of all terpenoids in plants (Chapter 17). In this route, called the direct pathway, the 15-carbon ABA molecule is formed from mevalonic acid and a 15-carbon precursor, farnesyl pyrophosphate (Fig. 19.10). In the alternative, or indirect, pathway ABA is formed from the cleavage of a 40-carbon carotenoid (Fig. 19.10). The indirect pathway involves xanthoxin, a neutral growth inhibitor with physiological properties similar to those of ABA. Xanthoxin is converted to ABA when it is supplied to shoots (Taylor and Burden, 1973) and in a cell-free system (Sindhu and Walton, 1987). Xanthoxin is a breakdown product of carotenoids such as violaxanthin. The uncertainties about the biosynthetic pathway of ABA are due to the facts that mevalonic acid can also be a precursor of carotenoids and that xanthoxin can also participate in the direct pathway. Experiments with ^{18}O suggest that ABA could be formed by different pathways in turgid leaves and leaves subjected to water stress (Zeevaart and Creelman, 1988).

The ABA concentration in tissues is determined by the balance between biosynthesis and catabolism. A major cause of inactivation of free ABA is oxidation, which results in the formation of an unstable intermediate, 6′-hydroxymethyl ABA, which is converted to **phaseic acid** (PA) and **dihydrophaseic acid** (DPA) (Fig. 19.10). Free ABA is also inactivated by conjugation, in which the hormone forms a covalent link with another molecule, such as a monosaccharide. A common example of an ABA conjugate is ABA β-D-glucosyl ester (ABAGE).

Metabolism of ABA can inactivate the hormone by changing an essential structural feature or by affecting its polarity. For example, ABAGE is found predominantly in vacuoles, and measurements in wilting tomato leaves showed an increase in free ABA but not in ABAGE. Thus, the increase in free ABA is not caused by the hydrolysis of ABAGE. The activity of PA and DPA is usually less than that of ABA, although PA was found to induce stomatal closure in some species.

ABA is transported by both the xylem and the phloem but is much more abundant in the phloem sap. When radioactive ABA is applied to a leaf, it is transported both up the stem and down toward the roots. Most of the radioactive ABA is found in the roots within 24 h. Destruction of the phloem by a stem girdle prevents ABA accumulation in the roots, indicating that the hormone is transported in the phloem sap.

The distribution of ABA is also affected by compartmentation. ABA is a weak acid with a pK_a of 4.7, and its dissociation depends on the pH of each cellular compartment (Fig. 14.3). The protonated form of ABA permeates freely across membranes, but the dissociated anion does not. As a result, the distribution of ABA between different compartments depends on their pH values; the more alkaline a compartment, the more ABA it accumulates (Cowan et al., 1982). Because of these properties, ABA concentrations in the chloroplast increase in the light, whereas the apoplast concentrations increase in the dark (Kaiser et al., 1985). This redistribution of ABA could play a role in the regulation of stomatal apertures.

ABA Causes Many Physiological Responses in Higher Plants

Bud Dormancy. In woody species, dormancy is an important adaptive feature in cold climates. When a tree faces very cold temperatures in winter, it protects its meristems with bud scales and temporarily stops bud growth. This response to cold temperatures requires a sensory mechanism that detects the environmental changes and a control system that transduces the sensory signal(s) and triggers the developmental processes leading to bud dormancy. ABA was originally suggested as the dormancy-inducing hormone by P. Wareing in 1964, but the ABA content of buds does not always correlate with the degree of dormancy. This discrepancy could underscore interactions between ABA and other hormones as part of a process in which bud dormancy and growth are regulated by the balance between bud growth inhibitors such as ABA and growth-inducing substances such as cytokinins and gibberellins.

Seed Dormancy. A characteristic feature of seeds is their ability to suspend developmental processes until the conditions necessary for germination are met (Bewley and Black, 1982). ABA appears to act as the dormancy-inducing hormone, as indicated by an observed correlation between levels of endogenous ABA and the physiological state of the seed. Dry dormant seeds usually contain higher concentration of ABA than nondormant seeds. However, increases in the amounts of growth promoters such as cytokinins and gibberellins have also been observed during seed germination. As in the case of bud dormancy, it appears that hormonal balance, rather than fluctuations in the concentration of a single hormone, controls the transition from dormancy to germination.

A role of ABA in dormancy is also indicated by the observation that exogenous ABA prevents germination and induces dormancy. For example, lettuce seeds, which require red light to germinate, do not germinate when illuminated in the presence of ABA. Exogenous ABA also reimposes dormancy on embryos kept in culture, often causing them to continue to grow and to accumulate reserves. ABA also prevents precocious germination of maize kernels (Zeevaart and Creelman, 1988).

ABA inhibits the synthesis of hydrolytic enzymes that are essential for the breakdown of storage reserves in the seed. For example, during the germination of cereal seeds, GA_3 induces the synthesis of α-amylase in the aleurone layers, which catalyzes the breakdown of starch (Chapter 17). ABA inhibits this GA_3-dependent enzyme synthesis by inhibiting the transcription of α-amylase mRNA (Chapter 17). The inhibition of hydrolytic enzymes by ABA could be central to its role in dormancy maintenance.

Growth. Auxin-induced growth of seedlings is inhibited by ABA, and for this reason ABA is also called the growth inhibitor hormone. It is well established that auxin-induced H^+ secretion causes the cell wall loosening that increases the rate of cell elongation (Chapter 16). ABA blocks H^+ secretion, thereby preventing cell wall acidification and cell elongation.

Stress and Stomatal Closure. The role of ABA in freezing, salt, and water stress (Chapter 14) had led to the characterization of ABA as the stress hormone. Under drought conditions, leaf ABA concentrations can increase up to 40 times, the most dramatic change in concentration reported for any hormone in response to an environmental signal. ABA is very effective in causing stomatal closure, and its accumulation in stressed leaves plays an important role in the reduction of water loss by transpiration under water stress conditions (Fig. 19.11) (Raschke, 1987). On the other hand, several studies have shown decreases in stomatal apertures before any increases in the bulk leaf ABA content. This apparent inconsistency is explained by recent data showing that the initial stomatal closure is caused by ABA redistribution within the leaf in response to pH changes in mesophyll chloroplasts under stress (Chapter 14). Stomatal closing can also be caused by ABA produced in the roots, without any changes in leaf turgor or ABA content (Chapter 14).

Mutants that lack the ability to synthesize ABA show permanent wilting because of their inability to close stomata. Application of exogenous ABA to such mutants causes stomatal closure and turgor buildup (Tal and Imber, 1971).

ABA-induced stomatal closure is a rapid response that can be observed in a few minutes. Guard cells appear to have specific ABA receptors in the outer surface of their

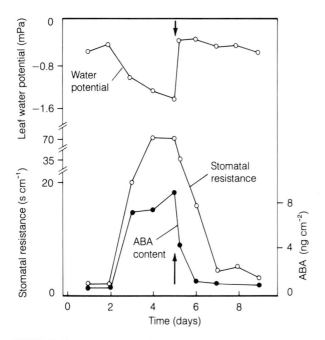

FIGURE 19.11. Changes in water potential, stomatal resistance (the inverse of stomatal conductance), and ABA content in maize. Watering was interrupted at the beginning of the treatment and restarted at the time indicated by the arrows. As the soil dried out, the water potential of the leaf decreased, and the ABA content and stomatal resistance increased. The process was reversed by rewatering. (From Beardsell and Cohen, 1975.)

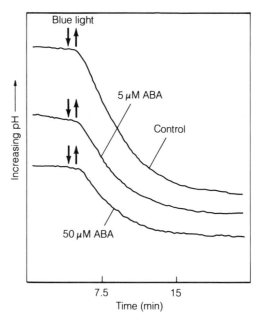

FIGURE 19.12. ABA inhibits the blue light–stimulated proton pumping by guard cell protoplasts. A suspension of guard cell protoplasts (Chapter 6) was incubated under red-light irradiation and the pH of the suspension medium was monitored with a pH electrode. At the time indicated by the arrows, a pulse of blue light was given. Blue light activates a proton pump at the guard cell plasma membrane, and proton extrusion from the pump acidifies the suspension medium, causing a decrease in pH. Addition of ABA to the suspension medium inhibits the acidification by 40%. These results indicate that either the rate of proton pumping or the permeability of the membrane has been affected by ABA. (From Shimazaki et al., 1986.)

plasma membrane, and ABA might cause stomatal closure by modulating the opening of ion channels and the activity of the proton pump (Fig. 19.12) (Zeevaart and Creelman, 1988; Mansfield et al., 1990). ABA also appears to modify membrane properties when causing root responses that increase water flux and ion uptake (Van Steveninck and Van Steveninck, 1983).

Water Uptake. Application of ABA to root tissue stimulates water flow and ion flux, suggesting that ABA regulates turgor not only by decreasing transpiration but also by increasing water influx into roots (Glinka and Reinhold, 1971). ABA increases the flow of water by increasing the hydraulic conductivity (decreasing the resistance to water) and by enhancing ion uptake, which causes an increase in the water potential gradient between the soil and the root (Chapter 4). In addition, ABA induces root growth and stimulates emergence of lateral roots, while suppressing leaf growth (Torrey, 1976). These antagonistic effects of ABA on roots and leaves cause a reduction in leaf area and an increase in the water-absorbing area of roots, which helps the plant cope with drought conditions.

Abscission and Senescence. Abscisic acid was originally isolated as an abscission-causing factor. However, it has since become evident that ABA stimulates abscission of organs in only a few species and that the primary hormone

causing abscission is ethylene. ABA, on the other hand, is clearly involved in senescence, and through its promotion of senescence it might indirectly increase ethylene formation and stimulate abscission. **Senescence** is the last developmental stage prior to the death of an organ or of the whole organism.

Senescence can be studied at the whole-plant level or at the level of excised organs such as leaf segments, flowers, and fruits. Studies with excised organs are less complicated and allow experimental analysis of the role of hormones in the senescence process. Because of these advantages, substantially more is known about the senescence of excised organs than that of the whole plant. Leaf senescence has been studied extensively, and the anatomical, physiological, and biochemical changes that take place during this process have been described (Nooden and Leopold, 1988). Leaf segments undergo senescence faster in darkness than in light, and they turn yellow as a result of chlorophyll breakdown. In addition, breakdown of proteins and nucleic acids is increased by the stimulation of several hydrolases. ABA greatly accelerates the senescence of both leaf segments and attached leaves. Cytokinins antagonize the action of ABA and delay senescence. Ethylene has been found to play a role in the senescence of oat leaf segments, but ABA

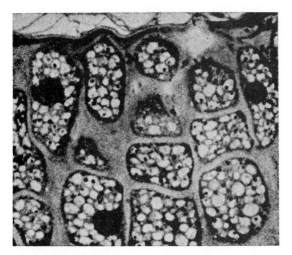

(a)

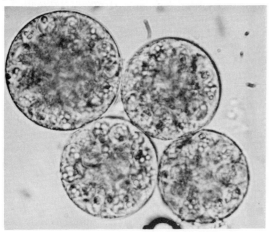

(b)

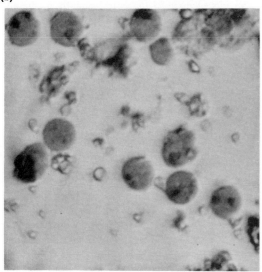

(c)

FIGURE 19.13. The isolation of nuclei from aleurone cells. Isolated nuclei are used to study signal transduction during responses to hormones such as ABA and GA. (a) Intact aleurone cells from a barley seed. (b) Protoplasts from aleurone cells obtained by cellulolytic digestion of the cell walls. (c) Isolated nuclei obtained after lysing the protoplasts. [(a) From R. L. Jones, (b), (c) from Jacobsen and Chandler, 1987.]

appears to be the initiating agent, whereas ethylene exerts its effect at a later stage (Gepstein and Thimann, 1981).

ABA Regulates Protein Synthesis

ABA has been shown to affect protein synthesis under several conditions, such as heat shock, adaptation to cold temperatures, and salt tolerance (Chapter 14). In these cases, ABA treatment results in transcriptional control leading to the synthesis of new proteins. In the case of inhibition of the GA_3-induced α-amylase, ABA caused a reduction in messenger RNA levels (Jacobsen and Chandler, 1987).

Organelle isolation (Fig. 19.13) and molecular studies of hormone-mediated gene regulation are providing new insights on the mode of action of growth regulators. For example, ABA induces a gene encoding glycine-rich protein (Gomez et al., 1988) and a gene promoter in wheat (Marcotte et al., 1988). Another study has characterized a rice gene that is induced by ABA and water stress (Mundy and Chua, 1988).

Summary

In contrast to auxins, cytokinins, and gibberellins, which act mainly as growth promoters, ethylene and ABA are the two primary plant hormones acting as inhibitors of growth and metabolic processes.

Ethylene is the only known gaseous plant hormone. It is formed in most organs of higher plants. Senescing tissues or ripening fruits produce more ethylene than young or mature tissues. The precursor of ethylene in vivo is the amino acid methionine, which is converted to S-adenosylmethionine (SAM), 1-aminocyclopropane-1-carboxylic acid (ACC), and ethylene. The rate-limiting step of this pathway is the conversion of SAM to ACC, which is catalyzed by ACC synthase. ACC can also be conjugated to malonic acid to form the biologically active compound N-malonyl ACC. Ethylene biosynthesis is triggered by various developmental processes, by auxins, and by environmental stresses. In all these cases the activity of ACC synthase increases. The physiological effects of ethylene can be blocked by biosynthetic inhibitors or by antagonists. AVG and AOA inhibit the synthesis of ethylene, whereas carbon dioxide and silver ions inhibit ethylene action. Ethylene can be detected by biological assays and by gas chromatography.

Ethylene regulates fruit ripening and other processes associated with leaf and flower senescence, leaf and fruit abscission, seedling growth, and hook opening. Ethylene also regulates ripening-associated gene expression and leads to accumulation of the mRNAs for cellulase and polygalacturonase, enzymes that contribute to cell wall solubilization and fruit softening.

Abscisic acid was discovered as a growth inhibitor that plays a major role in seed and bud dormancy. ABA is a 15-carbon compound that resembles the terminal portion of carotenoids. The endogenous level of ABA in tissues can be measured by bioassays based on effects such as stomatal closure. Gas chromatography and immunoassays are the most reliable and accurate methods available for measuring ABA levels.

There are two possible alternative pathways for ABA biosynthesis: a direct pathway from mevalonate and a 15-carbon precursor, and an indirect pathway that involves the cleavage of a 40-carbon carotenoid precursor. ABA is inactivated by oxidative degradation to phaseic acid and by conjugation to form the ABA glucose ester.

ABA is synthesized in almost all cells containing plastids and is transported to both the xylem and the phloem. In addition to mediating growth inhibition and dormancy, ABA regulates the water balance of plants under water stress by causing stomatal closure and maintaining water uptake by roots. ABA regulates cellular processes through effects on ion movements across membranes and on synthesis of specific proteins. ABA inhibits the synthesis of GA_3-induced α-amylase in the aleurone layer and also can cause the induction of other proteins under stress conditions.

GENERAL READING

Abeles, F. B. (1973) *Ethylene in Plant Biology.* Academic Press, New York.

Addicott, F. T. (1981) *Abscission.* University of California Press, Berkeley.

Addicott, F. T., ed. (1983) *Abscisic Acid.* Praeger, New York.

Addicott, F. T., and Carns, H. R. (1983) History and introduction. In: *Abscisic Acid*, Addicott, F. T., ed., pp. 1–21. Praeger, New York.

Bewley, J. D., and Black, M. (1982) *Physiology and Biochemistry of Seeds in relation to germination.* Springer-Verlag, Berlin.

Beyer, E. M., Jr., Morgan, P. W., and Yang, S. F. (1984) Ethylene. In: *Advanced Plant Physiology*, Wilkins, M. B., ed., pp. 111–126. Pitman, London.

Davies, P. J., ed. (1987) *Plant Hormones and Their Role in Plant Growth and Development.* Martinus Nijhoff, Dordrecht, The Netherlands.

Davies, W. J., and Mansfield, T. A. (1983) The role of abscisic acid in drought avoidance. In: *Abscisic Acid*, Addicott, F. T., ed., pp. 237–268. Praeger, New York.

Eisinger, W. (1983) Regulation of pea internode expansion by ethylene. *Annu. Rev. Plant Physiol.* 34:225–240.

Fuchs, Y., and Chalutz, E., eds. (1984) *Ethylene: Biochemical, Physiological and Applied Aspects.* Martinus Nijhoff, Dordrecht, The Hague.

Goldberg, R., ed. (1989) *The Molecular Basis of Plant Development.* Alan R. Liss, New York.

Mansfield, T. A., Hetherington, A. M., and Atkinson, C. J. (1990) Some current aspects of stomatal physiology. *Annu. Rev. Plant Physiol. Plant Mol. Biol.* 41:55–75.

Milborrow, B. V. (1983) Pathways to and from abscisic acid. In: *Abscisic Acid*, Addicott, F. T., ed., pp. 79–111. Praeger, New York.

Milborrow, B. V. (1984) Inhibitors. In: *Advanced Plant Physiology*, Wilkins, M. B., ed. pp. 76–110. Pitman, London.

Noodén, L. D., and Leopold, A. C., eds. (1988) *Senescence and Aging in Plants.* Academic Press, San Diego.

Sexton, R., and Woolhouse, H. W. (1984) Senescence and abscission. In: *Advanced Plant Physiology*, Wilkins, M. B., ed., pp. 469–497. Pitman, London.

Thimann, K. V., ed. (1980) *Senescence in Plants.* CRC Press, Boca Raton, Fla.

Torrey, J. G. (1976) Root hormones and plant growth. *Annu. Rev. Plant Physiol.* 27:435–459.

Van Steveninck, R. F. M., and Van Steveninck, M. E. (1983) Abscisic acid and membrane transport. In: *Abscisic Acid*, Addicott, F. T., ed., pp. 171–235. Praeger, New York.

Wareing, P. F., ed. (1982) *Plant Growth Substances 1982.* Academic Press, London.

Weiler, E. W. (1984) Immunoassay of plant growth regulators. *Annu. Rev. Plant Physiol.* 35:85–95.

Wilkins, M. B., ed. (1984) *Advanced Plant Physiology.* Pitman, London.

Yang, S. F., and Hoffman, N. E. (1984) Ethylene biosynthesis and its regulation in higher plants. *Annu. Rev. Plant Physiol.* 35:155–189.

Zeevaart, J. A. D., and Creelman, R. A. (1988) Metabolism and physiology of abscisic acid. *Annu. Rev. Plant Physiol. Plant Mol. Biol.* 39:439–473.

CHAPTER REFERENCES

Adams, D. O., and Yang, S. F. (1979) Ethylene biosynthesis: Identification of 1-aminocyclopropane-1-carboxylic acid as an intermediate in the conversion of methionine to ethylene. *Proc. Natl. Acad. Sci. USA* 76:170–174.

Amrhein, N., Breuing, F., Eberle, J., Skorupka, H., and Tophof, S. (1982) The metabolism of 1-aminocyclopropane-1-carboxylic acid. In: *Plant Growth Substances 1982*, Wareing, P. F., ed., pp. 249–258. Academic Press, London.

Beardsell, M. F., and Cohen, D. (1975) Relationships between leaf water status, abscisic acid levels, and stomatal resistance in maize and sorghum. *Plant Physiol.* 56:207–212.

Bennet-Clark, T. A., and Kefford, N. P. (1953) Chromatography of the growth substances in plant extracts. *Nature* 171:645–647.

Bengochea, T., Dodds, J. H., Evans, D. E., Jerie, P. H., Niepel, B., Shaari, A. R., and Hall, M. A. (1980) Studies on ethylene binding by cell-free preparations from cotyledons of *Phaseolus vulgaris* L. *Planta* 148:397–406.

Beyer, E. M., Jr. (1976) A potent inhibitor of ethylene action in plants. *Plant Physiol.* 58:268–271.

Beyer, E. M., Jr. (1979) Effect of silver ion, carbon dioxide, and oxygen on ethylene action and metabolism. *Plant Physiol.* 63:169–173.

Bradford, K. J., and Yang, S. F. (1980) Xylem transport of 1-aminocyclopropane-1-carboxylic acid, an ethylene precursor, in waterlogged tomato plants. *Plant Physiol.* 65:322–326.

Bradford, K., and Yang, S. F. (1981) Physiological responses of plants to waterlogging. *Hortic. Sci.* 16:25–30.

Burg, S. P., and Burg, E. A. (1965) Ethylene action and the ripening of fruits. *Science* 148:1190–1196.

Burg, S. P., and Burg, E. A. (1967) Molecular requirements for the biological activity of ethylene. *Plant Physiol.* 42:144–152.

Burg, S. P., and Clagett, C. O. (1967) Conversion of methionine to ethylene in vegetative tissue and fruits. *Biochem. Biophys. Res. Commun.* 27:125–130.

Burg, S. P., and Thimann, K. V. (1959) The physiology of ethylene formation in apples. *Proc. Natl. Acad. Sci. USA* 45:335–344.

Christoffersen, R. E. (1987) Cellulase gene expression during fruit ripening. In: *Plant Senescence: Its Biochemistry and Physiology*, Thomson, W. W., Nothnagel, E. A., and Huffaker, R. C., eds., pp. 89–97. American Society of Plant Physiologists, Rockville, Md.

Christoffersen, R. E., Tucker, M. L., and Laties, G. G. (1984) Cellulase gene expression in ripening avocado fruit: The accumulation of cellulase mRNA and protein as demonstrated by cDNA hybridization and immunodetection. *Plant Mol. Biol.* 3:385–391.

Cowan, I. R., Raven, J. A., Hartung, W., and Farquhar, G. D. (1982) A possible role for abscisic acid in coupling stomatal conductance and photosynthetic carbon metabolism in leaves. *Aust. J. Plant Physiol.* 9:489–498.

Gepstein, S., and Thimann, K. V. (1981) The role of ethylene in the senescence of oat leaves. *Plant Physiol.* 68:349–354.

Glinka, Z., and Reinhold, L. (1971) Abscisic acid raises the permeability of plant cells to water. *Plant Physiol.* 48:103–105.

Gómez, J., Sanchez-Martinez, D., Stiefel, V., Rigau, J., Puigdomenech, P., and Pages, M. (1988) A gene induced by the plant hormone abscisic acid in response to water stress encodes a glycine-rich protein. *Nature* 334:262–264.

Hoffman, N. E., and Yang, S. F. (1980) Changes of 1-aminocyclopropane-1-carboxylic acid content in ripening fruits in relation to their ethylene production rates. *J. Am. Soc. Hortic. Sci.* 105:492–495.

Hoffman, N. E., and Yang, S. F., and McKeon, T. (1982) Identification of 1-(malonylamino) cyclopropane-1-carboxylic acid as a major conjugate of 1-aminocyclopropane-1-carboxylic acid, an ethylene precursor in higher plants. *Biochem. Biophys. Res. Commun.* 104:765–770.

Imaseki, H., Yoshii, Y., and Todaka, I. (1982) Regulation of auxin-induced ethylene biosynthesis in plants: In: *Plant Growth Substances 1982*, Wareing, P. F., ed., pp. 259–268. Academic Press, London.

Jacobsen, J. V., and Chandler, P. M. (1987) Gibberellin and abscisic acid in germinating cereals. In: *Plant Hormones and Their Role in Plant Growth and Development*, Davies, P. J., ed., pp. 164–193. Martinus Nijhoff, Dordrecht, The Netherlands.

Kaiser, G., Weiler, E. W., and Hartung, W. (1985) The intracellular distribution of abscisic acid in mesophyll cells—the role of the vacuole. *J. Plant Physiol.* 119:237–245.

Kende, H., and Boller, T. (1981) Wound ethylene and 1-aminocyclopropane-1-carboxylate synthase in ripening tomato fruit. *Planta* 151:476–481.

Lieberman, M., and Kunishi, A., Mapson, L. W., and Wardale, D. A. (1966) Stimulation of ethylene production in apple tissue slices by methionine. *Plant Physiol.* 41:376–382.

Lieberman, M., and Mapson, L. W. (1964) Genesis and biogenesis of ethylene. *Nature* 204:343–345.

Lincoln, J. E., Cordes, S., Read, E., and Fischer, R. L. (1987) Regulation of gene expression by ethylene during *Lycopersicon esculentum* (tomato) fruit development. *Proc. Natl. Acad. Sci. USA* 84:2793–2797.

Marcotte, W. R., Jr., Bayley, C. C., and Quatrano, R. S. (1988) Regulation of a wheat promoter by abscisic acid in rice protoplasts. *Nature* 335:454–457.

Margossian, L. J., Federman, A. D., Giovannoni, J. J., and Fischer, R. L. (1988) Ethylene-regulated expression of a tomato fruit ripening gene encoding a proteinase inhibitor I with a glutamic residue at the reactive site. *Proc. Natl. Acad. Sci. USA* 85:8012–8016.

Morgan, P. W. (1984) Is ethylene the natural regulator of abscission? In: *Ethylene: Biochemical, Physiological and Applied Aspects*, Fuchs, Y., and Chalutz, E., eds., pp. 231–240. Martinus Nijhoff, The Hague.

Mundy, J., and Chua, N.-H. (1988) Abscisic acid and water stress induce the expression of a novel rice gene. *EMBO J.* 7:2279–2286.

Ohkuma, K., Lyon, J. L., Addicott, F. T., and Smith, O. E. (1963) Abscisin II, an abscission-accelerating substance from young cotton fruit. *Science* 142:1592–1593.

Osborne, D. J. (1982) The ethylene regulation of cell growth in specific target tissues of plants. In: *Plant Growth Substances 1982*, Wareing, P. F., ed., pp. 279–290. Academic Press, London.

Raschke, K. (1987) Action of abscisic acid on guard cells. In: *Stomatal Function*, Zeiger, E., Farquhar, G. D., and Cowan, I. R., eds., pp. 253–279. Stanford University Press, Stanford, Calif.

Sato, T., and Theologis, A. (1989) Cloning the mRNA encoding 1-aminocyclopropane-1-carboxylate synthase, the key enzyme for ethylene biosynthesis in plants. *Proc. Natl. Acad. Sci. USA* 86:6621–6625.

Shimazaki, K., Iino, M., and Zeiger, E. (1986) Blue light–dependent proton extrusion by guard-cell protoplasts of *Vicia faba*. *Nature* 319:324–326.

Sindhu, R. K., and Walton, D. C. (1987) Conversion of xanthoxin to abscisic acid by cell-free preparations from bean leaves. *Plant Physiol.* 85:916–921.

Sisler, E. C. (1979) Measurement of ethylene binding in plant tissue. *Plant Physiol.* 64:538–542.

Tal, M., and Imber, D. (1971) Abnormal stomatal behavior and hormonal imbalance in *flacca*, a wilty mutant of tomato. III. Hormonal effects on the water status in the plant. *Plant Physiol.* 47:849–850.

Taylor, H. F., and Burden, R. S. (1973) Preparation and metabolism of 2-[^{14}C]-*cis-trans*-xanthoxin. *J. Exp. Bot.* 24:873–880.

Yang, S. F. (1987) The role of ethylene and ethylene synthesis in fruit ripening. In: *Plant Senescence: Its Biochemistry and Physiology*, Thomson, W. W., Nothnagel, E. A., and Huffaker, R. C., eds., pp. 156–166. American Society of Plant Physiology, Rockville, Md.

Yang, S. F., Hoffman, N. E., McKeon, T., Riov, J., Kao, C. H., and Yung, K. H. (1982) Mechanism and regulation of ethylene biosynthesis. In: *Plant Growth Substances 1982*, Wareing, P. F., ed., pp. 239–248. Academic Press, London.

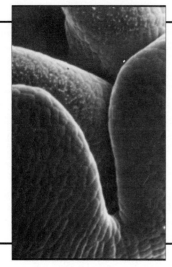

CHAPTER 20

Phytochrome and Photomorphogenesis

S EEDLINGS GROWN IN DARKNESS have a pale, almost ethereal appearance. This spindly, "etiolated" form of growth is dramatically different from the stockier, green appearance of light-grown seedlings (Fig. 20.1). Given the key role of photosynthesis in plant metabolism, one would be tempted to attribute much of this contrast to differences in the availability of light-derived metabolic energy. However, it takes very little light or time to initiate the transformation from the etiolated to the green appearance. Within 10 min after applying a single flash of relatively dim light to a dark-grown pea seedling, one can measure a decrease in the rate of stem extension, the beginning of apical hook straightening, and the initiation of the synthesis of pigments characteristic of green plants. Light has acted as a signal to induce a change in the form of the seedling from one that facilitates growth beneath the soil to one more adaptive to growth above ground in the light. Photosynthesis cannot be the driving force of this transformation because chlorophyll is not present during this time. Full de-etiolation does require some photosynthesis, but the initial rapid changes are induced by a distinctly different light response called *photomorphogenesis*.

Among the different pigments that can promote photomorphogenic responses in plants, the most important are those that absorb blue light and red light. The blue-light photoreceptors were discussed in relation to guard cells and phototropism in Chapters 6 and 16. The focus in this chapter will be the red-light photoreceptor, *phytochrome*.

The Photochemical and Biochemical Properties of Phytochrome

Although phytochrome was not identified as a unique chemical species and named until 1959, various red light–induced morphogenic responses in plants had been well documented since the 1930s. The list of such responses is now enormous and includes one or more responses at almost every stage in the life history of a wide range of different green plants (Table 20.1).

A key breakthrough in the history of phytochrome was the discovery that the effects of red light (650–680 nm) on morphogenesis could be reversed by a subsequent irradiation with light of longer wavelengths (710–740 nm), called far-red light. The first demonstrated case of red/far-red light photoreversibility was lettuce seed germination: red light promoted germination, and far-red light given immediately after the

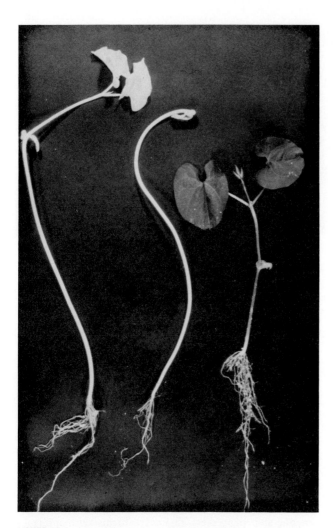

FIGURE 20.1. Bean (*Phaseolus vulgaris*) seedlings grown for six days either in continuous white light (right), in total darkness (center), or in darkness but with 5 min of dim red light given once a day (left). Note that 5 min of dim red light is sufficient to prevent some of the symptoms of etiolation, such as reduced leaf size and maintenance of the apical hook. (Photo courtesy of H. Smith.)

red light reversed its effects and inhibited germination (Fig. 20.2). It was predicted that the photoreceptor for this photoreversible system must itself be photoreversible; that is, it would change its own absorption properties (change color) reversibly after red and far-red light treatments.

An instrument to detect photoreversible color changes in plants was designed, and in 1959 this prediction was verified by observing such reversible changes in both plants and plant extracts (Butler et al., 1959). By using photoreversibility as a direct qualitative and quantitative assay of this pigment, scientists were soon able to purify and chemically characterize phytochrome as a protein pigment containing a specific light-absorbing molecule, or chromophore. Because knowledge of the photochemical and biochemical properties of purified phytochrome has shed light on how phytochrome functions in vivo, we will review these properties in this section.

The Two Forms of Phytochrome Are Interconvertible

In unirradiated plants, phytochrome is present in a red light–absorbing form referred to as **Pr**. This form, blue in color, is converted by red light to a far-red light–absorbing form called **Pfr**, which is blue-green in color. The Pfr form, in turn, can be converted back to Pr by far-red light. This quality of photoreversibility is the most distinctive property of phytochrome, and it may be expressed in an abbreviated form as follows:

$$Pr \underset{\text{Far-red light}}{\overset{\text{Red light}}{\rightleftharpoons}} Pfr$$

The interconversion of the Pr and Pfr forms can be measured in vivo or in vitro. In fact, most of the spectral

TABLE 20.1. Typical photoreversible responses induced by phytochrome in a variety of higher and lower plants.

Group	Genus	Stage of development	Pfr effects
Angiosperm	*Lactuca* (lettuce)	Seed	Promotes germination
	Avena (oat)	Seedling (etiolated)	Promotes de-etiolation (e.g., leaf unrolling)
	Sinapis (mustard)	Seedling	Promotes formation of leaf primordia, development of primary leaves
	Pisum (pea)	Adult	Inhibits internode elongation
	Xanthium (cocklebur)	Adult	Inhibits flowering (photoperiodic response)
Gymnosperm	*Pinus* (pine)	Seedling	Enhances rate of chlorophyll accumulation
Pteridophyta	*Onoclea* (sensitive fern)	Young gametophyte	Promotes growth
Bryophyta	*Polytrichum*	Germling	Promotes plastid replication
Chlorophyta	*Mougeotia*	Mature gameotophyte	Promotes chloroplast orientation to directional dim light

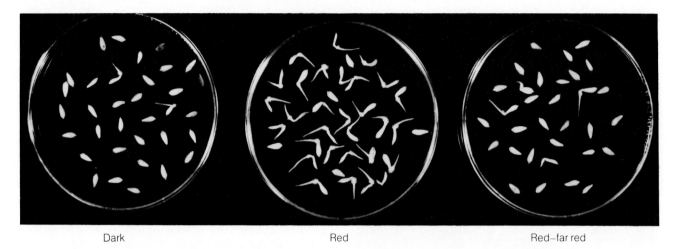

Dark   Red   Red–far red

FIGURE 20.2. The promotive effect of red light (R) on lettuce seed germination is reversed by far-red light (FR). Seeds irradiated by red light followed by far-red light (R-FR) have about the same germination rate as those that have been kept in darkness (D). Lettuce seed germination is a typical photoreversible response controlled by phytochrome. (From Kendrick and Frankland, 1976.)

properties of carefully purified phytochrome are the same as those observed in vivo.

The Pr form of phytochrome absorbs very little in the far-red region of the spectrum, but the spectra of Pr and Pfr have significant overlap in the red region (Fig. 20.3). Because red light is absorbed by both forms, when Pr molecules are given red light most of them will absorb it and be converted to Pfr, but then some of the Pfr will also absorb the red light and be converted back to Pr. The proportion of phytochrome in the Pfr form after saturating red-light irradiation is only 85%. Similarly, the very small amount of far-red light absorbed by Pr makes it impossible to convert Pfr entirely to Pr by broad-spectrum far-red light, though an equilibrium of 97% Pr and 3%

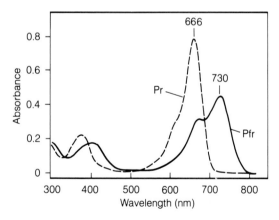

FIGURE 20.3. The absorption spectra of the Pr (dashed line) and Pfr (solid line) forms of phytochrome overlap. Because red light (640–690 nm) is absorbed by both Pr and Pfr, it does not generate pure Pfr but rather sets up a photoequilibrium of about 85% Pfr and 15% Pr. Similarly, far-red light (710–750 nm) does not convert all of phytochrome to Pr, but generates a photoequilibrium of about 97% Pr and 3% Pfr. (From Vierstra and Quail, 1983.)

Pfr can be achieved. This equilibrium is termed the **photostationary state**. Note that both forms of phytochrome also absorb in the blue region of the spectrum. However, most blue-light effects on plants are thought to be due to the action of a separate blue-light photoreceptor (Chapter 16) rather than phytochrome.

There Are Short-Lived Intermediates Between Pr and Pfr

The photoconversions of Pr to Pfr and of Pfr to Pr are not one-step processes. There are forms with intermediate absorbances between Pr and Pfr. By irradiating phytochrome with very brief flashes of light, absorption changes that occur in less than a millisecond can be observed. This technique shows that after Pr absorbs a flash of red light, several short-lived spectral forms are generated in a reproducible sequence before Pfr is formed. A different set of intermediate spectral forms occur in the photoconversion of Pfr to Pr (Rüdiger, 1986). Sunlight includes a mixture of all visible wavelengths. Under such white-light conditions, both Pr and Pfr are excited, and phytochrome cycles continuously between Pr and Pfr. In this situation, the intermediate forms of phytochrome accumulate and make up a significant fraction of the total phytochrome. It is possible that such intermediates could even play a role in initiating or amplifying phytochrome responses under natural sunlight conditions, but this question has as yet not been resolved.

In Most Cases Pfr Is the Physiologically Active Form

Because phytochrome responses are induced by red light, they could, in theory, result either from the appearance of

Pfr or from the disappearance of Pr. In most cases studied, a quantitative relationship holds between the magnitude of the response and the amount of Pfr generated by light, but no such relationship holds for the amount of Pr generated by response-inducing light. Evidence such as this has led to the conclusion that, in most cases, Pfr is the physiologically active form of phytochrome. In the cases in which it has been shown that a phytochrome response is not quantitatively related to the absolute amount of Pfr, it has been proposed that the ratio between Pfr and Pr determines the magnitude of the response.

The Amount of Pfr Can Be Regulated by Synthesis, Breakdown, and Dark Reversion

Light is not the only factor regulating the appearance or disappearance of Pfr. Both in vivo and in vitro, the Pfr of certain plants spontaneously changes to Pr in darkness by a reaction called **dark reversion**. The rate of this reaction is dependent on temperature and pH and can be greatly accelerated, even to a time course of seconds, by certain chemicals, especially reducing agents. Phytochrome extracted from seedlings of cereals, such as oat and rye, exhibits dark reversion only if it is partially fragmented by enzymes called proteases. It is possible that the dark reversion of Pfr observed in vivo in some plants is also dependent on partial proteolysis. Whatever the mechanism of dark reversion, the phenomenon of a timed disappearance of so potent an effector as Pfr has led to the hypothesis that reversion may have a regulatory function, particularly in relation to flowering (see Chapter 21). Unfortunately, the physiological significance of dark reversion remains obscure.

Both forms of phytochrome are targets of proteases present in cells and, like all other cellular proteins, undergo breakdown or destruction during the normal cellular turnover process. In most plants examined, the photoconversion of Pr to Pfr accelerates the destruction of phytochrome, suggesting that Pfr is more labile to proteolysis than Pr. There is evidence that Pfr may become marked or tagged for proteolytic destruction by the ubiquitin system (Shanklin et al., 1987). As discussed in Chapter 2, ubiquitin is a small polypeptide that binds covalently to proteins and serves as a recognition site for a large proteolytic complex. The disappearance of Pfr by this process would be quite distinct from dark reversion.

In addition to dark reversion and degradation, phytochrome levels are also regulated by the rate of synthesis. Phytochrome is synthesized in the Pr form in the dark. The same light stimulus that converts it to Pfr and thus accelerates its degradation also decreases its synthesis. As we shall see at the end of the chapter, the synthesis of phytochrome is itself phytochrome-regulated. The contributions of all these factors to Pr and Pfr levels are summarized in the diagram in Figure 20.4.

Two Types of Phytochrome Have Been Identified

The diversity and amount of biochemical information available on phytochrome are impressive. Obtaining this information required a method for purifying phytochrome from plant tissue, which was developed through research pioneered by H. W. Siegelman. After surveying many different plants and plant tissues, Siegelman found that the young etiolated seedlings of cereals, such as oats and rye, were particularly rich sources of phytochrome. Some initial success in purifying phytochrome from etiolated oats was reported in 1964 (Siegelman and Firer, 1964), and many workers subsequently used dark-grown oat seedlings for phytochrome extractions. As a result, most of the biochemical information on phytochrome is derived from studies on oat seedlings, although there is a considerable amount of data on phytochrome isolated from dark-grown rye and pea seedlings as well.

Although on a fresh-weight basis there is 50 times more phytochrome in etiolated tissue than in equivalent green tissue of the same age, the overall content of phytochrome in etiolated plants is low, about 0.2% of the total extractable protein. Thus, to purify even milligram quantities usually requires over a kilogram fresh weight of starting material. Etiolated seedlings are also a rich source of proteases, and phytochrome is especially susceptible to proteolysis. To minimize this problem,

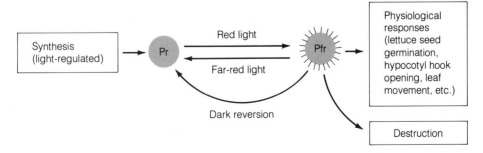

FIGURE 20.4. A summary of processes regulating the turnover of phytochrome in cells.

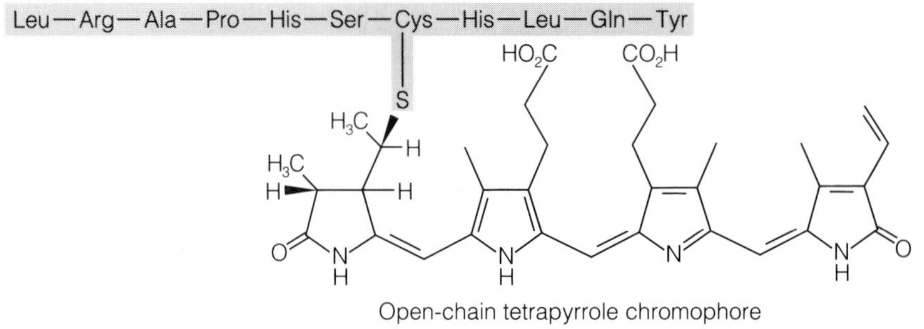

FIGURE 20.5. Structure of the Pr chromophore and the peptide region bound to the chromophore through a thioether linkage. Amino acid abbreviations: Leu, leucine; Arg, arginine; Ala, alanine; Pro, proline; His, histidine; Ser, serine; Cys, cysteine; Gln, glutamine; Tyr, tyrosine. (From Lagarias and Rapoport, 1980.)

Open-chain tetrapyrrole chromophore

extractions must proceed rapidly, at a low temperature, and in buffers containing protease inhibitors. The stability of phytochrome during its purification is also favored by using green-light illumination, which is absorbed only poorly by phytochrome, instead of white light.

Very little phytochrome is extractable from green plants, and some of it is different from the majority of the phytochrome present in etiolated plants. Recent research has shown that there are, in fact, two different phytochromes, with different (though similar) amino acid sequences. Current terminology refers to them as type I and type II phytochrome (Konomi et al., 1987). Type I is about nine times more prevalent in dark-grown pea seedlings than Type II, whereas in light-grown pea seedlings the amounts of the two types are about equal. There is evidence that the two types of phytochrome regulate distinctly different functions in plants. So far, almost all of the research on phytochrome has been done on type I, and very little is known about the structure and function of Type II. Thus, unless otherwise noted, most of the phytochrome referred to in the rest of this chapter can be assumed to be type I.

Two 124-kDa Monomers Make Up the Phytochrome Dimer

The chromophore of phytochrome is an open-chain tetrapyrrole and is covalently bound to the protein portion of phytochrome (phytochrome apoprotein) through a thioether linkage (Fig. 20.5). There appears to be only one chromophore per monomer of protein, and the primary sequence of amino acids immediately surrounding the thioether linkage to the chromatophore has been determined (Fig. 20.5). The Pr and Pfr chromophores differ mainly in the number of their resonating double bonds. In addition, the Pfr configuration has the methene bridge of its A ring bound to one or two amino acid residues in the protein structure.

Because phytochrome is unusually susceptible to proteolysis during purification, its published chemical properties have undergone several revisions as progressively larger versions of the molecule have been isolated. The currently accepted monomer molecular mass of intact type I oat phytochrome is 124,000 Da, and the value for phytochrome from other species is reported

to be close to this. The chemical characterizations of purified phytochrome reviewed here are all of the 124,000 Da version, which is apparently undegraded and most closely resembles the phytochrome prevalent in the living cells of dark-grown plants. The gene for phytochrome (type I) has been isolated, cloned, and sequenced, and the primary sequence of amino acids has been deduced from the DNA sequence (Hershey et al., 1985).

Native phytochrome has a molecular mass of about 240,000 Da and thus consists of a dimer of two equivalent subunits. The Pr form has been visualized by spraying a solution of the protein onto mica flakes, drying it under vacuum, and then "shadowing" the dried molecules by using platinum atoms to enhance its three-dimensional structure. After further treatment, the preparation can be viewed in an electron microscope. As shown in Figure 20.6a, phytochrome molecules appear as a collection of spots, the number depending on the orientation of the protein. Based on this information and more accurate measurements using x-ray scattering, the model shown in Figure 20.6b has been proposed (Tokutomi et al., 1989). Each of the subunits (I and II) is composed of two separate domains, an N-terminal chromophore binding domain (A) and a C-terminal nonchromophoric domain (B). Each domain is represented by a disk of the appropriate size and thickness. The shape of phytochrome may be important for its functions in the cell.

Phytochrome Undergoes Subtle Conformational Changes

The amino acid composition of phytochrome shows a higher percentage of hydrophobic amino acids than is typical of water-soluble proteins. Recent data indicate that the photoconversion of Pr to Pfr exposes more of the hydrophobic region of the molecule, consistent with the idea that Pfr, particularly the nonchromophoric domains, may function in association with membranes (Fig. 20.6b).

Amino acid side chains of the Pr and Pfr forms of phytochrome are differentially reactive to various reagents, suggesting that the protein undergoes some conformational change during the phototransformation. This shape change must be subtle because it cannot be detected by spectral assays, such as circular dichroism or nuclear magnetic resonance, that reveal general aspects of

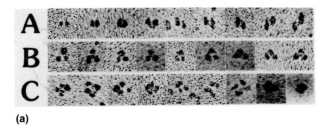

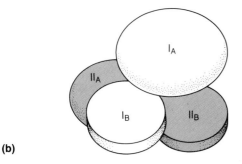

FIGURE 20.6. Visualization of phytochrome. (a) Phytochrome from peas, purified and then sprayed onto a mica flake. After drying, the specimen was shadowed with platinum at an angle of 6° to enhance the three-dimensional structure. A carbon replica was prepared and viewed in the electron microscope. Various images were produced, including pairs of spots (A), three spots (B), and clusters of spots (C). The variation in the pattern in (a) is believed to be caused by different orientations of the protein on the mica. (b) The structure of the phytochrome dimer based on a type of x-ray scattering that does not require crystallization. Each monomer consists of a chromophore binding domain (A) and a smaller nonchromophoric domain (B). The monomers are labeled I and II. The molecule as a whole has an ellipsoidal rather than a globular shape. Phytochrome may interact with membranes via conformational changes in the nonchromophoric B domains. (From Tokutomi et al., 1989.)

comparisons of phytochromes from different species. Such studies have revealed that, overall, phytochrome is not a highly conserved protein; that is, it does not have many sequences that have remained unchanged during its evolutionary history. However, the relatively rarer mAbs that do cross-react with phytochrome molecules derived from distantly related species are likely to be directed against the (presumably) few highly conserved domains of the molecule (Cordonnier et al., 1986).

Phytochrome Localization in Tissues and Cells

Valuable insights into the function of a protein can be gained from learning where it is localized in the organism—for example, whether it is present only in roots or, on the subcellular level, only in nuclei or mitochondria. It is not surprising, then, that significant effort has been expended to determine the location of phytochrome both throughout the plant and within individual cells. In this section we will review the results of these studies.

Phytochrome Can Be Detected in Tissues Spectrophotometrically

The photoreversible spectral changes that phytochrome undergoes after irradiations with red and far-red light uniquely identify the presence of phytochrome and permit its quantitation. These spectral changes can be detected in vivo; in fact, the initial discovery of phytochrome was made possible by the availability of a spectrophotometer that could quantitate changes in the red and far-red regions of the spectrum even against the background interference of living tissue.

The presence of chlorophyll makes it very difficult to detect phytochrome spectrally in green tissue. But in etiolated plants, phytochrome can easily be detected and measured by in vivo spectroscopy. Using this assay method, scientists have confirmed the presence of phytochrome in many different tissues of monocots and dicots and in gymnosperms, ferns, mosses, and algae.

Dark-grown seedlings have been studied in greatest detail. Here the highest phytochrome levels are usually found in meristematic regions or in regions that were recently meristematic, such as the bud and first node of peas and the tip and node regions of the coleoptile in oats (Fig. 20.7).

Immunocytochemical Methods Have Revealed Where Phytochrome Is Concentrated in Tissues and Cells

Physical and chemical fixation methods are available that can immobilize phytochrome in cells while preserving its cross-reactivity with antibodies. There are also many

a protein's shape. Some of the differential reactivity of Pr and Pfr may be due to a rearrangement of the chromophore relative to the protein rather than to a conformational change of the protein itself.

Another indication that Pr and Pfr differ conformationally comes from immunological studies. In this regard, monoclonal antibodies (mAbs) have been especially useful. These are antibodies produced by cells that have all been cloned from a single antibody-producing cell. Because each antibody-producing cell produces only one type of antibody molecule, each mAb raised against a protein is specific for a particular site on that protein. Thus, mAbs to phytochrome react only with specific regions of the molecule. Several workers who have raised mAbs to phytochrome have found that certain of these antibodies can discriminate between the Pr and Pfr forms. Such results support the hypothesis that conformational changes occur in defined regions of the phytochrome protein during its photoconversion.

Monoclonal antibodies have also been useful for

FIGURE 20.7. Distribution of phytochrome in an etiolated pea seedling, as measured spectrophotometrically. Notice that phytochrome is most heavily concentrated in the apical meristems of the root and epicotyl, where the most dramatic developmental changes are occurring. (Adapted from Kendrick and Frankland, 1983.)

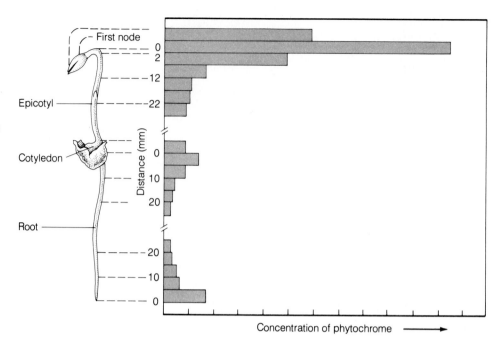

techniques for visualizing where antibodies have bound to phytochrome in fixed tissue. Such immunocytochemical methods have been used to localize phytochrome in plant cells and tissues. The early studies revealed a tissue-level distribution similar to that shown by spectrophotometric methods but added information on the location of phytochrome in roots (it is concentrated in the cap region). Moreover, they showed for the first time what spectrophotometric methods could not resolve: the subcellular distribution. Phytochrome was found both in the cytosol and in association with organelles, including plastids and nuclei.

One of the more interesting results from the early immunocytochemical studies was the finding that the photoactivation of phytochrome induced a dramatic change in its subcellular distribution. When phytochrome is converted to Pfr, its distribution in coleoptile parenchymal cells of oat and rice seedlings changes within minutes from mainly diffuse through the cytosol to mainly sequestered in discrete regions. When Pfr is photoconverted back to Pr by far-red light, phytochrome slowly returns to the original diffuse distribution. In pea seedlings, phytochrome seems to become inaccessible to antibodies immediately after the red-light treatment, and the sequestered distribution appears only after a subsequent 10-min incubation in the dark (Saunders et al., 1983). These results could indicate that Pfr associates tightly with some intracellular receptor that is important for its function soon after its production in the cell.

Recent ultrastructural studies have revealed that most of the sequestered Pfr in oat cells is associated with discrete structures that are globular to oval in shape, are about 300 nm in size, and have no morphologically

identifiable membranes surrounding them. The function or significance of these structures is as yet unknown, though one hypothesis is that they represent intracellular sites where the Pfr-dependent dark destruction of phytochrome occurs.

In summary, the immunocytochemical results indicate that phytochrome is most concentrated in young, undifferentiated tissue and, on the subcellular level, undergoes a distribution change from diffuse cytosolic to sequestered upon its conversion to Pfr. The strong correlation between phytochrome and cells that have the potential for dynamic developmental changes is certainly consistent with the important role of phytochrome in controlling such changes. In addition, the dramatic shift in the subcellular distribution of phytochrome upon conversion to Pfr is in accord with the idea that Pr and Pfr have radically different roles in the cell. Learning more about the distribution of Pfr may help to clarify how it alters plant cell development so significantly.

Photoreversible Responses Have Been Measured in Isolated Organelles

Because phytochrome regulates plastid development and modulates gene expression (see below), there is no doubt that it controls the function of at least some organelles in vivo. But does it *directly* associate with these organelles in order to alter their activity?

The most reliable data available on the intracellular sites of phytochrome are the in situ results reviewed above. These data, however, are equivocal on the question of whether phytochrome is associated with specific organelles. With some antibody preparations and some tissues, it appears that phytochrome is associated with plastids, mitochondria, and nuclei. With other antibody preparations and tissues, no evidence of phytochrome association with any subcellular membrane is evident.

A less direct way to address this problem is to investigate whether phytochrome copurifies with any organelles or regulates any organellar function in vitro. The answer in both cases is consistently yes. Photoreversible effects on organellar enzyme or other functions have been reported in isolated plastids, peroxisomes, nuclei, and mitochondria (Roux, 1987). Furthermore, phytochrome has been detected in association with these purified organelles by spectral and/or immunochemical methods. However, it has not been determined whether the association of phytochrome with these organelles occurred before they were isolated (i.e., in vivo) or only during (and only as a result of) the extraction procedure. Thus, the hypothesis that phytochrome directly associates with organelles as part of its cellular functioning must be tested further for confirmation or disproof.

Phytochrome-Induced Whole Plant Responses

The variety of different phytochrome responses in intact plants is great, in terms of both the *kinds* of responses and the quantity of light needed to induce the responses (Table 20.1). A review of this variety will show how diversely the effects of a single photoevent—the absorption of light by Pr—are amplified throughout the plant.

The kinds of phytochrome-induced responses may be logically grouped, for ease of discussion, into rapid biochemical events and slower morphological events, including movement and growth. Some of the early biochemical responses affect later morphological responses. These rapid responses will be treated in detail in a later section.

Phytochrome Responses Vary in Lag Time and Escape Time

Morphological responses to the photoactivation of phytochrome may be observed visually after a lag time as brief as a few minutes (e.g., the initiation of chloroplast rotation in *Mougeotia*) to as long as a few weeks (e.g., flowering in many photoperiodic species; see Chapter 21). The more rapid of these responses are usually reversible movements of organelles or reversible dimensional changes (swelling, shrinking) in cells, but even some growth responses occur remarkably fast. Red-light inhibition of the stem elongation rate of light-grown *Chenopodium album* is observed within 8 min after increasing its relative level of Pfr (Morgan and Smith, 1978). Information about the lag time for a phytochrome

response helps scientists to evaluate the kind of biochemical events that could precede and play a causal role in the induction of that response. The shorter the lag time, the more limited the range of biochemical events that could have been involved.

Variety in phytochrome responses can also be seen in the phenomenon called **escape from photoreversibility**. Red light–induced events are reversible by far-red light for only a limited period of time, after which the response is said to have "escaped" from reversal control by light. A currently favored model to explain this phenomenon assumes that phytochrome-controlled morphological responses are the result of a step-by-step sequence of linked biochemical reactions in the responding cells. Each of these sequences has a point of no return beyond which it proceeds irrevocably to the response. The escape time for different responses ranges from less than a minute to, remarkably, hours.

Phytochrome Effects on Plants Can Be Classified According to the Quantity of Light Required

Phytochrome responses may also be distinguished by the amount of red light required to induce them. This amount is referred to as fluence and is measured in moles of quanta per square meter. The total fluence of light given to a sample is a function of two factors: the photon flux density, or number of photons striking a given area per unit time (N), and the irradiation time (t). For every phytochrome response there is a characteristic range of light fluences over which the magnitude of the response is proportional to the fluence. This range varies greatly for different phytochrome responses. Some responses can be initiated by fluences as low as 0.1 nmol m^{-2} (one-tenth of the amount of light emitted from a firefly in a single flash). These are classified as **very low fluence (VLF) responses**. Red light can affect mesocotyl and coleoptile growth in etiolated oat seedlings at fluences this low. Other responses cannot be initiated until the fluences reach 1.0 μmol m^{-2}. These are referred to as **low fluence (LF) responses**, and they include most of the classical photoreversible responses, such as the promotion of lettuce seed germination, that are mentioned in Table 20.1.

The amount of light needed to induce VLF responses converts less than 0.02% of the total phytochrome to Pfr. Since the far-red light that would normally reverse a red-light effect converts 97% of the Pfr to Pr (see p. 492), about 3% of the phytochrome remains as Pfr—significantly more than is needed to induce VLF responses (Mandoli and Briggs, 1984). Thus, far-red light cannot reverse VLF responses. It is the red-light action spectrum, rather than photoreversibility, that suggests that phytochrome is the photoreceptor for VLF responses.

Still another group of phytochrome responses can be

TABLE 20.2. Some plant photomorphogenic responses induced by high irradiances.

Anthocyanin synthesis in various dicot seedlings and in apple skin
  segments
Inhibition of hypocotyl elongation in mustard, lettuce, and petunia
  seedlings
Induction of flowering in *Hyoscyamus* (henbane)
Plumular hook opening in lettuce
Enlargement of cotyledons in mustard
Ethylene production in sorghum

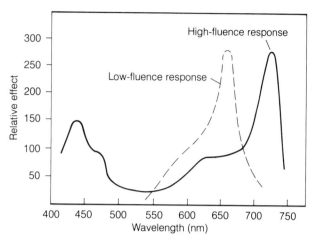

FIGURE 20.8. A comparison of the action spectra of typical high fluence (solid line) and low fluence (dashed line) responses. The high-fluence responses have peaks in the blue and far-red regions of the spectrum. Low fluence responses have a single peak at about 660 nm. High fluence responses are not photoreversible, although there is evidence that phytochrome is one of the pigments involved. (From Smith, 1974.)

distinguished by the quantity of light needed to promote them. These are the **high fluence (HF) responses,** several of which are listed in Table 20.2. These responses are elicited by prolonged or continuous irradiation, requiring exposures for hours, not minutes, and fluences in excess of 10 mmol m^{-2}. They have action spectra with peaks in the far-red and blue regions (Fig. 20.8), and they are not photoreversible. Experimental results suggest that phytochrome may be one of the photoreceptors that promote this response, although a blue-light photoreceptor and, in some cases, one or more of the chlorophylls in photosystem I or II may also be involved (Mancinelli, 1980). Perhaps it should not be surprising that a reaction which requires so much light over so long an exposure time would involve more than one photoreceptor.

Phytochrome Enables Plants to Adapt to Changes in Light Conditions

The presence of a red/far-red reversible pigment in all green plants, from algae to dicots, suggests that these wavelengths of light provide information that helps plants adjust to their environment. What environmental conditions result in changes in the relative levels of these two wavelengths of light in natural radiation?

The ratio of red light (R) to far-red light (FR) varies remarkably in different environments. This ratio can be defined as

$$R/FR = \frac{\text{photon fluence rate in 10-nm band centered on 660 nm}}{\text{photon fluence rate in 10-nm band centered on 730 nm}}$$

(Smith, 1982). Table 20.3 summarizes both the total photons (400–800 nm) and the R/FR values in six natural environments. Both parameters vary greatly in different environments. Compared with direct daylight, there is relatively more FR in sunset light, under 5 mm of soil, and especially under the canopy of other plants (as on the floor of a forest). The canopy phenomenon results from the fact that green leaves absorb red light because of their

high chlorophyll content but are relatively transparent to far-red light.

Phytochrome, then, can serve as an indicator of the degree of shading of a plant by other plants. As shading increases, R/FR decreases and the ratio of Pfr to the total phytochrome (Pfr/P$_{total}$) decreases. When simulated natural radiation was used to vary the FR content, it was found that the higher the FR content (i.e., the lower the Pfr/P$_{total}$) the higher the stem extension rate for "sun plants," that is, plants that normally grow in an open-field habitat (Fig. 20.9, solid line) (Morgan and Smith, 1979).

TABLE 20.3. Ecologically important light parameters. The light intensity factor (400–800 nm) is given as the photon flux density, and phytochrome-active light is given as the R/FR ratio.*

	Photon flux density ($\mu mol\ m^{-2}\ s^{-1}$)	R/FR
Daylight	1900	1.19
Sunset	26.5	0.96
Moonlight	0.005	0.94
Ivy canopy	17.7	0.13
Lakes, at a depth of 1 m		
Black Loch	680	17.2
Loch Leven	300	3.1
Loch Borralie	1200	1.2
At a depth of 5-mm soil	8.6	0.88

* Absolute values taken from spectroradiometer scans; the values should be taken to indicate the relationships between the various natural conditions and not as actual environmental means.
From Smith (1982); p. 493.

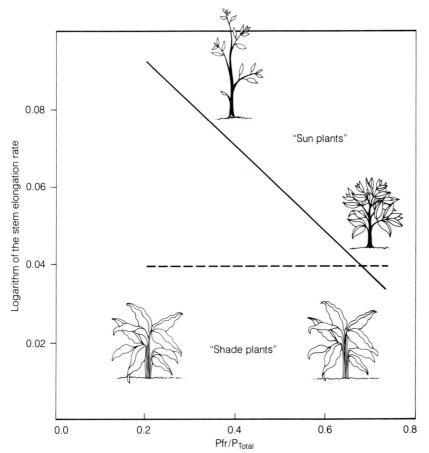

FIGURE 20.9. The relative ratio of Pfr to total phytochrome (Pfr/P$_{total}$) has an effect on the stem elongation rate of species that normally grow in a more sunlit environment (solid line) but not on species that normally grow in a more shaded environment (dashed line). This result supports the idea that phytochrome plays a role in shade perception for plants not adapted to a shaded environment. (Modified from Morgan and Smith, 1979.)

Simulated canopy shading induced these plants to allocate more of their resources growing taller. This correlation did not hold for the woodland understory plants that were tested. These "shade plants," which normally grow in a shaded environment, showed little or no alteration in their stem extension rate as they were exposed to higher R/FR values (Fig. 20.9, dashed line). Thus, there appeared to be a systematic relationship between phytochrome-controlled growth and species habitat. Such results are taken as an indication of the involvement of phytochrome in shade perception.

For a "shade-avoiding" plant there is a clear adaptive value in allocating its resources toward more rapid extension growth when it is shaded by another plant. In this way it can enhance its chances of growing above the canopy and acquiring a greater share of unfiltered, photosynthetically active light. The price for favoring internode expansion is usually reduced leaf area and reduced branching, but, at least in the short run, this evolutionary adaptation seems to work.

Light quality also plays a role in regulating the germination of some seeds. Large-seeded species, with ample food reserves to sustain prolonged seedling growth in darkness (e.g., underground), generally do not have a light requirement for germination. However, such a requirement is often observed in the small seeds of herbaceous and grassland species, many of which remain dormant, even while hydrated, if they are buried below the depth to which light penetrates. Even when such seeds are on or near the soil surface, their level of shading from the vegetation canopy (i.e., the R/FR they receive) is likely to affect their germination. There are well-documented instances in which the FR enrichment imparted by a leaf canopy inhibits germination in a range of small-seeded species.

For seeds of the tropical species *Cecropia obtusifolia* and *Piper auritum* planted on the floor of a deeply shaded forest, this inhibition can be reversed by placing a narrow-band red filter immediately above the seeds. This filter permits the red component of the canopy-shaded light to pass through while almost completely blocking the far-red component. Although the canopy transmits very little red light, it is enough to stimulate the seeds to germinate, probably because most of the inhibitory far-red light is excluded by the filter and the R/FR ratio is very high. These seeds would also be more likely to germinate in spaces receiving sunlight through gaps in the canopy than in densely shaded spaces. The sunlight would help to ensure that the seedlings became photosynthetically self-sustaining before their seed food reserves were exhausted.

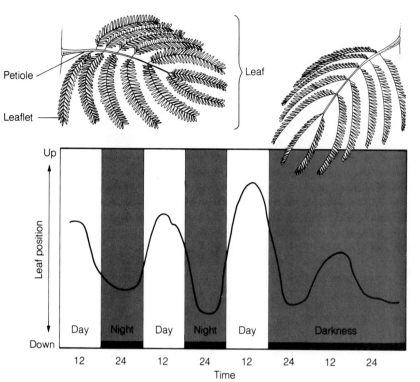

FIGURE 20.10. Circadian rhythm in the diurnal movements of *Albizzia* leaves. The leaves are elevated in the morning and lowered in the evening. The rhythm persists at a lower amplitude for a limited time in total darkness. In addition to the raising and lowering of leaves, the leaflets undergo opening and closing responses.

Phytochrome Regulates Certain Daily Rhythms

Various metabolic processes in plants, such as O_2 evolution and respiration, cycle alternately through high-activity and low-activity phases with a regular periodicity of about 24 h. These rhythmic changes are referred to as **circadian rhythms**, from the Latin *circa diem*, meaning "approximately a day." The **period** of a rhythm is the time elapsed between successive peaks or troughs in the cycle, and because the rhythm persists in the absence of external controlling factors it is considered to be **endogenous**. Not surprisingly, metabolic circadian rhythms are often expressed in visible forms that affect the whole plant or large parts of it. A well-described example is the leaf "sleep movements" of certain plants in the legume family (Fig. 20.10). These movements are the rhythmic expression of circadian ion fluxes into and out of motor cells at the base of the leaves. The swelling or shrinking of these cells in a structure called the **pulvinus** controls the up-and-down movement of the leaves.

There is abundant evidence that the molecular basis of rhythms involves, at some level, the regular cycle of ions or metabolites into and out of organelles and/or the cytosol of cells. The pulsed application of agents, such as lipid solvents, that alter membrane permeability shifts the phase of circadian rhythms. Such agents either advance or delay the daily cycle, depending on the point in the cycle at which they are applied. Phase-response curves, in which the magnitude of the advance or delay of a cycle is plotted against the time at which an agent was applied, are often used to illustrate this phenomenon (Fig. 20.11).

Light is a strong modulator of rhythms in both plants and animals. Although circadian rhythms that persist under controlled laboratory conditions usually have periods one or more hours longer or shorter than 24 h, in nature their periods tend to be uniformly closer to 24 h because of the synchronizing effects of light at daybreak. Pulses of light, like pulses of membrane-active agents, generate characteristic phase-response curves (Fig. 20.11c). A key photoreceptor for these light responses is phytochrome (see Chapter 21). It is no coincidence that phase-response curves generated by pulses of light resemble those generated by membrane reagents. As indicated below, phytochrome itself appears to modulate the permeability of membranes.

Cellular and Molecular Mode of Action

Phytochrome-regulated changes in plants all begin with absorption of light by the pigment. The absorbed light somehow alters the molecular properties of phytochrome so that it can induce a sequence of cellular reactions that ultimately cause a change in growth, development, or position of an organ (Table 20.1). Such a progression of causally linked events is termed a transduction sequence. Every environmental stimulus that affects plant or animal metabolism does so through one or more such sequences.

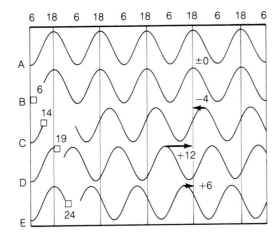

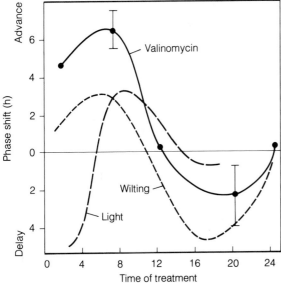

(a)

(c)

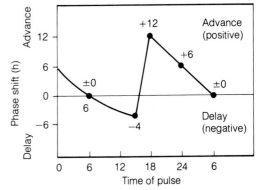

(b)

FIGURE 20.11. The phase-response curve of circadian rhythms. (a) Method for obtaining data for phase-response curves. The control response is curve A. The numbers at the top of the figure indicate the clock time, which is arbitrary. At various times during the cycle a pulse is given (shown by the boxes and adjacent numbers indicating the time), resulting in a shift of the curve by the amount indicated at the right of each curve. The rhythm may either be delayed (curve C) or advanced (curves D and E) in response to the pulse. (b) A plot of the data in (a). (From Sweeney, 1987.) (c) Effects of two environmental stimuli (light and water stress) and one agent that affects membrane permeability (the potassium ionophore valinomycin) on phase shifts of the circadian leaf movements of the bean *Phaseolus coccineus*. All the agents advance or delay the rhythm, depending on when during the rhythm they were applied. Zero indicates the last night peak (when the leaves are lowered) of the circadian cycle before treatment. These results are consistent with the idea that ion movements are an important factor in circadian timekeeping. (From Bunning and Moser, 1972.)

Identifying even the first few steps of such a sequence is a daunting task. Although it may be easy enough to document an early biochemical change induced by a stimulus, it is far more difficult to show that this change is causally related to a specific visible alteration in the plant. Furthermore, though we should not complicate our hypotheses more than necessary, there is no reason to propose that all phytochrome phenomena result from the same early sequence of biochemical reactions.

From the numerous explanations for phytochrome effects, certain hypotheses have emerged as particularly plausible. According to one of these hypotheses, many responses result from changes that phytochrome induces in membranes. Since other well-known light transduction pathways, such as photosynthesis and, in animals, vision, involve the mediation of membranes, it would be surprising if the phytochrome system were an exception. Phytochrome has repeatedly been shown to modulate membrane functions, though how it alters membrane proper-

ties is still controversial. The evidence that phytochrome-mediated changes in membrane properties may regulate photomorphogenesis is summarized in the next two subsections.

Phytochrome Can Associate with Membranes

As discussed earlier, immunocytochemical results give equivocal answers to the question of whether phytochrome is associated with organelles or other cellular membranes in situ. The evidence in support of the hypothesis that phytochrome can bind to membranes is indirect, but it comes from both in vivo (see the box on *Mougeotia*) and in vitro studies.

In vitro, the Pfr form of phytochrome binds tightly to pure lipid bilayers. Unlike Pr, it also facilitates ion movements across these bilayers. Although high salt concentrations can inhibit the binding of Pfr to spherical lipid bilayers, called liposomes, they cannot dissociate Pfr

Mougeotia: A Chloroplast with a Twist

MOUGEOTIA IS A filamentous green alga whose cells contain a single ribbonlike chloroplast nestled between two large vacuoles and surrounded by a layer of cytoplasm (Fig. 20.A, part a). The chloroplast can rotate about its long axis and can respond to incident radiation by orienting *perpendicular* to the direction of light (part b). Phytochrome is implicated in the response because red light elicits the greatest rotation and the effect of red light can be reversed by far-red light. Far-red light can also induce rotation, except that the chloroplast orients *parallel* to a source of far-red light (part b).

Using microbeams of red and far-red light, Wolfgang Haupt and his co-workers in Germany showed that the photoreceptor for chloroplast rotation is localized near the periphery of the cytoplasm in the vicinity of the plasma membrane. When a microbeam of red light was directed at the cell surface under a microscope, the edge of the chloroplast adjacent to the microbeam rotated 90°, even though the chloroplast itself was not illuminated.

The plasma membrane location of phytochrome in *Mougeotia* was deduced from studies using microbeams of plane-polarized red and far-red light. Phytochrome is a dichroic pigment; that is, it has a preferred electrical vector orientation for the absorption of light. The greatest absorption occurs when the electric vector is parallel to the absorbing surface of the pigment. If we think of the pigment as a rod and the electric vector as a narrow slit of light, more light will be absorbed when the light slit is oriented parallel to the long axis of the pigment than in any other orientation (part c). Haupt and his co-workers discovered that the ability of plane-polarized red light to cause rotation depended on the plane of polarization relative to the long axis of the cell. This phenomenon is depicted in the experiment shown in part d. Half of a cell was illuminated with plane-polarized red light vibrating transverse to the cell axis, while the other half was illuminated with light vibrating parallel to the cell axis. The chloroplast rotated from the profile to the face position only in response to red light vibrating in the transverse direction, producing a twist in the chloroplast. It was concluded that phytochrome has a defined orientation in the cell (Haupt, 1982). Later studies have specifically shown that Pr molecules are arranged in left-handed spirals around the cell.

What happens to the orientation of phytochrome after it is converted from Pr to Pfr? Part e shows the results of a photoreversibility experiment. The chloroplast is first treated with red light vibrating in a plane that induces rotation from a profile to a face position. Half of the cell is then immediately treated with far-red light vibrating in the same plane, while the other half is treated with far-red light vibrating perpendicular to the plane of the red light. The red-light effect is reversed only by the far-red light vibrating in a plane perpendicular to that of the red light, suggesting that phytochrome molecules change their orientation by about 90° during photoconversion (Kraml, 1986).

The effects described above are best accounted for by a model in which phytochrome is located on the plasma membrane of *Mougeotia* (part f). It is difficult to imagine how Pr and Pfr could maintain their respective orientations without being associated with the membrane. Rotation occurs because the chloroplast moves away from a localized region of Pfr and toward a region of Pr (part b). Thus, chloroplast movement occurs in response to a gradient of Pfr to Pr around the periphery of the cell. Does the reorientation of phytochrome molecules during photoconversion play a role in higher plants as well? The diffuse distribution of phytochrome in higher plant cells is inconsistent with such a model, but only further research will tell.

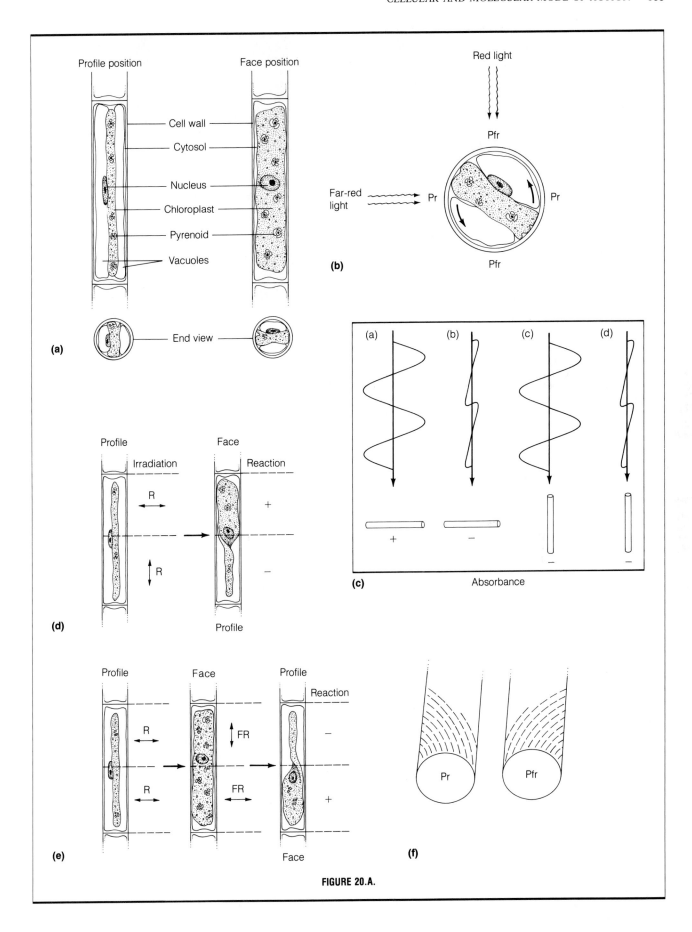

FIGURE 20.A.

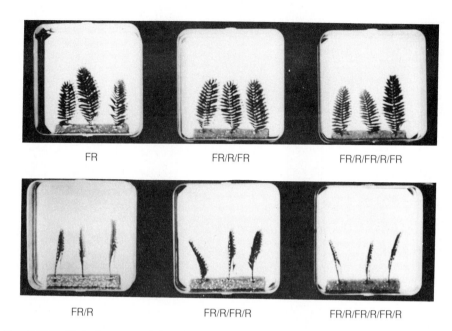

FIGURE 20.12. Phytochrome regulation of leaflet movement in *Mimosa pudica* L. The leaflets were first opened in white light and then transferred to darkness. Just before transfer to darkness they were exposed to a succession of red (R) or far-red (FR) pulses of 2-min duration. The leaflets on the top row all received a pulse of far-red light as the final treatment and remained open for 30 min after transfer to darkness. In contrast, the leaflets on the bottom row received red light as their final treatment and were fully closed after 30 min in the dark. The results show the requirement for Pfr for leaf closure in the dark and the photoreversibility of the effect. (From Fondeville et al., 1966.)

from liposomes once it has bound to them; detergents are required to produce this dissociation. These results suggest that the initial attachments of Pfr to membranes may be through ionic bonds but the final association is maintained by hydrophobic forces.

There is also in vitro evidence that Pfr binds to biological membranes. Mitochondrial preparations isolated from red light–irradiated oat seedlings have significant levels of phytochrome (Pfr) associated with them, whereas those from unirradiated plants contain no detectable phytochrome. The association of Pfr with isolated mitochondria does not appear to be a random surface-sticking phenomenon, because the Pfr is not accessible to digestion by enzymes (proteases) that would be expected to remove proteins bound only on organelle surfaces. Furthermore, the phytochrome associated with purified mitochondria can photoreversibly regulate some of their functions.

Distinct from the question of whether phytochrome binds directly to membranes is the hypothesis that phytochrome regulates membrane functions. Abundant in vivo data support this contention. The most convincing results are those that demonstrate rapid photoreversible modulation of the electrical potential across the plasma membrane. Such modulation has been directly measured in individual cells and has also been inferred from the rapid effects of red and far-red light on the surface potential of roots and oat (*Avena*) coleoptiles. The lag between the production of Pfr and the onset of mea-

surable potential changes varies from organism to organism; for example, it is about 1.7 s for depolarization (inside becomes less negative relative to outside the plasma membrane) in the giant alga *Nitella* and 4.5 s for hyperpolarization in *Avena*. The longer lag in *Avena* suggests that some secondary reaction (rather than Pfr itself) may induce the potential changes. As yet, this question has not been resolved.

Changes in the bioelectric potential of cells imply changes in the flux of ions across the plasma membrane. Phytochrome rapidly modulates fluxes of Ca^{2+} into and out of oat cells and, with a longer lag time, induces the flux of other ions in other cells. There is evidence linking the rapid red light–induced membrane depolarization in *Nitella* with a Ca^{2+} influx at the plasma membrane.

Other in vivo data indicating that phytochrome regulates membrane functions are those demonstrating the photoregulation of leaf and leaflet movements in *Mimosa*, *Albizzia*, *Samanea*, and other legumes (Fig. 20.12). As we have seen in the section on circadian rhythms, these movements result from the selective swelling and shrinking of pulvinar motor cells located at the bases of the leaves and leaflets. The swelling and shrinking, in turn, are regulated by the rapid entry and exit of ions, especially K^+. Phytochrome modulates these salt movements, presumably by affecting the permeability of the membranes across which these ions are transported (Satter and Galston, 1981).

Whether phytochrome directly or indirectly modulates the permeability of membranes remains an open question. A more important question is, what sort of permeability change could amplify a dim light signal into major changes in growth and development? If we interpret the photoregulated bioelectric potential changes as indicating that light induces membrane permeability changes, we should ask what ion fluxes are modulated by phytochrome and what developmental impact such fluxes could have.

Considering that the different ion transport systems in cells are closely coupled in their functions, it seems unlikely that phytochrome could affect the transport of any one ion without affecting others simultaneously. In fact, there are many reports of phytochrome-regulated proton fluxes, K^+ fluxes, and Ca^{2+} fluxes. Among these ions, Ca^{2+} has been considered a good candidate for transducing the photoactivation of phytochrome into significant physiological changes in plants. Since the Ca^{2+} studies are at an early stage, their discussion in the next section should be taken as only one example of how phytochrome-regulated membrane changes could affect plant growth and development.

Calcium and Calmodulin May Mediate Some Pfr Responses

As discussed in previous chapters on the plant hormones, calcium ions have long been known to be important for the regulation of sensory responses in animal cells. More recently, their role in regulating the responses of plants to environmental changes has also come to be appreciated. Hence, it was logical to examine the role of calcium in the response of plants to light.

The Ca^{2+}-regulated responses in both plant and animal cells generally require a Ca^{2+}-binding protein, and it is this complex of Ca^{2+} and a binding protein that triggers the response. There are several such proteins in plant and animal cells, but the one that is best known is calmodulin (Chapter 16).

Calmodulin is now known to function in almost all plant and animal cells as a transducer of small Ca^{2+} concentration changes into major physiological responses. The concentration of unbound Ca^{2+} in cells is usually very low (near 10^{-7} M), but calmodulin has such a high affinity for Ca^{2+} that it can begin binding the ion when its concentration rises to only 10^{-6} M. When one or more of its four high-affinity sites for Ca^{2+} is occupied by the ion, calmodulin changes conformation to an activated form that can bind to and stimulate a wide range of regulatory enzymes in cells. More than a half-dozen calmodulin-regulated enzymes have been found in plants and more than a dozen have been documented in animal cells. These include plasma membrane–localized Ca^{2+} pumps (Ca^{2+}-ATPases), NAD kinases, and enzymes (kinases and phosphatases) that regulate the phosphorylation (and thus the activity) of other enzymes. All of

these enzymes play central roles in the regulation of various metabolic activities in plants. Calmodulin is present both in the cytosol and in organelles, including plastids, mitochondria, and nuclei. Because of its high affinity for Ca^{2+}, calmodulin and the target enzymes it controls can be activated by very small changes in the Ca^{2+} concentration of the subcellular environment. This property of the calmodulin system makes it attractive for amplifying the photoconversion of a few phytochrome molecules into major cellular changes and for modulating other regulatory systems, such as hormonal responses, as well.

A current model of how Ca^{2+} could mediate phytochrome responses predicts that (1) Pfr should promote an increase in the Ca^{2+} concentration of one or more subcellular compartments, (2) some Pfr-stimulated enzymes will also be stimulated by calmodulin, and (3) calmodulin inhibitors should interfere with some red light–induced responses (Fig. 20.13). Indirectly, the model also predicts that some phytochrome responses should be inducible in darkness by chemical agents, known as *ionophores*, that promote Ca^{2+} uptake into cells. All of these predictions have been tested at some level, and the preliminary results, which appear to favor the model, are reviewed briefly here.

An obvious first question is, does phytochrome promote Ca^{2+} entry into plant cells? As assayed by autoradiography of $^{45}Ca^{2+}$ in *Mougeotia* (Fig. 20.14) and by quantitation of total Ca^{2+} in germinating fern spores, the answer is yes. To investigate whether the red light–stimulated entry of Ca^{2+} into cells played any role in modulating phytochrome responses, workers have tested whether chemically induced uptake of Ca^{2+} into cells could mimic Pfr effects. Pfr is a strong promoter of fern spore germination. Treatment of unirradiated fern spores with the Ca^{2+} ionophore A23187 can substitute for Pfr in inducing the germination of the spores (Wayne and Hepler, 1984). Although this ionophore can promote the entry of other divalent cations (such as Mg^{2+}), it stimulates germination only if Ca^{2+} is in the medium bathing the spores. In a similar experiment, ionophore A23187 can substitute for photoactivated phytochrome in promoting chloroplast rotation in *Mougeotia* (Fig. 20.15), but again only if Ca^{2+} is in the medium. In this experiment, the K^+ ionophore valinomycin is totally ineffective as an inducer, so the response is not merely due to nonspecific membrane disturbances resulting from ionophore incorporation. Finally, calcium and A23187 can substitute for red light in regulating leaflet closure in *Albizzia* (Moysset and Simon, 1989).

Is calmodulin the agent that transduces Ca^{2+} entry into physiological responses? This question has been addressed by testing whether certain chemicals that bind to Ca^{2+}-activated calmodulin and prevent it from activating target enzymes are able to block Ca^{2+}-inducible phytochrome responses. The phenothiazines and

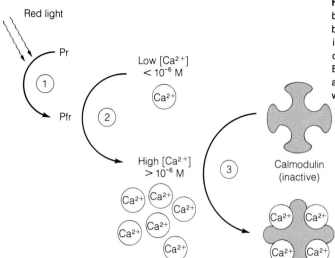

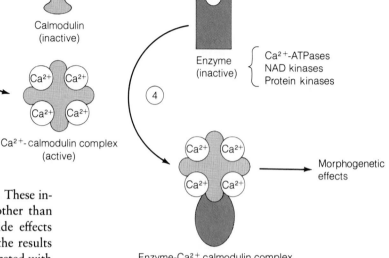

FIGURE 20.13. Proposed model for the cascade of events initiated by the activation of phytochrome. Step 1: Phytochrome activated by red light (650–680 nm) irradiation. Step 2: Pfr-mediated increase in the cytosolic Ca^{2+} concentration. Step 3: Activation of calmodulin by formation of a Ca^{2+}-calmodulin complex. Step 4: Binding of calcium-activated calmodulin to specific enzymes, thus activating them. These events could take place in the cytosol or within organelles, such as plastids, mitochondria, or nuclei.

calmidozolium are examples of such chemicals. These inhibitors can bind to Ca^{2+}-binding proteins other than calmodulin and they can have nonspecific side effects unrelated to calmodulin-binding proteins, so the results of studies using these inhibitors must be interpreted with caution. Nonetheless, both the fern spore germination and chloroplast rotation responses referred to in the preceding paragraphs are blocked by these calmodulin inhibitors (Wayne and Hepler, 1984). There are preliminary reports that other phytochrome responses are blocked by calmodulin inhibitors as well.

One of the more rapid visible responses to the photoactivation of phytochrome can be logically linked to preceding Ca^{2+}- and calmodulin-dependent cellular events. This response is the red-light inhibition of internode extension rates referred to earlier (Fig. 20.9). A major enzyme under calmodulin control is a plasma membrane Ca^{2+}-ATPase, which pumps Ca^{2+} out into the wall of plant cells. Pfr stimulates Ca^{2+} efflux from

oat cells (Hale and Roux, 1980). Increasing the Ca^{2+} concentration in plant cell walls inhibits some (as yet unidentified) biochemical process needed for cell wall loosening and cell extension (see Chapter 16). The induction of a transitory increase in cytosolic Ca^{2+} concentration by Pfr, then, could be expected to promote (1) the activation of calmodulin, (2) stimulation of the calmodulin-modulated Ca^{2+}-ATPase, (3) increased Ca^{2+} concentration in the walls of irradiated cells, and (4) decreased cell extension rates (Fig. 20.16). This model is obviously speculative, but it helps to illustrate how Pfr-induced Ca^{2+} fluxes could help lead to Pfr-induced growth changes.

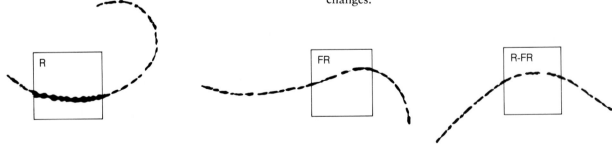

FIGURE 20.14. Photoreversible uptake of $^{45}Ca^{2+}$ by cells of the filamentous alga *Mougeotia*. Filaments of *Mougeotia* were transferred to a medium containing $^{45}Ca^{2+}$ after a portion of their cells (enclosed by the box) was irradiated with R, FR, or R followed by FR. The relative amount of $^{45}Ca^{2+}$ taken up by the irradiated cells was estimated by autoradiography; the darker the cell, the more $^{45}Ca^{2+}$ it took up. Cells irradiated with R took up more $^{45}Ca^{2+}$ than surrounding cells, whereas those irradiated with FR or with R-FR did not. These results indicated that phytochrome activation could result in calcium uptake by *Mougeotia* cells, consistent with step 2 of the model shown in Figure 20.13. (From Dreyer and Weisenseel, 1979.)

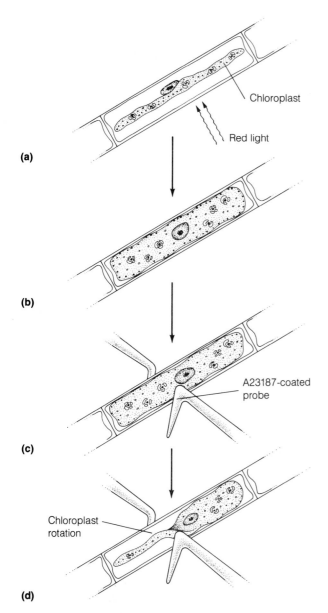

(a)

Chloroplast

Red light

(b)

(c)

A23187-coated probe

(d)

Chloroplast rotation

FIGURE 20.15. (left) The promotion of chloroplast rotation in *Mougeotia* by light or by positional application of a calcium ionophore, A23187. (a) A *Mougeotia* cell receives red-light irradiation through the edge surface of its chloroplast, which induces its chloroplast to turn to face view (b) as seen by an observer looking down from above. Chloroplast rotation can also be induced in the dark by placing A23187-coated probes against the two sides of the cell adjacent to the edge surface of the chloroplast (c, d'). This result supports the hypothesis that calcium functions as a chemical messenger to couple the stimulus of Pfr production with physiological responses in plants. (From Serlin and Roux, 1984.)

It should be noted that calmodulin is responsive only to increases in the concentration of *free* Ca^{2+}. Most cellular Ca^{2+} is bound to cell structures and is not free to react with calmodulin. To promote an increased Ca^{2+} concentration in the cytosol, where most of the cell's calmodulin is present, Pfr would have to promote either a Ca^{2+} influx from outside the cell or release of Ca^{2+} from organelles (e.g., the vacuole, mitochondria, or the endoplasmic reticulum) into the cytosol. Ruth Satter and her colleagues at the University of Connecticut have provided evidence that phosphoinositide turnover may be involved in the control of leaf movement by light in *Samanea saman*. White light caused a rise in the levels of inositol triphosphate (IP_3) and diacylglycerol (DAG) after less than a minute of exposure (Morse et al., 1989). IP_3 could then release free Ca^{2+} from organelles (see Chapter 16).

It is very difficult to measure the level of free Ca^{2+} in plant cells and even more difficult to determine whether a particular stimulus, such as light, has altered this level. Because the Ca^{2+} entering plant cells is so rapidly bound to or sequestered in organelles, the evidence that Pfr promotes Ca^{2+} uptake into cells or causes an increase in IP_3 does not indicate whether Pfr also promotes an increase of free Ca^{2+} in the cytosol. Answering this question experimentally will be crucial for assessing the

FIGURE 20.16. (right) A model for phytochrome regulation of wall extensibility based on calmodulin. Step 1: Red light converts Pr to Pfr. Step 2: Ca^{2+}-calmodulin complexes (Ca^{2+}-CaM) are formed. Step 3: The Ca^{2+}-CaM complex activates a calcium pump on the plasma membrane. Step 4: The calcium concentration of the cell wall increases, inhibiting cell wall–loosening processes.

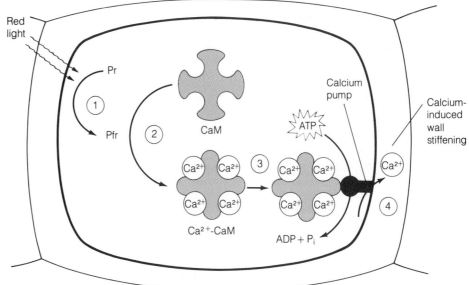

Red light

Pr

Pfr

CaM

Calcium pump

ATP

Calcium-induced wall stiffening

Ca^{2+}-CaM

ADP + P_i

validity of the hypothesis that Ca^{2+} and calcium-binding proteins such as calmodulin mediate some phytochrome responses. Until then, this idea must be considered only a plausible and testable hypothesis.

Phytochrome Regulates Gene Transcription

As the term photomorphogenesis implies, plant development is profoundly influenced by light. Etiolation symptoms include spindly stems, small leaves (in dicots), and the absence of chlorophyll. Complete reversal of these symptoms by light involves major long-term alterations in metabolism that can be brought about only by changes in gene expression. Historically, the hypothesis that phytochrome regulates morphogenesis primarily through its modulation of gene expression preceded the hypothesis that phytochrome affects primarily membrane properties. Whether the control of one is mediated by regulation of the other, or whether the two modes of action are independent, is not yet clear.

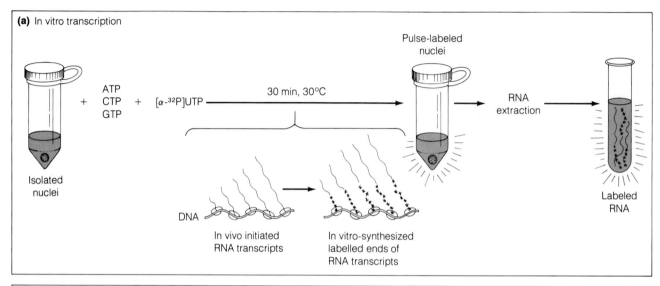

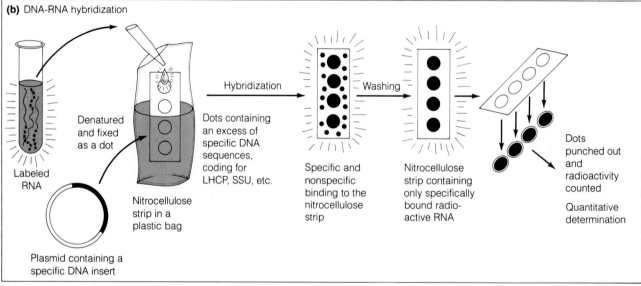

FIGURE 20.17. Nuclear run-off assay. (a) In vitro transcription. Isolated nuclei are incubated in a small tube with unlabeled ATP, CTP, and GTP, plus labeled UTP. During a 30-min incubation transcripts already initiated prior to nuclear isolation incorporate labeled UTP. The end-labeled RNA transcripts are purified. (b) DNA-RNA hybridization. Plasmids containing cDNA inserts for the gene of interest are spotted onto a strip of nitrocellulose paper. The strip is incubated in a plastic bag containing buffer and a drop of the labeled RNA. The mRNA complementary to the cDNA hybridizes to the spot, and the remainder of unbound RNA is removed by washing. The spots can then be punched out and the radioactivity quantified by counting. Alternatively, the radioactivity in each spot can be estimated by exposure to x-ray film. The size of the developed spot is roughly proportional to the radioactivity. (Adapted from Schäfer et al., 1986.)

FIGURE 20.18. Model for phytochrome regulation of *rbcS* and *cab* genes. Red light converts to Pr to Pfr, initiating a sequence of biochemical events that leads to the activation of one or more regulatory proteins in the cytosol. The regulatory proteins migrate to the nucleus, where they bind to specific light-regulated elements (LREs) in the promoter region of the *rbcS* and *cab* genes. Transcription is stimulated, leading to enhanced synthesis of the gene products, the small subunit (SSU) of Rubisco and the light harvesting chlorophyll *a/b* protein (LHCP). These proteins contain transit peptides that facilitate their entry into the chloroplast. Once inside the chloroplast, SSU combines with LSU (the large subunit of Rubisco) to form the holoenzyme. LHCP is incorporated into photosystem II on the thylakoid membrane. (Adapted from Schäfer et al., 1986.)

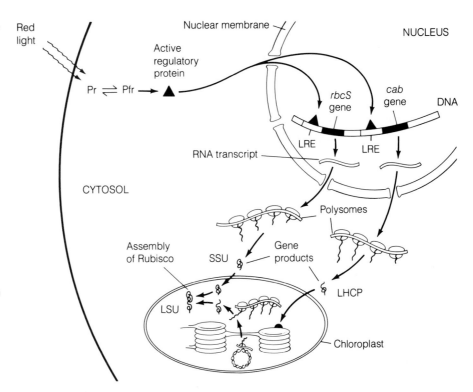

As in the case of cytokinin-regulated gene expression discussed in Chapter 18, the photoregulation of gene expression has focused on the nuclear genes encoding messages for chloroplast proteins, the small subunit (SSU) of ribulose-1,5-bisphosphate carboxylase/oxygenase (Rubisco) and the major light harvesting chlorophyll *a/b* protein (LHCP) associated with the light harvesting complex of photosystem II. These two proteins play an important role in chloroplast development and greening; hence, their regulation by phytochrome has been studied in detail. The genes for both of these proteins—*rbcS* and *cab*, respectively—are present in multiple copies in the genome; that is, they belong to multigene families.

Working with the duckweed *Lemna*, Tobin and her colleagues at the University of California, Los Angeles have shown by Northern blotting (see Chapters 16–18) that the abundance of the mRNAs for these two multigene families increases linearly as a function of time soon after exposure of dark-grown seedlings to light. Only a 1-min treatment is needed to increase the amount of mRNA, and the effect is reversed by far-red light. That phytochrome works at the transcriptional level was demonstrated through in vitro transcription in isolated nuclei (nuclear run-off experiments). Nuclei were isolated from tissue exposed to various light regimes and then incubated in the presence of radiolabeled ribonucleotide triphosphates. The labeled transcripts were hybridized to nitrocellulose filters containing cDNA copies of the *rbcS* or *cab* genes, and the bound radioactivity was counted (Fig. 20.17). Such assays have shown that phytochrome responding to red light enhances the rate of transcription of the *rbcS* and *cab* genes. Similar results have now been

obtained for a number of species, so the effect appears to be widespread.

Light-induced *rbcS* and *cab* mRNAs are translated on free polysomes (i.e., polysomes not attached to endoplasmic reticulum) as high-molecular-weight precursors containing a transit peptide at the amino-terminal end of the protein. The transit peptide contains hydrophobic sequences and serves as a targeting signal for transport into chloroplasts (Keegstra and Bauerle, 1988). That is, it facilitates binding to and transport across the double membrane of the chloroplast. Upon entry into the chloroplast, the transit peptides are cleaved by a specific processing protease, allowing the proteins to fold and assemble into their active conformations (see Fig. 20.18).

The Promoter Region of the *rbcS* Gene Is Light-Regulated

Promoters are DNA sequences located upstream from a gene—that is, on the 5′ flanking side—which function in the regulation of transcription. Eukaryotic promoters typically are composed of two functionally different regions: a short sequence that determines the transcription start site (the so-called TATA box, named for its most abundant nucleotides) and a more distant region that controls the developmental expression of the gene. Such regulatory sequences in the promoter are called cis-acting DNA elements (Chapter 2). Recently, the promoter regions of the *rbcS* gene from peas have been analyzed to identify cis-acting elements that might be involved in light regulation (Moses and Chua, 1988).

To study light-regulated elements (LREs), Chua and

colleagues at Rockefeller University transferred the *rbcS* gene (including the promoter region) of pea into petunia and tobacco plants, producing transgenic plants. This was accomplished with the tumor-inducing (Ti) plasmid of the soil bacterium *Agrobacterium tumefaciens*, which can be used as a vector for introducing foreign genes into plant cells (Chapter 18). Fortunately, the cDNA probe for the *rbcS* gene of peas did not hybridize very efficiently with the Rubisco small subunit mRNA of either petunia or tobacco. Thus, it was possible to use the cDNA from peas to monitor the expression of the pea gene in the transgenic plants.

The expression of the foreign *rbcS* gene in petunia and tobacco showed remarkable fidelity during development. That is, expression was highest in the leaves of the transgenic plant and was stimulated by red light and inhibited by far-red light. By selectively removing ("deleting") specific regions of the promoter region before transferring the gene to tobacco, it was possible to identify two short sequences of DNA in the 5′ upstream region that regulated both light induction and organ specificity of *rbcS* gene transcription. When a hybrid gene was constructed from the LRE region of the *rbcS* gene and the TATA box and coding sequence of a non-light-regulated foreign gene, the foreign gene became light-regulated when introduced into tobacco.

It is generally thought that cis-acting regulatory elements such as the LREs are regulated by proteins encoded elsewhere in the genome, called trans-acting factors. Trans-acting factors are DNA-binding proteins that bind to the cis-acting DNA sequences and regulate transcription of the gene. As discussed in relation to auxins (Chapter 16) and gibberellins (Chapter 17), DNA-binding proteins thought to play a role in hormone-regulated transcription have been identified using the gel retardation assay. A protein factor called GT-1 has been isolated from nuclear extracts that bind specifically to the LREs of the *rbcS* genes of several species (Kay et al., 1989). The binding site for the GT-1 protein has been identified. It is located about 150 bp upstream of the transcription start site of many *rbcS* genes. Recently, Lam and Chua (1990) demonstrated that the GT-1 sequence alone has sufficient information to turn a promoter that is not regulated by light into a light-regulated one. Thus, the GT-1 sequence "is likely to be a part of the molecular light switch for *rbcS* activation" (Lam and Chua, 1990). Other possible trans-acting factors have been identified, and it is likely that the regulation of transcription by light is exceedingly complex, involving a host of such factors.

Based on the foregoing results, the following model has been proposed. Red light causes the conversion of Pr to Pfr. Pfr either directly or indirectly (e.g., through Ca^{2+} and calmodulin) activates one or more regulatory proteins (phytochrome itself has no DNA-binding activity). The activated regulatory protein(s) then binds to the LRE sequences or other potential regulatory sites and stimulates transcription of the gene as illustrated in Fig. 20.19. It is assumed that most phytochrome-stimulated genes show this basic pattern of regulation.

Phytochrome Regulates the Expression of Its Own Gene

Phytochrome is encoded by a multigene family (e.g., the oat nuclear genome contains at least four phytochrome genes). When Pr is converted to Pfr by a pulse of red light, the steady-state phytochrome mRNA level decreases exponentially with a half-life of only 15 min, an extremely fast response by higher plant standards. This decrease has been shown by the nuclear run-off assay to be due to a sharp decrease in the rate of transcription (Kay et al., 1989). Thus phytochrome genes are **auto-regulated**. As noted earlier in the chapter, the repression of phytochrome gene expression is a contributing factor to the rapid decrease in total phytochrome when dark-grown plants are transferred to the light.

How does phytochrome repress the transcription of its own genes? Presumably, a trans-acting factor binds to a specific LRE in the promoter region, inhibiting transcription. A highly conserved sequence has now been identified in the promoter region of the phytochrome gene in rice (*Oryza sativa*) (Kay et al., 1989). The sequence is a potential LRE, although this function has not yet been demonstrated by deletion analysis, as described above for the *rbcS* gene. Surprisingly, GT-1, the factor that binds to the LRE of the *rbcS* genes, also appears to bind to the

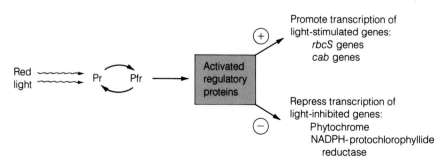

FIGURE 20.19. Scheme for phytochrome-regulated transcription of genes. Phytochrome in the Pfr form causes, directly or indirectly, the activation of one or more regulatory proteins. These regulatory proteins either promote the transcription of light-stimulated genes or repress the transcription of light-inhibited genes.

conserved sequence in the phytochrome gene. This raises the interesting possibility that the same DNA-binding protein that *promotes* the transcription of light-inducible genes may *repress* the transcription of light-inhibited genes, such as the gene for phytochrome. However, as noted earlier, the situation is undoubtedly much more complex, and much research remains to be done.

Summary

Photomorphogenesis refers to the dramatic effects of light on plant development and cellular metabolism. Red light exerts the strongest influence, and the effects of red light are often reversible by far-red light. Phytochrome is the pigment involved in most photomorphogenetic phenomena. Phytochrome is a large dimeric protein made up of two equivalent subunits. The monomer has a molecular mass of 124,000 Da and is covalently bound to a chromophore molecule, an open-chain tetrapyrrole. Phytochrome exists in two forms, a red light–absorbing form (Pr) and a far-red light–absorbing form (Pfr). Absorption of red light by Pr converts it to Pfr, and absorption of far-red light by Pfr converts it to Pr. However, the absorption spectra of the two forms overlap in the red region of the spectrum, leading to an equilibrium between the two forms called a photostationary state. Pfr is considered to be the active form in most cases; that is, it gives rise to the physiological response. Other factors in addition to light regulate the steady-state level of Pfr; these include dark reversion, degradation, and the rate of synthesis.

Spectroscopic and immunological studies have shown that phytochrome is most abundant in etiolated tissues and is concentrated in meristematic regions. During photoconversion to Pfr in higher plants, the phytochrome distribution within the cell changes from a diffuse to a sequestered pattern. Thus far, no organelle has been specifically associated with Pfr in higher plant cells. In the alga *Mougeotia*, phytochrome appears to be located on the plasma membrane, where it mediates chloroplast rotation by undergoing a 90° shift in orientation.

Phytochrome responses have been classified into very low fluence (VLF), low fluence (LF), and high fluence (HF) responses. These three types of responses differ not only in their fluence requirements but also in other parameters such as their escape times, action spectra, and photoreversibility. Phytochrome plays an important role in the detection of shade in plants adapted to high levels of sunlight. Phytochrome also regulates circadian rhythms, such as the sleep movements of leaves. Many of these phenomena can be traced to alterations of membrane properties. One way in which phytochrome might exert its effects is by increasing free calcium levels. The formation of a calcium-calmodulin complex might activate enzymes important for cell regulation. In addition, phytochrome is known to regulate the transcription of a number of genes. Many of the genes involved in greening, such as the nuclear-encoded genes for the small subunit of Rubisco and the chlorophyll *a/b* protein of the light harvesting complex, are activated by red light. Gene activation is thought to be mediated by regulatory proteins that bind to the promoter region of the gene. Red light also represses the transcription of a number of genes, including the gene for phytochrome. The repression of the gene for phytochrome is also presumed to be mediated by DNA-binding regulatory proteins.

GENERAL READING

Furuya, M., ed. (1987) *Phytochrome and Photoregulation in Plants.* Academic Press, Tokyo.

Haupt, W. (1982) Light-mediated movement of chloroplasts. *Annu. Rev. Plant Physiol.* 33:205–233.

Hendricks, S. B., and Van Der Woude, W. J. (1983) How phytochrome acts—perspectives on the continuing quest. In: *Encyclopedia of Plant Physiology*, Vol. 16A, Shropshire, W., Jr., and Mohr, H., pp. 3–23. Springer-Verlag, Berlin.

Kay, S. A., Keith, B., Shinozaki, K., Chye, M.-L., and Chua, N.-H. (1989) The rice phytochrome gene: Structure, auto-regulated expression, and binding of GT-1 to a conserved site in the 5′ upstream region. *Plant Cell* 1:351–360.

Keegstra, K., and Bauerle, C. (1988) Targeting of proteins into chloroplasts. *BioEssays* 9:15–19.

Kendrick, R. E., and Frankland, B. (1976) *Phytochrome and Plant Growth.* Edward Arnold, London. Also second edition (1983).

Kendrick, R. E., and Kronenberg, G. H. M., eds. (1986) *Photomorphogenesis in Plants.* Martinus Nijhoff, Dordrecht, The Netherlands.

Kraml, K. (1986) Light direction and polarization. In: *Photomorphogenesis in Plants*, Kendrick, R. E., and Kronenberg, G. H. M., eds., pp. 237–267. Martinus Nijhoff, Dordrecht, The Netherlands.

Leopold, A. C., and Kriedemann, P. (1975) *Plant Growth and Development*, 2d ed. McGraw-Hill, New York.

Mandoli, D. F., and Briggs, W. R. (1984) Fiber optics in plants. *Sci. Am.* 251 (February):90–98.

Moses, P. B., and Chua, N.-H. (1988) Light switches for plant genes. *Sci. Am.* 258 (April):88–93.

Roux, S. J. (1987) Phytochrome interactions with purified organelles. In: *Phytochrome and Photoregulation in Plants*, Furuya, M., ed., pp. 193–207. Academic Press, Tokyo.

Roux, S. J., Wayne, R. O., and Datta, N. (1986) Role of calcium ions in phytochrome responses: An update. *Physiol. Plant.* 66:344–348.

Rüdiger, W. (1986) The chromophore. In: *Photomorphogenesis in Plants*, Kendrick, R. E., and Kronenberg, G. H. M., eds., pp. 17–33. Martinus Nijhoff, Dordrecht, The Netherlands.

Satter, R. L., and Galston, A. W. (1981) Mechanisms of control of leaf movements. *Annu. Rev. Plant Physiol.* 32:83–110.

Smith, H. (1975) *Phytochrome and Photomorphogenesis: An Introduction to the Photocontrol of Plant Development.* McGraw-Hill, London.

Smith, H. (1982) Light quality, photoperception, and plant strategy. *Annu. Rev. Plant Physiol.* 33:481–518.

Sweeney, B. M. (1987) *Rhythmic Phenomena in Plants.* Academic Press, San Diego, Calif.

CHAPTER REFERENCES

Bunning, E., and Moser, I. (1972) Influence of valinomycin on circadian leaf movements in *Phaseolus. Proc. Natl. Acad. Sci. USA* 69:2732–2733.

Butler, W. L., Norris, K. H., Siegelman, H. W., and Hendricks, S. B. (1959) Detection, assay, and preliminary purification of the pigment controlling photoresponsive development of plants. *Proc. Natl. Acad. Sci. USA* 45:1703–1708.

Cordonnier, M.-M., Greppin, H., and Pratt, L. H. (1986) Identification of a highly conserved domain on phytochrome from angiosperms to algae. *Plant Physiol.* 80:982–987.

Dreyer, E. M., and Weinsenseel, M. H. (1979) Phytochrome-mediated uptake of calcium in *Mougeotia* cells. *Planta* 146:31–39.

Fondeville, J. C., Borthwick, H. A., and Hendricks, S. B. (1966) Leaflet movement of *Mimosa pudica* L. Identification of phytochrome action. *Planta* 69:357–364.

Hale, C. C., II, and Roux, S. J. (1980) Photoreversible calcium fluxes induced by phytochrome in oat coleoptile cells. *Plant Physiol.* 65:658–662.

Hershey, H. P., Barker, R. F., Idler, K. B., Lissemore, J. L., and Quail, P. H. (1985) Analysis of cloned cDNA and genomic sequences for phytochrome: Complete amino acid sequence for two gene products expressed in etiolated *Avena. Nucleic Acids Res.* 13:8543–8559.

Kaufman, L. S., Thompson, W. F., and Briggs, W. R. (1984) Different red light requirements for phytochrome-induced accumulation of *cab* RNA and *rbcS* RNA. *Science* 226:1447–1449.

Konomi, K., Abe, H., and Furuya, M. (1987) Changes in the content of phytochrome I and II apoproteins in embryonic axes of pea seeds during imbibition. *Plant Cell Physiol.* 28:1443–1451.

Lagarias, J. C., and Rapoport, H. (1980) Chromopeptides from phytochrome. The structure and linkage of the P_r form of the phytochrome chromophore. *J. Am. Chem. Soc.* 102:4821–4828.

Lam, E., and Chua, N.-H. (1990) GT-1 binding site confers light responsive expression in transgenic tobacco. *Science* 248:471–474.

Mancinelli, A. L. (1980) The photoreceptors of the high irradiance responses of plant photomorphogenesis. *Photochem. Photobiol.* 32:853–857.

Morgan, D. C., and Smith, H. (1978) Simulated sunflecks have large, rapid effects on plant stem extension. *Nature* 273:534–536.

Morgan, D. C., and Smith, H. (1979) A systematic relationship between phytochrome-controlled development and species habitat, for plants grown in simulated natural radiation. *Planta* 145:253–258.

Morse, M. J., Crain, R. C., Coté, G. G., and Satter, R. L (1989) Light-stimulated inositol phospholipid turnover in *Samanea saman pulvini. Plant Physiol.* 89:724–727.

Moysset, L., and Simon, E. (1989) Role of calcium in phytochrome-controlled nyctinastic movements of *Albizzia lophantha* leaflets. *Plant Physiol.* 90:1108–1114.

Saunders, M. J., Cordonnier, M.-M., Palevitz, B. A., and Pratt, L. H. (1983) Immunofluorescence visualization of phytochrome in *Pisum sativum* L. epicotyls using monoclonal antibodies. *Planta* 159:545–553.

Schäfer, E., Apel, K., Batschauer, A., and Mösinger, E. (1986) The molecular biology of action. In: *Photomorphogenesis in Plants*, Kendrick, R. E., and Kronenberg, G. H. M., eds., pp. 83–98. Martinus Nijhoff, Dordrecht, The Netherlands.

Serlin, B. S., and Roux, S. J. (1984) Modulation of chloroplast movement in the green alga *Mougeotia* by the Ca^{2+} ionophore A23187 and by calmodulin antagonists. *Proc. Natl. Acad. Sci. USA* 81:6368–6372.

Shanklin, J., Jabben, M., and Vierstra, R. D. (1987) Red light-induced formation of ubiquitin–phytochrome conjugates: Identification of possible intermediates of phytochrome degradation. *Proc. Natl. Acad. Sci. USA* 84:359–363.

Siegelman, H. W., and Firer, E. M. (1964) Purification of phytochrome from oat seedlings. *Biochemistry* 3:418–423.

Tokutomi, S., Nakasako, M., Sakai, J., Kataoka, M., Yamamoto, K. T., Wada, M., Tokunaga, F., and Furuya, M. (1989) A model for the dimeric molecular structure of phytochrome based on small-angle x-ray scattering. *FEBS Lett.* 247:139–142.

Vierstra, R. D., and Quail, P. H. (1983) Purification and structural characterization of 124-kDa phytochrome from *Avena. Biochem.* 22:2498–2505.

Wayne, R., and Hepler, P. K. (1984) The role of calcium ions in phytochrome-mediated germination of spores of *Onoclea sensibilis* L. *Planta* 160:12–20.

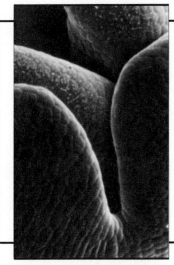

CHAPTER 21

The Control of Flowering

MOST PEOPLE LOOK FORWARD to the spring season and the profusion of flowers it brings. Many travelers carefully time their visits to the desert to coincide with the few days in which the ephemeral desert annuals bloom. In Washington, D.C., and throughout Japan, the cherry blossoms are received with spirited ceremonies. The strong correlation between flowering and seasons is common knowledge, but the phenomenon poses special questions for plant physiologists. Which environmental signals control flowering and how are those signals perceived? Can the environment be manipulated so that the flowering response is shifted in time? How do plants keep track of the seasons of the year and the time of the day? This chapter addresses these questions in connection with flowering and other time-dependent plant responses.

The transition from the vegetative to the reproductive stage is quite dramatic. In the vegetative state, plants continuously generate new leaves, stems, and roots. The switch to flowering involves a major change in the pattern of differentiation at the shoot apical meristems, leading to the development of the floral organs—sepals, petals, anthers, and carpels (Fig. 21.1). Through meiosis the anther gives rise to pollen grains containing microspores and the carpel gives rise to megaspores, which, upon fusion, will form the embryo of the next generation. The petals and sepals are usually colored to attract pollinators, and flowers often contain nectaries for the same purpose. Thus, flowering represents a complex array of highly specialized structures, which are completely different in form from those of the vegetative plant body and which also differ widely from species to species. Despite this complexity, flowering in all species is set in motion by internal and external factors that couple reproductive development with the plant environment.

Flowering may occur within a few weeks after germination, as in annual plants such as groundsel (*Senecio vulgaris*). In contrast, some perennial plants, especially many forest trees, may grow for 20 or more years before they begin to produce flowers. Thus, plants flower at widely different ages, indicating that age (or size) of the plant is one of the *internal* factors controlling the switch to reproductive development. A second major characteristic of flowering is that, in many species, it occurs only at one particular time of year and so must be under the control of *external* environmental conditions. Photoperiodism and vernalization are two of the most important mechanisms underlying such seasonal responses. **Photoperiodism** is a response to the length of day,

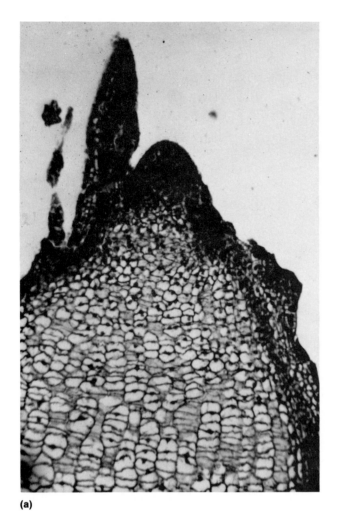
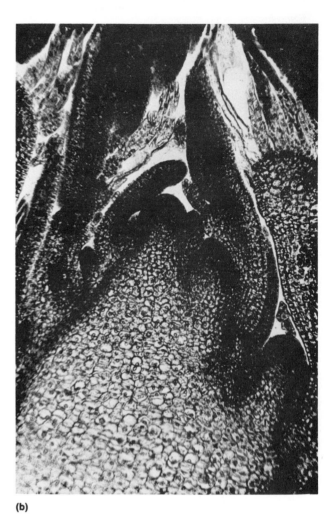

(a)

(b)

FIGURE 21.1. Vegetative (a) and reproductive (b) apices in the tropical legume *Stylosanthes guianensis* ("Stylo"). Ta, Terminal apex; AB, auxillary buds formed after a treatment of 12 short days. F_1 is the first floral primordium. L, leaf; LP, leaf primordium; B, bud. (From Ison and Hopkinson, 1985.)

while **vernalization** is an effect on flowering brought about by exposure to cold. The two control systems enable plants to synchronize reproduction. This synchrony has clear adaptive advantages, as it favors crossbreeding and adjusts flowering to coincide with favorable environments, particularly with respect to water and temperature.

The switch to reproductive development is clearly controlled by endogenous factors, such as plant age, and by several environmental conditions. The most important environmental signals are daylength and low temperature. Other signals, such as total light radiation and water stress, can be important modifiers of the responses to daylength and low temperature.

Effects of Plant Age

In the early, vegetative phase of growth, plants are said to be **juvenile**. Plants reach **maturity** when they achieve the *ability* to flower. In many plants flowering occurs without

responding to any particular environmental condition; this is called **autonomous induction** of flowering. In other cases, appropriate environmental treatments are necessary. Plants that have completed their juvenile phase but have not yet experienced the correct conditions for flowering are said to have achieved the state of **ripeness-to-flower.**

The transition from juvenility to maturity is often associated with changes in vegetative characteristics, such as leaf morphology, leaf arrangement on the stem, thorniness, rooting capacity, and leaf retention in deciduous plants (Fig. 21.2). Such changes occur particularly in woody species, although morphological differences between juvenile and mature plants are also found in herbaceous species. Lack of flowering in itself is not a clear-cut indication of juvenility because fully mature trees may not flower if growing very vigorously. Irregular flowering, such as biennial bearing in fruit trees, is also common. Because of these issues, a complete understanding of the effect of age in flowering must await the elucidation of its biochemical basis.

FIGURE 21.2. Juvenile and adult forms of ivy. The adult form (above) has entire ovate leaves arranged in spirals and an upright growth habit. Flowers are present. The juvenile form (below) has three- or five-lobed palmate leaves arranged alternately, a climbing growth habit, and no flowers. (Photo courtesy of D. Vince-Prue.)

Apical Changes Are Important in the Transition from Juvenility to Maturity

Attainment of a sufficiently large size appears to be more important than the plant's chronological age in determining the transition to maturity. Optimal conditions for the attainment of maturity are those that allow rapid growth to the minimum size at which the transition is possible. When this occurs, exposure to the correct flower-inducing treatment results in flowering. However, although size seems to be the most important factor, it is not yet clear what specific component associated with size is critical. Perhaps plants of a sufficient size transmit flower-inducing signals to the apex. Alternatively, the apex itself may undergo the transition to maturity.

Irrespective of whether flowering occurs in response to a signal arriving from elsewhere in the plant, the transition from juvenility to maturity is clearly associated with events occurring at the apex. For example, in mature plants of ivy (*Hedera helix*), cuttings taken from the basal region develop into juvenile plants, while those from the tip develop into mature plants. Results for grafted plants in which the upper part of the graft (the *scion*) originated from different parts of an old flowering specimen of silver birch (*Betula verrucosa*) also indicate that the basal part retains the juvenile condition (Longman, 1976). When scions were taken from the base of the flowering tree and grafted onto seedling rootstocks, there were no flowers on the grafts within the first 2 years. In contrast, the grafts flowered freely when scions were taken from the top of the flowering tree. Results of some experiments support the concept that some signal is transmitted to the apex when a plant becomes sufficiently large. For example, in mango (*Mango mangifera*), juvenile seedlings can be induced to flower when grafted to a mature tree. In other species, however, grafting does not induce flowering.

The Apex Receives Both Nutrients and Hormones from the Plant

The transition of the apex to maturity could be attributed to several transmissible factors. In many plants, exposure to low-light conditions causes rejuvenation or prolongs juvenility. A major consequence of the low-light regime is a reduction in the supply of carbohydrates to the apex; thus, carbohydrate supply may play a role in the transition between juvenility and maturity. One obvious connection between nutritional status and the maturity stage is the effect of nutrition on the size of the apex. For example, in the florist's chrysanthemum, (*Chrysanthemum mori-folium*), flower primordia are not initiated until a minimum apex size has been reached (Cockshull, 1985).

The apex receives a variety of hormonal factors from the rest of the plant in addition to carbohydrates and other nutrients. Could any one factor promote the change between juvenility and maturity? There is now a large body of experimental evidence showing that the application of gibberellins (*exogenous* GAs) is able to cause flowering in young, juvenile plants of several conifer families (Pharis and King, 1985). For example, although most conifers require several years to attain maturity, male cones have been induced to form in Arizona cypress (*Cupressus arizonica*) plants that were only 2 months old by spraying with GAs. The involvement of *endogenous* GAs in the control of flowering is indicated by the fact that other treatments that accelerate flowering in pines (e.g., root removal, water stress, and nitrogen starvation) often also result in a buildup of GAs in the plant. On the other hand, whereas gibberellins are a "maturity" factor in conifers, the application of GA_3 causes rejuvenation in *Hedera* and in several other woody angiosperms, so gibberellins can also be "juvenility" factors (Hackett and Srinivasan, 1985). Clearly, the role of gibberellins in the control of maturity is complex and probably underlies interactions with other factors.

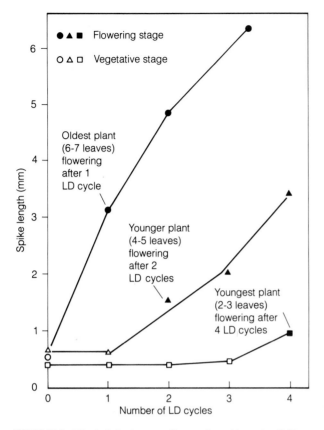

FIGURE 21.3. Effect of plant age on the number of long-day (LD) inductive cycles required for flowering in the long-day plant *Lolium temulentum*. The older the plant, the fewer cycles needed to produce flowering. An inductive long-day cycle consisted of 8 h of sunlight followed by 16 h of low-intensity incandescent light.

After maturity is attained, there is often an increasing tendency toward flowering as the plant ages. For example, in daylength-controlled plants, the number of short-day or long-day cycles necessary to achieve flowering is often fewer in older plants (Fig. 21.3). These observations suggest that the effect of daylength could be to accelerate a flowering process that is already occurring at a slow pace.

Photoperiodism

The ability of an organism to detect daylength makes it possible for an event to occur at a particular *time of year*, thus making possible a *seasonal* response. Organisms also possess a mechanism that can determine the *time of day* at which a particular event occurs.

Both processes have the common property of responding to cycles of light and darkness. These photoperiodic phenomena are common to both animals and plants. In the animal kingdom, daylength controls such seasonal activities as hibernation, development of summer or winter coats, and reproductive activity. Plant

responses controlled by daylength are numerous, including the initiation and further development of flowers, asexual reproduction, the formation of storage organs, and the onset of dormancy (Vince-Prue, 1975). There is considerable evidence that all photoperiodic responses share a common initial mechanism, with subsequent specific biochemical pathways regulating different responses. Since it is clear that monitoring the passage of time is essential to all photoperiodic responses, a time-keeping mechanism must underlie both the time-of-year and time-of-day responses. Before proceeding with our discussion on photoperiodism, we shall consider the time-keeping process that establishes the time of day when an event occurs.

Plants Can Show Daily Rhythms in the Absence of External Changes

Organisms are normally subjected to daily cycles of light and darkness, and both plants and animals often exhibit rhythmic behavior in association with these changes (Brady, 1982; Sweeney, 1987). Examples of such rhythms include leaf movements (day and night positions), stomatal and petal movements, growth and sporulation patterns in fungi (e.g., in *Pilobolus* and *Neurospora*), time of day of pupal emergence (e.g., in the *fruit fly, Drosophila*), and activity cycles in rodents, as well as metabolic processes such as photosynthetic capacity and respiration rate. When organisms are transferred from daily light/dark cycles to continuous darkness (or continuous dim light), many of these rhythms continue to be expressed, at least for some time. Under such uniform conditions the periodicity of the rhythm is then close to 24 h, and consequently the term **circadian** (from the Latin for "about one day") is applied. Because they continue in a constant light or dark environment, these circadian rhythms cannot be direct responses to the presence or absence of light but must be based on an endogenous pacemaker that is self-sustaining. This pacemaker is often called an **endogenous oscillator** (a clock).

Circadian rhythms exhibit a number of characteristic features (Fig. 21.4a). Rhythms derive from cyclic phenomena, and the pattern repeats over and over. The time between comparable points in the repeating cycle is known as the **period**. The term **phase** is used for any point in the cycle recognizable by its relationship to the rest of the cycle. The most obvious phase points are the maximum (peak) and minimum (trough) positions. The **amplitude** of a biological rhythm is usually considered to be the distance between peak and trough, and this can often vary while the period remains unchanged (Fig. 21.4d).

In constant light or darkness, rhythms depart from an exact 24-h periodicity. The rhythms then drift in relation to solar time, either gaining or losing time depending on

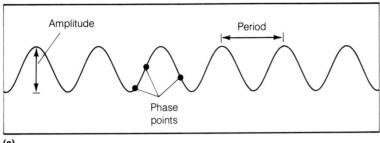

(a)

A typical circadian rhythm. The **period** is the time between comparable points in the repeating cycle; the **phase** is any point in the repeating cycle recognizable by its relationship with the rest of the cycle; the **amplitude** is the distance between peak and trough.

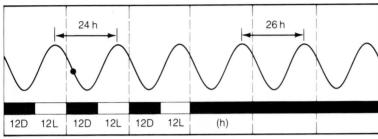

(b)

A circadian rhythm entrained to a 24-h light/dark cycle and its reversion to the free-running period (26 h in this example) following transfer to continuous darkness.

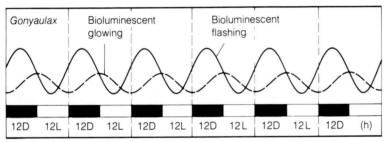

(c)

The phase relationship of two different rhythms in the unicellular alga, *Gonyaulax polyedra* entrained to 12-h light/12-h dark cycles.

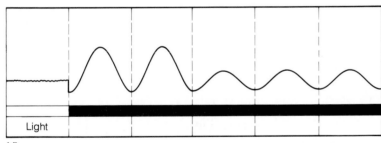

(d)

Suspension of a circadian rhythm in continuous bright light and the release or restarting of the rhythm following transfer to darkness.

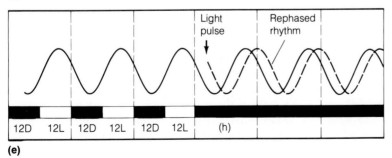

(e)

Typical phase-shifting response to a light pulse given shortly after transfer to darkness. The rhythm is rephased (delayed) without changing its period.

FIGURE 21.4. Some characteristics of circadian rhythms.

whether the period is shorter or longer than 24 h. Under natural conditions, the endogenous oscillator is **entrained** (synchronized) to a true 24-h periodicity (Fig. 21.4b) by environmental signals, the most important being the light-to-dark transition at dusk and the dark-to-light transition at dawn. Such signals are termed **zeitgebers** (from the German for "time-giver"). When the entraining signals are removed, for example, by transfer to continuous darkness, the rhythm is said to be free-running and reverts to the circadian period characteristic of the particular organism (Fig. 21.4b). Although the rhythms are innate, they normally require an environmental signal, such as exposure to light or a change in temperature, to start them. In addition, many rhythms damp out (i.e., the amplitude decreases) when the organism is in a constant environment for some time and then require an environmental zeitgeber, such as a light/dark transfer or change in temperature, to restart them (Fig. 21.4d). Temperature has little or no effect on the periodicity of the free-running rhythm. This insensitivity to temperature is an essential characteristic, as the clock would be of no value if it could not keep accurate time under the fluctuating temperatures experienced in natural conditions.

In these daily rhythms, the operation of the endogenous oscillator sets a response to occur at a particular time of day. For example, in the unicellular alga *Gonyaulax polyedra*, maximum "glowing" bioluminescence occurs toward the end of a 12-h night, whereas maximum "flashing" bioluminescence occurs near the middle (Fig. 21.4c). How do such responses remain on time when the daily durations of light and darkness are changing? The answer to this question lies in the fact that the phase of the rhythm can be changed by moving the whole cycle forward or backward in time without altering its period (Fig. 21.4e; see also Fig. 20.11). As an example, we can take a phase point that occurs 1 h after the onset of darkness in 12-h light/12-h dark cycles (Fig. 21.4b). If daylight is extended for 1 h, it must mean that this phase point has drifted forward relative to the dusk zeitgeber, and the appropriate response would be a correcting backward shift (or phase *delay*) such that this phase point still occurs 1 h after the end of the light period and the rhythm stays on local time. Similarly, the appropriate answer to light perceived at a phase point that would normally occur 1 h before dawn would be a phase *advance*. Adaptively, therefore, different *directions* of phase shifting are to be expected in response to light signals given at different phase points in the circadian cycle, and experimentally that is what is observed.

The response of the endogenous oscillator is usually tested experimentally by placing the organism in continuous darkness and examining the response to a short pulse of light (usually less than 1 h) given at different phase points in the free-running rhythm. When an organism is entrained to 12-h light/12-h dark cycles and then allowed to free-run in darkness, the phase of the rhythm coinciding with the light period of the previous entraining cycle is called the **subjective day** and that coinciding with the dark period is called the **subjective night**. If a light pulse is given during the first few hours of the subjective night, the rhythm is delayed; in contrast, a light pulse given toward the end of the subjective night advances the phase of the rhythm (Fig. 21.5). As already pointed out, this is precisely the pattern of response that would be expected if the rhythm is to stay on local time. These phase-shifting responses, therefore, enable the rhythm to be entrained to cycles with different durations of light and darkness and mean that the rhythm will run differently under different natural daylength conditions. It has been demonstrated that different photoperiodic rhythms show a very similar pattern of phase shifting when tested in the same way (Fig. 21.5).

The physiological mechanism whereby a light signal causes phase shifting is not yet known, but it is clear that the light response must be mediated by a photoreceptor. The role of phytochrome in phase shifting has been discussed in Chapter 20. Light reception by phytochrome in the rhythmic leaf movements in *Samanea*, a semitropical leguminous tree, is well characterized (Satter and Galston, 1981). Phytochrome is also involved in the phase shifting of photoperiodic rhythms. For example, the circadian rhythm in the flowering response of seedlings of Japanese morning glory (*Pharbitis nil*) can be phase-shifted by exposure to a few seconds of red light (Lumsden et al., 1986). On the other hand, phytochrome does not appear to be the universal photoreceptor for phase shifting, since it is not involved in the circadian rhythm of bioluminescence in *Gonyaulax*.

Daylength Is a Major Determinant of Flowering Time

For many years, plant physiologists thought that the correlation between long days and flowering was a consequence of the accumulation of photosynthetic products made on long days. The work of Wightmam Garner and Henry Allard, conducted in the 1920s at the U.S. Department of Agriculture laboratories in Beltsville, Maryland, provided solid evidence for the role of daylength in the induction of flowering in tobacco. They found that a new variety of tobacco, Maryland Mammoth, grew profusely to 3 to 5 m in height, but it failed to flower in the prevailing conditions of the summer months. On the other hand, the plants flowered when 1 m tall if grown under artificially shortened days. This requirement for short days was hard to reconcile with the explanation stating that higher total levels of radiation and photosynthesis underlie the long-day requirement. Garner and Allard concluded that the length of the day

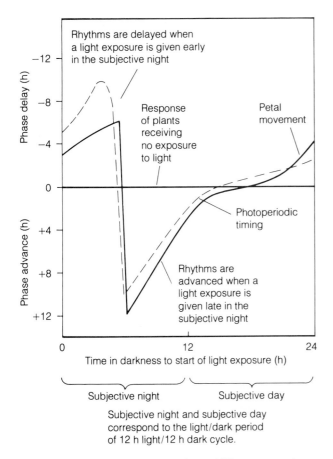

FIGURE 21.5. Characteristics of the phase-shifting response in circadian rhythms. The phase-shifting response of the petal movement rhythm in the succulent plant *Kalanchoe blossfeldiana* (solid line) was obtained by giving a 2-h light treatment at various times after transferring plants from cycles of 12-h light/12-h dark to continuous darkness. The curve shows the delay or advance of the petal movement rhythm compared with that of plants receiving no exposure to light. (Data of Zimmer, 1962; redrawn from Salisbury and Ross, 1978.) The response for photoperiodic timing in the short-day plant *Chenopodium rubrum* (dashed line) was obtained by giving a 6-h light treatment at different times after transfer to darkness. The curve shows the delay or advance of the rhythm of flowering response to a night-break. (Data of King and Cumming, 1972.)

was the determining factor in flowering and were able to confirm this hypothesis in a number of different species and conditions. This work laid the foundations for the extensive subsequent research on photoperiodic responses.

The classification of plants according to their photoperiodic responses is usually made on the basis of flowering, even though many other aspects of their development may also be affected by daylength. The two main photoperiodic response categories are **short-day plants** (SDPs), in which flowering occurs only in short days (qualitative SDP) or is accelerated by short days (quantitative SDP), and **long-day plants** (LDPs), in which flowering occurs only in long days (qualitative LDP) or is accelerated by long days (quantitative LDP) (Fig. 21.6). A few plants have more specialized daylength requirements. **Intermediate-day plants** flower only between quite narrow daylength limits—for example, between 12 and 14 h in one variety of sugarcane. Plants in another special category (**ambiphotoperiodic**) flower in long days or short days but not at intermediate daylengths. There are also many plants whose flowering is not regulated by daylengths; these are called **day-neutral plants**.

The essential distinction between long-day and short-day responses is that flowering in LDPs is promoted only when the daylength *exceeds* a certain duration in every 24-h cycle, whereas the promotion of flowering in SDPs requires a daylength that is *less than* a critical value. The actual value of the critical daylength varies widely between species, and only when flowering is examined for a range of daylengths can the correct photoperiodic classification be established (Fig. 21.6).

Long-day plants can effectively measure the lengthening days of spring or early summer and delay flowering until the critical daylength is reached. Many varieties of wheat (*Triticum aestivum*) behave in this way. In SDPs flowering often occurs in fall when the days shorten below the critical daylength, as in many varieties of *Chrysanthemum morifolium*. However, daylength alone is an ambiguous signal because it cannot distinguish between spring and fall. Plants exhibit several strategies that avoid

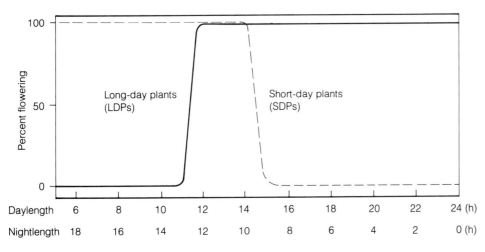

FIGURE 21.6. The photoperiodic response in long- and short-day plants. Short-day plants flower when the daylength is less than (or the nightlength exceeds) a certain critical duration in a 24-h cycle. Long-day plants flower when the daylength exceeds (or the nightlength is less than) a certain critical duration in a 24-h cycle. The critical duration varies between species; in this example, both the SDPs and the LDPs would flower in photoperiods between 12 and 14 h in duration.

this ambiguity. One is the coupling of a temperature requirement to a photoperiodic response. An example of this is found in some short-day varieties of strawberry (*Fragaria ananassa*) in which flowers are initiated when the critical daylength is reached in fall, but do not emerge until spring. The low temperatures of winter prevent the subsequent initiation of flowers in the spring even though the daylength conditions are appropriate. A different strategy for avoiding seasonal ambiguity would be a mechanism for distinguishing between *lengthening* and *shortening* days. Some animals appear to do this, but such a mechanism is not known to operate in plants.

Plants Track Daylength by Measuring the Length of the Night

Under natural conditions, day- and night-lengths configure a 24-h cycle of light and darkness. Perception of a critical daylength could, therefore, be achieved by measuring the duration of either light or darkness. Much experimental work in the early studies of photoperiodism was devoted to establishing which part of the light/dark cycle is the controlling factor in flowering. Results showed that flowering is determined primarily by the duration of darkness (Fig. 21.7a). Thus, it was possible

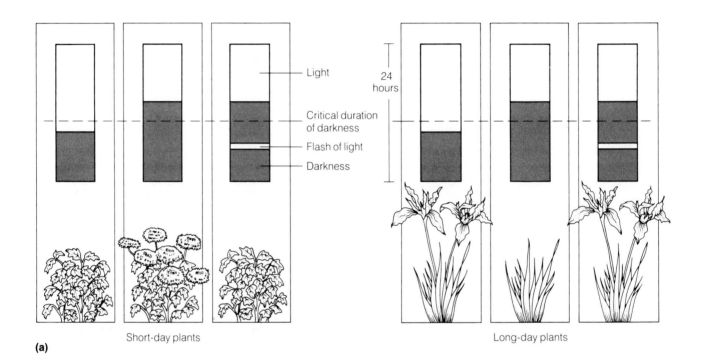

(a)

Short-day plants

Long-day plants

Light

Critical duration of darkness

Flash of light

Darkness

24 hours

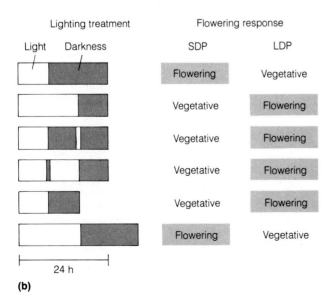

(b)

FIGURE 21.7. (a) The photoperiodic regulation of flowering. Short-day plants (long-night plants) flower when nightlength exceeds a critical dark period. Interruption of the dark period by a flash of light (a night-break) prevents flowering. Long-day plants (short-night plants) flower if the nightlength is shorter than a critical period. In some long-day plants, shortening the night with a flash of light induces flowering. (b) Effects of the duration of the dark period on flowering. Treating short- and long-day plants with different photoperiods, as illustrated , in the figure, clearly shows that the critical variable is the length of the dark period.

to induce flowering in SDPs with light periods longer than the critical value, provided that these were followed by sufficiently long nights (Fig. 21.7b). Similarly, flowering did not occur in SDPs when short days were followed by short nights. More detailed experiments demonstrated that photoperiodic timekeeping in SDPs is essentially a matter of measuring the duration of darkness (Hamner and Bonner, 1938; Hamner, 1940). For example, flowering occurred only when the dark period exceeded 8.5 h in cocklebur (*Xanthium strumarium*) or 10 h in soybean (*Glycine max*). The duration of darkness was also shown to be important in LDPs (Fig. 21.7a). These plants were found to flower in short days, provided that the accompanying nightlength was also short; however, long days followed by long nights were ineffective.

A feature that underlines the importance of the dark period is that it can be made ineffective by interruption with a short exposure to light, called a **night-break** (Fig. 21.7). In contrast, interrupting a long day with a brief dark period does not cancel the effect of the long day (Fig. 21.7b). Night-break treatments of only a few minutes are effective in *preventing* flowering in many SDPs, including *Xanthium* and *Pharbitis*, but longer exposures are often required to *promote* flowering in LDPs. Thus, the effect of a night-break varies greatly according to the time when it is given. For both LDPs and SDPs, a night-break was found to be most effective when given near the middle of a dark period of 16 h (Fig. 21.8). However, when considerably longer dark periods were given, it was established that the time of maximum sensitivity to a night-break is related to the time from the beginning of darkness. In *Xanthium*, for example, the maximum effect of a night-break still occurred about 8 h after the beginning of a 24-h dark period, even though the subsequent 16-h period of unbroken darkness considerably exceeds the critical nightlength of 8.5 h.

The discovery of the night-break effect and its time dependence had several important consequences. It established the central role of the dark period and provided a valuable probe for studying photoperiodic timekeeping. Because only small amounts of light are needed, it became possible to study the action and identity of the photoreceptor without the interfering effects of photosynthesis and other nonphotoperiodic phenomena. It has also led to the development of commercial methods for regulating the time of flowering in crop plants such as chrysanthemum and poinsettia (*Euphorbia pulcherrima*).

An Endogenous Oscillator Is Involved in Photoperiodic Timekeeping

The decisive effect of nightlength on flowering indicates that measurement of the passage of time in darkness is central to photoperiodic timekeeping. Two different hypotheses have been proposed to explain how plants measure nightlength. According to the **hourglass hypothesis**, time is measured by a unidirectional series of biochemical reactions that start at the beginning of the dark period. If not interrupted by light, this series of reactions would lead to induction of flowering in SDPs or to inhibition of flowering in LDPs. This type of mechanism can measure only one interval of time and, by ana-

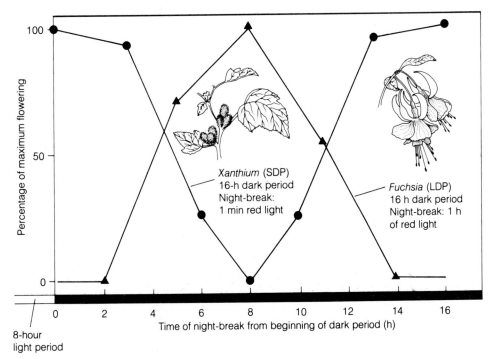

FIGURE 21.8. The time when a night-break is given determines the flowering response. When given during a long dark period, a night-break promotes flowering in LDPs and inhibits flowering in SDPs. In both cases, the greatest effect on flowering occurs when the night-break is given near the middle of a 16-h dark period. The LDP *Fuchsia* was given a 1-h exposure to red light in a 16-h dark period. (From Vince-Prue, 1975.) *Xanthium* was given a 1-min exposure to red light in a 16-h period. (From Salisbury, 1963; Papenfuss and Salisbury, 1967.)

Xanthium (SDP)
16-h dark period
Night-break:
1 min red light

Fuchsia (LDP)
16 h dark period
Night-break: 1 h
of red light

Percentage of maximum flowering

Time of night-break from beginning of dark period (h)

8-hour
light period

logy with an hourglass device, would be reset each day by the photoperiod.

Some of the characteristics of flowering induction are consistent with the accumulation of biochemical products, as predicted by the hourglass hypothesis (Thomas and Vince-Prue, 1984), but most of the available evidence favors a different mechanism, called the **clock hypothesis** (Bünning, 1960). This hypothesis proposes that photoperiodic timekeeping depends on an endogenous circadian oscillator of the type involved in the daily rhythms described earlier in the chapter. In this view, the role of the photoperiod is to set the phase of the rhythm in a way leading to the measurement of time in darkness and of the critical nightlength.

Measurements of the effect of the night-break on flowering may be used to investigate circadian rhythms in flowering, and many studies have shown that those rhythms exist. For example, when soybean plants are kept under a long dark cycle after 8 h of light, and the dark cycle is interrupted at different times by a 4-h-long night-break, the flowering response to the night-break shows a circadian rhythm (Fig. 21.9). Many other rhythmic responses of both LDPs and SDPs have been reported (Thomas and Vince-Prue, 1984). Further evidence for the presence of a circadian oscillator is the observation that the photoperiodic response can be phase-shifted by a light treatment (Fig. 21.5).

The involvement of a circadian oscillator in photoperiodism poses an important question. How does an oscillation with a 24-h periodicity measure a critical duration of darkness of, say, 8–9 h, as in the SDP

Xanthium? Erwing Bünning, working in Tübingen, Germany, proposed in 1936 that the control of flowering by photoperiodism is achieved by an oscillation of phases with different sensitivities to light. Bünning's hypothesis invokes two alternating phases, one requiring 12 h of darkness and the other requiring 12 h of light (Bünning, 1960)—hence, the involvement of both light and darkness in the photoperiodic response in flowering.

According to Bünning's hypothesis, light has two distinct roles in the flowering process. Light signals at dawn and dusk, acting as zeitgebers, set the phase of the photoperiodic rhythm. In addition, the rhythm has a light-sensitive phase called the **inducible phase**. When a light signal is received during the inducible phase of the rhythm, the effect is to promote or prevent floral induction. Thus, in nature, the phase of the rhythm is initially set by the light-on signal at dawn. The rhythm then continues to run until it reaches a specific phase point at which it is suspended while the plant remains in the light (Fig. 21.4d). At the transition to darkness at dusk, the rhythm is released and runs freely. This entrainment at dusk ensures that the rhythm is released at a specific phase point and that the inducible phase always occurs at a constant real time after transfer to darkness. In SDPs, exposure to light prevents flowering, so flowering would be induced only when dawn occurs after the completion of the inducible phase of the rhythm. In this way, the photoperiod establishes the time of sensitivity to light in the subsequent dark period and sets the conditions for the measurement of the critical nightlength.

Phytochrome Is Involved in Photoperiodism

Night-break experiments are optimally suited for studying the nature of the photoreceptors involved in the reception of the light signals during the photoperiodic response. Indeed, the night-break inhibition of flowering in SDPs was one of the first physiological processes shown to be under the control of phytochrome (Fig. 21.10) (Borthwick et al., 1952). In many SDPs, a night-break becomes effective only when the supplied light dose is sufficient to saturate the photoconversion of Pr to Pfr (Chapter 20). A subsequent exposure to far-red light, which photoconverts the pigment back to the physiologically inactive Pr form, restores the flowering response (Downs, 1956). Red/far-red reversibility has also been demonstrated in some LDPs, in which a red night-break promoted flowering and a subsequent far-red exposure prevented this response (Fig. 21.10).

Figure 21.11 shows action spectra for the inhibition and restoration of the flowering response in SDPs. The spectra for *Xanthium* provide an example of the response in green plants, in which the presence of chlorophyll can cause some discrepancy between the action spectrum and the absorption spectrum of Pr (Vince-Prue and Lumsden, 1987). Interference from chlorophyll can be avoided by

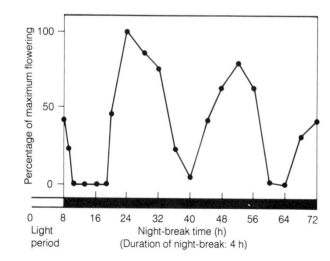

FIGURE 21.9. Rhythmic flowering in response to night-breaks. When plants are given a night-break at different times (indicated on the horizontal axis) in a long dark period, the flowering response shows a circadian rhythm. In this experiment, the SDP soybean (*Glycine max*) received cycles of an 8-h light period followed by a 64-h dark period. A 4-h night break was given at various times during this long inductive dark period. (Data of Coulter and Hamner, 1964.)

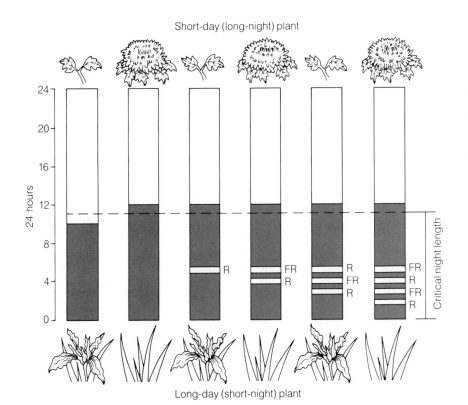

FIGURE 21.10. Control of flowering by red and far-red light. Red light is the most effective wavelength in night-breaks. A flash of red light during the dark period induces flowering in an LDP, and the effect is reversed by a flash of far-red light. This response indicates the involvement of phytochrome. In SDPs, a flash of red light prevents flowering.

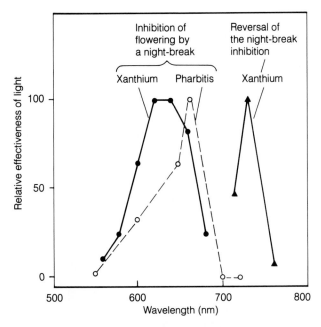

FIGURE 21.11. Action spectra for the control of flowering by night-breaks. Flowering in SDPs is inhibited by a short light treatment (night-break) given in an otherwise inductive period. In the SDP *Xanthium strumarium*, red night-breaks of 620–640 nm are the most effective. Reversal of the red-light effect is maximal at 725 nm. (Data of Hendricks and Siegelman, 1967.) In the dark-grown SDP *Pharbitis nil*, which is devoid of chlorophyll, night-breaks of 660 nm are the most effective. This maximum coincides with the absorption maximum of phytochrome. (From Saji et al., 1983.)

using dark-grown seedlings, as illustrated in Figure 21.11 with *Pharbitis*. In that case, the action spectrum shows a peak at 660 nm, the absorption maximum of Pr (Chapter 20). These action spectra confirm the role of phytochrome as the photoreceptor that absorbs the light signal during the inducible phase. It is of interest that the inductive event is completed fairly rapidly, despite the fact that the actual expression of floral induction does not occur for days or even weeks after reception of the light signal. The rapidity of the inductive event is evident from the fact that only 0.5 to 30 min of darkness can elapse between the initial photoconversion of Pr by red light and the reversal of the response by far-red light. Application of far-red light after that time interval no longer restores flowering in SDPs (Downs, 1956; Vince-Prue, 1975).

Far-Red Light Modifies Flowering in Some LDPs

Circadian rhythms have also been found in LDPs. A circadian periodicity in the promotion of flowering by far-red light has been observed in barley (*Hordeum vulgare*) (Deitzer et al., 1979) and in the darnel grass (*Lolium temulentum*) (Fig. 21.12). In both cases, when far-red light is added for 4–6 h, flowering is promoted compared with plants maintained under continuous white or red light. The rhythm clearly continues to run in the

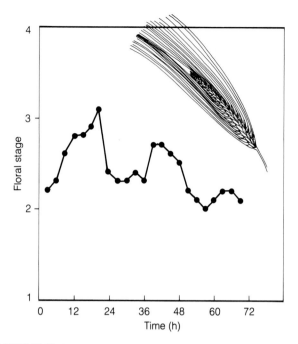

FIGURE 21.12. Effect of 6 h of far-red light on floral induction in barley. Far-red light was added for 6 h at the indicated times to a continuous 72-h daylight period. Data points in the graph are plotted at the centers of the 6-h treatments. Floral stage refers to the developmental stage of the apex. Stage 1 refers to an indeterminate vegetative apex, and stages 3 and 4 are distinct morphological steps in floral development. (From Deitzer et al., 1979.)

light. In SDPs, on the other hand, an essential feature of the circadian timing mechanism seems to be that the rhythm is suspended after a few hours in continuous light and released, or restarted, in transfer to darkness. The response to far-red light is not the only rhythmic feature in LDPs. Although relatively insensitive to a night-break of only a few minutes, many LDPs can be induced to flower with a longer night-break, usually of at least 1 h. A circadian oscillation in the flowering response to such a long night-break has been observed in LDPs, showing that a rhythm of responsivity to light continues to run in darkness.

Thus, circadian rhythms that modify the flowering response in LDPs have been shown to run both in the light (promotion by far-red) and in the dark (promotion by red or white light). However, we do not yet know how these rhythms affect the photoperiodic response.

Vernalization

Vernalization is the process whereby flowering is promoted by a cold treatment given to a seed that has imbibed water or to a growing plant. Dry seeds do not respond to the cold treatment. Temperature also has direct effects on floral initiation in some plants, but these effects can be distinguished from vernalization, which, like photoperiodism, is an inductive phenomenon leading to flowering some time after the low-temperature treatment. Without the cold treatment, vernalization-requiring plants show delayed flowering or remain vegetative. In many cases these plants grow as rosettes with no elongation of the stem (Fig. 21.13).

FIGURE 21.13. The control of flowering in the SDP *Campanula medium.* When grown in continuous long days, the plants grow as rosettes and the stem does not elongate. Eight weeks of short days followed by long days result in both elongation and flowering. The short days can be substituted by eight weeks of cold temperatures. Application of GA₃ results in stem elongation but not in flowering induction. (From Wellensiek, 1985.)

Continuous LD   SD → LD   Cold → LD   GA in LD

Plants differ considerably in the age at which they become sensitive to vernalization. Winter annuals, such as the winter forms of cereals (which are sown in the fall and flower in the following summer), respond to low temperature very early in their life cycle. They can be vernalized before germination if the seeds have been soaked. Other plants, including most biennials (which grow as rosettes during the first season after sowing and flower in the following summer), must reach a minimal size before they become sensitive to low temperature for vernalization.

The effective temperature range for vernalization is from just below freezing to about 10°C, with a broad optimum usually between about 1 and 7°C (Lang, 1965). The effect of cold increases with the duration of the cold treatment until the response is saturated. The response usually requires several weeks of exposure to low temperature, but the precise duration varies widely with species and variety. The vernalization effect can be lost as a result of exposure to devernalizing conditions such as high temperature (Fig. 21.14), but the longer the exposure to low temperatures, the more permanent the vernalization effect.

The vernalization process appears to take place primarily in the meristematic zones of the shoot apex (Thomas and Vince-Prue, 1984). Localized cooling treatments cause flowering when only the stem apex is chilled,

and this effect appears to be largely independent of the temperature experienced by the rest of the plant. Excised shoot tips have been successfully vernalized and, where seed vernalization is possible, fragments of embryos consisting essentially of the shoot tip are sensitive to low temperature.

Photoperiodism and Vernalization May Interact

A vernalization requirement is often linked with a requirement for a particular photoperiod (Napp-Zinn, 1984). The most common combination is a requirement for cold treatment followed by a requirement for long days—a combination that would lead to flowering in early summer at high latitudes. However, many other combinations have been recorded. In relation to the position of vernalization in the sequence of events leading to flowering, two types of interaction between vernalization and daylength requirements are particularly interesting. In the biennial form of *Hyoscyamus niger*, young plants can be photoperiodically induced by long days only *after* they have received at least 7 days of cold. Thus, in *Hyoscyamus*, vernalization at the shoot apex seems to be an earlier step than photoperiodic induction in the leaf in the sequence of events leading to flowering. In *Campanula medium* (Canterbury Bell), short days acting in the leaf can completely substitute for vernalization at the apex (Fig. 21.13). In this case vernalization and photoperiod appear to act as alternative pathways to flowering. The ability of short days to substitute for vernalization occurs in a number of species, but the relationship between the two processes is not known. Does vernalization at the shoot apex lead to the development of the induced state in the leaves without exposure to inductive photoperiods? Could floral initiation take place at the apex without the product(s) of photoperiodic induction in the leaves? Answers to these questions must await a better understanding of the biochemistry of the flowering process.

Protein Synthesis Appears to Be Required for Vernalization

Much remains to be learned about the metabolic processes associated with vernalization. Both sugars and oxygen are required for the effect of low temperature, so it appears unlikely that low temperatures simply arrest some metabolic reactions that are inhibiting flowering. The sugar and oxygen requirements are more consistent with activation of an aerobic metabolic reaction that is essential for flowering. On the other hand, the rates of most metabolic reactions *decrease* with temperature, so the induction of flowering by low temperatures is likely to be more complex than the simple activation of a metabolic pathway.

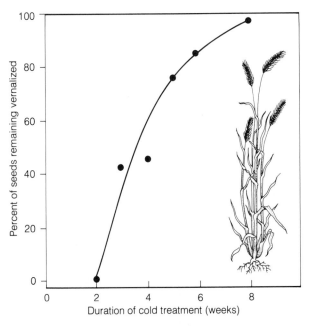

FIGURE 21.14. The duration of exposure to low temperature affects the stabilization of vernalization. The longer winter rye, *Secale cereale*, is exposed to a vernalizing cold treatment, the greater the number of plants that remain vernalized when the cold treatment is followed by a devernalizing treatment. In this experiment, seeds of rye that had imbibed water were exposed to 5°C for different lengths of time, immediately followed by a devernalizing treatment of 3 days at 35°C. (Data of Purvis and Gregory, 1952.)

One study in winter wheat (*Triticum aestivum*) resolved new proteins that appeared after vernalization (Teroaka, 1972). The protein pattern after vernalization resembled that typical of spring wheat, which does not require vernalization to flower. Further studies of these processes using the powerful approaches currently available in molecular biology should provide new insights into this important subject.

The Transition to Flowering

The preceding sections discussed the induction of flowering by environmental conditions, such as temperature and daylength, or by endogenous factors such as age. The morphological changes that give rise to the floral primordia occur at the apical meristems of the shoots. The events that result in the morphological changes appear to be triggered by biochemical signals arriving at the apex from other parts of the plant, especially from the leaves. In this section we shall consider the nature of these biochemical signals.

Photoperiodic Induction of a Single Leaf Can Cause Flowering

One well-established fact about the photoperiodic process is that it takes place in the leaves. Treatment of a single leaf of the SDP *Xanthium* with short photoperiods was sufficient to cause the formation of macroscopically visible flowers, even though the rest of the plant was exposed to long days (Hamner and Bonner, 1938). For both LDPs and SDPs, many experiments have confirmed that the photoperiod experienced by the leaves determines the response at the apex.

Photoperiodic induction can also take place in a leaf that has been separated from the plant. In the SDP *Perilla* (red-leaved perilla), Jan Zeevaart has shown that an excised leaf exposed to short days can induce flowering when subsequently grafted to a noninduced plant maintained in long days (Zeevaart, 1958). Thus, photoperiodic induction appears to depend on events that take place in the leaf. When several photoperiodic cycles are required, these must be given to the same leaf. The cumulative effect of repeated cycles appears to take place in the leaf and not to be a consequence of the accumulation of a metabolite at the apex.

Leaves have also been shown to play an important role in plants in which flowering is under endogenous control (Heinze et al., 1942). For example, grafting a single leaf of Agate, a day-neutral variety of *Glycine*, caused flowering of the short-day variety Biloxi when the latter was maintained in noninductive long days. Thus, induction in the leaf leads to the transmission of some kind of signal that regulates the transition to flowering at the shoot apex.

Available information indicates that the flowering signal is transported via the phloem and that it is chemical, rather than physical, in nature. For example, it appears that grafted leaves cause flowering only when a functional phloem connection is established between the donor and receptor plants. Treatments that restrict phloem transport, such as girdling (removing tissue external to the xylem) or localized heating, prevent the movement of the floral signal.

By manipulation of leaves it is possible to measure rates of transport of the flowering signal. This can be accomplished by removing an induced leaf at different times and comparing the time it takes for the signal to reach two buds located at different distances from the induced leaf. The rationale for this type of measurement is that a critical amount of the signaling compound has reached the bud when flowering takes place despite the removal of the leaf. Studies using this method (Evans and Wardlaw, 1966; King et al., 1968) have shown that the rate of transport of the flowering signal is comparable to or somewhat slower than the rate of translocation of sugars in the phloem (Chapter 7). These rates of transport are consistent with a chemical message and rule out the possibility of an electrical signal.

The role of phytochrome in floral induction provides some clues to the possible nature of the inductive signal. Phytochrome regulates gene expression (Chapter 20), and some aspects of photoperiodic induction are suggestive of the activation of specific genes in the leaf that could regulate floral induction. In addition, the phytochrome-mediated control of flowering could be partially associated with modification of membrane properties (Cleland, 1984; also see Chapter 20). Photoperiodic timekeeping depends on a circadian oscillator, and there is evidence that changes in membrane properties play an important role in circadian rhythms (Engelmann and Schrempf, 1980).

Identification of the Hypothetical Florigen Remains Elusive

The production at the leaf of a biochemical signal that acts in a remote place at the apex satisfies an important criterion for a hormonal effect. In the 1930s Michael Chailakhyan, working in Russia, postulated the existence of a flowering hormone and called it **florigen**. The many attempts to isolate and characterize this hypothetical hormone have been largely unsuccessful. The most common approach has been to make extracts from induced leaf tissue and test for their ability to elicit flowering in noninduced plants. In other experiments, investigators have extracted and analyzed phloem sap from induced plants. In some studies, extracts from one of these sources have induced flowering in test plants, but these results have not been consistently reproduced (Zeevaart and Boyer, 1987). Attempts to isolate a specific,

Gene Expression During Flower Development

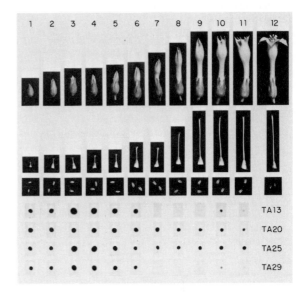

FIGURE 21.A. (From Goldberg, 1988)

THE FLOWERING PROCESS is attracting much attention from plant biologists using sophisticated new techniques of molecular biology to study molecular and cellular events involved in plant growth and development. Some of these studies have characterized molecular changes associated with the conversion of the vegetative apex into a floral meristem, including the regulation of gene expression during tobacco flower development (Goldberg, 1988). The development of a tobacco flower takes 1–2 weeks, depending on the growing conditions. By using the length of the floral bud and petals as markers, this developmental process was divided into 12 stages. Messenger RNAs extracted from the anthers of flowers in each stage were hybridized with cDNA clones isolated from the anthers. Results of the hybridization between mRNA isolated from each developmental stage and four cDNA clones are shown in the lower part of Figure 21.A. A large spot indicates a relative abundance of an mRNA that is complementary to the cDNA clone; blank spots indicate that the corresponding mRNA was absent at that developmental stage. The data show that the

TA20 and TA25 mRNAs persist throughout floral development, whereas the TA13 and TA29 mRNAs accumulate early in anther development and then decay prior to anther dehiscence and pollen release. Further characterization of the role of the proteins coded by these mRNAs in flower development should help us understand the molecular basis of flowering.

TABLE 21.1. Examples of the successful transfer of a flowering induction signal by grafting between plants of different photoperiodic response groups. Successful transfer of the flowering signal between long-day plants (LDPs), short-day plants (SDPs), and day-neutral plants (DNPs) indicates the existence of a transmissible floral hormone that is effective in plants of different photoperiodic classes.

Donor plants maintained under flower-inducing conditions	Photoperiod type	Vegetative receptor plant induced to flower	Photoperiod type
Helianthus annus	DNP in LD	*H. tuberosus*	SDP in LD
Nicotiana tabacum Delcrest	DNP in SD	*N. sylvestris*	LDP in SD
Nicotiana sylvestris	LDP in LD	*N. tabacum* Maryland Mammoth	SDP in LD
Nicotiana tabacum Maryland Mammoth	SDP in SD	*N. sylvestris*	LDP in SD

graft-transmissible inhibitor of flowering, or **antiflorigen,** have also been unsuccessful. Thus, despite unequivocal data showing that a transmissible factor regulates flowering (Table 21.1) (Lang et al., 1977; Zeevaart, 1984), the involved substances are yet to be characterized.

Grafting experiments have also been done to investigate possible signals from vernalized plants

(Table 21.2). Cold-requiring, untreated plants have been induced to flower when grafted with a vernalized donor, pointing to a transmissible stimulus. Whether the transmitted message is a specific compound involved in vernalization or is the same substance that induces flowering in plants that do not require vernalization remains to be determined (Thomas and Vince-Prue, 1984).

TABLE 21.2. Examples of the successful transfer of flowering signals by grafting in cold-requiring plants.

Flowering donor plant	Receptor plants induced to flower
Hyoscyamus niger, cold-requiring, vernalized	
Hyoscyamus niger, non-cold-requiring annual form	*H. niber*, cold-requiring, nonvernalized
Petunia hybrida, non-cold-requiring annual	
Beta vulgaris, cold-requiring, vernalized	
Beta vulgaris, non-cold-requiring annual form	*B. vulgaris*, cold-requiring, nonvernalized
Beta procumbens, non-cold-requiring annual	

TABLE 21.3. Examples of flowering induction by gibberellins in plants with different environmental requirements for flowering.

Flowering requirement	Plant	Effect of gibberellin
Long-day plants	*Lolium*	Promotes in SD
	Fuchsia	Inhibits in LD
	Anagallis	No effect
Short-day plants	*Zinnia*	Promotes in LD
	Fragaria	Inhibits in SD
	Xanthium	No effect
Dual-daylength plants	*Bryophyllum* (LSDP)	Promotes in SD
	Coreopsis (SLDP)	Promotes in SD
	Cestrum (LSDP)	Inhibits
Day-neutral plants	Many conifers	Promotes
	Many woody angiosperms	Inhibits
Plants requiring vernalization	*Daucus*	Promotes
	Oenothera	No effect

Gibberellins Can Induce Flowering in Some Plants

Of the major groups of naturally occurring growth hormones, gibberellins (Chapter 17) have an important role in flowering (Table 21.3). Exogenous gibberellin can substitute for photoperiodic induction in some plants (Lang, 1965; Vince-Prue, 1985). In particular, GAs cause flowering when applied to long-day plants that grow as rosettes in short days; in these plants, the flowering response (either to GAs or to long days) is accompanied by elongation of the flowering stem. However, it is important to note that flower formation and stem elongation are independent processes (Zeevaart, 1984). Also, application of GAs can evoke flowering in a few short-day plants in noninductive conditions and can substitute partially or completely for a low-temperature signal in several cold-requiring plants. Flowering can also be accelerated in juvenile plants of several gymnosperm families by addition of GAs (Pharis and King, 1985). Thus, GAs can substitute for the endogenous trigger of age in autonomous induction and for the primary environmental signals of daylength and low temperature.

Considerable attention has been given to effects of daylength on GA metabolism in the plant. Amounts of endogenous GAs are often higher in long days than in short days, and this seems to be independent of the photoperiodic response group of the plant. Furthermore, the metabolism of endogenous GAs differs considerably between plants growing in different photoperiods. For example, in the long-day plant spinach (*Spinacia oleracea*), GA_{19} accumulates in short days. Following transfer to long days, the rate of conversion from GA_{19} to GA_{20} and the further metabolism of GA_{20} are increased (Metzger and Zeevaart, 1982). In *Spinacia* GA_{20} appears to be involved in stem elongation rather than flowering, but in another long-day plant, pea (*Pisum sativum*), the GA present in short days appears to act primarily as an inhibitor of flowering. It has been suggested that in *P. sativum* daylength controls flowering via the production of an inhibitory GA in short days (Proebsting and Heftmann, 1980).

Only the less polar GAs cause flowering in seedlings of the Pinaceae, and these GAs may be endogenous flowering factors because cultural treatments, such as water stress, that accelerate flowering have in some cases increased the amount of less polar GAs in the plant (Ross et al., 1983). Different GAs also have markedly different effects on flowering and stem elongation in the long-day plant *Lolium* (Fig. 21.15). These observations suggest that the regulation of flowering may be associated with specific GAs.

In addition to GAs, the application of other growth hormones can either inhibit or promote flowering (Vince-Prue, 1985). Because of the possibility of commercial exploitation of flowering control, a great deal of information is available on flowering responses to applied growth hormones in a wide range of plant species. One commercially important example is the striking promotion of flowering in pineapple (*Ananas comosus*) by ethylene and ethylene-releasing substances, a response that seems common in members of the Bromeliaceae. In most plants, ethylene is either inhibitory or ineffective. The application of abscisic acid modifies the flowering response in some plants, but endogenous abscisic acid does not appear to play a major role in the control of flowering.

The Floral Stimulus May Have Several Components

Ultimately, the formation and development of floral primordia occur at the shoot apex as a result of biochemical and cellular changes leading to the production of the floral organs. The transition to flowering

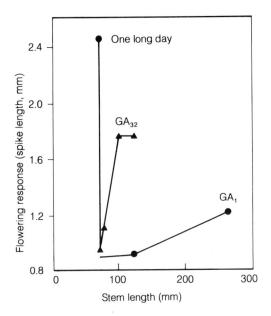

FIGURE 21.15. The relative effect of two different gibberellins on flowering (spike length) and elongation (stem length). In the long-day plant *Lolium temulentum*, GA_{32} strongly promotes flowering but has only a small effect on stem elongation. In contrast, GA_1 strongly promotes stem growth but has only a small effect on flowering. Gibberellin-treated plants were maintained in short days and, for comparison, the response to a single 24-h-long day is also shown. This long day strongly promotes flowering but has almost no effect on stem length. (Data of Pharis et al., 1987.)

may involve a complex system of interacting factors including, among others, carbohydrates, gibberellins, and cytokinins (Bernier, 1988). Some of these factors may act in sequence, since some of the cellular changes seem to occur *before* the floral stimulus arrives at the apex. Sometimes, changes normally associated with flowering can be caused by treatments that do not themselves bring about floral initiation. For example, one of the earliest events observed at the shoot apex following photoperiodic induction is a transitory increase in the number of cells undergoing mitosis. When cytokinins are applied to the stem apex of the long-day plant white mustard (*Sinapis alba*), they cause an increase in mitotic activity similar to that caused by exposure to a single long day. However, the long day induces flowering but the cytokinin treatment does not. Cytokinin may, therefore, be a component of the floral stimulus, although it is insufficient by itself to bring about flowering. If the floral stimulus does have more than one component, it is possible that all factors need not be absent in conditions that do not allow flowering. This could explain the wide range of substances and conditions that can cause flowering. However, many physiological experiments show that the stimulus exported from an induced leaf has some specific properties. It remains a major challenge to establish the biochemical basis of flower induction.

Summary

Flowering occurs at shoot apical meristems and is a complex morphological event. The ability to flower (i.e., the transition from juvenility to maturity) is attained when the plant has reached a certain age or size. In some plants, the transition to flowering then occurs independently of the environment (autonomous induction). Other plants require exposure to appropriate environmental conditions. The most important environmental controls for flowering are daylength and temperature; both are inductive signals and flowering occurs some time after exposure to the flower-inducing treatment. A response to daylength—photoperiodism—locates flowering at a particular time of year, and several different categories of response are known. The photoperiodic mechanism is located in the leaf. Exposure to low temperature—vernalization—is required for flowering in some plants, and this is often coupled with a daylength requirement. Vernalization occurs at the shoot apical meristem. There are several types of interaction between photoperiodism and vernalization.

Daily rhythms—circadian rhythms—can locate an event at a particular time of day. Timekeeping in these rhythms is based on an endogenous circadian oscillator. Keeping the rhythm on local time depends on the phase response of the rhythm to environmental signals. The most important signals are dawn and dusk. The timekeeping mechanism in photoperiodism also probably involves an endogenous circadian oscillator. Plants keep track of time by measuring the duration of darkness, and flowering depends on the duration of the right period. Short-day plants flower when a critical duration of darkness is exceeded. Long-day plants flower when the length of the dark period is less than a critical value. Light given at certain times in a dark period that is longer than the critical value—a night-break—prevents its effect. The response to a night-break is mediated through phytochrome: Pfr inhibits flowering in short-day plants but promotes flowering in long-day plants. In some instances, Pfr has been shown to be required for flowering in short-day plants. Light also acts on the circadian oscillator to entrain the photoperiodic rhythm, an effect that is important for dark timekeeping. The photoperiodic mechanism shows some variation in short-day and long-day responses, but both appear to involve phytochrome and a circadian oscillator.

When induced by exposure to appropriate daylengths or by autonomous induction, leaves send a chemical signal to the apex to bring about flowering. This transmissible signal is able to cause flowering in plants of different photoperiodic response groups. In noninductive daylengths, a transmissible inhibitor of flowering may be produced by the leaves. Although physiological experiments, especially grafting, indicate the existence of specific transmissible floral promoters, called florigens, and inhibitors, called antiflorigens, the chemical identity of these factors is not known. Plant growth hormones, especially the gibberellins, can modify flowering in many plants.

GENERAL READING

Atherton, J. G., ed. (1987) *Manipulation of Flowering*. Butterworths, London.

Bernier, G. (1988) The control of floral evocation and morphogenesis. *Annu. Rev. Plant Physiol. Plant Mol. Biol.* 39:175–219.

Brady, J., ed. (1982) *Biological Timekeeping*. Cambridge University Press, Cambridge, England.

Bünning, E. (1973) *The Physiological Clock: Circadian Rhythms and Biological Chronometry*, 3d rev. ed. Springer-Verlag, New York.

Bünning, E. (1979) Circadian rhythms, light and photoperiodism. A re-evaluation. *Bot. Mag.* 92:89–103.

Halevy, A. H., ed. (1985) *CRC Handbook of Flowering*. CRC Press, Boca Raton, Fla.

Napp-Zinn, K. (1984) Light and vernalization. In: *Light and the Flowering Process*, Vince-Prue, D., Thomas, B., and Cockshull, K. E., eds., pp. 75–88. Academic Press, London.

Pharis, R. P., and King, R. W. (1985) Gibberellins and reproductive development in seed plants. *Annu. Rev. Plant Physiol.* 36:517–568.

Ross, D., Pharis, P., and Binder, W. D. (1983) Growth regulators and conifers; their physiology and potential use in forestry. In: *Plant Growth Regulating Chemicals*, Nickell, L. G., ed., pp. 35–78. CRC Press, Boca Raton, Fla.

Salisbury, F. B., and Ross, C. (1978) *Plant Physiology*, 2d ed. Wadsworth, Belmont, Calif.

Satter, R. L., and Galston, A. W. (1981) Mechanisms of control of leaf movements. *Annu. Rev. Plant Physiol.* 32:83–110.

Sweeney, B. M. (1987) *Rhythmic Phenomena in Plants*. Academic Press, San Diego.

Thomas, B., and Vince-Prue, D. (1984) Juvenility, photoperiodism and vernalization. In: *Advanced Plant Physiology*, Wilkins, M. B., ed., pp. 408–439. Pitman, London.

Vince-Prue, D. (1975) *Photoperiodism in Plants*. McGraw-Hill, London.

Vince-Prue, D. (1985) Photoperiod and hormones. In: *Encyclopedia of Plant Physiology*, New Series, Vol. 11 (III), Pharis, R. P., and Reid, D. M., eds., pp. 308–364. Springer-Verlag, Berlin.

Vince-Prue, D. (1986) The duration of light and photoperiodic responses. In: *Photomorphogenesis in Plants*, Kendrick, R. E., and Kronenberg, G. H. M., eds., pp. 269–305. Martinus Nijhoff, Dordrecht, The Netherlands.

Zimmerman, R. H., Hackett, W. P., and Pharis, R. P. (1985) Hormonal aspects of phase change and precocious flowering. In: *Encyclopedia of Plant Physiology*, New Series, Vol. 11 (III), Pharis, R. P., and Reid, D. M., eds., pp. 79–115. Springer-Verlag, Berlin.

CHAPTER REFERENCES

Borthwick, H. A., Hendricks, S. B., and Parker, M. W. (1952) The reaction controlling floral initiation. *Proc. Natl. Acad. Sci. USA* 38:929–934.

Cleland, C. F. (1984) Biochemistry of induction—the immediate action of light. In: *Light and the Flowering Process*, Vince-Prue, D., Thomas, B., and Cockshull, K. E., eds., pp. 123–136. Academic Press, London.

Cockshull, K. E. (1985) *Chrysanthemum morifolium*. In: *Handbook of Flowering*, Halevy, A. H., ed., Vol. II, pp. 238–257. CRC Press, Boca Raton, Fla.

Coulter, M. W., and Hamner, K. C. (1964) Photoperiodic flowering response of Biloxi soybean in 72-hour cycles. *Plant Physiol.* 39:848–856.

Deitzer, G. F., Hayes, R., and Jabben, M. (1979) Kinetics and time dependence of the effect of far red light on the photoperiodic induction of flowering in Wintex barley. *Plant Physiol.* 64:1015–1021.

Downs, R. J. (1956) Photoreversibility of flower initiation. *Plant Physiol.* 31:279–284.

Engelmann, W., and Schrempf, M. (1980) Membrane models for circadian rhythms. *Photochem. Photobiol. Rev.* 5:49–86.

Evans, L. T., and Wardlaw, I. F. (1966) Independent translocation of ^{14}C-labelled assimilates and of the floral stimulus in *Lolium temulentum*. *Planta* 68:310–326.

Goldberg, R. B. (1988) Plants: Novel developmental processes. *Science* 240:1460–1467.

Hackett, W. P., and Srinivasan, C. (1985) *Hedera helix* and *Hedera canariensis*. In: *Handbook of Flowering*, Halevy, A. H., ed., Vol. III, pp. 89–97. CRC Press, Boca Raton, Fla.

Hamner, K. C. (1940) Interrelation of light and darkness in photoperiodic induction. *Bot. Gaz.* 101:658–687.

Hamner, K. C., and Bonner, J. (1938) Photoperiodism in relation to hormones as factors in floral initiation and development. *Bot. Gaz.* 100:388–431.

Heinze, P. H., Parker, M. W., and Borthwick, H. A. (1942) Floral initiation in Biloxi soybean as influenced by grafting. *Bot. Gaz.* 103:518–530.

Hendricks, S. B., and Siegelman, H. W. (1967) Phytochrome and photoperiodism in plants. *Comp. Biochem.* 27:211–235.

Ison, R. L., and Hopkinson, J. M. (1985) Pasture legumes and grasses of warm climate regions. In: *Handbook of Flowering*, Halevy, A. H., ed., Vol. I, pp. 203–251. CRC Press, Boca Raton, Fla.

King, R. W., and Cumming, B. (1972) Rhythms as photoperiodic timers in the control of flowering in *Chenopodium rubrum* L. *Planta* 103:281–301.

King, R. W., Evans, L. T., and Wardlaw, I. F. (1968) Translocation of the floral stimulus in *Pharbitis nil* in relation to that of assimilates. *Z. Pflanzenphysiol.* 59:377–388.

Lang, A. (1965) Physiology of flower initiation. In: *Encyclopedia of Plant Physiology* (old series), Ruhland, W., ed., Vol. 15, pp. 1380–1535. Springer-Verlag, Berlin.

Lang, A., Chailakhyan, M. Kh., and Frolova, I. A. (1977) Promotion and inhibition of flower formation in a day-neutral plant in grafts with a short-day plant and a long-day plant. *Proc. Natl. Acad. Sci. USA* 74:2412–2416.

Longman, K. A. (1976) Some experimental approaches to the problem of phase change in forest trees. *Acta Hortic.* 56:81–90.

Lumsden, P. J., Vince-Prue, D., and Furuya, M. (1986) Phase-shifting of the photoperiodic flowering response rhythm in *Pharbitis nil* by red-light pulses. *Physiol. Plant.* 67:604–607.

Metzger, J. D., and Zeevaart, J. A. D. (1982) Photoperiodic

control of gibberellin metabolism in spinach. *Plant Physiol.* 69:287–291.

Papenfuss, H. D., and Salisbury, F. B. (1967) Aspects of clock resetting in flowering of *Xanthium. Plant Physiol.* 42:1562–1568.

Pharis, R. P., Evans, L. T., King, R. W., and Mander, L. N. (1987) Gibberellins, endogenous and applied, in relation to flower induction in the long-day plant *Lolium temulentum. Plant Physiol.* 84:1132–1138.

Proebsting, W. M., and Heftmann, E. (1980) The relationship of (^{3}H) GA$_9$ metabolism to photoperiod-induced flowering in *Pisum sativum* L. *Z. Pflanzenphysiol.* 98:305–309.

Purvis, O. N., and Gregory, F. G. (1952) Studies in vernalization of cereals. XII. The reversibility by high temperature of the vernalized condition in Petkus winter rye. *Ann. Bot.* 16:1–21.

Saji, H., Vince-Prue, D., and Furuya, M. (1983) Studies on the photoreceptors for the promotion and inhibition of flowering in dark-grown seedlings of *Pharbitis nil* Choisy. *Plant Cell Physiol.* 67:1183–1189.

Salisbury, F. B. (1963) Biological timing and hormone synthesis in flowering of *Xanthium. Planta* 49:518–524.

Teroaka, H. (1972) Proteins of wheat embryos in the period of vernalization. *Plant Cell Physiol.* 8:87–95.

Vince-Prue, D., and Lumsden, P. J. (1987) Inductive events in the leaves: Time measurement and photoperception in the short-day plant, *Pharbitis nil.* In: *Manipulation of Flowering*, Atherton, J., ed., pp. 255–269. Butterworths, London.

Wellensiek, S. J. (1985) *Campanula medium.* In: *Handbook of Flowering*, Halevy, A. H., ed., Vol. II, pp. 123–126. CRC Press, Boca Raton, Fla.

Zeevaart, J. A. D. (1958) Flower formation as studied by grafting. *Meded. Landbouwhogesch. Wageningen* 58:1–88.

Zeevaart, J. A. D. (1984) Photoperiodic induction, the floral stimulus and flower-promoting substances. In: *Light and the Flowering Process*, Vince-Prue, D., Thomas, B., and Cockshull, K. E., eds., pp. 137–142. Academic Press, London.

Zeevaart, J. A. D., and Boyer, G. L. (1987) Photoperiodic induction and the floral stimulus in *Perilla.* In: *Manipulation of Flowering*, Atherton, J. G., ed., pp. 269–277. Butterworths, London.

Zimmer, R. (1962) Phasenverschiebung und andere Störlichtwirkungen auf die endogen tagesperiodischen Blütenblattbewegungen von *Kalanchoë blossfeldiana* (Phase shift and other light-interruption effects on endogenous diurnal petal movements). *Planta* 58:283–300.

PHOTOGRAPH AND ILLUSTRATION CREDITS

PHOTO CREDITS

Cover © Dennis Brokaw

Unit Openers Overview: © Verna R. Johnson/ Photo Researchers, Inc. Unit I and Unit II: © Biophoto Associates/Photo Researchers, Inc. Unit III: © Jeremy Burgess/Photo Researchers, Inc.

Chapter Openers Chapters 1 and 2: © Verna R. Johnson/Photo Researchers, Inc. Chapters 3–14: © Biophoto Associates/Photo Researchers, Inc. Chapters 15–21: © Jeremy Burgess/ Photo Researchers, Inc.

Box Essay Photographs Blade of grass: © G. F. Leedale/Photo Researchers, Inc. Cross section of birch wood: © H. E. Stork, National Audubon Society/Photo Researchers, Inc. *Ficus* leaf: © D. Newman/Visuals Unlimited.

Chapter 1 1.4a: © Jack Bostrack. 1.4b: © John D. Cunningham/Visuals Unlimited. 1.5a: T. P. O'Brien and M. E. McCully, *Plant Structure and Development*, Collier-Macmillan Ltd., 1969. 1.6: From *Botany*, 4th ed. by Carl L. Wilson and Walter E. Loomis, copyright © 1967 by Saunders College Publishing, a division of Holt, Reinhart and Winston, Inc., reprinted by permission of the publisher. 1.8b: Keith Porter and Myron Ledbetter, *Introduction to the Fine Structure of Plant Cells*, Springer-Verlag, 1970. 1.9a: © Biophoto Associates/Photo Researchers, Inc. 1.9b: Courtesy D. Branton. 1.11a: H. T. Bonnett and E. H. Newcomb, *J. Cell Biol.* 27 (1965):423. Reproduced by copyright permission of Rockefeller University Press. 1.11b: Courtesy R. Warmbrodt. 1.12: Dr. T. W. Fraser, from Gunning and Steer, *Ultrastructure and Biology of Plant Cells*, Edward Arnold Press, 1975. 1.13c: Courtesy H. Depta and D. G. Robinson. 1.14b: M. E. Doonan, courtesy of E. H. Newcomb, University of Wisconsin, Madison. 1.15a, b: Micrographs by W. P. Wergin, provided by E. H. Newcomb. 1.16: S. E. Frederick and E. H. Newcomb, *J. Cell Biol.* 43 (1969): 343. Reproduced by copyright permission of Rockefeller University Press. 1.19: W. T. Jackson. 1.20b: E. H. Newcomb and W. P. Wergin, University of Wisconsin, Madison/BPS. 1.24a: W. P. Wergin, provided by E. H. Newcomb. 1.24c: E. H. Newcomb.

Chapter 4 4.4: D. Cosgrove, *Planta* 171 (1987): 272. 4.5b: W. A. Cote, N. C. Brown Center for Ultrastructure Studies. 4.7a: W. A. Russin and R. Evert. 4.7b: © Jack Bostrack. 4.12a: B. A. Palevitz, *Protoplasma* 107 (1981): 115–125. 4.12b: L. Srivastava and J. Singh, *J. Ultrastructure Res.* 39 (1972): 345–363. 4.12c (top): E. Zeiger/BPS. 4.12c (bottom): E. Zeiger and P. Hepler, *Plant Physiology* 58 (1976): 492–498. 4.13: F. Sack.

Chapter 6 6.15: E. Zeiger. 6.17: E. Zeiger and G. Tallman.

Chapter 7 7.1: © CBS/Visuals Unlimited. 7.2: © Richard Gross. 7.4a,b: A. L. Christy and D. B. Fisher, *Plant Physiology* 61 (1978): 283–290. 7.6: R. Warmbrodt, *Amer. J. Bot.* 72:414–

432. 7.7: R. F. Evert, *BioScience* 32 (1982): 789–795. 7.8: D. S. Neuberger and R. F. Evert, *Protoplasma* 84 (1975): 109–125. 7.9: Susan A. Sovonick-Dunford. 7.10: B. Bentwood and J. Cronshaw, *Planta* 140 (1978): 111–120. 7.12a–c: M. H. Zimmerman and C. L. Brown, *Trees: Structure and Function,* Springer-Verlag, 1971. 7.14: R. F. Evert and R. J. Mierzwa, from J. Cronshaw, W. J. Lucas, and R. T. Giaquinta, eds., *Phloem Transport. Proceedings of an International Conference on Phloem Transport.* Alan R. Liss, Inc., 1986, pp. 419–432. B. R. Fondy and D. R. Geiger, *Plant Physiology* 59 (1977): 953–960. 7.19: R. Turgeon and J. A. Webb, *Planta* 113 (1973): 179–191.

Chapter 8   8.13: Courtesy of J. Swafford, Department of Botany, Arizona State University. 8.14: From L. A. Staehelin, D. P. Carter, and A. McDonnel, in *Membrane-Membrane Interactions,* N. B. Gilula, ed., p. 179, Raven Press, 1980. 8.24: E. Gantt, *BioScience* 25 (1975): 781–788.

Chapter 9   9.5: J. A. Bassam. 9.10a,b: S. E. Frederick and E. H. Newcomb, *Planta* 96 (1971): 152–174. 9.10c: E. H. Newcomb, University of Wisconsin, Madison. 9.10e: D. Goodchild and S. Craig, *Austr. J. Bot.* 25 (1977): 277–290. 9.17: S. E. Frederick, courtesy of E. H. Newcomb, University of Wisconsin, Madison.

Chapter 10   10.3a,b: D. McCain, *Plant Physiology* 86 (1988): 16–18. 10.4: A. Schwartz.

Chapter 11   11.2: M. E. Doonan, courtesy of E. H. Newcomb, University of Wisconsin, Madison. 11.14b: R. N. Trelase, P. J. Gruber, W. M. Becker, and E. H. Newcomb, *Plant Physiology* 48 (1971): 461.

Chapter 12   12.2a,b: W. D. P. Stewart, in *A Treatise on Nitrogen Fixation*, R. W. F. Hardy and W. S. Silver, eds., pp. 63-123, John Wiley & Sons, 1977. 12.3a: The Nitragin Company. 12.3b: H. J. Evans. 12.6a,b: A. Hirsch. 12.6c: E. H. Newcomb and S. R. Tandon, University of Wisconsin, Madison/BPS.

Chapter 13   13.2b: N. D. Hallam. 13.3: J. Troughton and L. A. Donaldson, *Probing Plant Structure,* McGraw-Hill, 1972. 13.8: Courtesy of R. Miller. 13.9: Courtesy of G. Kreitner and J. Gershenzon. 13.B: Courtesy of M. Levy

Chapter 14   14.2: Courtesy of B. L. McMichael. 14.5: I. M. Rao, R. E. Sharp, and J. S. Boyer, *Plant Physiology* 84 (1987): 1214–1219. 14.8a-c: Courtesy of D. M. Oosterhuis. 14.9: H. J. Bohnert, in *Environmental Stress in Plants*, J. H. Cherry, ed., pp. 159–171, Springer-Verlag, 1989. 14.13: Courtesy of J. L. Basq and M. C. Drew.

Chapter 15   15.2b: Courtesy of C. H. Busby. 15.2c: Courtesy of A. Hardham. 15.3a: © Jack Bostrack. 15.5: Courtesy of A. Hardham. 15.10a,b: Courtesy of A. Hardham. 15.11a,b: Courtesy of A. Hardham. 15.12: Courtesy of A. Hardham. 15.13a,b: Courtesy of D. J. Flanders. 15.14: Courtesy of A. Hardham. 15.15a,b: Courtesy of A. Hardham. 15.16a: A. Hardham and B. Gunning, *J. Cell Sci.* 37 (1979): 411–442. 15.16b,c: A. Hardham and B. Gunning, *Protoplasma* 102 (1980): 31–51. 15.20a: A. Hardham, *McGraw-Hill Yearbook of Science and Technology,* pp. 252–254, 1979. 15.20b: A. Hardham and B. Gunning, *J. Cell Sci.* 37 (1979): 411–442. 15.20c,d: Courtesy of A. Hardham. 15.22a,b: A. Hardham and M. McCully, *Protoplasma* 112 (1982): 143–151.

Chapter 16   16.11: M. Jacobs and S. F. Gilbert, *Science* 220 (1983): 1297–1300, © 1983 AAAS. 16.16a: A. Sievers.

Chapter 17   17.2: Courtesy of B. O. Phinney. 17.3: A. Lang. *Proc. Nat. Acad. Science* 43 (1957): 709–717. 17.9: Peter Davies. 17.12: Courtesy of S. H. Wittwer. 17.13: C. Rogler and W. Hackett, *Physiol. Plant* 34 (1975): 141-152. 17.17: Courtesy of J. A. D. Zeevaart. 17.21b: P. A. Adams, et al, *Plant Physiology* 51 (1973): 1102–1108. 17.22b: R. L. Jones, *Planta* 85 (1969): 359–375.

Chapter 18   18.1: USDA. 18.10: F. Skoog and C. O. Miller, *Symposia for the Society of Experimental Biology XI,* 118–140, Cambridge University Press, 1957. 18.11: Courtesy of Roy Morris. 18:13: Courtesy of A. K. Huff and C. W. Ross, based on findings in *Plant Physiology* 56 (1975): 429–433.

Chapter 19   19.3a–d: Courtesy of S. Gepstein. 19.13a: R. L. Jones, *Planta* 85 (1969): 359–375. 19.13b,c: J. V. Jacobsen and P. M. Chandler, in *Plant Hormones and Their Role in Plant Growth and Development*, P. J. Davies, ed., pp. 164–193, Martinus Nijhoff, 1987.

Chapter 20   20.1: Courtesy of H. Smith. 20.2: R. E. Kendrick, *Phytochrome and Plant Growth*, p. 78, Edward Arnold, 1976. 20.6a: S. Tokutomi, et al., *FEBS Letters* 247 (1989): 139–142. 20.12: J. C. Fondeville, *Planta* 69 (1966): 358.

Chapter 21   21.1: R. L. Ison and J. M. Hopkinson, in *Handbook of Flowering*, Vol. I, A. H. Halevy, ed., p. 206, CRC Press, 1985. 21.2: Courtesy of D. Vince-Prue. 21.13: S. J. Wellensiek, in *Handbook of Flowering*, Vol. II, A. H. Halevy, ed., p. 124, CRC Press, 1985. 21.A: R. B. Goldberg, *Science* 240 (1988): 1460, © AAAS.

ILLUSTRATION CREDITS

Chapter 1   1.5b: Adapted from Esau, K. *The Anatomy of Seed Plants 2e*, p. 376, John Wiley & Sons: NY, 1960. 1.6: From Wilson, C. L. & Loomis, W. E. *Botany 4e* © 1967 Saunders College Publishing, a division of Holt, Rinehart and Winston, Inc., reprinted by permission of the publisher. 1.23a,b: Reproduced from the *Journal of Cell Biology*, 84(1987):327 by copyright permission of the Rockefeller University Press.

Chapter 2   2.2: Modified from Nicholls, D. G. *Bioenergetics: An Introduction to the Chemiosmotic Theory*, Academic Press, New York, 1982. 2:23: Darnell, Lodish, Baltimore, *Molecular Cell Biology 2e*, fig. 11.19a, W. H. Freeman, 1990.

Chapter 3   3.8: From P. S. Nobel, *Biophysical Plant Physiology and Ecology*, p. 17, W. H. Freeman, 1983. Reprinted by permission. 3.12: *Agricultural and Forest Meterology*, T. C. Hsiao, 1974, 14:59–84. Elsevier Science Publishers. 3D: P. B. Green, *Plant Physiology* 43:1169–1184. Reprinted by permission of American Society of Plant Physiologists © 1968.

Chapter 4   4.11: From G. G. Bange, 1953, *Acta Botanica Neerlandica* 2:254–297.

Chapter 5   5.1, 5.2: Adapted from Weaver, J. E., *Root Development of Field Crops* McGraw-Hill: NY 1926 © McGraw-Hill, Inc. 5.5: Adapted from Huck, M. G. and Taylor, H. M. *Advances in Agronomy* 35:1–35, Academic Press: New York, 1984. 5.6: Adapted from R. E. Lucas and J. F. Davis, *Soil Science* 92:177–182 © by Williams & Wilkins, 1961. 5.7: From Rovira et al., *Encyclopedia of Plant Physiology* New Series Vol. 15A, pp. 61–86, Springer-Verlag: Heidelberg, 1983. T5.1: Reprinted with permission of Macmillan Publishing Company from *Nature and Properties of Soils 8e*, by Nyle C. Brody Copyright © 1974 by Macmillan Publishing Company. T5.2: Adapted from Epstein, *Mineral Nutrition of Plants: Principles and Perspectives*, John Wiley & Sons: NY, 1972. T5.3: From Mengel and Kirkby, 1979, *Principles of Plant Nutrition*. International Potash Institute, Berne, Switzerland. T5.4: Adapted from Epstein, *Mineral Nutrition of Plants: Principles and Perspectives* John Wiley & Sons: NY 1972. T5.5: Reproduced, with permission, from the *Annual Review of Plant Physiology*, Evans & Sorger, v. 17. © 1966 by Annual Reviews, Inc. T5A: Adapted from Huck, M. G. & Taylor, H. M., *Advances in Agronomy* 35:1–35 Academic Press, 1984.

Chapter 6   6.9: From Higenbotham, Graves, Davis, *Journal of Membrane Biology* 3:210–222, Springer-Verlag: Heidelberg, 1970. 6.12: From Novacky, A. et al., *Planta* 149:321–326 Springer-Verlag: Heidelberg. 6.21: W. Lin et al, *Plant Physiology* v.75. Reprinted by permission of American Society of Plant Physiologists © 1984.

Chapter 7   7.11: Adapted from Joy, K. W., *Journal of Experimental Botany* 15(1964): 485–494, Oxford University Press. 7.18: Giaquinta, R. T., W. Lin, N. L. Sadler, V. R. Francheschi, *Plant Physiology* 72:362–367. Reprinted by permission of American Society of Plant Physiologists © 1983. 7.21: Adapted from Geiger, D. R. & Sovonick, S. A. *Encyclopedia of*

Plant Physiology ?:256–286, Springer-Verlag, 1975. 7.22: Reproduced with permission from the *Annual Review of Plant Physiology*, Preiss, J. v. 33 p. 444, 1982 by Annual Reviews, Inc. T7.2: From Hall, S. M. and D. A. Baker, *Planta* 106:131–140 Springer-Verlag: Heidelberg 1972. Adapted from Richardson, P. T., D. A. Baker and L. C. Ho, *Journal of Experimental Botany* 33(1982):1239–47, Oxford University Press.

Chapter 8 8.1f: Adapted from Ke, B. *Biochemical and Biophysical Acta* 301(1973):1–33, Elsevier Science Publishers. 8.4a,b: Adapted from Sauer, K. *Bioenergetics of Photosynthesis* Govindgee, ed. pp. 116–181 Academic Press: NY, 1975. 8.6: C. G. Avers, *Molecular Cell Biology* © 1985, Addison-Wesley. 8.11: Reprinted by permission from *Nature* Vol. 190 pp. 510–11 Copyright © 1961 Macmillan Magazines Ltd. 8.14b: From Staehelin, L. A., D. P. Carter and A. McDonnel, 1980. *Membrane-Membrane Interactions*, N. B. Gilula, ed., p. 179 Raven Press: New York. 8.17: Adapted with permission from *Biochemistry* Kramer et al., 765:156–165. Copyright 1984 American Chemical Society. 8.18: Reprinted by permission from *Nature* Vol. 318 pp. 618–624 Copyright © 1985 Macmillan Magazines Ltd. 8.19: *On-Page Credit:* Reproduced, with permission, from the *Annual Review of Plant Physiology*, A. N. Glazer and A. Melis, 38:11–45 © 1987 Annual Reviews, Inc. 8.25: Adapted from Blankenship & Prince, *Trends in Biochemical Science* 10:382–383, 1985, Elsevier Science Publishers. 8.26: Adapted from Blankenship & Prince, *Trends in Biochemical Science* 10:382–383, 1985, Elsevier Science Publishers. 8.29: From N. Murata and M. Miyao, *Trends in Biochemical Sciences* 10:122–124, 1985, Elsevier Science Publishers. 8.36: From Curtis, *Photosynthesis Research* 18:223–244 Martinus Nijhoff © 1988 Kluwer Academic Publishers.

Chapter 9 9.10d: From Luttge and Hinginbotham, *Transport in Plants*, Springer-Verlag: Heidelberg, 1979.

Chapter 10 10.2: From Smith, H. *Photomorphogenesis in Plants* pp. 187–217. Martinus Nijhoff, © 1986 Kluwer Academic Publishers. 10.4a: From Haupt, W. *Photomorphogenesis in Plants* pp. 415–441. Martinus Nijhoff, © 1986 Kluwer Academic Publishers. 10.6: Adapted from Harvey, G. W., 1979 Carnegie Institution Yearbook 79:161–164, 1979. 10.8: From Bjorkman, U. *Encyclopedia of Plant Physiology*, New Series 12A:57–107 Springer-Verlag: Heidelberg, 1981. 10.10: Adapted from Berry, *Photosynthesis, Development, Carbon Metabolism and Productivity* Vol. II pp. 263–343 Academic Press: NY, 1982. 10.11: *Plant Physiology* 86:700–705. Reprinted by permission of American Society of Plant Physiologists © 1988. 10.12: Reprinted by permission of the publishers from *The Cactus Primer* by Arthur C. Gibson and Park S. Nobel, Cambridge, Mass.: Harvard University Press © 1986 by Arthur C. Gibson and Park S. Nobel. 10.13: *On-Page Credit:* Reproduced, with permission, from the *Annual Review of Plant Physiology*, Berry, J. & Bjorkman, O. v. 31 © 1980 by Annual Reviews, Inc. 10.14: Adapted from Berry, *Photosynthesis, Development, Carbon Metabolism and Productivity* Vol. II pp. 263–343, Academic Press: NY, 1982.

Chapter 12 12.1b: From Burns, R. C., Hardy, R. W. F., *Nitrogen Fixation in Bacteria and Plants*, Springer-Verlag: Heidelberg, 1975. 12.4a,b: *On-Page Credit:* Reproduced, with permission, from the *Annual Review of Plant Physiology*, Rolfe, B. & Gresshoff, P., v. 39. © 1988 by Annual Reviews, Inc. 12.7: Adap. from Dixon, Rod, C. T. Wheeler, *Nitrogen Fixation in Plants* (1986). 12.16: Adapted from D. A. Rees, *Polysaccharide Shapes* p. 52 Chapman and Hall, 1977.

Chapter 14 14.1: Adapted from Matthews, M. A., Volkenburgh, E. and Boyer, J. S. *Plant Cell and Environment* 7:199–206, 1984 Blackwell Scientific Publications. 14.4: Boyer, J. S. *Plant Physiology* 46:233–235. Reprinted by permission of American Society of Plant Physiologists © 1970. 14.5: Rao et al, *Plant Physiology* 84:1214–1219. Reprinted by permission of American Society of Plant Physiologists © 1987. 14.6: Sung and Kreig, *Plant Physiology* 64:852–856 Reprinted by permission of American Society of Plant Physiologists © 1979. 14.7: Adapted from *Crop Science*, v. 27; no. 3, May–June, 1987, pp. 539–543 by permission of Crop Science Society of America, Inc. 14.10: From Patterson, B. D., Paull, R. and Smillie, R. M. *Australian Journal of Plant Physiology* 5:609–617. 14.11: Adapted from Bjorkman, O., Badger, M. R. & Armond, P. A. *Adaptation of Plants to Water and High Temperature Stress*, N. C. Turner & P. J. Kramer, eds. pp. 233–249, John Wiley & Sons: NY, 1980. 14.12: Reproduced, with permission, from the *Annual Review of Plant Physiology*, Greenway & Munns, v. 31. © 1980 by Annual Reviews, Inc. 14A: Brown, Pereira, Finkle, *Plant Physiology* 53: 709–711. Reprinted by permission of American Society of Plant Physiologists © 1974.

Chapter 15 15.21: Adapted from Hardham et al. *Planta* 149:181–195, 1980, Elsevier Science Publishers.

Chapter 16 16.7: Cohen, J. D., 1983. *What's New in Plant Physiology* 14:41–44, p. 43. 16.10: Jacobs, N. 1983, *What's New in Plant Physiology* 14:17–20. 16.16a,b: Hasenstein, K. H., M. L. Evans *Plant Physiology* 86:890–894. Reprinted by permission of American Society of Plant Physiologists © 1988. 16.17: From Vanderhoef et al., *PNAS USA* 72:1822, 1975.

Chapter 17 17.4: From Frydman et al., *Planta* 118:123–132 Springer-Verlag: Heidelberg, 1974. 17.7: From Sponsel, *Plant Hormones and Their Role in Plant Growth and Development* pp. 43–75 P. J., Davies, ed. Kluwer, Boston, 1987. 17.14: Adapted from Davies, P. J. *Plant Hormones and Their Role in Plant Growth and Development* Kluwer: Boston, 1987. 17.19: Davies et al., *Plant Physiology* 81:991–996 Reprinted by permission of American Society of Plant Physiologists © 1986. 17.20: Reproduced, with permission, from the *Annual Review of Plant Physiology*, Sachs, R. M. v. 16 © 1965 by Annual Reviews, Inc. 17.21a,b: Adams et al, *Plant Physiology* 51:1102–1108. Reprinted by permission of Amer. Society of Plant Physiologists © 1983. 17.22a: From Jones, R. L. & MacMillan, J. *Advanced Plant Physiology* M. B. Wilkins, ed. Longman: London 1987 pp. 21–25. 17.22b: *Plant Phys.* 48:137–142. Reprinted by permission of American Society Plant Physiologists. 17.23a,b: Reprinted by permission from *Nature* Vol. 260 pp. 166–169 Copyright © 1976 Macmillan Magazines Ltd. 17.24: From Ou-Lee, T. M., Turgeon, R., Wu, R. *Proceedings of National Academy of Sciences USA* 85:6366–69 1988.

Chapter 18 18.6: From Leonard, N. J., Hecht, S. M., Skoog, F., Schmitz, R. Y. *PNAS USA* 59:15–21, 1968. 18.7: Esashi, T., A. C. Leopold, *Plant Physiology* 44:618–620. Reprinted by permission of American Society of Plant Physiologists © 1969. 18.10: From Chilton, M. D. *Scientific American* 248:50–59 1983. 18.12: Reproduced, with permission, from the *Annual Review of Plant Physiology*, Morris, v. 37. © 1986 by Annual Reviews, Inc. 18.15: Adapted from Tepfer, D. A. and Fosket, D. E. *Developmental Biology* 62:486–497 Academic Press 1978. 18.16: From Flores, S. and Tobin, E. M. *Plant Molecular Biology* 11:409–415 Kluwer Academic Publishers © 1988.

Chapter 19 19.1: Adapted from Beyer et al., *Advanced Plant Physiology* M. B. Wilkins, ed. pp. 111–126. Longman: London, 1987. 19.2b: Yang, *Plant Senescence* pp. 156–166. Reprinted by permission of American Society of Plant Physiologists © 1987. 19.4: Burg, S. *Science* 148:1190 © 1965 by American Association for the Advancement of Science. 19.5: Adapted from Morgan, *Ethylene, Biochemistry Physics and Applied Asp.*, Y. Fuchs and E. Chalutz, eds. © 1984 Kluwer Academic Publishers. 19.7: Christophersen, R. E. *Plant Senescence* p. 89–97. Reprinted by permission of American Society of Plant Physiologists © 1987. 19.12: Reprinted by permission from *Nature* Vol. 319 pp. 324–326 © 1986 Macmillan Magazines Ltd. 19.13: Adapted from Jacobsen & Chandler, *Plant Hormones and Their Role in Plant Growth and Development* pp. 164–193 © 1987 Kluwer Academic Publishers.

Chapter 20 20.3: Smith, H. *Phytochrome and Photomorphogenesis* McGraw-Hill, London, 1974. 20.5: Reprinted with permission from *Journal of the American Chemical Society*, 102: 4821–4828. Lagarias, J. C. and Rapaport, H. 1980 H. Copyright 1980 American Chemical Society. 20.7: After R. E. Kendrick and B. Frankland, *Phytochrome and Plant Growth*, Edward Arnold Publishers, London, 1976, 1983. 20.8: Smith, H. *Phytochrome and Photomorphogenesis* McGraw-Hill, London, 1974. 20.9: Reprinted by permission from *Nature* Vol. 273 pp. 534–536 Copyright © 1978 Macmillan Magazines Ltd. 20.10: From Leopold, A. C. & Kriedemann, P. *Plant Growth and Development 2e* p. 373 McGraw-Hill: NY, 1975 © McGraw-Hill, Inc. 20.11b: Adapted from Sweeney, B. M. *Rhythmic Phenomena in Plants* Academic Press: San Diego, 1987. 20.11c: From Bunning, E. & Moser, E. *PNAS USA* 69:2732–33, 1972. 20.14: From Dreyer, E. M. and J. H. Weisenseel *Planta* 146:31–39, Springer-Verlag: Heidelberg, 1979. 20.15: From Serlin, B. S. & Roux, S. J. *PNAS USA* 81:6368–6372, 1984. 20.17: From Shafer, E., Apel, K., Batschauer, A., and E. Mosinger. Martinus Nijhoff: © 1986 Kluwer Academic Publishers. 20.18: From Shafer, E., Apel, K. Batschauer, A. and E. Mosinger. Martinus Nijhoff: © 1986 Kluwer Academic Publishers. 20A a–f: From Kraml, K. *Photomorphogenesis in Plants* pp. 237–267 Martinus Nijhoff: © 1986 Kluwer Academic Publishers.

AUTHOR INDEX

Abe, H., 494
Abeles, F. B., 476
Adams, D. D., 475
Adams, P. A., 444
Adams, W. W., 39
Addicott, F. T., 482
Adesnik, M., 288
Ahmad, I., 118
Akiyoshi, D. E., 465
Albersheim, P., 336
Alberts, B., 50, 139
Allen, J. F., 199
Altman, S., 53
Ames, B. N., 331
Amesz, J., 190, 191, 205
Amhrein, N., 476
Anderson, J. M., 197
Andrews, T. J., 220
ap Rees, T., 279
Apel, K., 508, 509
Apostolakos, P., 388
Appleby, C. A., 295
Arkins, C. A., 156
Armond, P. A., 361, 362
Armstrong, D. J., 456, 469
Arnold, W., 188
Arnon, D. E., 108
Arnon, D. I., 212
Arny, D. C., 360
Artntzen, C. J., 199
Asher, C. J., 110, 111
Ashton, F. M., 211, 212
Assmann, S. M., 136, 138
Atchinson, M. L., 55
Atkinson, C. J., 486
Avers, C. G., 20, 187, 193
Avigne, W. T., 153

Babcock, G. T., 206
Bacastow, R. B., 256
Badger, M. R., 361, 362
Baker, D. A., 154, 173
Baliga, B. S., 312
Baltimore, D., 50, 54

Bandurski, R. S., 402, 403
Bange, G. G. J., 93, 94
Bar-Yosef, B., 102
Barber, J., 206
Barker, R. F., 494
Barnhart, M. C., 282
Bartholomew, G. A., 282
Bassham, J. A., 226
Batschauer, A., 508, 509
Bauer, W. D., 298
Bauerle, C., 509
Bayley, C. C., 487
Beach, L. R., 446
Beachy, R. N., 390
Beale, M. H., 448
Bearce, B. C., 73
Beard, W. A., 215
Beardsell, M. F., 486
Beck, E., 243
Becker, M., 196
Becker, W. M., 193
Becwar, M. R., 359
Beevers, H., 288
Beevers, L., 293, 302, 303, 305
Bellissimo, D., 303
Bendall, F., 190
Bengochea, T., 481
Bennet-Clark, T. A., 482
Bennett, A. B., 132
Bennett, J. H., 111, 311
Berenbaum, M., 331
Berger, S., 380
Bernier, G., 529
Berry, J. A., 254, 259, 261, 262, 361
Berryman, A. A., 324
Berthold, D. A., 274, 278
Bewley, J. D., 362, 485
Beyer, E. M., Jr., 475, 476, 481
Bienfait, H. F., 115, 311, 312
Binder, W. D., 528
Binns, A. N., 460
Birnberg, P. R., 435
Björkman, O., 254, 255, 261, 361, 362
Black, C. C., 269

Black, M., 485
Blackman, P. G., 351
Blankenship, R. E., 203, 204
Blatt, M. R., 252
Blizzard, W. E., 355
Blum, W., 418
Blumwald, E., 133
Bohnert, H. J., 356
Boller, T., 132, 475, 481
Bomhoff, G., 461
Bonham-Smith, P. C., 362
Bonner, J., 241, 521, 526
Bonner, W., 241
Bonnet, M., 459
Borisov, A. Y., 197
Borthwick, H. A., 504, 522, 526
Bottomley, W., 215, 227
Boudet, A., 330
Bouma, D., 115, 116
Bouton, J. H., 300
Bowen, C. D., 106
Bowling, D. J. F., 142
Bowman, B. J., 133
Bowman, E. J., 133
Bowman, R. L., 73
Boyer, G. L., 526
Boyer, J. S., 72, 171, 346, 347, 348, 351, 352, 355
Bozarth, C. S., 351
Bradford, K. J., 480
Bradstreet, E. D., 76
Brady, J., 516
Brady, N. C., 104, 295
Brain, R. D., 412
Brandon, D. L., 270, 457
Braun, A. C., 454
Bray, D., 50, 139
Bray, E. A., 351
Brenner, M. L., 402, 435
Brentwood, B., 152
Bresler, E., 102
Breuing, F., 476
Briggs, W. R., 252, 412, 497
Briskin, D. P., 133

539

SUBJECT INDEX

Supercooling, 359–60
  deep, 358–60
Superoxide anions, 312
Surface, cuticle, 320
Surface tension, air-water interface,
  63–64
Symbiotic relationships, *Alnus* with
  Frankia, 300
Symbols
  α, Bunsen absorption coefficient, 221
  ΔE_n, Nernst potential, 124
  $\Delta\mu$, electrochemical potential
    difference, 37
  ε, molar extinction coefficient, 181
  F, Faraday's constant, 121
  h, Planck's constant, 180
  L, hydraulic conductance, 78
  Lp, specific hydraulic conductivity, 409
  μ, chemical potential, 122
  υ, frequency, 180
  P, hydrostatic (gauge) pressure, 69,
    348
  P, turgor, 69, 348
  Φ, quantum yield
  π, osmotic pressure, 68
  ψ, water potential, 68
  m, wall extensibility, 410
  Y, yield threshold, 410
  z, electrostatic charge of an ion, 121
Sym plasmid, 297
Symplast, 84–85, 102
  communication by, 390
  ion movement to, 142
Symplast pathway
  blocking, 164
  for phloem loading, 160, 160f
  for water transport, 85
Symplocarpus foetidus, alternative
  respiration in, 278
Symport, 130, 132, 174
  hypothetical model for cotransport,
    131
  proton/sucrose, 129, 164
Synechococcus lividus, thylakoids with
  phycobilisomes attached, 201
Synthase
  ATP, 275–76
  ATP, and proton transport, 129
  ATPase, 213
  callose, 151–52, 154
  cellulose, 382, 384
  citrate, and concentration of ATP, 279
  cytokinin, 458–59, 470
  glutamine, 306
  octopine, 465
  starch, 170, 244
  succinyl CoA, 271
  sucrose-phosphate, 246
Synthetic cytokinins, 456

Takahashi, Nobutaka, 426
Tamarix species, salt glands in the leaves
  of, 364
Tamura, Saburo, 426
Tannins, 337–38
Taproot, 101
Taxodium, growth in flooded
  conditions, 283

TCA cycle, 49, 267, 270–72, 281
  manganese requirements, 115
  and respiration, 290
T-DNA, mutant, and crown gall
  development, 464–65, 465f
Temperature
  and available energy, 31–32
  control by transpiration, 61
  coupling to photoperiodic response,
    520
  and dark reversion, 493
  effect on water potential, 72–73
  and enzyme catalysis, 47, 47f
  leaf, and water loss, 92
  and photorespiration/photosynthesis
    balance, 232
  and photosynthesis, 261–62
  and photosynthetic efficiency, 241
  and respiration rate, 284, 290
  and Rubisco catalysis, 221
Temperature compensation point, 361
Tensile strength of water, 64–65
Tension, 65
  as a matric potential component, 76
  measuring, 76
  origin of leaf, 91
  of soil water, 82
  transport of water to the tops of trees
    by, 88–89
  and water movement in a plant, 69
Teratomas, and crown gall development,
  464–65, 465f
Terminal differentiation, 452
Terpenes, 322–28, 344
Terpenoids, 284, 429
Tetrapyrroles, 494, 494f
  in phycobilisomes, 187
Thenoyltrifluoroacetone, 274
The Power of Movement in Plants
  (Darwin), 399
Thermal agitation, and diffusion, 65
Thermodynamic efficiency
  of photosynthesis, 227
  and oxygenation/carboxylation
    competition, 232
Thermodynamics, 28
Thimann, Kenneth, 399
Thin-layer chromatography (TLC),
  402
  gibberellin assays, 433
Thioglucosidase, 341
Thiol proteases, papain, 43, 43f, 46
Thioredoxin system, regulation by light
  intensity, 240
Third law of thermodynamics, 31–32
Threonine in glycoproteins, 43
Thylakoid lumen, 207, 227
  proton accumulation in, 209–10
Thylakoid membranes
  effects of air pollution on, 368
  freeze-fractured, 193
  galactolipids in, 286–88
  organization of, 197
  photosynthesis in, 216
  synthesis of diacylglycerol and
    galactolipids for, 286–88
Thylakoids, 17, 18f, 179
  chlorophyll in, 192–95

integral membrane proteins in, 192,
  194
source, 19
Thylakoid stacking, 199
Tidestromia oblongifolia
  photosynthesis as a function of CO_2
    concentrations, 259
  response to heat stress, 361, 361f
Tilia americana, transverse section of
  stem, 146
Time-resolved optical absorbance
  measurements, to monitor the redox
  state of chlorophylls, 205
Timothy grass (*Phleum pratense*), 18, 26
Tip growth, 377–78
Ti plasmid, 461
  and genetic engineering, 462
  as a probe, 510
Tissue culture, 453, 462, 464
Tissue systems, 8–9, 26
Tobacco, mRNA sequences of nitrate
  reductase, 302–3
Tomato plants, nutrient deficiency
  symptoms of, 113
Tonoplast, 11f, 16–17
  proton ATPase in, 132–33
  proton gradients in, 129
  transport across, 163
  water transport across, 84
Tosyl-lysine chloromethyl ketone
  (TLCK), 46
Totipotency, 386
Toxicity
  of aglycone, 341–42
  of aluminum from acid rain, 368
  of nutrients in excess, 116
  phototoxicity of furanocoumarins, 331
  of terpenes, 324–27
Tracheary elements, 376–77
  cavitation detection using, 89
  differentiation of, 386–87
  in the xylem, 86, 87
Tracheids, 9, 86–87, 87f
Transacetylase, 51
Transaminase, tryptophan, 404
Transamination reactions, 305
Transcellular current, 378
Transcellular transport, 140–43
Transcription, 12–13
  blocking by furanocoumarins, 331
  and gene expression, 50, 50f
  of genes encoding the cell wall-
    digesting enzymes, 481
  regulation by DNA-binding proteins,
    54–55
Transduction, 29
Transduction sequence, 500–501
Transfer cells, 152–53
Transgenic plants, 510
  experimental use of, 390
Trans Golgi network, 16
Transition state (TS), 44
Transition to source from sink, 164
Transit peptide, 216
Transketolase
  in the C3 PCR cycle, 224, 225
  in the pentose phosphate pathway, 280